7th IGRSM International Remote Sensing and GIS Conference and Exhibition 2014

AA008908

IOP Conference Series: Earth and Environmental Science
Volume 20

Kuala Lumpur, Malaysia
22-23 April 2014

ISBN: 978-1-63439-961-6
ISSN: 1755-1307

Printed from e-media with permission by:

Curran Associates, Inc.
57 Morehouse Lane
Red Hook, NY 12571

Some format issues inherent in the e-media version may also appear in this print version.

Copyright© by the Institute of Physics
All rights reserved.

Printed by Curran Associates, Inc. (2015)

For permission requests, please contact the Institute of Physics at the address below.

Institute of Physics
Dirac House, Temple Back
Bristol BS1 6BE UK

Phone: 44 1 17 929 7481
Fax: 44 1 17 920 0979

techtracking@iop.org

Additional copies of this publication are available from:

Curran Associates, Inc.
57 Morehouse Lane
Red Hook, NY 12571 USA
Phone: 845-758-0400
Fax: 845-758-2634
Email: curran@proceedings.com
Web: www.proceedings.com

TABLE OF CONTENTS

SPACE APPLICATIONS FOR ENVIRONMENT (SAFE)

LAND USE/COVER MAPPING

IOP Conf. Series: Earth and Environmental Science **20** (2014) 011001 doi:10.1088/1755-1315/20/1/011001

Preface

This proceedings consists of the peer-reviewed papers from the *7th IGRSM International Conference and Exhibition on Remote Sensing & GIS* (IGRSM 2014), which was held on 21–22 April 2014 at Berjaya Times Square Hotel, Kuala Lumpur, Malaysia. The conference, with the theme *Geospatial Innovation for Nation Building* was aimed at disseminating knowledge, and sharing expertise and experiences in geospatial sciences in all aspects of applications. It also aimed to build linkages between local and international professionals in this field with industries.

Highlights of the conference included:

- Officiation by Y B Datuk Dr Abu Bakar bin Mohamad Diah, Deputy Minister of Minister of Science, Technology & Innovation
- Keynote presentations by:
  - Associate Professor Dr Francis Harvey, Chair of the Geographic Information Science Commission at the International Geographical Union (IGU) and Director of U-Spatial, University of Minnesota, US: *The Next Age of Discovery and a Future in a Post-GIS World.*
  - Professor Dr Naoshi Kondo, Bio-Sensing Engineering, University of Kyoto, Japan: *Mobile Fruit Grading Machine for Precision Agriculture.*
  - Datuk Ir Hj Ahmad Jamalluddin bin Shaaban, Director-General, National Hydraulic Research Institute of Malaysia (NAHRIM), Malaysia: *Remote Sensing & GIS in Climate Change Analyses.*

Content from this work may be used under the terms of the Creative Commons Attribution 3.0 licence. Any further distribution of this work must maintain attribution to the author(s) and the title of the work, journal citation and DOI.

- Oral and poster presentations from 69 speakers, from both Malaysia (35) and abroad (34), covering areas of water resources management, urban sprawl & social mobility, agriculture, land use/cover mapping, infrastructure planning, disaster management, technology trends, environmental monitoring, atmospheric/temperature monitoring, and space applications for the environment.
- Post-conference workshops on:
  - *Space Applications for Environment* (SAFE), which was be organised by the Japan Aerospace Exploration Agency (JAXA)
  - *Global Positioning System (GPS) Receiver Evaluation Using GPS Simulation*, which was be organised by the Science & Technology Research Institute for Defence (STRIDE), and sponsored by RFI Technologies Sdn. Bhd. and Aeroflex Inc.
- Two awards were presented by Dr Noordin Ahmad, Director-General of the National Space Agency during the conference's closing ceremony:
  - **Best Paper Award:** Dr Rizatus Shofiyati, Indonesian Center for Agricultural Land Resources Research and Development (ICALRD), Indonesia: *Indonesian Drought Monitoring from Space. A Report of SAFE Activity: Assessment of Drought Impact on Rice Production in Indonesia by Satellite Remote Sensing and Dissemination with Web-GIS*
  - **Best Student Paper Award:** Rosnani Rahman, Space Science Centre (ANGKASA), Institute of Climate Change, Universiti Kebangsaan Malaysia (UKM), Malaysia: *Monitoring the Variability of Precipitable Water Vapor Over the Klang Valley, Malaysia During Flash Flood*

The success of the IGRSM 2014 was due to commitments of many: authors, keynote speakers, session chairpersons, the organising and technical programme committees, student volunteers from Universiti Putra Malaysia (UPM), and many others of various roles. We acknowledge the sponsors of IGRSM 2014, namely Antaragrafik Systems Sdn. Bhd. and Geospatial Media and Communications Sdn. Bhd. We also thank all exhibitors and contributors: E J Motiwalla, Fajar Saintifik Sdn. Bhd., Bandwork GPS Solutions Sdn. Bhd., Tenaga Nasional Bhd., TSKAY Technology Sdn. Bhd., Geo Spatial Solutions Sdn. Bhd. and Accutac Sdn. Bhd.

Associate Professor Sr Dr Abdul Rashid Mohamed Shariff

Chairman
7th IGRSM International Remote Sensing & GIS Conference and Exhibition (IGRSM2014)

President
Institution of Geospatial and Remote Sensing Malaysia (IGRSM), 2012–2014

IOP Conf. Series: Earth and Environmental Science **20** (2014) 011002 doi:10.1088/1755-1315/20/1/011002

Organising Committee

Chairman

Associate Professor Sr Dr Abdul Rashid Mohamed Shariff

Secretariat:

Dr Ahmad Fikri Abdullah
Dr Farrah Melissa Muharram
Normalina Jamaluddin
Siti Nooradzah Adam

Members

Professor Dr Shattri Mansor, Universiti Putra Malaysia (UPM)
Professor Sr Dr Mazlan Bin Hashim, Universiti Teknologi Malaysia (UTM)
Professor Dr Ruslan Ruainis, Universiti Sains Malaysia (USM)
Associate Professor Dr Norzailawati Mohd Nor, International Islamic University Malaysia (IIUM)
Associate Professor Dr Biswajeet Pradhan, Universiti Putra Malaysia (UPM)
Associate Professor Dr Helmi Zulhaidi Mohd Shafri, Universiti Putra Malaysia (UPM)
Associate Professor Dr Siva K Balasundram, Universiti Putra Malaysia (UPM)
Dr Muhamad Radzali Mispan, Malaysian Agricultural Research and Development Institute (MARDI)
Dr Siti Khairunnisa Bejo, Universiti Putra Malaysia (UPM)
Dr Mohd Johari Mohd Yusof, Universiti Putra Malaysia (UPM)
Dr Ali Ariapour, Islamic Azad University
Dr Eran Sadek Said Md Sadek, Universiti Teknologi MARA (UiTM)
Dr Dinesh Sathyamoorthy, Science & Technology Research Institute for Defence (STRIDE)
Sr Mokhtar Azizi Mohd Din, Universiti Malaya
Noor Azawani Wahap, National Space Agency
Mohd Azahari Faidi, Forest Research Institute Malaysia (FRIM)
Abdul Munir Hafizy Ladoni, Universiti Malaysia Sabah (UMS)
Kay Thoan San, TSKAY Technology Sdn Bhd
Kumar Veliayudam, Geo Spatial Solutions Sdn Bhd
Muhammad Zakri Tarmidi, Universiti Putra Malaysia (UPM)
Zailani Khuzaimah, Universiti Putra Malaysia (UPM)
Saleh Abdullahi, Universiti Putra Malaysia (UPM)
Mohd Amiruddin, Universiti Putra Malaysia (UPM)
Siti Sarah Emran, Universiti Putra Malaysia (UPM)
Sarah Hanim Samsudin, Universiti Putra Malaysia (UPM)
Maizatul Akma Mohamad, Universiti Putra Malaysia (UPM)

Content from this work may be used under the terms of the Creative Commons Attribution 3.0 licence. Any further distribution of this work must maintain attribution to the author(s) and the title of the work, journal citation and DOI.
Published under licence by IOP Publishing Ltd

IOP Conf. Series: Earth and Environmental Science **20** (2014) 011003 doi:10.1088/1755-1315/20/1/011003

Technical Programme Committee

Chairpersons

Dr Dinesh Sathyamoorthy, Science & Technology Research Institute for Defence (STRIDE), Ministry of Defence, Malaysia
Associate Professor Sr Dr Abdul Rashid Mohamed Shariff, Universiti Putra Malaysia (UPM), Malaysia
Dr Ahmad Fikri Abdullah, Universiti Putra Malaysia (UPM), Malaysia
Dr Farrah Melissa Muharram, Universiti Putra Malaysia (UPM), Malaysia

Members

Professor Dr Li Jing, Beijing Normal University, China
Professor Dr Iyyanki Muralikrishna, Administrative Staff College of India (ASCI), India
Professor Dr Alias Abdul Rahman, Universiti Teknologi Malaysia (UTM), Malaysia
Professor Dr Ismat Mohamed El Hassan, King Saud University, Saudi Arabia
Professor Dr George Miliaresis, Open University of Cyprus, Cyprus
Professor Dr Christine Pohl, Universiti Teknologi Malaysia (UTM), Malaysia
Professor Dr Mahender Kotha, Goa University, India
Associate Professor Dr Paolo Gamba, University of Pavia, Italy
Associate Professor Dr Behara Seshadri Daya Sagar, Indian Statistical Institute (ISI), India
Associate Professor Sr Ranjit Singh, Infrastructure University Kuala Lumpur (IUKL), Malaysia
Associate Professor Dr Abdul Nasir Matori, Universiti Teknologi Petronas (UTP), Malaysia
Associate Dr Lucian Dragut, West University of Timişoara, Romania
Associate Professor Dr Saied Pirasteh, Islamic Azad University, Iran
Associate Professor Dr Peter Yuen, Cranfield University, United Kingdom
Associate Professor Dr Lim Hwee San, Universiti Sains Malaysia (USM), Malaysia
Associate Professor Dr Wayan Suparta, Universiti Kebangsaan Malaysia (UKM), Malaysia
Associate Professor Dr Tuong Thuy Vu, The University of Nottingham Malaysia Campus, Malaysia
Associate Professor Dr Maged Mahmoud Marghany, Universiti Teknologi Malaysia (UTM), Malaysia
Associate Professor Dr Rami Al-Ruzouq, University of Sharjah, UAE
Associate Professor Dr Biswajeet Pradhan, Universiti Putra Malaysia (UPM), Malaysia
Associate Professor Dr Helmi Zulhaidi Mohd Shafri, Universiti Putra Malaysia (UPM), Malaysia
Associate Professor Dr Benny Peter, Universiti Teknologi Malaysia (UTM), Malaysia
Associate Professor Dr Norzailawati Mohd Nor, International Islamic University Malaysia (IIUM), Malaysia
Dr Josée Lévesque, Defence Research and Development Canada (DRDC), Canada
Dr Ali Ariapour, Islamic Azad University, Iran
Dr Zulkiflee Abd Latif, Universiti Teknologi MARA (UiTM), Malaysia
Dr Lim Tien Sze, Multimedia University (MMU), Malaysia
Dr Ruzinoor Che Mat, Universiti Utara Malaysia (UUM), Malaysia

Content from this work may be used under the terms of the Creative Commons Attribution 3.0 licence. Any further distribution of this work must maintain attribution to the author(s) and the title of the work, journal citation and DOI.
Published under licence by IOP Publishing Ltd

Dr Eran Sadek Said Md Sadek, Universiti Teknologi MARA (UiTM), Malaysia
Dr Siti Khairunniza Bejo, Universiti Putra Malaysia (UPM), Malaysia
Dr Ramin Nourqolipour, National Organization of Forest, Range and Watershed Management, Iran
Sr Mohktar Azizi Mohd Din, Universiti Malaya, Malaysia
Col (Rt) Frederic Hernoust, Magellium, France
Marzieh Mokarram, University of Isfahan, Iran
Mohd Fadhil Abuhan, Royal Malaysian Police, Malaysia

Peer review statement

All papers published in this volume of *IOP Conference Series: Earth and Environmental Science* have been peer reviewed through processes administered by the proceedings Editors. Reviews were conducted by expert referees to the professional and scientific standards expected of a proceedings journal published by IOP Publishing.

Content from this work may be used under the terms of the Creative Commons Attribution 3.0 licence. Any further distribution of this work must maintain attribution to the author(s) and the title of the work, journal citation and DOI.
Published under licence by IOP Publishing Ltd

IOP Conf. Series: Earth and Environmental Science **20** (2014) 012001 doi:10.1088/1755-1315/20/1/012001

Assessment of water resource potential for common use of cow and goat by GIS (Case study: Boroujerd Rangeland, Sarab Sefid, Iran)

A Ariapour[1], K Karami[2], A Sadr[3]
[1]Department of Natural Resource, College of Range Management, Boroujerd Branch, Islamic Azad University, Boroujerd, Iran Iran (Corresponding author).
[2]M.Sc. Student, Department of Rangeland Management, Young Researchers Club, Boroujerd Branch, Islamic Azad University, Boroujerd, Iran.
[3]M.A. Student, Islamic Azad University, Boroujerd Branch, Boroujerd, Lorestan, Iran
E-mail: aariapour@yahoo.com

Abstract. One of the most important factors to sustainability utilization of natural potential by rangeland grazing suitability is water resources suitability. This study is a model for quantitative, qualitative and spatial distance assessment of water resource's propriety for goat and cow grazing based on geographic information systems (GIS) in Boroujerd Sarab Sefid rangeland, Lorestan province, Iran 2013. In this research from combining three factors such as quantity, quality and water resource's distances; the final model of degree of propriety of water resources for goat and cow grazing is characterized. Results showed that slope factor was the reason of limitation, and it is considered as a limiting factor in propriety of water resources, so in terms of access to water resources for goat grazing, 4856.4 ha (100%) located in S1 classes and for cow grazing, 4023.14 ha (68.6%) located in S1(suitability) classes, 1,187 ha (20.24%) in S2 classes and 654.8 ha (11.16%) located in S3 classes, respectively for both. So according to the results the rangelands in this region are most suitable for goat because of terrain and weather but this, in combination with, cow hasbandry will allow diversity of economic production and stability of incomes.

1. Introduction

Correct management in rangeland ecosystem needs recognizing water, soil and plant resources, which are basic resources for production. So, one of the most important factors in analysis and evaluation of rangelands is utilization based on those potential and abilities. Recognition of factors affects those, and has crucial importance to desirable suitable management of rangeland. One of the main problems of developing countries such as Iran is utilization of natural resources area without ecological situation and its potentials consequently destruction of soil, water and plant as the most significant basis of suitable production. Most areas of Zagros Mountains include rangelands which most of them have not suitable quality and quantity forage, also access to water resources, cause more erosion for requirement of providing water and forage of livestock. These factors affected by rangeland utilization history [25]. FAO guided a standard evaluation system for assessment of land. In 1972 its background prepared and in the next year the first format was written, then final format of land evaluation issued in 1976. Subsequent issued guides of land for different land usages such as dry farming [11], forestry [9], dry land farming cultivation [12] and expanded grazing evaluated [10]. First of all need to suitable uses and land resources balancing, more information and its utility to different systems relevant to the earth such as natural resources in the FAO guides then nature of dynamic and those changeable, cause human to use new methods and electronically instruments [22]. RS and GIS methods are the new

Content from this work may be used under the terms of the Creative Commons Attribution 3.0 licence. Any further distribution of this work must maintain attribution to the author(s) and the title of the work, journal citation and DOI.
Published under licence by IOP Publishing Ltd

sciences as well as nano technology that are vastly used in natural science [23]. Suitable utilization of rangeland need to recognize its parameters, therefore determination of rangeland suitability is one of the most significant factors and more difficult to rangeland analysis [26]. 70 percent of Iranian livestock rely on rangeland, so for better using of rangeland potential, it is needed to investigate the limitation and none limitation of animal grazing in rangeland suitable determination which one of them is water resource, therefore Boroujerd Sarab Sefid region is selected according to more number of livestock like cow and goat. Besides in using rangeland forage for grazing of livestock there must be enough water to drink during grazing season that causes maximizes using of forage by livestock for producing animal products [29]. In some cases there is enough water but its propagation is not good, therefore animal cannot use all of the forage in rangeland [28]. In fact it is important to provide enough water for drinking livestock and wildlife. As a general rule one of the effective factors in site selection of grazing is water distance [20]. Livestock and wildlife often more graze plants in the vicinity of water [7]. When water is the most significant limited factor for using forage, it affects on animal movement, so it is important for management of rangeland to find distance of animal to water resource which must cover the distance and amount of forage uses in the course of movement [30]. Generally access ability to water resource depends on maximum distance where livestock can be far away from water resource to forage grazing. The distance depends on topography, utilization season, age and kind of animal and plant coverage [3]. Several factors affect on water consumption such as kind, age and breed, topography, quality and quantity of forage accessibility, grazing season and distance from water [6]. Water with more salinity or with poisonous elements may endanger animal health or make meat and milk uneatable. Also water with unsuitable quality affects on forage productivity. To determine usable water we must consider area situation, age and breed of animal and nutrition composition [21]. In fact animals prefer to drink fresh water [25].
The aim of this study was conducted to preparing water resource sustainable model to cow and goat based on FAO method (1991).

2. Materials and Methods

2.1. Study area

Sarab sefid rangeland under study is located in west of Boroujerd city in Lorestan province, Iran (48° 27′ 46″ to 48° 36′ 30″ E and 31° 53′ 33″ to 33° 58′ 24″ N) comprises of 5,864 ha (58.6 Km2) of which 3.78% of Boroujerd township (Figure 1). The average mean of annual precipitation (20 years) of the area is 450.9 mm, falling mainly in the autumn and winter. The average minimum and maximum temperatures are 11.5ºC and 39.2ºC, respectively. Its climate based on Henry Pabo (Senior expert of FAO) is located in Iran-Toran (High Mountain Region). The mean height at the sea level is 2,744 m and the minimum and maximum are 1,947 and 3,451 m, respectively.

2.2. Water resources of the area

Permanent river in the basin is Sarab Sefid, that is categorized too many main secondary rivers. Its direction is from south west to north. Seasonal springs in the area are so many. To determine contour distance map from water resources first, distance from water resources map was prepared (Figure 2).

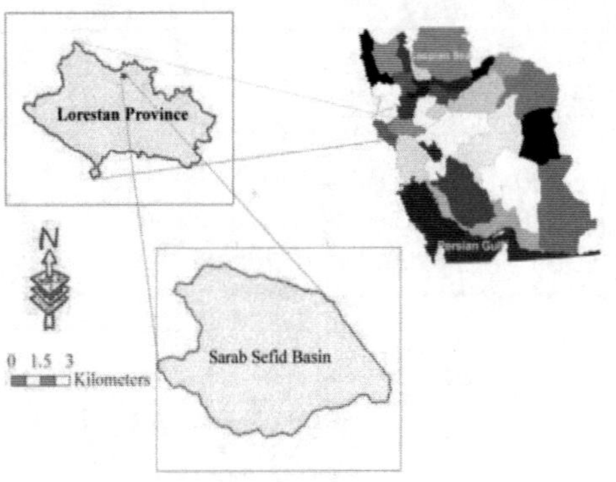

Figure 1. Location of the study area, Sarab Sefid basin.

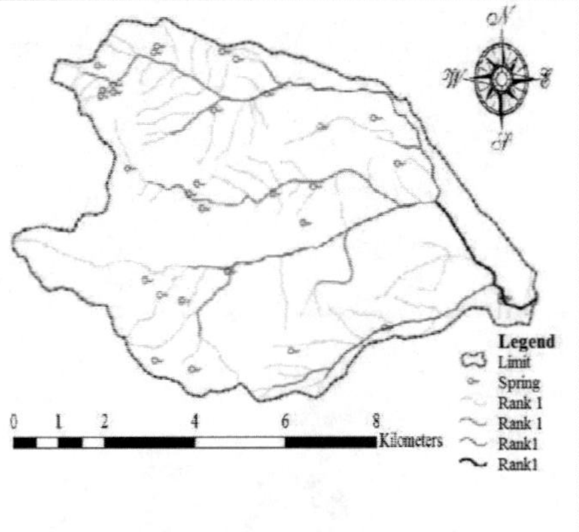

Figure 2. The map of basin hydrological networks and springs.

2.3. Criteria of water resources suitability

Three sub models such as: distance, quality and quantity were used (Figure 3). According to these sub models in each types of rangeland and combine them together, water resource suitability in the area for goats and cows grazing determined.

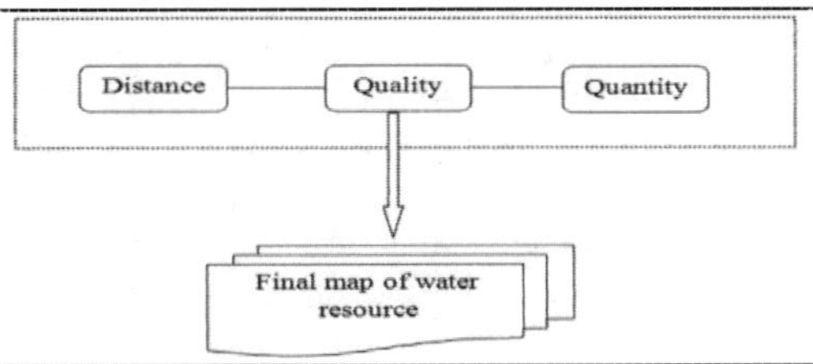

Figure 3. The final model of water resources suitability for goat and cow grazing.

2.3.1. Water accessibility sub-model

The slope map of the study areas was classified and extracted by using ArcGIS®9.3. Overlaying both maps (slope and distance) led to the final water accessibility model. The distance from water resources suitability classes in livestock (goats and cows) usage are illustrated in 'table 1-2 and (Figure 4-5).

Table 1. Water resources distances (m) and its suitability for grazing cow classes.

Suitability class	Slope class (%)			
	0-15	15-30	30-45	>45
S1	>3000	>2000	>800	N
S2	3000-4000	2000-3000	800-1500	N
S3	4000-5000	3000-4000	1500-2200	N
N	>5000	>4000	>2200	N

Table 2. Water Resources distances (m) and its suitability for grazing goats classes.

Suitability class	Slope class (%)			
	0-15	15-40	40-75	>75
S1	0-4200	0-3700	0-1200	N
S2	4200-6200	3700-6000	1200-4500	N
S3	6200-8000	6000-7500	4500-5100	N
N	>8000	>7500	>5100	N

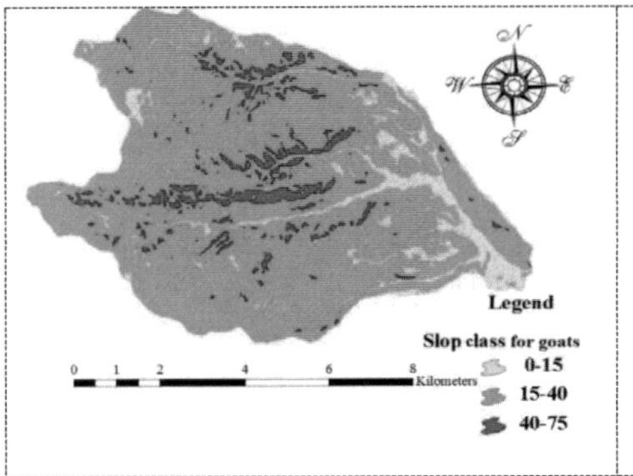

Figure 4. Slope class for goats.

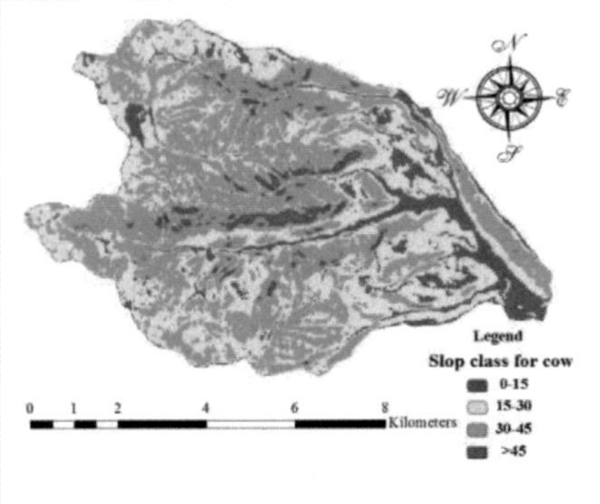

Figure 5. Slope class for cows.

2.3.2. Water quantity sub-model

Water resources were determined and summed up within each types of plant boundary to calculate water availability. For each livestock water demand with available water indicates the results in the water quantity suitability sub-model. Based on climatic conditions, vegetation characteristics, grazing season and animal type such as goat and cow water demand were estimated. Categories were determined by comparison of the available water with the water needed by the livestock (Table 3).

Table 3. Water resource suitability classes.

Available water in pasture ration to livestock need (%)	>76	51-75	26-50	<25
Suitability classes	S1	S2	S3	N

2.3.3. Water quality sub-model

Water quality data of water resources such as; [Acidity (pH), Electrically conductive (EC), Total Dissolved Salts (TDS), Sodium (Na), Claire (Cl), Bicarbonate (HCO3-), Magnesium (Mg), Sulphate (SO4), Calcium (Ca), Total Hardness (TH), Sodium absorption ratio (S.A.R), Potassium (K+)] were acquired from local offices, Boroujerd water management and other researches and compared with standards to determine water quality suitability.

3. Results

Based on previous studies and field experiences, three limiting conditions for grazing (FAO 1991) were taken into account.

Results show that based on the water resources quality and considering the water quality, there were no limitation in the range area in question, and the whole range area fell within the S1 suitability category. Also results show that Total Hardness (TH) based on standard is good for cows and goats. These results are similar to Cl, pH (7/8), NO3 (4/4), EC (235mimhos per cm), TDS (140/8 mg/lit) and other factors except SO4. For the last factor (SO4) there is a little limitation for drinking by goat and cow. The results revealed that there were no limitations for the amount of water in the region, so all of

them fell into the S1 suitability category, because of high precipitation of the region which is more than 700 mm per year (annually) and it has good intensity during year. In mountain and high elevations precipitations are most as snow and cause to save it, and its result is producing spring in the basin which is suitable water quantity for cows and goats local breed. In this basin there is enough water during grazing season for livestock and wildlife and it is more than water needed based on determined grazing capacity.
The results of the sub-model on the distance from water resources suitability for goats grazing revealed that 5864.8 ha (100%) located in S1 classes and about cow grazing, 4023.14 ha (68.6%) located in S1 classes, 1187 ha (20.24%) in S2 classes and 654.8 ha (11.16%) located in S3 classes, respectively (Figure 6-7) and Table 4'.

Table 4. Area of categorize of distance form water resources suitability for cows and goats.

Categorize of suitability		Area (ha)	Area (%)
For goats	S1	5864.8	100
For cows	S1	4023	68.6
	S2	1187	20.24
	S3	654.8	11.16

Final water resources suitability
Results show that there is no problems regarding the quantity and quality of the water resources; it was only the distance from the resources that mainly determined the suitability of the rangeland with respect to water resources for cow (Figure 7).

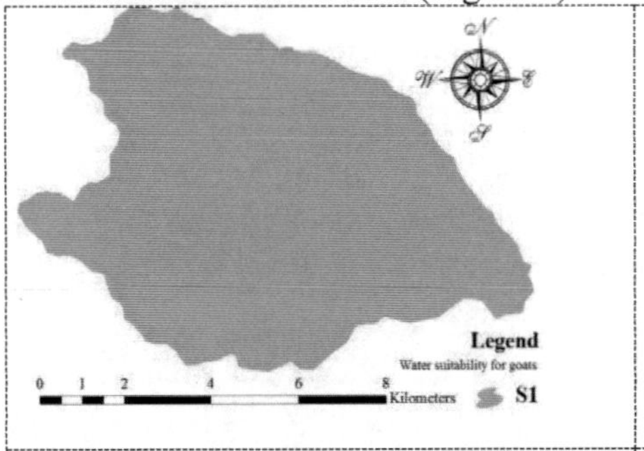

Figure 6. The model of water resource suitability for goats.

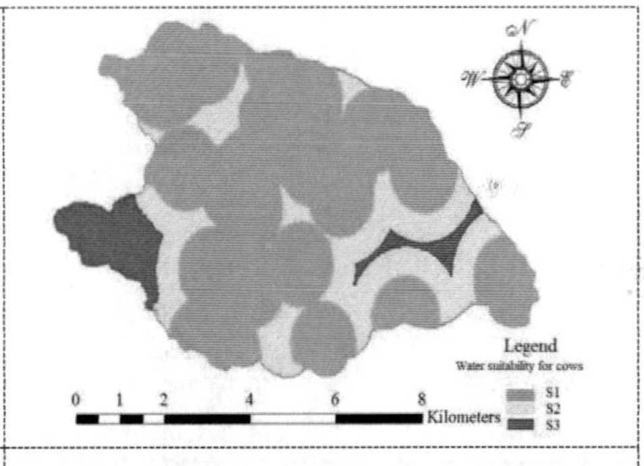

Figure 7. The model of water resource suitability for cows (final model).

4. Discussion
Iran is one of the dry countries that medium precipitation in annual is about 230mm, so water is important for agriculture and renewable resource management. Zagros Mountains chain is main resource for animal husbandry. Keit (2000) showed that for determining water resource suitability for cattle grazing assessment of two factors is needed such as slope, number of water resources, steep slope and suitable distance from water resource. Result of present study was similar to Keit research in which his study pointed out to significant factors to suitability. In studies of Jangjoo Borzelabadi (1996), Mohtashamnia (2000), Yoosefi khanghah (2004), Arzani and Yoosefi (2006), for determining animal grazing suitability carried out three factors such as forage production, water resource and soil sensibility based on FAO (1991) method that the present research path same method and showed the same results. Minor (2002) for determining rangeland grazing capacity of Fergosen-California region used RS and GIS abilities. He applied three sub-models such as plant coverage, slope and precipitation

for determination of final model of grazing capacity and declared that the results by GIS had acceptable accuracy for management of rangeland. The results are based on Paul Tueller and Reno Nevada (2001) opinion in for using RS technique to investigate rangeland forage production. The results of the study showed that the quantity (number of permanent water resources), quality and the distance from the water resources did not impose many limitations on the rangelands suitability for grazing livestock. However, the steep slopes along the livestock path to the water resources resulted in the formation of an 'unsuitability' category for livestock. Valentine (2001) reported on the importance of the slope factor in reaching the water resources, and declared that by increasing the slope the ability to graze decreases and increases the livestock demand to expend lots of energy. The quality and quantity of the water resources in the rangeland did not impose any limitations. The result of the research indicates the slope as the reducing and sometimes limiting factor in the range suitability. Hence, the slope factor is of considerable importance in determining the suitability of the pasture for grazing. As slope increases the water retention time on the ground decreases, the rate of penetration decreases, and the amount of water run-off increases. Cook (1954) explained that on slopes of more than 60 degrees little forage is grazed. Amiri (2009) and Gavili *et al.,* (2011) defined the slopes with more than 60 percent as useless for all kinds of livestock, while [17] reported slopes of more than 60 percent, and Arzani *et al.,* (2006) defined slopes of more than 60 percent as useless for livestock grazing. On such steep slopes wild animals would graze better than livestock. Guenther *et al.,* (2000) in determining the suitability of a rangeland in Australia noted the two factors of slope and water resources as the suitability limiting factors of rangeland for grazing cattle. Due to the existence of numerous permanent water resources in Sarab Sefid rangelands, the water resources factor does not impose many limitations on the suitability of the rangeland. However, the slope factor in reaching the water resources in limited areas of the rangeland was a suitability limiting factor. It should be noted that the present study was similar to results reported by [16]. Finally should told that water is first factor to development of livestock husbandry that the study area is most suitable for goat.

5. Reference

[1] Aghafashami Mohseni M 2002 Determination of rangeland suitability in Lare region by GIS M.Sc. Thesis, Natural resource faculty, Tarrbiat Moddares university (In Persian)

[2] Amiri F 2009 A GIS model for determination of water resources suitability for goats grazing *African Jour. of Agri. Re.* **4(1):** 014-020

[3] Arnold G W and Dudzinaski M L 1995 *Ethology of Free Ranging Domestic Animals 1edn* Amsterdam: Elsevier Sci. Publ. Co, USA. 51-55 pp

[4] Arzani H, Jangjo M, Shams H, Mohtashamnia S, Fashami M, Ahmadi H, Jafari M, Darvishsefat A and Shahriary E 2006 A model for classification of range suitability for sheep grazing in Central Alborz, Ardestan and Zagros regions *Jour. Sci. and Tech. of Agri. and Natural Res.* **10(1):** 273-290

[5] Arzani H and Yousefi S 2006 A GIS model of range suitability assessment for sheep grazing (Case study Taleghan Region in Tehran Province). 8th International Conference on: Information Systems in Sustainable Agriculture, Agro-environment and Food Technology, 20-23 September Thessaly 911-918 pp

[6] Bagley C V, Amacher J K and Kitt F P 1997 *Analysis of water Quality for livestock. Utah state Extension, Animal Helth Fact sheet, Utah State University,* Logan UT 84322-5600 Electronic Publishing 7 pp

[7] Bailey D M 2004 Management Strategies for Optimal Grazing Distribution and Use of Arid Rangelands *Jour. Anim. Sci.* 82 (E. Suppl.): E147–E153

[8] Cook CW 1954 Common use of summer range by sheep and cattle *Jour. Range Manag.* **7(1):**10-13

[9] FAO 1976 *A framework for land evaluation* Food and Agriculture Organization of the United Nations, Soils Bulletin 32 FAO Rome

[10] FAO 1991 *Gudelines: Land evaluation for extensive grazing* FAO Soils Bulletin No. 58 FAO Rome, Italy 170 pp

[11] FAO 1993 *Gudeline for land use planning* FAO Development Series No: 1 FAO, Rome, 96 pp
[12] FAO 2002 *Agricultural drainage water management in arid and semi-arid areas* Food and Agriculture Organization of United Nations, Rome
[13] FAO 2007 *Land evaluation Towards a revised framework Food and Agriculture Organization of the United Nations* Rome, Italy
[14] FAO-UNDP and UNEP 2007 *Land degradation in south Asia: its severity, causes and effects upon the pepole* World Soil Resources Report No. 78, FAO, Rome, Italy 100 pp
[15] Gavili E, Ghasriani F, Arzani H, Vahabi M and Amiri F 2011 Determine water resources accessibility for sheep grazing by GIS technology (Case study: Feraidun Shahr rangeland in Isfahan Province) *Jour. Applied RS and GIS Techno. in Natural Resource Sci.* **1(1):** 89-99
[16] Guenther KS, Guenther GE and Redick PS 2000 *Expected-use GIS maps* Rangelands **22(2):**18-20.
[17] Holechek J L, Pieper RD and Herbel CH 1995 *Range management* principles and practices 2 nd edn. Prentice-Hall, USA
[18] Jangjoo Borzolabad M 1996 *Determination of rangeland suitability by using GIS* M.Sc. Thesis, natural resource faculty, Tehran university, Iran
[19] Kiet S 2000 *Expected use GIS map*, Rangeland, **22(2)**: 18-20
[20] Lotfollahzadeh D 1999 *Relation between the spatial and temporal patterns of rangeland condition / flock movement and soil properties* MSc thesis, ITC, Enschede, The Netherlands
[21] Mahdavi M 1999 *Application hydrology* Second edition, Tehran university press, Iran Pages: 400
[22] Makhdom M 2001 *Evaluation of environment by GIS* First edition, Tehran university press, Iran, Pages 298
[23] Malczwski J 1990 *GIS and Multi criteria decision analysis,* John Wiley and sons Ins. Canada
[24] Minor T 2002 *Range Land Evaluation in Freson County, California, based on livestock carrying capacity modeling using GIS*, Proje : County of freson, California and Agricast, Inc.
[25] Moghadam M R 1998 *Range management* First edition, Tehran University Press, Pages 470
[26] Mohtashamnia S 2000 *Determination of rangeland suitability in semi steppe region of Fars province by GIS.* M.Sc. Thesis, natural resource faculty, Tarrbiat Moddares university, Tehran, Iran
[27] Paul T, Tueller and Reno Nevada 2001 Remote Sensing of Range Production and Utilization. *Jour. Range Manage.* **54:** A77-A89
[28] Scarlett T 2002 *Drought feeding of goats* New South Wales Agriculture: Sydney. http//www. agric. nsw. gov. u/reader/5337 22/11/02
[29] Schacht W H, Volesky J D and Waller S S 2003 *Proper Livestock Grazing Distribution on Range Land. Published by Cooperative Extension*, Institute of Agriculture and Natural Resources, University of Nebraska-Lincoln
[30] Vallentine J F 2001 *Grazing Management* Academic Press, San Diego, CA.

Modeling groundwater vulnerability to pollution using Optimized DRASTIC model

Kehinde Anthony Mogaji[1, 2], Hwee San Lim[2], Khiruddin Abdullar[2]

[1]Department of Applied Geophysics, Federal University of Technology, P.M.B. 704, Akure, Nigeria.
[2] School of Physics, Universiti Sains Malaysia, 11800 Penang, Malaysia.

E-mails: mogakeh@yahoo.com, hslim@usm.my, khirudd@usm.my

Abstract. The prediction accuracy of the conventional DRASTIC model (CDM) algorithm for groundwater vulnerability assessment is severely limited by the inherent subjectivity and uncertainty in the integration of data obtained from various sources. This study attempts to overcome these problems by exploring the potential of the analytic hierarchy process (AHP) technique as a decision support model to optimize the CDM algorithm. The AHP technique was utilized to compute the normalized weights for the seven parameters of the CDM to generate an optimized DRASTIC model (ODM) algorithm. The DRASTIC parameters integrated with the ODM algorithm predicted which among the study areas is more likely to become contaminated as a result of activities at or near the land surface potential. Five vulnerability zones, namely: no vulnerable(NV), very low vulnerable (VLV), low vulnerable (LV), moderate vulnerable (MV) and high vulnerable (HV) were identified based on the vulnerability index values estimated with the ODM algorithm. Results show that more than 50 % of the area belongs to both moderate and high vulnerable zones on the account of the spatial analysis of the produced ODM-based groundwater vulnerability prediction map (GVPM).The prediction accuracy of the ODM-based - GVPM with the groundwater pH and manganese (Mn) concentrations established correlation factors (CRs) result of 90 % and 86 % compared to the CRs result of 62 % and 50 % obtained for the validation accuracy of the CDM - based GVPM. The comparative results, indicated that the ODM-based produced GVPM is more reliable than the CDM - based produced GVPM in the study area. The study established the efficacy of AHP as a spatial decision support technique in enhancing environmental decision making with particular reference to future groundwater vulnerability assessment.

1. Introduction

Vulnerability is the degree to which human or environmental systems are likely to experience harm because of perturbation or stress. According to Pospecu *et al* [1] vulnerability can be identified for a specific system, hazard, or group of hazards. Vulnerability mapping entails quantifying the sensitivity of resources to the environment and is a practical visualization tool for decision making Rahman [2], Norlan [3]. In hydrogeology, vulnerability assessment is typically utilized to assess the susceptibility of a water table, an aquifer, or water well to contaminants that can reduce groundwater quality. Groundwater pollution is the artificially induced degradation of natural groundwater quality. It is primarily caused by anthropogenic and agricultural activities [4], Nobre *et al* [5], AtqurRahman [6]. These potential sources of pollution are responsible for the deterioration of groundwater quality, consequently increasing vulnerability. According to Thirumalaivasan *et al* [7], groundwater contamination has become one of the most serious global environmental problems because once an aquifer is polluted; its remediation is

Content from this work may be used under the terms of the Creative Commons Attribution 3.0 licence. Any further distribution of this work must maintain attribution to the author(s) and the title of the work, journal citation and DOI.
Published under licence by IOP Publishing Ltd

prohibitively costly and often impractical. The development of sustainable plans geared toward groundwater protection against contamination is therefore critical in water resources planning and management Chen, *et al* [8], Said Ettazarini [9]. The existing methods of assessing groundwater vulnerability are classified into three categories: i) overlay and index method; ii) process-based methods that apply deterministic models based on physical processes; and iii) statistical models (National Research Council, 1993). However, this article focuses on overlay and index methods for groundwater vulnerability assessment. The overlay and index methods which emphasized on the combination of different regional maps by allocating a numerical index includes GOD, AVI and DRATIC. The most widely used among is the DRASTIC method Aller *et al* [10], Durnford *et al* [11], [12-13], Rundquist *et al* [14], Fritch *et al* [15]. However, the DRASTIC modeling technique disregards the effect of regional characteristic; hence uniform weights and rating are mostly used. The inappropriate handling of parameters rating and weighting assignment in this model often result into series of subjectivity and uncertainty in prediction. Thus several researchers have continued to enhance the model with different methods. However, fewer attempts have been made to enhance this model performance through a modification that is based upon simple statistical procedure involving the revision of the rating scale of each parameter and revision of the factor weights as proposed by Panagopoulos *et al* [16]. For reliability and precision of the assessment of pollution potential in this study, the conventional DRASTIC model (CDM) is optimized by incorporating a multi-criteria decision analysis technique.

An optimized DRASTIC model (ODM) is proposed in this study to establish a groundwater vulnerability model with high reliability and precision. Optimization is performed by normalizing the rating scale of each parameter and revising factor weights via a hybrid approach of the DRASTIC model and Analytical hierarchy process (AHP). The AHP technique is employed as a decision support model. This technique is utilized for the elicitation of attributes weights of the model parameters for the purpose of resolving the limitations of the convectional DRASTIC modeling approach in terms of the inherent subjectivity and uncertainty in the integration of data obtained from various sources.
Though, there is no record of groundwater pollution in the area investigated in this study, however, assessing groundwater quality and developing strategies to protect aquifers from future contamination are necessary in planning and designing water resources in the area. Therefore, the objectives of the study are: (1) to produce a groundwater vulnerability prediction map with high reliability and precision through application of the proposed optimized DRASTIC model (CDM) algorithm to the DRASTIC model parameters; and (2) to compare the results of the proposed CDM algorithm with the CDM algorithm via validation of the predictive models with the analyzed results of groundwater quality obtained from the study area.

2. Materials and methods

2.1. *Geography, hydrology, and hydrogeology of the study area*
The proposed methodology was applied in a site situated between the boundary of Perak and Selango in Peninsular Malaysia. Figure 1 presents the 2863 km^2 study area and shows the borehole wells and other important features. The site lies between longitudes 101° 0' E and 101° 40'E and between latitudes 3°37' N and 4°18'N in the southern part of Perak. The area is having four major underlying rock types defined by geological age, namely, Quaternary (mainly recent alluvium), Devonian (sedimentary rocks), Silurian (sedimentary rocks with lava and tuff), and acidic and undifferentiated granitoids as depicted in figure 2. The region is characterized by equatorial maritime climate with nearly uniform air temperatures throughout the year. The average daily temperature is approximately 27 °C, and the relative humidity has a monthly mean value of 62% and 78% for the dry period and peak of the rainy season, respectively. The

general annual precipitation in Perak state ranges from 830 mm to 3,000 mm. However, 10years daily records of precipitation between 2000 and 2010 for the study area were acquired at quarter-degree resolution (0.25°) from the Tropical Rainfall Measuring Station (*TRMM*) database.

2.2. *Analytic Hierarchy Process (AHP)*

AHP is one of the most commonly utilized multi-criteria decision methods reported in literature [17]. Multi-criteria decision methods involve the assessment of a given alternatives by a group of decision makers based on a selected set of criteria. AHP is a decision support model that requires the summation of certain weights on a given level of the decision Beynon *et al* [18]. AHP involves the construction of a series of pair-wise comparison matrices, which compare the criteria to one another. Comparison is performed to estimate a rating or weight for each criteria; this rating describes the extent of the contribution of each criteria to the overall objective. Saaty's scale of 1 to 9 [19] plays a key role in the implementation of AHP. The scale is coupled with the experience and knowledge of experts or users to determine the factors or criteria affecting the decision process HO [20], [21]. AHP is capable of capturing both subjective and objective evaluation measures and providing a useful mechanism to check the consistency of the evaluation measures and alternatives suggested by experts or decision makers, thereby reducing bias in decision making Ariff *et al* [22].

2.3 *DRASTIC model*

The DRASTIC model was developed for the US Environmental Protection Agency by Aller *et a*l [10] from the National Water Well Association. The model was designed as an easy-to-use model that would allow a user with basic knowledge of hydrogeology to assess the relative potential for groundwater contamination. The term "DRASTIC" was derived from the seven variables in the model, namely, depth-to-water table, net recharge, aquifer media, soil media, topography, impact of vadose zone, and conductivity (hydraulic). These variables or parameters are utilized to define the hydrogeologic setting of an area. These parameters are further subdivided into ranges or zones that represent various hydrogeologic settings and are assigned different ratings on a scale of 1 to 10. The rating assigned to each range or zone indicates the relative importance of each parameter within the zone in contributing to aquifer vulnerability. The linear additive of the above mentioned variables is usually combined with ratings and weights to determine the DRASTIC index Aller *et a*l 10]. DRASTIC index (DI) is expressed as follows:

$$\mathrm{DI} = D_r D_w + R_r R_w + A_r A_w + S_r S_w + T_r T_w + I_r I_w + C_r C_w \tag{1}$$

Where *D*, *R*, *A*, *S*, *T*, *I*, and *C* represent the seven parameters, r is the rating, and w is the weight assigned to each parameter. The resulting index is a relative measure of vulnerability to contamination. Shown in figure 3 is the DRASTIC model evaluation approach and the optimization procedures

2.4 *Data preparation for the DRASTIC model*

The DRASTIC model with its seven parameters was derived from several data sets. This study utilized several data types, and the sources are presented in Table 1. The depth-to-water table (D) is one of the most important factors in evaluating pollution potential because it provides a greater chance for natural attenuation in terms of dispersion, absorption, and biodegradation to occur as the depth to water increases [23]. This study considered data determined based on water level measurements performed in several bore wells drilled in the year 2009 and uniformly spread in the entire study area to produce a depth-to-water table thematic map (Figure 4a). This map was evaluated for pollution potential mapping based on the ranges of depth-to-water table distribution in the area. The hydrological implication of the map indicated that at a shallower water table, the more susceptibility is the groundwater reservoir to pollution and vice-versa [4]. Hence, the pollution potential weights and ratings of depth to water thematic map classes are in accordance with Delphi technique Atiqur Rahman [6].

This study adopted 10 years daily records of precipitation between 2000 and 2010 to estimate net recharge in the area using the modified version of the Chaturvedi [24] formula (Equation 2)

$$R = 6.75(P-14)^{0.5} \tag{2}$$

where R is the net recharge from precipitation during the year, and P is annual precipitation.

The obtainable net recharge values were interpolated to generate net recharge (R) map (Figure 4b) in GIS environment. The net recharge represents the annual average amount of water that infiltrates the vadose zone and reaches the water table Aller *et al* [10]. The higher the net recharge, the more vulnerable is the groundwater reservoir. This was supported in Samake *et al* [25] who reported that recharge controls the volume of water available for dispersion and dilution of the contaminant in the vadose zone. The aquifer medium (A) is a water-bearing formation that yields useable quantities of water from a well or spring Heath [25]. The nature of soil composition constituting this medium determines the attenuation capacity of pollutants. The aquifer layer delineated from 2D resistivity imaging section was interpreted and for geoelctrical parameter determination (sees Figure 5). The estimated mean resistivity values of the delineated aquifer media was evaluated for the nature, texture, porosity, permeability properties. These properties play important roles in the filtering ability of this media. The attenuation capacity of pollutants tends be lower with porous and high permeable aquifer media and vice-versa Anwar *et al* [27]. Hence, the pollution potential rating of the aquifer medium map produced (Figure 4c) was evaluated accordingly (see Table 4). Furthermore, the impact of soil media (S) on pollution potential mapping cannot be overemphasized. The amount of recharge infiltrating the ground surface, the amount of potential dispersion, and the purifying process of contaminants are strongly influenced by the underlying soil type in the area Anbazhagan *et al* [28]. The soil media (Figure 4d) in this study was prepared based on the 1962 district soil type map of Malaysia. The varying properties of the different soil types were used in the rating pollution potential of this layer (Table 4).The topography (T) for this evaluation was derived based on the sloping factor of the area which is in accordance with [25] who reported that topography is the slope variability of an area. The degree of slope, which varies from one place to another controls the likelihood that rainfall and the contained pollutants have to run off or be retained in one area long enough to infiltrate it. The slope layer (Figure 4e) is derived from the processed ASTER DEM image of 30 m resolution (downloaded from NASA's land processes distributed active archive center (LP DAAC)). This layer was interpreted by adopting a similar approach discussed in prior studies of Adiat *et al* [29], Doumouya *et al* [30]. An area with a low slope degree (flat) tends to retain water for longer time, hence providing greater chance for the infiltration of recharge water, which may contain a considerable amount of pollutants. Meanwhile, an area with a steeper slope has little potential for recharge, thus allowing pollutants little opportunity to reach the groundwater table (see Table 4). Meanwhile, the standard descriptions of varying types of slope used in the evaluation of the slope/topography layer are presented in Table 2. The role of vadose zone in evaluating pollution potential rating has been documented in literatures Chen *et al* [8], Samake *et al* [25]. It influences the aquifer pollution potential depending on the permeability and attenuation characteristics of its soil cover media. Hence, if the vadose zone is vastly permeable, then the probability of such zones/area aquifer being vulnerable to pollution is very high [28].The vadose zone thematic layer (Fig. 4f) was generated from the estimated mean resistivity values of the delineated vadose layer (see Figure 5) and evaluated accordingly (Table 4).The hydraulic conductivity parameter (C) is an important factor that controls contaminant migration and dispersion from the injection point within the saturated zone, and consequently, within the plume concentration in the aquifer AtiqurRahman [6]. The transmission rate of water in subsurface formation is determined by this parameter, which in turn controls contaminant movement rate. Therefore, an area underlain with water bearing formation (aquifer) with high hydraulic conductivity is more vulnerable to potential contamination, relative to other formations with low hydraulic conductivity. The hydraulic conductivity layer (Figure 4g) was prepared from borehole data and evaluated for pollution potential mapping.

IOP Conf. Series: Earth and Environmental Science **20** (2014) 012002 doi:10.1088/1755-1315/20/1/012002

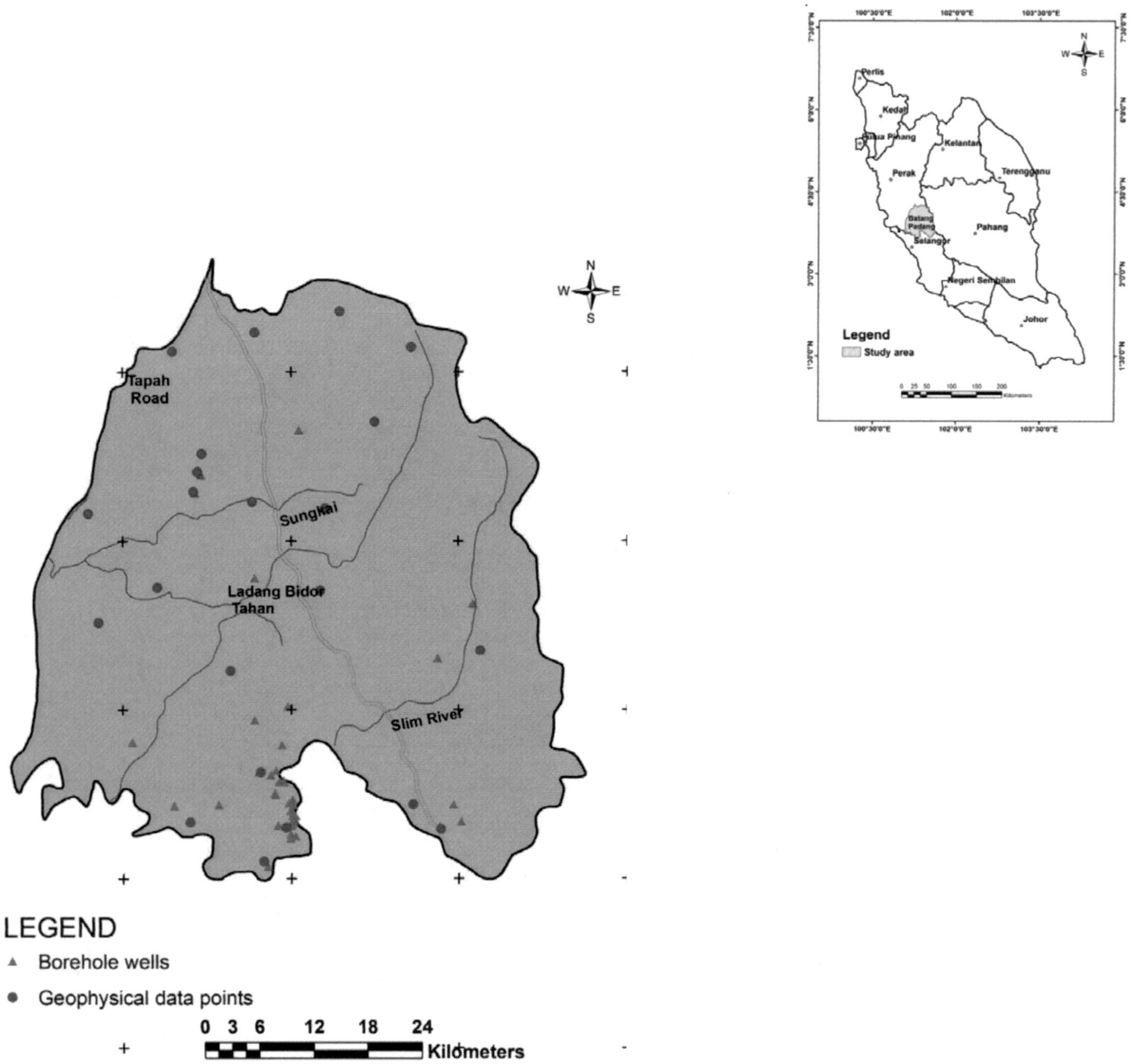

Figure 1. Study area location

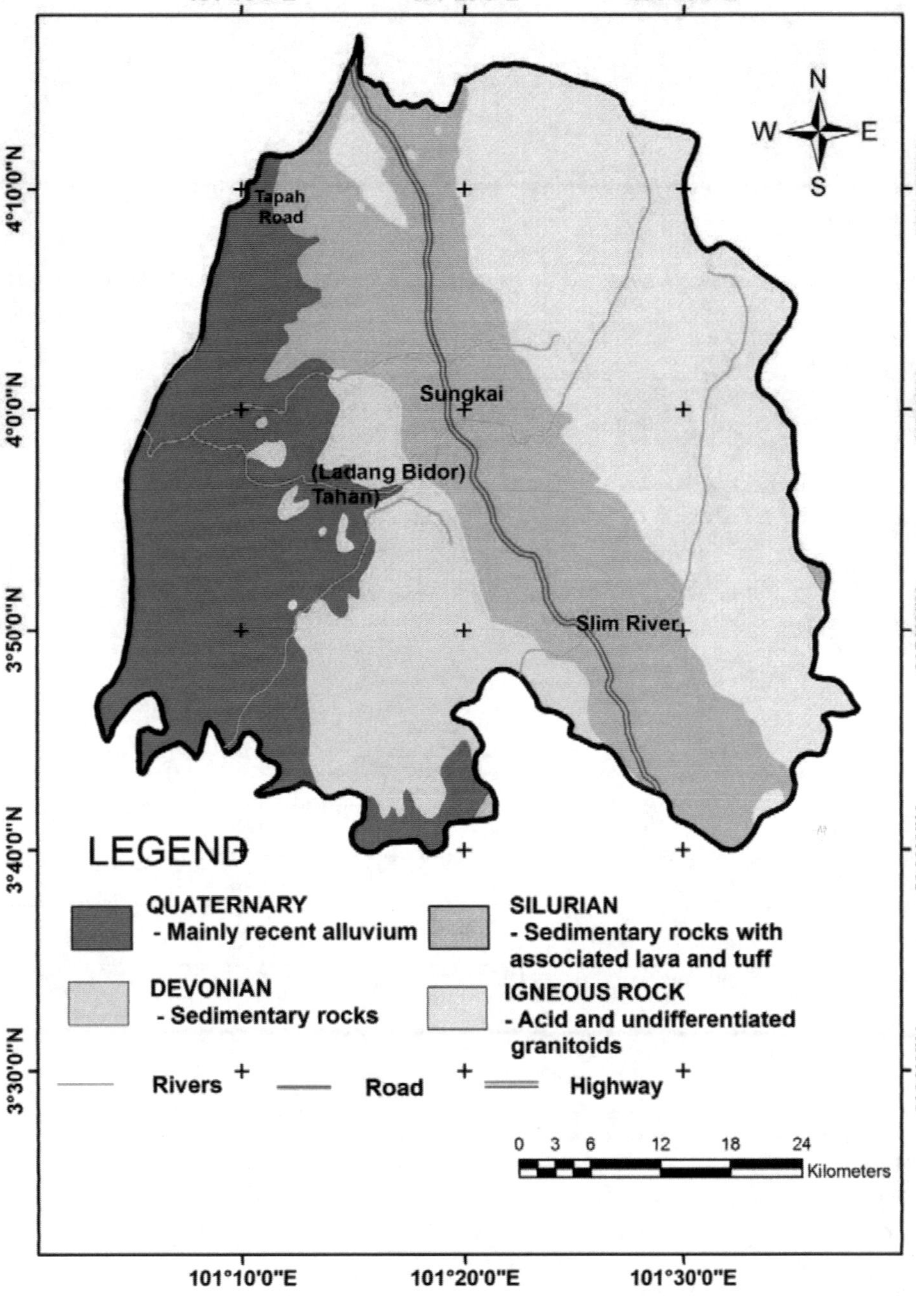

Figure 2. Geological map of the area

Table 1. Information and sources of data used in this study

Data type	Detail of data	Format	Output layer
Borehole data	Malaysian Department of Mineral and Geosciences Date acquired: 12 Feb, 2009	Table and lithology log	Depth to water table(D)
Average annual rainfall	Climatic Research Unit database CRU 2.1 Resolution: 0.5^0	Table	Recharge (R)
Geophysical data	Batang Padang, Perak	Point	Aquifer (A)
Soil map	Ministry of agriculture Kuala Lumpur, Malaysia(1962) Scale: 24 mile : 1 inch	Map	Soil type (S)
Remote sensing imagery	ASTER DEM data NASA((LP DAAC) Resolution: 30 m	Satellite image	Topography/Slope (T)
Geophysical data	Batang Padang, Perak	Point	Impact of vadoze zone(I)
Borehole data	Malaysian Department of Mineral and Geosciences Date acquired: 12 Feb, 2009	Table and lithology log	Hydraulic conductivity(C)

Table 2. Regional slope classification based on Soil Terrain (SOTER) Model

Slope (%)	Classification	Groundwater storage potentiality
0 - 2	Flat	Very high
2 - 8	Undulating	High
8 - 15	Rolling	Moderate
15 - 30	Moderately steep	Low
30 - 60	Steep	Very low

Table 3. Saaty scale for pair-wise comparisons

Score	Judgment	Explanation
1	Equally	Two factors contribute equally to the objective
3	Slightly favour	Slightly favour one attribute over another
5	Strongly favour	Strongly favour one attribute over another
7	Very strongly	Strongly favour one attribute with demonstrated importance over another
9	Extremely	Evidence favouring one attribute over another is one of the highest possible order of affirmation
2,4,6,8	Intermediate	The intermediate values are used when compromise is needed

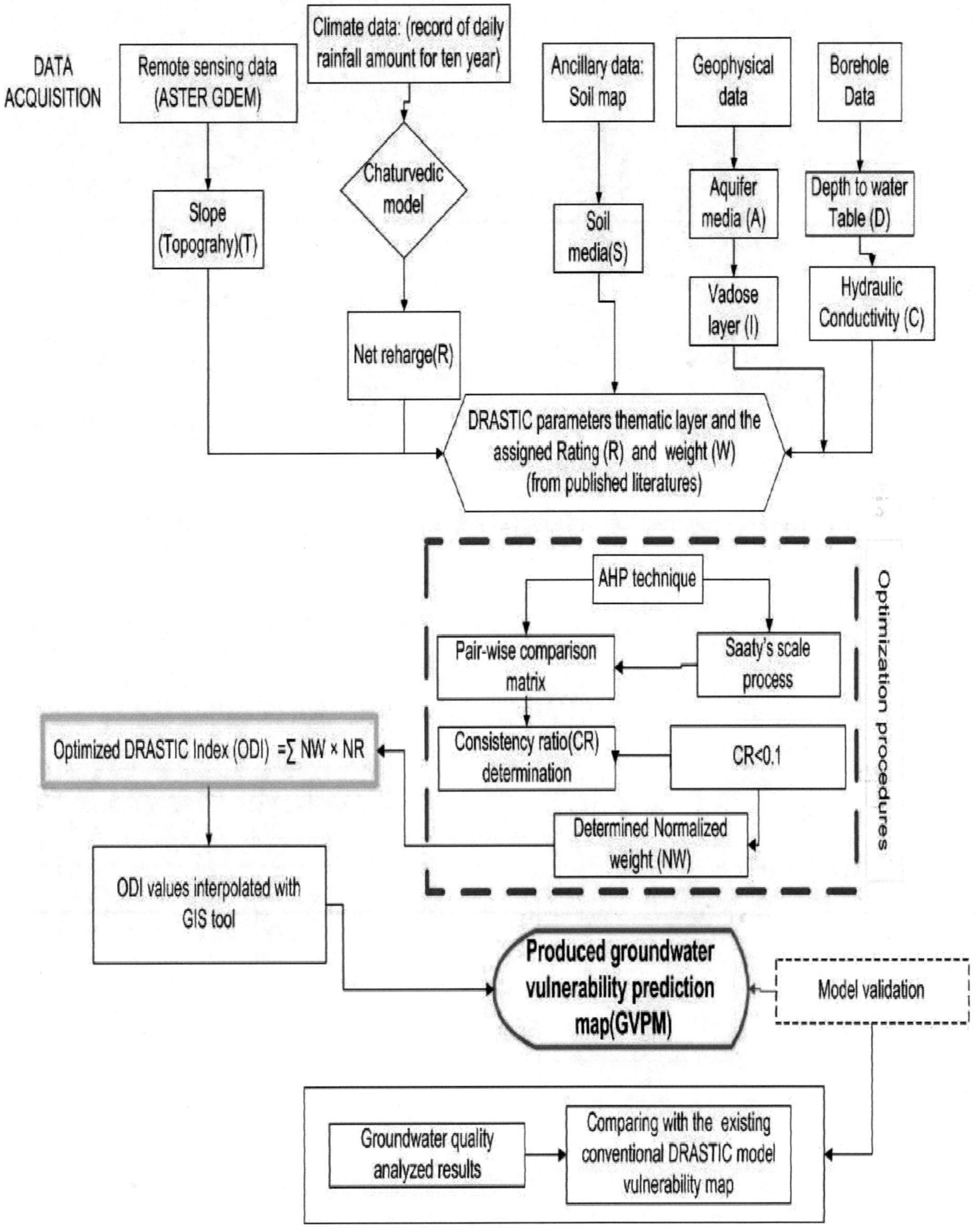

Figure 3. The DRASTIC model evaluation approach and the optimization procedures methodology

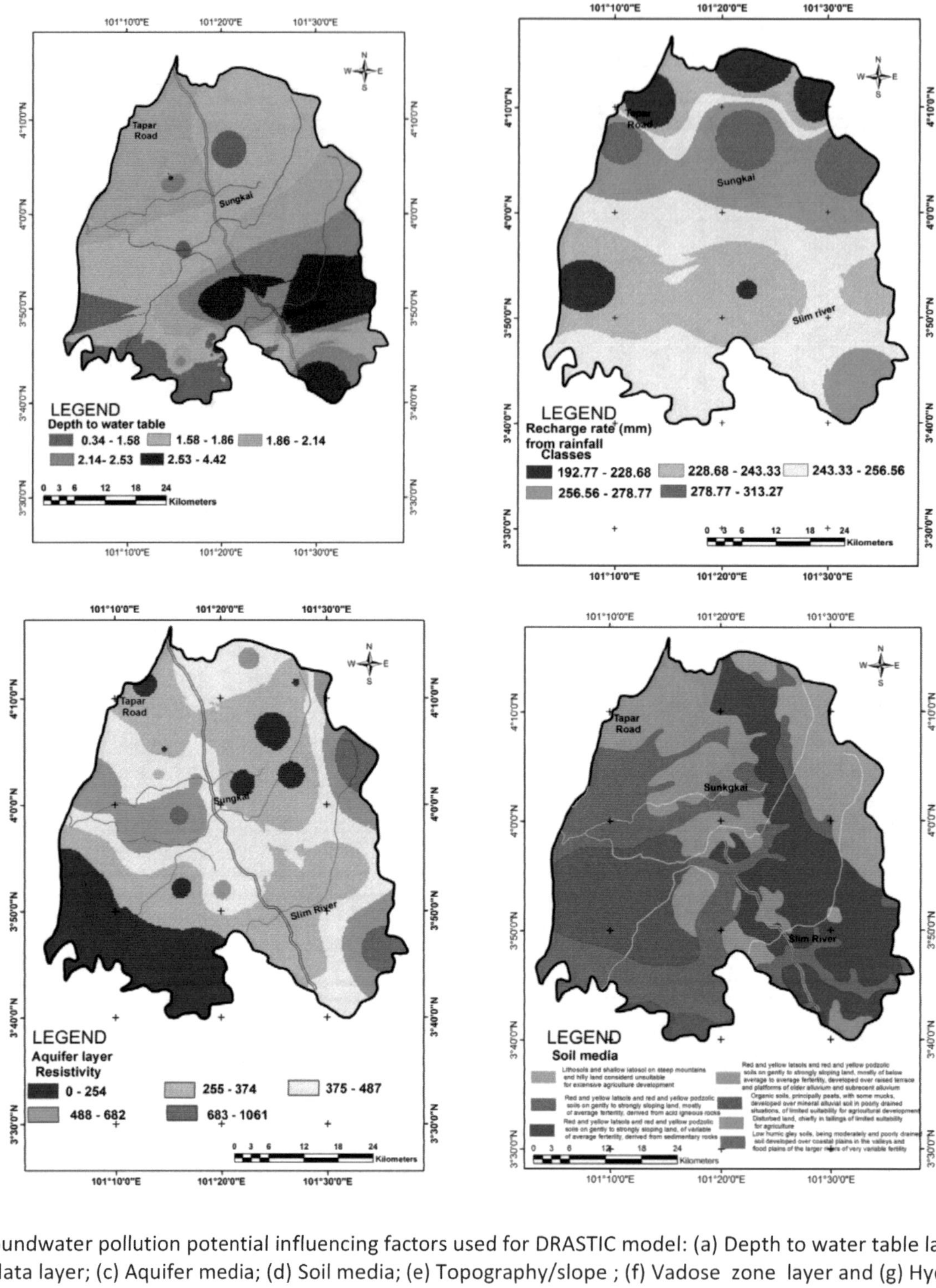

Figure 4. Groundwater pollution potential influencing factors used for DRASTIC model: (a) Depth to water table layer; (b) Rainfall data layer; (c) Aquifer media; (d) Soil media; (e) Topography/slope ; (f) Vadose zone layer and (g) Hydraulic conductivitv

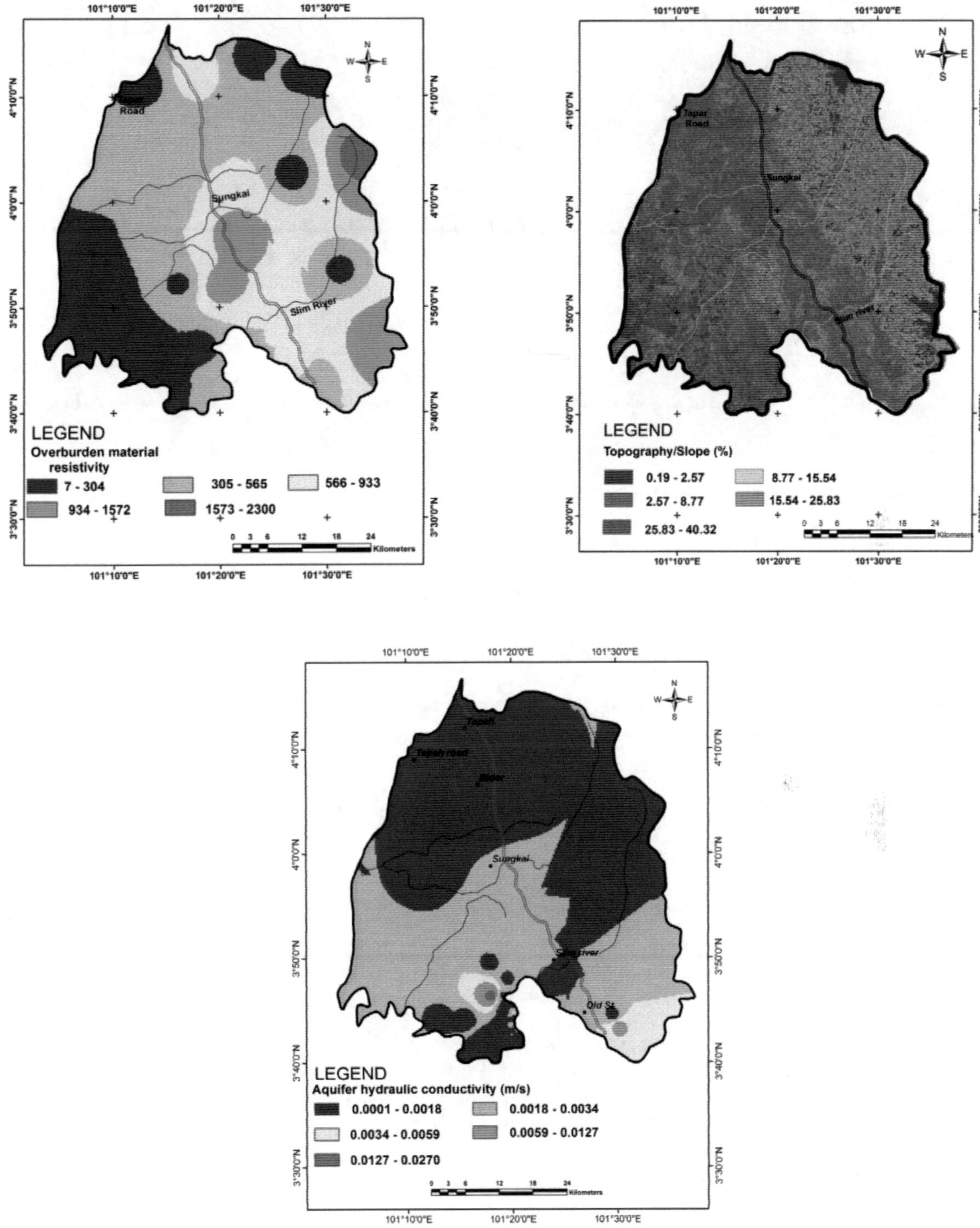

Figure 4. Continued

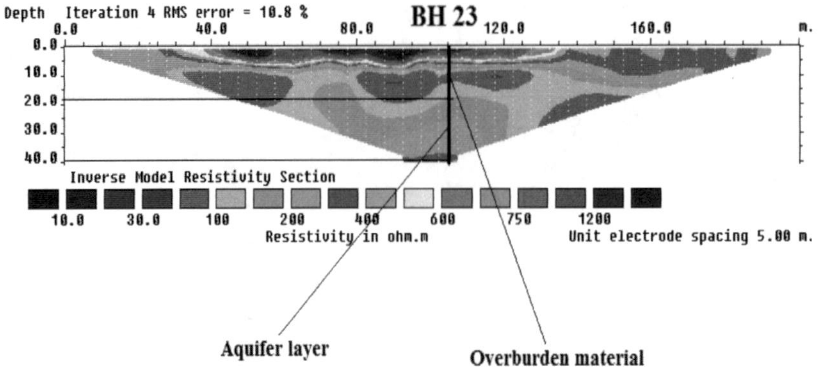

Figure 5. Example of 2D section showing how aquifer media and vadose layer were delineated

2.5. DRASTIC model and decision model optimization procedures

2. 5. 1. Pairwise construction and comparison

The major limitation of the DRASTIC method is its inherent subjectivity in determining rating scales and weighting coefficients [4]. The method is optimized in the present study with a decision model technique. AHP helps in reducing the complexity of the decision problem by considering two parameters/factors at a time, with each parameter being scored according to its relative influence/importance on the pollution potential evaluation. The comparison ratings are based on a scale of 1 to 9 (see Table 3). Construction of the AHP matrix with Saaty's scale in this study was however guided by the use of published expert weights previously assigned to the DRASTIC parameters where parameters are weighted from "1" to "5" according to their relative contribution to potential contamination as reported in the works of (Chen *et al* [8]; AtiqurRahman[6]; Pradhan *et al* [31]; Aller *et al*[10]) coupled with the several field observation in the area. The process adopted for the used Saaty's scale in this study is presented in Table 5, where A, B and C are used for indicating a more preference, less preference and equality in between two factors at a time, respectively. Thereafter, the pair-wise comparison matrix shown in Table 6 was developed. The pair-wise comparison was normalized, where each cell of the matrix was divided by its column total. The normalized weight (NW) was obtained by computing the average value of the normalized pair-wise comparisons. The consistency ratio (CR) obtained for the pair-wise comparison matrix was 0.039 (see Table 6). According to [32] CR=0.039 indicates that the comparisons of the parameters characteristics are perfectly consistent and that the determined normalized weights can be utilized (see Table 4).

2. 5. 2. Estimation of DRASTIC vulnerability index

Using the given equation (1), the conventional DRASTIC vulnerability index based on the rating (R) and the assigned relative weigh (see Table 4) was estimated. Thereafter, the R and W of each factors in equation 1 were substituted with the normalized rating (NR) and the normalized weight (NW) in Table 4. Hence, the conventional DRASTIC model algorithm in equation (1) becomes an optimized DRASTIC model algorithm for estimating optimized DRASTIC index (ODI) in equation (3):

$$ODI = D_{NW}D_{NR} + R_{NW}R_{NR} + A_{NW}A_{NR} + S_{NW}S_{NR} + T_{NW}T_{NR} + I_{NW}I_{NR} + C_{NW}C_{NR} \quad (3)$$

The DI and ODI vulnerability indices were estimated using equations (1) and (3), respectively. The, DI and ODI values obtained for all the grids are presented in Table 7 and Table 8, respectively.

Table 4. Probability ratings (R) and weight (W) assignment for classes of the DRASTIC parameters (Modified after (Atiqur Rahman [6] and Nobre et al [5])

DRASTIC parameters	Ranges(classes)	Pollution Potentiality for groundwater vulnerability	Rating (R)	Normalized Rating(NR)	Relative Weight	Normalized Weight (NW)
Depth to water (D)	0.34 -1.58	High	5	0.33		
	1.58 -1.86	Medium high	4	0.26		
	1.86 – 2.14	Medium	3	0.2	5	0.31
	2.14 -2.53	Low	2	0.13		
	2.53 – 4.42	Very low	1	0.07		
Recharge(R)	1,082 – 1,292	Very low	1	0.07	4	0.16
	1,292 – 1,476	Low	2	0.13		
	1,476 – 1,681	Medium	3	0.2		
	1,681 - 1,887	Medium high	4	0.26		
	1,887 – 2,161	High	5	0.33		
Aquifer media (A)	0 –254(silty-clay with fine sand)	Very low	1	0.07		
	255-374(silty-clay with gravel/siltstone)	Low	2	0.13		
	375-487(claysand with gravel)	Medium	3	0.2	3	0.08
	488–682(weathered sandstone/coarse gravel)	Medium high	4	0.26		
	683–1061(coarse sand/gravel/Fractured/ jointed sandstone)	High	5	0.33		
Soil media(S)	Lithosol&Shallow latosol(Steep mountail)	Very low	1	0.07		
	Red &Yellow podozolic soil from acid igneous rock	Low	2	0.13		
	Red &Yellow latsols podozolic soil from sedimentary rocks	Medium	3	0.2		
	Red &Yellow latsols podozolic soil from older and sub recent alluvium	Medium high	4	0.26	2	0.04
	Organic soil- peat and poorly drained	Very low	1	0.07		
	Low humic gley soil developed in the valley and flood plain	High	5	0.33		

	Agricultural land	Medium high	4	0.26		
Topography (T)	0 - 2.57	(Flat)High	5	0.33		
	2.57 – 8.77	(Undulating) Medium high	4	0.26		
	8.77 – 15.54	(Rolling)Medium	3	0.2		
	15.54 – 25.83	(Moderately steep)Low	2	0.13	1	0.02
	25.83 – 40.32	(Steep)Very low	1	0.07		
Impact of vadose zone (I)	7 –304 clay/silty sand/consolidated shale)	Very Low	1	0.07		
	305 -565 (silty sand with gravel)	medium	3	0.2		
	566-933 (fracture /weathered sandstone With coarse grave	High	5	0.33	5	0.31
	934 –572(Sandstone with gravel)	Medium high	4	0.26		
	1573 -300(consolidated sandstone/fresh)	Low	2	0.13		
Hydraulic Conductivity(C)	0.0001 – 0.0018	Very low	1	0.07		
	0.0018 – 0.0034	Low	2	0.13		
	0.0034 – 0.0059	Medium	3	0.2	3	0.08
	0.0059 – 0.0127	Medium high	4	0.26		
	0.0127 – 0.0270	High	5	0.33		

Table 5. The expert process adopted for the used Saaty's scale based on published literatures expert Assigned weights (After: Chen *et al* [8]; AtiqurRahman[6]; Pradhan *et al* [31])

Factor 1/(weight)	Extremely favors (9)	Very strong Favors (7)	Strongly favor (5)	Slightly favor (3)	Equal (1)	Slightly favor (3)	Strongly favor (5)	Very strong Favors (7)	Extreme favors (9)	Factor 2/(weight)
Depth to water table/(5)				A						Recharge/ (4)
Depth to water table/(5)			A							Aquifer media/ (3)
Depth to water table/(5)		A								Soil media/ (2)
Depth to water table/(5)	A									Topography/ Slope/ (1)
Depth to water table/(5)					C					Impact of Vadose zone/(5)
Depth to water table/(5)			A							Hydraulic conductivity /(3)
Recharge/ (4				A						Aquifer media/(3)
Recharge/ (4)			A							Soil media/(2)
Recharge/ (4)	A									Topography/ Slope/(1)
Recharge/ (4)						B				Impact of Vadose zone/(5)
Recharge/ (4)				A						Hydraulic conductivity /(3)
Aquifer media/(3)				A						Soil media/(2)
Aquifer media/(3)			A							Topography/ Slope/(1)
Aquifer media/(3)							B			Impact of Vadose zone/(5
Aquifer media/(3)					C					Hydraulic conductivity /(3)
Soil media/(2)				A						Topography/ Slope/(1)
Soil								B		Impact of

media/(2)										Vadose zone/(5)
Soil media/(2)						B				Hydraulic conductivity /(3)
Topograph y/Slope/(1)									B	Impact of Vadose zone/(5)
Topograph y/Slope/(1)							B			Hydraulic conductivity /(3)
Impact of Vadose zone/(5)			A							Hydraulic conductivity /(3
										Hydraulic conductivity /(3

Table 6. A pairwise comparison matrix for calculation criteria weights for pollution potential mapping using DRASTIC parameters

	D	**R**	**A**	**S**	**T**	**I**	**C**	**Weights**	**CR**
D	1	3	5	7	9	1	5	0.31	0.039
R	1/3	1	3	5	9	1/3	3	0.16	
A	1/5	1/3	1	3	5	1/5	1	0.08	
S	1/7	1/5	1/3	1	3	1/7	1/3	0.04	
T	1/9	1/9	1/5	1/3	1	1/9	1/5	0.02	
I	1	3	5	7	9	1	5	0.31	
C	1/5	1/3	1	3	5	1/5	1	0.08	

IOP Conf. Series: Earth and Environmental Science **20** (2014) 012002 doi:10.1088/1755-1315/20/1/012002

Table 7. Conventional DRASTIC index (DI) calculations

Grid Nos	Grid Center's coordinates		D(W=5)		R(W=4)		A(W=3)		S(W=2)		T(W=1)		I(W=5)		C(W=3)		DI
	LONG	LAT	R	R*W	R	R*W	R	R*W	R	R*W	R	R*W	R	R*W	R	R*W	$\sum W*R$
1	754386.3	468233.4	3	15	3	12	2	6	3	6	2	2	3	15	1	3	59
2	765042.5	468233.4	4	20	3	12	5	15	1	2	2	2	1	5	1	3	59
3	743977.9	456833.7	3	15	4	16	3	9	4	8	1	1	4	20	1	3	71
4	754386.3	456833.7	4	20	4	16	4	12	4	8	2	2	5	25	1	3	84
5	765042.5	456833.7	4	20	5	20	1	3	3	6	4	4	5	25	1	3	80
6	775450.8	456833.7	4	20	4	16	3	9	1	2	2	2	5	25	2	6	79
7	733321.8	445434.1	3	15	3	12	5	15	5	10	1	1	4	20	1	3	75
8	743977.9	445434.1	3	15	3	12	4	12	5	10	1	1	4	20	1	3	72
9	754386.3	445434.1	3	15	3	12	5	15	5	10	1	1	5	25	1	3	81
10	765042.5	445434.1	4	20	4	16	5	15	4	8	2	2	3	15	1	3	78
11	775450.8	445434.1	3	15	3	12	3	9	1	2	3	3	5	25	2	6	72
12	733321.8	434034.5	4	20	1	4	1	3	1	2	1	1	1	5	2	6	41
13	743977.9	434034.5	4	20	1	4	1	3	1	2	4	4	4	20	1	3	56
14	754386.3	434034.5	4	20	2	8	4	12	4	8	2	2	5	25	2	6	79
15	765042.5	434034.5	2	10	1	4	3	9	3	6	4	4	3	15	2	6	54
16	775450.8	434034.5	1	5	2	8	4	12	3	6	2	2	1	5	3	9	47
17	786107	434034.5	1	5	2	8	4	12	1	2	4	4	1	5	4	12	48
18	733321.8	422882.7	5	25	1	4	3	9	3	6	4	4	4	20	3	9	76
19	743977.9	422882.7	4	20	1	4	3	9	3	6	5	5	5	25	2	6	74
20	754386.3	422882.7	2	10	2	8	3	9	3	6	2	2	5	25	1	3	62
21	765042.5	422882.7	2	10	1	4	4	12	4	8	1	1	5	25	2	6	65
22	775450.8	422882.7	1	5	2	8	2	6	4	8	2	2	4	20	4	12	60
23	786107	422882.7	2	10	2	8	4	12	3	6	4	4	1	5	4	12	56
24	743977.9	411235.2	5	25	2	8	3	9	3	6	5	5	3	15	2	6	73
25	754386.3	411235.2	5	25	2	8	3	9	4	8	3	3	3	15	4	12	79
26	775450.8	411235.2	1	5	2	8	1	3	3	6	1	1	1	5	5	15	43
27	786107	411235.2	2	10	2	8	4	12	3	6	2	2	4	20	4	12	68

IOP Conf. Series: Earth and Environmental Science **20** (2014) 012002 doi:10.1088/1755-1315/20/1/012002

Table 8. Optimized DRASTIC index (ODI) calculations.

Grid No	Grid Center's coordinates		D(NW = 0.31)		R(NW = 0.16)		A(NW = 0.08)		S(NW = 0.04)		T(NW=0.02)		I(NW=0.31)		C(NW=0.08)		ODI
	LONG	LAT	NR	NR*NW	NR	NR*NW	NR	NR*NW	NR	NR*NW	NR	NR*NW	NR	NR*NW	NR	NR*NW	
1	754386.3	468233.4	0.2	0.062	0.2	0.032	0.1	0.0104	0.2	0.008	0.1	0.0026	0.2	0.062	0.1	0.0056	0.183
2	765042.5	468233.4	0.3	0.0806	0.2	0.032	0.3	0.0264	0.07	0.0028	0.1	0.0026	0.1	0.0217	0.1	0.0056	0.172
3	743977.9	456833.7	0.2	0.062	0.3	0.0416	0.2	0.016	0.26	0.0104	0.1	0.0014	0.3	0.0806	0.1	0.0056	0.218
4	754386.3	456833.7	0.3	0.0806	0.3	0.0416	0.3	0.0208	0.26	0.0104	0.1	0.0026	0.3	0.1023	0.1	0.0056	0.264
5	765042.5	456833.7	0.3	0.0806	0.3	0.0528	0.1	0.0056	0.2	0.008	0.3	0.0052	0.3	0.1023	0.1	0.0056	0.260
6	775450.8	456833.7	0.3	0.0806	0.3	0.0416	0.2	0.016	0.07	0.0028	0.1	0.0026	0.3	0.1023	0.1	0.0104	0.256
7	733321.8	445434.1	0.2	0.062	0.2	0.032	0.3	0.0264	0.33	0.0132	0.1	0.0014	0.3	0.0806	0.1	0.0056	0.221
8	743977.9	445434.1	0.2	0.062	0.2	0.032	0.3	0.0208	0.33	0.0132	0.1	0.0014	0.3	0.0806	0.1	0.0056	0.216
9	754386.3	445434.1	0.2	0.062	0.2	0.032	0.3	0.0264	0.33	0.0132	0.1	0.0014	0.3	0.1023	0.1	0.0056	0.243
10	765042.5	445434.1	0.3	0.0806	0.3	0.0416	0.3	0.0264	0.26	0.0104	0.1	0.0026	0.2	0.062	0.1	0.0056	0.229
11	775450.8	445434.1	0.2	0.062	0.2	0.032	0.2	0.016	0.07	0.0028	0.2	0.004	0.3	0.1023	0.1	0.0104	0.230
12	733321.8	434034.5	0.3	0.0806	0.1	0.0112	0.1	0.0056	0.07	0.0028	0.1	0.0014	0.1	0.0217	0.1	0.0104	0.134
13	743977.9	434034.5	0.3	0.0806	0.1	0.0112	0.1	0.0056	0.07	0.0028	0.3	0.0052	0.3	0.0806	0.1	0.0056	0.192
14	754386.3	434034.5	0.3	0.0806	0.1	0.0208	0.3	0.0208	0.26	0.0104	0.1	0.0026	0.3	0.1023	0.1	0.0104	0.248
15	765042.5	434034.5	0.1	0.0403	0.1	0.0112	0.2	0.016	0.2	0.008	0.3	0.0052	0.2	0.062	0.1	0.0104	0.153
16	775450.8	434034.5	0.1	0.0217	0.1	0.0208	0.3	0.0208	0.2	0.008	0.1	0.0026	0.1	0.0217	0.2	0.016	0.112
17	786107.0	434034.5	0.1	0.0217	0.1	0.0208	0.3	0.0208	0.07	0.0028	0.3	0.0052	0.1	0.0217	0.3	0.0208	0.114
18	733321.8	422882.7	0.3	0.1023	0.1	0.0112	0.2	0.016	0.2	0.008	0.3	0.0052	0.3	0.0806	0.2	0.016	0.239
19	743977.9	422882.7	0.3	0.0806	0.1	0.0112	0.2	0.016	0.2	0.008	0.3	0.0066	0.3	0.1023	0.1	0.0104	0.235
20	754386.3	422882.7	0.1	0.0403	0.1	0.0208	0.2	0.016	0.2	0.008	0.1	0.0026	0.3	0.1023	0.1	0.0056	0.196
21	765042.5	422882.7	0.1	0.0403	0.1	0.0112	0.3	0.0208	0.26	0.0104	0.1	0.0014	0.3	0.1023	0.1	0.0104	0.197
22	775450.8	422882.7	0.1	0.0217	0.1	0.0208	0.1	0.0104	0.26	0.0104	0.1	0.0026	0.3	0.0806	0.3	0.0208	0.167
23	786107.0	422882.7	0.1	0.0403	0.1	0.0208	0.3	0.0208	0.2	0.008	0.3	0.0052	0.1	0.0217	0.3	0.0208	0.138
24	743977.9	411235.2	0.3	0.1023	0.1	0.0208	0.2	0.016	0.2	0.008	0.3	0.0066	0.2	0.062	0.1	0.0104	0.226
25	754386.3	411235.2	0.3	0.1023	0.1	0.0208	0.2	0.016	0.26	0.0104	0.2	0.004	0.2	0.062	0.3	0.0208	0.236
26	775450.8	411235.2	0.1	0.0217	0.1	0.0208	0.1	0.0056	0.2	0.008	0.1	0.0014	0.1	0.0217	0.3	0.0264	0.106
27	786107.0	411235.2	0.1	0.0403	0.1	0.0208	0.3	0.0208	0.2	0.008	0.1	0.0026	0.3	0.0806	0.3	0.0208	0.194

2.6. Preparation of the groundwater vulnerability model maps

The DI and ODI vulnerability indices values for each grid in Table 7 and Table 8, respectively were plotted at the center of the grids. Using the coordinates of the grids' center, the obtained DI and ODI values estimated were utilized for modeling the groundwater vulnerability prediction maps (GVPM) shown in Figures 6 and 7 using the kriging interpolation technique in GIS environment. Thereafter, it was observed from the model maps that the DI and ODI values for the study area are in the range 5.67 to 84.45 and 0.0178 to 0.2639. Employing the natural break classification method according to [33] in GIS platform, the DI and ODI values in Tables 7 and 8 were classified into possible five levels of groundwater vulnerability zones namely; No vulnerable (NV), Very low vulnerable (VLV), Low vulnerable (LV), Moderate vulnerable (MV) and High vulnerable (HV) as presented in Table 9

.

Table 9. The possible groundwater vulnerable zone classification

DI Values	ODI Values	Classifications
5.67 – 36.88	0.0178 – 0.1047	No vulnerable (NV)
36.88 – 51.70	0.1047 – 0.1491	Very low vulnerable (VLV)
51.70 – 61.28	0.1491 – 0.1857	Low vulnerable (LV)
61.28 – 70.55	0.1857 – 0.2176	Moderate vulnerable (MV)
70.55 -84.45	0.2176 – 0.2639	High vulnerable (HV)

3. Results and Discussion

3.1. Distribution of vulnerable zones in the area

The GVPM models (Figures 6 and 7) were evaluated for the distribution of the different groundwater vulnerable zones in GIS environment. Figure 8 shows the area percentages distribution covering the predicted vulnerable zones namely NV, VLV, LV, MV and HV estimated from the produced groundwater vulnerability prediction maps (Figures 6 and 7). It was deduced from figure 8 that more than 50 % of the area are under the moderate to high vulnerable zones which are areas that are highly susceptible to contamination AtiqurRahman [6].

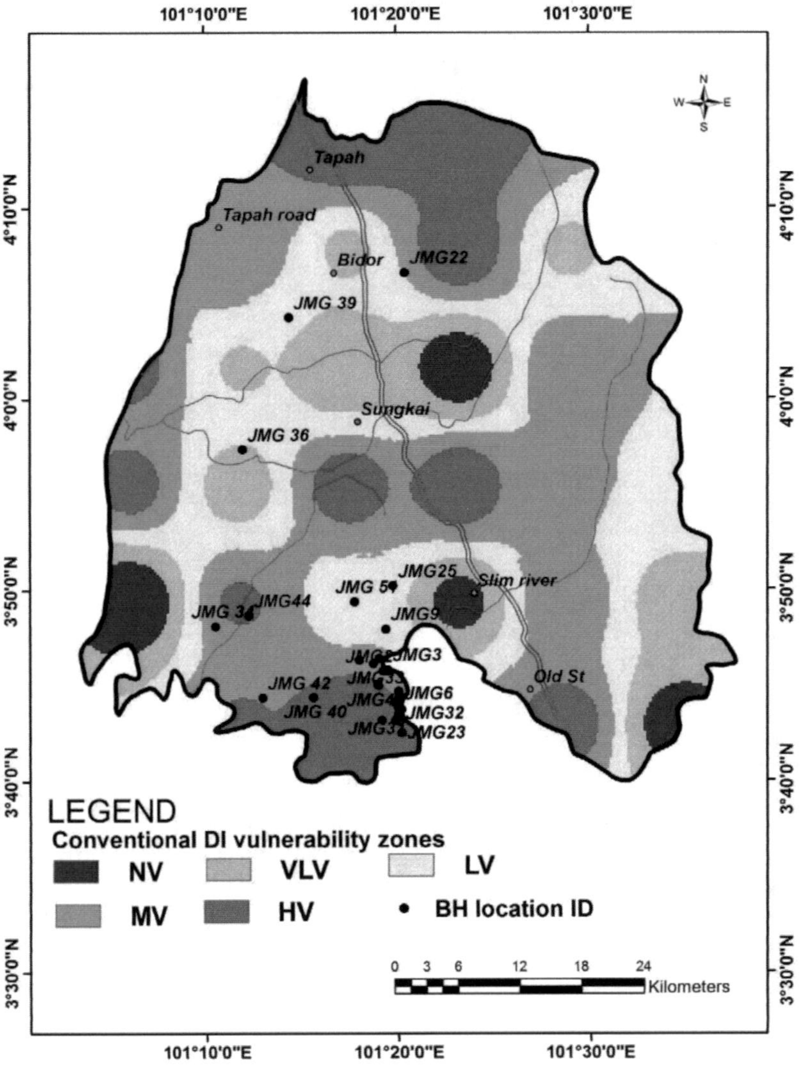

Figure 6. GVPM produced by conventional DRASTIC model

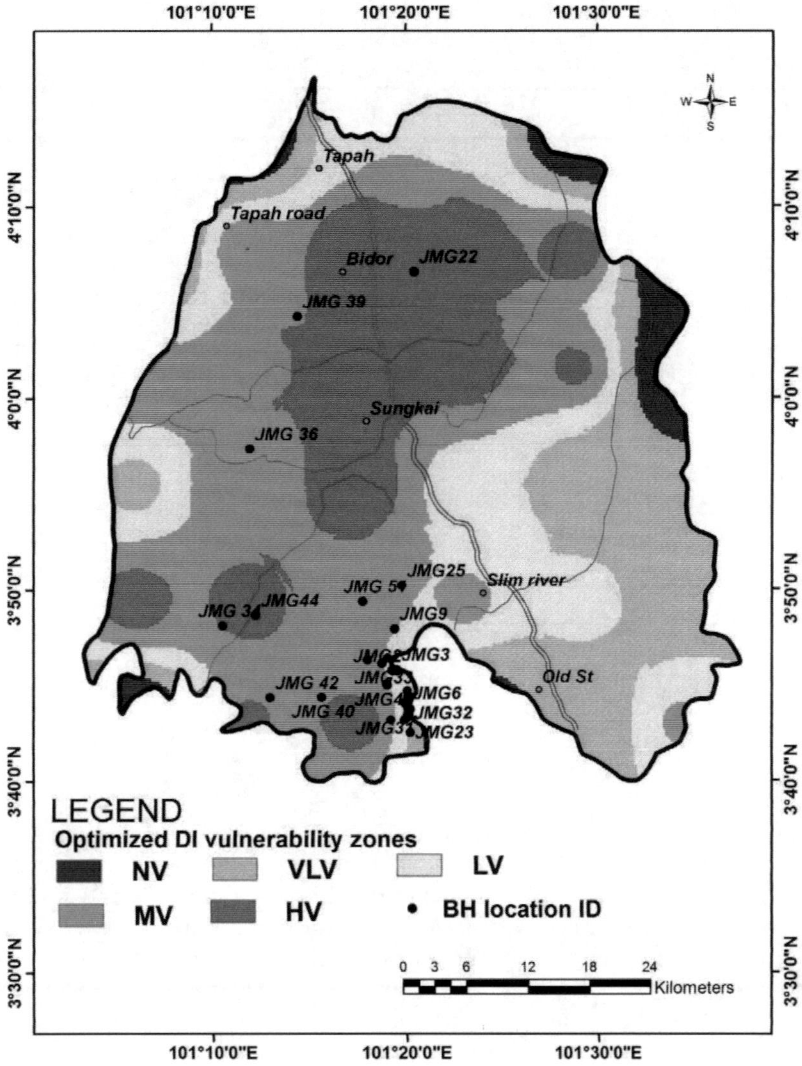

Figure 7. GVPM produced by Optimized DRASTIC model

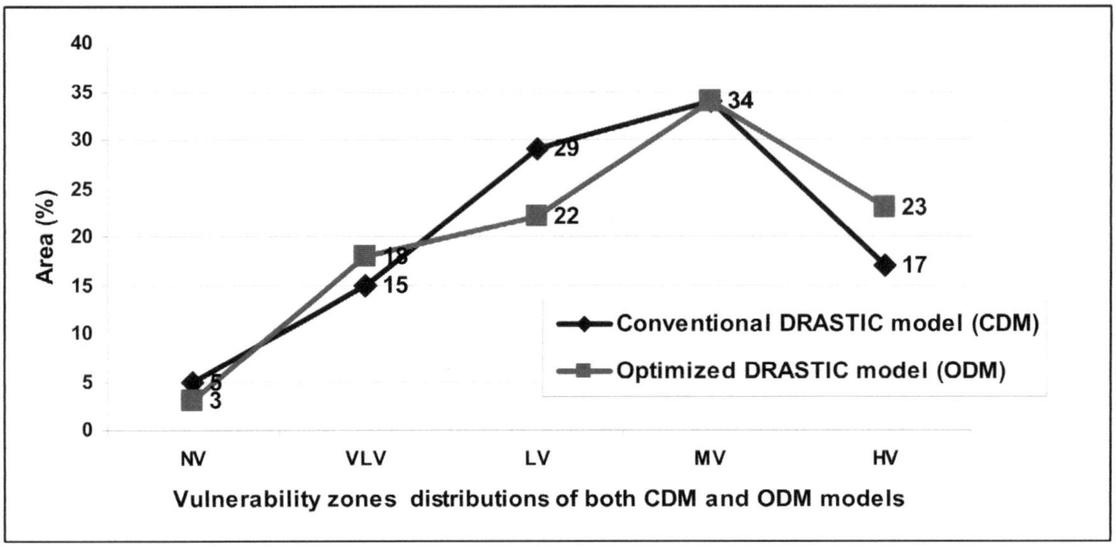

Figure 8. Percentage of vulnerable zones distribution between CDM and ODM approach

3.2. Model validation

The produced groundwater vulnerability prediction maps (Figures 6 and 7) were subjected to validation assessment to establish their reliability and validity for efficient groundwater resources quality preservation in the area. The approach according to Kalinski *et al* [33] and McLay *et al* [34] where chemical and contaminant parameters concentrations in groundwater samples were used to correlate vulnerability index were adopted for validation technique in this study. The groundwater quality analysis obtained in the area enabled determination of physio-chemical parameters and heavy metals element concentration records in the groundwater samples obtained from the drilled boreholes in the area. However, it should note that, a rare reference on groundwater contamination problem was present in the area despite being an area covered with agricultural lands. Hence, this study aimed to assess impending contamination to be model from estimated vulnerability index in the area. The determined physio-chemical parameters and heavy metals that were assumed to be completely soluble in water include pH, TDS, nitrate (NO_3), sulfate (SO_4), calcium (Ca), iron (Fe), zinc (Zn) and manganese (Mn). The records of the concentrations of these chemical parameters and elements were compared with the WHO (World Health Organization) and Food and Agriculture Organization (FAO) standard levels to assess the water quality status of the boreholes sampled. Hence, the water quality in the area has been classified into good or bad depending on the permissible limit of these standards. Y and X are used for indicating good and bad water qualities, respectively depending on the measured concentrations of each chemical elements/parameters relatively to the permissible limit based on WHO and FAO standards. Therefore, for each determined parameters where the concentration (mg/l) is within the permissible limit of these standards, "Y" is indicated (see Table 10). On the other hand, where we have the measured concentration of the chemical elements exceeded the permissible limit, "X" is indicated. The results of this water quality assessment from each boreholes drilled across the entire study area were used to assess the accuracy of the vulnerable maps produced. The locations of these boreholes and their respective names are displayed on the prediction maps shown in figures 6 and 7. The comparison results based on Table 11 general showed that water quality in the area is relatively of good status. However, the level of concentration of pH and Manganese chemical elements which exceeded the permissible limit of FAO and WHO standards have shown traceable groundwater contamination or pollution in the area. Hence, these two chemical parameters were used as groundwater contamination indices for evaluating prediction accuracy of the GVPM models. The DRASTIC vulnerability index of the most vulnerable zones i.e the moderate and the high (MV and HV) of the vulnerability maps were spatially correlated with concentration records of pH and Mn in each of the boreholes located in those zones. The considered MV and HV are in accordance with Artiqur Rahman [6] and Thirumalaivasan *et al* [7] where the moderate to high vulnerable zones in a predicted vulnerability map are typified areas that are highly susceptible to pollution compare to the NV, VLV and LV zones that are areas that are characterized with attributes of high resistance to pollution.

The results of the spatial relationship of the conventional DRASTIC vulnerability index model (CDM) and the optimized DRASTIC index model (ODM) to groundwater pH and Manganese (Mn) concentration in the area are shown in figure 9.

Table 10 Groundwater quality evaluation criteria based on FAO and WHO standards (After FAO [35] and WHO [36])

Standards	Physio-chemical Parameters					Heavy Metals			Water quality remark	
	pH	TDS	NO_3	Ca	SO_4	Fe	Zn	Mn	Good	Bad
WHO	6.5 – 8.5	500	50	75	20	0.5 - 50	3	0.1	Y	
FAO	6.5 – 8.5	2000	10	20	20	-	0.01	-		X

The results in Figure 9 present the comparative analysis of the prediction accuracy of GVPM models both for the conventional DRASTIC model and the optimized DRASTIC model(Figures 6 and 7), respectively. The results clarified the prediction accuracy of the produced GVPM based on optimized DRASTIC model (ODM) compare to GVPM based on conventional DRASTIC model (CDM) in the study area. The computed correlation factor results obtained from the spatial relationship between the vulnerability index of MV and HV zones of ODM - based model map to pH and Mn concentrations yielded 90 % and 86 %, respectively while the CDM - based model map have correlation factor of 62 % and 50 %, respectively. These results established the reliability and validity of the ODM based groundwater vulnerability prediction map as decision support tool that can provide a satisfactory assessment of intrinsic vulnerability of groundwater to pollution in the study area.

4. **Conclusions**

This study developed an optimized DRASTIC model (ODM) algorithm through application of AHP as a decision support model to conventional DRASTIC model (CDM) algorithm. With the sound mathematical basis of AHP application to the weights of the CDM parameters, namely, depth-to-water table, recharge, aquifer media, soil media, topography, impact of the vadose zone, and hydraulic conductivity, obtainable in the vicinity of oil palm plantation and other agricultural activities in southern Perak, Malaysia, the optimized DRASTIC index for groundwater vulnerability modeling in the area was obtained.. Five vulnerable zones, namely, NV, VLV, LV, MV, and HV where more than 50 % of the area belong to MV and HV zones were identified based on application of the ODM algorithm. The prediction accuracy of the ODM - based groundwater vulnerability prediction map with the groundwater pH and Mn concentrations yielded correlation factor of 90 % and 86 %, respectively compared to 62 % and 50 % correlation factors computed for CDM - based prediction map. These results indicated that the optimized DRASTIC model (ODM) could provide better results as compared with the conventional DRASTIC model (CDM). An advantage of optimized DRASTIC model (ODM) algorithm is that it can provide an excellent insight into the reduction of bias in decision making in vulnerability mapping through appropriate management of the uncertainty and subjectivity of the CDM parameters. The ODM-based vulnerability prediction model map produced in this study is recommended for evaluating groundwater vulnerability to pollution in the study area. Therefore, the ODM–based groundwater vulnerability prediction map (GVPM) can be utilized as an economical tool to identify the zones of concern and as a tool to overcome problems regarding the haphazard and uncontrolled development of land and undesirable activities that can affect groundwater quality in the study area. However, the produced GVPM can only be utilized as a screening tool and not as a replacement for detailed site-specific analysis.

5. Acknowledgments

This project was carried out with financial support from RUI, Investigation Of The Impacts Of Summertime Monsoon Circulation To The Aerosols Transportation And Distribution In Southeast Asia Which Can Lead To Global Climate Change, 1001/PFIZIK/811228 and Science Fund, Environmental Effects And Its Influence of Increased Green House Gasses In Peninsular Malaysia, 305/PFIZIK/613615.

IOP Conf. Series: Earth and Environmental Science **20** (2014) 012002 doi:10.1088/1755-1315/20/1/012002

Table 11 Groundwater quality analysis and the comparison results with the FAO and WHO standards

BH NAMES	BH location coordinates		Physio-chemical Parameters										Heavy Metals						General BH Water quality Remark
	LAT	LONG	pH	B	TDS	B	NO_3	B	Ca	B	SO_4	B	Fe	B	Zn	B	Mn	B	
JMG2	416182	758308	6.5	X	338	Y	3.90	Y	3.6	Y	3.8	Y	53.8	Y	0.02	Y	0.50	X	Y
JMG3	416304	758263	5.6	X	102	Y	0.09	Y	4.7	Y	0.6	Y	53.7	Y	0.12	Y	0.54	X	Y
JMG 4	416214	757808	5.5	X	100	Y	0.05	Y	3.2	Y	0.2	Y	59.9	Y	0.10	Y	0.26	X	Y
JMG6	412612	759329	5.5	X	41	Y	0.10	Y	5.3	Y	0.3	Y	39.0	Y	0.10	Y	0.51	X	Y
JMG9	420230	758064	5.9	X	29	Y	0.00	Y	1.6	Y	7.1	Y	28.2	Y	0.10	Y	0.38	X	Y
JMG16	411468	757687	5.4	X	33	Y	0.25	Y	1.4	Y	2.4	Y	23.2	Y	0.10	Y	0.45	X	Y
JMG17	411450	759032	5.5	X	73	Y	0.10	Y	3.8	Y	18.4	Y	16.9	Y	0.03	Y	0.71	X	Y
JMG1 8	411716	759298	5.8	X	39	Y	0.05	Y	4.5	Y	1.9	Y	44.8	Y	0.35	Y	1.16	X	Y
JMG 19	411616	759198	6.8	X	59	Y	0.01	Y	5.0	Y	1.9	Y	37.1	Y	0.00	Y	1.21	X	Y
JMG21	416181	757886	6	X	38	Y	0.68	Y	2.7	Y	0.2	Y	51.3	Y	0.10	Y	0.24	X	Y
JMG22	454638	759856	5.8	X	46	Y	0.05	Y	2.7	Y	0.2	Y	52.0	Y	0.09	Y	0.54	X	Y
JMG23	410246	759591	5	X	68	Y	0.20	Y	4.2	Y	62.7	Y	36.3	Y	0.04	Y	1.20	X	Y
JMG25	424414	758753	5.1	X	38	Y	0.00	Y	1.8	Y	1.5	Y	20.2	Y	0.10	Y	0.43	X	Y
JMG27	414271	759280	5.1	X	27	Y	0.00	Y	1.2	Y	0.6	Y	20.9	Y	0.10	Y	0.19	X	Y
JMG28	413718	759359	5.7	X	58	Y	0.10	Y	2.8	Y	1.5	Y	42.5	Y	0.03	Y	0.52	X	Y
JMG29	412513	759574	5.5	X	82	Y	6.90	Y	3.8	Y	3.8	Y	38.5	Y	0.10	Y	0.59	X	Y
JMG31	412037	759330	5.4	X	10	Y	0.10	Y	0.5	Y	1.9	Y	4.1	Y	0.10	Y	0.13	X	Y
JMG32	411971	759431	5.4	X	19	Y	0.30	Y	2.2	Y	0.3	Y	14.9	Y	0.10	Y	0.39	X	Y
JMG33	416909	756862	5.4	X	21	Y	0.00	Y	1.6	Y	1.0	Y	17.9	Y	0.10	Y	0.28	X	Y
JMG 34	420497	741653	5.5	X	30	Y	0.20	Y	1.8	Y	0.7	Y	24.5	Y	0.10	Y	0.48	X	Y
JMG 35	414952	757311	6.1	X	50	Y	0.10	Y	3.8	Y	3.4	Y	17.1	Y	0.03	Y	0.91	X	Y
JMG 36	437560	744242	5.3	X	31	Y	0.10	Y	2.5	Y	0.2	Y	46.3	Y	0.10	Y	0.37	X	Y
JMG 38	413076	759094	5.4	X	29	Y	0.10	Y	2.8	Y	0.3	Y	25.1	Y	0.01	Y	0.51	X	Y
JMG 39	450292	748761	5.5	X	90	Y	0.05	Y	3.2	Y	0.5	Y	30.7	Y	0.07	Y	0.48	X	Y
JMG 40	413664	751115	6.2	X	58	Y	0.05	Y	5.2	Y	3.1	Y	7.3	Y	0.06	Y	0.20	X	Y
JMG 41	413109	759150	5.5	X	43	Y	0.03	Y	2.5	Y	0.2	Y	30.7	Y	0.10	Y	0.35	X	Y
JMG 42	413618	746226	5.4	X	5	Y	0.20	Y	2.4	Y	0.2	Y	32.8	Y	0.10	Y	0.29	X	Y
JMG 43	413220	759138	5.9	X	56	Y	0.00	Y	2.4	Y	0.4	Y	31.5	Y	0.10	Y	0.32	X	Y
JMG45	414808	757378	5.4	X	57	Y	0.20	Y	3.1	Y	1.9	Y	26.2	Y	0.01	Y	0.67	X	Y
JMG 47	417237	755516	5.2	X	63	Y	0.00	Y	3.2	Y	0.5	Y	34.7	Y	0.01	Y	0.76	X	Y
JMG 50	417408	757405	5.9	X	3	Y	0.10	Y	1.7	Y	2.9	Y	25.0	Y	0.10	Y	0.36	X	Y
JMG 51	422877	755057	4.9	X	28	Y	0.10	Y	3.1	Y	1.0	Y	34.0	Y	0.07	Y	0.52	X	Y
JMG44	421523	744839	4.7	X	43	Y	0.10	Y	1.2	Y	3.3	Y	19.2	Y	0.01	Y	0.25	X	Y

B:Water quality measurement level; Y: Good water quality and X: Bad water quality

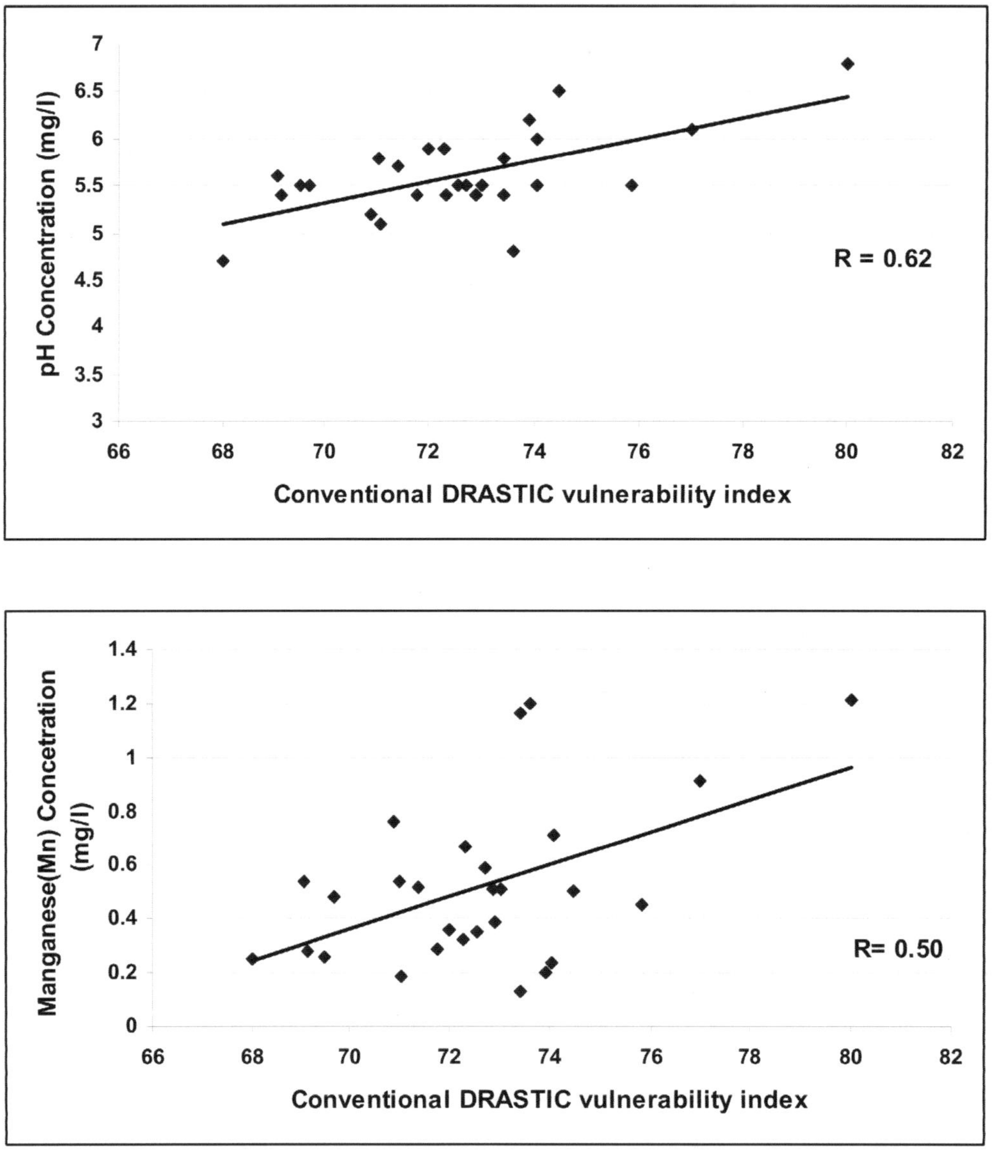

Figure 9. Spatial relationship of conventional DRASTIC vulnerability index and optimized DRASTIC vulnerability index to groundwater pH and Mn concentrations in the study area.

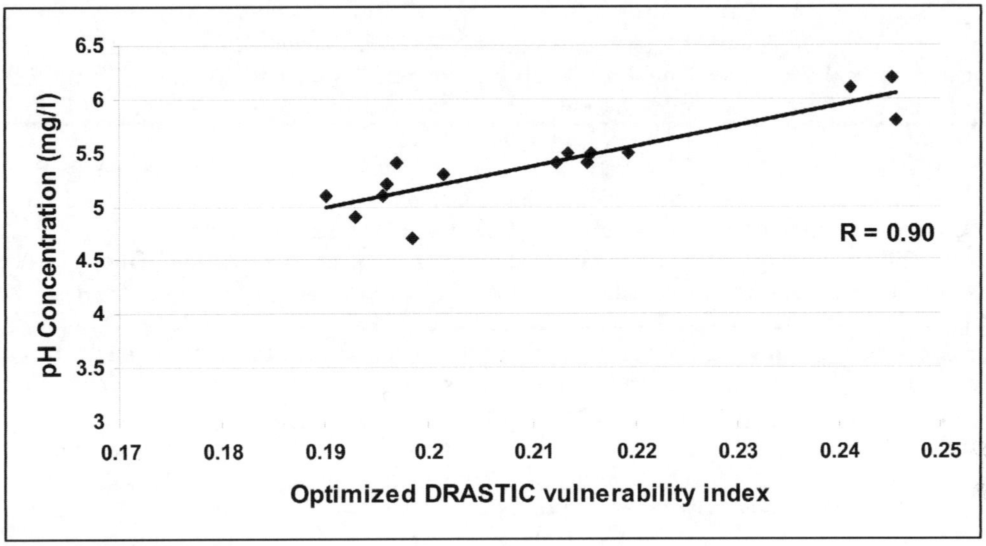

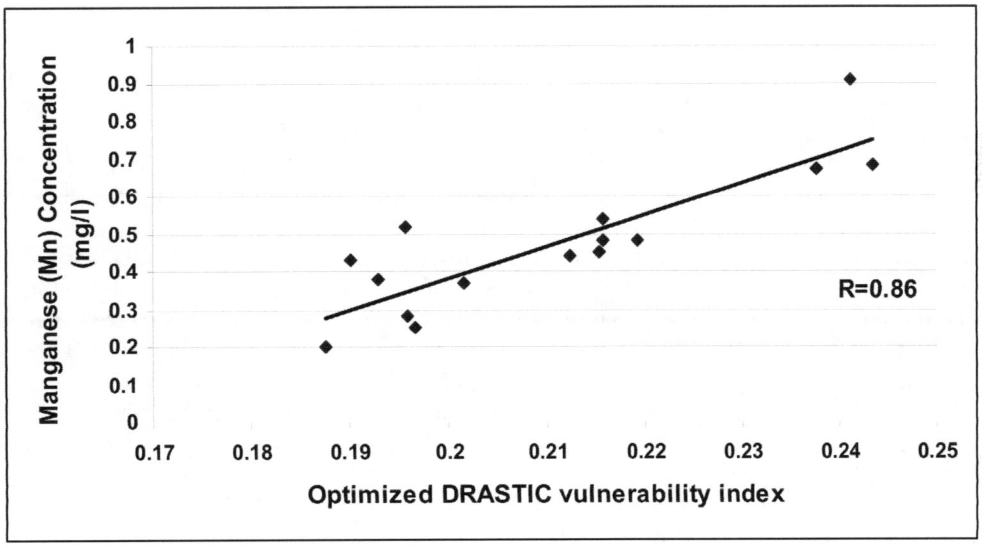

Figure 9. Continued

6. References

[1] Pospecu I C, Gardin N, Brouyere S and Dassargues A 2008 Groundwater vulnerability assessment using physical-based modeling: From challenges to pragmatic solutions. In modelCARE 2007 proceedings, Calibiration and Reliability in Groundwater modeling.J.C. Refsgaard, K. Kovar, E. Haarder, and E. Nygaard (editors), Denmark. IAHS.

[2] Rahman MA 2009 Coastal Vulnerabilities and its Integrated Management along Bangladesh Coast". *Proceedings* of the International conference on coastal environment and management– for the future human lives in coastal regions, pp. 68-78, held in Mie, Japan.more.

[3] Nowlan L 2005 Buried treasure: Groundwater permitting and pricing in Canada.Walter and Duncan Gordan Foundation, with case studies by Geological Survey of Canada, West Coast Environmental Law, and Sierra Legal Defence Fund.

[4] Antonakos A K and Lambrakis N J 2007 Development and testing of three hybrid methods for the assessment of aquifer vulnerability to nitrates, based on the drastic model, an example from NE Korinthia, Greece, *J of Hydrology*, 288– 304 doi:10.1016/j.jhydrol.2006.08.014

[5] Nobre R C M, Rotunno Filho O C, Mansur W J, Cosenza C A N and Nobre M M M 2007 Groundwater vulnerability and risk mapping using GIS,modeling and a fuzzy logic tool . *Journal of Contaminant Hydrology* **94** (2007) 277–292.

[6] AtiquRahman 2008 A GIS based DRASTIC model for assessing groundwater vulnerability in shallow aquifer in Aligarh, India. *Applied Geography* **28** (2008) 32–53.

[7] Thirumalaivasan D, Karmegam M and Venugopal K 2003 AHP-DRASTIC: software for specific aquifer vulnerability assessment using DRASTIC model and GIS. *Environmental Modelling & Software* **18** 645–656 doi:10.1016/S1364-8152(03)00051-3.

[8] Chen S K, Jang C S and Peng Y H 2013 Developing a probability-based model of aquifer vulnerability in an agricultural region *Jhydrol.***486** 494–504 doi.org/10.1016/j.jhydrol.2013.02.019

[9] Said Ettazarini 2006 Groundwater pollution risk mapping for the Eocene aquifer of Oum Er-Rabia basin, Morocco, *Springer Journal* Doi 10. 1007/s00254-006-0330-1.

[10] Aller L, Bennett T, Lehr J H, Pretty R J and Hackett G 1987 DRASTIC: A Standardized System for Evaluating Ground Water Pollution Potential Using Hydrogeologic Settings. US Environmental Protection Agency, Ada, Oklahoma (EPA-600/2-87-035)

[11] Durnford D S, Thompson K R, Ellerbrook D A, Loftis J C and Davies G S 1990 Screening methods for ground water pollution potential from pesticide use in Colorado agriculture: Colorado Water Resources Research Institute, Fort Collins.Completion Report, 157–165

[12] Evans B M and Myers W L 1990 A CIS-based approach to evaluating regional groundwater pollution potential with DRASTIC. Journal of Soil and Water Conservation, 242–245.

[13] Halliday S L and Wolfe M L 1991 Assessing ground water pollution potential from nitrogen fertilizer using a geographic information system.*Water Resources Bulletin* **27**, 237–245.

[14] Rundquist D C, Peters A J, Di L, Rodekohr D A, Ehrman R L and Murray G 1991 Statewide groundwater-vulnerability assessment in Nebraska using the DRASTIC/ GISmodel. *Geocarto international* **6** (2), 51–57.

[15] F ritch T G, McKnight C L, YeldermanJr J C and Arnold J G 2000 An aquifer vulnerability assessment of the paluxy aquifer, central Texas, USA, using GIS and a modified DRASTIC approach. *Environmental Management*, **25**, 337–345.

[16] Panagopoulos G, Antonakos A., Lambrakis N., Panagopoulos G, Antonako A., Lambrakis, N. 2006 Optimization of DRASTIC model for groundwater vulnerability assessment, by the use of simple statistical methods and GIS. *Hydrogeology Journal* .

[17] Awasthi A and Satyaveer C S 2011 Using AHP and Dempster–Shafer theory for evaluating sustainable transport solutions. *Environmental Modelling & Software*, **26** (6), 787-796. doi: 10.1016/j.envsoft.2010.11.010.

[18] Beynon M, Curry B and Morgan P 2000. The Dempster–Shafer theory of evidence: an alternative approach to multicriteria decision modelling. *Omega,* **28** (1), 37-50.

[19] Saaty T L and Vargas G L 1991 Prediction, Projection and Forecasting. Kluwer Academic Publishers, Dordrecht

[20] HO W 2008 Integrated Analytic Hierarchy Process and Its Applications–A Literature Review. Journal of European Journal of Operation Research. 186: 211–228.

[21] Dweiri F and Al-Oqla F M 2006 Material Selection Using Analytical Hierarchy Process *International Journal of Computer Applications in Technology*. **26**(4):82–189

[22] Ariff H, Salit M S, Ismail N and Nukman Y 2008 Use of Analytical Hierarchy Process (AHP) for selecting the best design concept.*JurnalTeknologi*, **49**(A):1–18.

[23] Ckakraborty S 2007 Assessing aquifer vulnerability to arsenic pollution using DRASTIC and GIS of North Bengal Plain: A case study of English Bazar Block, Malda District, West Bengal, India, **Vol**.7, No.1, Springer 2007.

[24]Chaturvedi RS 1973 A Note on the Investigation of Ground Water Resources in Western Districts of Uttar Pradesh". *Annual Report*, U. P. Irrigation Research Institute, 1973, pp. 86-122

[25]Samake M Tang Z, Hlaing W, Ndoh M I, Kasereka K and Waheed O B 2011 Groundwater Vulnerability Assessment in Shallow Aquifer in Linfen Basin, Shanxi Province, China Using DRASTIC Model. Journal of Sustainable Development, Vol. 4,No.1

[26] Heath RC 1987 Basic Ground-Water Hydrology. US Geol Surv Water Suppl Pap 2220, 84 pp.

[27] Anwar M, Prem C C and Rao V B (2003) Evaluation of groundwater potential of Musi River catchment using DRASTIC index model. In B. R. Venkateshwar, M. K. Ram, C. S. Sarala, & C. Raju (Eds.), Hydrology and watershed management. *Proceedings* of the international conference 18– 20, 2002 (pp. 399–409).

[28] Anbazhagan S, Ramasamy S M and Das G S 2005 Remote sensing and GIS for artificial recharge study, runoff estimation and planning in Ayyar basin, Tamil Nadu, India. *Environ Geol* **48**:158–170. doi:10.1007/s00254-005-1284-4.

[29] Adiat K A N, Nawawi M N M and Abdullah K 2012 Assessing the accuracy of GIS-based elementary multi criteria decision analysis as a spatial prediction tool – A case of predicting potentia zones of sustainable groundwater resources. *Journal of Hydrology*, Volume **440**, p.75-89.

[30] Doumouya I, Dibi B, Kouame K I, Saley B, Jourda J P, Savane I and Biemi J 2012 Modeling of favourable zones for the establishment of water points by geographical information system (GIS) and multicriteria analysis (MCA) in the Aboisso area (South –east of Cote d'Ivoire) Environmental earth sciences 08/2013; **67**(6). DOI: 10. 1007/s12665-012-1622-2.

[31]Pradhan B, Neshat A, Pirasteh S, Mohd Shafri H Z 2013 Estimating groundwater vulnerability to pollution using a modified DRASTIC model in the Kerman agricultural area, Iran. Environ Earth Sci.DOI 10.1007/s12665-013-2690-7.

[32] Feizizadeh B and Blaschke T 2013 GIS-multicriteria decision analysis for landslide susceptibility mapping: Comparing three methods for the Urmia lake basin, iran *Nat Hazards* (2013) **65**:2105–2128, DOI 10.1007/s11069-012-0463-1.

[33] Kalinski R J, Kelly WE, Bogardi I, Ehrman R L and Yamamoto P D 1994 Correlation between DRASTIC vulnerabilities and Incidents of VOC contamination of municipal wells in Nebraska. Ground Water Journal 32 (1), 31–34.

[34] McLay C D A, Dragden R, Sparling G and Selvarajah N 2001 Predicting groundwater nitrate concentrations in a region of mixed agricultural land use: a comparison of three approaches. Environmental Pollution 115, 191– 204.

[35] FAO 1995 Food and Agriculture Organization(FAO) . Environmental Impact Assessment of Irrigation and Drainage Projects, FAO, Rome.

[36] WHO 2004 Guidelines for Drinking Water Quality, 3rd Edn. Vol. 1, Recommendation, Geneva, 515 pp.

IOP Conf. Series: Earth and Environmental Science **20** (2014) 012003 doi:10.1088/1755-1315/20/1/012003

Use of morphometric analysis and self-organizing maps for alluvial fan classification: case study on Ostorankooh altitudes, Iran

Marzieh Mokarram[1*], Abdollah Seif[2], Dinesh Sathyamoorthy[3]
[1*]Deparment of Geographic Sciences and Planning, University of Isfahan
[2]Deparment of Geographic Sciences and Planning, University of Isfahan
[3]Science & Technology Research Institute for Defence (STRIDE), Ministry of Defence, Malaysia

E-mail: m.mokarram.313@gmail.com

Abstract. The aim of this study is to classify alluvial fans formed by high-gradient braided streams and torrents that discharge into the Oshtorankook altitudes in the Lorestan province, Iran. The morphology of the fans and their watershed is quantitatively described through estimated morphometric parameters. For relationships between geomorphological features of the fans and their drainage basins, self-organizingmaps (SOM) were used. In SOM, according to both qualitative data and morphometric variables, the clustering tendency of alluvial fans was investigated using 15 alluvial fans parameters. The results of the analysis showed that several morphologically different fan types were recognizedbased on their geomorphological characteristics in the study area. A strong positive relationship was found between the drainage basin area and size of the fan with a simple power function. In addition, the relationship between fan slope and drainage area was found to be negative and moderately strong with a simple power function.

1. Introduction

Alluvial fans are created on steep high power channels where reduced stream power and serve as a transitional environment between a degrading upland area and adjacent lowland [1]. The morphology of a fan resembles a cone segment with concave slopes that typically range from less than 25° at the apex to less than 1° at the toe [2]. The analysis of the main controlling factors on past and present fan processes (debris flows and stream flows) is a major concern in order to distinguish between the two dominant sedimentary processes on alluvial fan formation and evolution [3].

Alluvial fans occur in any climatic environment, such as temperate mountain areas [4], humid temperate [5] and humid tropical environments [6]. Several studies have explored the relationship between the size of fans and their contributing basins to understand the mechanisms of fan construction [7]. Other studies on fan-basin morphometric focus on differentiation between debris flow and fluvial dominated fans [4-8].

Alluvial fan morphometry has been studied since the early 1960s, with a number well-researched relationships developed, such as a drainage basin area to alluvial fan area. Denny [9] and Hooke [10] have described classic alluvial fan form process variables using morphometry. More recent morphometry

Content from this work may be used under the terms of the Creative Commons Attribution 3.0 licence. Any further distribution of this work must maintain attribution to the author(s) and the title of the work, journal citation and DOI.
Published under licence by IOP Publishing Ltd

research used geography information system (GIS) to explore alluvial fan morphology for determination of fans location and fast and accurate analysis [11,12]. Volker et al. [12] investigated formative surface processes using morphometry to identify formative surface processes generated by alluvial fan flows. Fan morphometry typically evaluates a number of alluvial fan control variables to demonstrate the impact of each of these variables on fans [12]. They used topographic maps, aerial photographs and derived equations, but were seldom paired up with a field analysis of landforms [13].

Self-organizing maps (SOM) have been applied as a clustering and projection algorithm for high dimensional data [14]. Modeling utilizing SOM has recently been applied to a wide variety of geomorphology fields. It was employed to cluster volcanic ash arising from different fragmentation mechanisms [15], assess sediment quality, and define mortality index on different sampling sites [16].In addition, Ferentinou et al. [17] used SOM to classify marine sediments. Karymbalis et al. [18] also used SOM to investigate the clustering tendency of alluvial fans.

The aim of this study is to perform the classification of alluvial fans, located in the Oshtorankooh altitudes in the Lorestan province, Iran, according to their morphological features, using morphometry and SOM. The investigation will use 15 morphometric variables for each fan as described in Table 1.

Table 1. Morphometric parameters of the fan deltas and their corresponding drainage basins measured for this study [19].

Symbol	Morphometric parameter
A_b	Drainage basin area
P_b	Perimeter of the drainage basin
ΣL_c	Total length of 20 m contour lines within the drainage basin
C_b	Basin crest
R_b	Basin relief
L_c	Total length of the channels within the drainage basin
Ap_f	Fan apex
D_b	Drainage basin density
S_b	Drainage basin slope($S_b=e\Sigma L_c/A_b$,wheree is the equidistance)
Cir_b	Drainage basin circularity ($Cir_b = 4\pi A_b/P_b^2$)
M	Melton's ruggedness number ($M= R_b A_b^{-0.5}$)
L_f	Fan length
A_f	Fan area
S_f	Fan slope
p_f	Fan concavity

2. Material and methods

2.1. Study area

The study area is the Oshtorankooh altitudes in the Lorestan province, Iran. This area is located at 33° 9′ to 33° 22′ N and 49° 12′ to 49° 23′ E with an area of 3,260.62 km^2 (Figure 1).The lowest and highest elevations in this area are 885 and 4,049 m respectively. Rainfall displays high inter-annual and seasonal variability, with the annual long-term precipitation of the study area being 425 mm.

This study is based on quantitative and qualitative data depicting the morphology and morphometry of fans and their catchments derived from fieldwork. In order to determine the role of the fluvial sediment supply for the evolution of the fan deltas, the correlation between geomorphological features of the drainage basins and features of their fan deltas was estimated.

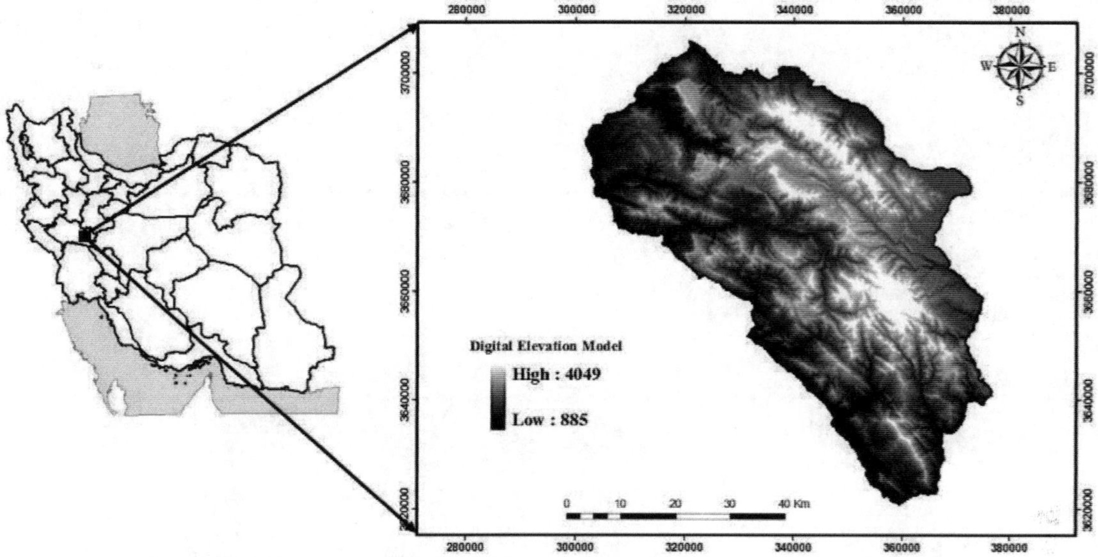

Figure 1. Location of the study area.

Drainage networks were delineated from aerial photograph interpretation and topographic maps with a scale of 1:50,000. The same maps, with 20 m contour lines, were used for the measurement of the morphometric parameters of the drainage basins.

In the study area, there are several sub-basins, with each having an alluvial fan(Figure 2). Their respective measured morphometric parameters are shown in Table 2. Maximum, minimum and mean elevations were calculated for each drainage basin using zonal statistics with the mapped basins and filled DEM as inputs.

Table 2. Morphometric parameters measured for the alluvial fans.

Fan	$A_b(km^2)$	$P_b(km)$	$\sum L_c(km)$	C_b	R_b	$L_c(km)$	Apf	D_b	S_b	Cir_b	M	L_f	$A_f(km^2)$	P_f	S_f
1	23.66	21.03	530.07	3006	1159	22.15	2006	0.94	0.45	0.67	0.24	1.47	0.44	2.95	0.32
2	43.61	27.09	1140.38	3509	1750	34.46	2509	0.79	0.52	0.75	0.26	1.51	0.42	3.03	0.23
3	32.90	23.46	601.72	2644	947	32.38	1644	0.98	0.37	0.75	0.17	0.81	0.16	1.62	0.25
4	6.09	10.64	113.97	2086	666	5.58	1086	0.92	0.37	0.68	0.27	1.44	0.40	2.88	0.46
5	20.26	19.50	389.26	2774	1310	18.37	1774	0.91	0.38	0.67	0.29	1.14	0.25	2.28	0.36
6	9.26	13.53	135.61	2362	718	12.05	1362	1.30	0.29	0.64	0.24	0.92	0.17	1.85	0.43
7	13.52	14.38	255.82	2893	1255	14.37	1893	1.06	0.38	0.82	0.34	0.93	0.19	1.86	0.41
8	18.74	18.90	374.16	2844	1522	19.42	1844	1.04	0.40	0.66	0.35	1.55	0.44	3.10	0.39
9	25.54	20.03	593.39	3200	1843	20.21	2200	0.79	0.46	0.80	0.36	1.51	0.46	3.02	0.27

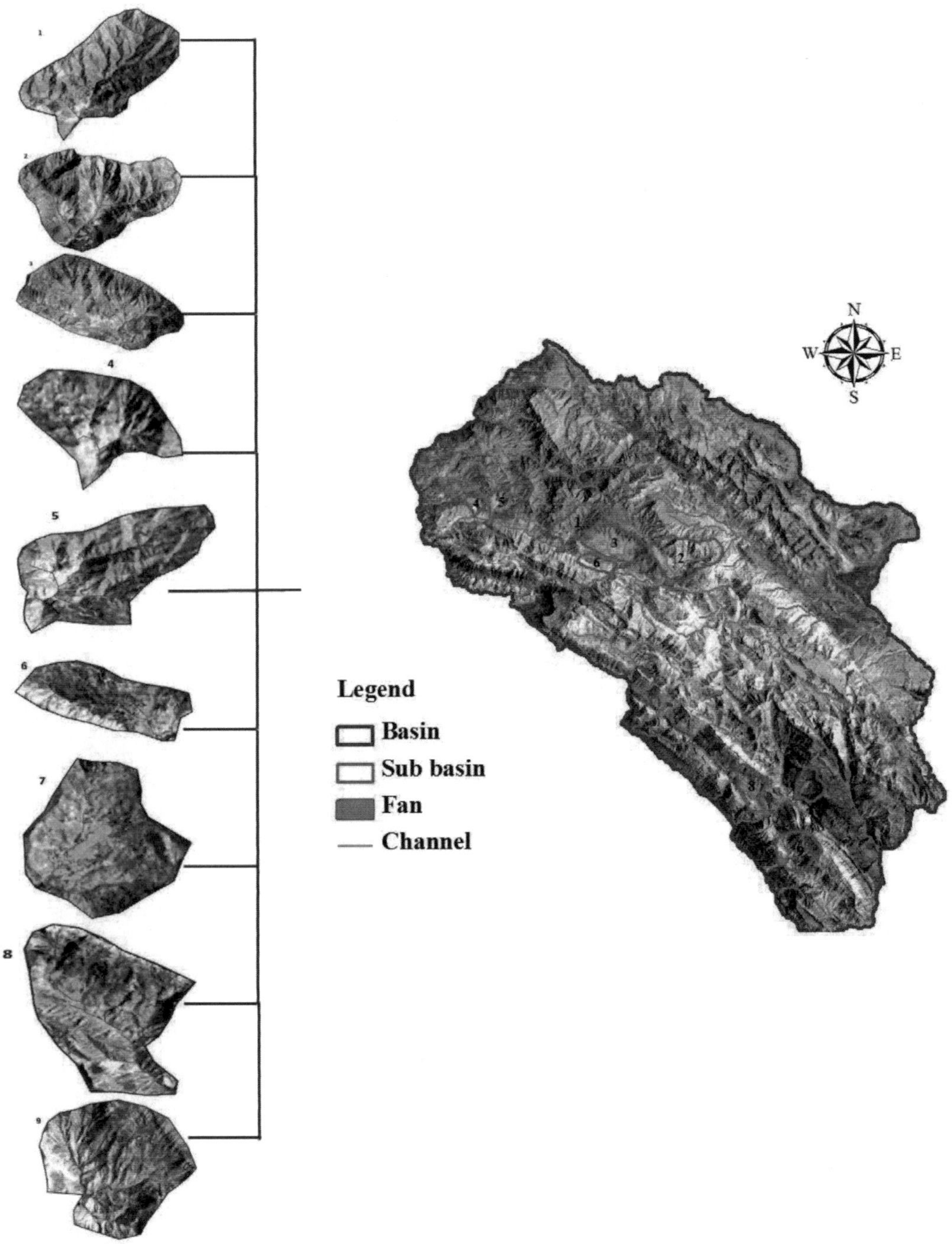

Figure 2. The sub-basins in the study area.

2.2. SOM

Artificial neural networks (ANNs) are non-linear mapping structures based on the function of the human brain, and are powerful tools for modelling when the underlying data relationship is unknown. ANNs can identify and learn correlated patterns between input data sets and corresponding target values, after which can be used to predict the outcome of new independent input data. ANNs imitate the learning process of the human brain and complex data even if the data is imprecise and hence, are ideally suited for the modeling of alluvial fan data, which are known to be complex and often non-linear. ANNs have great capacity in predictive modeling, whereby all the parameters describing the unknown situation can be presented to the trained ANNs [20].

SOM are unsupervised ANNs formed from neurons located on a regular, usually, two-dimensional regular planar array grid (Figure 3). It is based on unsupervised learning, which means that no human intervention is needed during the learning and little needs to be known about the characteristics of the input data.SOM offers a solution to use a number of visualizations linked together [21]. When several visualizations are linked together, scanning through them is very efficient because they are interpreted in a similar way. The U-matrix produced from SOM visualizes distances between neighboring map units and thus shows the cluster structure of the map. Samples within the same cluster will be the most similar according to the variables considered.

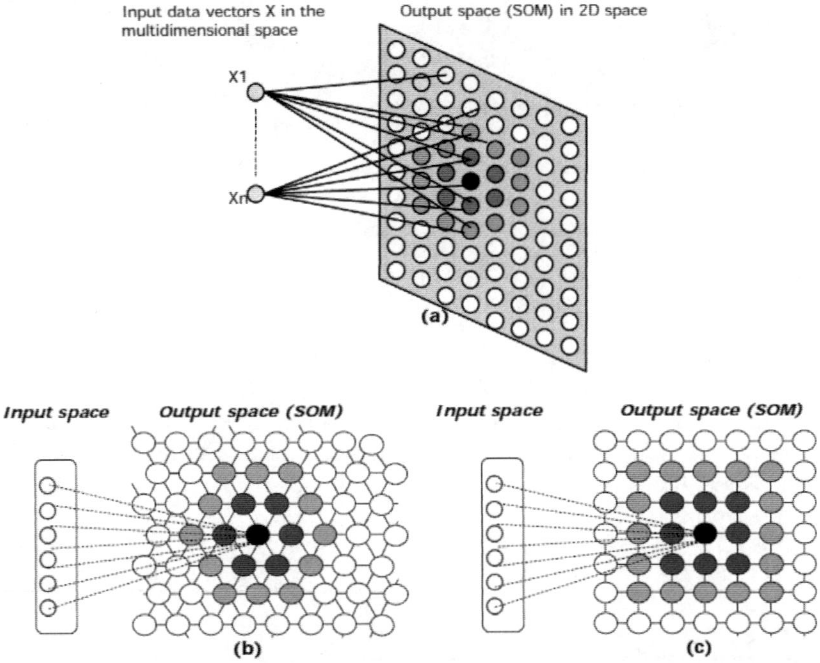

Figure 3. The structure of a SOM network: (a) Selection of a node and adaptation of neighboring nodes to the input data. The SOM grid can be (b) hexagonal or (c) rectangular. The black object indicates the node that was selected as the best match for the input pattern[22].

3. Results and discussion

3.1. SOM

SOM was applied for the study area to describe the alluvial fan morphometry. The visualization in Figure 4 consists of 16 hexagonal grids, with the U-matrix in the upper left, along with the 15 component layers (one layer for each morphometric parameter examined in this study).

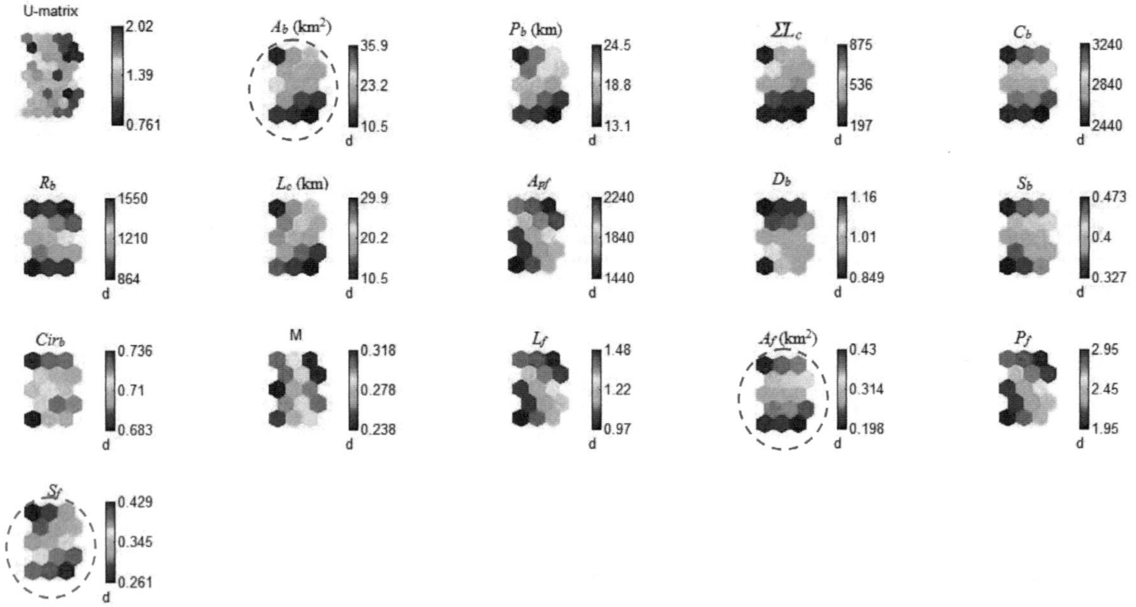

Figure 4. SOM visualization through U-matrix (top left) and 15 component layers (one layer for each morphometric parameter examined).

The warm colours in the layers represent the boundaries of the clusters, while cold colours represent the clusters themselves. In Figure 5, the same visualization is presented through hit numbers and post-it labels. The hit numbers in the polygons represent the record number of the dataset that belongs to the same neighborhood (cluster). Based on this, two clusters were generated. The records that belong to the same cluster are mapped closer and have the same colour.

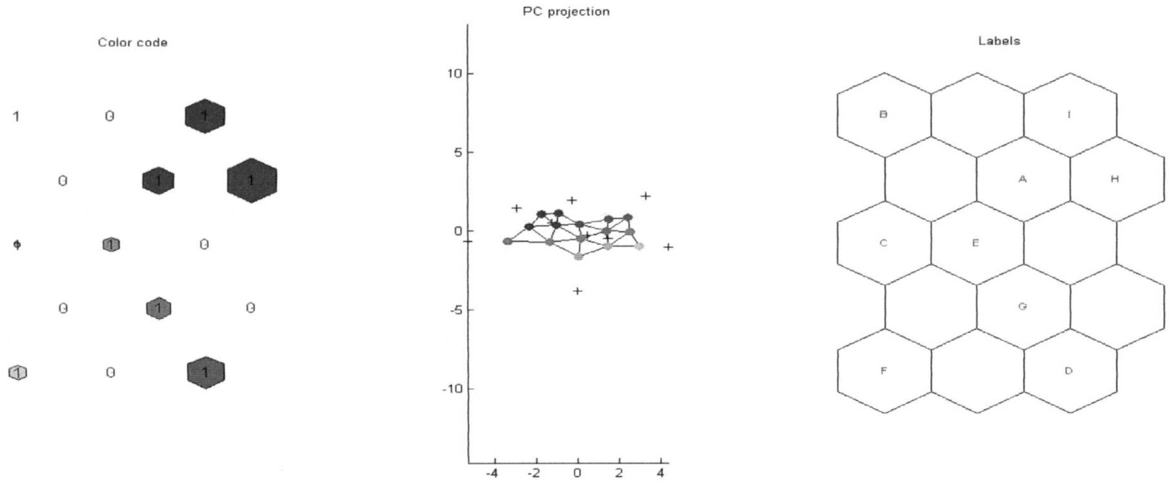

Figure 5. Different visualizations of the clusters obtained from the classification of the morphological variation through SOM: (a) Color code. (b) Principal component projection. (c) Label map with the names of the alluvial fans.

Based on the clusters for relationship between drainage basins and alluvial fans, it is found that:

- Drainage basin area A_b is correlated with fan area A_f.
- Fan slope S_f and drainage basin area A_b are inversely correlated.

Analysis of each cluster is then carried out to extract the rules that best describe each cluster by comparison of the component layers. The rules to predict the generation of alluvial fans are extracted by mapping the clusters presented in Figure 4 using the input morphometric parameters (component layers) in Figure 5. Using the SOM method, the fans in the study area are divided into two groups (Figure 6). The characteristics of each group of fans are provided in the Table 3.

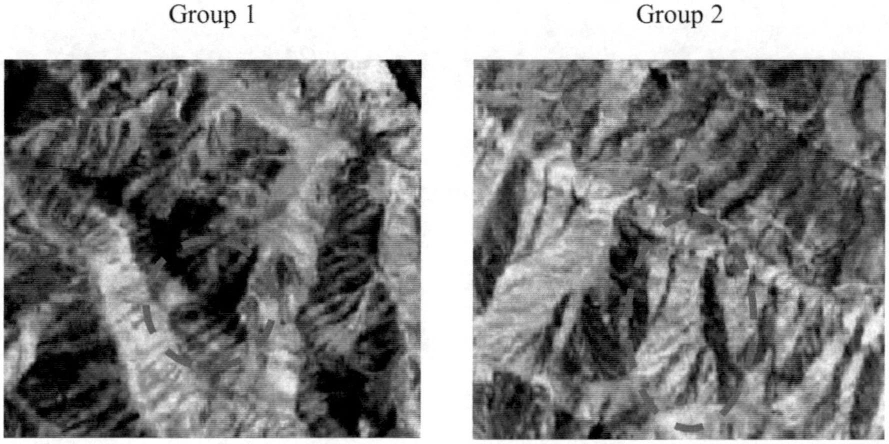

Figure 6. The two groups of fans in the study area.

Table 3. Characteristics of each group of fans in the study area.

Parameters	Group 1	Group 2
Fan area A_f(km^2)	0.19	0.43
Fan concavity P_f	1.9	3
Fan length L_f	0.95	1.5
Fan slope S_f	0.36	0.33

3.2 Morphometric relationships

The correlations between the geomorphic features of the studied fans and drainage basins include relationships between drainage basin area and fan area, drainage basin area and fan slope.

3.2.1. Drainage Area A_b and Fan Area A_f

Bull [23]was the first to describe quantitatively that as drainage basin area increases, the size of the alluvial fan also increases. He quantified the relationship with a simple power function of the form:

$$A_f = cA_b^k \tag{1}$$

where A_f is alluvial fan area, A_b is drainage basin area and c is an empirical derived coefficient representing the area of an alluvial fan with a drainage basin area of 1.0. The exponent k is the slope of the regression line and measures the rate of change in the fan area with increasing drainage basin area. By representing the relationship between these two parameters (Figure 7), it becomes clear that the data fits into a single exponential function:

$$A_f = 0.21A_b^{0.1096} \tag{2}$$

with a high correlation coefficient of 0.83, so there is a strong positive relation between A_b and A_f.

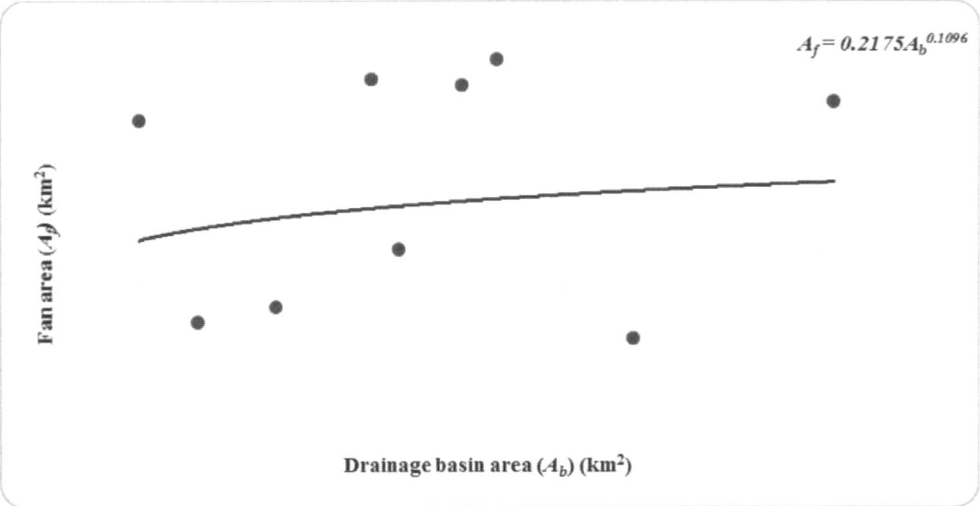

Figure 7. Correlation between drainage basin area $A_b$and fan area A_f for the studied fans and drainage basins.

The coefficient c, which according to Harvey [24] ranges from 0.1 to 2.2, has the value of 0.21 for the study area, similar with the value calculated for the coast of Gulf of Corinth, Greece [18]. In the literature, the different values of c and k for humid and sub humid area fans have been reported [25]. Table 4 includes values of c and k in the power law relationships between A_b and A_f from research works in various morphoclimatic environments. The value of the exponent for the study area agrees with the typical value for arid regions (about 0.9) [10] and 0.88 [18], and is higher than those derived from humid (0.58) and polar (0.65) regions. This value shows that the fans increase little in extension when the drainage area increases.

3.2.2. Drainage Area A_b and Fan Slope S_f

The relationship between the drainage basin area A_b and fan gradient S_f was investigated by Hooke [10] and Harvey *et al.*[1]. This relation is quantified with a simple power function of the form:

$$S_f = cA_b^k \tag{3}$$

Table 4. Values of coefficient c and exponent k in the power law relationship (A_f=cA_b^k) between drainage area and fan area from previous research works performed in various environments.

Location	*c*	*k*	Reference
Dellwood,North Carolina,USA	0.23	0.53	26
Roan Mountain, North Carolina, USA	0.38	0.76	27
General River Valley, Costa Rica	0.92	1.01	6
Banff, Alberta, Canada, fluvial fans,	0.48	0.32	28
debris-flow fans	0.17	0.48	28
single-group of fans	0.15	0.77	4
Japan	2.23	0.40	29
Central Alps, Northen Italy	0.29	0.33	25
Gulf of Corinth, Greece	0.13	0.88	18

The coefficient *k* is negative, indicating that increase of A_b results in decrease of S_f. In the study area, this relationship (Figure 8) can fit into the following function:

$$S_f = 1.0142A_b^{-0.377}$$

(4)$S_f = cA_b^k S_f = 1.0142A_b^{-0.377}$

The value of the exponent *k*(-0.377) falls within the range of values of -0.40 to 0, as given by Harvey[24]. In the coast of Gulf of Corinth, Greece, the calculated values of *c* and *k* are 0.14 and -0.45 respectively [18].

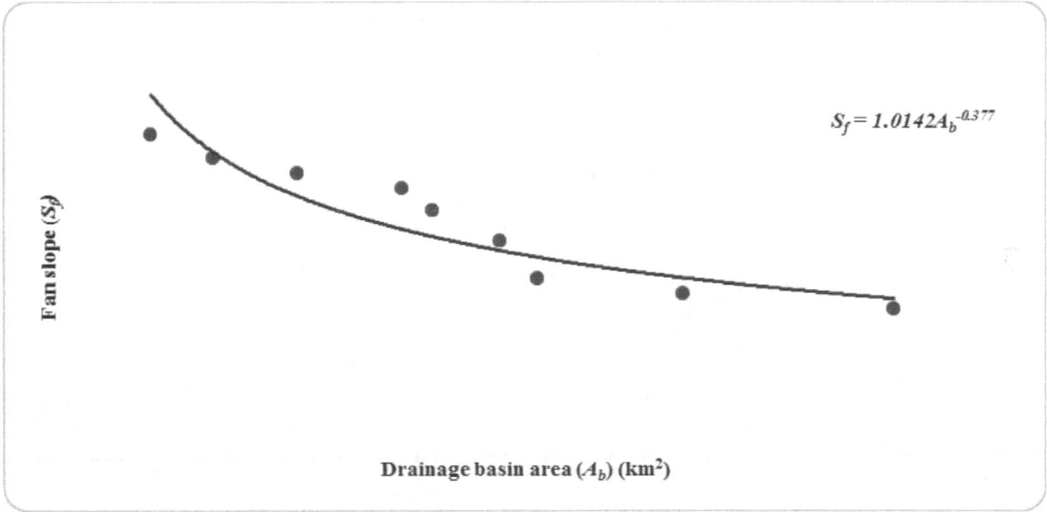

Figure 8. Correlation between drainage basin area A_b and fan slope S_f for the studied fans and drainage basins.

The values of exponent *k* from previous studies are 1.00 [18] and hence, Church & Mark [7] postulated that a general linear relationship exists for S_f versus *M*. For the study area, the value of *k* (0.26) indicates that the fan slope increases more rapidly than the ruggedness of the basin. According to De Scally & Owens [8] and Karymbalis et al. [18], debris-flow fans are supplied by basins with a higher value of *M* while the fans are steeper, smaller and less concave. This relationship provides an initial assessment of debris-flow potential for the broader study area, but this needs to be supported by field investigation of debris flows and sediment supply conditions in the basin.

4. Conclusion

The aim of this study is to determine the effectiveness of SOM as a clustering tool in the field of applied geomorphology. In this research, the geomorphological characteristics of alluvial fans in the Oshtorankooh altitudes were investigated. Qualitative observations, quantitative geomorphological analysis and application of SOM led to the definition of two main types of fans with different morphological features and the identification of certain correlation schemes between the studied parameters. Comparison of the geomorphological map and SOM showed that the applied methodology is a promising method for the mapping of drainage network channels. This method could be applied as a generic tool of alluvial fan classifier to larger datasets, in order to assess and interpret dominant formation processes through the study of multiple morphometric features describing alluvial fans and corresponding drainage basins. SOM provides an efficient scalable tool for the analysis of geomorphometric features as meaningful landform elements, leading to better understanding complex geomorphological systems.

5. References

[1] Harvey A M, Silva P G, Mather A E, Goy, J L, Stokes M and Zazo C 1999 The impact of Quaternary sea-level and climate change on coastal alluvial fans in the Cabo de Gata ranges, southeast Spain. *Geomorphology***28**, 1-22.

[2] Mwasiagi J I 2011 Self Organizing Maps - Applications and Novel Algorithm Design. ISBN 978-953-307-546-4Hard cover, 702 pages. Publisher InTech. Published online 21, January, 2011 Published in print edition January, 2011.

[3] Vandine D F 1996 Debris Flow Control Structures for Forest Engineering. Ministry of Forests, Research Program. *British Columbia.* 75pp.

[4] Giles P T 2010 Investigating the use of alluvial fan volume to represent fan size in morphometric studies. *Geomorphology***121**, 317-328.

[5] Kochel R C 1990 *Humid alluvial fans of the Apalachian Mountains.* In: Rachocki, A.H. & M. Church (Eds), Alluvial Fans: Field Approach,Wiley, Chichester, 109-129.

[6] Kesel R H and Spicer B E 1985 Geomorphic relationships and ages of soils on alluvial fans in the Rio General valley. *Costa Rica. Catena***12**, 149-166.

[7] Church M and Mark D M 1980 On size and scale in geomorphology. *Progress in Physical Geography***4**, 342-390.

[8] De Scally F A and Owens I 2004 Morphometric controls on fans in the SouthernAlps, New Zealand. *Earth Surface Processes and Landforms*, **29**, 311-322.

[9] Denny C S 1965 Alluvial fans in the Death valley region, California and Nevada.*US Geological Survey Professional Paper*, 466.

[10] Hooke R L B 1968 Steady-state relationships on arid region alluvial fans in closed basins. *American Journal of Science***266**, 609- 629.

[11] Whipple K X, Dunne L 1992 The influence of debris-flow rheology on fan morphology, Owens Valley California. *Geological Society of American Bulletin*, **104**: 887 - 900.

[12] Volker H X, Wasklewicz T A and Ellis M A 2007 A topographic Fingerprint to Distinguish Alluvial Fan Formative processes. *Geomorphology,* **88**: 34 - 45.

[13] Bobbins KL 2011 *Developing a form-process framework to describe the functioning of semi-arid alluvial fans in the Baviaanskloof valley, South Africa*, Rhodes University. 174.

[14] Kohonen T 1995 *Self - Organizing Maps* . Springer, Berlin.Lambeck, K., 1996. Sea-level change and shore-line evolution in Aegean Greece since UpperPalaeolithic Time. Antiquity **70**, 588-611.

[15] Ersoya O, Aydara E, Gourgaudb A, Artunerc H and Bayhan H 2007 Clustering ofVolcanic Ash Arising from Different Fragmentation Mechanisms Using KohonenSelf-Organizing Maps. Computers and Geosciences **33**, 821-828.

[16] Tsakovski S, Kudlakb B, Simeonova V, Wolskab L and Namiesnikb J 2009 Ecotoxicity and chemical sediment data classification by the use of self-Organizing maps. *Analytica Chimica Acta***631**, 142–152.

[17] Ferentinou M, Hasiotis T and Sakellariou M 2010 *Clustering of geotechnical properties ofmarine sediments through self – organizing maps: An example from the ZakynthosCanyon–Valley system,*

Greece. In: Submarine Mass Movements and their consequencesIV , Mosher, D., Shipp, C., Moscardelli, Chaytor J, Baxter C, Lee H and Urgeles R, (Ed.), Advances in Natural and Technological Hazards Research, v. 28, 43-54,Springer, The Netherlands.

[18] Karymbalis E, Gaki- Papanastasiou K, Ferentinou M 2011 Coastal Fan DeltasClassification Along the NW Coast of Gulf Of Corinth, Greece CouplingMorphometric Analysis and Artificial Neural Networks. *Hellenic Journal of Geosciences* , vol. 45, (in press).

[19] Karymbalis E, Gaki-Papanastassiou K and Maroukian H 2007 Recent geomorphic evolution of the fan delta of the Mornos river, Greece: natural processes and human impacts. *Bulletin of the Geological Society of Greece*, XXXX, 1538-1551.

[20] Dhubkarya D C, Nagariya D, Kapoor R 2010 Implementation Of A Radial Basis Function Using VHDL. *Global Journal of Computer Science and Technology,* 16-19.

[21] Buza A, Mcdonald JA, Michalak J and Stuetzle W 1991 Interactive data visualization using focusing and linking. Proceedings of IEEE conference of visualization, 156-63.

[22] Dykes J 2005 *Exploring Geovisualization*. Elsevier.

[23] Bull W B 1962 Relations of alluvial fan size and slope to drainage basin size and lithology in western Fresno Country. CA. *US Geo- logical Survey Professional Paper*, 450-B, 51-53.

[24] Harvey A M 1997 *The role of alluvial fans in arid zone fluvial systems*. In: Thomas D.S.G. (Ed.), Arid Zone Geomorphology: Process, Form and Change in Dry lands, John Wiley and Sons Limited, Chichester, 231-259 pp.

[25] Crosta G and Frattini P 2004 Controls on modern alluvial fan processes in the central Alps, Northern Italy. *Earth Surface Processes and Landforms***29**, 267-293.

[26] Mills H H 1982 Piedmont-cove deposits of the Delwood quadrangle, Great Smoky Mountains, North Carolina, USA: Morphometry. *Zeitschrift fur Geomorphologie Neue Folge***26**, 163-178.

[27] Mills H H 1983 Piedmont evolution at Roan Mountain, North Carolina, USA. *Geografiska Annaler A***65**, 111-126.

[28] Kostaschuck R A, MacDonald G M and Putnam P E 1986 Depositional processes andalluvial fan-drainage basin morphometric relationships near Banff, Alberta,Canada. *Earth Surf. Proc***11** , 471–484.

[29] Oguchi T and Ohmori H 1994 Analysis of the relationships among alluvial fan area, source basin area, basin slope, and sediment yield. *Zeitschrift fur Geomorphologie Neue Folge***38(4)**, 405-420.

IOP Conf. Series: Earth and Environmental Science **20** (2014) 012004 doi:10.1088/1755-1315/20/1/012004

Leveraging of remote sensing and GIS on mapping in urban and regional planning applications

Norzailawati Mohd Noor[1], **Alias Abdullah and Nur Aulia Rosni**
Urban and Regional Planning Dept, Kulliyah Architecture and Environment Design, International Islamic University Malaysia, P.O Box 10, 50278, Kuala Lumpur

E-mail: norzailawati@iium.edu.my

Abstract. While remote sensing applications represent a major though still underused source of urban data, the proposed combination between remote sensing and Geo-information System (GIS) in urban and regional planning is not fully explored. In order to measure changes in land use, the need of platform in monitoring, recording, and predicting the changes is necessary for planners and developers. In advance technology of mapping process, remote sensing and GIS as tools for urban planning are already recognised. But, due to lack of implementation and awareness about the benefits of these tools, these terms look unusual. Therefore, this paper reviews the history of remote sensing and GIS in urban applications, technical skills and the challenges, and future development of remote sensing and GIS especially for urban development particularly in developing countries.

1. Introduction

In urban planning, mapping plays an important role in the development of the city. An urbanised, well-developed city has industrial, commercial, and residential areas evolved from time to time. [1] stated that urban and regional planning is the art and science of ordering the use of land and the character and sitting of buildings and communication routes so as to secure the maximum practicable degree of economy, convenience, and beauty. Nowadays, remote sensing and Geo-information System (GIS) have been introduced as an initiative to map the urban area. According to [2], urban remote sensing presents renewed opportunities for analysis of the city as a system. Urban planning is essential especially in the developing country to ensure that the city is well-planned and to preserve the sustainability of the environment. Urban planning spatial database analysis is a subset of analytic method in which the objects are repositioned within it [3]. This analysis gives detailed information related to cadastral plan, such as area, perimeter, and lot numbering [5]. However, urban mapping is quite complicated when dealing with remote sensing imagery data sources because of the atmospheric disturbance in high altitude that affects the spectral response of an object such as rooftop. In addition, the rooftop has different type of materials and colours, so the spectral response of the object varies too [6]. Thus, to overcome the problem of mixing pixels, object-oriented classification is applied. Pixels are combined and represent one object. Therefore, this paper reviews the leveraging of remote sensing and GIS for urban mapping in the context of history, techniques, and challenges, as well as the future of remote sensing and GIS in urban mapping.

[1] To whom any correspondence should be addressed.

Content from this work may be used under the terms of the Creative Commons Attribution 3.0 licence. Any further distribution of this work must maintain attribution to the author(s) and the title of the work, journal citation and DOI.
Published under licence by IOP Publishing Ltd

2. History of remote sensing and GIS in urban mapping

GIS was originally developed as an environment technology. According to [7], the term "Geographic Information System" (GIS) started in early 1960s in the project of mapping Canada's natural resource. A scholar in civil engineering and planning, Prof. Edgar Harwood from the University of Washington worked on some of the earliest computer mapping software. The finding stated that the Urban and Regional Information System Association has conducted high numbers of influenced conferences and short courses [7-8]. In the early 1990s, GIS began to expand into business market as it became available in personal computer. Since then, it had become more valuable for a broader spectrum of business users [9] and GIS in urban planning trends was identified mostly after that [7][10][11]. According to [12-14][16], GIS is also used in specific planning in non-American countries. Land use is one of the most important criteria in urban development, by using GIS, land uses are being recorded, mapped, monitored, and changes has been updated, planned, and predicted [15-16].

The first Earth observation data used in urban system application were carried out in a map scale of 1:100 000 (80 to 30 m spatial resolution) on the city of North America [17-18]. Later on, the higher spatial resolution data were available with 10m to 20 m enable mapping with more detail with the scale 1:100 000 to 1:25 000 for instance to assess the compact urban area on European cities [4][19]. The higher spatial resolution is capable to provide a notable contribution on rural-to-urban land conversion [18-20], including in estimating housing and population density [21-24].

In 1990s, the use of GIS in town planning in Malaysia started, the same year in which GIS had become popular all over the world. The application of GIS is expected to increase the rationality in generating better planning decision. The success of GIS implementation in urban planning depends on complete and updated data. According to Act 172: Town Planning Act, in urban planning, there are three major levels that encompass the development plans namely National Physical Plan (Federal Level), Structure Plan (State Level), Local Plan, and Special Area Plan (Local Level). In Malaysia, there are certain areas that have been tested using GIS technology for urban planning, such as Kuala Lumpur City Council, City Development Management System and Safe City Monitoring System. GIS applications are not only used in planning process, but they are also used for monitoring purposes, thus the data in the applications should be as recent as possible. In addition, the remote sensing applications are also being used in order to get the most recent data in the fastest ways.

3. Remote sensing as a platform to urban applications and its challenges

Generally, remote sensing activities include recording, observing, and perceiving (sensing) object or targets far away from that place (remote). By using remote sensing technology, the sensor scans the object on Earth and the output will be represented in digital images that contain digital number and spectral reflectance of the object of interest. Rooftop from brick material, an example of an object of interest, has different digital numbers and reflectance. Digital number, radiance, and reflectance can each be defined in various definitions. Digital number is the value of pixel that describes the image that has not yet been calibrated into physically meaningful units. Radiance is the amount of radiation emitted from an area including radiation from neighbouring surface, cloud, and the target itself. Reflectance is the proportion of the radiation striking a surface to the radiation reflected off of it. Therefore, materials can be identified by reflectance spectra. So, these materials are required to correct an image to reflectance as a primary step. According to [25], the spectral library for urban area acquired from 400–1100 nm is the range from visible band to infrared band. Other than that, urban built-up materials (e.g., roofs and roads) are best identified and separated by wavelength [26-28]. Hyper spectral data can give good results in separation of different materials of rooftops because the bandwidth is specific and the range is very small. Roofs and roads have specific absorption features associated with mineral or material properties that make them up. However, the interaction of atmospheric condition disturbs the reflectance of materials from rooftop, so the spectral reflectance of target does not represent the actual reflectance. In order to overcome these problems, atmospheric correction is crucial to remove the presence of spectral reflectance aerosol. There are many types of atmospheric correction methods such as MOTRAN [29] and 6s [30]. However, in remote sensing, the assumption of flat ground is used to model the signal at ground and sensor level even though in cities with very high spatial resolution, this assumption is no longer valid according to Thomas *et al* [31] because of the complexity introduced by relief. Therefore, it requires a new formalism adapted in this case namely AMARTIS v2 [31].

The interpretation of different urban objects is not only subject to radiometric factor as the spectral reflectance values can correspond to too many objects. Confusing features of extraction can be made. For instance, a tiled roof does not reflect the same way as a concrete low roof or terrace. On the other hand, geometric factor only arises more problems because a similar geometric shape can refer to too many objects. Therefore, geometric and radiometric factors should be combined so that urban structure can be identified more easily [32]. In Puissant and Weber's research, they found that 5 m resolution was adapted to identify wide road of approximately 15 m, while 2.5 m resolution allowed secondary road network to be identified. In order to study urban morphology and to distinguish three main constitutive objects of urban structure, normally resolution that is less than 0.80 m is necessary. If the cases is the houses is urban smaller object less 5 m inside a functional of 0.80 m resolution is necessary. Puissant and Weber also found that automatic extraction feature can be done using lower than 1 m spatial resolution. However, some adjustments still need to be done. Finally, simple visual interpretation remains the most accurate method because human operator can extract the information from very high spatial resolution relatively easily by considering radiometric and geometric factors [33].

4. Future GIS in urban mapping

Remote sensing methods have been widely applied in mapping land surface features in urban areas. Several recent developments in remote sensing have the potential to significantly improve the mapping of urban areas [2]. A new technology of GIS namely hybrid GIS solution combines spatial database and server-based geospatial processing system with a consumer-oriented geospatial visualisation components. The hybrid GIS can support the urban development by creating opportunities for technology communities and planners to embellish the process of plan development and management through modelling. Before this, urban planners are facing a problem with maps that are not geometrically correct and unscaled and they only rely on traditional GIS tools for urban development planning. The evolution of GIS into an enterprise-wide system has led to the identification of potential problems in urban development. Indeed, this technology enables many agencies to respond to those problems efficiently and the results are rapidly shared with the public. These Internet-based mapping platforms are able to create public-facing portal that leverages digital database of land parcel, whereas the land parcel is created by using aerial imagery and satellite as well as surveying with GPS. In the presence of GPS, the location of an area is determined so that better planning for urban growth can be decided and made. With this hybrid GIS, planners can develop 3D city model, terrain for interactive mapping, and Internet-based urban information system featuring aerial view. This powerful tool is capable to view city for example zoning, parcels, building outlines, address, land contour, and slope. It also includes areas prone to flooding as a permit for construction, demolition, and land use changes. The benefits are not just for planning and development purposes but also for sustaining the urban development in order to avoid anthropogenic disaster in the city.

5. Conclusion

Remote sensing and GIS really help the planners to monitor the changes in urban land use. Besides, the invention of high spatial resolution imagery allows the automatic extraction of urban features to be done easily, added with prior knowledge of image interpretation. Hyper spectral imagery is able to distinguish the materials of rooftop because of the sensitivity of the sensor and the range of bandwidth is small and precise. The evolution of future GIS requires a high capacity of processors in the computer, in order to process the data efficiently.

6. Reference

[1] Keeble, L. (1969). Principles and Practice of Town and Country Planning. London: Estate Gazette.

[2] Herold, M.,Couclelis, H and Clarke, K.C (2005). The role of spatial metrics in the analysis and modeling of urban land use change. Computers, Environment and urban systems 29, pp369-399.

[3] Goodchild, M. F, and Longley, P.A (1999). The Future of GIS and Spatial Analysis, In Geographic Information Systems: Principle, Techniques, Management and Applications.p.p. 567-580.

[4] Welch, R (1982). Spatial Resolution Requirements for Urban Studies. International Journal of Remote Sensing 3.pp. 139-146.

[5] Donnay, J.P, Barnsley, M.J and Longley, P.A (2001). Remote Sensing and Urban Analysis. New York.

[6] Herold, M., Gardner, M. and Roberts, D. (2003). Spectral Resolution Requirements for mapping urban areas, IEEE Trans. Geoscience Remote Sensing 41(9), pp. 1907-1919.

[7] Tomlinson, R. (1998). The Canadian geographic information system. In T.W.Foresman (Ed), The history of geographic information system: Perspective From Pioneers (pp.21-32). Upper Saddle River, NJ:Prentice Hall.

[8] Chrisman, N.R (1998). Academics origins of GIS. In T. W. Foresman (ed.). The history of geographic information system: Perspective From Pioneers (pp.33-43). Upper Saddle River, NJ:Prentice Hall.

[9] Castle, G. H (Ed.)(1993).Profiting from a geographic information system. Ft.Collins, CO:GIS World.

[10] Deuker , K.J (1979) Land Resource Information System; A review of Fifteen Years Experience. Geo-Processing ,1, 105-128.

[11] Wellar, B.S (1975) Urban data management system. In Canada: A federal perspective.Paper from the 12th Annual URISA Conference, Urban and Regional Information Systems; Resources and Results, Vol. 1,pp.377-391. Montreal.

[12] Warnecke, L. (1992) State geographic information activities compendium. Lexington:Council for State Governments.

[13] Vonderohe, A. P. and Saleh, R. (1991) Status of GIS within state transportation agencies (abstract). Technical Papers 1991 ACSM-ASPRS Annual Convention, 2 Cartography and GISLIS, 324, Baltimore.

[14] Huxhold, W. E. (1990). An introduction to urban geographic information systems. Oxford: Oxford University Press.

[15] Engelen, G., Geertman, S. Smits, P. & Wessels, C. (1999).Dynamic GIS and strategic physical planning support; a practical application. Geographical Information and Planning, London, Springer, pp.87-111.

[16] Stillwell, J, Geertman, S. & Openshaw, S (1999). Developments in Geographical Information Systems and Planning Geographical Information and Planning, London, Springer.pp 3-22.

[17] Forster, B.C. (1980) Urban residential ground cover using Landsat digital data, Photogrammetric Engineering and Remote Sensing46: 547-558.

[18] Jensen, J.R. (1983) Urban / suburban land use analysis, in R.N. Colwell (Ed.) Manual of remote sensing, Second Edition, pp15711666.

[19] Forster, B.C. (1985) An examination of some problems and solutions in monitoring urban areas from satellite platforms, of Environment, 6:139-151.

[20] Gomascara, M.A., Brivio, P.A, Pagnogni, F. and Galli, A. (1993) One century of land use changes in the metropolitan Milan (Italy), International Journal of Remote Sensing, 14:211-223.

[21] Lo, C.P. (1993) A GIS approach to population estimation in a complex urban environment using SPOT multispectral images, International Archives of Photogrammetry and Remote Sensing29:935-941.

[22] Dureau, F. and Weber, C. (1995) Télédétection et système d'informations urbains, Anthropos, Paris, 357p.

[23] Baudot, Y. (1994) TélÔnÆuÔs•{on aérospatiale en analysis géographique de la population des

villes dans les pays en développement, Ph.D Catholic University of Louvain, Belgium, 350p.

[24] Weber, C. (2000) Urban agglomeration delimitation using remote sensing data, GISDATA 9, Taylor & Francis, Great Britain, 8:145159.

[25] Ben-Dor, E., Levin, N., and Saaroni, H., (2001). "A spectral based recognition of the urban environment using the visible and near-infrared spectral region (0.4–1.1 m). A case study over Tel-Aviv," Int. J. Remote Sens., vol. 22, no. 11, pp. 2193–2218, 2001.

[26] Heiden, U., Roessner, S., Segl, K., and Kaufmann, H., (2001). "Analysis of spectral signatures of urban surfaces for their area-wide identification using hyper spectral Hy-Map data," I Proc. IEEE-ISPRS Joint Workshop on Remote Sensing and Data Fusion over Urban Areas, Rome, Italy, 2001

[27] Hepner, G.F., Houshmand, B., Kulikov, I., and Bryant, N., (1998). Investigation of the integration of AVIRIS and IFSAR for urban analysis, Photogrammetric. Eng. Remote Sens., vol. 64, no. 8, pp. 813–820, 1998.

[28] Hepner, G.F., and Chen, J., (2002) "Investigation of imaging spectroscopy for discriminating urban land covers and surface materials," in Proc. AVIRIS Earth Science and Applications Workshop, Palo Alto,CA, 2002, Available: [Online] http://popo.jpl.nasa.gov/docs/workshops/01_docs/2001Chen_web.pdf.

[29] Berk, A; Anderson, G.P.; Bernstein, L.S.; Acharya, P.K.; Dothe, H.; Matthew, M.W. MODTRAN4 radiative transfer modelling for atmospheric correction. Proc. SPIE 1999, 3756, 348-353.

[30] Vermote, E.F.; Tanre, D.; Deuzé, J.L.; Herman, M.; Morcrette, J.J. Second simulation of the satellite signal in the solar spectrum, 6S: An overview. IEEE Trans. Geosci. Remote Sens. 1997, 35, 675-686.

[31] Thomas, C., Doz, S., Briottet, X., Lachérade, S., (2011), AMARTIS v2: 3D Radiative Transfer Code in the Spectral Domain Dedicated to Urban Areas. ISSN 2072-4292 article Remote Sensing www.mdpi.com/journal/remotesensing.

[32] Puissant, A., and Weber, C., (2002). The utility of very high spatial resolution images to identify urban object. Geocarto International, Vol. 17, No. 1, March 2002.Published by Geocarto International Centre, G.P.O. Box 4122, Hong Kong.

[33] Bähr, H.P. (2000). Image segmentation for change detection in urban environments, GISDATA 9, Taylor & Francis, Great Britain,6:95-113.

IOP Conf. Series: Earth and Environmental Science **20** (2014) 012005 doi:10.1088/1755-1315/20/1/012005

Study for urbanization corresponding to socio-economic activities in Savannaket, Laos using satellite remote sensing

S Kimijiama and M Nagai

School of Engineering and Technology, Asian Institute of Technology, Moo 9 km 42 Phaholyohin Rd, Klong Nueng, Klong Luang, Pathumthani 12120, Thailand

E-mail: st114307@ait.ac.th

Abstract. In Greater Mekong Sub-region (GMS), economic liberalization and deregulation facilitated by GMS Regional Economic Corporation Program (GMS-ECP) has triggered urbanization in the region. However, the urbanization rate and its linkage to socio-economic activities are ambiguous. The objectives of this paper are to: (a) determine the changes in urban area from 1972 to 2013 using remote sensing data, and (b) analyse the relationships between urbanization with respect to socio-economic activities in central Laos. The study employed supervised classification and human visible interpretation to determine changes in urbanization rate. Regression analysis was used to analyze the correlation between the urbanization rate and socio-economic variables. The result shows that the urban area increased significantly from 1972 to 2013. The socio-economic variables such as school enrollment, labour force, mortality rate, water source and sanitation highly correlated with the rate of urbanization during the period. The study concluded that identifying the highly correlated socio-economic variables with urbanization rate could enable us to conduct a further urbanization simulation. The simulation helps in designing policies for sustainable development.

1. Introduction

Asian Development Bank (ADB) initiated the Economic Corporation Program (ECP) in the Greater Mekong Sub-region (GMS) in 1992 and is known as GMS-ECP. The GMS-ECP was started to promote mutual economic and social development by strengthening corporation among the GMS basin countries. The GMS basin countries include Cambodia, Lao People's Democratic Republic (Laos), Myanmar, Thailand, Vietnam and Yunnan Province in the People's Republic of China (PRC). The GMS-ECP facilitated the market-liberal policies which provide excellent opportunities to develop the sectors of 1) transportation; 2) telecommunication; 3) energy; 4) human resources; 5) environment; 6) trade; 7) investment; 8) tourism; and 9) agriculture in the region.

The impact of GMS-ECP, for example in case of Laos, can be indicated by the amount of investment, number of projects implemented in different sectors, and the growth of GDP mainly due to the program. Indeed, the total amount invested during 2000-2010 by the top 10 investing countries is 10,660 million US dollar. These donor countries are Australia, China, France, India, Japan, Korea, Malaysia, Norway, Thailand and Vietnam. Among these donor countries, 76.6 percent consists of investment from Vietnam, China and Thailand. Furthermore, the total number of projects implemented during the period is 1,280. Among them, major sectors invested are electricity generation, mining, service, agriculture and industry/handicraft (Investment Promotion Department, 2010). Additionally, the growth of GDP in Laos from 2006 to 2012 shows 6.5 to 8.5 percent by year and this higher growth rates are resulted from the development of resource sectors such as electricity generation, mining, agriculture and service (JETRO, 2013). These developments largely due to the GMS-ECP also

Content from this work may be used under the terms of the Creative Commons Attribution 3.0 licence. Any further distribution of this work must maintain attribution to the author(s) and the title of the work, journal citation and DOI.
Published under licence by IOP Publishing Ltd

accelerated urbanization by providing basic needs such as infrastructure.

Although the GMS-ECP contributed significantly to the rate of urbanization over two decades, time series change of its rate is not clearly known in least developing country in the region like Laos. In case of developed country like USA, relationships between vegetation amount represented by Normalized Spectral Mixture Analysis (NDVI) and socio-economic variables were studied (Lo, *et al*, 1997; Lo, 1998; and Mennis, 2006). Similar case might be true for countries like Laos; however, literatures linking such changes in Laos including other developing countries in the region is sparsely documented.

Although many researches have been conducted in developing countries relating urbanization affecting land cover but its linkage to the socio-economic activities in sub-urban areas in developing country like Laos is not clearly known. One of the tools to measure the rate of urbanization over time is satellite remote sensing technology. Remote sensing technology is a powerful tool to observe land cover and detect amount of area by assigning classes without direct interaction. This would also help understanding the trend of urban development and interaction between the urbanization and socio-economic variables. Therefore this paper (a) determines the rate of urbanization from 1972 to 2013 by using Landsat MSS, TM and ETM+ remote sensing data and (b) analyzes the relationships between the rate of urbanization with respect to socio-economic activities such as education, literature rate, mortality, environmental facilities, labour force, telecommunication and FDIs in central Laos. This paper is expected to contribute to rank area-specific variables for a next step that is a simulation of urbanization.

2. Data and methods

The date used for this study comprises of Landsat MMS, TM and ETM+ images and information from the ADB that includes education, literature rate, mortality, environmental facilities, labour force, land use, telecommunication and FDIs. The set of images included cloud-free parts acquired from January 22, 1973; December 22, 2000; November 22, 2001; November 10, 2002, December 15, 2003; January 2, 2005; December 15, 2006; November 18, 2007; January 29, 2009; December 15, 2009, December 26, 2010; January 6, 2012, and January 24, 2013. For image interpretation, the study used Environment for Visualizing Images (ENVI) and Earth Resources Data Analysis System (ERDAS) software. The complete process of the study can be envisioned as four major steps: 1) image preparation, 2) image classification, 3) area calculation and 4) analysis of urbanization corresponding to socio-economic activities.

2.1. Image preparation

The first step was to conduct gap filling of the Landsat 7 ETM+ Scan Line Corrector off imageries taken in 2003, 2005, 2007, 2009, 2010, 2012 and 2013 by using ERDAS. After the process, the study created subsets of central Savannaket.

2.2. Image classification

The study applied supervised classification and human visible interpretation in order to identify urban area in time series. For the supervised classification, the study created 30 training data (ROIs) and assigned different colors on each ROIs and conducted Maximum Likelihood classification. The study selected ground points from the latest image available from Google Earth captured in February 2012. The study selected totally 40 ground truth points for the accuracy assessment. The supervised classification showed 79.8 percent of accuracy. As visualization of the images was limited due to the Scan Line Corrector (SLC) problem of the Landsat ETM+, the study considered these accuracy rates acceptable. At the same time, for the human visible interpretation, digitization was done on images visualized by R:G:B (4:3:2) with the reference of GoogleMap and online data source in the ArcMap.

2.3. Area calculation
After checking the accuracy, the study calculated the urban areas generated from supervised classification by ENVI and human visible interpretation by ArcGIS in time series and visualize them.

2.4. Analysis of urbanization corresponding to socio-economic activities
The study obtained socio-economic variables from ADB. The selected socio-economic variables and available time period of data were 1) telephone lines (1960-2011), 2) literacy rate, youth total (% of people ages 15-24) (1995-2005), 3) school enrolment (primary) (1971-2011), 4) morality rate under 5 (per 1,000 live births) (1966-2011), 5) improved water source (% of population with access) (1994-2010), 6) improved sanitation facilities (% of population with access) (1994-2010), 7) labour force (1990- 2011) (1990-2011) and 8) FDI, net inflows (US$) (1985-2011). The data is available only at country level. However the Savannaket province is the second largest province in Laos; therefore, the study considered that the country level data is applicable to the study area to find out correlations of parameters to the rate of urbanization. Furthermore, it is difficult to find out information of parameter-relating project implementations; therefore, this study considered the yearly data as an input activity. In order to find out scientific correlations between socio-economic variables and the rate of urbanization, the study employed regression model by using Microsoft Excel.

3. Result and discussion
The results are obtained using two methods (Fig1). Different methods showed different rate of urbanization. Such differences might have been occurred due to four technical errors and limitations; namely accuracy error during the gap-filling process; limited spatial resolutions; mixture of pixels; and misclassification of objects. USGU (2004) noted that the gap-filling technique performs well in homogenous landscape such as agricultural fields. However, it is difficult in heterogeneous landscape where feature size on surface is smaller than the local window size (normally 19x19 pixel). One multispectral pixel size of Landsat ETM+ is 30 meter; therefore, 19x19 pixel means 570m x 570m. Objects in the study area can be smaller than this size; therefore, the gap-filling process can be a factor which created errors. Furthermore, Weng (2012) also notes that results from image classifications are not satisfactory in terms of accuracy due to the limited spatial resolution in medium resolution satellite imagery and the heterogeneity of urban landscape mixture of trees, grassed, and soils. Moreover, Bauer *et al.* (2004) noted that as bare soil is spectrally similar to the impervious surface; therefore, these bare were misclassified as impervious surface in the case of the State Minnesota. In comparison to the case of Minnesota, the study area comprises of much bare land. Addition to that, many non-paved roads which are largely observed in the developing countries like Laos can be misclassified as a non-urban area. In this regard, the study found the human visible interpretation is the most accurate. The result from the human visible interpretation showed increase in the urban area from 9,721 square meters in 1973 to 5258,030 square meters in 2013. The urban area identified in 2013 is 541 times larger in comparison to that in 1973. The study utilized this data for the further analysis.

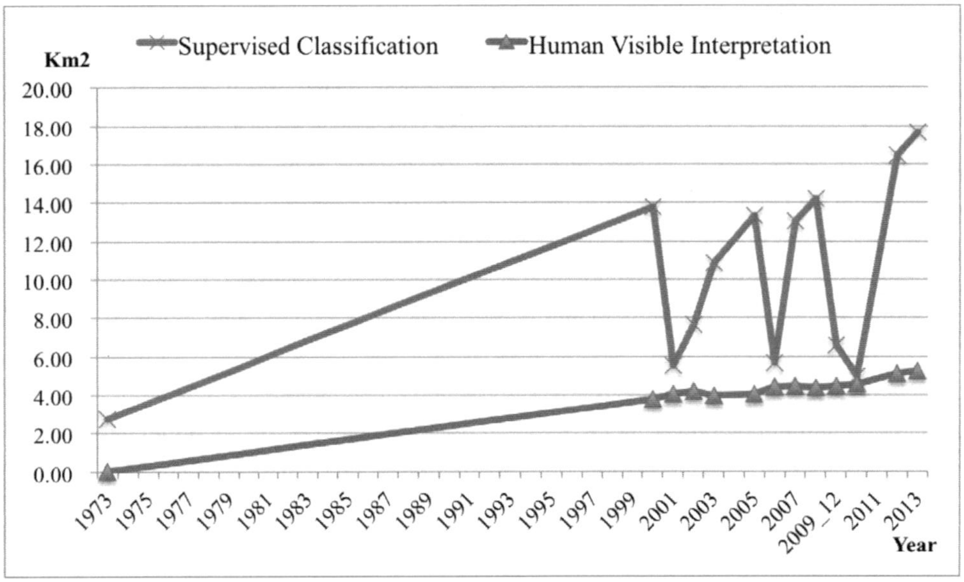

Figure 1. Comparison of urban area calculated by two methods.

Table 1 shows relationships between the rate of urbanization and socio-economic variables resulting from the regression model. As the accessibility of satellite images are limited and the availability of socio-economic data are different by variables, the regression equations were calculated from different range of time period. The study categorized each relationship based on the size of r as 1) very strong (0.9 to 1.0), 2) strong (0.8 to 0.9), 3) moderate (0.6 to 0.8), 4) weak (0.2 to 0.6) and 5) very weak or no relationship (0.0 to 0.2).

Table 1. Relationship between urbanization and socio-economic variables.

	Variables	R
1	School enrolment (Primary)	0.9664
2	Labour force	0.8877
3	Mortality rate, under-5 (per 1,000 live births)	0.8797
4	Improved water source (% of population with access)	0.8717
5	Improved sanitation facilities (% of population with access)	0.8702
6	Telephone lines	0.7497
7	Foreign direct investment, net inflows (US$)	0.7075
8	Literacy rate, youth total (people ages 15-24)	0.0141

Very strong relationships were found among the variables namely school enrolment (R=0.9664). For the relationship between the rete of urbanization and school enrolment, it can be assumed that due to the population growth under the urbanization, provision of basic education to the increase population was highly demanded. As a result, proper educational programs were implemented in order to meet a basic educational demand.

Strong relationships were found among the variables namely labour force (R=0.8877), mortality rate under 5 (R=0.8797), improved water source (R=0.8717), and improved sanitation facility (R=0.8712). For labour force, it is obvious that many projects established by donor countries will require huge number of labour force to maintain their productivities. Additionally, Elgin and Oyvat (2013) studied relationships between the level of urbanization and size of informal economy with cross country dataset. The authors found out an inverted-U relationship between urbanization and the share of informal sector explained by the pull and push factors. The pull factor is a factor attracting many individuals into the urban informal sector due to better employment opportunities and working conditions in comparison to rural area. On the other hand, the push factor is a factor facilitating many small-scale producers in rural sectors to engage in urban informal sectors due to economic damage caused by technical development under urban industrialization. These pull and push factors play an important role during the early stage of industrialization for meeting the demand of labour force. However, the impact of these factors tends to be reduced due to economic development of rural dwellers. As Laos may be considered in the early stage of industrialization, it can be said that labour force in informal sector will be continuously increased in the next a few decades. Thus, in addition to the formal employment opportunities, development of informal sectors in urban area is also a significant factor which increases labour force. For mortality rate under 5, Gong et al. (2012)'s study on urbanization and health conducted in China also pointed out a positive relationship between them. This is explained by 1) difficulty in access to health care, vaccination coverage, and accidents and insured among migrant population; 2) worse urban environmental quality such as air and water pollution; and 3) traffic-related accidents. The result of this study can be also explained by the circumstance observed in China. For provision of basic facility such as water source and sanitation facilities, it can be rationally assumed that urbanization with economic growth may become an opportunity to largely pay attention to basic human needs such as health and environmental facility developments.

Moderate relationships were identified among the variables namely telephone line (R=0.7497) and FDI (R=0.7075). For the telephone line, it can be assumed that the relationship with telephone line is not so significant due to the higher distribution of mobile phone in comparison to telephone line in Laos. In fact, the mobile cellular subscription (per 100 people) in Laos was reported at 3.6 in 2004 and 64.56 in 2010 (Trading Economics, 2013), which is approximately 18 times higher than it was in 2004. For the FDI, Sabu (2013) found a positive relationship between FDI growth and urbanization; however, it is not so significant in this study. This difference can be caused because of difference in targeted investment sectors among China and Laos. Chinese FDI is highly related to export-oriented industries which require establishment of many manufacturing industries. On the other hand, the main sectors largely invested in Laos are electricity generation, mining, service, agriculture and industry/handicraft which have not been largely carried out in the study area. Therefore, the relation between two were not so significant in this study.

Very weak or no relationship was identified among the variable namely literature rate (total youth) (R=0.0141). As data from the regression model was calculated by the data from 2000 to 2011, the study interpreted that literacy rate among youth has been developed already several decades before. Indeed, literacy rate of youth showed 71.13 percent in 1995; 80.6 percent in 2000; 78.46 percent in 2001; and 83.93 percent in 2005 (Index Mundi, 2013).

4. Conclusion

This study examined the change in the rate of urbanization in time series utilizing Landsat MSS, TM and ETM+ remote sensing data obtained from 1973 to 2013. The result from the human visible interpretation showed increase in urban area from 9,721 square meters in 1973 to 5,258,030 square meters in 2013 that is 541 times larger in comparison to that in 1973. The socio-economic variables such as school enrollment, labour force, mortality rate, water source and sanitation were highly correlated with the rate of urbanization during the period. Identifying the highly correlated socio-economic variables with urbanization rate could enable us to conduct a further urbanization simulation. The simulation helps in designing policies for sustainable development.

Reference

[1] Bauer M E, Loffelholz B C, Wilson B, and Bauer M 2004 Estimating and Mapping Impervious Surface Area by Regression Analysis of Landsat Imagery 612–625.
[2] Elgin C and Oyvat C 2013 Lurking in the cities: Urbanization and the informal economy. *Structural Change and Economic Dynamics* **27** 36–47

[3] Gong P, Liang S, Carlton E J, Jiang Q, Wu J, Wang L and Remais J V 2012 Urbanisation and health in China *The Lancet* **379**(9818) 843–52
[4] Index Mundi 2013 *Lao PDR-Literacy rate:* http://www.indexmundi.com/ (Accessed on 9th January 2014)
[5] Investment Promotion Department (IPD) 2010 LAOS: Land of Ample Opportunities & Success
[6] Japan External Trade Organization 2008 2009 2010 2011 2012 2013 *Laos* Japan External Trade Organization
[7] Lo C P 1998 Application of Landsat TM data for quality of life assessment in an urban environment *Environment and Urban System* **21**(3) 259–276
[8] Lo C P and Faber B J 1997 Integration of Landsat Thematic Mapper and Census Data for Quality of Life Assessment *Remote Sensing of Environment* **62** 143–157
[9] Mennis J 2006 Socioeconomic-Vegetation Relationships in Urban, Residential Land : The Case of Denver, Colorado *Photogrammetric Engineering & Remote Sensing* **72**(8) 911–921
[10] Trading Economics 2013 *Mobile cellular subscriptions (per 100 people) in Laos:* http://www.tradingeconomics.com/ (Accessed on 9th January 2014)
[11] Sahu K K 2013 Does Urbanisation Promote Foreign Direct Investment? Lessons and Evidences from *International Journal of Arts and Commerce* **2**(3) 17–26
[12] USGS 2004 Phase 2 gap-fill algorithm: SLC-off gap-filled products gap-fill algorithm methodology 1–12: http://landsat.usgs.gov/documents/L7SLCGapFilledMethod.pdf (Accessed on 10th January 2014)
[13] Weng Q 2012 Remote sensing of impervious surfaces in the urban areas: Requirements, methods, and trends *Remote Sensing of Environment* **117** 34–49

IOP Conf. Series: Earth and Environmental Science **20** (2014) 012006 doi:10.1088/1755-1315/20/1/012006

Quantitative analysis of urban sprawl in Tripoli using Pearson's Chi-Square statistics and urban expansion intensity index

Abubakr A. A. Al-sharif, Biswajeet Pradhan, Helmi Zulhaidi Mohd Shafri, Shattri Mansor

Department of Civil Engineering, Faculty of Engineering, University Putra Malaysia, 43400,

Email: biswajeet24@gmail.com or biswajeet@lycos.com

Abstract. Urban expansion is a spatial phenomenon that reflects the increased level of importance of metropolises. The remotely sensed data and GIS have been widely used to study and analyze the process of urban expansions and their patterns. The capital of Libya (Tripoli) was selected to perform this study and to examine its urban growth patterns. Four satellite imageries of the study area in different dates (1984, 1996, 2002 and 2010) were used to conduct this research. The main goal of this work is identification and analyzes the urban sprawl of Tripoli metropolitan area. Urban expansion intensity index (UEII) and degree of freedom test were used to analyze and assess urban expansions in the area of study. The results show that Tripoli has sprawled urban expansion patterns; high urban expansion intensity index; and its urban development had high degree of freedom according to its urban expansion history during the time period (1984-2010). However, the novel proposed hypothesis used for zones division resulted in very good insight understanding of urban expansion direction and the effect of the distance from central business of district (CBD).

1. Introduction

Metropolises in the world are one of human civilization aspects, their complex structures is a result of its complicated land use patterns. The urbanized land cover class changes considerably in short time as compared to other land cover classes [1]. The rapid urban land use change results in undesired urban sprawl. The urban sprawl is one of major significant problems in urban land use change and urban expansions management [2, 3]. However, the simple definition of urban sprawl is the amount of built up area and its dispersion level in the landscape [4, 5]. Furthermore, the urban sprawl level in the landscape in specific time represents the urban sprawl as a pattern. Whereas the change of urban sprawl patterns in a landscape in different time periods represents the urban sprawl as a process [5]. Nevertheless, both definitions of urban sprawl pattern and urban sprawl process can be included in one term ''urban sprawl'' [6]. In urban planning the accurate mapping of urban lands and monitoring urban expansions are significant for optimum urban analysis and planning [7]. The use of conventional urban analysis methods is really hard, time consuming, and expensive. Though, the urban expansions can be measured simply by calculating the percentage of extents covered by concrete and asphalt, i.e. impervious surfaces. Moreover, those extents which are covered by impervious surfaces are detectable and interpretable using remote sensing technique [8-11]. Hence statistical methods along with GIS and remotely sensed data can be used as efficient and cost effective option for urban growth analysis and detecting urban sprawl [5, 12, 13]. The main objectives of this work are analysis the urban expansions and their spatial patterns in Tripoli, Libya, from 1984 to 2010, by using GIS, remote sensing tools. This study employed statistical models of Pearson's Chi-Square and (UEII) analysis. Specifically, this paper identifies and assesses the urban sprawl occurrence in Tripoli and the effect of growth direction, and the distance from CBD on urban sprawl in three aspects namely; in each urban zone; in each temporal span; in overall study area.

Content from this work may be used under the terms of the Creative Commons Attribution 3.0 licence. Any further distribution of this work must maintain attribution to the author(s) and the title of the work, journal citation and DOI.
Published under licence by IOP Publishing Ltd

2. Study area

Tripoli is the capital of the Libyan state, it is thousands years old. Tripoli is selected as study area in this work. It is the main financial, political, commercial, economical and business centre of Libyan state. Tripoli is located at (N 32° 53′ and E 13° 10′) along the Mediterranean coast in the north western part of Libya. The Tripoli urban agglomeration covers about 1143.73km^2 with population more than 1.3 million individuals.

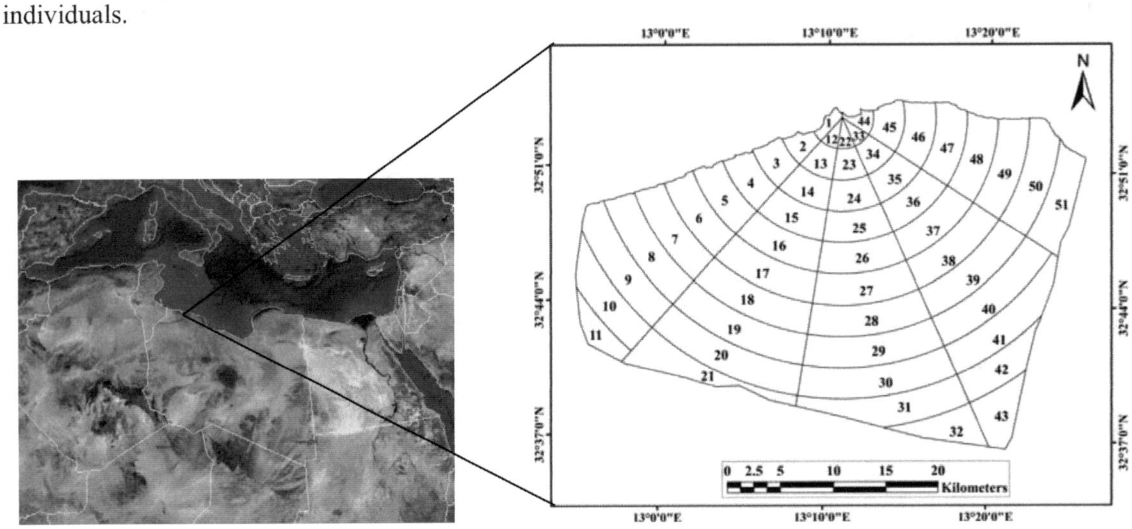

Figure1. Study area and used zones division.

3. Data used and methodology

Four remotely sensed images cover the study area of LANDSAT–TM 1984; LANDSAT–TM1996; SPOT-5 2002 and SPOT-5 2010 were classified to determine the built-up area extents of Tripoli metropolitan area. The used imageries have different spatial resolutions. A simple technique to solve this trouble is resampling the imageries which have large resolution. So that their spatial resolution equals the smallest resolution imagery. However, images resampling using bi-cubic or bi-linear resampling methods will change pixels values, i.e. average the adjacent pixel values. Resampling using nearest neighbour method will results in pixel dropout or pixel duplication. For that, imageries were used without any altering of the pixel size.

Consequently, vector map of the Tripoli metropolitan area was used for clipping used imageries. Thereafter, a supervised maximum likelihood method was used to classify all clipped images to extract the built-up areas. Since in this study only urban growth is assessed, for this reason only two classes were considered, namely: urban class and non-urban class as shown in Figure 2. Subsequently, classified imageries were clipped further by the vector map of study area which divided into 51 zones as shown in Figure 1. Then central point of Tripoli is the central business district which represents starting point of urbanization process along study area history was determined. The urban growth and built up area for all zones and for every temporal point were calculated with respective zone border by multiplying quantity of pixels in each zone by pixel size. However, the idea used in this paper based on dividing the area into five pie sections to consider and assess urban sprawl direction. Then, by further dividing the five sections into more zones to take into account of the effect of distance from central business district CBD; and to determine the sprawl in each zone. This approach will give more details about urban growth process and its patterns in whole study area and in each individual zone. Lastly, the theoretical expected and observed urban growths were calculated. Thereafter, both calculated growths were used to compute and assess the urban expansions using Pearson's chi-square technique and urban expansion intensity index.

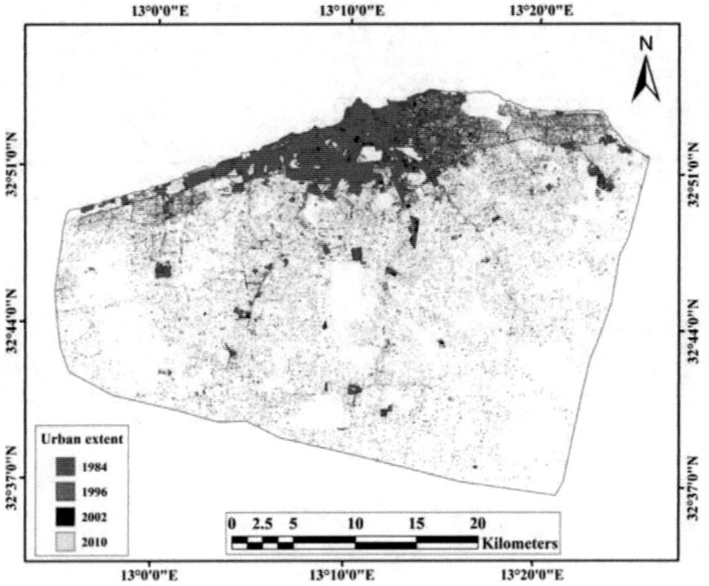

Figure 2. Urban extent in different years.

4. Results and discussions

4.1. Observed and theoretical expected built up area growth

To understand the divergence of urban growth, the observed growth should be compared with the forecasted theoretical urban growth. To calculate theoretical expected urban area growth statistically for all zones in each time period, equation 1 was used [14].

$$G_i^e = \frac{G_t^z \times G_t^t}{G} \quad (1)$$

G_i^e is expected urban growth in zone.
G_t^t is total growth in the three time periods in one zone.
G_t^z is total growth in the 51zones in one time period.
G is total growth of study area in all time periods.

Now, by subtracting theoretical expected growth from observed growth the divergence of urban growth for each zone and each temporal period can be easily identified as shown in Figure 3. Positive values confirm more growth than the anticipations, while negative values indicate less amount of growth, level of variation can also be identified by the magnitudes of differences.

The results clearly demonstrate that the observed urban expansion in several zones (especially at built up area fringes) is very much deviated from the expectations. In addition, the deviation is continuing, associating urban growth, and increasing with time. Those higher deviations reflect the freedom and independency of urban expansion; i.e. if the deviation is high means that the studied variable is independent to other similar type of variables. Based on these findings we can say there is a clear urban sprawl occurrence in most zones far from CBD of study area in all directions in time period 1996 to 2010.

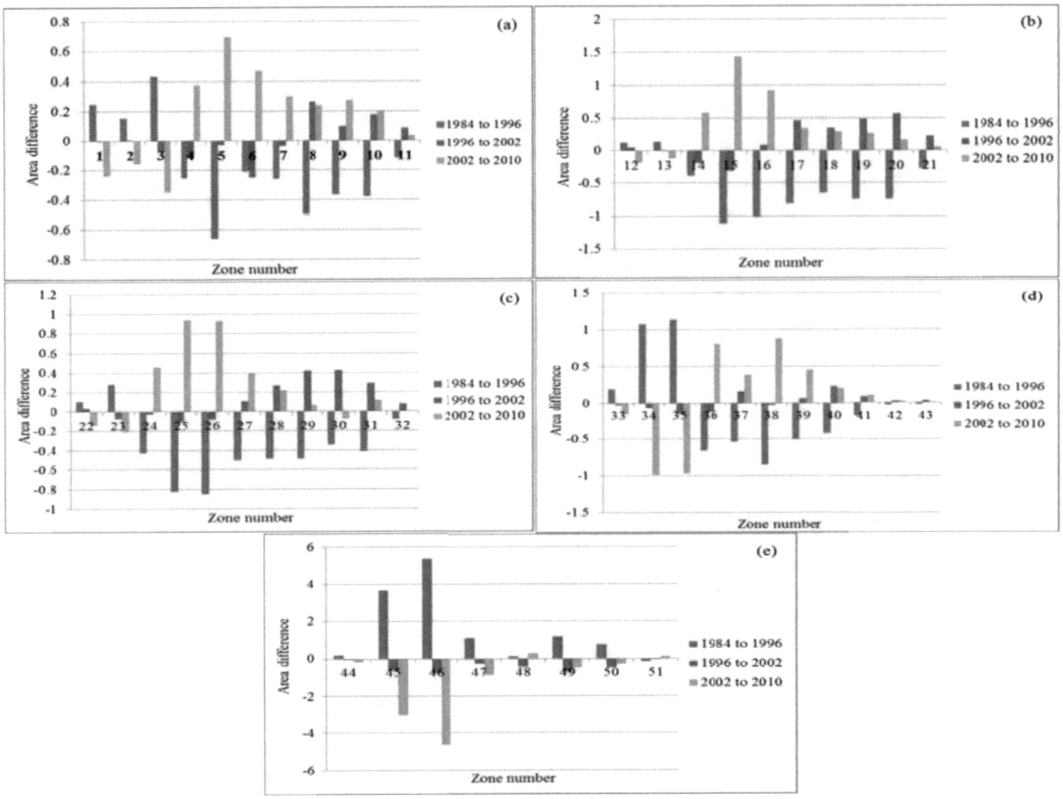

Figure 3. Difference between observed and expected built up area growth (in km^2); (a) in zones within 1st direction; (b) in zones within 2nd direction; (3) in zones within 3rd direction; (4) in zones within 4th direction; (e) in zones within 5th direction.

4.2. Degree of freedom of urban expansion

The statistics Pearson's chi square technique was used for checking the degree of freedom between pairs of variables that selected to describe the same class of land cover change [6, 14]. In this research, to find out the degree of freedom for the urban growth, Pearson's chi square analysis was carried out using following formula: $D =$ [(Observed Growth - Expected Growth)^2/ Expected Growth]. Pearson's chi square statistics estimates the freedom or degree of variation for the observed urban growth over the expected urban growth. The chi square statistics for each time period was computed using following equation 2:

$$D_i^t = \sum_i^n D_i^z \quad (2)$$

D_i^t is degree of freedom of growth in i^{th} time period.
D_i^z is degrees of freedom of growth for i^{th} zone in same time period.

To determine the degree of freedom for each zone same formula can be used as following:

$$D_i^z = \sum_i^n D_i^t \quad (3)$$

Moreover, the overall degree of freedom of study area can be calculated by summation of degrees of freedom of all time periods or by summation of degrees of freedom of all zones. The lower limit of chi square is 0, which means that observed growth value exactly equals the expected growth value.

Table 1 shows the degree of freedom of the three time periods which is considered very high. The overall degree of freedom of study area is 52.41 which is considered extremely high; i.e. high difference in observed and expected values. In addition, Figure 4 shows the degree of freedom in each zone specifically which is varying from zone to other. The highest degree of freedom value was recorded in zone number 46 and the lowest was in zone number 42. Generally speaking, a higher degree of freedom refers to the needs of consistency in planning, managing and controlling urban growth of the entire study area. Higher degree of freedom for a zone is a warning of unbalanced growth within the zone with the change of time. And higher degrees of freedom for a time period can be regarded as higher inter zone inconsistency in urban growth. Nonetheless, we cannot consider higher degree of freedom as sprawl, but we should consider it as disparity in urban growth.

Table 1. Degree of freedom and urban expansion intensity index of time periods

Time period	Degree of freedom (D_i^t)	UEII
1984 to 1996	10.77	0.35
1996 to 2002	5.22	0.45
2002 to 2010	36.43	1.28

4.3. Urban expansion intensity index (UEII)

In the urbanization process, due to the rule of urban deriving factors (such as road network, population density, slope, economics and so on) and their spatial impacts, the urban expansions will be different in each region and in each direction this phenomena named preference of urban growth [7]. UEII can be employed to assess and analyze the urban spatial expansion differences quantitatively. Moreover UEII can be used to recognize the preference of urban growth in a certain period [15]. The UEII reflects the probable future direction and potentials of urban expansions, and it compares the speed or intensity of urban land use change in different time periods.
The UEII standard is divided as follow: 0 to 0.28 is slow development; 0.28 to 0.59 is low-speed development; 0.59-1.05 is medium-speed development; 1.05-1.92 is high-speed development; and >1.92 is very high-speed development. The UEII for overall study area, each temporal span and each zone is calculated using formula as following:

$$UEII_{it} = \left[\frac{ULA_{i,b} - ULA_{i,a}}{t}\right] / TLA_i \times 100 \quad (4)$$

Where, $UEII_{it}$ is the annual average urban expansion intensity index of (i^{th}) zone in time period (t); $ULA_{i,a}$ and $ULA_{i,b}$ are the quantity of built-up area at time periods a and b in (i^{th}) spatial zone respectively; and TLA_i is the total area of (i^{th}) spatial zone.

The results showed that overall study area has expansion intensity index of 0.66 which is considered moderate urban expansion speed. But on the other hand the UEII has increased dramatically from 0.35 in time period 1984-1996 to reach 1.28 in time period 2002-2010; this high increase of UEII gives an alarm of increased urban sprawl occurrence. By reading Figure 4 further one can notice that the lowest UEIIs are recorded in zones near to CBD, which means these zones relatively have stable rate of expansion. The highest UEIIs were occurred in zones that include built up area fringes, i.e. high sprawled urban expansion occurrence. However, in most urban expansion directions in the study area, the values of UEIIs are declined after exceeding the urban area fringes, i.e. low urban expansion happened.

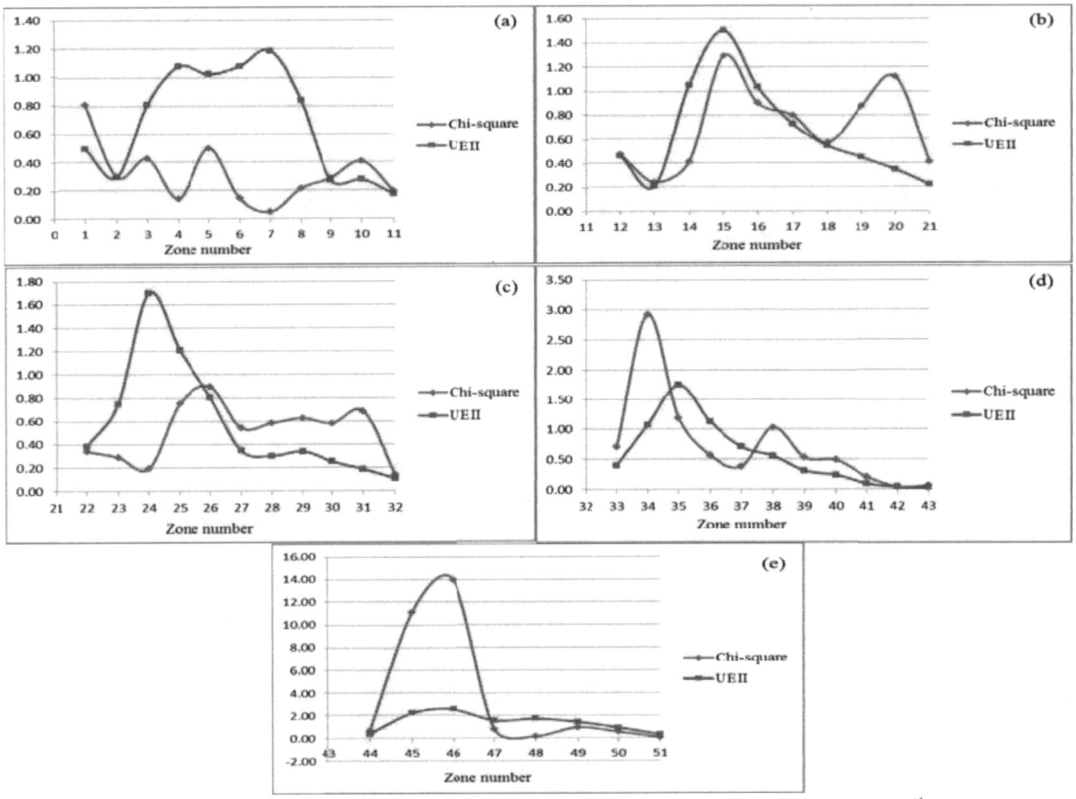

Figure 4. Variation of the two applied measures in different growth directions; (a) in 1st direction; (b) in 2nd direction; (c) in 3rd direction; (d) in 4th direction; (e) in 5th direction.

5. Conclusion

This study has investigated the urban expansion and its spatial patterns in Tripoli metropolis area from 1984 until 2010. The findings are very useful in terms of directing prospect urban plans and urbanization policies for Tripoli. The analysis results demonstrated that there is a clear signs of urban dispersion and urban sprawl for Tripoli city, i.e. sprawl is increasing with time. The techniques used in this paper can be employed to provide some guidance to recognize and assess the change likely to occur if the tendency of urban history persists. However, the proposed division technique gave very good insight into urban expansion of the study area. The expansion direction and distance to CBD which are considered in applied methods are very supportive for decision makers and urban planners. The study outcomes clearly demonstrate that Tripoli has increasing urban sprawl especially in last decade. Moreover, the observed urban expansion in most zones (especially at built up area fringes) is very much deviated from the theoretical expectations. In addition, very high sprawl was recorded at urban fringes whereas decreased sprawl was noticed in regions nearer to CBD. Finally, the sprawl level variation along all five directions offered another view of sprawl variation.

6. References

[1] Al-sharif, A.A. and B. Pradhan, *Monitoring and predicting land use change in Tripoli Metropolitan City using an integrated Markov chain and cellular automata models in GIS.* Arabian Journal of Geosciences, 2013b: p. 1-11.

[2] Barredo, J.I. and L. Demicheli, *Urban sustainability in developing countries' megacities: modelling and predicting future urban growth in Lagos.* Cities, 2003. **20**(5): p. 297-310.

[3] Weng, Q., *A remote sensing? GIS evaluation of urban expansion and its impact on surface temperature in the Zhujiang Delta, China.* International Journal of Remote Sensing, 2001. **22**(10): p. 1999-2014.

[4] Jaeger, J.A., et al., *Suitability criteria for measures of urban sprawl.* Ecological Indicators, 2010. **10**(2): p. 397-406.

[5] Al-sharif, A.A., et al., *Spatio-temporal analysis of urban and population growths in Tripoli using remotely sensed data and GIS.* Indian Journal of Science and Technology, 2013. **6**(8): p. 5134-5142.

[6] Bhatta, B., S. Saraswati, and D. Bandyopadhyay, *Quantifying the degree-of-freedom, degree-of-sprawl, and degree-of-goodness of urban growth from remote sensing data.* Applied Geography, 2010. **30**(1): p. 96-111.

[7] Alsharif, A.A. and B. Pradhan, *Urban sprawl analysis of Tripoli Metropolitan city (Libya) using remote sensing data and multivariate logistic regression model.* Journal of the Indian Society of Remote Sensing, 2013a: p. 1-15.

[8] Barnes, K.B., et al., *Sprawl development: its patterns, consequences, and measurement.* Towson University, Towson, 2001: p. 1-24.

[9] Sudhira, H., et al., *Urban growth analysis using spatial and temporal data.* Journal of the Indian Society of Remote Sensing, 2003. **31**(4): p. 299-311.

[10] Bhatta, B., *Analysis of urban growth pattern using remote sensing and GIS: a case study of Kolkata, India.* International Journal of Remote Sensing, 2009. **30**(18): p. 4733-4746.

[11] Al-Sharif, A.A., et al. *Revisiting Methods and Potentials of SAR Change Detection.* in *Proceedings of the World Congress on Engineering.* 2013c.

[12] Kumar, J.A.V., S. Pathan, and R. Bhanderi, *Spatio-temporal analysis for monitoring urban growth–a case study of Indore city.* Journal of the Indian Society of Remote Sensing, 2007. **35**(1): p. 11-20.

[13] Punia, M. and L. Singh, *Entropy Approach for Assessment of Urban Growth: A Case Study of Jaipur, INDIA.* Journal of the Indian Society of Remote Sensing, 2012. **40**(2): p. 231-244.

[14] Almeida, C.M.D., et al., *GIS and remote sensing as tools for the simulation of urban land-use change.* International Journal of Remote Sensing, 2005. **26**(4): p. 759-774.

[15] Ren, P., et al., *Spatial Expansion and Sprawl Quantitative Analysis of Mountain City Built-Up Area*, in *Geo-Informatics in Resource Management and Sustainable Ecosystem*. 2013, Springer. p. 166-176.

Six decades of urban growth using remote sensing and GIS in the city of Bandar Abbas, Iran

Mohsen Dadras*[1,2], Helmi Zulhaidi Mohd Shafri[1], Noordin Ahmad[1], Biswajeet Pradhan[1], Sahabeh Safarpour[3]
[1]Faculty of Engineering, Department of Civil Engineering, Universiti Putra Malaysia, 43400 UPM Serdang, Selangor, Malaysia
[2]Faculty of Engineering, Department of Civil Engineering, Islamic Azad University, Bandar Abbas Iran
[3]School of Physics, Universiti Sains Malaysia, 11800 Penang, Malaysia

*Corresponding author. Tel: +60 17 354 2063
E-mail addresses: dadras.mohsen@yahoo.com or m.dadras@iauba.ac.ir

Abstract. Bandar Abbas is the capital city of Hormozgan province, is the south of Iran. The city has witnessed rapid growth in the last three decades, mostly because of its economic, commercial and social attractions. However, forms and operations of urban sprawl may vary in important manners according to determine geographical and historical characteristics, and these difference need to be reviewed with creation geodatabase of spatial and attribute data during past periods until now of urban formation and expansion. We implemented this research to understand Bandar Abbas city growth dynamic during last six decades using aerial photo, Remote Sensing (RS) data and Geographical Information System (GIS), to investigate its sprawl for the during six decades and to prepare a basis for urban planning and management. We calibrated it with geospatial data derived from a time series of aerial photos and satellite images. Treated remote sensing data covering the six decades were used to calculate land use/cover and urban growth. The application of classification techniques to the remote sensing data enabled the extraction of eight main types of land use: agricultural, barren, coastal, hole, river, rocky hill, urban, and built-up. Growth was calculated through Shannon's entropy model. The urbanized area increased from 403.77 ha to 4959.59 ha from 1956 to 2012, a rate almost five times that of the population growth observed in the same period. Such findings make the case of Bandar Abbas important for several reasons. First, Bandar Abbas has undergone a rapid increase in urban sprawl according to urban growth indicators. Second, the urban sprawl quickly grew from medium-sized to large a process considered inappropriate according to physical and structural limitations on urban growth. Lastly, the excessive extension of the built-up boundary in the city resulted in the loss of coastal land and open space, two main sources of tourist attraction and economic sustainable development.

1. Introduction

Today, urban sprawl is introduced as one of the urban development processes. The event of urban development is comprehensive and general process and has a wide range of concepts. The urban development process should be discussed based on its history and effective factors such as social and economic factors over time. During the past two centuries, we have been witnessing the expansion of large cities and the growth of the space under their spatial influence. This process in many countries has been this way: the change in the model of life of residents caused new urban areas. Nowadays, a wide section of people live on the margins of cities, towns, and residential areas adjacent to cities. The people who live adjacent to the urban areas are completely under the influence of urban life styles, values, and expectations. In fact, urban centers are considered as center for the employment of non-agricultural activities and include employment activities such as industry and services. Also, the cities have been regarded over time as the centers of cultural, social, and intellectual development. Urban development is a process brought about by the emergence of a world dominated by cities and urban values. Here, it is very important that to distinguish between two main processes of urban development including: the sprawl of the city and urban growth [4].

Content from this work may be used under the terms of the Creative Commons Attribution 3.0 licence. Any further distribution of this work must maintain attribution to the author(s) and the title of the work, journal citation and DOI.
Published under licence by IOP Publishing Ltd

Urban sprawl is the shape of land use representing low levels of eight separate dimensions: density, continuity, concentration, clustering, centrality, nuclearity, mixed uses, and proximity. Density is the average number of residential units per square mile of developable land in an urban region. Continuity is the degree to which developable land has been built upon at urban densities in a continuous style. Concentration is the degree to which development is positioned disproportionately in a few square miles of the total urban area rather than spread evenly. Clustering is the degree to which development has been firmly bunched to minimize the amount of land in each mile of developable land occupied for residential or nonresidential use. Centrality is the degree to which residential or nonresidential development (or both) is located near the central business district of an urban area. Nuclearity is the extent to which an urban area is characterized by a mononuclear pattern of development (as opposed to a polynuclear pattern). Mixed uses refer to the degree to which a small area is used for multiple purposes and the degree to which this practice is prevalent throughout a region. Proximity is the degree to which lands used for various purposes are located near each other across an urban region [10]. Given these characteristics, urban sprawl is a hazard to sustainable urban development because it implies growth in the use of land, water, energy, and other resources as well as in the amount of contaminants and waste. Monitoring and forecasting urban sprawl are the basic sources of information required in long-term planning and development. For balanced development, urban authorities need tools to monitor how the land is used, to estimate future demand, and to take steps to guarantee the capability of future supply. For better planning of urban growth and infrastructure, urban authorities need to understand the urban sprawl phenomenon and the directions it may take in years to come.

Conventional surveying and mapping techniques for estimating urban sprawl are costly and time consuming, and the information obtained from them is inaccessible for most urban areas, particularly in undeveloped countries. Consequently, increasing research is directed toward the mapping and monitoring of urban sprawl by using geographic information systems (GIS) and remote sensing techniques [19].Population growth primarily affects urban sprawl, which is a significant feature of the urbanization process in developed countries, such as the US, China, India, Australia, Canada, and several European countries [9]. Urban sprawl is expanding to countries with large populations and to several developing countries, such as China ([22],[23], [24]), India ([3], [13]), and Turkey [17]. However, the different aspects of and reasons behind urban sprawl are studied still less extensively in the developing world. More examples of how this process occurs in specific areas are needed to discover methods, causes, and insights that improve our understanding of the urbanization process in areas where the process is more important.

This study presents the case of the city of Bandar Abbas in southern Iran. Both Bandar Abbas and Iran as a whole are interesting and relevant cases of explosive urban growth. According to the first Iranian census (in 1956), the number of Iranian cities was 199, and the proportion of the urban population was 31.8% of the total. In 2010, the number of cities had increased to 1200, and the proportion of the urban population exceeded 68.37% of the total ([8], [11]). This study aims to investigate the urban sprawl of Bandar Abbas in the past six decades and to analyze how various features have affected the spatial pattern of this urban sprawl. Section 2 introduces the study area. Section 3 discusses the materials and methods used in generating data through different aerial photos and satellite images. Section 4 considers the correlation between population and urban growth variables. Section 5 discusses and presents more specific land transformation processes. Section 6 concludes the paper and summarizes the observed characteristics of the urban sprawl of Bandar Abbas.

2. Study Area

Bandar Abbas, the capital of Hormozgan Province, is a commercial port city along the southern coast of Iran facing the Persian Gulf. The study area is located at a latitude of 27°8' N to 27°15' N and at a longitude of 56°13' to 56°22'. It has a land area of approximately 100 km² (Figure 1) and includes four regions and 70 districts. The city occupies a strategic position on the narrow Strait of Hormuz, and it is the site of the main base of the Iranian Navy. Bandar Abbas lies on flat ground with an average altitude of 9 m above sea level. The nearest elevated areas are Geno Mountain 17 km to the north and Pooladi Mountain 16 km northwest of the city. The population of Bandar Abbas was 0.72 million in 2012 and, given the present growth rate, is expected to rise to 0.95 million in 2030. For better future urban growth and infrastructure planning, urban authorities need to understand the nature of the urban sprawl in Bandar Abbas, its distribution, and the directions it is likely to take in years to come. The most important economic activities in Bandar Abbas include heavy industries (commercial ports, fishing ports, oil and gas refinery, and other industries), which employ about 74% of the active population. Thus, the city experiences rapid growth to the point that, among Iranian cities with a population larger than 300,000, it has the highest growth in terms of urban land development.

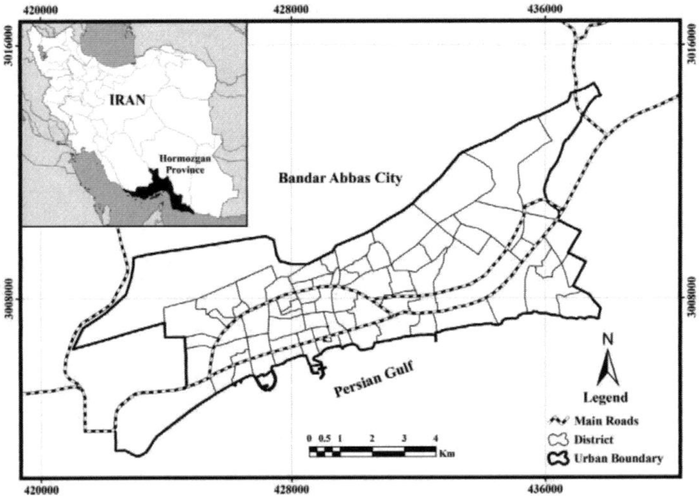

Figure 1. Bandar Abbas location in Iran.

3. Materials and methods

Understanding the dynamic phenomenon of urban sprawl/growth requires the identification of urban sprawl patterns, the computation of landscape criteria, and the analysis of built-up area change. The identification of built-up area change is a fundamental input in urban planning, management, and environmental analysis. The specific aerial photos and satellite images used were obtained from aerial photos (1956, 1965, 1957, 1987, and 2001), Landsat Multi-Spectral Scanner from 1979 (one of the first satellite image acquisitions over Bandar Abbas), Landsat Thematic Mapper from 1990, Landsat ETM+ from 2001, ASTER from 2005, IRS LISS-III from 2008, and GeoEye-1 from 2012, an image captured by a different sensor. Some ancillary data were also used. Three-dimensional digital topographical maps with a scale of 1:2000 were used for image georeferencing; digital elevation modeling (DEM), the basis for the land use/cover map; and the increased accuracy of the overall estimation. These nationwide maps were prepared in 2001 through the photogrammetric processing of 1:2000 scale aerial photographs. Other supporting data included the demographic censuses of Bandar Abbas. All satellite and supporting data used in this study are presented in Table 1and Figure 2 briefly illustrates the flowchart of entire methodology.

The aerial photos taken of the city of Bandar Abbas, at 5 time periods are vertical and panchromatic. After taking aerial photos, the photogrammetric operations were done for mosaicking of the photos and geo-referencing them in order to extract topographic maps. The geo-referencing of the aerial photos is done by using triangulating operations, benchmark points prepared by the mapping organization, and topographic maps. It's worth mentioning that geometric corrections required are done during the operation of the photogrammetry on the photos in order to reduce error and increase the accuracy of interpretation. After doing the process of photogrammetry and geo-referencing of aerial photos the stereoscopy of complications and 3D vision are made possible and by using this method, the range of sprawl and growth of cities can be identified and determined.

Table 1. Data used in this study.

Data	Pass&Row	Year	Spatial Resolution/Scale(m)
Aerial Photos			
	56-302	1956	1:20000
	68-438	1965	1:5000
	74-665	1975	1:15000
	88-459	1987	1:10000
	2793	2001	1:10000
Satellite Images			
Landsat MSS	172/41	05-Feb-1979	79
Landsat ETM+	160/41	07-Apr-2001	15-30
Aster	2653487	17-May-2005	15-90
IRS (Panchromatic Band)	75-52-D	21-Jul-2008	5.8
GeoEye-1	-	31-Dec-2012	0.5
Secondary Data			
3-D topographic map	-	2001	1:2000
DEM	-	2001	1m vertical
Iranian Censuses	-	1976–1986–1991–1996–2006	
UN Iran censuses	-	1995-2025	

To convert satellite images the method of nearest-neighbor resampling is used so that the original value of each pixel is maintained with the use of this procedure. The satellite images and various sensors are different in terms of the spatial resolution. The way to solve this problem is to replace the images with high resolution for comparison with the images with low resolution. Topographic maps with scale 1:25000 (prepared by Iran's mapping organization 1987-1989) and 1:500 (prepared by geography organization of the country 2002-2004) are used in order to control and analyze the accuracy of the accuracy of classification of aerial photos and satellite images in the present study. Also, the land use map of Bandar Abbas with a scale of 1:2000 (prepared by the Ministry of Roads and Urbanization Urban Development Master Plan 2004-2006) in order to check the accuracy of aerial photos and satellite images in 2001 and 2012. It should be noted that 1300 sampling points in the city of Bandar Abbas were used to compare the uses of existing situation and increasing the accuracy of the geo-reference of aerial photos and satellite images.

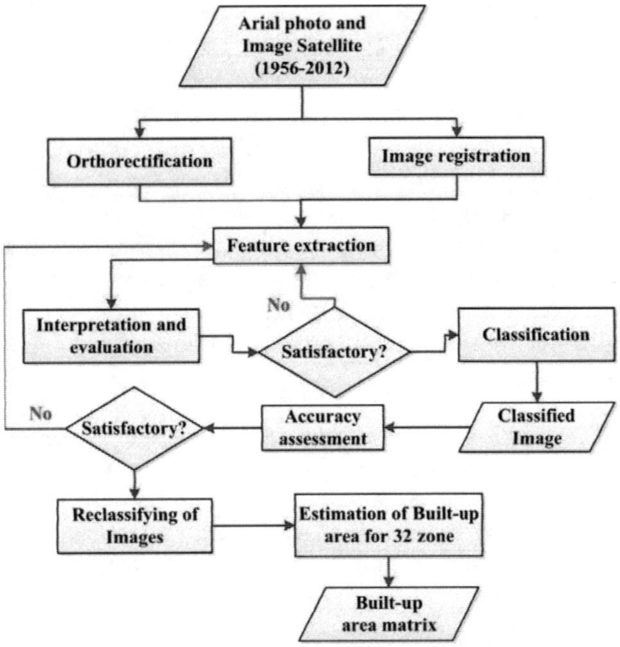

Figure 2. Flowchart of image analysis.

The central point of the city is considered as the central place of growth of city, which is in fact, the initial nucleus of formation of the city of Bandar Abbas located currently in the southwest of the city (the big market of city). The drawn circles are concentric and include the entire scope of the study from the center of city (nucleus of formation). Hence, the circles have been drawn in a way which include the regions constructed based on the radius of 500 meters from each other and different geographic directions. This division has been made so that the process of changes in construction in different parts and direction could be statistically compared. It should be noted that the radius of circle must be so large that it include all of urban boundary and constructed lands within it. Essentially, the structure of urban boundary is a dynamic process and greatly changes in different directions by the passing of time. However, in this study, the largest boundary is considered in the last period of time. Thus, the circle drawn is based on the last photo taken related to the GeoEye-1 satellite image in 2012.

Table 2. Assessment of satellite images classification.

Year	Overall Accuracy (%)	Kappa coefficient
1979	97.59	0.94
2001	98.26	0.951
2005	97.11	0.93
2008	98.12	0.89
2012	99.86	0.98

We use Shannon's entropy, a method widely applied in studies of urban sprawl. Shannon's entropy is an indicator of spatial concentration or distribution and can be applied to any geographical variable. The entropy method, another urban sprawl metric, is the most widely used technique for measuring the extent of urban sprawl with the integration of remote sensing and GIS ([2], [3], [12], [13], [14], [19], [23]).

Shannon's entropy (Hn) can be used to measure the degree of spatial concentration or dispersion of a geographical variable (xi) among n zones. Entropy is calculated according to:

$$_n = -\sum_{i=1}^{n} P_i log_e(P_i) \quad (1)$$

where Pi is the probability or proportion of a phenomenon (variable) occurring in the ith zone.

$$P_i = X_i / \sum_k^n X_k \quad (2)$$

X_i is the observed value of the phenomenon in the ith zone. For the purposes of this study, zones are defined as concentric circles around the city center or around a selected intersection of roads [23]. The value of entropy ranges from 0 to$\log_e(n)$. A value of 0 indicates that the distribution of the built-up areas is compact, whereas values close to $\log_e(n)$ indicate that the distribution of built-up areas is dispersed. High values of entropy indicate the occurrence of sprawl. The halfway mark of $\log_e(n)$ is generally considered as the threshold. If the entropy value crosses this threshold, the city is considered to be sprawling. Determining the threshold for this metric is robust and can be considered for any area under study.

For the purposes of the current study, 24 concentric circles and 32 geographic directions were determined, with a radius of 500 m between every two consecutive circles. The center of the circles was the historic district and the main core of the city. A zone was defined as the intersection of the city shape with the ring specified by two consecutive concentric circles. The operation was done in ArcGIS software covering a 10-year urban sprawl of the city (Figure 3).

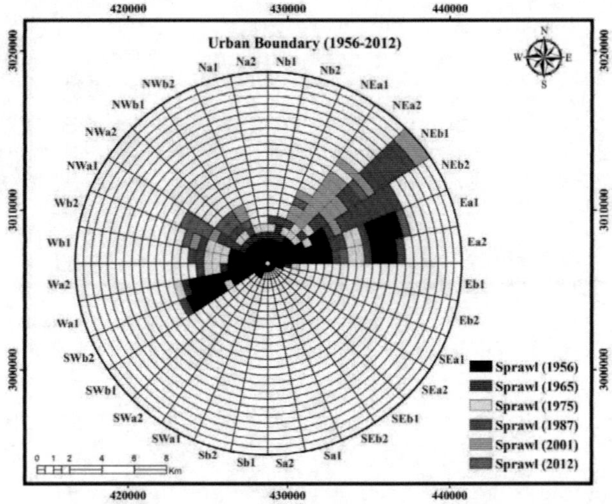

Figure 3. Radar chart showing built-up area in different directions, Bandar Abbas (1956-2012).

4. Result and discussion

Data on urban sprawl in 1956, 1965, 1975, 1987, 2001, and 2012 were estimated from data on impervious (built-up) areas captured from aerial photos and satellite images. Map statistics created from remote sensing images show that the built-up area encompassed some 403.77 ha in the first image (1956). In this year, the population density was 43.86 people per ha (Figure 4). As shown in Figure 5, the physical growth and development of Bandar Abbas proceeded around the historical center and the main core in a continuous pattern. However, this tendency changed in the next periods. In the second image (1965), the urban area grew to 753.78 ha and followed an easterly, northerly, and northwesterly direction. Between 1965 and 1987, the growth of urban land increased to 125.45%, with 2914.64 ha of urban land in 1987, more than twice larger than that in the previous image. During this period, population grew by 98.51%. Between 1987 and 2001, the built-up area increased, and the urban spaces filled were the eastern and northern boundaries. The growth in urban land increased to 43.51%, with 4182.88 ha of urban land in 2001 (Table 3 and Figure 7).

Table 3. Changes in all classes of land use/cover in six periods studied.

	Agricultural Land	Barren Land	Coastal Zone	Hole Land	Military Land	River	Rocky Hills	Urban Land
1956	155.62	2536.59	210.96	4.71	48.13	162.43	2097.24	403.77
1965	133.26	2232.09	192.82	6.45	49.42	153.86	2122.15	753.78
1975	49.94	2061.80	111.52	12.64	154.86	131.90	1834.29	1292.81
1987	115.35	541.45	111.33	16.64	337.93	122.59	1511.96	2914.64
2001	30.48	237.11	101.15	9.33	340.07	109.61	661.28	4182.88
2012	6.92	47.46	60.31	8.65	296.63	72.05	262.16	4959.59
Total Change	-148.70	-2489.14	-150.66	+3.94	+248.50	-90.39	-1835.08	+4555.82

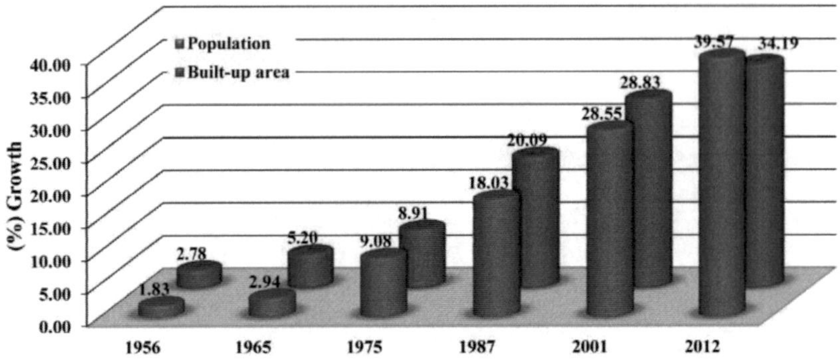

Figure 4. Growth of population and built-up area of Bandar Abbas in last 60 years.

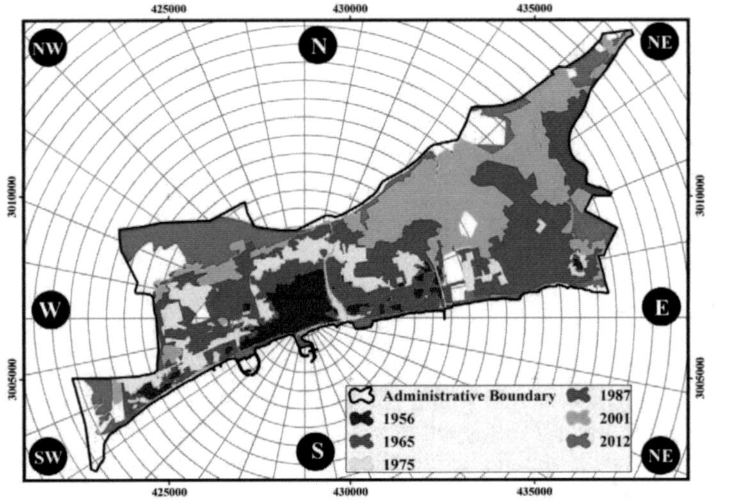

Figure 5. Urban growth and sprawl in Bandar Abbas (1956-2012).

In the last studied period, Bandar Abbas presented major growth in the built-up area. The increase of urban land between 2000 and 2012 led to the occupation of much of the remaining vacant area by new construction. In 2012, urban uses occupied 4959.59 ha of the land, and the population density increased to 77.27 persons per ha. This expansion shows a chaotic pattern that, among other effects, reduces spatial harmonies [20]. Since 1956, urban sprawl has continued in a disproportionate, dispersed, and leapfrogging manner in addition to rapid growth. In other words, it has followed the classic features of sprawl (Figure 6).

As previously supposed, the analysis of changes in land use/cover through remotely sensed data facilitates the detection of the history of the relationships between human activities and the environment as well as the assessment of possible trends for landscape conversions, thereby creating accurate and multidimensional scenarios for developing environmentally sustainable strategies and management practices [7]. Furthermore, the analysis of the spatiotemporal characteristics of land use/cover change is necessary in the evaluation and understanding of the ecological repercussions of urbanization [6]. The overall accuracy and Kappa coefficient of all the images are presented in Table 2. According to [5], more than 50 samples are required to certify the classification accuracy assessment for each of the classes, which are dispersed across the image in a random stratified pattern. The high accuracy obtained in the present study may be the result of the rough classification, as only eight classes are used. Moreover, the created maps had different classification accuracies because of the different multispectral bands between the images and the differences in ground resolution. Lastly, no thoroughly spatial matching was conducted among the maps. After complete land use/cover maps were created, the complete coverage of the different classes was determined. The results of the changes in land use/cover are shown in Figure 6 and Table 3. The most substantial change was the transformation of barren land to impervious built-up areas. Within the time period of study, 2489.13

ha of barren land was transformed into urban land use (e.g., residential, commercial, educational, green space, and transportation). The second main change is the conversion of barren land into military land. This change is also a significant deviation from the common direction of land replacement in many countries, in which military land is usually considered a limiting factor in urbanization. In Bandar Abbas, however, about 296.63 ha of barren land was converted for military use between 1965 and 2012. The third major change is the conversion of coastal zones into urban land, affecting some 150 ha. Coastal areas are important to economic development and tourism. Unfortunately, the coastal areas of Bandar Abbas have significantly changed as a result of the increase in growth of urban built-up areas since the 1950s. Irregular urban development in the coastal zones has violated regulations. The lack of adequate supervision and development planning has intensified the problem. Only the easternmost coastal city of Bandar Abbas has a concrete development plan. Other places located to the south and west of the city have been exposed to serious disasters because of irregular construction and the lack of supervision.

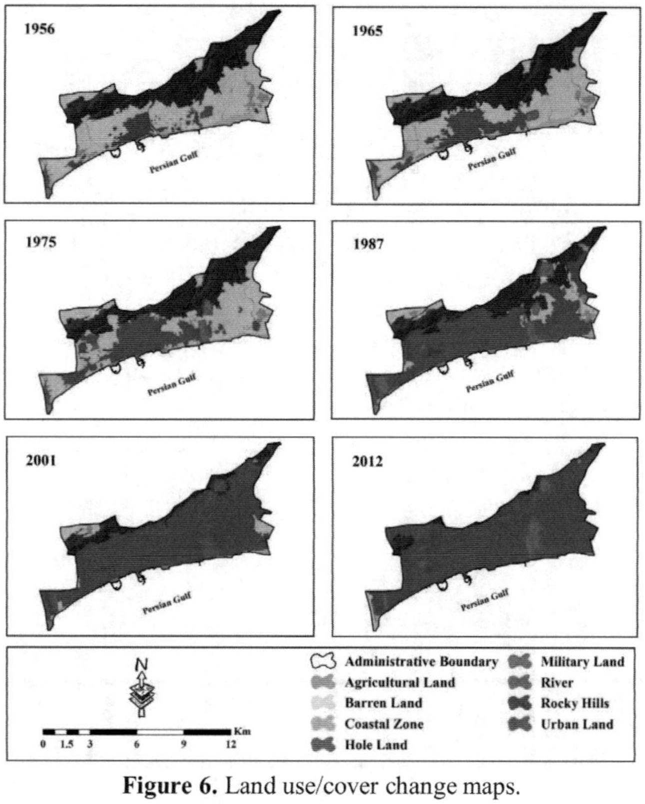

Figure 6. Land use/cover change maps.

Table 4 presents the land-use and land-cover change matrix in the eight categories of land use selected for Bandar Abbas between 1956 and 2012. Urban or built-up land increased by 1028.32% (4152 ha), and military lands surrounding the city increased by 516.30% (248.5 ha). By contrast, barren land decreased by 1.87% (47.46 ha). In these results, the problems related to different pixel sizes were minimized, particularly in the case of the unlikely change of barren land into built-up areas. Based on the data from the maps and tables, urban sprawl is the main cause of the changes in land use in the city. The land-use map of 2006, which is related to the master plan of Bandar Abbas, illustrates that only 68.4% of the land is actively used for residential, commercial, educational, cultural, tourism, and transportation purposes, whereas about 31.6% is undeveloped and vacant [15]. This undeveloped land, nearly one third of the city area, has mostly three uses: vacant land and desolate homes (23.3 %), military land (5.46%), and coastal zone (2.84). These uses circumvent to a certain extent the irregular nature of urban sprawl and the alternative uses of abundant spaces left behind by the urbanization process, such as stores, workshops, and green spaces. Vacant land also offers an opportunity for developing infilling policies.

Table 4. Land use/cover matrix, 1956-2012 (ha).

		1956								
		Urban Land	Agricultural Land	Hole Land	River	Coastal Zone	Military Land	Rocky Hills	Barren Land	Total
2012	Hole Land	0	0	1.66	0	0	0	0	6.99	8.65
	River	0	0.49	0	39.97	0.60	0	4.34	26.56	71.96
	Rocky Hills	0	0	0	3.04	0	0	225.04	34.06	262.14
	Agricultural Land	0	0	0	0.00	0	0	6.92	0	6.92
	Military Land	0	0	0	0.17	0.56	47.51	78.72	169.54	296.49
	Coastal Zone	0	0	0	12.72	46.89	0	0	0.60	60.21
	Barren Land	0	0	0	30.81	0	0	0	16.64	47.45
	Urban Land	403.74	155.11	3.05	75.68	162.87	0.62	1782.06	2329.03	4912.17
	Total	403.74	155.60	4.71	162.38	210.91	48.13	2097.08	2583.43	5665.99

As discussed earlier, Shannon's entropy is a well-known and functional method of identifying the growth of urban sprawl [14].Table 5 show the absolute and relative Shannon's entropy of Bandar Abbas. The growth of Bandar Abbas cannot be characterized as a complete sprawl because of the significant difference between the calculated absolute Shannon's entropy and the maximum possible value of log(24), which represents dispersed development. During the study period, Shannon's entropy established a moderately increasing rate, which indicates that Bandar Abbas grew toward dispersed development or sprawl. The rate of change of the built-up area per capita considerably increased from 1956 to 1987, and the slight increase in 2012 has subsided, indicating that the development has changed from a compact to a dispersed or sprawled pattern (Figure 7). The results present excellent harmonization between Shannon's entropy and built-up per capita as two fast and simple criteria for assessing growth patterns.

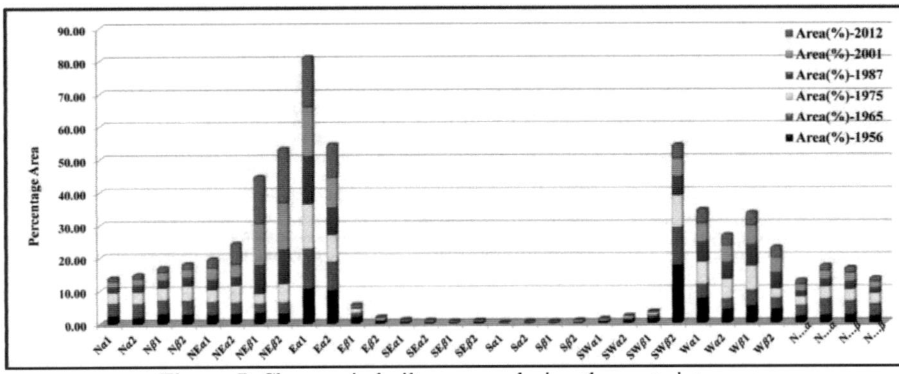

Figure 7. Changes in built-up area during the past sixty year.

Table 5. Calculated values for absolute and relative Shannon's entropy.

Year	Absolute Shannon's entropy	log(24)	Relative Shannon's entropy
1956	0.742541332	1.380211242	0.651245781
1965	0.794124236		0.672511408
1975	0.853845671		0.697413781
1987	0.952147319		0.701423211
2001	0.970421045		0.714589453
2012	0.996340724		0.723251987

5. Conclusion

This study examined the urban sprawl of Bandar Abbas, one of the most important tourism, industrial, and commercial centers in southern Iran. Remote sensing and GIS techniques were used to monitor the dynamic phenomenon of sprawl with the help of aerial photos, satellite images, and census data. Bandar Abbas is a medium-sized city to the south of Iran that has experienced rapid urban sprawl since the mid-1970s. Since then, the physical size of the city has increased every 10 years or so. In particular, between 1987 and 2001, the new urban area added to Bandar Abbas was as large as the entire urban area developed in more than 2500 years of its history. As in many cities in developing countries, the sprawl in Bandar Abbas occurs in disorderly and unplanned patterns, influenced by the proximity of highways, towns, and villages and the differences in communication networks, industrial locations, and land rents. Moreover, irregular rapid and unplanned sprawl produces important problems, such as lack of security, sufficient service, and facilities for residents; lack of public transport systems; increase in commuting length; increase in energy consumption; and increase in air pollution. Lastly, the irregular sprawl of the urban area entails the loss of coastal tourism attractions, a factor that hinders economic growth and minimizes employment opportunities. As previously discussed, the total growth rates of the city and of its population from 1956 to 2012 are similar. Thus, based on the classification of sprawl proposed by [1], the growth of Bandar Abbas cannot be considered as a classic model of sprawl growth. Instead, if the study period is divided into smaller periods, the period between 1986 and 2001 clearly emerges as the period during which Bandar Abbas experienced sprawl accompanied with the conversion of large plots of coastal zone, barren land, and rocky hills into built-up areas. This pattern of growth is also supported by calculations of Shannon's entropy, a functional method for identifying sprawl The cases presented in this paper may instead be interpreted as presaging a possible sprawl in the future.

On methodological grounds, this study strengthens the high value of remote sensing in urban studies and environmental monitoring. Although many studies on a global scale have used aerial photos, satellite images, and remote sensing in the study of urban sprawl, few have studied Iranian cities ([16], [21], [18]). In light of the increase of urbanization and population growth and the decrease of suitable land for urban sprawl, the importance of the study of built-up area change and the preparation of development plans based on remote sensing and geospatial data is crucial for other cities in Iran, as well as other cities around the world, facing the threat of limited land for development.

References

[1] Barnes K B, Morgan J M, Roberge, M C, Lowe S (2001) Sprawl development: Its patterns, consequences, and measurement, Towson University. <http:// chesapeake.towson.edu/landscape/urbansprawl/download/Sprawl_ white_paper.pdf>.

[2] Bhatta B (2009b) Modeling of urban growth boundary using geoinformatics. Int. J. Digital Earth, 2(4), 359-381.

[3] Bhatta B, Saraswati S, Bandyopadhyay D (2010) Quantifying the degree-of-freedom, degree-of-sprawl, and degree-of-goodness of urban growth from remote sensing data. Appl Geogr, 30(1), 96-111.

[4] Clark, D. (1982). Urban geography: An introductory guide. Taylor & Francis. p. 231.

[5] Congalton, R. G. (1991). A review of assessing the accuracy of classification of remotely sensed data. Remote Sens Environ, 37, 35-46.

[6] Deng J, Wang K, Hong Y, Jia G (2009) Spatio-temporal dynamics and evolution of land use change and landscape pattern in response to rapid urbanization. Landscape Urban Plan, 92, 187-198.

[7] Doygun H (2009) Effects of urban sprawl on agricultural land: a case study of Kahramanmaras, Turkey. Environ Monit Assess, 158, 471-478.

[8] Farhoudi R, Zanganeh Shahraki S, Saed Moucheshi R (2009) Spatial distribution of population in Iranian urban system. Quat. Geo. Res., 68, 55-68.

[9] Gill J (2008) The effect of urban sprawl on Sydney's peri-urban agricultural region. Soc. Environ Pol Sus 8, 219-416.

[10] Glaeser E, Kahn M, Chu C (2001) Job sprawl: Employment location in U.S. metropolitan areas. Washington, DC: Survey Series of Center for Urban & Metropolitan Policy, The Brookings Institution.

[11] Iranian Statistic Center (2012) Census data. Tehran, Iran.

[12] Kumar J A V, Pathan S K, Bhanderi R J (2007) Spatio-temporal analysis for monitoring urban growth – a case study of Indore city. J Indian Soc Remote Sens 35(1), 11-20.

[13] Lata K M, Rao C H S, Prasad V K, Badarianth K V S, Rahgavasamy V (2001) Measuring urban sprawl: a case study of Hyderabad. GIS Dev 5(12), 26-29.

[14] Li, X., & Yeh, A. G. O. (2004). Analyzing spatial restructuring of landuse patterns in a fast growing region remote sensing and GIS. Landscape Urban Plan, 69, 335-354.

[15] Ministry of Roads and Urban Development. (2009). Iranian cities and their spatial distribution in different periods. Tehran, Iran.

[16] Mahdi Sabet Sarvestani, Ab Latif Ibrahim, Pavlos Kanaroglou (2011) Three decades of urban growth in the city of Shiraz, Iran: A remote sensing and geographic information systems application. Habitat Int 35: 521-528.

[17] Onur, I., Maktav, D., Sari, M., & Sonmez, N. K. (2009). Change detection of land cover and land use using remote sensing and GIS: a case study in Kemer, Turkey. Int. J. Remote Sens., 30(7), 1749-1757.

[18] Rafiee, R., Salman Mahiny, A., Khorasani, N., Darvishsefat, A., & Danekar, A. (2009). Simulating urban growth in Mashad City, Iran through the SLEUTH model (UGM). Cities, 26, 19-26.

[19] Sudhira H S, Ramachandra . V, Jagdish K S (2004) Urban sprawl: metrics, dynamics and modeling using GIS. Iint J Apll Earth Obs 5, 29-39.

[20] Saraie M H (2008) Multiple patterns of Yazd growth. Quat. Geo. Res, 84, 75-98.

[21] Saeed Zanganeh Shahraki, David Sauri, Pere Serra, Sirio Modugno, Faranak Seifolddini, Ahmad Pourahmad (2011) Urban sprawl pattern and land-use change detection in Yazd, Iran. Habitat Int 35, 521-528.

[22] Yang L, Xian G, Klaver JM, Deal B (2003) Urban land-cover change detection through sub-pixel imperviousness mapping using remotely sensed data. Photogramm Eng Rem S 69:1003-1010.

[23] Yeh A G O, Li X (2001) Measurement and monitoring of urban sprawl in a rapidly growing region using entropy. Photogramm Eng Rem S 67(1), 83-90.

[24] Yuan F, Sawaya KE, Loeffelholz BC, Bauer ME (2005) Land cover classification and change analysis of theTwin Cities (Minnesota) Metropolitan Area by multitemporal Landsat remote sensing. Remote Sens Environ 98:317-328.

IOP Conf. Series: Earth and Environmental Science **20** (2014) 012008 doi:10.1088/1755-1315/20/1/012008

Markov-CA model using analytical hierarchy process and multi-regression technique

N.Q. Omar[1], S. A. M.Sanusi[2], W.M.W.Hussin[3], N Samat[4],K.S.Mohammed[5]

[1]Geomatic Engineering Unit, School of Civil Engineering, Engineering Campus, Universiti Sains Malaysia
Civil Engineering Unit, College of Engineering, Kirkuk University, Kirkuk, Iraq
[2,3]Geomatic Engineering Unit, School of Civil Engineering, Engineering Campus, Universiti Sains Malaysia, 14300 Nibong Tebal, Pulau Pinang. Tel: 04-5937788,
[4,5]Geography section, School of Humanities, Main Campus, University Sains Malaysia, 11800, Penang, Malaysia Phone: 604-6532854 Fax: 604-6563707

E-mail: [1]nqo11_civ078@student.usm.my, [2]cesanusi@eng.usm.my,
[3]cewan@eng.usm.my, [4]narimah@usm.my , [5]khamaj7582@yahoo.com

Abstract. The unprecedented increase in population and rapid rate of urbanisation has led to extensive land use changes. Cellular automata (*CA*) are increasingly used to simulate a variety of urban dynamics. This paper introduces a new *CA* based on an integration model built-in multi regression and multi-criteria evaluation to improve the representation of *CA* transition rule. This multi-criteria evaluation is implemented by utilising data relating to the environmental and socioeconomic factors in the study area in order to produce suitability maps (*SMs*) using an analytical hierarchical process, which is a well-known method. Before being integrated to generate suitability maps for the periods from 1984 to 2010 based on the different decision makings, which have become conditioned for the next step of *CA* generation. The suitability maps are compared in order to find the best maps based on the values of the root equation (R^2). This comparison can help the stakeholders make better decisions. Thus, the resultant suitability map derives a predefined transition rule for the last step for *CA* model. The approach used in this study highlights a mechanism for monitoring and evaluating land-use and land-cover changes in Kirkuk city, Iraq owing changes in the structures of governments, wars, and an economic blockade over the past decades. The present study asserts the high applicability and flexibility of *Markov-CA* model. The results have shown that the model and its interrelated concepts are performing rather well.

1. Introduction

In recent times, there has been a rapid increase in urbanization all over the globe. The approximate world population in 2000 was 2.9 billion and predicted to be 5.0 billion in 2030 (United Nations, 2007). This ratio will increase to over 72% by 2050 (United Nations, 2012). The unprecedented increase in population and rapid rate of urbanisation can lead to extensive land-use changes.

The rapid urbanization in Asia and West Asia, both has exhibited a process distinctively different in late 20th Century [1]. Many researchers have studied land-use and land-cover change (*LUCC*) since the end of the 20th century [2]. The input issues outlined by *LUCC* agenda include the study of integrated global and regional models and the evaluation of databases land-use and land-cover change is a complex process involving both environmental and socioeconomic factors. Land-use and land-cover change research increasingly takes the form of integrated land-use change research, important to this change is human activity that impinges on land-use systems and surface features, known as land use and land cover change (*LUCC*) [3].

Content from this work may be used under the terms of the Creative Commons Attribution 3.0 licence. Any further distribution of this work must maintain attribution to the author(s) and the title of the work, journal citation and DOI.
Published under licence by IOP Publishing Ltd

Monitoring land-use and land-cover change help to develop an understanding of past trends, while simulation based modelling can provide insights into potential future developments. Both complementary approaches are necessary strategies for implementing appropriate actions including formulating better land use policies, providing infrastructure, identifying future development pressure points, and developing previous visions of urbanization process implications. [4].

Therefore, knowledge concerning past, current, and future growth plays an important role in the decision-making process [5, 6]. The simulation and prediction of urbanization can give input to various environmental and planning models. In the last three decades, urban models based on *CA* techniques were developed for improved understanding of urban evolution .The core of a *CA* is a transition rule that has been represented either using weight matrices [7], *SLEUTH* model [8], multi-criterion evaluation (*MCE*) [9], logistic regression [10], neural networks [11], and decision catalogue trees [12]. In these methods, many variables are mixed up for defining transition rules. Every variable is usually connected with a criterion that indicates its importance in the simulation process. These criteria significantly have an effect on outcomes of urban simulation [10, 11].

The cellular automata (*CA*) model is regularly employed in land utilities research. The grid cell in *CA* is similar in its spatial description to remote sensing images and raster data of geographic information systems (*GIS*), which frequently support the land utilities data [13]. The *CA* model also open managing of dynamic spatial models with time [14]. The setting of *CA* is in the simulation processes of adjacent interaction cells. This method is similar to land use activities that are frequently characterized by the interaction of adjoining landscape patterns.

The integration of various geospatial technologies in the form of technical support for data collection, processing, analysis, and production [15]. With developments in remote sensing (*RS*) and *GIS*, comprehensive computing, and prophecy techniques, simulation models have been developed to better understand urban progress dynamics and to be hopeful of urban planning activities [16]. Significant contributions in the urban field have been established thanks to the advancement of *GIS* and *RS* [6, 17] both of which have been used to relate land-use and land-cover change to urban growth models [18] as well as techniques that are in the front position of the advances in the land-use suitability analysis such as multi-criteria decision analysis (*MCDA*) [19]. RS and GIS, given their cost-usefulness and technical reliability, have been used progressively to build up positive sources of information supporting decision making in various urban applications [20]. *MCE* is based on the combination of several measures to make a conclusion. Criteria can be of two kinds: factors and constraints [21]. The AHP is a mathematical method of analysing complex decisions problem with multiple criteria. It is based on three principles: decomposition, comparative judgment, and synthesis of priorities. For this purpose, (Saaty, 1980) has used and developed the Par wise Comparison Method (*PCM*).

The *PCM* is based on forming judgments between two particular elements rather than attempting to prioritize an entire list of elements [22]. It allows users to estimate the relative weight of multiple criteria against giving criteria in an intuitive method. If the quantitative evaluations are not available, decision makers can nevertheless realize whether one criterion is (equally important, extremely important, or less important) than another depending on the scale. A spreadsheet package, EXPERT CHOICE is based on *AHP* and used in calculating weight for respective layering. It has an ability to compute weights for multiple criteria with par wise comparisons. Weighted linear combination (*WLC*) or simple additive weighting is a procedure that multiplies normalised criteria values of relative criteria weights for each alternative [23-25].

IOP Conf. Series: Earth and Environmental Science **20** (2014) 012008 doi:10.1088/1755-1315/20/1/012008

Fuzzy Membership Functions are used to standardize the criterion scores. Decision makers have to decide based on their knowledge and fair judgment the function to be used for each criterion. Regression technique determines the empirical correlations between binary dependent and several independent categorical and continuous variables [26].

Kirkuk is one of the ancient provinces in Iraq inhabited by different ethnicities and characterized by a rich culture with a distinguished history It is heritage of about 5,000 years and significant geographic location linking between central and northern Iraq. It is an oil-rich province having one of the top quality oils in the world .The last official master plan for the City of Kirkuk was completed by *Doxiadis* in 1974 and last updated on 1986. Since then, the Master Plan and the development of the City of Kirkuk were politically influenced, and since 2003, the city's growth and current situation have dramatically affected and altered the city to the challenge by many difficulties such as unplanned environmental pollution and other socioeconomic problems. (Kirkuk Government, 2001).In addition owing to changes in government structures, political, wars, and economic blockade over the past three decades. In this respect, it is a good deal needed to study the land-use and land-cover changes and anticipate the future maps of Kirkuk city. In this paper, an incorporated model was developed based on multi-regression model and multi-criteria evaluation to improve the representation of *CA* transition rules as a new contribution framework . The objective of this article is to present analyses that are complete by *GIS* and *RS* of the period from 1984 to 2010 and depict the implementation of a validation procedure, which show how uncertainty affects the simulation results then predictive for future land use land cover change model in Kirkuk city in 2020,2030 and 2040 based on the past trend (from 1984 to 2010) using a method called *Markov-CA* along with transition probability maps taken into account by *MCE*.

2 . Material and Dataset

Kirkuk city in Iraq is the capital of the Kirkuk Governorate (formerly known as Tameem) located along Khasa River, within the geographical coordinates (Lat 35° 28' 5"N, Long 44° 23' 31"E) at 350m above sea level as shown in 'figure 1'. It is situated in the northern part of Iraq, 236 km in northwest of the capital Baghdad city, 83 km south of Erbil, the capital of Kurdistan Region, 149 km southeast of Mosul, 97 km west of Sulaymaniyah, and 116 km northeast Tikrit.

The topography of Kirkuk is generally very flat with common terrain heights in the northern part (786km from the Arabian Gulf). The Kirkuk region lies between the Zagros Mountains (northeast), Zap and Tigris River (west), Hamrin Mountains (south), and Sirwan (Diyala) River (southeast) (Google Maps Distance Calculator). All the geographic data for the study area were collected in two stages: primary and secondary phase.

2.1. Primary Data

Group discussions, interviews, and reviews of existing studies conducted during the two field works (from the 15th of November 2011 to 01st of February 2012, and from the 21th of July 2012 to the 27th of September 2012), which provided the key information for identifying the driving criteria of urban growth in the last decade. The selections of urban growth criteria accorded through literature review [9, 27], academic knowledge, engineering expert, and members of the city planner. Synthesis of these data led to the derivation of five representative factors out of the 11 criteria identified, (Distance to road accessibility, Distance to railway, Distance to airport, Distance to urban centres, Social services, Slope, Elevation, Environmental factors, Hazard lands, Agricultural value/soil type, Urban suitability, Zoning, Population density, Land value, Water supply, river, water bodies, Social housing… etc.).

Five factors closely associated with land-use and land-cover change forces were selected and incorporated into the transition potential calculation, and three constraints.

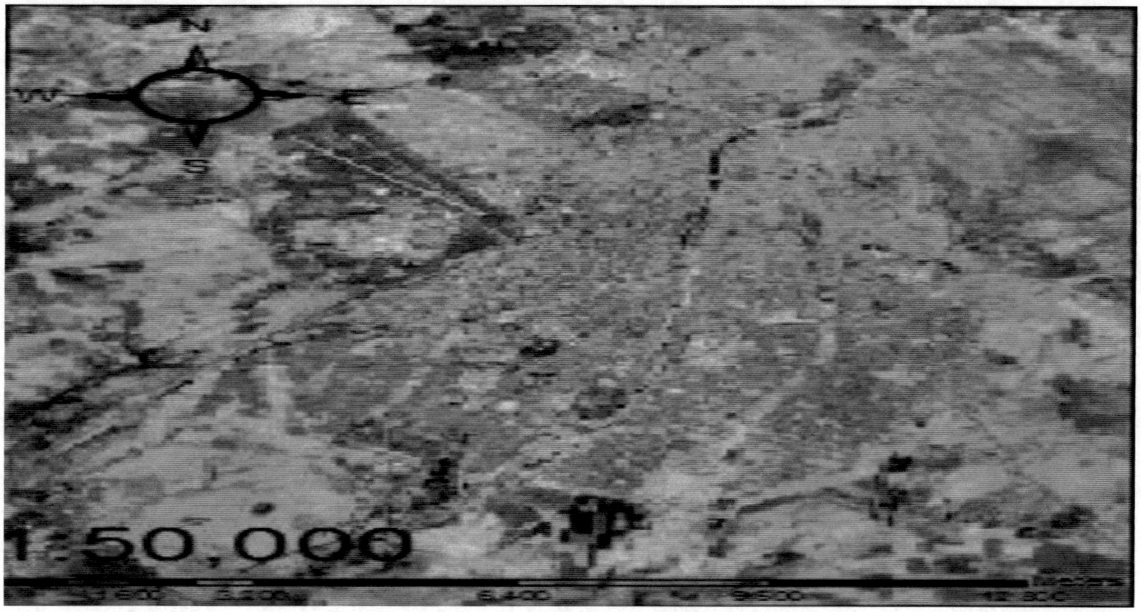

Figure 1. Location of the study area.

Ranging from socioeconomic to environmental, for further evaluation (table 1). These factors are populated , topography, accessibility to the Central Business District (*CBD*), Distance to the road, and Distance to the Khasa Rivers. Consequently, data pertaining to urban growth factors were obtained through a set of questionnaires was developed based on *AHP* technique, interviewed, in which the respondents stated the relative importance of each factor with respect to the others. The group of the decision making process uses 27 professionals where from their expert opinions weights to the genes that best define their preferences are obtained and assigned.

Table 1. The relative weights for different decision making groups.

Criteria	Normalized Weight Total	Normalized Weight PL	Normalized Weight EE	Normalized Weight AC
Population DENSITY	0.4650	0.4535	0.4431	0.4926
Topographic(Slope)	0.1438	0.1721	0.1220	0.1515
Accessibility To (C.B.D)	0.1101	0.1313	0.0860	0.1320
Distance To Road	0.1210	0.1004	0.1425	0.1075
Distance To River	0.1600	0.1426	0.2064	0.1165
Total Weight	1.0000	1.0000	1.0000	1.0000

2.2. Secondary Datasets

In this study four scenes of the land satellite "*LANDSAT*" images LANDSAT -5 Thematic Mapper (*TM*) and LANDSAT-7 Enhance Thematic Mapper (*ETM*) for path 169, row 35 covers the Kirkuk and the surrounding area , images were acquired on 24 June 1984, 30 April 1990 ,16 April 2000, and 18 February 2010. These images were downloaded one by one from the official website of the United States Geological Survey (USGS).

More details concerning the spectral ranges and spatial resolution of LANDSAT-5 can be referred to the *USGS* web site. The 30m spatial resolution of "*LANDSAT*" satellite images is sufficient to capture the characteristic scales of human development, and the spectral range of the tool is capable to recognize land -use of other types of land-cover change. The digital elevation model (*DEM*) data were obtained from the project Shuttle Radar Topography Mission www.glcf.umiacs.umd.edu/data/srtm (*SRTM*) operated by the US Geological Survey (*USGS*) Earth Resources Observation Systems (*EROS*) Data Centre (2006). *DEM* data were part of the 1 arc second (30m) SRTM Digital Terrain Elevation Data (*DTED*).

2.3. Software and Processing tools used

The work utilised various software for all the analyses, data preparation , image processing and analysis performed during the creation of the model of Kirkuk city expansion was accomplished using a combination of the following software sets:

Expert choice (*EC*) to determine the relative weight of the factors (Expert Choice , 1982-2004). *AutoCAD* merged and edited the *CAD* (dwg extent file) data (Autodesk, 1982-2009). Microsoft excel in the arrangement of relative importance weights. *ERDAS* Imagine-8.5 (ERDAS, 1999) to register the satellite data, and then cut the interest area of studies. *ARC GIS 9.2* (*ESRI*) to set up the database and digitize all layers for criteria and convert to raster. *GIS IDRISI* (Eastman, 1987-2012) to resample enhancement of images, standardization of all strata image criteria, map calculation (*MCE*), statistical classification, Markova change, Markova cellular automata and validation. Lastly the Microsoft word for writing of the paper.

2.4. Base map

At first, all "*LANDSAT*" images and *GIS* data needed for use in model, converted to the same projection and raster maps, standardized to the same cell size and grid dimensions (number of rows and columns), under a Universal Transverse Mercator map (*UTM*) projection system , referenced to the world geodetic system (*WGS*) 1984 coordinates system, Zone 38N.Secondly, for the purpose of ground-treating the base maps, they were registered them geometrically in the map projection system by using the image in 2005 as a basis. The Land sat images contain massive numbers of numerical values that signify information such as the region.

The region was encompassed by the resolution and the number of spatial band used. The effort in reducing supply and treatment operations in the computer was accomplished by partitioning the image (ERDAS, 1999), that covers the area of study to Minimum X and Y coordinate (3933427.13539o, 432502.457664o), Maximum X and Y coordinate (3903144.38623o, 459357.992169o) in decimal degrees using the WGS84 datum. The images used to create land cover maps for the Kirkuk area covers a surface area of 827 km2 representing the administrative boundary of the city. TM image has seven bands with a spatial resolution of 30m-120m, while ETM image has eight spectral bands with a resolution of 15m-30m. Images from different sensors will have a different spatial resolution.

Therefore one method of resolving this problem is to resample higher resolution i.e. 120m, 90m, and 60m to lower resolution of 30m, which is enough to capture spatial details and small enough to reduce computation time (Bhatta, 2009).The mean square error (*MSE*) is equal to 0.010734.

Thirdly, by utilizing the widely used supervised maximum likelihood classification method, we separate the three images under a Nine-class (*C1-C9*), for each pixel equal probabilities which includes urban (consists the building's residential commercial industrial public serves road network and parking lots.), mountainous area, oil field area, mineral land, wetland, pastures water bodies, cultivated land and vacant land respectively [21]. Digitized training sites for each class were delimited by 1) comparing the visual interpretation of composites true colour imagery for each year with supervised classification images, 2) multi temporal imagery through Google earth, 3) field observation and 4) topographic map of 1987 (Department of survey and mapping, Kirkuk, 1987) at the scale of 1:50,000 before the signature of all grades are compiled. We calculate the overall accuracy and Kappa statistics were calculated for assessing the classification accuracy [28]. The results show that overall accuracies of the classification are 74 %, 78 %, 77 % and 82 % in 1984, 1990 , 2000and 2010 respectively.

2.5. Factors specification

The series of factors are prepared using *Arc GIS*, and they are transformed into a normalized scale before being input into the model. The "*LANDSAT*" imagery of 1984, 1990, 2000 and 2010, and topographic map of Kirkuk (Department of Surveying and Mapping, Ministry of Agriculture, Kirkuk, 2000) were used for digitised to creating vector layers by digitizing the road network, the Khasa Rivers and the Central Business District (*CBD*) for the mentioned years consecutively. The population density vector layer was digitized based on the entire population of the respective zones in Kirkuk for all the years and then the estimated population data was compared, with the last census population data in 1997. The Demographic Data details from Primary depended on the Census abstracts for 1957, 1977, 1987 and 1997 (National Census Bureau of Statistics, Kirkuk, 2005). The *SRTM DEM* data were re-sampled from its original resolution of 30m. The slope layer (in percentage) map is derived from the *DEM* data. Additionally, constraint vector data files such as water bodies, restricted military area, and oil producing area were created, which are stored in Shape file format.

3 . Modelling approach

3.1. Suitability maps

These suitability maps are organized based on Multi Criteria Evaluation (*MCE*) process. At present, the obtained weights can be computed automatically in Expert choice computer package software. The indicator score weights of factors in three aspects: First the Academic represent the theoretical aspect, second the Planner as the planning aspect, and the Expert Engineer as applied aspects of land-use and land-cover change is shown in (table 1) based on factors to compute the relative indicator score weights of decision makers. Thus, par wise comparison matrixes are calculated into Expert Choice determining priority weight. However, the normalisation of factors weight is (divide by the sums of the pillars, and average across rows to find the relative weights of each component).

Where, the vector layers converted to raster's fuzzy set membership is used for standardising criteria. The factors or land cover constraints which have been standardised to a continuous scale to minimum and maximum standard numerical range from 0 signify the least suitable to 255 signify the most suitable, the Boolean map for each land-cover type has been prepared. Then the images 'figure 2' for Boolean land-cover map were produced and compared with a model in a range from 0 to 1 [29]. The measurement unit was used is an international system (IS).

The distance images are created using a simple Euclidean distance function in *Arc GIS* which measures the distance between each cell from the featured image such as a road network, Khasa rivers, and *CBD* respectively. The lowest and highest values obtained from the distance image are used as the input for fuzzy set membership analysis. Then, the distance images have been standardised to the same continuous suitability scale (0-255) using fuzzy set membership analysis process. 'figure 2' provides a summary of the main input digital maps data (factors and constraints)to the *MCE* model.

We thus draw the following relative weights for the Five indicators. Factor as in (table 1) and for the three decision maker (Planner, Engineer, and academicians). Then, these will be entered in *Arc GIS* for spatial analysis to ascertain the suitability for development maps. These weights are used in the *WLC* as explained in 'figure 3'. In the overlay process, each factor map will be multiplied by its weight and the result is then summed up producing a suitable maps for consecutive years (2000 and 2010) as shown in 'figure 4'. The process involved constraint maps, where the suitability maps will be multiplied by each of the Boolean constraints to zero out the unsuitable area.

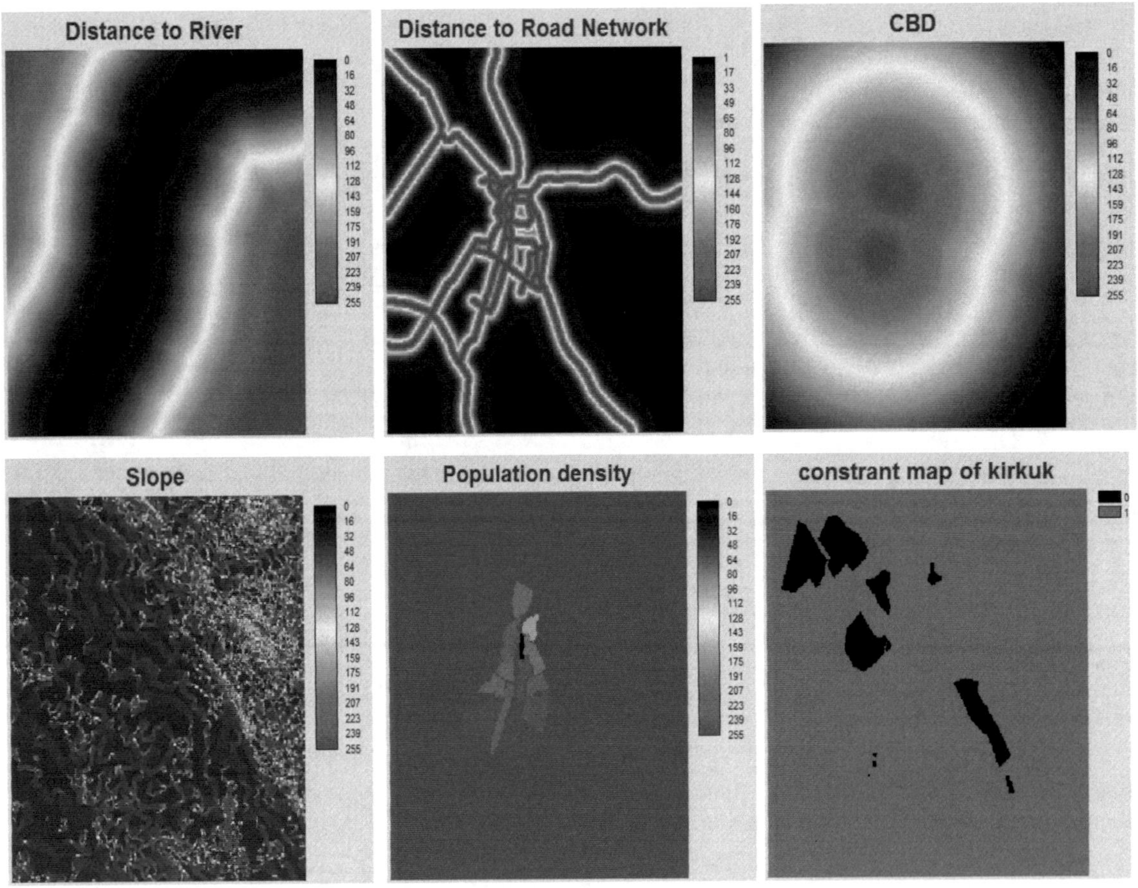

Figure 2 . Criterion as a raster maps.

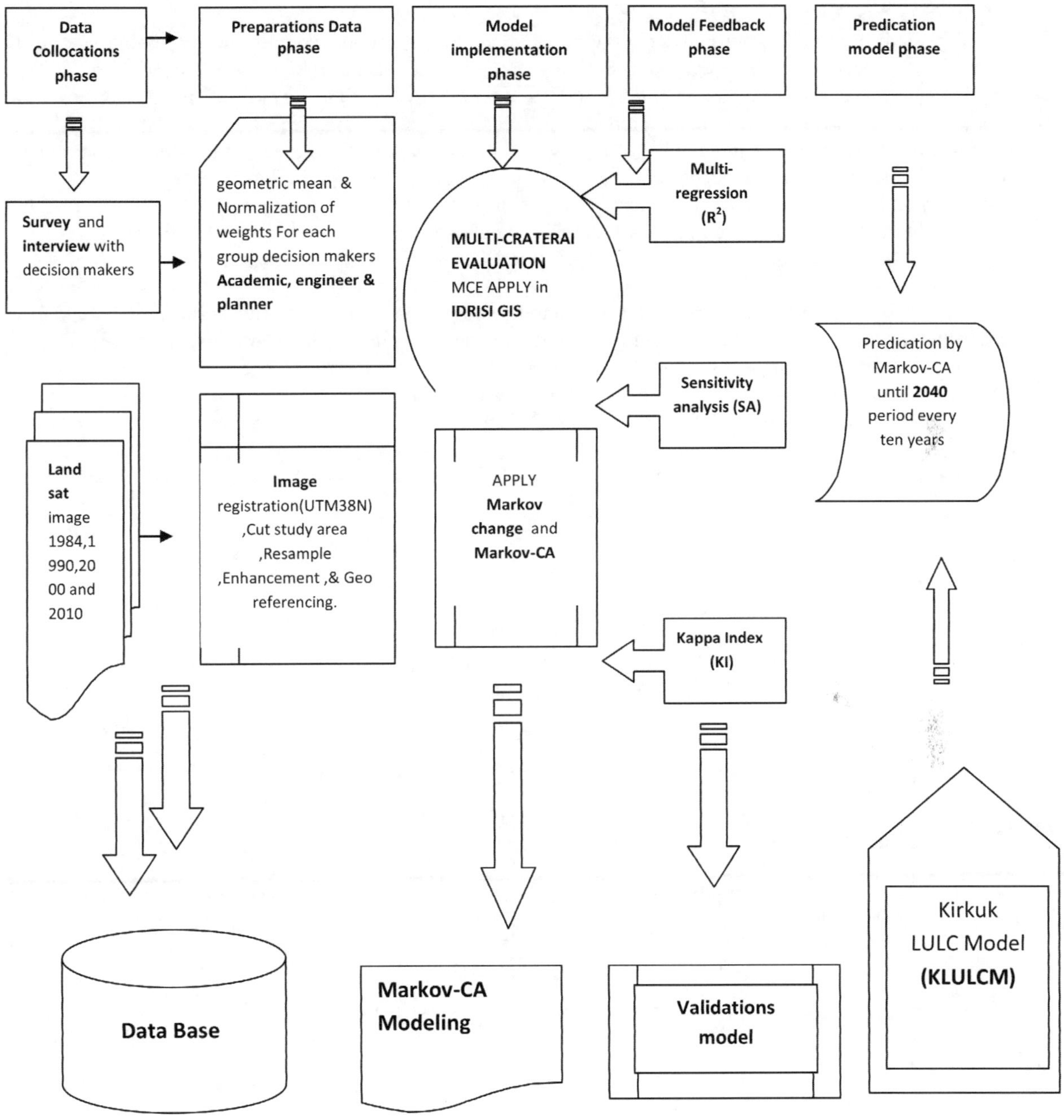

Figure 3. Flow chart modelling process for land-use and land-cover changes in Kirkuk city.

3.2. Multi-regression techniques

The suitability maps were analysed statistically in the form of a multiple regression model in *IDRISI* software. A stepwise regression building approach was used, which starts by including all the independent maps and then discarding those that do not have a significant role in determining the goodness or best fit suitability map. Hence, only maps that are significant in explaining variations in the dependent map are included. To deal with the problems of the potential non-linear relationship between the dependent and independent 16 maps in the models, the suitability change was transformed into common in *Markov-CA*. To show these changes the multi-regression analysis could be the best way to interpret the relationship between this map related to land use and land cover changes in the study area. Multi regression techniques were applied to the spatial statistical analysis variance (*ANOVA*).

This approach is based on the comparison of suitability maps obtained from active states in the simulation and in the actual city map through a set of spatial statistics. *ANOVA* tables show that the weights of the three groups are different. It is possible to better understand the similarities between suitability maps at different decision makers [30].By reference to the (table 2), the adjusted R^2 coefficient range (0.735 - 0.891) for the *AHP* method at different decision makers is a consequence of the high similarity in all the maps.

Table 2. Assessing R^2 for suitability maps model at different decision making for AHP method.

DC	years	Equations	R^2
AC	1984	AHP84AC = 0.1998 + 0.1014*AHP84EE + 0.7662*AHP84PL	0.823
EE		AHP84EE = -0.0586 + 0.1802*AHP84AC + 0.8765*AHP84PL	0.787
PL		AHP84PL = 0.1477 + 0.5720*AHP84AC + 0.3681*AHP84EE	0.878
AC	1990	AHPAC90 = 0.1843 + 0.1061*AHPEE90 + 0.7701*AHPPL90	0.829
EE		AHPPL90 = 0.1478 + 0.3568*AHPEE90 + 0.5830*AHPAC90	0.879
PL		AHPEE90 = -0.0454 + 0.8583*AHPPL90 + 0.1932*AHPAC90	0.785
AC	2000	AHP99AC = 0.1962 + 0.7252*AHP99PL + 0.1393*AHP99EE	0.827
EE		AHP99EE = -0.0360 + 0.8509*AHP99PL + 0.1910*AHP99AC	0.827
PL		AHP99PL = 0.1324 + 0.5116*AHP99AC + 0.4378*AHP99EE	0.888
AC	2010	AHPAC10 = 0.2347 + 0.5099*AHPEE10 + 0.3302*AHPPL90	0.753
EE		AHPEE10 = -0.0036 + 0.7491*AHPAC10 + 0.3085*AHPPL90	0.735
PL		AHPPL10 = 0.1092 + 0.4291*AHPEE10 + 0.5266*AHPAC10	0.891

Where the AC is university academic staff , EE is expert engineer, PL is urban planner and DM is decision maker.

3.3. Markov Change model

To model the urban expansion with spatial models is a useful way to understand the land-use and land-cover process, with the results offering support to urban planning and management policies. Markov Change model (*MCM*) is clear as a set of land-use and land-cover states where the process begins with one of the states and moves consecutively from one state to another [31].

Each transition matrix is defined as a step for land-use and land-cover change, a transition areas matrix and a set of conditional probability images where analysing a number of qualitative land-use and land-cover changes from two different dates (1990 and 2000). It is a random process (illustrated in table 3) which shows how likely one state is to change to another state through the transition probability matrix [32]. In the tables , the rows represent the older land cover categories and the pillars represent the newer land cover categories. Markov Chain model Analysis also produces related conditional probability images with the help of transition probability matrices.

3.4. Markov-CA model

The *Markov-CA* model used to simulate and predict land-use and cover-maps for all classes are presented in 'figure 5'. These can be very effectively modelled using Cellular Automata (*CA*). A cellular automaton is a cellular entity that independently varies its new state based on its previous state and that of its immediate neighbours according to a specific rule.

Cellular automata consider the composition of associations of pixels around one pixel. Clearly there is a similarity between *MCM* and *CA* process. Both depend on the previous state, *CA* also upon the province of the local neighbourhood. So, combining both Markov Chain and CA (*MARKOV-CA*) will be much more accurate and logical for predicting the future land cover change for a certain domain.

Markov-CA is a powerful model for the state of several categories of a cell based on transition areas matrix; transitional suitability images and a user defined contiguity filter default a 5 x 5 mean have been utilised . The estimated years are 2020 (time 2), 2030 (time 3) and 2040 (time 4) that is based on the transitional years 2010 (time 1), 2000 (time 1) , 1990 (time 1) and 1984 (time 0). Therefore, for this research purpose, 10 iterations are considered for future prediction depending on years. In the final step , the transition area matrix, all the suitability maps, the 5×5 *CA* contiguity filter and the base map of Kirkuk city (2010) are used to predict the future land-use and land-cover images for 2020,2030, and 2040.

Table 3. Markov transitions probabilities matrix for 1984-1990.

Classes	Cl. 1	Cl. 2	Cl. 3	Cl. 4	Cl. 5	Cl. 6	Cl. 7	Cl. 8	Cl. 9
Class 1	0.4238	0.1804	0.0465	0.0682	0.0254	0.0347	0.1386	0.0378	0.0445
Class 2	0.0781	0.3207	0.0598	0.2241	0.0559	0.1018	0.0477	0.0722	0.0399
Class 3	0.0365	0.2187	0.0633	0.2851	0.0587	0.1495	0.0317	0.1277	0.0288
Class 4	0.0229	0.2247	0.0599	0.2817	0.0731	0.1668	0.0236	0.1169	0.0304
Class 5	0.0221	0.0784	0.0398	0.1615	0.1778	0.2752	0.0361	0.1491	0.0601
Class 6	0.0223	0.1329	0.0474	0.2148	0.1396	0.2391	0.0302	0.1251	0.0486
Class 7	0.1313	0.1709	0.0606	0.1515	0.1003	0.1312	0.1051	0.0787	0.0705
Class 8	0.0560	0.1732	0.0563	0.1981	0.0828	0.1736	0.0529	0.1649	0.0421
Class 9	0.0362	0.0969	0.0433	0.1645	0.1729	0.2491	0.0469	0.1255	0.0648

Where *C1, C2,C3,C4,C5,C6,C7,C8,and C9* to equal be the urban (consists the buildings residential commercial industrial public serves road network and parking lots.), mountainous area, oil field area, mineral land, wetland, pastures water bodies, cultivated land and vacant land respectively.

3.5. Kappa statistic index

Validation techniques by [33] were used to determine the agreement between the 1990, 2000 and 2010 urban land use reference map with the1990, 2000 and 2010 simulated urban land use map. Furthermore, the techniques were used to compare the agreement between the reference map and the simulated map with the null model such as, agreement between the 1990 reference map and the 2000 reference map. Specifically, the validation technique described modest sources of agreement and disagreement between the simulated map and the reference map with higher accuracy as shown in (table 4).

The precision of the simulation or classification image results on a pixel by pixel beginning was assessed via the Kappa statistic index see (table 4). This statistic measures the goodness of fit or the best value between two model predication and reality, corrected for accuracy by chance [34]. Since land use maps are categorical maps, Kappa can be used to assess the goodness of fit between the simulation maps and the real land use map at the end of the simulation period [35, 36]. Kappa values range from 1 to -1, where positive values show sign of improved agreement than usual by chance, and negative values is agreement that is not good [21].

Table 4. Validation between actual and simulated maps in AHP for different decision makers.

Simulation Actual Maps	Methods	K no	K location	K location Strata	K standard
1990	AHP	0.9782	0.9865	0.9865	0.9739
	AC AHP	0.9775	0.9856	0.9856	0.9730
	EE AHP	0.9784	0.9867	0.9867	0.9740
	PL AHP	0.9778	0.9860	0.9860	0.9734
2000	AHP	0.9827	0.9867	0.9867	0.9782
	AC AHP	0.9819	0.9856	0.9856	0.9772
	EE AHP	0.9824	0.9863	0.9863	0.9779
	PL AHP	0.9822	0.9860	0.9860	0.9775
2010	AHP	0.9911	0.9893	0.9893	0.9893
	AC AHP	0.9915	0.9898	0.9898	0.9898
	EE AHP	0.9918	0.9902	0.9902	0.9902
	PL AHP	0.9912	0.9895	0.9895	0.9895

Where the K no is Kappa information, K location is kappa location, K location Strata is kappa location strata, and K standard is kappa standard.

4. Results and Discussion

The multi-regression analyses were carried out following the same methodology for all the metrics. Initially the value for each adjusted R^2 coefficient was calculated for each suitability map in all years at different decision makers. The first remarkable result of this analysis is the overall high adjusted R^2 coefficient obtained for most of the studied Suitability map.

The first sign most important suitability maps in explaining maps variations of urban planner decision maker in 2010 at *AHP* is 0.891 indicators of good fit models between 1984,1990,200, and 2010, although while comparing the weight of different decision making groups the same decision maker for 2000 the adjusted R-squared coefficient is 0.888 reasonably good degree correspondence between the evaluated Suitability map in mention yours. The result is shown in (table 2) and 'figure 4' respectively.

In this study, it is now possible to predict the land cover change map of Kirkuk city using *Markov-CA* modelling. The *Markov-CA* module in *IDRISI Andes* calculates the transition probabilities and outputs a text file with a transition probability matrix, a text file with the number of transitioning cells from one land use, land cover class to another, and raster lattice representing Markov transition areas. For example, (table 3) indicates that during 1984–1990 there was a 4.0 % chance the total pixels would transition.

The Markov transition probability matrix is calculated from a cross-tabulation of earlier and later land use, land cover images. Markov transition areas are derived by multiplying each column representing a land cover category in Markov probability matrix (table 3) by the number of cells of the same land cover class in the subsequent image [21]. As a *Markov-CA* model set in motion, a standard 5×5 contiguity filter re-weights the suitability maps during each iteration increasing the suitability of pixels in close nearness to contiguous areas of the same land use land cover class[21]. During each iteration, pixels with the maximum transition probability and the maximum suitability score for an exacting class transition to a new class while pixels with minimum probabilities and minimum suitability scores remain unchanged. If the input consists of 10 iterations[21].

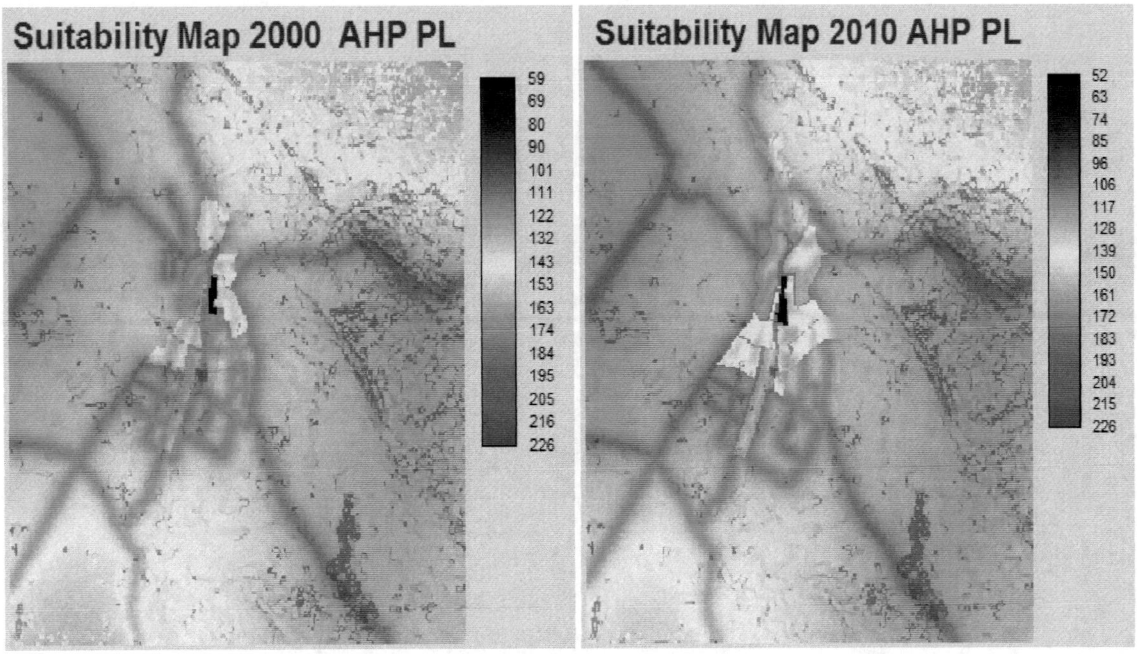

Figure 4. Suitability maps.

IOP Conf. Series: Earth and Environmental Science **20** (2014) 012008 doi:10.1088/1755-1315/20/1/012008

This Kappa index will assess the changes in the kappa statistics with various image resolutions and cell sizes. The prediction accuracy of the model was highly stable, with less than 1 % change in accuracy between the filters from 3*3 to 13*13. Furthermore, the model was suitable at the resolution image from 30 m to 120 m, which is adequate for the remote sensing resolution of *TM*, allocating more time to simulate land use change [37]. The projections specify that approximately 15 % of the study area will be urbanized by the year 2020,2030, and 2040, though the extent of urban development under academics and engineers decision is slightly less than that under the planners' decision. The validity of the model results has been evaluated by comparing the projected land cover image with the existing land cover map. The overall Kappa standard in different years was 0.97–0.99, which indicated a very good agreement between observed and projected land-use and land-cover see (table 4).

A separate Kappa statistic was calculated for the correspondence between observed and projected built-up areas [38]. Both kappa location and kappa location strata scores indicate that the model has improved performance over the multi-regression model. The relatively low kappa information scores for academics, engineers, and planners in 1990 are caused by the appearance of certain large patches of this particular change in the same year. These are the results of one multi- decision maker. Therefore, the simulated using a bottom-up technique, as used in *CA* very usefully. Nevertheless, Kappa values are taken as relative to the results of the multi-regression model. For instance, the suitability map in (table 2) shows the R^2 coefficient (2010) obtained for both maps is high as the previous as mentions 0.891 , although (table 4) representing Kappa Index was used to validate the simulation model, and the best kappa standard value for (2010) is 0.990. Thus, this model can be regarded as good results indicator of similarity between both maps. Kappa was used statistically as a linear distance decay role to account for slightly unfavourable pixels [39].

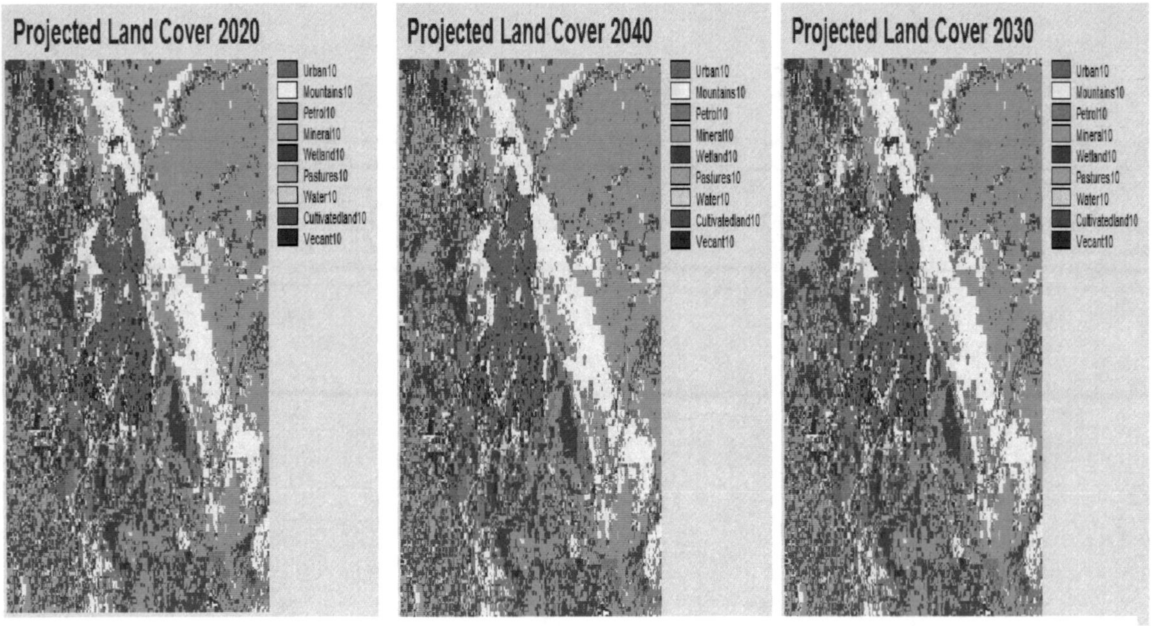

Figure 5. CA-Markov Projected Land-Use and Land-Cover Map of Kirkuk City (2020,2030 and 2040).

In the context , the result are summarized as follows: First, the integrated model has proven to be competent in monitoring and projecting the land-use and land-cover changes. Second, the future land use land cover map of Additionally , the value 0.990 for this index implies a high degree of similarity between both maps. From the point of view of their simulation model usually measure spatial configuration of land-use and land-cover , these techniques a powerful instrument for the comparison of suitability maps and simulated in the land use land cover change modelling.

The city will extend on the north-eastern, north-western and a few bit south- eastern parts in 2020,2030, and 2040. See (figure 5). Over the years land-use will increase intensively, while the land-cover will be diminished. Kirkuk city for the years of 2020,2030 and 2040 has been is predicted to increase at an annual rate of 3%. In spatially case , the newly increased land-use is most likely to expand around the vicinity of the city and mainly along a west–east axis and a north–south axis along the *CBD*.

5. Conclusions

The model is just a generalization of the reality. A goal of urban expansion modelling is evaluating potential for future paths improvement [40]. This paper, develops and illustrate new ways of assessing and managing the land-use and land-cover changes. In particular, it investigate a sampling method of criterion weight under three groups of decision makers: academic staff decision, expert engineer's decision, and planner decision makers, for five factors and three constraints. It is feasible and effective to build up a model to predict urban expansion trend through *Markov-CA*, Markov probability , multi-criteria evaluation and multi-regression which allowed the simulation of nine land cover classes simultaneously and thus furnished a basis for a significant examination of the land-use and land-cover changes under different decision makers. The analysis in this case has testified that the sample results in multi-regression to choose the best suitability map and kappa statistical index of the validation model indicates similarity between the choices of three decision makers. The transition rules were based on the score achieved by multi-regression.

MCE could not resolve this situation and therefore, some more is to be done to resolve this conflict. Setting transition rules in the form of suitable maps based on *MCE* allows for consideration of various components that are commonly used in land use planning and decision-making such as changes in population density, topography (slope), proximity to network roads, proximity to the *CBD* as well as protection riparian corridors of Kasha river, water bodes resources, military restricted area, and oil sensitive areas. The Group (*AHP*) is easy for usage , a powerful and useful decision-making tool that allows group decision makers to compare and select options as part of the group decision-making process However, the process for controlling and simplifying urban expansion model of integration with geographical information systems is mentioned in more researcher works [8, 9, 14, 41, 42]. In addition, the performance of the Kirkuk model indicates its output and evaluation via Investigational application of the model to an imitation city which produced realistic patterns of development, supporting the modeling approach. Iraq is one of the fastest growing country. After the year 1972, the unplanned appeared in many parts especially in Kirkuk . This research will add to the urban form of the city at first step in planning method. The decision makers: academic staff , expert engineer's as well the city planners can initiate appropriate plans based on the results. That will make the Kirkuk city a much more habitable and planned in the future.

Further studies can describe the results obtained from applying the model to Kirkuk, Iraq. A more appropriate and rigorous test of the model is required to validate its approach and output. For example the paper shows Kirkuk model the optimum suitability map used as a transition rule in *CA* model. This emphasizes the intensity of our work, the ability to address multiple criteria with multiple decision makers in choosing the best model.

References

[1] W. Bing Sheng and D. Z. Sui, "CA-based Simulation of Asian Urban Dynamics: A Case Study of Taipei Metropolitan Area, Taiwan," in *Geoscience and Remote Sensing Symposium, 2008. IGARSS 2008. IEEE International,* 2008, pp. V - 13-V - 16.

[2] E. F. Lambin, M. D. A. Rounsevell, and H. J. Geist, "Are agricultural land-use models able to predict changes in land-use intensity?," *Agriculture, Ecosystems & Environment,* vol. 82, pp. 321-331, 2000.

[3] D. R. Foley JA, Asner GP, Barford C, Bonan G, Carpenter SR, Chapin FS, Coe, D. G. MT, Gibbs HK, Helkowski JH, Holloway T, Howard EA, Kucharik CJ,, and P. J. Monfreda C, Prentice IC, Ramankutty N, Snyder PK., "Global consequences of land use.Science. 2005 Jul 22;309(5734):570-4," 2005.

[4] H. Taubenböck, M. Wegmann, A. Roth, H. Mehl, and S. Dech, "Urbanization in India - Spatiotemporal analysis using remote sensing data," *Computers, Environment and Urban Systems,* vol. 33, pp. 179-188, 2009.

[5] H. Shafizadeh Moghadam and M. Helbich, "Spatiotemporal urbanization processes in the megacity of Mumbai, India: A Markov chains-cellular automata urban growth model," *Applied Geography,* vol. 40, pp. 140-149, 2013.

[6] J. E. Patino and J. C. Duque, "A review of regional science applications of satellite remote sensing in urban settings," *Computers, Environment and Urban Systems,* vol. 37, pp. 1-17, 2013.

[7] R. White and G. Engelen, "Cellular automata and fractal urban form: a cellular modelling approach to the evolution of urban land-use patterns," *Environment and planning A,* vol. 25, pp. 1175-1199, 1993.

[8] K. C. Clarke and L. J. Gaydos, "Loose-coupling a cellular automaton model and GIS: long-term urban growth prediction for San Francisco and Washington/Baltimore," *International Journal of Geographical Information Science,* vol. 12, pp. 699-714, 1998.

[9] F. Wu and C. J. Webster, "Simulation of natural land use zoning under free-market and incremental development control regimes," *Computers, Environment and Urban Systems,* vol. 22, pp. 241-256, 1998.

[10] F. Wu, "Calibration of stochastic cellular automata: the application to rural-urban land conversions," *International Journal of Geographical Information Science,* vol. 16, pp. 795-818, 2002.

[11] X. Li and A. G.-O. Yeh, "Neural-network-based cellular automata for simulating multiple land use changes using GIS," *International Journal of Geographical Information Science,* vol. 16, pp. 323-343, 2002.

[12] X. Li and A. Gar-On Yeh, "Data mining of cellular automata's transition rules," *International Journal of Geographical Information Science,* vol. 18, pp. 723-744, 2004.

[13] A. M,nard and D. J. Marceau, "Exploration of spatial scale sensitivity in geographic cellular automata," *Environment and Planning B: Planning and Design,* vol. 32, pp. 693-714, 2005.

[14] D. F. Wagner, "Cellular automata and geographic information systems," *environment and Planning B,* vol. 24, pp. 219-234, 1997.

[15] Z. Yang, "PAML 4: phylogenetic analysis by maximum likelihood," *Molecular biology and evolution,* vol. 24, pp. 1586-1591, 2007.
[16] I. Benenson and P. M. Torrens 2004, " Geosimulation: object-based modeling of urban phenomena," *Computers, Environment and Urban Systems,* vol. 28, pp. 1-8.
[17] B. Bhatta, S. Saraswati, and D. Bandyopadhyay, "Quantifying the degree-of-freedom, degree-of-sprawl, and degree-of-goodness of urban growth from remote sensing data," *Applied Geography,* vol. 30, pp. 96-111, 2010.
[18] J. Jokar Arsanjani, M. Helbich, W. Kainz, and A. Darvishi Boloorani, "Integration of logistic regression, Markov chain and cellular automata models to simulate urban expansion," *International Journal of Applied Earth Observation and Geoinformation,* vol. 21, pp. 265-275, 2013.
[19] J. Malczewski, *GIS and Multicriteria Decision Analysis*: Wiley, 1999.
[20] Q. Weng and D. A. Quattrochi, *Urban remote sensing*: CRC Press, 2007.
[21] J. R. Eastman, *IDRISI Kilimanjaro: guide to GIS and image processing*: Clark Labs, Clark University Worcester, 2003.
[22] T. L. Saaty, "How to make a decision: the analytic hierarchy process," *European journal of operational research,* vol. 48, pp. 9-26, 1990.
[23] J. Geldermann, M. Treitz, and O. Rentz, "Towards sustainable production networks," *International Journal of Production Research,* vol. 45, pp. 4207-4224, 2007.
[24] M. Treitz, H. Schollenberger, B. Schrader, J. Geldermann, and O. Rentz, "Multi-criteria decision support for integrated technique assessment," *RADTECH Europe 05: UV/EB-Join the Winning Technology,* pp. 153-160, 2005.
[25] T. L. Nyerges and P. Jankowski, "regional and urban Gis," *A Decision Support Approach. New York,* 2010.
[26] P. McCullagh and J. A. Nelder, "Generalized linear models (Monographs on statistics and applied probability 37)," *Chapman Hall, London,* 1989.
[27] C. He, N. Okada, Q. Zhang, P. Shi, and J. Zhang, "Modeling urban expansion scenarios by coupling cellular automata model and system dynamic model in Beijing, China," *Applied Geography,* vol. 26, pp. 323-345, 2006.
[28] S. Stehman, "Estimating the kappa coefficient and its variance under stratified random sampling," *Photogrammetric Engineering and Remote Sensing,* vol. 62, pp. 401-407, 1996.
[29] T. L. Saaty, *The Analytic Hierarchy Process: Planning, Priority Setting, Resource Allocation*: McGraw-Hill, 1980.
[30] K. McGarigal and S. A. Cushman, "Comparative evaluation of experimental approaches to the study of habitat fragmentation effects," *Ecological applications,* vol. 12, pp. 335-345, 2002.
[31] X. Zhang, T. Kang, H. Wang, and Y. Sun, "Analysis on spatial structure of landuse change based on remote sensing and geographical information system," *International Journal of Applied Earth Observation and Geoinformation,* vol. 12, pp. S145-S150, 2010.
[32] A. J. Mousivand, Alimohammadi Sarab, A., Shayan, S. , "A new approach of predicting land use and land cover changes by satellite imagery and Markov chain model (Case study: Tehran)." 2007.
[33] R. G. Pontius, E. Shusas, and M. McEachern, "Detecting important categorical land changes while accounting for persistence," *Agriculture, Ecosystems & Environment,* vol. 101, pp. 251-268, 2004.
[34] J. v. Vliet, R. White, and S. Dragicevic, "Modeling urban growth using a variable grid cellular automaton," *Computers, Environment and Urban Systems,* vol. 33, pp. 35-43, 2009.
[35] G. M. Foody, "Status of land cover classification accuracy assessment," *Remote Sensing of Environment,* vol. 80, pp. 185-201, 2002.

[36] R. G. Pontius and L. C. Schneider, "Land-cover change model validation by an ROC method for the Ipswich watershed, Massachusetts, USA," *Agriculture, Ecosystems & Environment,* vol. 85, pp. 239-248, 2001.

[37] G. Wang, J. Song, B. L. Xue, X. Xu, and K. Otsuki, "Land use and land cover Cchange of hulun lake nature reserve in inner Mongolia,China: A Modeling analysis," *Journal of the Faculty of Agriculture, Kyushu University,* vol. 57, pp. 219-225, 2012.

[38] R. G. Pontius Jr and M. Millones, "Death to Kappa: birth of quantity disagreement and allocation disagreement for accuracy assessment," *International Journal of Remote Sensing,* vol. 32, pp. 4407-4429, 2011.

[39] A. Hagen-Zanker, B. Straatman, and I. Uljee, "Further developments of a fuzzy set map comparison approach," *International Journal of Geographical Information Science,* vol. 19, pp. 769-785, Aug 2005.

[40] M. Herold, H. Couclelis, and K. C. Clarke, "The role of spatial metrics in the analysis and modeling of urban land use change," *Computers, Environment and Urban Systems,* vol. 29, pp. 369-399, 2005.

[41] F. Wu, "SimLand: a prototype to simulate land conversion through the integrated GIS and CA with AHP-derived transition rules," *International Journal of Geographical Information Science,* vol. 12, pp. 63-82, 1998/01/01 1998.

[42] D. P. Ward, A. T. Murray, and S. R. Phinn, "A stochastically constrained cellular model of urban growth," *Computers, Environment and Urban Systems,* vol. 24, pp. 539-558, 2000.

IOP Conf. Series: Earth and Environmental Science **20** (2014) 012009 doi:10.1088/1755-1315/20/1/012009

Using Arc GIS to analyse urban growth towards torrent risk areas (Aswan city as a case study)

Omar Hamdy[1], Shichen Zhao[2], Mohamed A. Salheen[3], Y. Y. Eid[4]

[1] Research Visitor, Department of Architecture and Urban Design, Faculty of Human-Environment Studies, Kyushu University, Japan

[1] Assistant Lecturer & PhD Student, Architectural Engineering Dept., Faculty of Engineering, Aswan University, Aswan, Egypt

[2] Associate Professor, Department of Architecture and Urban Design, Faculty of Human-Environment Studies, Kyushu University, Japan

[3] Associate Professor, Urban Planning & Design Dept., Faculty of Engineering, Ain Shams University, 1 Sarayat St., Abassia, Cairo, Egypt

[4] Professor, Urban Planning & Design Dept., Faculty of Engineering, Ain Shams University, 1 Sarayat St., Abassia, Cairo, Egypt

E-mail: omargem56@yahoo.com

Abstract: Areas suffering from storm water drains are considered to be the places most at risk, water torrents have an effect on urban areas and can cause a lot of damage to buildings and infrastructure. Moreover, there is dangerous situation whereby urban growth is occuring towards at-risk areas. The urban growth rate in risk areas rose up to 24.9% in 2001, and reached 48.8% in 2013. Urban growth in "Abouelreesh" village had been influenced by the construction of larger buildings, because most people were looking forward to live in bigger houses. We can discover the previous problem by observing the average size increase of the buildings' areas from 2001 until 2013, especially in risky areas where the average building's area had grown from 254 m2 in 2001 to 411 m2 in 2013. This Phenomenon is considered to be very important factor which attracts the urban growth towards the risky areas in spite of the danger surrounding them.

1. Introduction

Urban growth usually leads to the change of land use and the form of land cover at many areas around the world, especially in developing countries [1]. Moreover, urban areas have a natural extension in multiple directions, even sometimes they tend to grow towards places threatened by natural disasters. Hence, the governments' decision makers and planners should solve this problem through risk management and prevention of urban growth in that direction with the provision of different alternatives. Moreover, changing the urban areas is considered to be an indicator of urban growth, in terms of growth rate and direction. The urban areas include urban clusters of buildings, streets, areas of services, and land space. Urban sprawl which had become an issue for many rapidly developing areas refers to the uncontrolled growth of an urban area resulting from poorly or totally unplanned urbanization [2].

For example, the growth of urban areas depends on the population growth and buildings in a city. This type of metrics measures what is present and the householders' relative amounts or properties

Content from this work may be used under the terms of the Creative Commons Attribution 3.0 licence. Any further distribution of this work must maintain attribution to the author(s) and the title of the work, journal citation and DOI.
Published under licence by IOP Publishing Ltd

regardless to where on the landscape they may be located. Therefore, the analyst should be aware about the spatial / non-spatial nature of the selected metrics. [3]

Torrents are one of the most important natural disasters that threaten with destruction of urban areas, especially those which are located directly in the areas of storm water drains, where the rush of water is very large and the concentrated amounts makes huge destruction for the buildings, streets, and infrastructure of the region. The danger degree is depending on torrents' power and the layout readiness of the city or village to face and manage that risk. It is found that many scholars focused on using indicators to measure urban sprawl by establishing multi-dimensional indicators by GIS analysis or descriptive statistical analysis [2]. Growing of urban sprawl is a serious concern worldwide for a number of environmental and economic reasons [4]. Rapid increase of urban sprawl in many countries worldwide had become a major concern because of its detrimental effects on the environment. Existing measures of urban sprawl suffer from a confusing variety of differing and sometimes contradictory interpretations of the term ''urban sprawl'' [5].

2. The impact of natural disasters on urban growth

The world had witnessed an alarming increase in the frequency and severity of disasters; 240 million people on average were affected by natural disasters world-wide each year between 2000 and 2005. During each of these six years these disasters claimed an average of 80,000 lives and caused damage of an estimated 80 billion US$. Disaster losses are rising throughout the world due to a number of factors that include more frequent extreme weather events associated with increasing climate change. Population growth accompanied with demographic change and movements are leading to unplanned urbanization.

Catchments which are located in the east of River Nile at the region between Edfu and Aswan cities are dangerous, particularly at the area of Kom Ombo and the east of Aswan. In May of 1979 the runoff spate had resulted in disrupting the rail lines, besides affecting the central of Edfu, Kom-Ombo and Aswan which had led to the collapse of the 200 houses. Moreover, the torrents flooding had resulted in killing three children under the rubble and around 300 families had been displaced, besides the falling of boulders on some parts of the agricultural road and cutting the rail lines. These floods were repeated in October from the same year which led to the collapse of more than 300 buildings and cracking a large number of buildings. Moreover, torrents had accompanied with these dust storms, thunderstorm, cyclonic rain and frequent floods in the years 1980, 1987, 2005 and 2010 [6] [7].

Figure 1. Buildings are destroyed in Abouelreesh village after a torrent in 2010.

Severe hurricane shocked the people of the province of Aswan on the evening of 17th Jan. 2010, which had followed directly by a torrential rain for half an hour continuously. About 50 electricity pylons of high voltage had dropped on the ground which led to the cutting of Aswan province power. After that, rains which accumulated on the eastern mountains had turned into a torrent and swept the home furnishings and people, leaving behind hundred houses destruction, which caused a sudden catastrophe that was new bad experience for people. In addition, hundreds of houses had been cracked because of directly rainfall and threatening to collapse at any moment. As rains and severe storms caused damage to thousands of acres of agricultural and uprooting of trees from the root, as well as damaging tons of harvest dates which people had put them under the sun to dry out. That catastrophe happened at the village of northern "Abouelreesh" at a distance of 5 kilometers northern Aswan adjacent to the Red Sea Mountains at a length of 500 meters north-south and at the depth of 1300 meters east-west. The most vulnerable places to torrents were houses that had been built along the east of the railway, until the mid of the mountain. The competent administrative authorities had alerted the danger of this region and its surroundings for construction on it, because of the water storm risk (Fig.1).

Because of lacking of other available areas; people have been forced to build in this region and settle there. This time was not big damage, because there were not too much people at that time [8].

Around 20,000 people are living in this region. Therefore, torrents caused in the demolition of a total of about 25 houses and partially demolished for about 50 home mostly built of stone and mortar clay addition water storm Abu Spirh, It is away from the city of Aswan about 15 kilometers to the north and then eastward about 4 kilometers and inhabited by tribes Ababdh numbering about 3000 people can assess the damage, which suffered the destroy of 10 houses completely and about 30 homes in part also draws large tracts of agricultural land and the uprooting of many trees (Fig. 2) [7].

Figure 2. Roads and buildings affected by torrents in Egypt.

3. Study area

The study area (Northern Abouelreesh village) is located in the city of Aswan in southern Egypt Where it lies between longitudes 32°52'E to 32°55'E and latitudes 24°10'N to 24°14'N. Limiting the study area to the east of the mountain and from the west was agricultural area and then River Nile and in the village there are three areas at risk torrents (Fig. 3).

The area of urban space approved for the village of northern Abouelreesh approximately 453.01 acres in 2010 and that the total area of roads and pathways and urban spaces (106.11) acres, land space (private property) about (67.33) acres, desert land of approximately (126.25) acres and farmland of about (4.73) acres.

The population of the village was about 9896 inhabitants in 1986 rose to 12206 inhabitants in 1996, Then to 13189 in 2006, The growth rate was 2.12% per annum in the period 1986/1996, decreased to 0.78% per annum in the period 1996/2006. The average of growth rate was 1.45% per annum in the period 1986/2006. The total population of the village in 2010 about 14202 and the average family size is about 4.16 and 3204 the number of families [8].

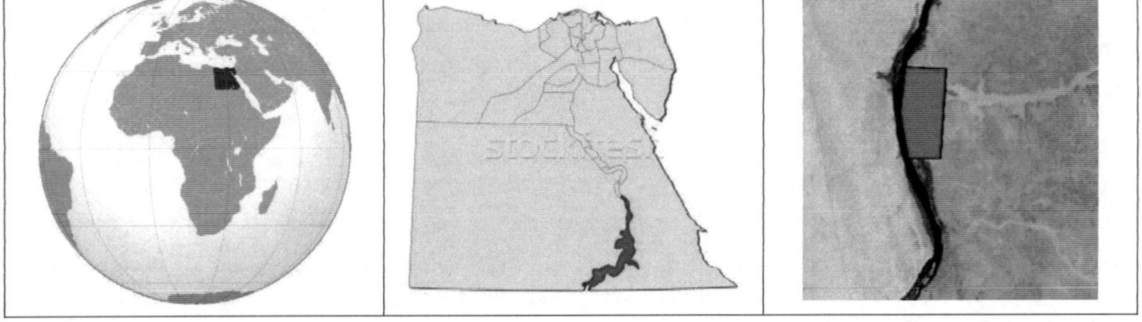

Figure 3. The study area related the world.

4. Methodology

Research methodology is based on the study of the region in different time periods and identifies urban growth especially in risky areas which are facing natural hazards; it was conducted in several steps.

It started by discussing the detailed outline of the study area and updating it by using Google Earth to determine the approximate periods for buildings that have been where construction by observing changes during the display of historical images in Google Earth.

Followed by using Arc GIS software to analyze the urban areas results, by studying the land use and the average area of the building for each time period and determine the ratio of the urban growth rate from period to another.

Then, we were identifying risky areas through locating the storm water drains from the official bodies of government in addition to using the Digital Elevation Model (DEM), and converting it into a contour map using GIS to get the lowest possible elevations which affected by torrents water.

Finally, the study area was divided into two region types according to their exposure to the flooding risk; safe areas, and far from the scope of the water floods.

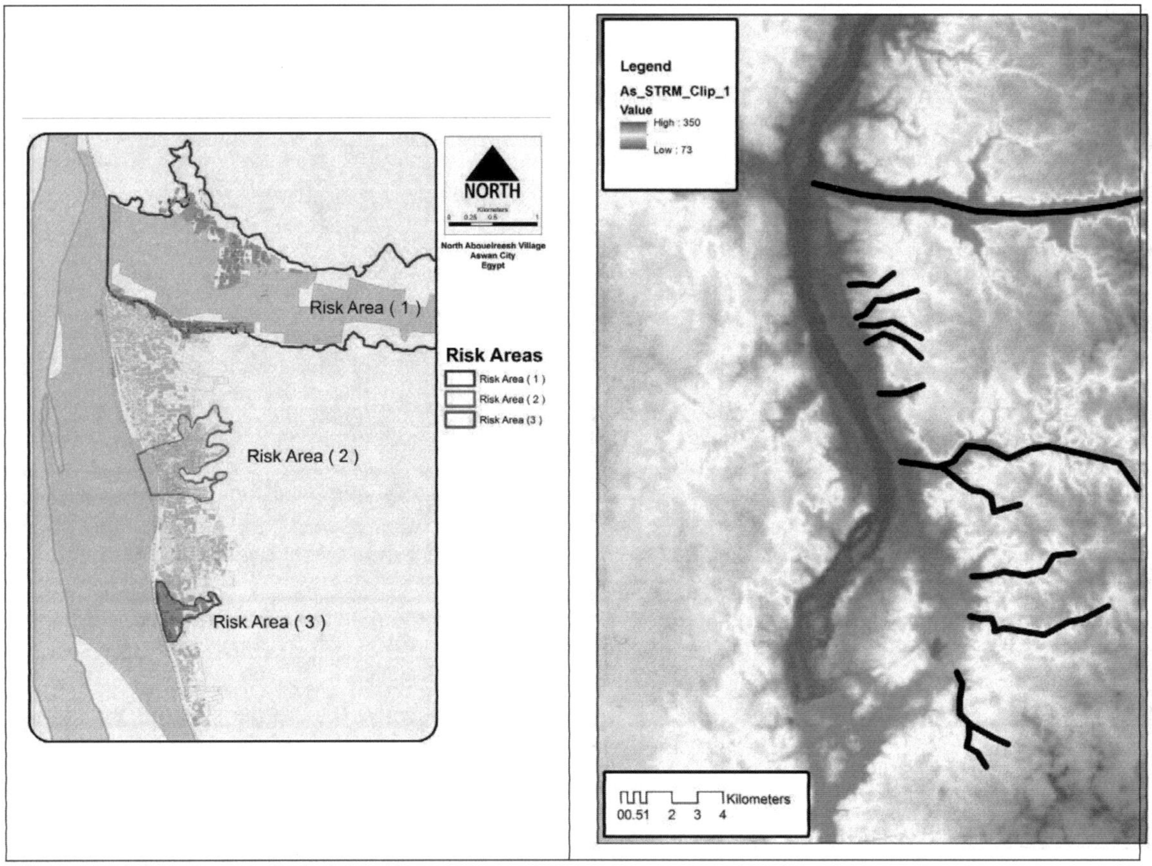

Figure 4. Identify the risk area by digital elevation model and convert it to contour map.

5. Data Analysis

We studied and analyzed the buildings ages using Google Earth and land use in the study area according to (Fig. 4), It appears the distribution of land uses inside the village which is divided into a small part in the north and big part in south which separated by agricultural land, The greater part has mountainous area of rugged terrain from the east, There is agricultural land and the Nile River in the west, In south located the village of southern Abouelreesh as shown in (Fig.5).

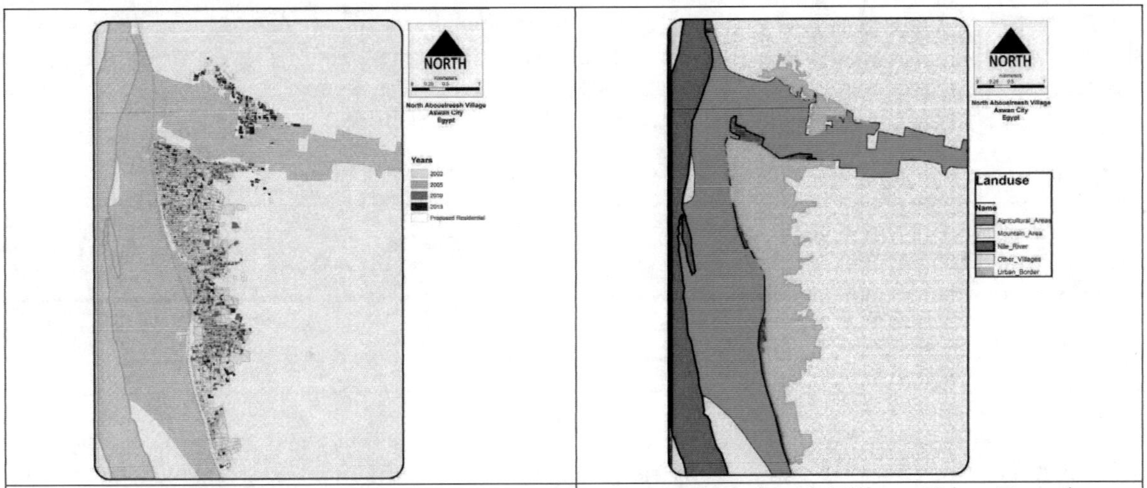

North Abouelreesh Village, Years | North Abouelreesh Village, Urban Boundary

Figure 5. Study the current status of the village of Abouelreesh (land use – years of construction - Urban Structure).

We used Arc GIS to analyze the data about the time periods for buildings and obtaining the total area of buildings for each period (Fig. 5), Also calculate the average area of buildings for each year to compare between them and find the urban growth, Then we made the same analysis with the study of safe areas and risky areas to determine the total area and the average area of the buildings for each year separately in each region according to table (1).

Table 1. This table shows the area of urban areas and the average area units.

Year	Sum Total Area (m^2)			Percentage Area (%)		Ave Total Area (m^2)		
	Total Areas	Risk Areas	Safe Areas	Risk Areas	Safe Areas	Total Areas	Risk Areas	Safe Areas
2001	1180287	293909	926074	24.9 %	78.5 %	260	254	216
2005	46075	18751	30326	40.7 %	65.8 %	268	254	221
2009	328853	124667	212978	37.9 %	64.8 %	317	307	241
2013	291523	142260	156462	48.8 %	53.7 %	351	411	236

Table (1) shows the percentage of urbanization in each year of the study years of 2001 to 2013 excluding the first year, also shows the highest rate of growth that was between 2005 and 2009 amounted to 17.81% of the total area of urbanism at the public level, also table shows the level of safe areas where amounted to 16.06%, while the highest rate was in risky areas around 24.54% between 2009 and 2013, But less than percentage between 2001-2005 about 2.49%.

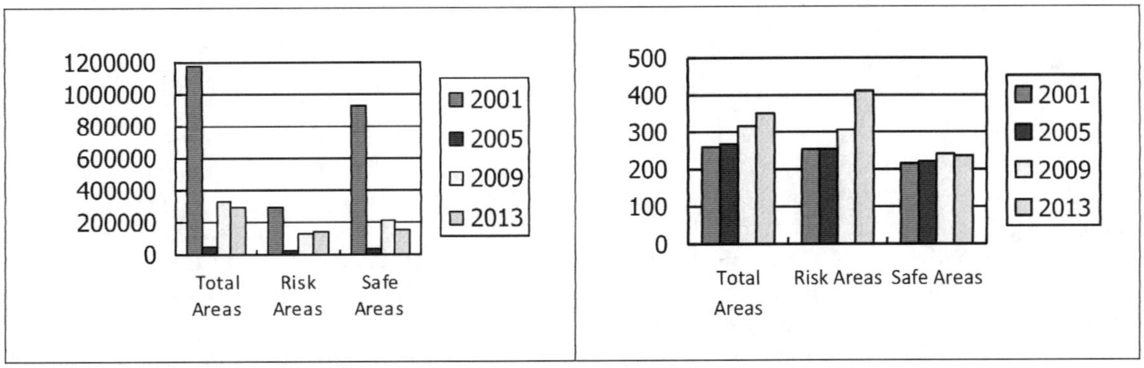

Figure 6. It is show the relationship between the Urban Area and average area for units in safe areas and risky areas.

As shown in Table (1) urbanization rate increase during the years of the study in all areas and showing that the highest increasing rate between the two years was 26.32% in total urban areas, but risky areas and safe areas were 39.87% and 22.27% respectively, It happened between 2005 and 2009 while it was less than the increasing rate between 2001 and 2005. (Fig. 6)

Table 2. The amount of urban growth each year and the percentage increase.

Year	Annual Extend_%			Increase_%		
	Total Areas	Risk Areas	Safe Areas	Total Areas	Risk Areas	Safe Areas
2001	63.91 %	50.71 %	69.85 %	0.00 %	0.00 %	0.00 %
2005	2.49 %	3.24 %	2.29 %	3.90 %	6.38 %	3.27 %
2009	17.81 %	21.51 %	16.06 %	26.82 %	39.87 %	22.27 %
2013	15.79 %	24.54 %	11.80 %	18.74 %	32.53 %	13.38 %

When we studied the relationship between the average area of building and urban growth we have noted that the average area is getting big in risk areas at about 411 m2 in 2013, while the average area of building in safe areas reached 236 m2 in the same year that It is an indication to the trends of the population in the urban sprawl around risky areas in search of big space table (2) and (Fig. 7).

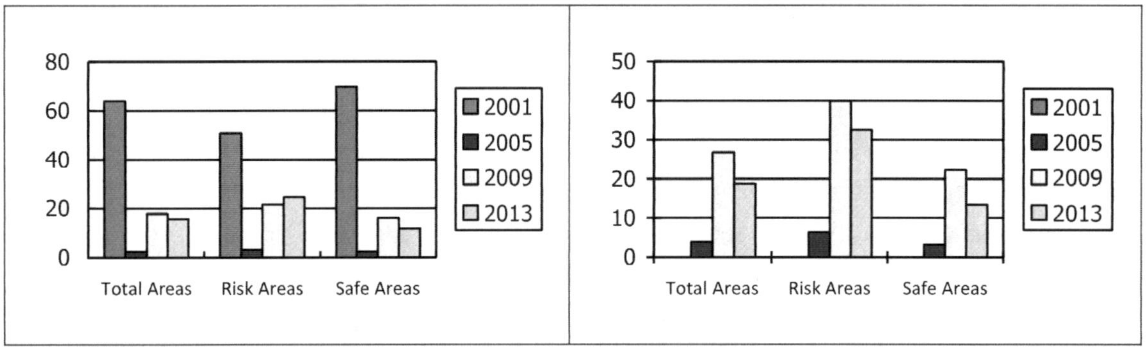

Figure 7. Graphic shows the amount of urban growth each year and the percentage increase in the safe areas and risk areas.

6. Results

After studying urban growth in Abouelreesh village in each of the following years 2001, 2005, 2009 and 2013 we did that by selecting all of area urban and determine the ratio in each of the safe areas and risky areas, Research found that Abouelreesh village grows in risky areas that's where the percentage urban areas in risky area was 48.8 %, Which threatens the future of the village, the government requires to develop a plan to manage the risk in the village with the provision of alternatives for the population to avoid construction in risky areas.

Building area is one of very important factors that affecting on attract people to certain areas in urban growth areas, If it was one of the dangerous areas and explained that the average area of buildings in the study area that increases in risky areas for safe areas, where the advantage urban areas in risky areas terrain easy and vast areas which differ from the mountainous overcome the difficult nature of the village.

7. Conclusion

Some of the towns and villages exposed to the risk of natural disasters such as some villages in Aswan to the risk of torrents which requires the study of the behavior of urbanization and urban growth, moreover It should be study the relationship between the urban areas and risks and find the solutions to risk management and the protection of urban structure, in spite of we should study the causes of urban sprawl into risky areas and try to avoid it.

Abouelreesh village is facing a big problem that is the significant growth of the village in risky areas at risk of torrents so we must study the solutions to redirect the urbanization toward the safe areas with providing the means to manage the risk of storm water drains in the village.

IOP Conf. Series: Earth and Environmental Science **20** (2014) 012009 doi:10.1088/1755-1315/20/1/012009

References

[1] Belal, A. (2011). Detecting urban growth using remote sensing and GIS techniques in Al Gharbiya governorate, Egypt. elsevier.

[2] Noor, N. M. (2013). Determination of Spatial Factors in Measuring Urban Sprawl in Kuantan Using Remote Sensing and GIS. ASEAN Conference on Environment-Behaviour Studies. Hanoi, Vietnam.

[3] Bhatta, B. (2010). Urban sprawl measurement from remote sensing data. elsevier.

[4] Jaegera, J. A. (2014). Improving the measurement of urban sprawl: Weighted UrbanProliferation (WUP) and its application to SwitzerlandJochen oliferation (WUP) and its application to Switzerland. elsevier.

[5] Jaeger, J. A. (2010). Suitability criteria for measures of urban sprawl. elsevier.

[6] Nassar, W. M. (2001). The Impact of the Natural Disasters on Re-planning of the Disastered Zones.

[7] Hisham Mubarak , C. (2010). A preliminary report on the torrent damage in Aswan. Cairo.

[8] GOPP, G. O. (2010). Detailed planning for the village of Abouelreesh North - Aswan Governorate. Cairo.

IOP Conf. Series: Earth and Environmental Science **20** (2014) 012010 doi:10.1088/1755-1315/20/1/012010

Multi-Temporal Satellite Imagery for Urban Expansion Assessment at Sharjah City /UAE

R Al-Ruzouq and A Shanableh

Department of Civil and Environmental Engineering, University of Sharjah, United Arab Emirates.

E-mail: ralruzouq@sharjah.ac.ae

Abstract. Change detection is the process of identifying differences in land cover over time. As human and natural forces continue to alter the landscape, it is important to develop monitoring methods to assess and quantify these changes. Recent advances in satellite imagery, in terms of improved spatial and temporal resolutions, are allowing for efficient identification of change patterns and the prediction of areas of growth. Sharjah is the third largest and most populous city in the United Arab Emirates (UAE). It is located along the northern coast of the Persian Gulf on the Arabian Peninsula. After the discovery of oil and its export in the last four decades at UAE, it has experienced very rapid growth in industry, economy and population. The main purpose of this study is to detect urban development in Sharjah city by detecting and registering linear features in multi-temporal Landsat images. This paper used linear features for image registration that were chosen since they can be reliably extracted from imagery with significantly different geometric and radiometric properties. Derived edges from the registered images are used as the basis for change detection. Image registration and pixel–pixel subtraction has been implement using multi- temporal Landsat images for Sharjah City. Straight-line segments have been used for accurate co-registration as well as main element for a reliable change detection procedure. Results illustrate that highest range of growth that represented by linear features (building and roads) have been accrued during 1976 – 1987 and stand for 36.24 % of the total urban features inside Sharjah city. Moreover, result shows that since 1976 to 2010, the cumulative urban expansion inside Sharjah city is 71.9 %.

1. Introduction

The demand for up-to-date geographic data is increasing due to fast changes to the landscape that are taking place as a result of nature and/or human actions. Such changes have to be accurately and reliably inventoried to fully understand the physical and human processes at work [1]. Change detection is the process of identifying differences in the state of an object or phenomenon by observing it at different times [2]. It involves the ability to quantify changes using multi-resolution, multi-spectral, and/or multi-source imagery captured at different times. Traditional change detection studies are based on visual/manual comparison of temporal datasets (such as satellite scenes, aerial images, maps, etc.). However, the huge flux of imagery that is being captured by an ever increasing number of earth observing satellites necessitates the development of automatic, reliable, and fast change detection techniques. Such techniques are essential to reduce the high cost associated with spatial data updating activities. Several change detection methods have been developed and reported in the literature [2],

Content from this work may be used under the terms of the Creative Commons Attribution 3.0 licence. Any further distribution of this work must maintain attribution to the author(s) and the title of the work, journal citation and DOI.
Published under licence by IOP Publishing Ltd

[3],[4] and [5]. These procedures are based on image subtraction, image ratio, change vector analysis, principle component analysis, neural network, or morphological mathematics.
The main purpose of this study is to detect urban development in Sharjah city by detecting and registering linear features in multi-temporal Lnadsat images. This paper used a registration methodology that is based on the Modified Iterated Hough Transform (MIHT) [6], [7] and [8] for simultaneously estimating the parameters of the registration transformation function while establishing correspondence between conjugate straight-line segments in an image pair. Derived edges from the registered images are used as the basis for change detection. Image registration and pixel–pixel subtraction has been implement using multi- temporal Landsat images for Sharjah City.

2. Study Area

The city of Sharjah, United Arab Emirates (UAE) is located along the northern coast of the Persian Gulf on the Arabian Peninsula with a central coordinate of 25.3° N 55.5° E , Figure 1(a), (b) and (c) shows in sequence UAE map , Landsat image of Sharjah city and the study area within Sharjah city.

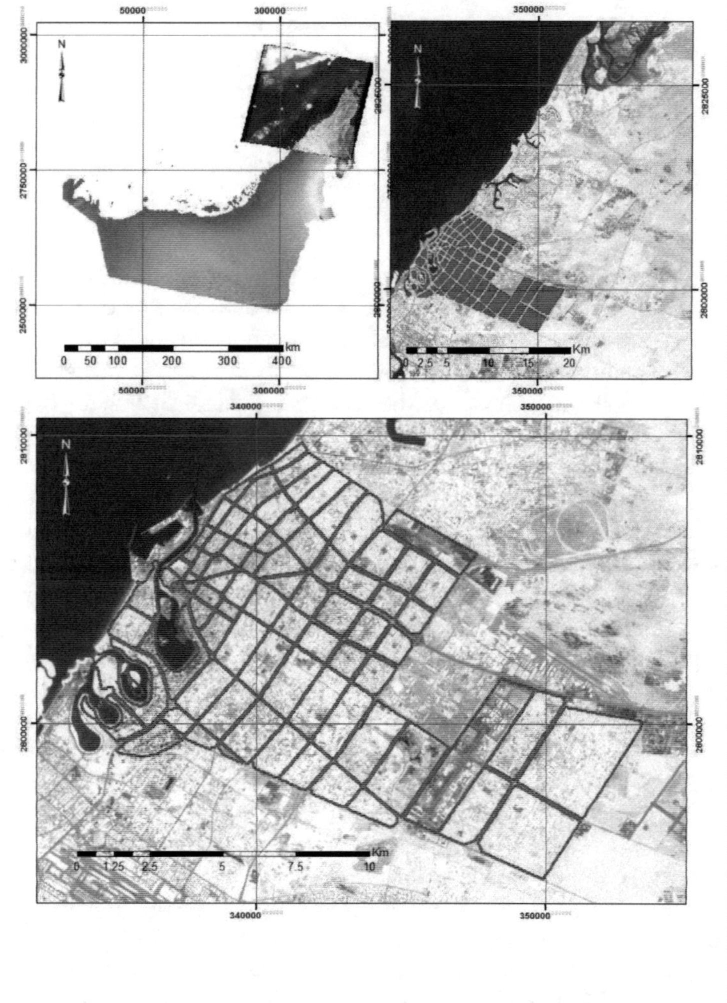

Figure 1. Location of the study area: UAE map, Landsat image of Sharjah city and the study area within Sharjah city.

Sharjah City is the capital and largest city in Emirates of Sharjah located in a dry hot region, with daily high temperatures of 24-42°C and daily main temperature of 18-34°C. The rainy month is typically December to February, with average rainfall of approximately 100 mm/year. The rainfall in dry years doesn't generally exceed 40 mm/year and 200mm/year in heavy rain years.

3. Geospatial Data

The data used in this research include diversity of satellite images with different radiometric and geometric resolution captured in different years. These data will be used for image registration and change detection. Table 1 lists the source of each image including capturing data and the spatial resolution in meters. Figure 2-a shows sample of the Landsat images for the study area for years 1976 and 2000.

Table 1. Landsat images captured at various years.

Satellite	**Year**	**Resolution**
Landsat Multispectral Scanner (MSS)	1976	79 m
Landsat Thematic Mapper (TM)	1987	60 m
Landsat Thematic Mapper (TM)	2000	30 m
Landsat Thematic Mapper (TM)	2007	30 m
Landsat Enhanced Thematic Mapper (ETM)	2010	15 m

4. Methodology

In this paper linear feature, image registration and pixel–pixel subtraction has been implementing using Landsat images that have different spatial and temporal resolution. Linear features (straight-line segments) have high semantics and can be reliably extracted from the images. These linear features (streets and buildings in urban areas) were used as the base for change detection. Research has been conducted using multi-temporal imagery over the city of Sharjah. All images have been registered to the Landsat image from 2010. Interest and well distributed points (main intersections and corners of buildings) have been manually digitized in these images to perform the registration process where 10 - 15 points (end points of the linear Feature) have been used to drive the transformation parameters needed for the image registration. It has to be mentioned that accurate registration of multi-temporal imagery can be considered as the most important component of an accurate and reliable change detection procedure, [5].

Having established the transformation function between the images, the input image can be resampled into the reference frame associated with the reference image, Figure 2-a shows an example of Landsat clipped image after image registration and resampling process. The resampling is followed by applying canny edge detection and majority filter to both images. Then, the resulting images are subtracted to produce a change image, which is enhanced by re-application of the majority filter. Section 5 will discuss the theory and principle of the image registration process. Change detection criteria and stages will be discussed in section 6.

5. Image Registration Paradigm

An image registration process aims at combining data and/or information from multiple sensors in order to achieve improved accuracies and better inference about the environment than could be attained through the use of a single sensor. An effective automated image registration methodology must deal with four issues; namely registration primitives, transformation function, similarity measure, and matching strategy

IOP Conf. Series: Earth and Environmental Science **20** (2014) 012010 doi:10.1088/1755-1315/20/1/012010

To carry out the registration process, a decision has to be made regarding the choice of the appropriate primitives (for example, distinct points, linear features, or homogeneous regions). In this research, straight-line segments have been used as the registration primitives, Figure 1.
The second issue in a registration procedure is concerned with establishing the transformation function that mathematically describes the mapping function between the imagery in question. In other words, given a pair of images, *reference and input images*, the transformation function attempts to properly overlay these images. Habib et al, 2001a, 2001b showed that affine transformation, Equation 1, could be used as the registration transformation function for imagery captured by satellite imaging systems with narrow angular field of view over relatively flat terrain as compared to the flying height.

$$\begin{bmatrix} x' \\ y' \end{bmatrix} = \begin{bmatrix} a_0 \\ b_0 \end{bmatrix} + \begin{bmatrix} a_1 & a_2 \\ b_1 & b_2 \end{bmatrix} \begin{bmatrix} x \\ y \end{bmatrix} \quad (1)$$

Where: (x, y) Denotes the coordinates of a point in the reference image, and
(x', y') Denotes the coordinates of the conjugate point in the input image.

The next step in the registration paradigm is the selection of the similarity measure, which mathematically describes the necessary constraints for ensuring the correspondence of conjugate primitives. The similarity measure formulation depends on the selected registration primitives and their respective attributes. In this work, the registration primitives, straight-lines, will be represented by their end points, which need not be conjugate.
Let's assume that a line segment **A** in the reference image corresponds to the line segment **B** in the input image. The similarity measure should mathematically describe the fact that the line segment **A** will coincide with the corresponding line segment **B** after applying the transformation function relating the reference and input images. Such a measure can be derived by forcing the normal distances between the end points of a line segment in the reference image, after applying the transformation function, and the corresponding line segment in the input image to be zero. Equation 2 mathematically describes such a constraint for one of the end points of the line segment in the reference image.

$$x_1' \cdot \cos\theta + y_1' \cdot \sin\theta - \rho = 0 \quad (2)$$

Where: (ρ, θ) Are the polar coordinates representing the line segment **B** in the input image.
(x_1', y_1') Are the transformed coordinates of point 1 in the reference image after applying the registration transformation function.

Another constraint of the form in Equation 2 can be written for point 2 along the line-segment in the reference image. To automate the solution of the registration problem, a controlling framework that utilizes the primitives, similarity measure, and transformation function must be established. This framework is usually referred to as the matching strategy. In this research, the (MIHT) is used as the matching strategy. Such a methodology is attractive since it allows for simultaneous matching and parameter estimation. Moreover, it does not require complete correspondence between the primitives in the reference and input images. MIHT has been successfully implemented in several photogrammetric operations such as automatic single photo resection and relative orientation [5], [9] and [10].

6. Change Detection

After deriving the parameters of the registration transformation function, one of the images can be resampled into the reference frame associated with the other one. Within the resampled images, corresponding pixels are assumed to point at the same object space feature. Therefore, a simple pixel-by-pixel comparison/differencing between the resampled images could be used to highlight object space changes. However, radiometric differences should be expected as a result of varying atmospheric conditions and/or different sensor types. The effect of these differences can be reduced by applying intensity normalization techniques to the images in question (e.g., to ensure that they have the same intensity mean and variance values). One can argue that this procedure would be still unreliable since it would be affected by the noise and the spectral properties associated with the input imagery [11], [12] and [13]. Instead of working with the original images, after intensity modification, one can use derived edge images as a basis for the change detection procedure. Utilizing the edge images has two advantages. First, derived edges are robust to possible radiometric differences between the registered images (e.g., due to noise and/or different spectral properties). Also, the edges would correspond to interesting features (e.g., building boundaries, roads, trails, etc.). Therefore, comparing edge images will be useful in outlining the amount of urbanization activities, which is one of the most important objectives of change detection exercises. The suggested change detection methodology starts by extracting edge cells using canny edge detector (Canny, 1986), Figure 2-b shows sample of derived edges for the study area at different years. To compensate for small geometric differences between the imagery (in the order of few pixels), a majority filer is used to fill small gaps within an area with numerous edges, as well as eliminate isolated edges, Figure 2-c.

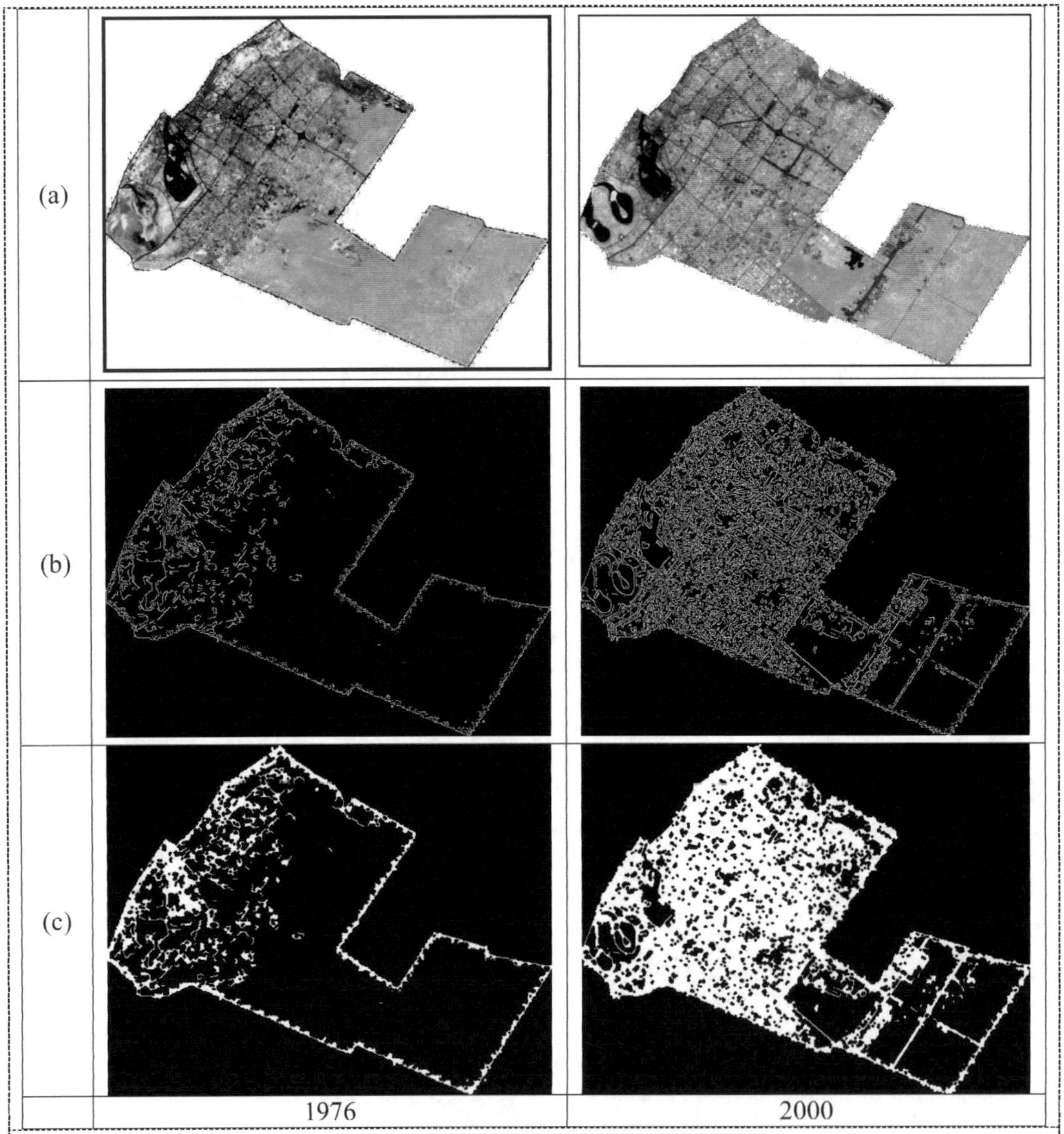

Figure 2. (a) Cropped Landsat Image for years 1976 and 2000 (b) Derived Edges. (c) Derived Edges after applying the majority filter.

The filtered images will highlight areas with interesting features since they would lead to a dense distribution of edge cells. Afterwards, the filtered images are subtracted to highlight areas of change. Finally, a majority filter can be applied to the difference image to eliminate small areas (since changes/no-changes are expected to be locally spread – i.e., they are not isolated), Figure 3 shows the difference image from 1987 until 2010. In the difference image, white areas indicate changes while black areas indicate parts with no change. Visual inspection for figure 3, it can be clearly noticed the areas where changed have been accrued and represented by white pixels.

Figure 3. Difference image for the study area for years 2010 and 1987.

Figure 4 shows the clipped Landsat image for the study area for years 1976, 1987 and 2000 and the change images where white pixels represent changes. Simple statistics show that highest range of growth that represented by linear features (building and roads) have been accrued during 1976 – 1987 and stand for 36.24 % of the total urban features inside Sharjah city. Moreover, result shows that since 1976 to 2010, the cumulative urban expansion inside Sharjah city is 71.9 %.

7. Conclusion and future work

This paper presents an image registration together with a suggested procedure for detecting changes between the involved images. The approach has been tested on real datasets, which showed its effectiveness in registering and detecting changes among multi-temporal and multi-resolution imagery.

The MIHT procedure has been used for automatic registration of multi-source imagery with varying geometric and radiometric properties. The presented approach used linear features (straight-line segments) as the registration primitives since they can be reliably extracted from the images.To avoid the effect of possible radiometric differences between the registered images, due to different

IOP Conf. Series: Earth and Environmental Science **20** (2014) 012010 doi:10.1088/1755-1315/20/1/012010

atmospheric conditions, noise, and/or different spectral properties, the change detection is based on derived edge images. The use of edge images is attractive since it would lead to an effective detection of urbanization activities since they would lead to a dens distribution of edge cells. Also, a majority filter has been applied to compensate for small registration errors as well as eliminating small gaps and isolated edges. The images are then subtracted to produce a change image, which could be enhanced by applying a majority filter to remove small regions. The change detection results are found to be consistent with these visually identified.

Future research will concentrate on using high resolution images for change detection at the same time establishing ground truth for quantitative evaluation of the suggested approach.

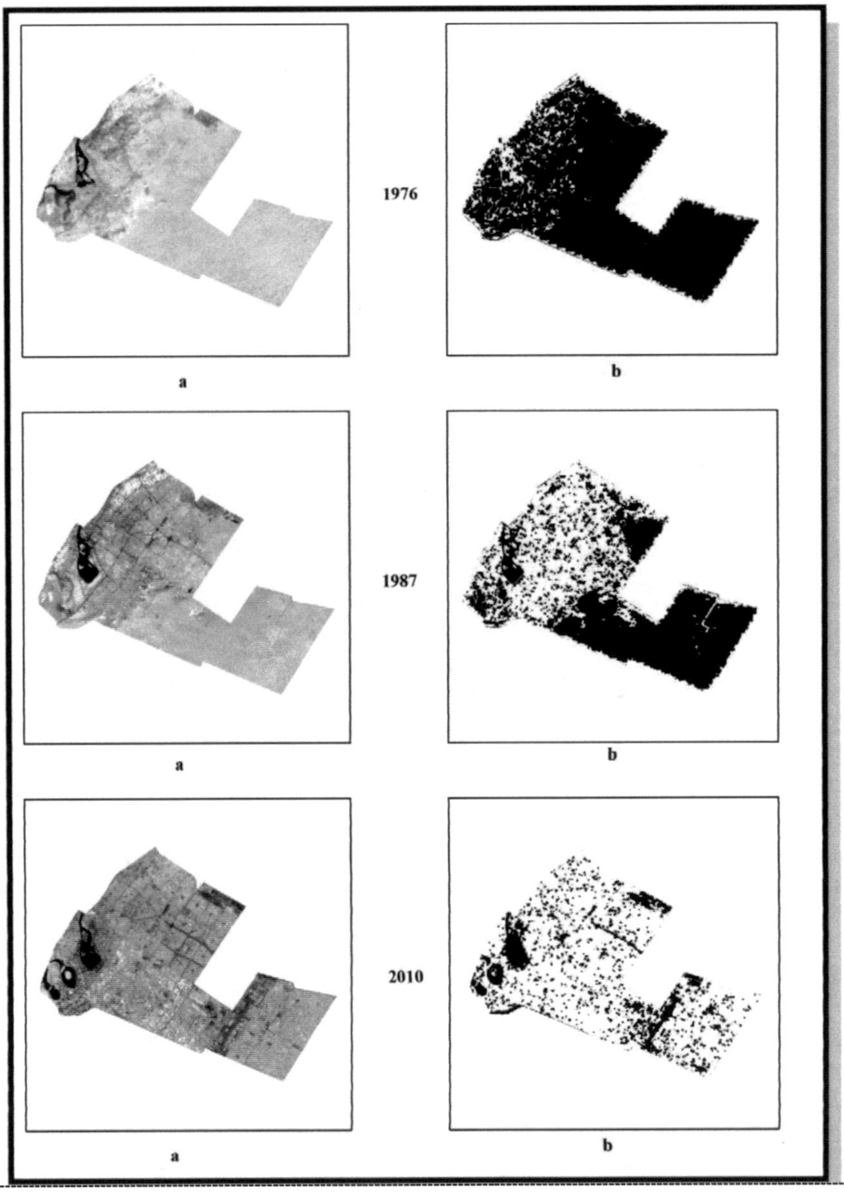

Figure 4. (a) Clipped Landsat image for the study area for years 1976 ,1987 and 2000, (b)Change detection image where white pixels represent changes.

8. References

[1] Estes, J., 1992. *Technology and Policy Issues Impact Global Monitoring*. GIS world, 5(10): 52-55

[2] Singh, A., 1989. Digital Change Detection Techniques Using Remotely Sensed Data. International journal of remote sensing, 10(6): 989-1003.

[3] Dowman, I. 1998. *Automated Procedures for Integration of Satellite Images and Map Data for Change Detection:* The Archangel Project. GIS-between visions and applications. IAPRS, 32(4): 162-169.

[4] Bruzzone, L. and D. Prieto, 2000. *Automatic Analysis of the Difference Image for Unsupervised Change Detection.* IEEE Transactions on Geoscience and Remote Sensing, 38(3): 1171-1182.

[5] Al-Ruzouq and Habib, 2012 . *Linear Features for Automatic Registration and Reliable Change Detection of Multi-Source Imagery.* Journal of Spatial Science, Taylor & Francis,Volume 57, Issue 1, 2012 ,pages 51-64.

[6] Habib A., R. Al-ruzouq, 2004. *Line-Based Modified Iterated Hough Transformation For Automatic Registration Of Multi-Source Imagery*. Photogrammetric Record, 19(105), pp. 5-21.

[7] Habib, A. and D. Kelley, 2001a. Automatic Relative Orientation of Large Scale Imagery Over Urban Areas Using Modified Iterated Hough Transform. International journal of photogrammetry and remote sensing, 56 (2001): 29-41.

[8] Habib, A. and D. Kelley, 2001b. Single Photo Resection Using the Modified Hough Transform. Journal of Photogrammetric Engineering and Remote Sensing, 67(2001): 909-914.

[9] Habib, A., M. Morgan and Y. Lee, 2001b. Integrating Data from Terrestrial Mobile Mapping Systems and Aerial Imagery for Change Detection Purposes. Third Mobile Mapping Symposium, Cairo, Egypt, 3-5 January, 2001.

[10] Habib, A., Y. Lee and M. Morgan, 2001a. Surface Matching and Change Detection Using Modified Iterative Hough Transform for Robust Parameter Estimation. Photogrammetric Record, 17 (98): 303-315.

[11] Agouris, P., G. Mountrakis, and A. Stefanidis, 2000. *Automated Spatiotemporal Change Detection in Digital Aerial Imagery*. SPIE Proceedings, 4054: 2-12.

[12] Cavallaro, A. and E. Touradj, 2001. *Change Detection Based on Color Edges.* Proc. of IEEE International Symposium on Circuits and Systems (ISCAS-2001).

[13] Canny, J., 1986. *A computational approach to edge detection*. IEEE Transactions on Pattern Analysis and Machine Intelligence, 6(6):679-698.

IOP Conf. Series: Earth and Environmental Science **20** (2014) 012011 doi:10.1088/1755-1315/20/1/012011

The influence of urban development and social mobility on socioeconomic level: The application of GIS on urban ecosystems

Nur Suhaili Mansor *, Helmi Zulhaidi Mohd Shafri, Shattri Mansor, Biswajeet Paradhan
Civil Engineering Department, Faculty of Engineering, Universiti Putra Malaysia, 43400 Serdang, Selangor, Malaysia

*E-mail. nursuhailimansor@gmail.com (Corresponding author)

Abstract. Specifically, the integration between social sciences and natural science are fundamental in our understanding of the economic, social and technological transformations that have drastically changed the society. This study will be based on the municipality of Sungai Petani, Kedah as it has been most influenced by urbanization and urban development. Urban development in Sungai Petani is closely associated with a tremendous increase in demand for land, which is highly related to population growth, human movement and their social mobility. The qualitative case study taken will rely on the visual interpretation technique that would allow the researcher to develop a map of urban changes detection. The potential application of GIS information to estimate socioeconomic indicators and the modelling of socio-economic activities that are explored in this study is hoped to increase further our understanding of the impacts of development and urbanization on social life.

1. Introduction

Nowadays, urbanization on land uses had been increasingly subjected to changes of different forms, sorts and types since the urban explosion of the 1970s that effected on environmental [1]. Most people moved to cities from rural area who expected to get benefit from the urban economy to improve social life. Mobility of the people typically moves to urban areas in search of economic opportunities. This phenomenon is particularly significant in developing countries, where rural poverty and the rural-urban income gap are worrying aspects for development. Therefore, the agriculture sector has been greatly modified by human activity that gives more economic opportunities to the locals.

Since environmental changes are going on in association with changing patterns of economy, society, culture and civilization. Multifaceted changes, transformation and conversions are the results of multilevel pressures on ecosystems and the human dimensions of global environmental changes happen through a subset of multilevel driving forces which directly or indirectly alter aspects of the physical, economic, social and cultural environment in ways that have global as well as regional effects.

Typically, in Malaysia, development of the urban for the last two decades due to industrialization, an increase in economic prosperity and urban population has opened opportunities for new developments in urban areas and their periphery. This growth has significantly changed the landscape of many cities. In the context of urbanization, a large amount of agricultural land has been converted to built-up or urban land uses. The dynamics of urban landscape systems of Malaysian cities are driven by complex political, social and economic systems [2].

Content from this work may be used under the terms of the Creative Commons Attribution 3.0 licence. Any further distribution of this work must maintain attribution to the author(s) and the title of the work, journal citation and DOI.
Published under licence by IOP Publishing Ltd

From the above literature, this is an exciting research area is emerging around the integration of detailed social and spatial networks that expect that upcoming cohort of demographers will increasingly utilize new forms of geospatial data and analyze those using spatially explicit methods [3]. Also, clearly the research that applied remote sensing information and GIS application on the social mobility and socioeconomic research that link them to the economy, and the urban ecosystem is very limited in Malaysia. There is some similar research on social and economic such as, cultural factors are evident in mobility patterns, even when many economic and social factors are controlled [4]. It is also clear that much has been done to analysis aimed at improving the Malaysia social wellbeing through a related study done by previous research which analysis of social, economic, and political changes that took place between 1966 and 1978 in the rural Malay community.

Further to improving Malaysian social mobility and their socioeconomic, various studies show that lack of study about related land use and social mobility, also the lack of study, which uses GIS and remote sensing to see this relationship. In levels within a developing country, i.e. Malaysia needs a study about a social mobility concept to ensure continuation of stability and development across the country, especially in Sungai Petani. Therefore, it is to full fill structural plan of Sungai Petani and implementation.

2. Data collection

This research will employ the survey research methodology by using primary and secondary data to map the land use and socioeconomic activity. Land use map will be generated from SPOT-5 covering the periods of 2000 and 2012 will be used to prepare land use maps and monitor the land use changes. For the change detection, time-series of images of SPOT-5 2000/2012 will be compared.

Then, use statistical reports, periodicals or annual for economic indicator reports and structure plan and household survey secondary data. The structure questionnaire will be distributed to the targeted respondents. Most especially farmers were asked about their activity related rubber plantation, paddy field, including the pattern of land use over the past years surrounding Sungai Petani. On the local stakeholders such as rubber plantation owner, private sector and district, administrative office such as district planning and investment office, district agriculture and forestry office will be interviewed.

3. Conceptual framework of study

In this research study, data preparation and pre-processing are a phase to achieve the objective of the study. The data preparation task is to see the change detection over a period; also a statistical analysis model for growth based on their population and socioeconomic indicators.

For satellite imagery should be used in each time, corresponding to two different periods of time. The adopted interpretation technique, which is preferred to the standard classification, is based on visual interpretation assisted by computer to see the change detection.

The process consists of displaying the rectified or registered imagery on the screen which representing different land use categories, based on image characteristics such as tone, texture, colour and pattern, which are translated into land use attributes.

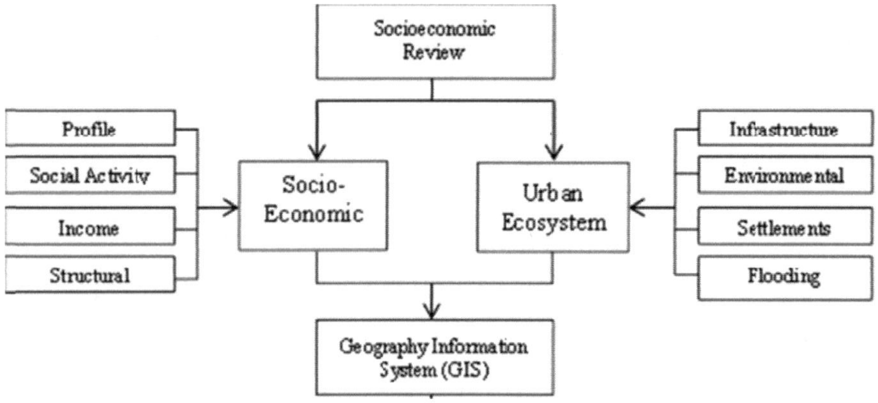

Figure 1. Conceptual framework of the research.

3.1. Linking of urban ecosystem (RS) and GIS with social mobility

Selected socioeconomic characteristics may be extracted directly from remote sensor data or by using surrogate information derived from the imagery. Two of the most important attributes are a population and quality of life indicators [5]. Remote sensing techniques may provide a population estimate that approaches the accuracy of traditional census methods if sufficiently accurate in situ data are available to calibrate the remote sensing model.

The most accurate remote sensing method for estimating the population of a local area is to count individual dwelling units based on the following assumptions [6,7] (i) The imagery must be of a sufficient spatial resolution to identify individual structures, even though tree cover and whether they are residential, commercial, or industrial buildings; (ii) Some estimation of the average number of persons per dwelling unit must be available; (iii) Some estimate of the number of homeless, seasonal, and migratory workers are required; and (iv) It is assumed that all dwelling units are occupied, and that only n family lives in each unit (calibrated using in situ investigation).

Besides that, [8] that adequate income, decent housing, education and health services, and good physical environment are important indicators of social wellbeing and quality of life. Evaluating the quality of life of a population on a continuing basis is important because it helps planners and government agencies involved with the delivery of human services to be aware of problem areas. In the past, most quality of life studies made use of census data to extract socioeconomic indicators.

The previous research on the mobility experiences of very low-income families, expectation patterns of mobility among low income families that emphasize the role of personal safety nets in searching for and securing housing, instability (that is, frequent moves), and a prevalence of informal housing solutions. These factors produce mobility dynamics that are mostly independent of neighborhood concerns. Therefore, several more attempts have been made to use GIS for measurements and analysis of individual mobility under special space, time constraints [9].

There many current models of human activity are fundamentally stochastic from Erlang's formula used in telephony to L´evy-walk models describing human mobility[10]. Gonz´alez found out that, in contrast with the random trajectories predicted by the prevailing L´evywalk models, human trajectories show a high degree of temporal and spatial regularity, each individual being characterized by a time-independent characteristic travel distance and a significant probability to return to a few highly frequented locations. Similarly, by measuring the entropy of each individual's trajectory, [11] demonstrated that a 93% potential predictability in user mobility across the whole user base.

Indeed, linking GIS and RS with social movement has long been a critical issue [12]. GIS and RS are used in urban growth models as spatial data and link them to social mobility data as non-spatial data including questionnaires to provide functions such as data processing and visualization. The combination both data, spatial and non-spatial data will be processed in GIS software and to model them. The GIS and RS are mainly used for providing data inputs for the purpose built model program and for presenting the model results. It has been far easier to produce purpose built programs and special graphics and interface functions for the model than to embed these into the proprietary GIS software. The core model functions are implemented through mainstream statistic or computing software, including SPSS, SEM, and Matlab. GIS and RS have also been used mainly for data inputs and display of model results.

3.2. Socioeconomic level

Changes in land development are often associated with policy implementation and socioeconomic factors. The growing population or social mobility and the increasing socioeconomic necessities have created pressures on land use and land covers. Therefore, economic development strategies have focused on labor force training, business recruitment, and tax incentives, rather than seeking a better spatial arrangement of the socioeconomic activities within a region. Moreover, in many planning studies and models, the interactions have been considered for a 'top down' perspective without consideration of 'bottom up' connections; that is, regional economic growth and transformation are only assumed to have significant effects on land use and the spatial structure within the region.

Addition to that, it contributes to a more systematic coordination of economic development and land-use planning, which is essential for achieving a more complete realization of urban development that is ecologically and socio-economically sustainable and prosperous. Former researcher agreed that processes were based on the interrelationships between the social and ecological processes and notably how changes in the landscape induced changes in the community [13].

Figure 2 below represents the first attempt to identify land development and social mobility and their impacts in an urban setting also the development as well as some other factors in previous literature such as institution, geography, mobility and human capital are significant factors in current socioeconomic development rather than urban growth. Then, there exists an optimal secondary ratio to see the socioeconomic level status and social mobility.

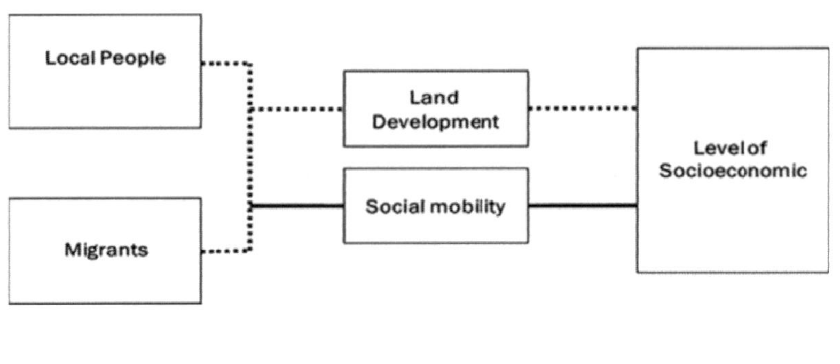

Figure 2. The relationship land development and social mobility.

4. Results and discussion

By applying an integrated framework, this paper has analyzed and compared the urban changes detection, which are expected to occur due to urbanization strategies in Sungai Petani. In the assessment acquire the data about the urban changes detection in Sungai Petani in 2000 and 2012 respectively (Figure 3), the accuracy assessment of urban change detection between 2000 (Table I) and 2012 (Table II). According to figure 3, the massive increase of urban area (including residential, industrial) and green area (including paddy field, dry land), and bare land (including transportation, land) and, forestland (including rubber estate) contributed to the main features of urban changes in Sungai Petani between 2000 and 2012. Moreover, the change amplitude of green area around urban area which include residential and industrial was significant, population year 2000 about 339,898 people and increase to 1,993,000 people year 2012 (source: Local Authority, 2013).

Urban changes detection map of Sungai Petani

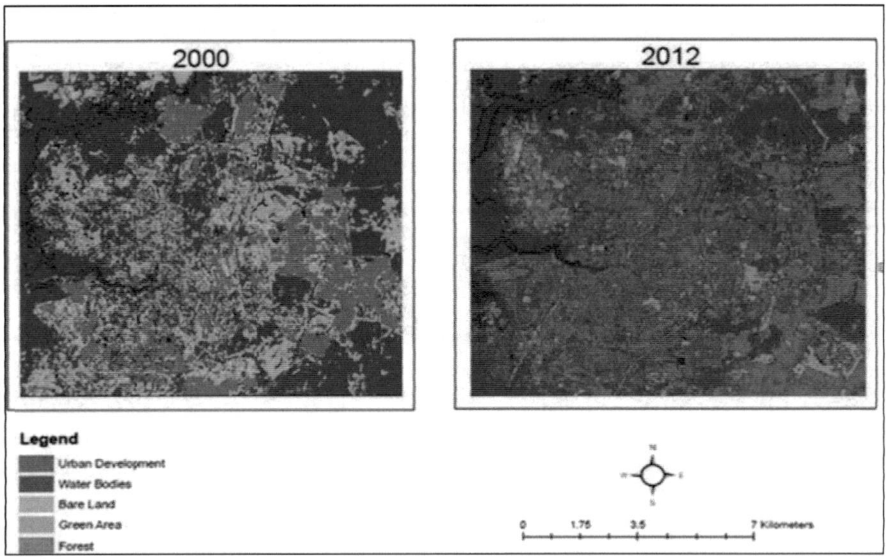

Figure 3. Urban changes detection type in 2000 and 2012 in Sungai Petani.

According to the results obtained, this process of modification of the land can improve the efficiency of the economy and the income level of the population. Although the indicators may not represent all the socioeconomic and environmental aspects of the problem at hand, they have been useful in order to highlight some important links between urbanization strategies and the agriculture change process in the area of study. It may be concluded that the most significant features of the change of urban development type in Sungai Petani between 2000 and 2012 were construction, land occupied, massive agricultural land, land-use type of agricultural land changed internally to that of relatively high economic benefits. Therefore, it is shown the economic opportunities and living conditions are the main reason the mobility occurred.

Table 1. Accuracy assessment for urban changes detection types year 2000.

	Reference Total	Classified Total	Number Correct	Producers Accuracy	Users Accuracy
Water bodies	10	10	9	90.00%	90.00%
Forest	42	40	39	92.86%	97.50%
Bare land	13	16	12	92.31%	75.00%
Green area	16	17	14	87.50%	82.35%
Urban area	19	17	17	89.47%	100.00%

Kappa 97.35, Accuracy 98.00%

Table II. Accuracy assessment for urban changes detection types year 2012.

	Reference Total	Classified Total	Number Correct	Producers Accuracy	Users Accuracy
Water bodies	11	10	10	90.91%	100.00%
Forest	31	31	30	96.77%	96.77%
Bare land	11	11	11	100.00%	100.00%
Green area	15	16	16	100.00%	93.75%
Urban area	32	32	32	100.00%	100.00%

Kappa 87.87%,Accuracy 91.00%

5. Conclusion

In the social sciences, primary research themes span problems at local through global scales. They range from the sense of place associated with daily life to the interdependences associated with regional and global interconnections. They also reflect the needs for descriptive and predictive tools that enhance insights on the meanings of spatial patterns and how they relate to societal processes that impinge on all aspects of social wellbeing [14].

Using GIS as a method for modeling the urban development change observed against the environment with socioeconomic variables. In this research, the relation between science and social will produce the map of urban changes detection. This study attempts to embark on a science based social research that studies social wellbeing and the application of remote sensing and Geography Information Science (GIS). Specifically, the integration between social sciences and natural science are fundamental in our understanding of the economic, social and technological transformations that have drastically changed the society.

Acknowledgements
I would like to acknowledge to Assoc. Prof Dr. Helmi Zulhaidi Mohd, Prof Dr. Shafri Mansor and Assoc. Prof Dr. Biswajeet Paradhan for helpful discussion, I would also like to thank Mrs Suziyanna Arshad for Remote Sensing Agency. A special thanks also goes to the anonymous reviewers for their comments and suggestions that helped to improve an earlier version of this paper.

References

[1] G. Siciliono, "Urbanization strategies, rural development and land use changes in China: A multiple-level integrated assessment," *Land Use Policy 29,* p165-178, 2011.

[2] M. N. Sharif and R. Ruslan, "Modelling Urban Spatial Structure Using Geographically Weighted Regression" At 18th World IMACS / MODSIM Congress, 2009.

[3] S. A. Matthews and D. M. Parker, "Progress in Spatial Demography," Vol.28, page271-312. http://www.demographic-research.org/Volumes/Vol28/10 ,2013.

[4] R. W. Johnson and J .Davanzo, "Economic and Cultural Influences On The Decision To Leave Home In Peninsular Malaysia," *Demography*, *35*(1), 97–114. Retrieved From Http://Www.Ncbi.Nlm.Nih.Gov/Pubmed/9512913 , 1998.

[5] J. R. Jensen and D. C. Cowen, "Remote Sensing of Urban/Suburban Socio-economic Attributes," *Proceedings Land Satellite Information in the Next Decade II,* American Society for Photogrammetry & Remote Sensing, Bethesda, Maryland, Compact Disk, 1997 .

[6] C. P. Lo, "Automated Population and Dwelling Unit Estimation from High-Resolution Satellite Images: A GIS Approach," *International Journal of Remote Sensing, 16:17-34,* 1995.

[7] B. Haack, S. Guptill, R.Holz, S. Jampoler, J. Jensen, and R.Welch, "Chapter 15: Urban Analysis and Planning," *Manual of Photographic Interpretation,* American Society for Photogrammetry & Remote Sensing, Bethesda, Maryland, pp. 517-553, 1997.

[8] C. P. Lo and B. J. Faber, "Interpretation of Landsat Thematic Mapper and Census Data for Quality of Life Assessment," *Remote Sensing of Environment,* in press, 1998.

[9] S. L. Shaw and H. Yu, "A GIS-based Time-geographic Approach of Studying Individual Activities and Interactions in A Hybrid Physicalvirtual Space," *Journal of Transport Geography.* 2009, vol.17, pp.141-149, 2009.

[10] M. C. Gonz´alez, C. A. Hidalgo and A. L. Barab´asi, "Understanding individual human mobility patterns," NATURE. vol.453, pp.779-782, 2008.

[11] C. Song, Z. Qu, N. Blumm and A. L. Barab´asi, "Limits of Predictability in Human Mobility," Science. vol.327, no.1018, pp.1018-1021, 2010.

[12] M. J. Mlotha, "Remote Sensing And Gis Linked To Socio-Analysis For Land Cover Change Assessment," 0-7803-7031-7/01/ 2001 IEEE, 2001.

[13] C. M. Raboanarielina, "The forgotten resource: Community perspectives on conservation and well - being in Zahamena National Park, Madagascar," *Madagascar Conservation & Development* Volume 7 | Issue 2s — November 2012,2012.

[14] D. G. Janelle and M. F. Goodchild, "Concepts, Principles, Tools, and Challenges in Spatially Integrated Social Science," *The Sage Handbook Of GIS And Society Research,* 2011.

IOP Conf. Series: Earth and Environmental Science **20** (2014) 012012 doi:10.1088/1755-1315/20/1/012012

Mapping palm oil expansion using SAR to study the impact on the CO_2 cycle

Christine Pohl[1]

Professor, *Institute of Geospatial Science and Technology (INSTeG)*,
UniversitiTeknologi Malaysia (UTM), 81310 Johor Bahru, Malaysia

Email: c.pohl@utm.my

Abstract. With Malaysia being the second largest palm oil producer in the world and the fact that palm oil ranks first in vegetable oil production on the world market the palm oil industry became an important factor in the country. Along with the expansion of palm oil across the nation causing deforestation of natural rain forest and conversion of peat land into plantation land there are several factors causing a tremendous increase in carbon dioxide (CO_2) emissions. Main causes of CO_2 emission apart from deforestation and peat-land conversion are the fires to create plantation land plus the fires burning waste products of the plantations itself. This paper describes a project that aims at the development of a remote sensing monitoring system to allow a continuous observation of oil palm plantation activities and expansion in order to be able to quantify CO_2 emissions. The research concentrates on developing a spaceborne synthetic aperture radar information extraction system for palm oil plantations in the Tropics. This will lead to objective figures that can be used internationally to create a policy implementation plan to sustainably reduce CO_2 emission in the future.

1. Background

In the past two decades the international community became aware of impacting factors that influence changes in our environment and the global climate. One of the identified global climate warming factors is the increasing green house gas (GHG) emission. Deforestation and forest degradation is one major contributing aspect to the global emission balance. Its share in total global anthropogenic CO_2emissions accounts for 7 – 14% in the period of 2000 – 2005 [1].

Palm Oil became the most used vegetable oil in the world when it passed the consumption of soybean oil in 2004/2005 according to the United States Department of Agriculture [2]. This is no surprise due to the fact that it is the highest yielding crop with a rising demand based on increased income, population and biodiesel production. It takes only 1/10 of the size of the land needed for soybean to produce the same amount of yield [3].

With the increasing success of palm oil on the global market large areas of primary forest were converted into palm oil plantations. In 2012-2013 85% of the crude oil originates from Indonesia (52%) and Malaysia (34%) [4]. From Malaysia's 32 million ha land area 19% (5.4 million ha) contain agricultural crops. Of these four million hectare were covered by oil palms in 2005 [5]. Malaysia is taking efforts to reduce deforestation of primary forest looking for alternatives to benefit from the popularity of palm oil, forming the second largest exported product of the nation. A large part of the

[1] To whom any correspondence should be addressed.

Content from this work may be used under the terms of the Creative Commons Attribution 3.0 licence. Any further distribution of this work must maintain attribution to the author(s) and the title of the work, journal citation and DOI.
Published under licence by IOP Publishing Ltd

increasing wealth of the population and the fast economic growth is based on the fossil fuel and palm oil industry. Many of the newly created oil palm plantations emerged by converting less beneficial crops, such as cacao, rubber and coconut into oil palms. Other efforts deal with the improvement of management practices to increase the yield from currently four tons per ha to a feasible ten tons per ha to reduce the area impact [3].

Apart from these issues the international agreement on Reducing Emissions from Deforestation and Forest Degradation (REDD+) under the United Nations Framework Convention on Climate Change initiated in 2005 provides compensation to tropical forest countries to protect forested areas and reduce CO_2 emission by deforestation. The successful implementation requires regularly updated spatio-temporal information on land use / land cover changes [6, 7]. Since monitoring entails continuous data acquisition it is not possible to rely on optical remote sensing (VIR) alone because especially in the Tropics persistent cloud coverage hinders cloud-free image acquisition. Examples show that it takes between one and seven years in average to get cloud-free scene [4]. A minimum of 12 months coverage contains the chance of one scene with less than 10% cloud cover [8]. Another issue in optical remote sensing is the degradation of the imagery through haze which is a very common phenomenon in Malaysia and other countries in South East Asia [9]. Synthetic aperture radar (SAR) sensors do not only provide complementary information to VIR images. The SAR is an active sensor that operates weather and daylight independent delivering information all year round at the time that it is needed. Clouds, haze and smoke do not prevent the sensor from providing images. The challenge in this respect is the different nature of radar data in terms of information content and interpretation. The intensity backscattered energy of microwaves is sensitive to texture, size and orientation of structural objects, moisture content and ground conditions. For the application discussed in this paper the interaction of microwaves with canopies, trees and soil is of interest. The longer wavelengths (e.g. L-band 15-20 cm) penetrate the canopy and deliver information on leaves, branches, stems and soil conditions; the longer the wavelength the greater the penetration [10]. It has been proven that radar is sensitive to the structure of the canopy. The received backscatter intensity represented in the image is a composition of interactions with the crown, the trunk and the ground surface. Using fully polarimetric SAR it is possible to derive a relationship between backscatter, texture and crop status [11]. If we consider an oil palm as crop this is very interesting. It would help to derive certain patterns for different growth stages of the oil palm. This is very relevant information for optimized and sustainable oil palm plantation management.

Using radar for land cover mapping is a recent technological development that has not been exploited in its full potential [7]. That is why a joint initiative between the Institute of Geospatial Science and Technology (INSTeG), UniversitiTeknologi Malaysia and SkyMap GlobalPtd. Ltd. aims at investigating the benefits of using radar satellite remote sensing to monitor palm oil plantations. The first step is to conduct a feasibility study to understand which parameters that are relevant to a sustainable palm oil plantation management can be derived from SAR data. For a proper, successful and sustainable oil palm plantation management it is necessary to monitor the plantation on a regular basis. From preparing the ground, through planting the trees, growing the trees plus their accompanying plants to prevent erosion etc., to nutrition and health status of the trees plus yield prediction there is a palm tree life long process to be managed. Therefore, the plantation owner needs regular and continuous access to a set of parameters that are vital to take decisions about fertilizer, pest control, soil moisture regulation, environmental impact, water quality, and yield.

The initiative builds on existing experience in connecting to efforts taken by the Federal Land Agency (FELDA) research centre for palm oil. An established cooperation with Felda Agricultural Services Sdn. Bhd. ensures exchange of information and access to oil palm plantations and laboratories to provide ground truth data for validation of developed procedures. The entire project is embedded in the UTM CO_2 Initiative using Palm Oil as application for the research focus. The project headed by Sune Hansen, Director of the Palm Oil Research Centre of UTM focuses on CO_2 in the

IOP Conf. Series: Earth and Environmental Science **20** (2014) 012012 doi:10.1088/1755-1315/20/1/012012

palm oil industry within four major components: The Science of CO_2, Technologies and Methods, Regulatory & Policies and Training & Capacity Building.

The following sections of this paper will describe achieved results using spaceborne SAR for forest monitoring in general (Section 2) and oil palm plantations in particular (Section 3). Then Section 4 elaborates on the research methodology, identified data sources and planned processing flow. At the end the conclusions and outlook are summarized in Section 5.

2. SAR for forest monitoring

The use of radar remote sensing for forestry applications in the Tropics started with sensors on-board of aircrafts. In 1993 the usefulness of fully polarimetric airborne SAR to derive relationships between backscatter and tropical forest structures was investigated [12]. The research team was able to quantify an empirical relationship between biomass, forest structure and backscatter properties. Another study conducted by [13] showed the feasibility of using high resolution radar at a scale of 1:50,000 for the monitoring of a sustainable forest management for timber production in 1997. In parallel researchers started to exploit spaceborne SAR with the launch of the European Remote Sensing Satellite-1 (ERS-1) making SAR images more widely available for forestry applications. Today we have access to three different wavelengths, increased resolution and multiple polarizations which increases the information that can be derived to describe the status and useful parameters of the forest. The wavelength determines the depth of penetration into the canopy. The polarization is sensitive to different orientations of the tree and forest structures (stems, branches, twigs, leaves or needles) [14].

2.1. Spaceborne SAR sensors for forestry

From the literature commonly used radar satellite sensors during the past six years are

- European Remote Sensing (ERS-1/2) satellites: [14-16]
- Japanese Earth Resources Satellite (JERS-1): [14, 16-18]
- Canadian Remote Sensing Satellite Radarsat-2: [19-23]
- Advanced Land-Observing Satellite Phased Array L-band Synthetic Aperture Radar (ALOS PALSAR) (only 2013): [6, 24-26]
- High-resolution radar satellites TerraSAR-X/TanDEM-X: [10, 27-30]
- Environmental Satellite (Envisat) Advanced Synthetic Aperture Radar (ASAR): [26, 31-33]

The Kyoto & Carbon (K&C) Initiative by the Japan Aerospace Exploration Agency (JAXA) Earth Observation Research and Applications Centre (EORC) in 2000, to support environmental conventions, carbon cycle science and natural conservation, produced during the years a lot of useful results using ALOS PALSAR (L-band) in forestry. Lessons learned from all the research that has gone into this data have led to tremendous advancement in knowledge for information extraction from SAR in forestry applications. With the increase in available high-resolution multipolarization satellite SAR sensors the possibilities in up-to-date mapping of forest related parameters has become more and more feasible.

2.2. Change detection

The most straight forward application of SAR in forestry is the identification of changes, in particular the mapping of deforestation or clear-cuts. Recent findings underline the possibility of using L-band radar to detect losses and gains in above-ground biomass woodland [34]. Also C-band radar has proven its suitability to continuously detect large-scale, clear-cut and selective logging in a tropical rainforest irrespective of weather conditions [35]. [36] used multi-sensor C-band SAR to map forest changes over large areas in China in a ten year period with great success. They applied interferometric and classification methods to derive the thematic maps. X-band radar has been used for storm damage detection fusing multi-aspect SAR images [37]. The authors point out the advantage of increasing spatial resolution of the new sensors. Open research questions relate to the precise quantification of uncertainties in identified changes [34].

2.3. Biomass estimation

Backscatter from SAR is correlated to stem volume which means that biomass estimates can be derived. Again the information quality increases with wavelength since shorter wavelengths interact strongly with the canopy of the trees. For stem volume estimates L-band (3.75 – 7.5 cm wavelength) or P-band (> 1 m wavelength) are suitable [38]. P-band so far is only available on airborne platforms.

Other researchers succeeded with Radarsat-2 C-band SAR because they applied advanced processing techniques to estimate biomass making use of polarization, incidence angle, high spatial resolution and texture, especially from ratio images [39].

[10] presented a method for biomass estimation using SAR and derived parameters. They perform the process in two steps:

1. Analysis of selected Haralick textural parameters (HTP), i.e. entropy, contract, correlation, energy and homogeneity
2. Variogram analysis

In a case study they identified forest in VV (vertical transmitted, vertical received) polarized TerraSAR-X data (X-band, wavelength 3cm) of 0.5m ground sampling distance (GSD). Applying the support vector machine (SVM) supervised classifier to the image plus the derived texture parameters they were able to improve the classification accuracy by more than 50%.

A sub-parameter in biomass estimation is the derivation of tree heights that can be performed using polarimetric interferometric radar. The idea behind this concept is applying the knowledge that an HH microwave signal has a double bounce from the ground and the tree canopy. HV polarized signals have a smaller penetration into the canopy but a large diffuse scattering. VV backscatters most from the canopy and less from the tree trunk [40].

3. Oil palm plantation monitoring with SAR

Having understood the scientific achievements in using active microwave remote sensing for forestry, it takes a short step to convert the findings to applications in tree plantations and for oil palms in particular. In a way the applications for oil palm plantations is somewhat more straight forward since we already know what we are looking at. The trees are planted systematically. The plantation is covered with one species only. Each plantation by itself contains trees of the same age. The practices and activities to grow oil palms, to nurture and to harvest are specified. There is very little variation.

Similarly to applications in forestry in general, researchers have used SAR to estimate above-ground biomass (AGB) of oil palms. Using the backscatter coefficient of fully polarized ALOS PALSAR data [41] produced biomass estimates for an oil palm plantation in the state of Perak, Malaysia. Best results were obtained from VH polarization data using a window size of 5x5 for the Laplace filter to detect large changes in pixel values. Another study used a regression analysis of HV-polarized PALSAR data to estimate biomass of oil palm plantations in Sabah (Borneo), Malaysia. The authors found a precipitation significant effect on the backscatter. Also scattering from large leaves of mature palm trees significantly changed the estimates [42]. This definitely needs more investigation.

The measurement of changes in oil palm plantation expansion does not seem to be a major problem using multimodal SAR imagery [16].

However, the establishment of a correct relationships between backscatter and reality parameters of oil palm trees needs to consider multimodal data, ranging from multitemporal acquisitions, through multimodal SAR images, up to verification data from optical imagery or fieldwork [43].

4. Methodology, data sources and processing

Based on the collected information from the literature, discussion amongst oil palm plantation researchers and managers, as well as SAR image information extraction capabilities, the project starts with the design of a processing flow based on state-of-the art approaches. After having decided on the final test site which is most probably going to be a 'super' test site near the Malaysia National Park Taman Negara in the area of Jengka, Malaysia we will start the remote sensing data collection. The test data package will consist of COSMO-SkyMed SAR, ALOS PALSAR, TerraSAR-X, IKONOS, WorldView-2, and other available scenes.

Regarding the optical data the selection will depend on existing data from the archive with < 10% cloud coverage. The radar images will be taken at different times in coordination with ground truth data collection at the test site.

The data will be pre-processed to correct for atmospheric effects (optical), sensor-induced errors, speckle reduction (SAR) and geometry. All images will be geocoded to create a joint reference which enables multimodal image processing and mapping. The optical data and the ground truth will be taken as reference data to derive oil palm plantation relevant parameters from the SAR data.

The following SAR image processing techniques, for oil palm plantation parameter extraction are taken into account:

- multifrequency approach and use of polarimetric information for oil palm tree age determination using normalized backscatter coefficients and segmentation [44]
- classification of polarimetric and interferometric data to estimate forest heights (biomass); comparison to biomass from empirical BioMass Index (BMI) [28, 45]
- extraction of biomass from HV polarized L-band SAR through adapted Malaysian above ground biomass index of oil palm plantation [41]
- oil palm plantation analysis using the Bayesian approached based on mixture modelling followed by Markov Random Field (MRF) classification [7]

Of special interest will be the investigation of the potential of high-resolution multifrequency, interferometric and polarimetric SAR to provide information on oil palm frond shapes and changes as well as moisture parameters of plants and soil. The establishment of oil palm plantation parameter extraction models will be supported by optical image exploitation. For the multimodal image processing various analysis methods will be investigated amongst which joint classification of stacked multimodal images as well as fused remote sensing images are two options. Further image enhancements prior to analysis will be considered. Of high interest is the recent trend in integrating Light Detection and Ranging (LiDAR) and hyperspectral data for forest analysis [46].

5. Conclusions and Outlook

From the literature it becomes obvious that a stand-alone SAR system will not provide the necessary information needed for sustainable oil palm plantation management. The system has to be seen in the context of multimodal imaging. While the radar signal is mainly sensitive related to the woody biomass optical imagery delivers information in respect to seasonally variable plant leaves. However once a relationship between SAR and optical remote sensing data in terms of parameter extraction has been established, it is anticipated that an operation SAR image acquisition for the timely information extraction is sufficient to draw conclusions. Another important advantage is the integration of different radar sensors. With the availability of various operational high resolution fully polarized systems, such as Constellation of small Satellites for the Mediterranean basin Observation (COSMO-SkyMed), Environmental Satellite Advanced Synthetic Aperture Radar (ENVISAT ASAR), Advanced Land Observing Satellite Phased Array type L-band Synthetic Aperture Radar (ALOS PALSAR), TerraSAR-X, its twin-satellite TanDEM-X, add-on for Digital Elevation Measurements and upcoming systems like Sentinel-1 additional, valuable and regularly accessible information can be derived to the benefit of sustainable palm oil production. It should be mentioned that the trend in forest monitoring is to use LiDAR data in combination with polarimetric SAR and optical data. In this context hyperspectral imaging plays an important role. With the upcoming hyperspectral German satellite sensor onboard of the Environmental Mapping and Analysis Programme (EnMap) it is hoped to have access to additional valuable information. The satellite has been re-scheduled to be launched in 2017. It cannot be neglected that a multimodal approach always leads to improved, more reliable and trustworthy results.

Acknowledgement
This research is supported by the Ministry of Higher Education in Malaysia and the UniversitiTeknologi Malaysia under the research grant Q.J130000.2709.00K71. The support is gratefully acknowledged.

References

[1] Harris N L, Brown S, Hagen S C, Saatchi S S, Petrova S, Salas W, Hansen M, Potapov P V and Lotsch A 2012 Baseline Map of Carbon Emissions from Deforestation in Tropical Regions *Science* **336**(6088) 1573-76

[2] United States Department of Agriculture 2005 Oilseeds: World Markets and Trade *Cirular Series FOP* **10 - 05** 28

[3] Sulaiman F, Abdullah N, Gerhauser H and Shariff A 2011 An outlook of Malaysian energy, oil palm industry and its utilization of wastes as useful resources *Biomass and Bioenergy* **35**(9) 3775-86

[4] McDonald Pittman A Carlson K, Curran L M and Ponette-Gonzalez 2013 A NASA satellite data used to study the impact of oil palm expansion across Indonesian Borneo *The Earth Observer* **25**(5) 12-16

[5] Tan K T, Lee K T, Mohamed A R and Bhatia S 2009 Palm oil: Addressing issues and towards sustainable development *Renewable and Sustainable Energy Reviews* **13**(2) 420-27

[6] Thapa R B, Shimada M, Watanabe M, Motohka, T and Shiraishi T 2013 The tropical forest in south east Asia: Monitoring and scenario modeling using synthetic aperture radar data *Applied Geography* **41** 168-78

[7] Hoekman D H, Vissers M and Wielaard N 2010 PALSAR wide-area mapping of Borneo: Methodology and map validation *IEEE Journal of Selected Topics in Applied Earth Observations and Remote Sensing* **3**(4 PART 2) 605-17

[8] Sano E E, Ferreira L G, Anser G P and Steinke E T 2007 Spatial and temporal probabilities of obtaining cloud-free Landsat images over the Brazilian tropical savanna *International Journal of Remote Sensing* **28**(12) 2739-52

[9] Wielaard N 2011 Impact of oil palm plantations on peatland conversion in Sarawak 2005-2010 *Report SarVision,* 14 pages http://www.wetlands.org/Portals/0/publications/Report/Malaysia%20Sarvision.pdf (last accessed 18 March 2014)

[10] Benelcadi H, Frison P, Lardeux C, Capel A, Routier J and Rudant J 2012 Contribution of TerraSAR-X radar images texture for forest monitoring *IEEE Geoscience and Remote Sensing Symposium (IGARSS)* DOI 10.1109/IGARSS.2012.6352130 6427-30

[11] Izzawati, Lewis P and McMorrow J 1998 3D model simulation of polarimetric radar backscatter and texture of an oil-palm plantation *IEEE International Geoscience and Remote Sensing Symposium (IGARSS) Part 1* Seattle USA 1502-04

[12] Hoekman D H, Amar F and Quinones M J 1995 Biomass and structure estimation of primary and secondary tropical rain forests using AirSAR data *IEEE Geoscience and Remote Sensing Symposium (IGARSS) 'Quantitative Remote Sensing for Science and Applications' **1** 706-7*

[13] Hoekman D H 1997 Radar monitoring system for sustainable forest management in Indonesia *IEEE Geoscience and Remote Sensing (IGARSS) 'Remote Sensing - A Scientific Vision for Sustainable Development'* ***4*** 1731-33

[14] Koch B and Dees M 2008 Forestry applications, in *Advances in Photogrammetry, Remote Sensing and Spatial Information Sciences* CRC Press 439-465

[15] Einzmann K, Haarpaintner J and Larsen Y 2012 Forest monitoring in Congo Basin with combined use of SAR C- & L-band *IEEE Geoscience and Remote Sensing Symposium (IGARSS)* 6573-76

[16] Hartzell D, Seelan S K and Bolick L 2009 Monitoring oil-palm and other threats to Tanjung Puting National Park, Indonesia, with synthetic aperture radar *60th International Astronautical Congress*

[17] Guo Z, Ni W and Sun G 2009 Analysis of the effect of crown structure changes on backscattering coefficient using modeling and SAR data *IEEE Geoscience and Remote Sensing Symposium* Cape Town, South Africa **4** IV-386-89

[18] Shimada M, Watanabe M, Motooka T, Shiraishi T, Isoguchi O, Mukaida A, Okumura H, Otaki T and Itoh T 2011 Generation of 10m resolution PALSAR and JERS-1 SAR MOSAIC and forest/non-forest maps for forest carbon tracking *IEEE Geoscience and Remote Sensing Symposium (IGARSS)* 256-59

[19] Kovacs J M, Jiao X, Flores-de Santiago F, Zhang C and Flores-Verdugo F 2013 Assessing relationships between Radarsat-2 C-band and structural parameters of a degraded mangrove forest *International Journal of Remote Sensing* **34**(20) 7002-19

[20] Kumar T and Patnaik C 2013 Discrimination of mangrove forests and characterization of adjoining land cover classes using temporal C-band Synthetic Aperture Radar data: A case study of Sundarbans *International Journal of Applied Earth Observation and Geoinformation* **23**(1) 119-31

[21] Maghsoudi Y, Collins M J and Leckie D J 2013 Radarsat-2 Polarimetric SAR Data for Boreal Forest Classification Using SVM and a Wrapper Feature Selector *IEEE Journal of Selected Topics in Applied Earth Observations and Remote Sensing* **6**(3) 1531-38

[22] Mitchell A L, Williams M, Tapley I and Milne A K 2012 Interoperability of multi-frequency SAR data for forest information extraction in support of national MRV systems *IEEE Geoscience and Remote Sensing Symposium (IGARSS)* 3166-69

[23] Sarker M L R and Nichol J 2013 Forest biomass estimation from the fusion of C-band SAR and optical data using wavelet transform *SPIE Remote Sensing for Agriculture, Ecosystems, and Hydrology XV* Christopher M. U. Neale, Antonino Maltese (eds) **8887** 88870S-1

[24] Dong J, Xiao X, Chen B, Torbick N, Jin C, Zhang G and Biradar C 2013 Mapping deciduous rubber plantations through integration of PALSAR and multi-temporal Landsat imagery *Remote Sensing of Environment* **134** 392-402

[25] Peregon A and Yamagata Y 2013 The use of ALOS/PALSAR backscatter to estimate above-ground forest biomass: A case study in Western Siberia *Remote Sensing of Environment* **137** 139-46

[26] Wang X, Ge L and Li X 2013 Pasture Monitoring Using SAR with COSMO-SkyMed, ENVISAT ASAR, and ALOS PALSAR in Otway, Australia *Remote Sensing* **5**(7) 3611-36

[27] Baldauf T 2013 Monitoring Reduced Emissions from Deforestation and Forest Degradation (REDD+) : Capabilities of High- Resolution Active Remote Sensing *Staats- und Universitätsbibliothek Hamburg: Germany*

[28] Boldt M, Thiele A, Schulz K and Hinz S 2011 Structural analysis of forest areas in high-resolution SAR images *Earth Resources and Environmental Remote Sensing/GIS Applications II, Ulrich Michel & Daniel L. Civco (Eds)* ***81810S***

[29] Englhart S, Keuck V and Siegert F 2011 Aboveground biomass retrieval in tropical forests — The potential of combined X- and L-band SAR data use *Remote Sensing of Environment* **115**(5) 1260-71

[30] Solberg S, Astrup R, Breidenbach J, Nilsen B and Weydahl D 2013 Monitoring spruce volume and biomass with InSAR data from TanDEM-X *Remote Sensing of Environment* **139** 60-67

[31] Erasmi S and Twele A 2009 Regional land cover mapping in the humid tropics using combined optical and SAR satellite data—a case study from Central Sulawesi, Indonesia *International Journal of Remote Sensing* **30**(10) 2465-78

[32] Holecz F Barbieri M, Cantone A, Pasquali P and Monaco S 2010 Synergetic Use of ALOS PALSAR, ENVISAT ASAR and Landsat TM ETM+ Data for Land Cover and Change Mapping in *K&C Phase 2 (2008-2009) Science Team Reports*

[33] Santoro M, Cartus O, Fransson J, Svidenko A, McCallum I, Hall R, Beaudoin A, Beer C and Schmullius C 2013 Estimates of Forest Growing Stock Volume for Sweden, Central Siberia, and Québec Using Envisat Advanced Synthetic Aperture Radar Backscatter Data *Remote Sensing* **5**(9) 4503-32

[34] Mitchard E T A, Meir P, Ryan C M, Woollen E S, Williams M, Goodman L E, Mucavele J A, Watts P, Woodhouse I H and Saatchi S S 2012 A novel application of satellite radar data: measuring carbon sequestration and detecting degradation in a community forestry project in Mozambique *Plant Ecology & Diversity* **6**(1) 159-70

[35] Van Der Kooij M Radarsat-2: A powerful solution for global forest monitoring needs *GEO: connexion* **12**(2) 39-41

[36] Ling F, Leiterer R, Huang Y, Reiche J and Li Z 2011 Forest change mapping in northeast China using SAR and INSAR data *34th International Symposium on Remote Sensing of Environment (ISRSE), Sydney (AUS), 10 April 2011 - 15 April 2011*

[37] Thiele A, Boldt M and Hinz S 2012 Automated detection of storm damage in forest areas by analyzing TerraSAR-X data *IEEE Geoscience and Remote Sensing Symposium (IGARSS) 1672-75*

[38] Thiel C, Schmullius C and Santoro M 2010 Boreal Forest Mapping in Siberia, *K&C Phase 1 (2006-2008) Science Team Reports*

[39] Sarker M L R, Nichol J, Iz H B, Ahmad B B and Rahman A A 2013 Forest Biomass Estimation Using Texture Measurements of High-Resolution Dual-Polarization C-Band SAR Data *Geoscience and Remote Sensing, IEEE Transactions on Geoscience and Remote Sensing* **51**(6) 3371-84

[40] Shimada M 2007 Coherence dependency of the PALSAR POLinSAR on forest in japan and amazon *IEEE Geoscience and Remote Sensing Symposium (IGARSS)2636-39*

[41] Shashikant V, Shariff A R M, Nordin L and Pradhan B 2012 Estimation of above ground biomass of oil palm trees by PALSAR *IEEE Colloquium on Humanities, Science & Engineering Research (CHUSER 2012)* December 3-4 Kota Kinabalu, Sabah, Malaysia 838-41

[42] Morel A C, Saatchi S S, Malhi Y, Berry N J, Banin L, Burslem D, Nilus R and Ong R C 2011 Estimating aboveground biomass in forest and oil palm plantation in Sabah, Malaysian Borneo using ALOS PALSAR data *Forest Ecology and Management* **262**(9) 1786-98

[43] Santos C and Messina J P 2008 Multi-Sensor Data Fusion for Modeling African Palm in the Ecuadorian Amazon *Photogrammetric Engineering & Remote Sensing* **74**(6) 711-23

[44] Lee J S, Papathanassiou K P, Hajnsek I, Mette T, Grunes M R, Ainsworth T and Ferro-Famil L 2005 Applying polarimetric SAR interferometric data for forest classification *ESA PolInSAR Workshop* ESRIN, Frascati, Italy on 17-21 January, **7** 4848-51

[45] Nordin L 1996 Application of AirSAR data to oil palm tree characterization *Malaysian Centre for Remote Sensing (MACRES)* 7 pages http://citeseerx.ist.psu.edu/viewdoc/download?doi=10.1.1.124.7147&rep=rep1&type=pdf (last accessed 18 March 2014)

[46] Koch B 2010 Status and future of laser scanning, synthetic aperture radar and hyperspectral remote sensing data for forest biomass assessment *ISPRS Journal of Photogrammetry and Remote Sensing* **65**(6) 581-90

IOP Conf. Series: Earth and Environmental Science **20** (2014) 012013 doi:10.1088/1755-1315/20/1/012013

Auto Guided Oil Palm Planter by using multi-GNSS

I Nur Aini[1], Aimrun W [1,2], M S M Amin[1,2], M H Ezrin[1,2] and H Z Shafri[3]
[1] Department of Biological and Agricultural Engineering
[2] Smart Farming Technology Research Center
[3] Department of Civil Engineering
Faculty of Engineering
Universiti Putra Malaysia
Serdang, Selangor, 43400, Malaysia

e-mail: aimrun@upm.edu.my

Abstract. Planting is one of the most important operations in plantation because it could affect the total area of productivity since it is the starting point in cultivation. In oil palm plantation, lining and spacing of oil palm shall be laid out and coincided with the topographic area and a system of drains. Conventionally, planting of oil palm will require the polarization process in order to prevent and overcome the lack of influence of the sun rise and get a regular crop row. Polarization is done after the completion of the opening area by using the spike wood with 1 m length painted at the top and 100 m length of wire. This process will generally require at least five persons at a time to pull the wire and carry the spikes while the other two persons will act as observer and spikes craftsmen respectively with the ability of the team is 3ha/day. Therefore, the aim of this project is to develop the oil palm planting technique by using multi-GNSS (Global Navigation Satellite System). Generally, this project will involve five main steps mainly; design of planting pattern by using SOLIDWORKS software, determine the boundary coordinate of planting area, georeference process with ArcGIS, stakeout process with Tracy software and finally marking up the location with the wooden spikes. The results proved that the multi- GNSS is capable to provide the high accuracy with less than 1 m in precise positioning system without augmentation data. With the ability of one person, time taken to complete 70 m x 50 m planting area is 290 min, which is 25 min faster than using GPS (Global Positioning System) only.

1. Introduction

Oil palm cultivation in Malaysia began in 1917 and accelerated during the last 50 years through large-scale investments to become one of the approved crops for diversifying the country's agricultural development. As reported by MPOB [1], Malaysia has become one of the largest producers and exporters of oil palm in the world with the total of 4.69 million ha oil palm planted area in 2009. Due to this high demand and preference of oil palm, new technology should be implemented to facilitate the planting activities within a huge area of plantation.

Turner and Gillbanks [2] mentioned that there are several operational activities that involve in oil palm transplanting which are preparations of planting holes, placement of seedling in the prepared hole, covering the seedling in prepared hole and compaction of the soil around the planted seedling. Prior to the planting operation, planting points need to be identified and marked precisely in the plantation field before preparing the planting hole by using a hoe or mechanized drill. Darius et al. [3] stated that these present practices are very laborious and impose extensive stress to the field workers

Content from this work may be used under the terms of the Creative Commons Attribution 3.0 licence. Any further distribution of this work must maintain attribution to the author(s) and the title of the work, journal citation and DOI.
Published under licence by IOP Publishing Ltd

and thus become an unattractive job for others to pursuit. According to Basiron [4], almost all field operations in the plantation could be mechanized with the exception to the cutting operation of fresh fruit bunches. Hence, efforts to promote mechanization have to be made in order to combat the problem of labor shortages in plantation and also to maintain production.

1.1. Oil palm planting pattern

Desa et al. [5] suggested that planting pattern must make optimum use of space per hectare and at the same time be of optimal economic density which would minimize the competition between palms for nutrients to ensure high productivity. Generally, oil palm is planted in triangular system at spacing of 9 x 9 x 9 m accommodating 143 plants per ha. Hartley [6] proposed a palm spacing of 7.5 to 10 m in a triangular arrangement to give total number of 115 and 205 palms per ha.

The planting rows run north to south and the palm in one row will not be opposite the palm in adjacent row. The line from each palm to the nearest palm in the adjacent row will make an angle of 60° with the row and the distance between rows will be less that planting distance. According to Siamak [7], this triangular orientation and planting arrangement ensures maximum sunlight falls on the individual palms.

1.2. Conventional method in planting the oil palm

Contemporary techniques and technologies employed by the plantation industries in the planting and replanting of oil palm seedlings are very laborious and time consuming. Referring to Basiron [8], oil palm planting is preceded by preparation of the land, which may be previously logged-over areas alienated for agricultural use, old oil palms stands or areas once planted with rubber or cocoa. The places to plant the oil palm seedlings need to be pegged out carefully in order to provide an evenly spaced of oil palms and to achieve the optimum stand per ha for systematic irrigation and further harvesting.

Presently, the pegging out process is usually commenced soon after the completion of land preparation or one to two months prior to seedlings being planted in the plantation area by using wooden spike with 1 m length painted at the top and 100 m length of wire. This process will generally require at least five persons at a time to pull the wire and carry the spikes while the other two persons will act as observers with the ability of the team is 3 ha/day. For a huge plantation area, the theodolite is used to determine the angle, horizontal distance and to fix the planting points. This method has created a lot of problems at which it will consume a lot of time and also the high cost is required for the labour wages, equipment and overall activity.

Therefore, the purpose of the present study was to develop a standardised procedure of planting the oil palm by reducing the time for planning, lining, alignment and labour workforce. Field testing has been carried out to evaluate the accuracy, availability and also the reliability of multi-GNSS which comprise of GPS (Global Positioning System) from United States and GLONASS (Global Navigation Satellite System) from Russia.

2. Materials and method

2.1. Equipment

The study area was conducted at Plot Industrial Crops Plot, Universiti Putra Malaysia at latitude 2°59''N and longitude of 101°43''E which covers an area of about 70 m x 50 m. The main equipment that was used throughout this research study was JAVAD GrAnt Antenna, JAVAD Sigma Receiver and also JAVAD Victor with Tracy software installed.

JAVAD GrAnt antenna is a versatile high performance antenna that has the capability to track not only GPS but also GLONASS, Galieo, Compass, WAAS (Wide Area Augmentation System), EGNOS (European Geostationary Navigation Overlay Service), MSAS (Multi- Functional Satellite Augmentation System) and also QZSS (Quasi-Zenith Satellite System) signals. JAVAD Victor on the other hand is a handheld controller that is equipped with Tracy software, comprehensive field software to control the receiver, automate GNSS post processing surveying task, performing RTK (Real-Time Kinematic) survey and to perform the stakeout tasks. For this study this controller was used locate the oil palm planting point precisely. So far, only two satellites systems are available in this Tracy software which is GPS L1 + L2 and also GLONASS L1 + L2.

2.2. Procedure

To start the experiment, the oil palm planting pattern was sketched predominantly by using SOLIDWORKS software. For this case, triangular planting pattern was chosen as it can provide the most optimum density per hectare with the distance between each oil palm was 9 m. The boundary coordinates of the planting area were then determined by using JAVAD Victor for further georeferencing and rectifying process in ArcGIS software. Georeferencing process will enable us to obtain the coordinates of each planting points by only adding the boundary point coordinates of the planting area.

After each coordinates of the planting point was identified in ArcGis software, it was then exported to JAVAD Victor for navigating to a certain planting point. Once the target point has been determined, the survey process was carried out to specify the coordinates of the new point and the wooden spikes were staked out onto the ground to mark out the surveyed point. The distance difference between the surveyed points and designed points were then computed by using CoGo tab (Direct Transverse) in Tracy software that gave the readings of horizontal, vertical and also the slant distance between both coordinates.

Figure 1. Survey process to the target planting point by using JAVAD Victor.

Figure 2. Wooden spikes that were staked out onto the ground to mark the surveyed points.

3. Results and discussion

Table 1 below shows the summarize result for the availability, accuracy and time taken during the field test. The results showed that the number of satellite available during the field test for GPS only was about 5-6 SVs while, multi-GNSS obtained about 13-14 SVs, which is more than double. The standard deviation of the point by using GPS was about 1.013 m and 0.715 m when using multi-GNSS. This really proved that multi-GNSS is capable enough to provide more satellite availability compared to GPS only. For this study, the accuracy of the satellite system is determined by the average offline distance deviated from the designed points. Without any correction, the average offline distance for GPS was 0.9952 m while, multi-GNSS was 0.5375 m. This shows that multi-GNSS will provide higher accuracy than GPS. Time to complete the navigation process was 5 hrs 15 min when using GPS and 4hrs 50 min when using multi-GNSS. The longer time taken might be due to the low availability of GPS to acquire satellite signals.

Table 1. Summarize results of the accuracy, availability and time taken for Multi-GNSS and GPS.

	Multi-GNSS	GPS
Number of satellite	13-14	5-6
Standard deviation (m)	0.715	1.103
Average offline distance (m)	0.995	0.538
Time taken (minutes)	290	315

Figure 3 below on the other hand shows the oil palm planting pattern designed with SOLIDWORKS software and the target points navigated with Multi-GNSS and GPS. The pattern produced shows that the navigation with multi-GNSS could still produce the planting point in straight line even though the average offline distance was still high (0.538 m). Compared to GPS, the planting points are scattered within the planting area due to the loss of continuity. This is because most of the time GPS simply stops working and cannot broadcast the signals to compute the planting position and hence, will lead to longer time taken to complete all 40 surveyed points.

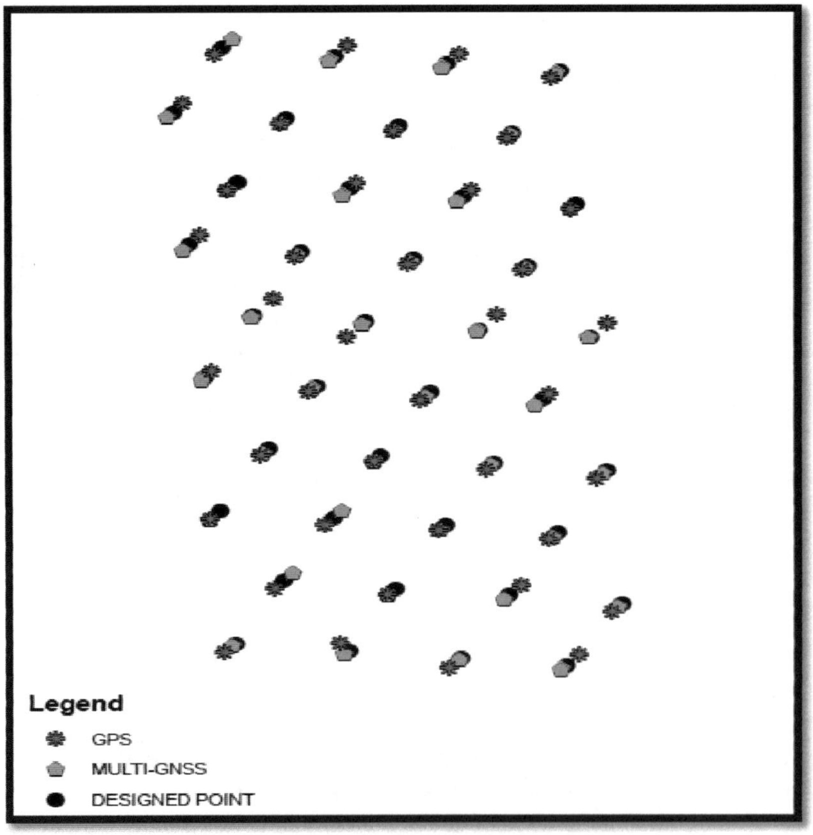

Figure 3. Comparison of designed and target planting point navigated by using GPS and multi-GNSS.

4. Conclusion

Overall, it can be concluded that Multi-GNSS could be adapted as a new technology in planting the oil palm without the requirement for planning, lining and alignment activities. It can also be proven that the multi-GNSS is capable to provide a high accuracy with less than 1 m in precise positioning system without any augmentation data with the minimum labour workforce. This result could be improved further if more satellites systems are available in Tracy software and configuration to JAVAD Victor is made to obtain the correction from the base station. However, at the huge plantation area, adapting this standard procedure would promise a great achievement in time management, proceeding benefit and as well as the efficiency of agricultural activities. Using multi-GNSS positioning is very current and will continue to grow in intensity as it will become a crucial element in increasing profit of agriculture field.

Acknowledgements

This project is supported by Multi-GNSS Joint Experiment Matching Grant (9300414) from Universiti Putra Malaysia. The authors are very grateful to Japanese Aerospace Exploration Agency (JAXA) for lending the JAVAD Sigma Receiver and GrAnt Antenna throughout this research study.

References

[1] MPOB 2010 *Oil Palm Statistics 2009* (Economic and Industry Division MPOB)
[2] Turner P D and R A Gillbanks 1974 *Oil Palm Cultivation and Management* (The Incorporated Society of Planter, Kuala Lumpur, Malaysia)
[3] D EL Pebrian and Azmi Y 2002 Design and development of a prototype trailed type oil palm seedling transplanter *J. Oil Palm Research* **15** 32-40
[4] Basiron Y 1998 *Role Mechanisation in the Oil Palm Industry (*National Seminar On Mechanisation in Oil Palm Plantation. PORIM, Bangi, Malaysia)
[5] Desa A Jayaselan and Anusia H 2011 Development of a mechanization selection system for oil palm plantations with alternative planting patterns *J. Oil Palm Research* **23** 990-998
[6] Hartley C W S 1977 *The Oil Palm (Elaeis guineensis gacq.)* (London, Longman Group Limited)
[7] Siamak M R, Azmi Y and SA Aziz 2011 Field performance of a single chassis integrated machine system in planting oil palm seedlings *J. Agricultural Machinery Science* **7** 185-190
[8] Basiron Y 2007 Palm oil production through sustainable plantations *Eur. J. Lipid Science Technology* **109** 289-295

Developing web-based data analysis tools for precision farming using R and Shiny

Ebrahim Jahanshiri, Abdul Rashid Mohd. Shariff

Geospatial Information Science Research Centre (GIS RC)
Faculty of Engineering
Universiti Putra Malaysia (UPM)
43400 Serdang, Selangor
Tel:603-89466424; Fax:603-89466425

e.jahanshiri@gmail.com; rashid@eng.upm.edu.my

Abstract. Technologies that are set to increase the productivity of agricultural practices require more and more data. Nevertheless, farming data is also being increasingly cheap to collect and maintain. Bulk of data that are collected by the sensors and samples need to be analysed in an efficient and transparent manner. Web technologies have long being used to develop applications that can assist the farmers and managers. However until recently, analysing the data in an online environment has not been an easy task especially in the eyes of data analysts. This barrier is now overcome by the availability of new application programming interfaces that can provide real-time web based data analysis. In this paper developing a prototype web based application for data analysis using new facilities in R statistical package and its web development facility, Shiny is explored. The pros and cons of this type of data analysis environment for precision farming are enumerated and future directions in web application development for agricultural data are discussed.

1. Introduction

World Wide Web has given new opportunities for making static and dynamic websites. With the advent of new programming languages like HTML5 [1] and JavaScript [2], making user-friendly web applications [3] [4] [5] is getting easier than ever. These advancements lead to the availability of many web applications that has truly influenced real time decision making. Data driven applications have also been recently used to facilitate data analysis online. Web Geo-porocessing applications [6] however have recently gained momentum in the business and research community. True geo-processing online which consists of accepting user data, processing it the way user requests and showing the results in a coherent and easy to understand manner has not been widely available to the developers due to complexity of data, process and visualization. Note that web mapping is different from web geo-processing that focuses on static online map delivery. In this case the maps pre-produced and the user can request them online. Most of the online mapping website including OpenGIS Web Map Service Interface Standard (WMS) is also based on web mapping concept [7] [8].

ESRI ArcGIS has been one of the pioneers in providing the online web mapping services based on Microsoft technologies [9]. Other companies have followed the lead and have provided facilities that are mainly tied to huge commercial data base systems and windows web application interfaces.

Content from this work may be used under the terms of the Creative Commons Attribution 3.0 licence. Any further distribution of this work must maintain attribution to the author(s) and the title of the work, journal citation and DOI.
Published under licence by IOP Publishing Ltd

Google has recently provided its web mapping interfaces that will let the developers and users to map their own events and applications over the already existent Google maps and earth [10]. Most of these applications however lack the complexity that is needed by the geo-processing operations.

In the open sources world, there have been attempts to provide web geo-processing services using web clients and Python language (as web server) but that initiative did not become widespread mainly because Python does not have dedicated facilities for the level of statistical analysis of geographic data. With the advent of R open source statistical programming language [11] and gaining popularity among the spatial analysis community, and also the introduction of very useful "Shiny" package [12] that lets the programmers to create applications online, a new opportunity has been shown itself for creating web geo-processing services .

Precision farming or precision agriculture is highly reliant on the accurate and easy to access maps for management of variability in the field. Interpolation techniques used in the map making for precision farming are mostly using complex mythologies and therefore any application that attempts to provide the users such facility is limited to only a few options that are available within the software [13][14]. For example low-level ArcGIS interpolation elements are not available to the users to tweak. The rich data analysis facilities in R gives freedom to the user to access low level parameters that are necessary in order to create accurate maps.

This paper introduces an attempt to make an online web geo-processing service and mapping for precision farming using common interpolation and visualization techniques available in R.

2. Methodology

2.1. System Architecture

The interface is consisted of two parts; web interface and web server. Both of these components are controlled by the code that is written within the frame work of Shiny application in R. The building block of Shiny package is based on reactive programming [15]. Since the major task of the web application is to get the inputs and produce outputs, the whole programming language is designed so that a change in any input whether it is input data or method parameters will change the end result. The change will immediately be reflected into the form of texts, tables or figures. This is an advantage for the web applications that rely on user inputs. The shiny procedure can provide different outputs without the need to refresh the web page.

Within the shiny package, ordinary controllers or widgets are provided for ease of use of application programmers. Many of the procedures like uploading files and refreshing the page for drawing new plots and tables are provided automatically. These tasks are done based on the pre-built output widgets. The communication between the web interface and R is done with fast websockets package [16]. Websockets are exclusively important in situations where there are constant back and forth dialogue or data exchange between the clients and servers. Figure 1 provides general overview of the system components.

The communication between the client and server is done over the normal TCP connection. The bulk of live data traffic that is needed for many of web applications (i.e. online games) between the browser and the server is facilitated over the websockets protocol. This protocol operates separately and only handshake between the client and server is done over the HTTP protocol. The duplex connection is open all the time and therefore the authentication is not needed when exchange is done. The websocket is currently supported by many modern browsers.

2.2. Web Interface

The web interface provide means for the user to upload files, change the file reading format and provide necessary parameters to the R engine to produce the results and then reflecting the results in a consistent manner.

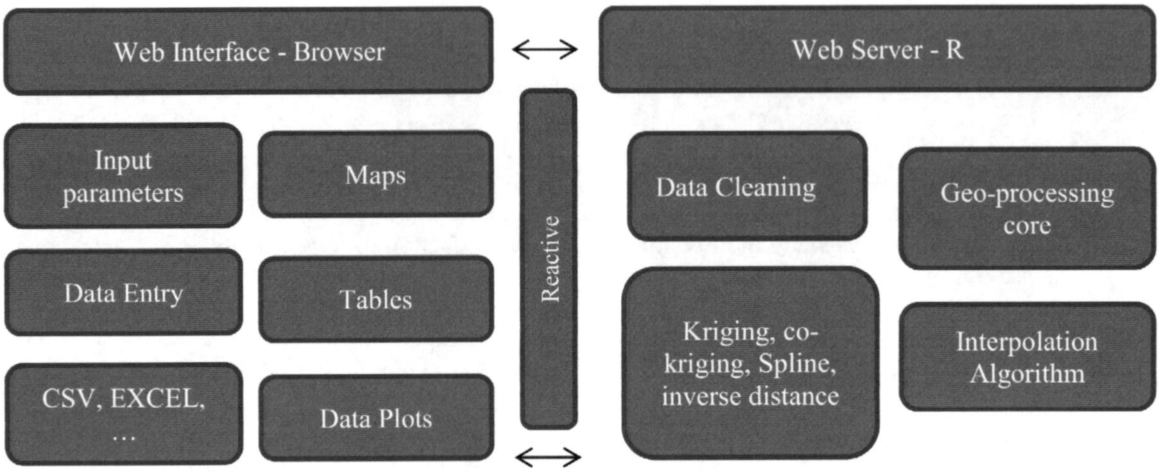

Figure 1: System architecture for the shiny application.

2.2.1 Data

The mapping procedure will start by providing the input data. Since the data is geographic by nature, the coordinates (x and y or Northing and Easting) should be part of the data. Also at least one target variable should be present in the data so that the interpolation can be done for that variable. Currently the input format is comma separated values (CSV). This is a very common format for storing the tabular data and is recognized by R in addition to the commercial data analysis software like Microsoft Excel. The support for other types of data inputs can be added in the future given the flexibility of R to accept many types of data.

Some elementary data screening like checking variable names, checking for duplicates in the coordinate points are done in this stage. In this stage the data also will be converted to specific spatial point data or SpatialPointsDataframe object [17], [18] so that spatial methods in R can take advantage of these formats. Most of these activities are done on the server side of the application and therefore hidden from the user.

2.2.2 Exploratory Analysis.

In this section normal exploratory spatial data analysis is done on the target of interest. As can be seen in Figure 2, the user must choose a target variable and the histogram, boxplot and distribution of variable (bubble plot) is drawn in the web page. The plot visualizations are done using elegant methods in the ggplot2 package [19]. Using these plots the user can decide whether to use any specific measure in the interpolation like transformation of target variable.

2.2.3 Mapping. In this section the result of interpolation using the standardized kriging for the selected variable is shown in the form of a map with a legend showing different levels of concentrations. The interpolation algorithm is currently limited to inverse distance weighting and geostatistical kriging (ordinary kriging). Other interpolation algorithms like co-kriging, universal and regression kriging will be implemented in the future. Due to extensive methods and procedures that are available in R, the program can become a very sophisticated interpolation web app for everyone who is willing to make a map for agricultural or geological variables whenever the data is available.

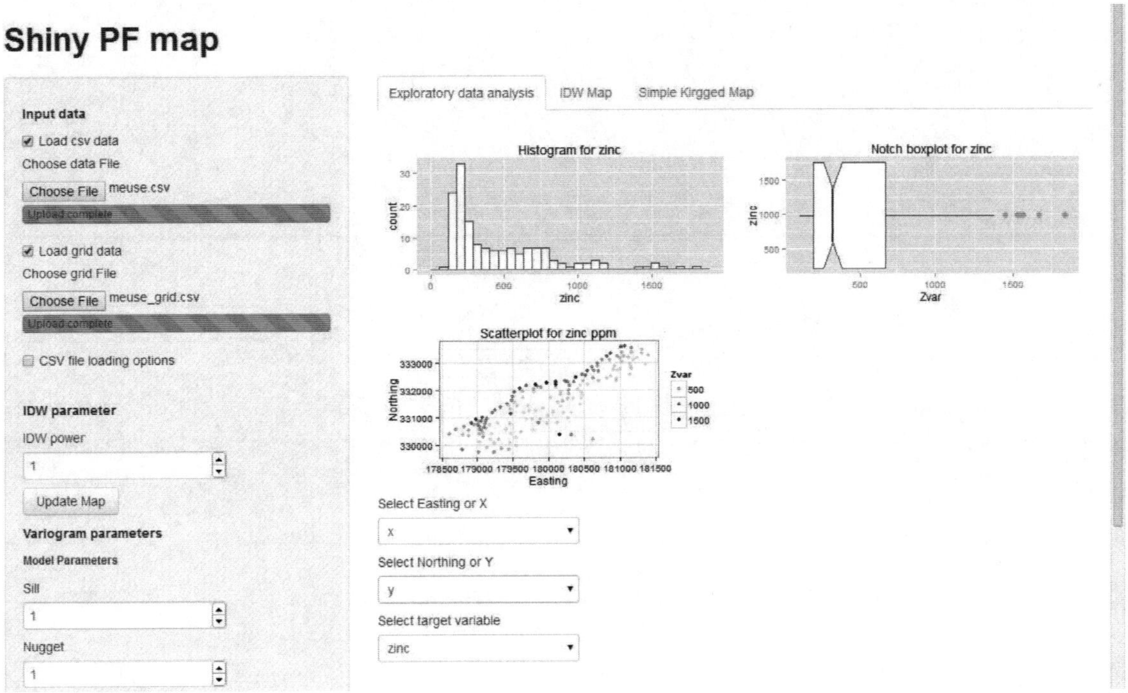

Figure 2: Exploratory spatial data analysis for the given target variable.

Figure 3 shows the mapping panel for kriging method. The variogram of the target variable in conjunction with the thematic map is shown in this panel. The advantage of this system is that the user can easily change the variogram and kriging variable and almost instantly see the results on both variogram fit and the resulting map. This will give an invaluable tool to analysts to fine tune the variogram and therefore the map. Given the vast potential of R, extending this application to encompass even minute parameters in modelling and mapping is also possible.

3. Future Direction

R as an extensive statistical programming language has provided suitable means for the spatial analysis. These methods are created and enhanced by the researchers and are subject to improvement and enhancement by a global network of researchers who test these methods. Many of the methods that are available in R for analysis are not implemented in the proprietary GIS software world. The implementation of shiny package unleashed this potential for web geo-processing applications. This web mapping application will provide basis to further enhance the analysis methods. The interpolation and mapping will be extended to encompass many new and older methods for spatial analysis of point data. New data screening facilities will be added in the future so that using different data methods (from SPSS files to ESRI shapefiles and geodatabases) will be possible with the Shiny PF app. Also providing on-the-fly projection for the data and then projecting the data and the resulting map on the Google maps or Open Street maps will be considered in the future so that the user will have a better understanding of the data and map. Further visualization plots for the data can also be added to this web app to enrich its utility everywhere.

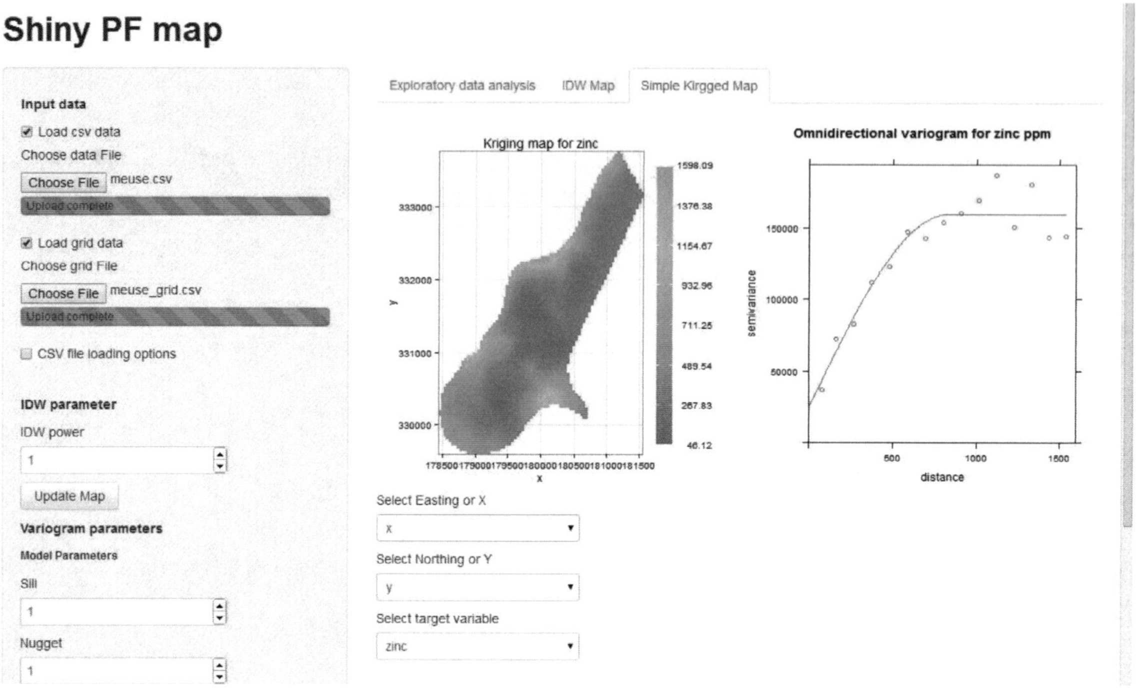

Figure 3: Mapping panel for ordinary kriging method.

4. Conclusion

In this paper we set out to present the feasibility of creating a web mapping application for precision farming data. Because of the variability in the field, it is important for the farmers and managers to be able to view the data visually. Because of the nature of agricultural data, that is geographical, any mapping and statistical analysis and has an added complexity. Given the vastness of methods that are available in R statistical package for analysis of spatial data and freedom of the analyst to analyse and map theses data, developing based upon R has tremendous advantage. The newly available shiny package from RStudio that lets the developer inter-connect R statistical system to the web interface has opened new horizons. Developing a complete data analysis and mapping solution however does not come without challenges. Coordinate systems transformations, variety of data formats, inaccurate and missing data and complexity of some of the interpolation algorithms are among some of them. These challenges can be overcome as data collections formats are becoming standardized and as new upgrades on the statistical methods and web interface are becoming available to the mainstream data analysts.

References

[1] B. Lawson and R. Sharp, *Introducing HTML5 (Google eBook)*. New Riders, 2011, p. 312.
[2] D. Flanagan, *JavaScript Pocket Reference (Google eBook)*. "O'Reilly Media, Inc.," 1998, p. 89.
[3] D. Nations, "Web Applications," *About.com*.
[4] M. Calore, "How Do Native Apps and Web Apps Compare?," *Wired.com*.
[5] P. Fraternali, "Tools and approaches for developing data-intensive Web applications: a survey," *ACM Comput. Surv.*, vol. 31, no. 3, pp. 227–263, Sep. 1999.

[6] P. Zhao, F. Lu, T. Foerster, and P. Yue, "The Geo-processingWeb," *Comput. Geosci.*, vol. 47, pp. 3–12, 2012.
[7] "Web mapping - Wikipedia, the free encyclopedia." [Online]. Available: https://en.wikipedia.org/wiki/Web_mapping.
[8] "Web Processing Service | OGC®." [Online]. Available: http://www.opengeospatial.org/standards/wps. [Accessed: 05-Oct-2009].
[9] ESRI, "ArcGIS Help 10.1 - What is a geo-processingservice?" ESRI, 2014.
[10] "Google Maps API Web Services - Google Maps API Web Services — Google Developers." [Online]. Available: https://developers.google.com/maps/documentation/webservices/.
[11] R Development Core Team, "R: A Language and Environment for Statistical Computing." Vienna, Austria, 2010.
[12] RStudio and Inc., "shiny: Web Application Framework for R." 2013.
[13] A. Fauzul, A. Aziz, A. Rashid, M. Shariff, M. Amin, A. A. Rahim, E. Jahanshiri, N. Norasma, C. Ya, and F. Azhan, "GIS based System for Paddy Precision Farming," in *World Conference on Agricultural Information and IT*, 2008, no. Did, pp. 417–422.
[14] C. Y. N. Norasma, A. R. M. Shariff, E. Jahanshiri, M. S. M. Amin, S. Khairunniza-Bejo, and A. R. Mahmud, "Web-Based Decision Support System for Paddy Planting Management," *Pertanika J. Sci. Technol.*, vol. 21, no. 2, pp. 343–364, 2013.
[15] Z. Wan and P. Hudak, "Functional reactive programming from first principles," *ACM SIGPLAN Not.*, vol. 35, no. 5, pp. 242–252, May 2000.
[16] "Introducing Shiny: Easy web applications in R | RStudio Blog." [Online]. Available: http://blog.rstudio.org/2012/11/08/introducing-shiny/.
[17] R. S. Bivand, E. J. Pebesma, and V. Gómez-Rubio, *Applied Spatial Data Analysis with R*, 1st ed. Springer, 2008, p. 390.
[18] E. Pebesma and R. Bivand, "Classes and methods for spatial data in R," *R News*, vol. 5, no. 2, pp. 9 – 13, 2005.
[19] H. Wickham, *ggplot2: elegant graphics for data analysis*. Springer New York, 2009.

Using SPOT-5 images in rice farming for detecting BPH (Brown Plant Hopper)

F Ghobadifar [1], A Wayayok [2], M Shattri [3] ,H Shafri[4]

[1] PhD student, Geospatial Information Science Research Centre (GIS RC), Faculty of Engineering, Universiti Putra Malaysia, Selangor, Malaysia (Corresponding author).

[2] Senior Lecturer, Department of Biological and Agricultural Engineering Faculty of Engineering, Universiti Putra Malaysia, Selangor, Malaysia.

[3] Professor, Department of Civil Engineering, Faculty of Engineering, Universiti Putra Malaysia, Malaysia.

[4] Associate Prof, Department of Civil Engineering, Faculty of Engineering, Universiti Putra Malaysia, Malaysia.

E-mail: faranakghobadi@gmail.com

Abstract. Infestation of rice plant-hopper such as Brown Plant Hopper (BPH) (*Nilaparvata lugens)* is one of the most notable risk in rice yield in tropical areas especially in Asia. In order to use visible and infrared images to detect stress in rice production caused by BPH infestation, several remote sensing techniques have been developed. Initial recognition of pest infestation by means of remote sensing will spreads, for precision farming practice. To address this issue, detection of sheath blight in rice farming was examined by using SPOT-5 images. Specific image indices such as Normalized decrease food production costs, limit environmental hazards, and enhance natural pest control before the problem Normalized Difference Vegetation Index (NDVI), Standard difference indices (SDI) and Ratio Vegetation Index (RVI) were used for analyses using ENVI 4.8 and SPSS software. Results showed that all the indices to recognize infected plants are significant at $\alpha = 0.01$. Examination of the association between the disease indices indicated that band 3 (near infrared) and band 4 (mid infrared) have a relatively high correlation. The selected indices declared better association for detecting healthy plants from diseased ones. Consequently, these sorts of indices especially NDVI could be valued as indicators for developing techniques for detecting the sheath blight of rice by using remote sensing. This infers that they are useful for crop disease detection but the spectral resolution is probably not sufficient to distinguish plants with light infections (low severity level). Using the index as an indicator can clarify the threshold for zoning the outbreaks. Quick assessment information is very useful in precision farming to practice site specific management such as pesticide application.

1. Introduction

Rice is considered as an important crop globally: more than half of the world's population depends on it for supplying food. In aim of maintaining the crop, several parameters might affect sustainable rice farming which consist of both well-timed and appropriate pest control. The potential of remote sensing makes it useful as a proficient and low-priced system to recognize unhealthy plants in a ground scale, primarily for the reason that infected plants contain various spectral reactions in comparison to healthy ones (18). Few investigations have been undertaken on remote sensing applications for pest control in rice (19). Sensing issues have been discussed in terms of remote sensing methods (2). Most of the research on rice remote sensing was conducted using field zone mapping and production estimating, nevertheless, in order to evaluate plant greenery for estimating the growth stage of plants, almost entire works were undertaken by means of hyperspectral remote sensing data. (1, 3, 4, 5, 6, 8 and 9). Several researches (13, 14, and 15) evaluated rice canopy spectra regarding to parameters such as leaf area index (LAI) and airborne phytomass in visible, near infrared and mid- infrared areas. Assessment of plant growth could be achieved by low-cost hyperspectral proximal sensing which evaluate detected data by purpose of measuring rice canopies (6, 7).

Content from this work may be used under the terms of the Creative Commons Attribution 3.0 licence. Any further distribution of this work must maintain attribution to the author(s) and the title of the work, journal citation and DOI.
Published under licence by IOP Publishing Ltd

However visible light (solar radiation) is absorbed in live green vegetation in the photosynthesis process, but plants scatter (reflect) solar energy at the near infrared waveband simultaneously. This difference in absorption is relatively distinctive in live vegetation and measures the greenness of the plants.

Symptoms of BPH will not be visible from outside in the early stages, but if we enter the field and tap the plants many times, this insect can be seen. They are visible only when the damage is severe, indeed, the plants present a burned up appearance - so called hopper burn - in circular patches. Symptoms of BPH appear more in the later stages of plant growth, when they turn from a green colour to brownish very quickly. Although, nutrient deficiency caused by various reasons are recognized with the plant colour changing, first to dark green and then to yellow or brown, but this mostly happen at an early stage in different parts of the plant, while BPH often affects leaves and chlorophyll content. Morphological plant issues which indicating infection with pests are accessible in two ways: the appearance of the canopy caused by the inner destruction of chlorophyll pigments and then, the tissue construction for photosynthesis and metabolism. Therefore, the unhealthy plants will possess different spectral structures compared to healthy ones. Remote sensing can be used for both discriminating the spectral contrast and recognising the unhealthy plants or patches in the field (19). Rice cultivation around the world could be undesirably affected by pests and diseases. There is a requirement for an agricultural study in both spatial and temporal terms to demonstrate the features and variability of the above mentioned problems. Hence, to address this issue, the existing investigation prepared a pest and disease database for rice by means of satellite imagery and remote sensing analysis in Malaysia. Furthermore, database administration and network knowledge, to detect sheath blight of BPH in the field, will be applied.

2. Detection of crop condition by remote sensing

Functional changes inside vegetation may be seen throughout drought periods. Satellite devices are capable to identify such changes using spectral radiance measurements and applying these records to vegetation indices that are sensitive to the proportion of plant growth, amount of growth and variations of vegetation in terms of moisture stress.

Satellite multispectral sensors that monitor the greenness of vegetation are generated by means of the visible and near infrared (NIR) bands. Reflection of the NIR band in stressed vegetation, which in visible band has no energy absorption, is less than in non-stressed ones. Consequently, for observing the influence of drought on vegetation in moisture stressed and standard crops, wavelengths are more appropriate.

For identifying vegetation situations and moisture stress in vegetation the temporal configuration of NDVI is beneficial. For example, a poor NDVI value reflects rainfall deficiency.
The most critical process and time-consuming scheme in information management-related studies is data collection. Based on a systematic analysis of the current materials, including published and unpublished literature in crop diseases and pests, relevant information was collected for crop diseases and pest classifications from different sources.

3. Methodology

3.1 Study area

Tanjung Karang consists of a small agricultural town sited 7 km north of Kuala Selangor, beyond Pasir Penambang. Most Malays are living in rural areas and are involved in agricultural activities, mainly rice cultivation.

Tanjung Karang Rice Irrigation Scheme is positioned at about 3°25′ to 3°45′N and 100°58′ to 101°15′E in the state of Selangor Malaysia. Rice is grown twice a year mostly from August to January (main or wet season) and February to July (off or dry season).

3.2 Image data collection

SPOT-5 images were collected within 101°12′42.82″N and 3° 28′49.88″E at upper left and also 101°13′49.71″N and 3°27′35.57″E at the lower right side. The images had four bands, in green: band 1 (530–600 nm); red: band 2 (610–680 nm), NIR: band 3 (780–1000 nm) and mid-infrared: band 4 (1000-2500nm). Most of the time, at the late growth stage BPH infection is greater. By this means, these dates of imagery were selected which are in both growth stages for comparison. The images were kept as 8-bit digital number (DN) values alternating from 0 to 255. Spatial resolution of the images was 2.5 m.

3.3 Image processing to extract data

Image data were extracted by help of ENVI 4.8 for producing subset captures of the field from the SPOT-5. Initially, BPH sheath blight through the paddy field was detected to identify unhealthy areas; the ENVI spectral utility was applied for data extraction from the subset images. To reduce probable bias, 4-pixel and 8-pixel were verified for arrangements of data extraction for evaluation. Additionally, the 8-pixel scheme was used in the study to extract data.

To apply change detection and threshold methods on multi temporal remotely sensed images, these must first be geometrically registered and radiometrically normalised. Converting DN to radiance was completed and SPOT gain and offset was determined. Among the processes, masking the clouds through subset images was performed.

Figure 1. SPOT-5 Subset Image.

3.4 Ground Observation

The ground orientation points were estimated through the field trip to the survey zones in the late season, by visiting the farms. Ground observation data was collected on 26 March in the Tanjung Karang area. These factors were used to address the locations where crops had been detected as attacked by BPH, due to SPOT image comparisons.

3.5 Process development to discover disease

Useful and appropriate indicators were identified to improve the technique for image processing. A number of indices such as Normalized Difference Vegetation Index (NDVI), Standard difference indices (SDI) and Ratio Vegetation Index (RVI) were used to isolate infected plants from healthy ones.

3.6 Image index computation

3.6.1 Normalized Difference Vegetation Index (NDVI)

NDVI is an index which provides a measure of vegetation density and condition. It is influenced by the fractional cover of the ground by vegetation, the vegetation density and greenness, and it indicates the photosynthetic capacity of the land surface cover. Its value is always between -1 and +1. Vegetation NDVI in Australia typically ranges from 0.1 to 0.7, with higher values associated with greater density and greenness of the plant canopy. NDVI decreases as leaves come under water stress, become diseased or die. Bare soil and snow values are close to zero, while water bodies have negative values.

The calculation of NDVI for an expected pixel always results in a number that ranges from minus one (-1) to plus one (+1); nevertheless, no green leaves provide a value near to zero; it means that if there is no vegetation the NDVI is shown at zero and if the area holds the maximum likely density of green leaves it is close to +1 (0.8 - 0.9).
The formula for NDVI is:

$$NDVI = (NIR - RED)/(NIR + RED) \quad (1)$$

3.6.2 Ratio Vegetation Index (RVI)

The RVI is a modest vegetation index, and is more than 30 years old. The definition of RVI is:

$$RVI = NIR/RED \quad (2)$$

RVI distributes the near infrared reflectance values by the visible red reflectance values. The ratio vegetation index is least influenced by soil brightness at LAI greater than three.

3.6.3 Standard difference indices (SDI)

The standard difference index is based on the fact that vegetation condition is closely related to environmental conditions. It displays the effects of climate and insects on vegetation over relatively long periods of time. Low SDI values show poor vegetation condition which could be the result of environmental conditions. Moisture shortage, pests and extreme temperature can cause low values.

The SDI can be expressed as follows:

$$SDI = (b_2 - b_p)/(b_2 + b_p) \quad (3)$$

- Where P is the range of bands 1, 3 and 4, and b2 refers to band 2; SDI is the standard difference index. The aim of this part of the study was to look for improved indicators for detecting rice sheath blight via remote sensing, by means of these indices.

3.7 Method development to detect disease

Appropriate indicators were designated to improve the technique for image processing. Various indices such as RI14, SDI14 and SDI24 show better correlations for detecting sheath blight in rice (20).

The following procedures have been used to develop the method:

Step 1: Processing the images to recover indices RI14, SDI14 and SDI24.

Step 2: The indices have dissimilar measurement components; it is necessary to convert their values into the same scheme for statistical analysis.

In this study the first stage for the conversion was to apply the following formulae:

$$DRI14 = \frac{RI\,14 - MRI\,14}{SRI14} \quad (4)$$

$$DSDI14 = \frac{SDI14 - MSDI\,14}{S\,SDI14} \quad (5)$$

$$DSDI24 = \frac{SDI24 - MSDI\,24}{S\,SDI24} \quad (6)$$

$$DNDVI = \frac{NDVI - MNDVI}{S\,NDVI} \quad (7)$$

- In which DSDI14, DRI14, DNDVI and DSDI24 are standardizations of RI14, SDI14, NDVI and SDI24; MRI14, MSDI14, MNDVI and MSDI24 are average of the indices; SSDI14, SRI14, SNDVI and SSDI24 are the standard deviations.

Calculating a new index as a mean of the three designated indices relevant to the formulae is shown below:

$$SRI = \frac{DRI14 + DSDI14 + DSDI24 + NDVI}{4} \quad (8)$$

- Where SRI refers to Significant Ratio Index. The upper value of SRI shows the greater disease infection rate of the pixel.

For testing the correlation between indices we computed T-test values, using SPSS software.

4. Results & Discussion

Unhealthy plants and soil apparently have a spectrum that is noticeably different from the healthy ones; the soil had the highest reflectance in the visible range, and the healthy plants had the lowest. Reflectance for diseased plants was among the reflectance interpretations of soil and healthy ones. Nevertheless, a spectral variance between the infected plants and the healthy ones was minor in identical ranges of the wavelengths. (19).

Table 1 shows the results of statistical analysis using indices, which have been used for differentiating healthy plants from unhealthy ones.

Table 1. Correlation analysis of indices.

Image indices	T-test[a]
SDI_{14}	-7.47^{***}
SDI_{24}	-6.29^{***}
RVI_{14}	-7.44^{***}
NDVI	8.89^{***}

*** Significant at $\alpha = 0.001$ ^{ns} non-significant

IOP Conf. Series: Earth and Environmental Science **20** (2014) 012015 doi:10.1088/1755-1315/20/1/012015

(Table 1) shows that all the indices are significant at $\alpha = 0.01$ and specified definitions to recognize infected plants were assigned. By considering the table results, the NDVI index indicates better correlation compared to the rest.

Statistical analysis were conducted for testing sample based on another set of data derived from ten points on the three different dates; Results were illustrated in following table which shows that NDVI index proves better correlation compared to the others, moreover, similar finding to the training data set was found.

Table2. Correlation analysis of indices for 10 points.

Image indices	T-test[a]
SDI_{14}	-5.76***
SDI_{24}	- 6.89***
RVI_{14}	-5.80***
NDVI	13.24***

*** Significant at α =0.001 ns non-significant

Examination of the association between the disease indices indicated that both band 3 (near infrared) and band 4 (mid infrared) have a relatively high correlation. A similar analysis was extended to a further three images, which generated identical results. Consequently, it was necessary to carefully observe the image indices for further investigation. Four types of image indices were inspected here: ratio vegetation indices, standard difference indices and a normalized difference vegetation index. Rice spectral signs would be closer to the spectra of healthy plants while it had low levels of infection. However, differences between unhealthy plants and healthy ones could be detected, but recognition of infected ones from the others is only possible while the infection levels are above a certain value with larger spectral reactions.

4.1Association examination for indices

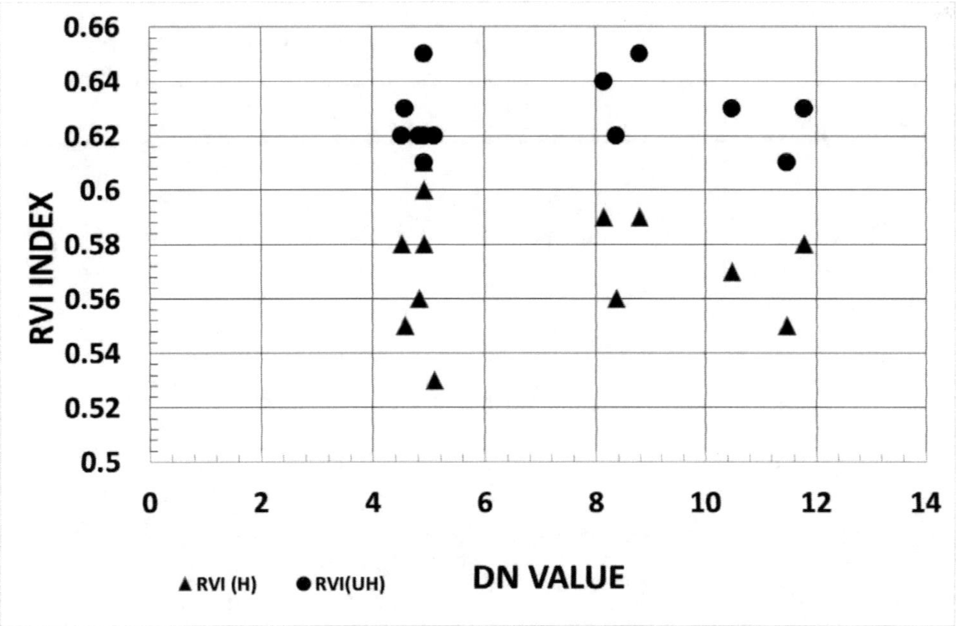

Figure 2. Correlation between ratio vegetation index and DN.

(Figure 2) demonstrates superior relationships among the ratio indices, healthy and diseased plants for the image of 9 March, 2012. It indicates that the amount of RVI_{14} for unhealthy plants tend to increase.

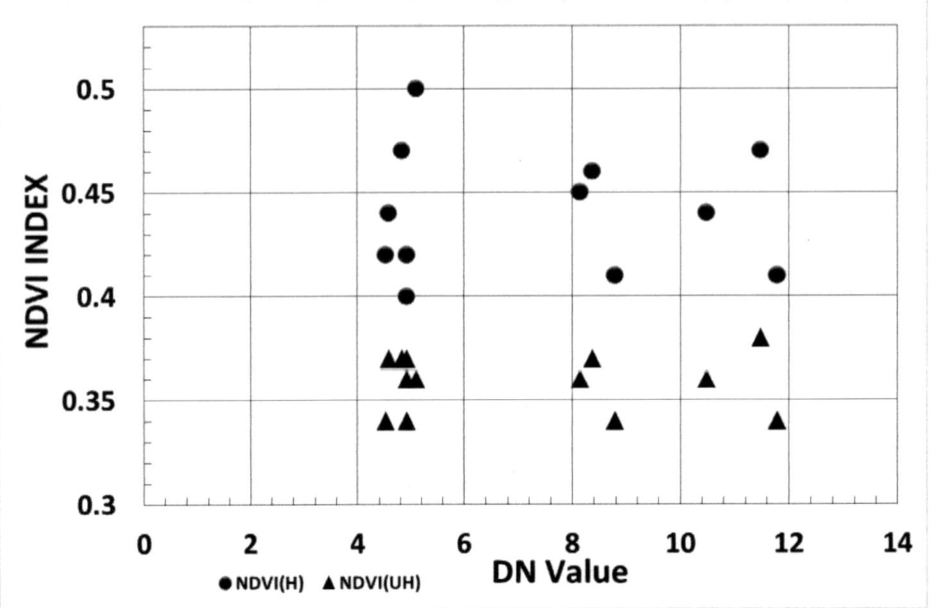

Figure 3. Correlation between normalized differences vegetation index and DN.

(Figure 3) indicates a significant correlation between infected and uninfected plants for the image from 9 March, 2012. It shows that a higher amount of NDVI belongs to healthy plants.

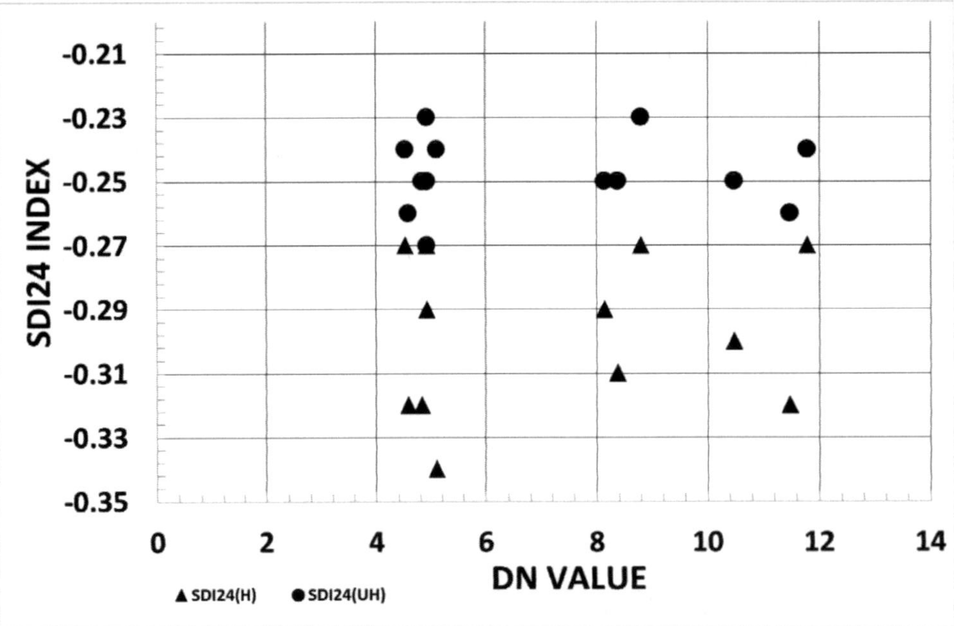

Figure 4. Correlation between standard differences index and DN.

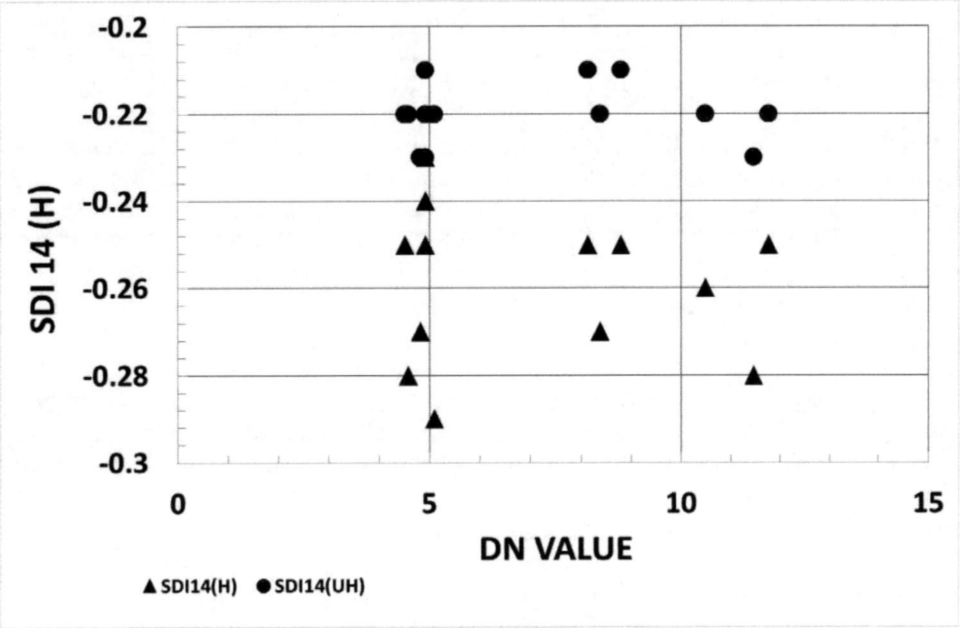

Figure 5. Correlation between standard differences index and DN.

(Figures 4 & 5) display connections between diseased plants and the standard differences index for the image from 9 March 2012.

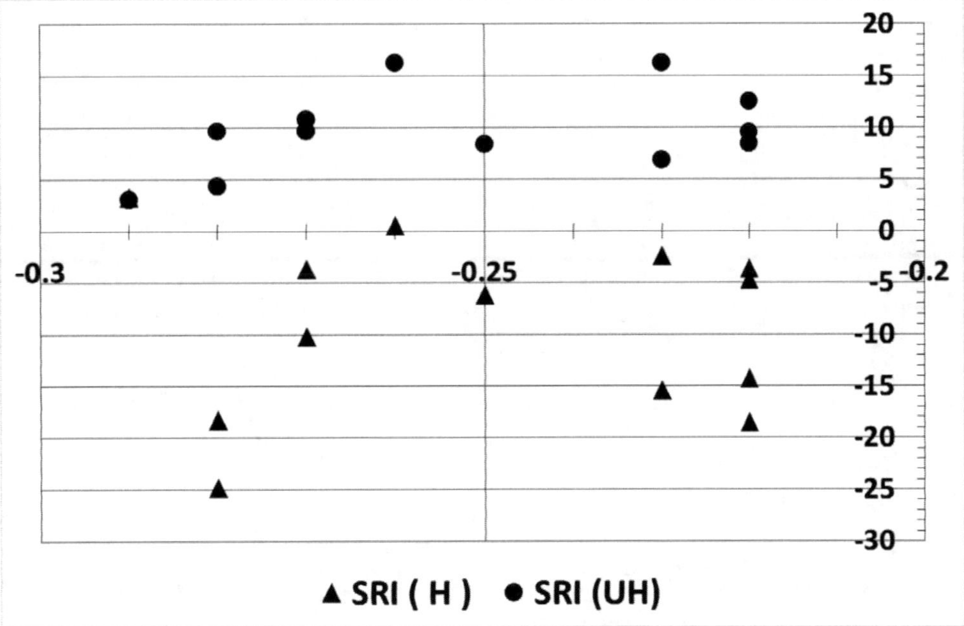

Figure 6. Computing a new index as an average of the four selected indices.

An extensive image index was established on the foundation of the selected indices SDI24, RI14, NDVI and SDI14 to organize the images (Figure 6). The images were classified into three types:

healthy to low infection is indicated by the point with RSI < 0, Class 2 refers to low to medium with RSI between 0 and 5, and Class 3 refers to severe infection with RSI > 5. The accuracy of the organization was ~76 % which was equivalent to other's investigations (20).

This misclassification was basically from class 2, containing some low infection samples, which in action may profit pest management in fields to control the disease at early stage. Results showed that there were dissimilarities among the images between early and late growing seasons. Observing particular image indices can conclude that NDVI and SDI_{24} indicated better correlations for detecting healthy plants from diseased ones. These types of index could be valuable indicators for a developed technique to detect the sheath blight of rice via remote sensing.

The image demonstrated the spatial spreading of the disease and displayed the likely infection enlargement in the ground. The eastern part of the rice ground was considerably more severely infected by sheath blight than its western half, but overall, infection is spread across the field. Moreover, we observed that the plants beside the levees had the highest levels of infection. This conclusion specifies a variation of the infection and disease growth beside levees. Irrigation and scheme channels for the field are served by means of levees. These probably lead to better conditions for the disease to spread spatially in the field, due to the fact that water constructive organization is suitable for the BPH improvement (11, 16).

With the knowledge that the late growth stage of plant has more severe BPH attacks, and by means of T-test values, the results concluded that 9/10 March had a more severe infestation of the pest.

5. Conclusion

Diseased plants do not display an identifiable spectral variance from healthy plants over a wide range of visible wavelength (18); the SPOT images might not be able to precisely identify the severity of infection at these wavelengths. This infers that they are useful for remote sensing to detect crop disease but the spectral resolution is probably not sufficient to distinguish plants with light infections. In Japan, Yamamoto et al. (1995) utilized infrared thermal images for recognition of rice blast disease by visual examination. The existence of rice blast disease was noticeable in thermal images, but their study did not include quantitative analysis using appropriate indices for more precise identification. Our study showed that SPOT images could be used to demonstrate the coarse scattering of infected plants in the field by recognizing chlorophyll content and the degree of green colour in the leaves. Using the indices as an indicator can clarify the threshold for identifying healthy plants from unhealthy ones. However, a better spectral resolution imagery system is needed (narrower bandwidth and more bands particularly in the infrared range) and in particular, higher spatial resolution.

6. References

[1] Bailey K et al. (2001) Remote sensing analysis of wild rice production using Landsat 7 for the Leech Lake band of Chippewa in Minnesota. Photogram. Eng. Remote Sensing. 67 (2), 189–192.
[2] Cline B (1970) New eyes for epidemiologists, aerial photography and other remote sensing techniques. Am. Journal of Epidemiol. 92, 85– 89.
[3] Fang H (1998) Rice crop area estimation of an administrative division in China using remote sensing data. International Journal of Remote Sensing. 19 (17), 3411–3419.
[4] Fang H (1998) Using NOAA AVHRR and Landsat TM to estimate rice area year-by-year. International Journal of Remote Sensing. 19 (3), 521–525.
[5] Hay S (1997) Remote sensing and disease control: past, present and future. Trans. R. Soc. Trop. Med. Hygiene 91, 105–106.
[6] Hong S et al (1997) Radiometric estimates of grain yields related to crop aboveground net production (ANP) in paddy rice. In: Proceedings of the International Geoscience and Remote Sensing Symposium, 4. pp. 1793–1795.

[7] Hong S et al (1997) Remote sensing for estimating chlorophyll amount in rice canopies. In: Proceedings of the International Geoscience and Remote Sensing Symposium, 1. pp. 89–91.

[8] Inoue Y (2000) Hyperspectral and directional remote sensing measurements of rice canopies for estimation of plant growth variables. In: Proceedings of the International Geoscience and Remote Sensing Symposium, 4. pp. 1471–1473.

[9] Kurosu T et al (1997) Identification of rice fields using multi-temporal ERS-1 C band SAR data. International Journal of Remote Sensing. 18 (14), 2953–2965.

[10] Maskova Z (2008) Normalized Difference Vegetation Index (NDVI) in the management of mountain meadows. Boreal Environmental Resort. 13: 417 – 432.

[11] Nandakumar R (2001) Induction of systemic resistance in rice against sheath blight disease by Pseudomonas fluorescens. Soil Biol. Biochem. 33, 603–612.

[12] Okamoto K, Kawashima H (1999) Estimation of rice-planted area in the tropical zone using a combination of optical and microwave satellite sensor data. International Journal of Remote Sensing. 20 (5), 1045–1048.

[13] Ribbes F, Le Toan T (1999) Rice field mapping and monitoring with RADARSAT data. International Journal of Remote Sensing. 20 (4), 745–765.

[14] Shao S (2001) Rice monitoring and production estimation using multitemporal RADARSAT. Remote Sensing of Environment. 76 (3), 310–325.

[15] Shibayama M (1993) Canopy water deficit detection in paddy rice using a high resolution field spectroradiometer. Remote Sensing of Environment. 45 (2), 117–126.

[16] Vidhyasekaran P (1997) Host specific toxin production by Rhizoctonia solani, the rice sheath blight pathogen. Phytopathology 87, 1258– 1263.

[17] Yamamoto, H., Suzuki, Y., Iwano, M., Hayakawa, S., 1995. Remote sensing of occurrence place of rice blast disease by infrared thermal image. Japanese Journal of Crop Science. 64 (3), 467–474.

[18] Zhang M (2002) Spectral discrimination of phytophthora infestans infection on tomatoes based on principle component and cluster analyses. International Journal of Remote Sensing. 23 (6), 1095– 1107.

[19] Zhang M (2003) Detection of stress in tomatoes induced by late blight disease in California, USA, using hyperspectral remote sensing. International Journal of Apply Earth Observation and Geoinformation. 4, 295–310

[20] Zhihao Q, Minghua Z (2005) Detection of rice sheath blight for in-season disease management using multispectral remote sensing, USA, using hyperspectral remote sensing. International Journal of Apply Earth Observation and Geoinformation. 7 (2005) 115–128.

IOP Conf. Series: Earth and Environmental Science **20** (2014) 012016 doi:10.1088/1755-1315/20/1/012016

Spot-5 multispectral image for 60-75 days of rice mapping

Mohd Amiruddin Ramli[1], Abdul Rashid Mohamed Shariff [2], and Siti Khairunniza Bejo[2]

[1]Department of Agriculture Peninsular Malaysia
[2]Department Of Biological and Agicultural Engineering, Faculty Of Engineering,University Putra Malaysia, 43400, Serdang, Selangor, Malaysia

Email:ultimatekeylogger2010@gmail.com

Abstract. The objective of this study is to investigate the potential application of Spot-5 multispectral satellite data in monitoring rice cultivation areas in IADA (Integrated Agriculture Development Area) located at Kerian District, Perak Malaysia. Information of the rice cultivation areas is a global economic and environmental significance. Multi-spectral images acquired at high spatial resolution are an important tool, especially in agricultural applications. This paper addresses the relationship between normalize difference vegetation index (NDVI) and ancillary data acquired from Farmers Organization Authority (PPK) for 217 farmer's field in IADA Kerian. The results indicated that NDVI range 0.62 - 0.75 has a strong positive relationship with the ground survey area estimation with ($r = 0.85$; $p < 0.01$) ($r^2 = 0.722$). The r^2 value of 0.722 indicated a statistically significant linear relationship between the rice area estimate using NDVI range 0.62 - 0.75 and on the ground surveyed data for 217 farmers' fields. The equation of unstandardized distribution can be described as $\hat{Y}=0.\ 0197+0.852x$. The equation for standardized regression formula for this distribution is $\hat{Y}= 0.850x$. Thus, the results indicate that 60-75 days of rice area can be estimated from the following equation $\hat{Y}=0.\ 197+0.852x$, where $\hat{Y}$ is the predicted rice area and x is area calculated using NDVI range 0.62-0.75 in IADA Kerian Perak Malaysia. The results appear promising and rice mapping operations using SPOT-5 multispectral image data can be foreseen.

1. Introduction

Rice (*Oryza Sativa L.*) attains second place among cereal crop, both for acreage and production in the world. Rice forms the economic, cultural and nutritional basis for many countries, making this crop's successful annual production of key significance to the world community. Recent projections made by the International Rice Research Institute (1998) show that demand for rice will increase by about 1.8% per year until 2025 [1]. Malaysia has experienced, in recent years, achieving self-sufficiency in rice production at about 65% -70% of the local consumption. Since Malaysia does not have the comparative advantage in rice production, it implements a wide range of market interventions to achieve the intended level of rice production [2]. In Malaysia, Granary areas refer to major irrigation scheme which is the area is greater than 4,000 hectares and recognized by the Government in the National Agricultural Policy as the main paddy producing areas. There are 8 Granary areas in Malaysia, namely MADA, KADA, IADA Kerian, IADA Barat Laut Selangor, IADA Pulau Pinang, IADA Seberang Perak, IADA Ketara and IADA Kemasin Semerak [17].

The main season is the period when the paddy is grown without depending wholly on any irrigation system. For administrative purpose, the main season is defined as the commencement date of paddy

Content from this work may be used under the terms of the Creative Commons Attribution 3.0 licence. Any further distribution of this work must maintain attribution to the author(s) and the title of the work, journal citation and DOI.
Published under licence by IOP Publishing Ltd

planting between 1st August to 28/29 February of the following year. Off season is the dry period and paddy planting normally depends upon the irrigation system. For administrative purposes, Off season is defined as the commencement date of paddy planting between 1st March to 31st July of the year. The definition of paddy planting is sowing for direct seeding system and transplanting for nursery system. The definition of a paddy parcel is a piece of land, which is usually used for planting Paddy meanwhile planted paddy area is a portion or might be full of the paddy parcel. The planted rice area in Malaysia was reduce -9.34% from 680,650 ha in 1990 decrease to 617,028 ha in 2011 [16].

Remote sensing has long been recognized as a useful tool for acquiring rice information. Leaf pigments, particularly chlorophyll, absorb mainly the blue and red portions of the incoming solar energy spectrum, which can result in very low canopy reflectance in the visible region. Increase reflectance in the near infrared (NIR) region is related to increased crop biomass and leaf area Index (LAI) [3]. Thus, changes in canopy reflectance spectrum in the visible and near infrared regions should be associated with crop yield and area estimation [4]. Normalized difference vegetation index (NDVI), can reflect the growth of crops, and it has difference between difference crops in different phenology period [5]. The objective of this study is to investigate the potential application of Spot-5 multispectral satellite data in monitoring rice cultivation areas in IADA (Integrated Agriculture Development Area) located at Kerian District, Perak Malaysia.

2. Previous studies

In the recent years, many methods have been developed and applied in rice mapping. Some studies have used optical remote sensing data, such as the NOAA Advanced Very High Resolution Radiometer (AVHRR), Landsat Thematic Mapper (TM) data and SPOT data to identify paddy rice fields. Some studies use NDVI data derived from optical remote sensing data to analyze and develop an algorithm for mapping paddy rice fields such as previous research by [6]. A number of studies have shown that vegetation index (VI) can be used as an indicator of the status of vegetation and crops. There are many vegetation indices such as NDVI, EVI (enhanced vegetation index) and SAVI (soil – adjusted vegetation index), which have been used to discriminate crops and monitor crop growth conditions. Among these vegetation indices, the NDVI is the most widely used VI and other indices are its refined form. There is a consensus that NDVI can be used as an effective measure of photosynthetically active biomass and chlorophyll activity of vegetation and crop.

The optical multispectral image has been used as an ancillary data for rice parameter retrieval. In a previous research by [7], NDVI range 0.6152, 0.6748, 0.6852, 0.7102 and 0.7322 have been found as an average parameter for rice mapping in the middle of Jiangsu, Province China. However, studies by [6] found that the pixel with greater value than 0.65 in NDVI can be identified as 60 – 75 days of rice in Guangdong Province, China. Like most other crops, planting date varies among the rice fields. Mapping rice fields are a fundamental step for growth monitoring and yield prediction. The rice calendar is significantly affected by temperature differences due to latitude [1].

The originally normalized difference vegetation index (NDVI) is an effective vegetation measure since it is sufficiently stable to permit meaningful comparisons between seasonal and inter annual changes in vegetation growth and activity. This is because it can reduce different forms of multiplicative noise (illumination differences, cloud shadows, atmospheric attenuation and certain topographic variations) [8].

Traditionally, rice growth monitoring methods based on field checking are still used for collection or rice paddy production information. However, the result is that accuracy of the collected data is in doubt and limited in local scale, particularly that concerning planted or harvested area [7]. This research aims to perform further analysis to see the relationship between NDVI range > 0.62 and the rice age 60-75 days on the ground and also the number of NPK fertilizer use by rice farmers in every rice parcel.

2.1. NPK fertilizer

Different rates of Nitrogen (N), phosphorus (P) and potassium (K) (NPK), in intensive rice monoculture do have the effect of healthiest rice growth. Soil and fertilizer management is very complex and dynamic in nature. Fertilizer recommendation for crops in a cropping pattern needs change over a certain period of time. With the advancement of technology and with a progress fertility and fertilizer management research in the country, there has been a continuous need for updating the Fertilizer Recommendation guide. The application of fertilizer in a proper amounts must be done to boost up agricultural production to an economically describe level [9].

The result of mean data of two years showed that paddy yield, increase in yield, number of grains per panicle, 1000 grain weight, gross return (GR), value cost ratio (VCR) and a grain nutrient ratio (GNR) were significantly influenced by different NPK levels [9]. Improved management of nitrogen (N) in low fertile soil is critical for increased land productivity and economic sustainability [10].

It is known that fertilization is crucial for maintaining soil available nutrient levels, because fertilization ensure the largely constant presence of active microorganisms and the regular dynamic of biomass carbon [11]. The increase in plant height, number of tillers per hill, spikelet number per panicle, grain yield and 1000-grain weight in response to the application of organic and chemical fertilizers is probably due to enhanced availability of nutrients. The variation in plant height due to nutrient sources was considered to be due to variation in the availability of nutrients [12].

In a previous previous research by [12] have found that using 1.5 ton/ha organic fertilizer treatments along with 50% recommended chemical fertilizer like NPK could give a similar yield of rice. From the economic point of view farmers can use the combination of organic fertilizer and reduce the rate of inorganic fertilizers to boost the yield of rice as well as maintain and improve soil health.

To sustain food security, the land-scarce Asian countries would need to increase fertilizer intake to over 250 kg of NPK/ha by the year 2020 from the present level of less than 100kg/ha [13]. The increase in demand for fertilizer is, however, constrained by the efficiency of nutrient uptake of the crop, unstable supply of soil moisture, the risk of investments due to fluctuation in yield from climatic variations, and fertilizer-grain price ratio. The intensity fertilizer-use is higher for modern varieties of cereals compared to traditional ones, on irrigated farming compared to the rain-fed, and on well-drained land with medium elevation than on lands that are subjected to droughts and floods.

3. Methodology

Rice mapping with such imagery is often challenging where paddy rice grows in a warm, humid season with frequent cloud cover and heavy precipitation. Although cloud-free optical imagery could be more easily acquired in the dry season, the complicated farming practices make it less optimal in rice mapping [14]. In this research, statistical regression and correlation analysis were chosen, both area estimation for 217 farmer's fields, using threshold expert knowledge classification NDVI range > 0.62 and on the ground area estimation for rice age 60-75 days, also a number of NPK fertilizer (17:3:25:2MgO) use by farmers were used to evaluate their relationship in rice mapping. Statistical software SPSS v21 was used in this study.

3.1. Study area

The study area is located at Kerian district in Perak, Malaysia approximately (5° 2'20.86"N 100°34'0.71"E, 5° 2'18.79"N 100°42'13.96"E, 4°55'35.32"N 100°42'12.09"E, 4°55'39.83"N 100°34'8.02"E). It is one of the rice growing areas in the state. Its local government is Majlis Daerah Kerian or Kerian District Council, which is based at Parit Buntar. From the 2010 population and housing census, Kerian has a population of 173,625.

The major town is Parit Buntar, which is located near the border of Penang and Kedah States. Other towns in Kerian are Bagan Serai, Kuala Kurau, Kamula Sarji and Tanjung Piandang. Sungai

Kerian is the Major river flowing through this district. The majority of the people here are Malays, with a sizeable population of Banjar People. Integrated Agriculture Development Area (IADA Kerian) has 25 Sections. In this study, we will focus on 217 farmer's field at Kerian Section 2 which is reported planted by rice age 60-75 days of rice age during 05/06/2012.

3.2. Existing cadastral maps

This includes the existing 217 farmer's field cadastral maps for extracting information on land parcel and their boundaries. These maps were requested from department survey and mapping to analyze with image analysis and integrate with ancillary data from farmers organization authority.

3.3. Remote sensing (RS) imagery

This includes the existing 217 farmer's field cadastral maps for extracting information on land parcel and their boundaries. The NDVI is the best indication factor for vegetation cover and grow stage. Traditional study mainly introduces vegetation index, especially normalize different vegetation index (NDVI) to extract, rice from remote sensing image [15]. There is a consensus that the NDVI can be used as an effective measure of photosynthetically active biomass and chlorophyll activity of vegetation and crop. In addition to serving as an indication of the "greenness" of the vegetation, the NDVI is also able to offer valuable information on the dynamic changes of crop groups. The principle of applying NDVI in crop mapping is that the crop is highly reflective in the near infrared and highly absorbed in the visible green. The contrast between these channels can be used as an indicator of the status of vegetation and crop. Consider the other vegetation indices has different calculation based on certain crop and soil conditions, the NDVI and soil conditions, the NDVI is employed to analyze spectral of rice and other crops in the study area. The NDVI can be calculated as follows:

$$NDVI = R_{nir} - R_{red} / R_{nir} + R_{red} \quad (1)$$

The equation (1), where R_{nir} and R_{red} represent reflectance of near infrared and red bands, respectively, which correspond to one and two bands for multispectral SPOT-5 level 1A 5/6/2012 images for Kerian district in Perak Malaysia. Before transplanting, NDVI has low values of 0.11 for bare soil. During the transplanting period, rice fields are flooded, which makes the NDVI value decrease to about 0.02 after 10 days after transplanting. After transplanting, the NDVI values increase rapidly with the growth of rice and reaches a peak of 0.73 during the ear differentiation period about 63 days after transplanting.The NDVI values decrease after the holding period with decrease in water content and photosynthesis activities. After harvesting, rice field turned into bare soil with an NDVI value of 0.09. For other ground types, the NDVI of water and fish pond varied little over the whole period with values about – 0.43 and - 0.26 because of the absorption in the near infrared [6].

3.4. Rice field mapping algorithm

SPOT-5 multispectral high resolution visible level 1A data at study area has been requested from Malaysian Remote Sensing Agency with radiometric and geometric correction. The key to mapping rice fields is to find proper remote sensing data combinations. The above NDVI profiles can suggest that the NDVI can be used as a possible factor to discriminate rice fields from other ground types in the study area though their unique genealogy and dynamic profiles of NDVI. Based on the above analysis, an algorithm for mapping rice fields was proposed as follows.

(i) First, SPOT-5 HRV multispectral images are processed for removing cloudy areas, if the study areas infected with cloud cover. There are some cloud removal methods which have been used for TM, MODIS and SPOT Images.
(ii) The mask is made to exclude water areas and fish pond with NDVI values below 0 in the NDVI images.

(iii) Other ground types are detected and excluded using masks with the NDVI values more than 0.0 – 0.62
(iv) The pixels with a value greater than 0.62 in the images are identified as rice fields.

Remote Sensing can help predict the 60-75 days of rice field area by using the NDVI images integrated with cadastral data of this study area. This is by means, all of the rice field area can be integrated with cadastral information and thus information on every pixel classification in NDVI can help discriminate rice with other ground types. In this study, the result of the NDVI value > 0.62 will be classified as rice field and will then be intersected and clip with the cadastral and thus can be extracted to calculate the geometry of the rice granary area in the cadastral. Threshold and expert knowledge classification for rice age 60-75 days were then having performed using threshold and expert knowledge classification. At last, accuracy assessment will then be performed using statistical correlation and regression analysis with on the ground data surveyed by the farmers organization authority.

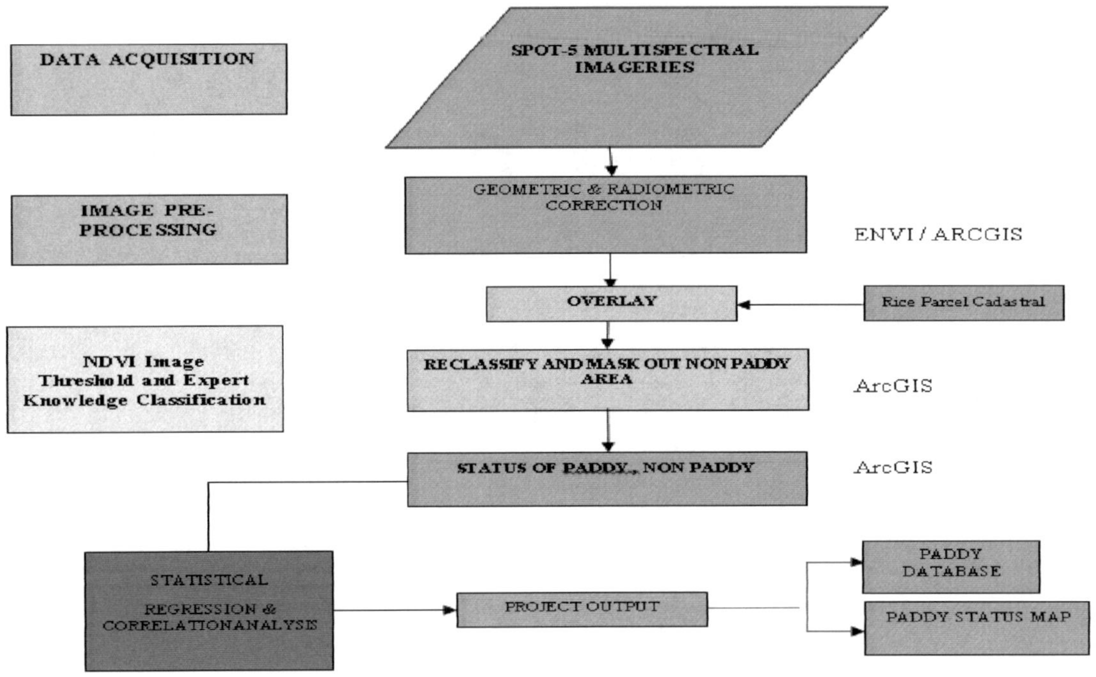

Figure 1. Rice field mapping algorithm.

4. Results and discussion

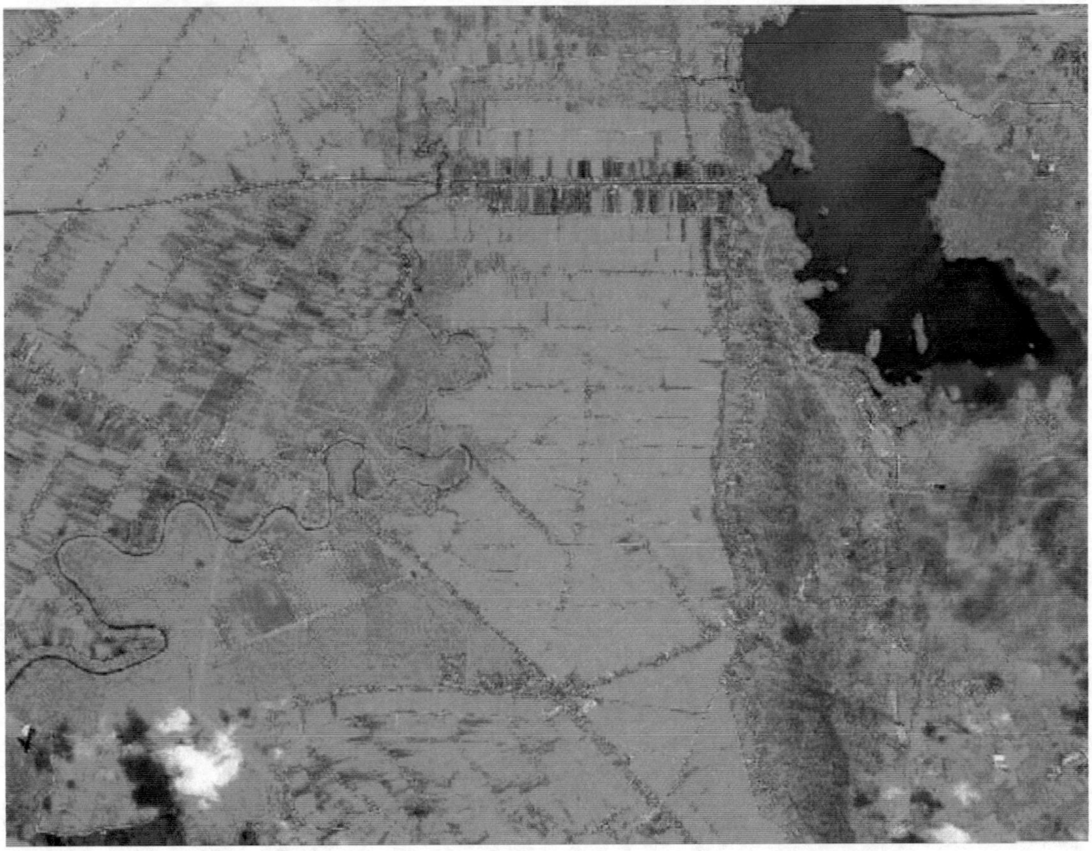

Figure 2. SPOT-5 image acquired 05/06/2012 3:1:2 band combination.

The offer of high spectral and high spatial resolution images has grown in the last decade. It is now possible to obtain data from different sources with different spatial and spectral resolution. In this study SPOT-5 image acquired from the Malaysian remote sensing agency. This data product name is level 1A and the acquisition date is 5/6/2012. The Sun Azimuth degrees is 46.34 Sun Elevation degrees is 63.61 and off-nadir degrees is 24.64. Wavelength (NM) is contained in this data is near infrared is 780-890, Red is 610-680, green is 500-590 and short wave Infrared is 1580-1750 nm. The pixel size is 3.627173 x 3.627173 meters as confirmed with Envi version 5 software in pixel information.

Figure 3. NDVI image.

The NDVI is the best indication factor for vegetation cover and grow stage. Traditional study mainly introduces vegetation index, especially normalize different vegetation index (NDVI) to extract, rice from remote sensing image [15]. Reflectance of near infrared and red bands, respectively, which correspond to one and two bands for multispectral SPOT-5 level 1A 5/6/2012 images for Kerian district in Perak Malaysia.

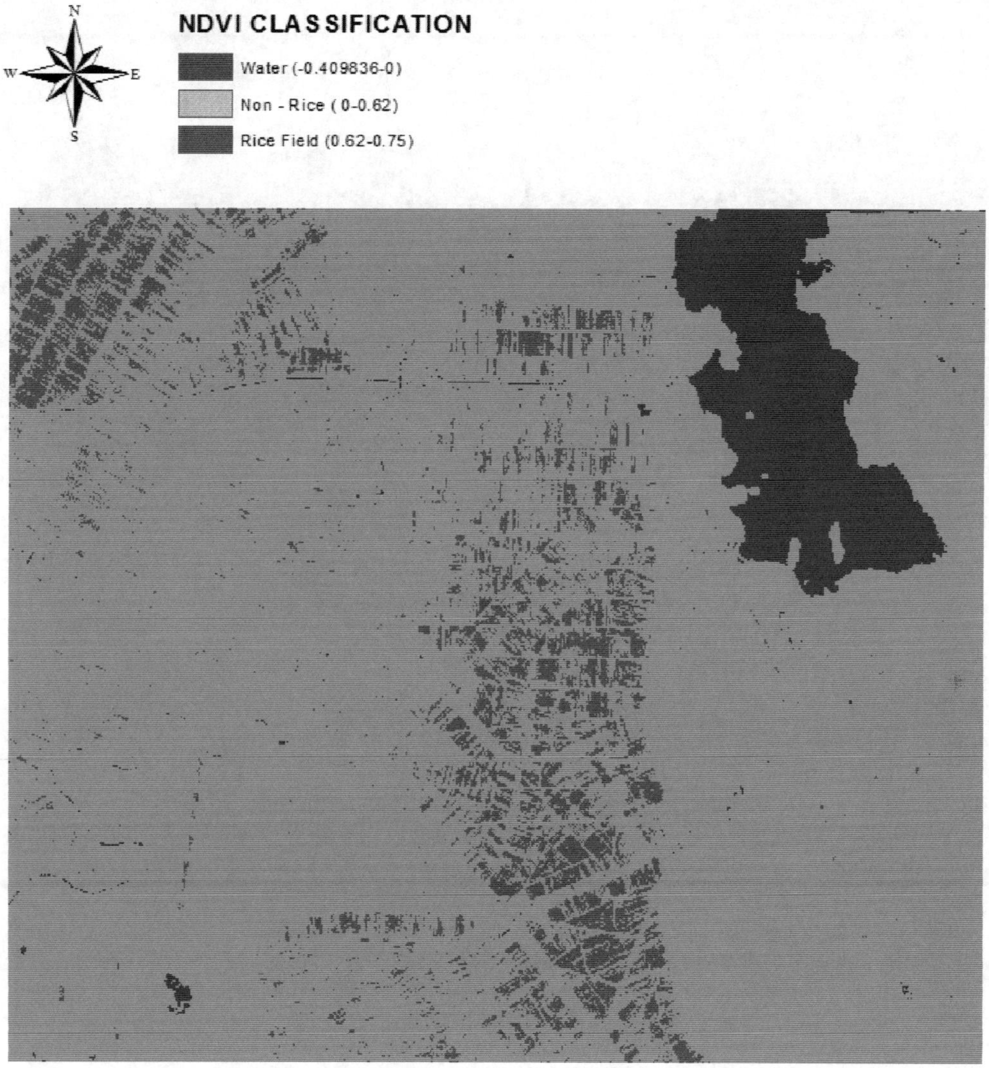

Figure 4. Rice age 60-75 days / non-rice map.

A good quality image of the study area has been developed using threshold and expert knowledge classification in NDVI images. Rice field age 60-75 days were categorized as (0.62 – 0.75), Non- Rice is (0-0.62) and Water is (-0.409836 – 0).

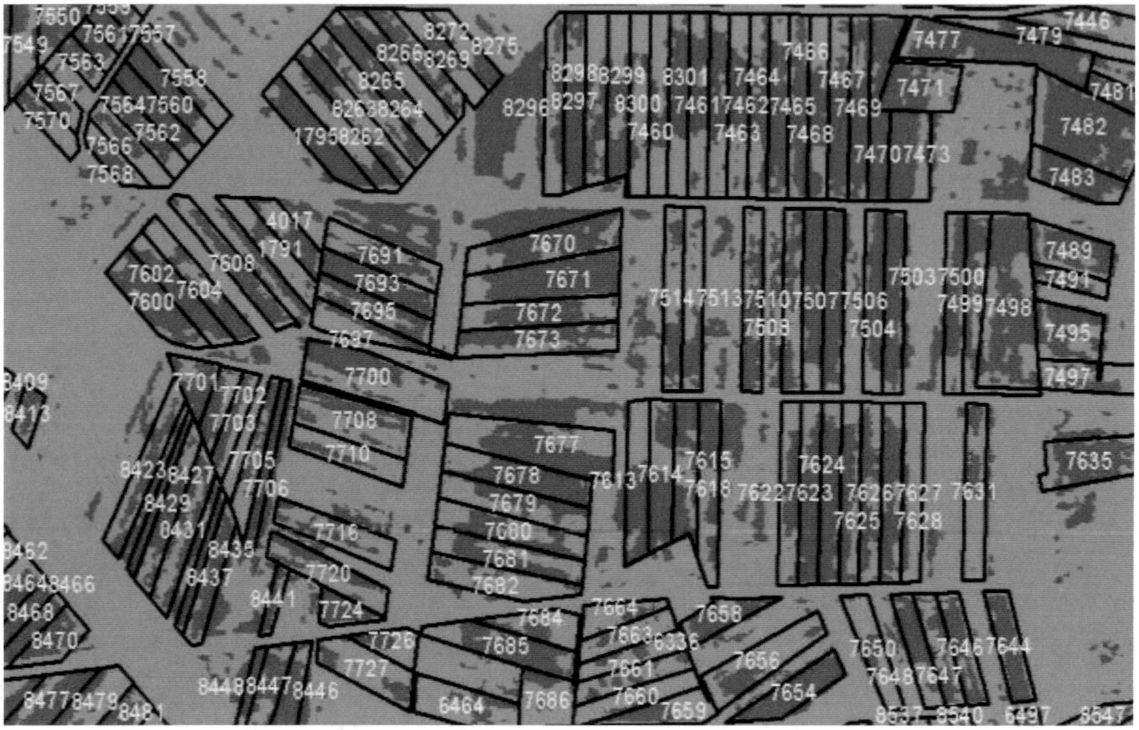

Figure 5. Rice cadastral parcel.

A good quality of 217 cadastral shapefile for the study area was acquired from department of survey and mapping. Cadastral shapefile were then overlaid with the NDVI classification image and the raster split tool were performed to split the raster for 217 parcels. The raster split tool is a new Arc Toolbox command implemented by importing a toolbox (. tbx) file into the users' standard Arc Toolbox. This is accomplished by right clicking on the standard Arc Toolbox and selecting "Add Toolbox". The purpose of the tool is to automate the process of breaking up large rasters into a small pieces which may be desired such as countries, watershed, etc. and allowing the multiple output rasters to be named according to an attribute in the split shapefile. Without automation this process can be quite time-consuming and tedious. After this process 217 split raster classification was then calculated for the number of rice pixel which is valued 0.62 - 0.75 to calculate the hectares of rice in every parcel. Statistical analysis was then being performed using SPSS v21.

4.1. Correlation analysis

Ho: There is no positive relationship between the Rice area estimate using NDVI range 0.62-0.75 for Rice Age 60-75 days and Rice area estimate on the ground of Farmers Organization Authority and Number of NPK Fertilizer use by farmers

Ha: There is a positive relationship between the Rice area estimate using NDVI range 0.62-0.75 for Rice Age 60-75 days Rice area estimate on the ground by Farmers Organization Authority Number of NPK Fertilizer use by farmers

Correlations

		Rice Planted Area Estimation using NDVI range 0.62-0.75 (Ha)	Rice Planted Area Estimation On The Ground by Farmers Organization Authority (Ha)	Number Of NPK 17:3:25:2MgO Used by Farmers 25Kg/ (Bag)
Rice Planted Area Estimation using NDVI range 0.62-0.75 (Ha)	Pearson Correlation	1	.850**	.851**
	Sig. (2-tailed)		.000	.000
	N	217	217	217
Rice Planted Area Estimation On The Ground by Farmers Organization Authority (Ha)	Pearson Correlation	.850**	1	.994**
	Sig. (2-tailed)	.000		.000
	N	217	217	217
Number Of NPK 17:3:25:2MgO Used by Farmers 25Kg/ (Bag)	Pearson Correlation	.851**	.994**	1
	Sig. (2-tailed)	.000	.000	
	N	217	217	217

**. Correlation is significant at the 0.01 level (2-tailed).

Figure 6. Correlation table rice area estimation using NDVI vs rice area estimation on the ground by farmers organization authority.

If $p>0.01$ we fail to reject Ho. Therefore, the NULL stands and the Ha are rejected. In this case we got $p<0.01$, therefore we reject Ho and we accept Ha. There is therefore a strong positive relationship between the Rice area estimate using NDVI range 0.62-0.75 for Rice Age 60-75 days and Rice area estimate on the ground of Farmers Organization Authority and Number of NPK Fertilizer use by farmers 25Kg/Bag/Ha.

4.2. Regression analysis

Ho: There is no significant relationship between the rice area estimate using NDVI range 0.62 -0.75 for rice age 60-75 days and rice area estimate on the ground by farmers organization authority.

Ha: There is a significant relationship between the rice area estimate using NDVI range 0.62-0.75 for rice Age 60-75 days rice area estimate on the ground by farmers organization authority.

Model summary

Model	R	R Square	Adjusted R Square	Std. Error of the Estimate
1	.850[a]	.722	.721	.23496

a. Predictors: (Constant), Rice planted area estimation using NDVI range 0.62-0.75 (Ha)

Figure 7. Regression model 1.

Coefficients[a]

Model		Unstandardized Coefficients		Standardized Coefficients	t	Sig.
		B	Std. Error	Beta		
1	(Constant)	.197	.032		6.107	.000
	Rice Planted Area Estimation using NDVI range 0.62-0.75 (Ha)	.852	.036	.850	23.658	.000

a. Dependent variable: Rice planted area estimation on the ground by farmers organization authority (Ha)

Figure 8. Coefficients model 1.

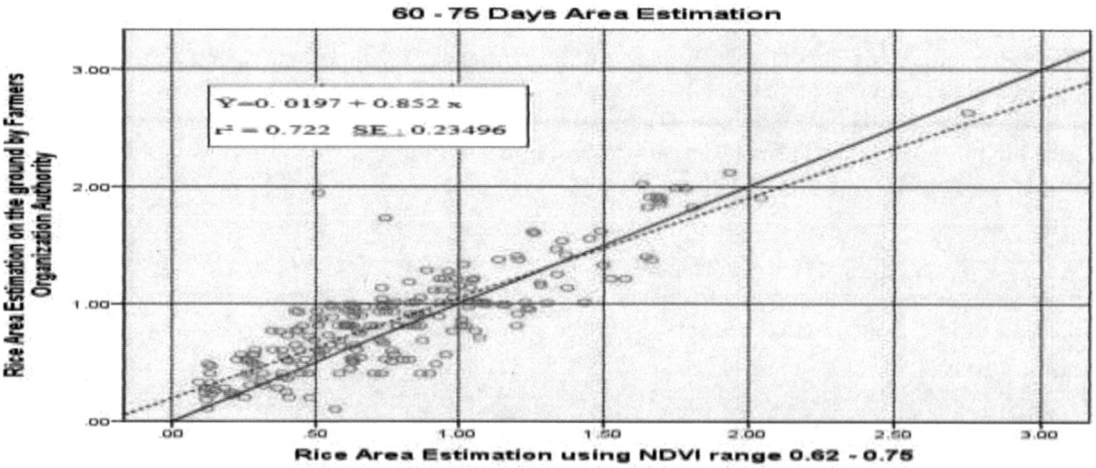

Figure 9. Scatter plot model 1.

If $p>0.05$ we fail to reject Ho. Therefore, the NULL stands and the Ha is rejected. In this case we got $p<0.05$, therefore we reject Ho and we accept Ha. Therefore, there is a significant relationship between the Rice area estimate using NDVI range 0.62-0.75 for Rice Age 60-75 days and Rice area estimate on the ground by farmers organization authority.

4.3. Regression analysis 2

Ho: There is no significant relationship between the Rice area estimate using NDVI range 0.62-0.75 for Rice Age 60-75 days and number of NPK Fertilizer 17:3:25:2 MgO use by rice farmers 25Kg/Ha

Ha: There is a significant relationship between the Rice area estimate using NDVI range 0.62-0.75 for Rice Age 60-75 days and number of NPK Fertilizer 17:3:25:2 MgO use by rice farmers 25Kg/Ha

Model Summary[b]

Model	R	R Square	Adjusted R Square	Std. Error of the Estimate
1	.851[a]	.725	.724	.23328

a. Predictors: (Constant), Number Of NPK 17:3:25:2MgO Used by Farmers 25Kg/ (Bag)

b. Dependent variable: Rice planted area estimation using NDVI range 0.62-0.75 (Ha)

Figure 10. Regression model 2.

Coefficients[a]

Model		Unstandardized Coefficients		Standardized Coefficients	t	Sig.
		B	Std. Error	Beta		
1	(Constant)	-.023	.037		-.630	.529
	Number Of NPK 17:3:25:2MgO Used by Farmers 25Kg/ (Bag)	.142	.006	.851	23.801	.000

a. Dependent variable: Rice planted area estimation using NDVI range 0.62-0.75 (Ha)

Figure 11. Coefficients model 2.

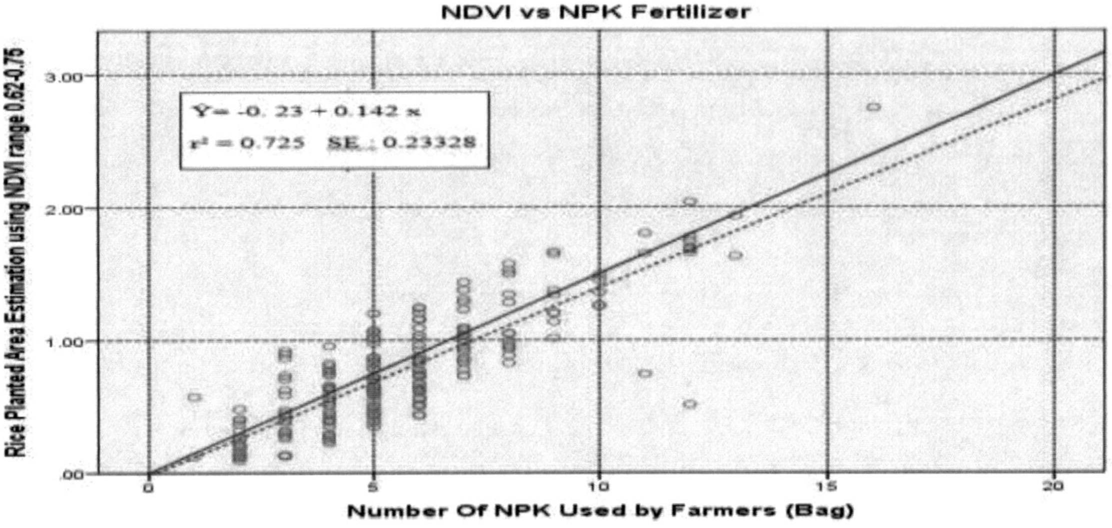

Figure 12. Scatter plot model 2.

If p>0.05 we fail to reject Ho. Therefore, the NULL stands and the Ha is rejected. In this case we got p<0.05, therefore we reject Ho and we accept Ha. There is therefore a significant relationship between the rice area estimate using NDVI range 0.62-0.75 for Rice Age 60-75 days and the number of NPK fertilizer use by farmers 25Kg / Bag / Ha.

5. Conclusion

Remote Sensing can help predict the 60-75 days of rice field area by using the threshold and expert knowledge classification in NDVI images integrated with cadastral data of this study area. This is by means, all of the rice field area can be integrated with cadastral information and thus information on every parcel and pixel classification in NDVI can help discriminate rice with other ground types.

Traditionally, rice growth monitoring methods based on field checking are still used for collection or rice paddy production information. However, the result is that accuracy of the collected data is in doubt and limited in local scale, particularly that concerning planted or harvested area [7]. Therefore, in this study regression equation can be used to predict the rice area in every rice parcel which is age 60-75 days.

The results indicated that NDVI range 0.62 - 0.75 has a strong positive relationship with the ground survey area estimation with (r = 0.85; p <0.01) (r^2 = 0.722). The r^2 value of 0.722 indicated a statistically significant linear relationship between the rice area estimate using NDVI range 0.62 - 0.75 and on the ground survey data for 217 farmers's fields.

The equation of unstandardized distribution can be described as $\hat{Y}=0.0197+0.852x$ (2). The equation for standardized regression formula for this distribution is $\hat{Y}= 0.850x$ (3). Thus, the results indicate that 60-75 days of rice area can be estimated from the following equation (2), where $\hat{Y}$ is the predicted rice area and x is area calculated using NDVI range 0.62-0.75 in IADA Kerian Perak Malaysia.

There is also a significant relationship between the rice area estimate using NDVI range 0.62-0.75 for Rice Age 60-75 days and the number of NPK Fertilizer use by Farmers 25Kg/Bag/ Ha. The result indicated that NDVI range 0.62 – 0.75 has a strong positive relationship with a number of NPK fertilizer use by farmers with (r=0.851; p<0.01) (r^2 = 0.725). The r^2 value of 0.725 indicated a statistically linear relationship between the rice area estimate using NDVI range 0.62-0.75 and the number of NPK 17:3:25:2MgO fertilizer use by farmers.

The equation of unstandardarized distribution can be described as $\hat{Y}=-0.23+0.142x$ (4). The equation for standardized regression formula for this distribution is $\hat{Y}= 0.851x$ (5). Thus, the results indicate that number of NPK fertilizer 17:3:25:2MgO can be estimated from the following equation (4), where $\hat{Y}$ is the predicted number of NPK Fertilizer use by farmers and x is area calculated using NDVI range 0.62-0.75 in IADA Kerian Perak Malaysia. The results appear promising and rice mapping operations using SPOT-5 multispectral image data can be foreseen.

Acknowledgements
The author would like to acknowledge and express our appreciation to Department of Agriculture, Integrated Agriculture Development Area Kerian, Farmers Organization Authority Kerian, Department of Survey and Mapping and Malaysian Remote Sensing Agency for their support and contribution towards our project.

References
[1] C. Chen and H. Mcnairn, "A neural network integrated approach for rice crop monitoring," *Int. J. Remote Sens.*, vol. 27, pp. 1367–1393, 2006.
[2] E. F. Alias, F. Mohamed Arshad, K. Mohd Noh, and M. Tasrif, "Food security: Self sufficiency of rice in Malaysia," in *National Conference on the Tenth Malaysia Plan: Transformation Towards a High Income Advanced Economy*, 2011.
[3] Y.-P. Wang, S.-H. Chen, K.-W. Chang, and Y. Shen, "Identifying and characterizing yield limiting factors in paddy rice using remote sensing yield maps," *Precis. Agric.*, vol. 13, no.5, pp. 553–567, Apr. 2012.
[4] C. . Wiegand, J. . Rhoades, D. . Escobar, and J. . Everitt, "Photographic and videographic observations for determining and mapping the response of cotton to soil salinity," *Remote Sens. Environ.*, vol. 49, no. 3, pp. 212–223, Sep. 1994.
[5] E. YueSheng, "Research on remote sensing monitoring methods of single-crop rice planting area," *2011 Int. Conf. Remote Sensing, Environ. Transp. Eng.*, pp. 664–667, Jun. 2011.
[6] J. Chen, J. Huang, and J. Hu, "Mapping rice planting areas in southern China using the China Environment Satellite data," *Math. Comput. Model.*, vol. 54, pp. 1037–1043, 2011.
[7] S. Y. S. Yang, S. S. S. Shen, B. L. B. Li, T. L. T. T. Le Toan, and W. H. W. He, "Rice Mapping and Monitoring Using ENVISAT ASAR Data," *IEEE Geosci. Remote Sens. Lett.*, vol. 5, pp. 108–112, 2008.
[8] C. Qian and W. Xiuju, "Mapping paddy rice yield in Zhejiang Province using MODIS spectral index," *Turkish J. Agric. For.*, vol. 35, no. 6, pp. 579–589, 2011.
[9] S. Siddiq, A. Anayat, A. Sattar, A. Ali, and M. Yaseen, "RESPONSE OF DIFFERENT RICE (ORYZA SATIVA L.) CULTIVARS TO DIFFERENT NPK LEVELS IN THE CENTRAL CROPPING ZONE OF PUNJAB.," *Agric. Sci. Dig.*, vol. 31, no. 3, pp. 155– 160, Sep. 2011.
[10] M. F. El-kramany, B. A. Bakry, T. A. Elewa, and M. F. Zeidan, "Effect of Composted Rice Residues and NPK on Yield and Yield Components of Lentil Grown in Sandy Soil.," *Res. J. Agric. Biol. Sci.*, vol. 6, no. 6, pp. 976–980, Nov. 2010.
[11] W. Dong, X. Zhang, H. Wang, X. Dai, X. Sun, W. Qiu, and F. Yang, "Effect of different fertilizer application on the soil fertility of paddy soils in red soil region of southern China.," *PLoS One*, vol. 7, no. 9, p. e44504, Jan. 2012.
[12] M. Siavoshi, S. L. Laware, and S. L. Laware, "Effect of Organic Fertilizer on Growth And Yield Components in Rice (Oryza sativa L.)," *J. Agric. Sci.*, vol. 3, no. 3, pp.

217–225, Sep. 2011.
[13] M. Hossain and V. P. Singh, "Fertilizer use in Asian agriculture : implications for sustaining food security and the environment," pp. 155–169, 2000.
[14] C. Wang, W. A. Salas, Y. Zhang, J. Qi, and J. Wu, "Mapping paddy rice with multitemporal ALOS/PALSAR imagery in southeast China," *International Journal of Remote Sensing*, vol. 30, no. 23. pp. 6301–6315, 2009.
[15] C. X. C. Xu, X. Z. X. Zhu, Y. P. Y. Pan, W. Z. W. Zhu, and Y. L. Y. Lei, "Comparison study on NDII and NDVI based on rice extraction from rice and ginkgo mixed area," *IGARSS 2008 2008 IEEE Int. Geosci. Remote Sens. Symp.*, vol. 3, 2008.
[16] DOA (2012). Buku Perangkaan Tahunan Padi. Seksyen Teknologi dan Pengurusan Maklumat. Jabatan Pertanian Putrajaya
[17] MOA (2013). Integrated Agriculture Development Area (IADA) Kementerian Pertanian Malaysia

IOP Conf. Series: Earth and Environmental Science **20** (2014) 012017 doi:10.1088/1755-1315/20/1/012017

Oil palm fruit grading using a hyperspectral device and machine learning algorithm

O.M. Bensaeed , A.M. Shariff *, A. B. Mahmud , H. Shafri, M. Alfatni

[1] Faculty of Engineering, Universiti Putra Malaysia, 43400 Serdang, Selangor, Malaysia

* Author to whom correspondence should be addressed; E-Mail: rashid@eng.upm.edu.my, abuhajar1035@gmail.com;

Tel.: +601-23025723; +601-8481141; Fax: +603-89466425.

Abstract. In this paper, a hyperspectral-based system was introduced to detect the ripeness of oil palm fresh fruit bunches (FFB). The FFBs were scanned using a hyperspectral device, and reflectance was recorded at different wavelengths. A total of 469 fruits from oil palm FFBs (nigrescens, virescens, oleifera) were categorized as overripe, ripe, and underripe. Fruit attributes in the visible and near-infrared (400 nm to1000 nm) wavelength range regions were measured. Artificial neural network (ANN), classified the different wavelength regions on oil palm fruit through pixel-wise processing. The developed ANN model successfully classified oil palm fruits into the three ripeness categories (ripe, underripe, and overripe). The accuracy achieved by our approach was compared against that of the conventional system employing manual classification based on the observations of a human grader. Our classification approach had an accuracy of more than 95% for all three types of oil palm fruits. The research findings will help increase the quality harvesting and grading efficiency of FFBs.

1. Introduction

Oil palm fruit is typically grown in the tropical areas in South East Asia, South Africa and South America [1]. Palm oil provides vital food for millions of people and has been found to be very healthy for the human diet. Palm oil has good resistance to oxidation and prolonged exposure to high temperatures, thus making it ideal for frying. A high percentage of palm oil is usually added to frying oils because of its structure and other economic advantages.

Content from this work may be used under the terms of the Creative Commons Attribution 3.0 licence. Any further distribution of this work must maintain attribution to the author(s) and the title of the work, journal citation and DOI.
Published under licence by IOP Publishing Ltd

The Malaysian palm oil industry is considered to be highly regulated. A major problem faced by oil palm exporters and producers is the accurate grading of fresh oil palm fruits according to their ripeness levels before processing. The maturity or ripeness of the oil palm fruits dictates the quality as well as overall marketability of the palm oil produced [2].Oil palm fruits can generally be categorized into four ripeness grades: ripe, underripe, unripe, and overripe [3]. Bunches range in color from yellow, reddish orange, red, to black. A fruit before maturity is typically yellow at the base and dark purple to black at the apex. A young palm has 50 to 100 red-violet ripe fruits per bunch. The ratio of oil palm fruit pigments, such as carotenoids and chlorophylls, affects the color of the oil palm fruit. For example, unripe fruits have a higher proportion of chlorophyll that gradually decreases upon maturity [4]. Similarly, carotenoids increase as oil palm fruits mature [2]. Color changes resulting from biochemical reactions can likewise be related to fruit maturity [5].

Among the challenges during oil production is the grading of oil palm fresh fruit bunches (FFBs) in terms of maturity. Workers still employ the conventional method that requires the use of their experience to assess oil palm fruit bunch condition visually by making a small cut in the fruits to see the mesocarp color and counting the number of loosened fruits per bunch [6, 7].The manual grading of oil palm FFBs is a time-consuming and labor-intensive process that is prone to biased appraisal and human error, drastically affecting the growers' profitability [4]. Therefore, a rapid, reliable, and accurate grading technique for the detection of oil palm FFB ripeness is necessary.

Successful automation of this process requires a system that can yield results that are comparable with human grading. The application of a color vision camera system to differentiate the classes of palm oil has recently been studied [8]. Researchers have developed a maturity color index based on different color intensities. Moisture measurements are efficient indicators of the internal features and characteristics of fruit and can thus be used in different applications to obtain valuable information on fruit [9].

A method used to design a non-destructive machine for fruit inspection is the optical imaging technique. Fruits and vegetables have previously been categorized based on physical characteristics [10, 11]. Optical sensors have more recently been used for fruit quality detection in various horticultural crops [12-15]. Cameras used for detecting the maturity of the fruit have optical sensors for multi-spectrum imaging. A number of the techniques used include near-infrared (NIR) reflectance spectroscopy, laser photon counting spectroscopy, and image analysis. García-Ramos et al. [14] reviewed several non-destructive techniques used for determining post-harvest fruit firmness. Multispectral imaging techniques collect spectral information in two or three selected spectral bands. Different features of the target are determined based on the band inspected. The success of a multispectral machine vision system depends on the accuracy of spectral band selection from a range of probable inspection spectra. In infrared color composites, the colors associated with those bands in the 0.7 - 1.1 μm interval are normally richer in hue and bright from tree leaves. The representative original spectra bands of leaves in the high-, middle-degree water stress and no water stress, respectively is shown in figure 1.

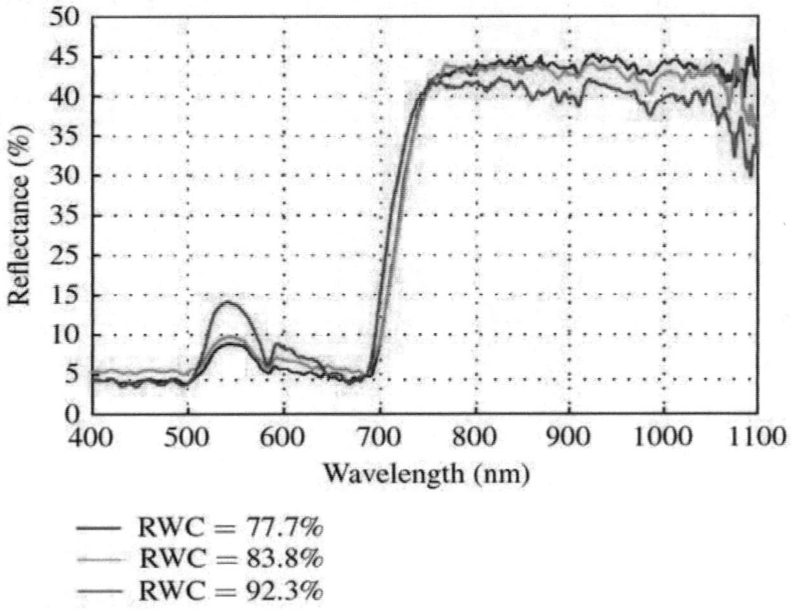

Figure 1. Original spectra of leaves with different relative water content (RWC).

A visual system is designed to produce high-resolution spectral bands called imaging systems or hyperspectral sensing. Sensor systems that have been investigated for vision-based fruit bunch grading include optical RGB cameras and hyperspectral imaging cameras [2, 5, 8, 16-19].

Hyperspectral imaging has also been widely used (mostly on apples) to measure the internal quality attributes of fruits, such as sugar or SSC, flesh and skin color, firmness, acidity, and starch index [20]. These techniques provide evidence of the potential of using optical sensors for FFB maturity determination. This approach provides more useful details for determining the most important spectral bands that can be used to differentiate normal and abnormal apples [21].

Non-parametric machine-classification algorithms are among the simplest and oldest methods of pattern recognition and are suitable for determining the ripeness of fruits [22].However; most previous studies were conducted under laboratory conditions. In this study, selective visible NIR bands in a portable optical sensor system were used for the determination of oil palm FFB maturity. More information on the quality of agriculture can be provided by the more detailed reflectance data of hyperspectral images compared with RGB [23].To evaluate the internal qualities of oil palm fruit in a laboratory, the use of a non-destructive technique such as hyperspectral imaging enables the testing of a larger number of oil palm bunches. [2] used hyperspectral imaging for FFB maturity/ripeness estimation and achieved a classification accuracy of up to 100%.

In such a system, a hyperspectral sensor is used to acquire reflectance data, and artificial neural network (ANN) algorithms are employed to classify three classes of oil palm FFBs (underripe, ripe, and overripe FFBs). ANN is application is ranging from data classification to data prediction and data visualization [24-26]. The ANN model used in this work was the multilayer feed forward network with three layers (30 input layers, 15 hidden layers, and one output), as shown in figure 2.

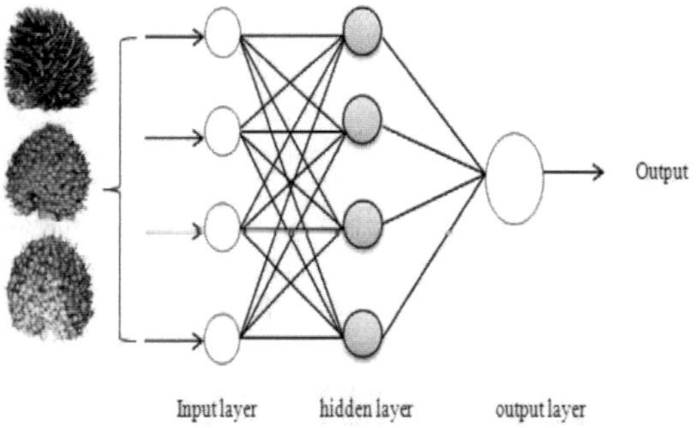

Figure 2. Simple feed-forward network.

A training dataset was used to train the algorithm, whereas testing datasets were used to test the developed (trained) algorithm in terms of predicting the class of the test dataset samples. The classification accuracies were determined by the reflectance data, which were directly used as input to the classification algorithm.

The statistical analysis method is important for determining the difference between the categories in this study, such as the receiver operating characteristic (ROC). ROC analysis offers a more robust evaluation of the relative prediction performance of alternative models compared with traditional comparisons of relative error [27, 28]. ROC is considered a statistical measure for studying the performance of an imaging or diagnostic system with respect to its capability to detect abnormality accurately and reliably [true positive (TP) without providing false detection [29]. In other words, ROC analysis provides a systematic analysis of the sensitivity and specificity of a diagnosis [30-32]. The true negative (TN) and TP indexes represent agreement with the classification of a human expert classifier. The false negative (FN) and false positive (FP) indexes represent disagreement in classification. At the end of each epoch (e), when all validation patterns are presented to the ANN classifier, the statistical indexes of such epoch are calculated for each threshold (t), including sensibility and specificity [29]. Sensitivity refers to the capacity of the classifier to identify a positive pattern among truly positive patterns, as shown in figure 3.

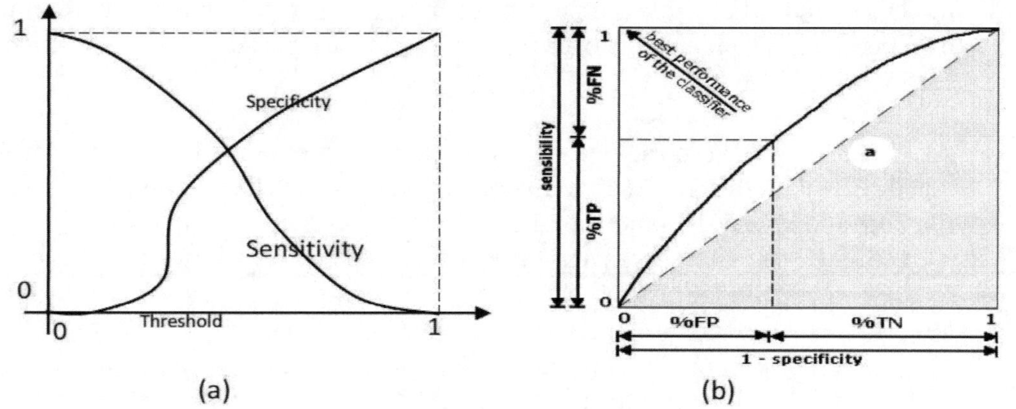

Figure 3. Sensibility versus threshold and specificity versus threshold (a) and the ROC curve with a hypothetical example (b).

The value of the sensitivity varies between 0 (when FN $\neq$ 0 and TP = 0) and 1 (when TP $\neq$ 0 and FN = 0).A smaller number of FNs denotes a higher sensitivity of the test. The sensitivity values are given along the abscissa axis and are given by:

$$\text{Sensitivity } (e,t) = \frac{\sum \mathrm{TP(e,t)}}{\sum TP(e,t) + \sum FN(e,t)} \tag{1}$$

Specificity denotes the capability of the classifier to identify negative patterns among the truly negative patterns. The specificity varies between 0 and 1. The (1-specificity) values are arranged along the axis and are given by:

$$\text{Sensitivity } (e,t) = \frac{\sum TN(e,t)}{\sum TN(e,t) + \sum FP(e,t)} \tag{2}$$

The ROC curve is a Cartesian graph that represents the dependency of the sensitivity and specificity of a classification system. An ideal classifier has a process sensitivity = 1 (TP = 1) and specificity = 1 (FP = 0) [30]. The ROC curve is an alternative approach to achieve accuracy in the evaluation of learning algorithms on natural datasets. The key assumption of ROC analysis is that TP and FP rates describe the performance of the model independent of class distribution. This analysis is conducted to provide a more robust comparative evaluation of the expected performance based on target data compared with a simple comparison of error, which assumes that the observed class distribution does not reflect any differences in the cost of different types of error. ROC analysis is of value in the evaluation of expected classifier performance under varying class distributions.

ROC curves describe the predictive behavior of a classifier independent of class distributions or error costs, thus enabling the decoupling of classification performance from these factors. ROC analysis is often called the ROC accuracy ratio, a common technique used to determine the accuracy of default probability models. A classifier has a higher probability of ranking a randomly chosen positive instance than a randomly chosen negative instance based on the area under ROC curve (AUC) [33].

AUC is a significant measure of the accuracy of ripeness determination. If AUC is equal to 1, then the ROC curve comprises two straight lines: one line vertical from (0, 0) to (0, 1) and the other line horizontal from (0, 1) to (1, 1). This test is 100% accurate because both sensitivity and specificity are 1.0, thus yielding no FPs and FNs. In other words, a test that cannot distinguish between what is normal and abnormal corresponds to an ROC curve that has a diagonal line from (0, 0) to (1, 1). The ROC area for this line is 0.6. ROC curve areas are typically between 0.6 and 1.0. Consequently, the value of AUC will always satisfy the following inequalities: $0 \leq \text{AUC} \leq 1$ [42]. An AUC of close to 1 (area of unit square) indicates very reliable diagnostic test.

Based the literature review there are still limitation on use indoor hyperspectral scanner device for oil palm ripeness different. This study was carrying out to develop an automated system for oil palm fruits bunch grading by using hyperspectral scanner technique. In addition, several wavelengths are investigated to distinguish between the three categories of oil palm fruit ripeness. The detailed objectives of this research are as follows:

i. To identify the relevant technologies to ensure the only ripe palm oil bunches can be collected.
ii. To design and build an intelligent prototype for a real time grading system using hyperspectral scanner.
iii. To test and validate the developed system through actual palm oil bunch collection. From the statistical pattern recognition of view, three band selection methods will apply.

From the statistical pattern recognition of view, three band selection methods will apply. The main aim of the study described in this paper also was to develop a hyperspectral technique that can assist the quality evaluation and classification for oil palm fruit.

2. Materials and methods

The flow chart of the method logy is given in figure 4. Details for each step are given in the following section.

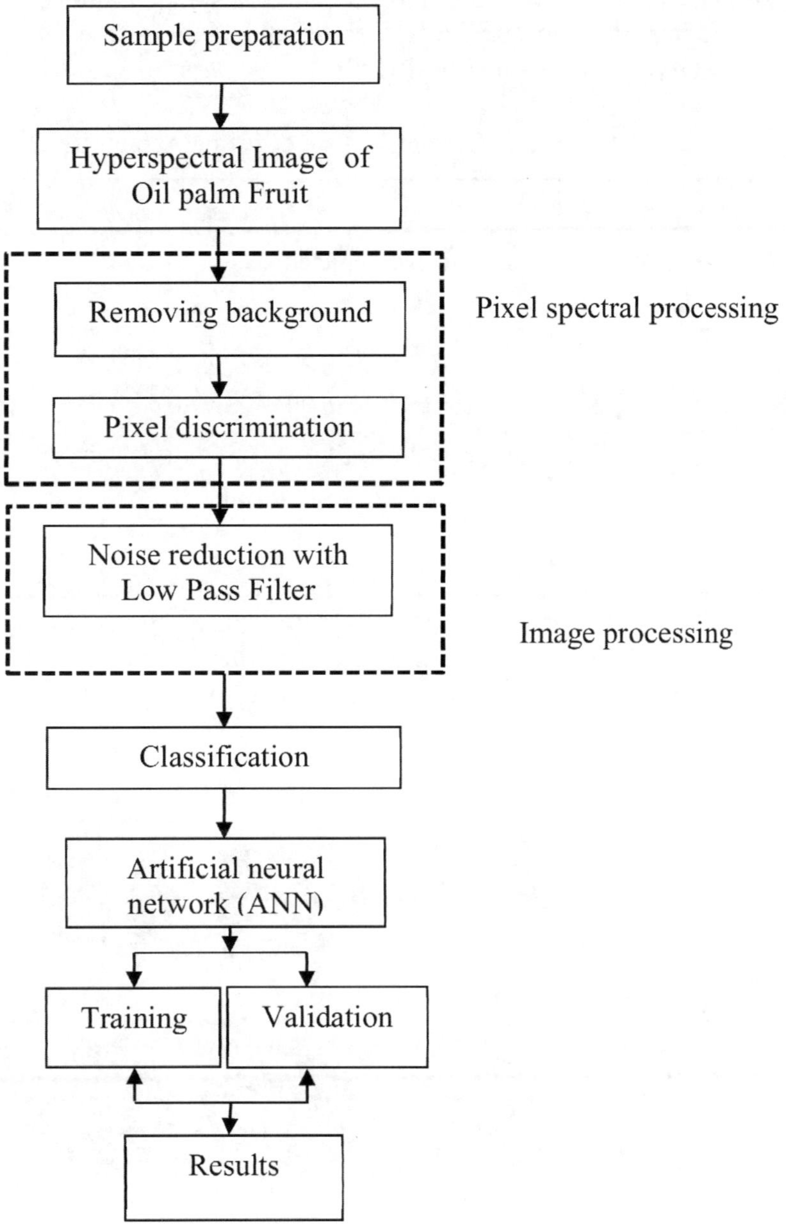

Figure 4. Flowchart of methodology.

IOP Conf. Series: Earth and Environmental Science **20** (2014) 012017 doi:10.1088/1755-1315/20/1/012017

2.1 Samples preparation

Oil palm fruits can be classified into six categories: black, underripe, ripe, overripe, empty, and rotten. The standards were established by the Malaysian Palm Oil Board (MPOB), as shown in Table 1.

Table 1. Categorization of oil palm bunches.

Category	Description
Black (Unripe)	Bunch with complete fruits
Hard (Underripe)	Bunch with 1 to 9 fruits detached
Ripe	Bunch with 10 %to 50 % fruits detached
Over ripe	Bunch with 50% to 90% fruits detached
Empty bunch	Bunch with more than 90% fruits detached
Rotten	Bunch with all or part having turned black

In this paper, total of 469 bunches evaluated by inspectors were allocated, tested, and divided into three types (nigrescens, virescens, oleifera). Each type of oil palm fruit bunches has three categories of ripeness (underripe, overripe, and ripe) were qualitatively determined by a human expert as shown in figure 5.

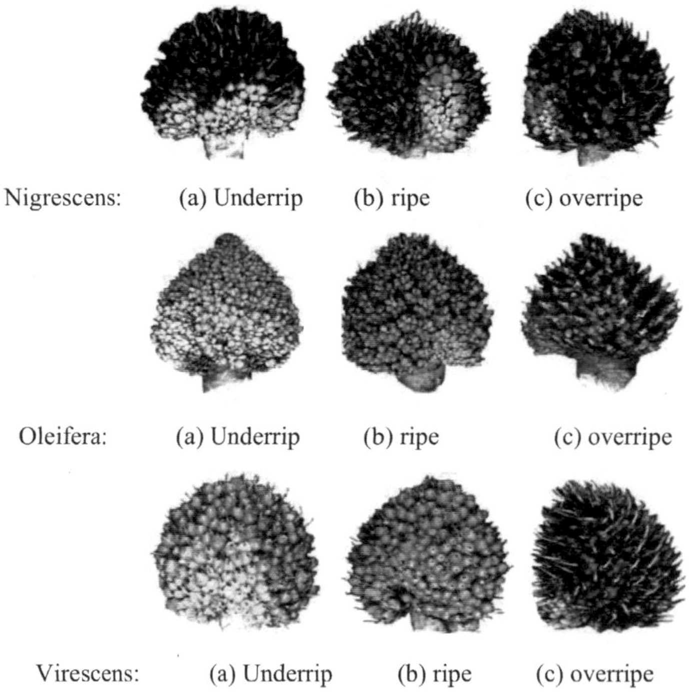

Figure 5. Typical image of oil palm fruits.

All samples were freshly taken from the MPOB farm area at Kluang, Johor, Malaysia. All fruits from the same bunch were in a similar status of ripeness despite the fact that their colors and size may vary with location in the bunch.

2.2 Hyperspectral device preparation

The hyperspectral active sensor system used for data collection is shown in figure 6. The image acquisition device utilized for this study has a high resolution of 1600 × 1200 pixels and pixel depth of 12 bits/pixel with 824 spectral bands from 400 nm to 1000nm.

Figure 6. Schematic view of the hyperspectral imaging system showing a bunch on the conveyor.

In this study, only reflectance measurements were analyzed. The hyperspectral imaging system employed in this research enabled different configurations for imaging in the visible NIR range setting. The hyperspectral system is very important to achieve good accuracy. The system set up of the spectral camera (Imperx IPX-2 M 30) was performed by connecting the x-stage USB connector and PCMCIA card to the assigned laptop and X-stage scanner. The camera was controlled by the software (Remote Sensing Cube, Specim Ltd. and Auto Vision Inc. Finland).

A white reference tile and calibration sheet was installed on the x-stage scanning platform. Illumination lights (type 4 Lamp/A 128932 and 4 ASD Pro Lamp, 14.5 V and 50 W) that were tripod mountable are especially designed for indoor laboratory diffuse reflectance measurements over a 400 nm to 2500 nm region were switched on to ensure even light distribution. SpectralDAQ was launched from the laptop. Spectral imaging properties were launched automatically upon selection of desired band. Using the camera controls, the frame and exposure time was adjusted to 15 ms. through monitoring, we ensured that the profile plot peak did saturate nor exceed 4096.The camera height was accordingly adjusted to 1.1m from object height. For the OLE23 objective lens, focus was adjusted using the calibration sheet, such that the image was sharp and not blurred. The x-stage scanner of spectralDAQwas used to control the scan mirror control for the determination of the start and end points along the sample area. Scan rate was adjusted by visually comparing the ratio of actual length to width based on the viewed image. These steps ensured that the spectral camera was ready for the assessment of the oil palm fruit bunch. The oil palm fruit and the white reference were placed on the x-stage platform, and the start and end positions were defined.

The hyperspectral imaging system was calibrated both spectrally and spatially by using the following procedures: Spectral calibrations employed eight lamps. For the spatial calibration, a white paper printed with thin parallel lines that were 2 mm apart was placed at the sample holder. The calibration results showed that the system was highly linear and that the distortion of spectral and spatial information was within one pixel on the charge-coupled device detector. Thus, no spectral and spatial corrections were needed for the system. The camera and spectrograph were used to scan the oil palm fruit line by line as the conveyer moved the oil palm fruit through the field of view of the optical system. The oil palm fruits were manually placed on the translational plate, which was covered by rubber mat to prevent the reflection of light, with region of interest facing the camera. After finishing the scans on one entire fruit bunch, the spatial-by-spectral matrices were combined to construct a three dimensional (3D) spatial and spectral data space. The scanning time for one oil palm was dependent on the integration time used for the camera, which was fixed to 200 ms and 5 s, and on the size of the oil palm.
The hyperspectral data must be normalized with a standard reference under the same illumination system setup to establish a reflectance coefficient at each pixel location. The normalization process can be expressed as:

$$R_{factor}(x, y, z) = \frac{I(x, y, z) \_ I_{dark}(x, y, z)}{I_{ref}(x, y, z)} \times 1000 \quad (3)$$

where I(x,y,z) is the reflectance radiation intensity of spectral band z at pixel location (x,y), Iref (x,y,z) is the radiation intensity of a white reference tile with a known reflectance coefficient at band z and pixel location (x,y) under the same illumination light state, Idark (x,y,z) is the noise of the sensor at band z and pixel location (x,y) under no light reflectance, andRfactor (x,y,z) is the normalized reflectance factor used for every pixel (x,y) at band z. A scaling factor of 10000 was used to increase the dynamic range of the reflectance, as shown in above equation.

The normalization and rearrangement processes comprised the calibration process of real hyperspectral data. After the calibration process, the 3D data cube is considered a 3D image cube.

The acquired hyperspectral images were processed using Environment for visualizing Images (ENVI 4.7) to do classification such as subseting image, image resizing, filtering the image to remove the noise from the image, region of interest ROI. About 35000 pixels were manually selected from every corrected image as a region of interest (ROI). The average spectrum from ROI of normal surface of each fruit was calculated by averaging the reflectance spectral value of all pixels in the ROI. The denoised image is used with data processing software such as Matlab® to do the analysis of the FFB classification.

11×11 low pass filter to remove the image noise. Low pass filter maintains the low-frequency components of the image, which smooth it. Low pass filter have the same weights in every kernel element, also changing the center pixel value with an average of the neighboring values. The practical kernel size is 11×11.

3. Data analysis

A total of 469 fruits were inspected and distributed into three types of oil palm fruit which have three classes (underripe, ripe, and overripe). These samples were analyzed on two stages, one stage is every type of oil palm fruits individually analyzing to get the reflectance for each class of fruit and which wavelength can distinguish between the three categories, and second stage all the types of oil palm fruit is together at specific wavelength which can selected by the percentage of high reflectance for each type of oil palm fruit. These samples were randomized and separated into independent training and testing datasets (75:25) to evaluate different classification algorithms. The replicates for each sample were averaged prior to further analysis. Matlab® was used for the analysis of the spectral data. This section analysis the most generally used statistical method found in the scientific literature for this purpose: Artificial neural network (ANN).

4. Results and discussion

4.1 Spectral reflectance of the Nigrescens fruit

Figure 7 presents the typical shapes of ripe, under ripe and overripe oil palm fruit reflectance spectra. The most important difference is the chlorophyll absorption hole (around 675 nm), which disappears as the fruit ripens. Nigrescens oil palm fruits had three broad band absorptions in the ripeness around 460, 675, and 970 nm regions. The relative reflectance increases steadily over the wavelengths 680-900 nm. The spectral bands from 750–910 nm differentiate between the three different categories of the fruit, which clearly demarcates the ripeness between three categories of oil palm fruit based on the wavelength.

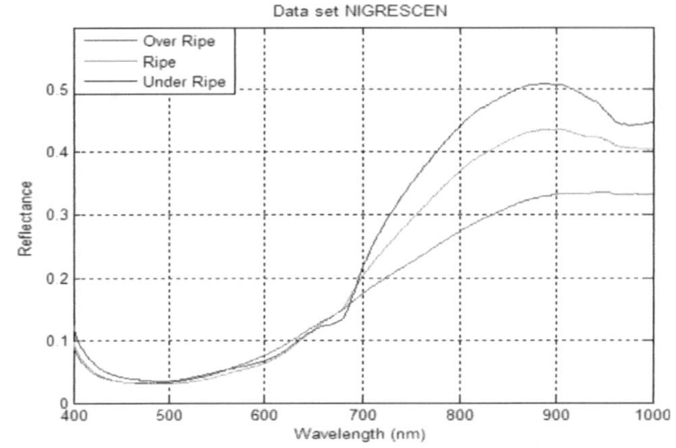

Figure 7. Spectral profile at different ripeness stages of nigrescens oil palm fruits.

4.2 Spectral reflectance of the Virescens fruit

The typical shapes of ripe, under ripe and overripe oil palm fruit reflectance spectra was present in figure 8. The most important difference is the chlorophyll absorption hole (around 675 nm), which disappears as the fruit ripens. Virescens oil palm fruits had three broad band absorptions in the ripeness around 520, 670, and 970 nm regions. The relative reflectance increases steadily over the wavelengths 700-880 nm. The spectral bands from 750–910 nm differentiate between the three different categories of the fruit, which clearly demarcates the ripeness between three categories of oil palm fruit based on the wavelength.

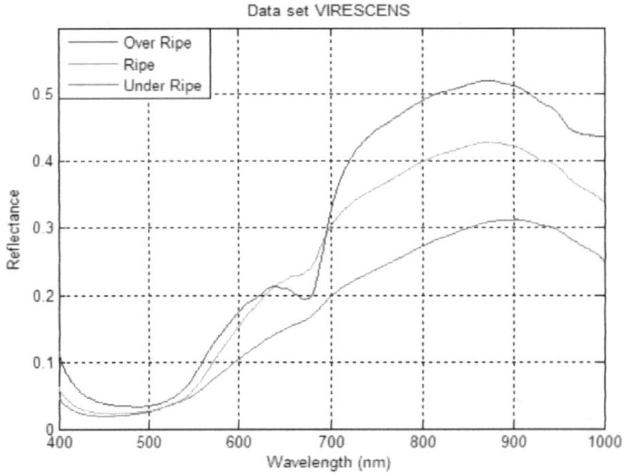

Figure 8. Spectral profile at different ripeness stages of virescens oil palm fruits.

4.3 Spectral reflectance of the Oleifera fruit

The typical shapes of ripe, under ripe and overripe oil palm fruit reflectance spectra was present in figure 9. The most important difference is the chlorophyll absorption hole (around 675 nm), which disappears as the fruit ripens. Oleifera oil palm fruits had three broad band absorptions in the ripeness around 520, 670, and 970 nm regions. The relative reflectance increases steadily over the wavelengths 680-900 nm. The spectral bands from 750–910 nm differentiate between the three different categories of the fruit, which clearly demarcates the ripeness between three categories of oil palm fruit based on the wavelength.

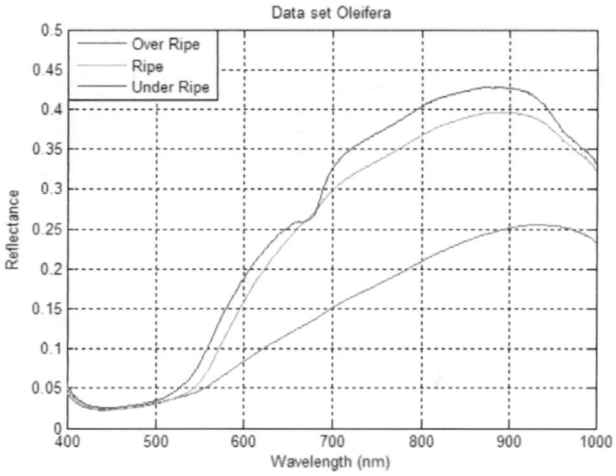

Figure 9. Spectral profile at different ripeness stages of oleifera oil palm fruits.

4.4 CHAID growing method results

151 samples for virescens (36 under-ripe, 41 ripe and 74 over ripe oil palm images), 159 samples for nigrescens (46 under-ripe, 47 ripe and 66 over ripe oil palm images) and 159 samples for oleifera (52 under-ripe, 44 ripe and 63 over ripe oil palm images) 30 samples were used as test set (10 samples per categories) and others used for training set. The wavelength used was interval of 10. The results show the overall classification accuracy using classification tree (CHAID growing method) as shown in figure 10.

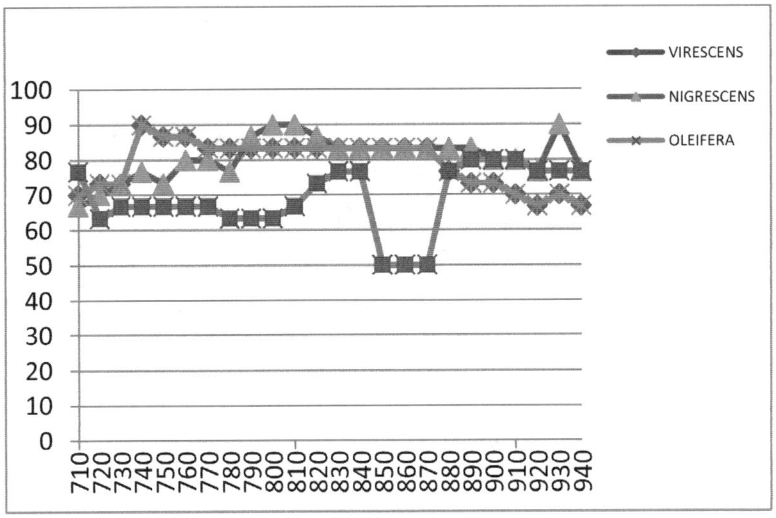

Figure 10. Overall classification accuracy using classification tree (CHAID growing method) for three types of oil palm fruits.

The spectral bands from 710–940 nm showed the best test accuracy between the three different types of the oil palm fruit as shown in Table 2. Virescens type has stable range from 770 nm to 870 nm 83.3 %, also nigrescens type has same accuracy 83.3 % in the range 830 nm to 890 nm, and there is overlap between from 830 nm to 870 nm, 830 nm it consider the best wavelengths to differentiate between three categories of oil palm fruits. Compared with oleifera type, it has accuracy medium at special region and the accuracy is limited, so there is no overlap between oleifera and another two types of oil palm fruit, 880 nm is consider the best wavelength for oleifera type.

Table 2. Overall classification accuracy for three types of oil palm fruits by using (CHAID growing method).

Wavelength (nm)	Test set for Virescens	Test set for Nigrescens	Test set for Oleifera
710	70	66.7	76.7
720	73.3	70	63.3
730	73.3	73.3	66.7
740	90	76.7	66.7
750	86.7	73.3	66.7
760	86.7	80	66.7
770	83.3	80	66.7
780	83.3	76.7	63.3
790	83.3	86.7	63.3
800	83.3	90	63.3
810	83.3	90	66.7
820	83.3	86.7	73.3
830	83.3	83.3	76.7
840	83.3	83.3	76.7
850	83.3	83.3	50
860	83.3	83.3	50
870	83.3	83.3	50
880	76.7	83.3	76.7
890	73.3	83.3	80
900	73.3	80	80
910	70	80	80
920	66.7	76.7	76.7
930	70	90	76.7
940	66.7	76.7	76.7

4.5 Spectral reflectance of the all data (Nigrescens. Virescens, and Oleifera) Fruits

The typical shapes of ripe, under ripe and overripe oil palm fruit reflectance was presented in figure 11. The first part of the graph (0 – 46 samples) present the nigrescens type which has three categories (underripe, ripe, and overripe), second part of the graph is virescens type between (46-93 samples); the last part of the graph is oleifera between (94 -140 samples). During three parts of the graph, the differentiation between three categories is clearly in nigrescens and virescens but difficult in oleifera type especially underripe and ripe samples.

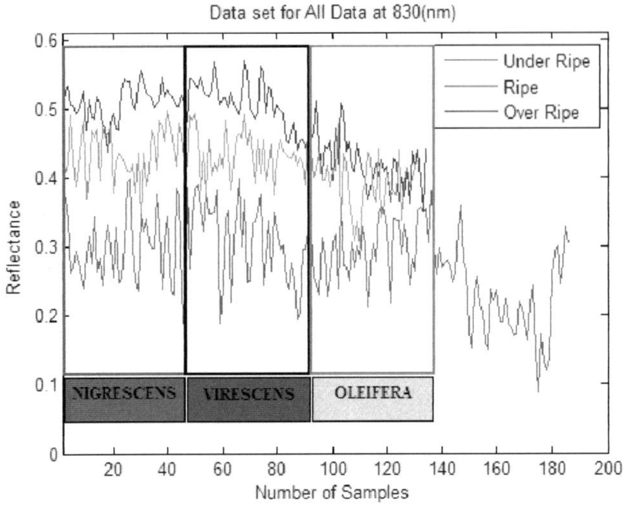

Figure 11. Reflectance values for all data samples (nigrescens, virescens, and oleifera) at best wavelength 830 nm.

4.6 ANN-MLP evaluation using receiver operating characteristic

The real performances of the ANN—MLP network is evaluated by the receiver operating characteristic (ROC) analysis and area under curve (AUC). The ROC analysis is related in a direct and natural way to the cost/benefit analysis of decision making.

4.6.1 Optimal neural network classifier nesults. During the training stage, a total of 469 samples of each dataset (nigrescens, virescens, oleifera), the dataset were presented as the full set of input samples to the ANN-MLP. Each dataset is graded as under-ripe (-1), ripe (0) and over-ripe (1), the maximum sum squared error (SSE) was empirically set at 10e-4, and the process was carried out at 10e6 epochs. After examining the efficiency and mean square error (MSE), the architectures of three layers comprising [50 30 1] [60 20 1] and [35 15 1] for nigrescens, virescens, oleifera, respectively. Those structures for each dataset were found to provide suitable performance result for each the dataset model. The "logsig" logarithmic sigmoid transfer function was used.

After several trainings, the ANN-MLP model for ripeness classification succeeded in learning and matching the target appropriately with MSE equal to $9.7276e^{-004}$, $1.09x10e^{-6}$, and $9.7476e^{-005}$ with model efficiency of 0.9999 in full training datasets. Based on the learning of the ANN-MLP, the full testing dataset (30% for each type type) was presented as input samples through the model. In the testing stage, viable performance results were obtained with MSEs of 0.00184, 0.00156 and 0.00173 and efficiency rates of 0.8574, 0.7812 and 0.8245 in each full testing datasets (nigescens, virescens, oleifera), respectively. This result indicated the excellent performance of the method, as well as the capability of the ANN-MLP to

IOP Conf. Series: Earth and Environmental Science **20** (2014) 012017 doi:10.1088/1755-1315/20/1/012017

classify the ripeness of the three types of oil palm three individually as shown in figure 12. As well as is the case all data together.

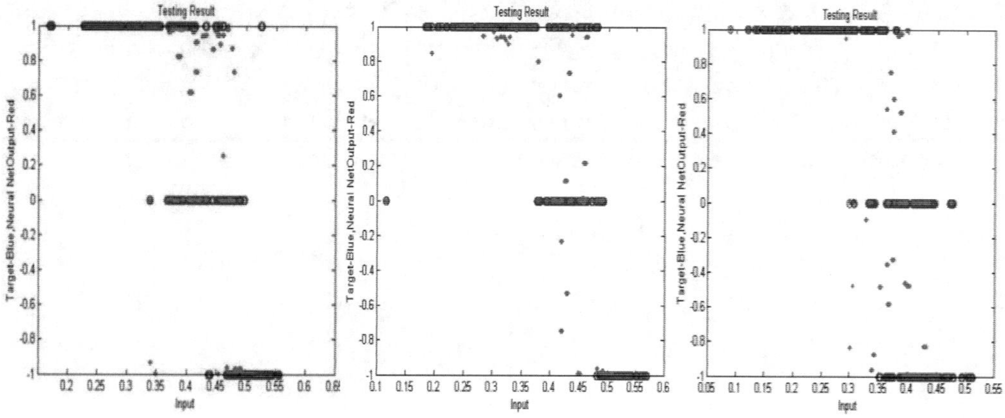

(a) (b) (c)

Figure 12. ANN-MLP model for ripeness classification for (a) Nigresence (b) Virecsens (c) Oleifera

To evaluate the ANN model for ripeness classification that distinguishes between the under-ripe, ripe and over-ripe grades for three types of oil palm fruits, the same sets are examined. The ripeness classification and evaluation performed by the score of the AUC is measured from the ROC curve as shown in figure 13.

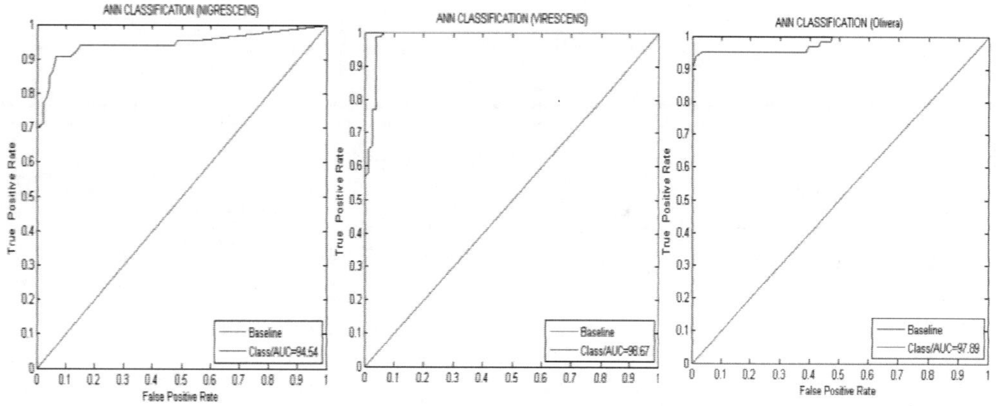

(a) (b) (c)

Figure 13. AUC measured from the ROC curve for (a) Nigresence (b) Virecsens (c) Oleifera.

Figure 13 (a, b, c) depict the ROC graphs of c for nigrescens, virescens andoleifera performed by the ANN-MLP with the higher AUC score obtained to recognize grade in the three datasets. The obtained results in figure 13 (a) indicate that the selected band features with ANN-MLP classifier performs perfect classification with AUC test accuracy of nigrescens ripeness (94.54). In addition, the ANN-MLP classifier could perform at a higher rate with AUC test accuracy of virescens ripeness (98.67) as shown in figure 13 (b). Meanwhile, figure 12 (c) shows the ROC graphs of oleifera ripeness classification performed by the ANN-MLP to recognize selected band features with AUC score (97.89).

Figure 14 shows the ROC graphs all data ripeness classification performed by the ANN-MLP to recognize selected band features with AUC score (95.73).

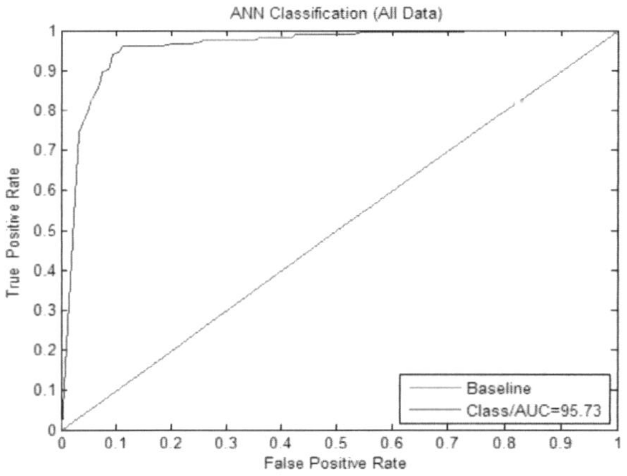

Figure 14. AUC measured from the ROC curve for all data.

Overall, the AUC score is an indicator of classification rate performance, which is high when the AUC test accuracy is higher than 94 in all cases.

The ripeness classification performed by the ANN-MLP decision system to recognize the selected bands features using all datasets showed great performance results.

5. Conclusions

This paper proposed a framework for (oil palm fruits bunch grading by using hyperspectral scanner technique). The ripeness detection of oil palm fruit bunch suffers from the lack of productivity, efficiency problems, and still graded by human visualization. Research studies have shown that the major cause of problems can be traced back to the lack of automating inspection system. One commonly cited means to overcome this problem is through advanced technologies. However, the challenge arises when a decision has to be made to choose the best technology to fulfill the present needs, since each system has its own technical, economic, and risk considerations. The goal of this paper was to develop a prototype that can help palm oil companies to choose the best investigation system that fits their needs.

After enhancing the images, the resulted wavelength and reflectance was taken to be processed for classification. After analyzing the region of spectral bands between 750 to 900 nm, the best wavelength to distinguish all data of oil palm fruit (nigrescens, virescens) was 830 nm as show in figure (10), also has same reflectance values in three categories of nigrescens and virescens but oleifera type difficult to differentiate between two categories (underripe and ripe), that is meaning that nigrescens and virescens has almost same characteristic and the properties but oleifera does not has same characteristic with nigrescens and virescens. The classification was performed by ANN. The developed system shows high classification results on accuracy of the ripeness detection for the three types of oil palm fruits separately (nigrescens, virescens and oleifera) with rates of 94.54 %, 98.67 % and 97.89 % respectively using the ANN-MLP classifier, A high classification results on accuracy of the ripeness detection for all data (nigrescens, virescens and oleifera) was 98.67 %.

The solutions discussed here are oil palm detection system. By using a set of carefully selected hardware that used in indoor, the degree of ripeness can be detected in both with high accuracy. This developed system can be used to enhance and optimize the entire grading process. The proposed prototype is not only practical for detecting the ripeness of the bunches in terms of design and development, but also very affordable since the components of the proposed system are not very expensive. Therefore, further research in the area of hyperspectral has a great potential and can provide technological advancement for ripeness of oil palm fruits.

In summary, the application framework improves the harvesting operation and presents the potential for improved workflow reliability and grading performance and for effortless derivation of performance indicators in project management. This study will be useful to the oil palm industry, oil palm engineers, oil palm harvesters, graders, mill operators, plantation managers, small holders and to the research community.

Acknowledgments

This study was conducted by a research team from University Putra Malaysia (UPM), jointly supported by spatial and numerical modeling lab (SNML), spatial research group (SRG) and the Malaysian Palm Oil Board (MPOB). The authors would like to thank the Malaysian Palm Oil Board (MPOB) for the valuable support rendered in carrying out this research

References

[1] Pantzaris, T. P.; Ahmad, M. J., Properties and utilization of palm kernel oil. Palm Oil Developments 2001, 35, 11-23.

[2] Junkwon, P.; Takigawa, T.; Okamoto, H.; Hasegawa, H.; Koike, M.; Sakai, K.; Siruntawineti, J.; Chaeychomsri, W.; Sanevas, N.; Tittinuchanon, P., Potential Application of Color and Hyperspectral Images for Estimation of Weight and Ripeness of Oil Palm (Elaeis guineensis Jacq. var. tenera). Agricultural Information Research 2009b, 18, (2), 72-81.

[3] Fadilah, N.; Mohamad-Saleh, J.; Abdul Halim, Z.; Ibrahim, H.; Syed Ali, S. S., Intelligent Color Vision System for Ripeness Classification of Oil Palm Fresh Fruit Bunch. Sensors 2012, 12, (10), 14179-14195.

[4] Saeed, O. M. B.; Sankaran, S.; Shariff, A. R. M.; Shafri, H. Z. M.; Ehsani, R.; Alfatni, M. S.; Hazir, M. H. M., Classification of oil palm fresh fruit bunches based on their maturity using portable four-band sensor system. Computers and Electronics in Agriculture 2012, 82, 55-60.

[5] Choong, T. S. Y.; Abbas, S.; Shariff, A. R.; Halim, R.; Ismail, M. H. S.; Yunus, R.; Ali, S.; Ahmadun, F. R., Digital image processing of palm oil fruits. International Journal of Food Engineering 2006, 2, (2), 7.

[6] Jamil, N.; Mohamed, A.; Abdullah, S. In Automated grading of palm oil fresh fruit bunches (FFB) using neuro-fuzzy technique, Soft Computing and Pattern Recognition, 2009. SOCPAR'09. International Conference of, 2009; IEEE: 2009; pp 245-249.

[7] Shaarani, S. M.; Cárdenas-Blanco, A.; Amin, M. G.; Soon, N. G.; Hall, L. D., Monitoring development and ripeness of oil palm fruit (Elaeis guneensis) by MRI and bulk NMR. International Journal of Agriculture and Biology 2010, 12, (1), 101-105.

[8] Ismail, W. I. W.; Bardaie, M. Z.; Hamid, A. M. A., Optical properties for mechanical harvesting of oil palm FFB. Journal of Oil Palm Research 2000, 12, 38-45.

[9] Alfatni, M. S. M.; Shariff, A. R. M.; Abdullah, M. Z.; Marhaban, M. H. B.; Saaed, O. M. B., The application of internal grading system technologies for agricultural products -Review. Journal of food engineering 2013, 116, (3), 703-725.

[10] Butz, P.; Hofmann, C.; Tauscher, B., Recent developments in noninvasive techniques for fresh fruit and vegetable internal quality analysis. Journal of food science 2005, 70, (9), R131-R141.

[11] Narendra, V. G.; Hareesh, K. S., Quality inspection and grading of agricultural and food products by computer vision-A Review. International Journal of Computer Applications IJCA 2010, 2, (1), 43-65.

[12] Chen, P.; Sun, Z., A review of non-destructive methods for quality evaluation and sorting of agricultural products. Journal of Agricultural Engineering Research 1991, 49, 85-98.

[13] Menesatti, P.; Zanella, A.; DâAndrea, S.; Costa, C.; Paglia, G.; Pallottino, F., Supervised multivariate analysis of hyper-spectral NIR images to evaluate the starch index of apples. Food and Bioprocess Technology 2009, 2, (3), 308-314.

[14] García-Ramos, F. J.; Valero, C.; Homer, I.; Ortiz-Cañavate, J.; Ruiz-Altisent, M., Non-destructive fruit firmness sensors: a review. Spanish Journal of Agricultural Research 2005, 3, (1), 61-73.

[15] Ruiz Altisent, M.; Ruiz-Garcia, L.; Moreda, G. P.; Lu, R.; Hernandez-Sanchez, N.; Correa, E. C.; Diezma, B.; Nicola, B.; Garcىa-Ramos, J., Sensors for product characterization and quality of specialty crops--A review. Computers and Electronics in Agriculture 2010.

[16] Abdullah, M. Z.; Guan, L. C.; Mohd Azemi, B. M. N., Stepwise discriminant analysis for colour grading of oil palm using machine vision system. Food and bioproducts processing 2001, 79, (4), 223-231.

[17] Abdullah, M. Z.; Guan, L. I. M. C.; Mohamed, A.; Noor, M. A. M., Color vision system for ripeness inspection of oil palm Elaeis guineensis. Journal of food processing and preservation 2002, 26, (3), 213-235.

[18] Alfatni, M. S. M.; Shariff, A. R. M.; Shafri, H. Z. M.; Saaed, O. M. B.; Eshanta, O. M., Oil Palm Fruit Bunch Grading System Using Red, Green and Blue Digital Number. Journal of Applied Sciences 2008, 8, (8), 1444-1452.

[19] Alfatni, M. S. M.; Shariff, A. R. M.; Shafri, H. Z. M.; Saaed, O. M. B.; Eshanta, O. M.; Abuzaed, a., Automated oil palm fruit bunch grading system using density of color (RGB). In 7th Saudi Engineering Conference, King Saud University, Riyadh, 2007; pp 299-313.

[20] Lorente, D.; Aleixos, N.; Gomez-Sanchis, J.; Cubero, S.; Garcia-Navarrete, O. L.; Blasco, J., Recent advances and applications of hyperspectral imaging for fruit and vegetable quality assessment. Food and Bioprocess Technology 2012, 1-22.

[21] Mehl, P.; Chao, K.; Kim, M.; Chen, Y., Detection of defects on selected apple cultivars using hyperspectral and multispectral image analysis. Applied engineering in agriculture 2002, 18, (2), 219-226.

[22] Altman, N., An introduction to kernel and nearest-neighbor nonparametric regression. American Statistician 1992, 175-185.

[23] Abdullah, M. Z.; Guan, L. C.; Lim, K. C.; Karim, A. A., The applications of computer vision system and tomographic radar imaging for assessing physical properties of food. Journal of Food Engineering 2004, 61, (1), 125-135.

[24] Ayrulu-Erdem, B.; Barshan, B., Leg motion classification with artificial neural networks using wavelet-based features of gyroscope signals. Sensors 2011, 11, (2), 1721-1743.

[25] BaladrÃn, C.; Aguiar, J. M.; Calavia, L.; Carro, B. n.; SÃnchez-Esguevillas, A.; HernÃ¡ndez, L., Performance Study of the Application of Artificial Neural Networks to the Completion and Prediction of Data Retrieved by Underwater Sensors. Sensors 2012, 12, (2), 1468-1481.

[26] Jeon, H. Y.; Tian, L. F.; Zhu, H., Robust crop and weed segmentation under uncontrolled outdoor illumination. Sensors 2011, 11, (6), 6270-6283.

[27] Bradley, A. P., The use of the area under the ROC curve in the evaluation of machine learning algorithms. Pattern Recognition 1997, 30, (7), 1145-1159.

[28] Hart, P. E.; Stork, D. G., Pattern classification. John Willey & Sons 2001.

[29] Mustapha, A.; Hussain, A.; Samad, S. A., A new approach for noise reduction in spine radiograph images using a non-linear contrast

adjustment scheme based adaptive factor. Scientific Research and Essays 2011, 6, (20), 4246-4258.

[30] Brown, C. D.; Davis, H. T., Receiver operating characteristics curves and related decision measures: A tutorial. Chemometrics and Intelligent Laboratory Systems 2006, 80, (1), 24-38.

[31] Sovierzoski, M. A.; Argoud, F. I. M.; de Azevedo, F. M. In Evaluation of ANN Classifiers During Supervised Training with ROC Analysis and Cross Validation, BioMedical Engineering and Informatics, 2008. BMEI 2008. International Conference on, 2008; IEEE: 2008; pp 274-278.

[32] Tilbury, J. B.; Van Eetvelt, W. J.; Garibaldi, J. M.; Curnsw, J. S. H.; Ifeachor, E. C., Receiver operating characteristic analysis for intelligent medical systems-a new approach for finding confidence intervals. Biomedical Engineering, IEEE Transactions on 2000, 47, (7), 952-963.

[33] Fawcett, T., An introduction to ROC analysis. Pattern recognition letters 2006, 27, (8), 861-874.

IOP Conf. Series: Earth and Environmental Science **20** (2014) 012018 doi:10.1088/1755-1315/20/1/012018

Oil palm fresh fruit bunch ripeness classification based on rule-based expert system of ROI image processing technique results

M S M Alfatni[1], A R M Shariff[1,*],M Z Abdullah[2], M H Marhaban[1], S B Shafie[1], M D BAmiruddin[3], O M B Saaed[1]

[1]Faculty of Engineering, Universiti Putra Malaysia, 43400, Serdang, Selangor, Malaysia
[2]School of Electrical and Electronic of Engineering, Universiti Sains Malaysia Engineering Campus, 14300, Seri Ampangan, Penang, Malaysia
[3]MPOB, Persiaran Institusi, Bandar, Baru, Bangi, 43000 Kajang, Selangor, Malaysia

E-mail:rashid@eng.upm.edu.my

Abstract. There is a processing need for a fast, easy and accurate classification system for oil palm fruit ripeness. Such a system will be invaluable to farmers and plantation managers who need to sell their oil palm fresh fruit bunch (FFB) for the mill as this will avoid disputes. In this paper,a new approach was developed under the name of expert rules-based systembased on the image processing techniques results of thethree different oil palm FFB region of interests (ROIs), namely; ROI1 (300x300 pixels), ROI2 (50x50 pixels) and ROI3 (100x100 pixels). The results show that the best rule-based ROIs for statistical colour feature extraction with k-nearest neighbors (KNN) classifier at 94% were chosen as well as the ROIs that indicated results higher than the rule-based outcome, such as the ROIs of statistical colour feature extraction with artificial neural network (ANN) classifier at 94%, were selected for further FFB ripeness inspection system.

1. Introduction

Agricultural product quality conventionally plays a fundamental role in nearly all food industry quality assessments. The traditional method of agricultural product quality assessment is tedious and costly. It is easily influenced by physiological factors, inducing subjective and inconsistent evaluation results[1]. Machine vision provides innovative solutions toward industrial automation [2]. Innovative computer technologies have been utilized in numerous research projects around the world in order to construct new machines for agricultural product quality assessment such as automatic grading systems [3]. The utilization of automatic grading systems for inspection has increased in recent years. Essentially, two inspection stages of the automatic grading system can be identified, viz. external and internal [2,4].

The external grading device is a combination of software and hardware, namely a programming language (PL) and an operating system (OS) with image processing capabilities. The hardware is comprised of a computer, a Charge coupled device (CCD) camera and a data acquisition and conveying system [5-10]. In this paper, rule based expert system (RBES)as a prediction system is proposed and

Corresponding author: Assoc. Prof.Dr. Abdul Rashid Mohamed Shariff. Mailing Address: Geospatial Information Science Research Centre (GISRC), Faculty of Engineering, Universiti Putra Malaysia, 43400, Serdang, Selangor, Malaysia. Telephones: +60389467543 (Office), +60123025723 (Mobile). Fax: +60389466425. E-mail: rashidpls@gmail.com

Content from this work may be used under the terms of the Creative Commons Attribution 3.0 licence. Any further distribution of this work must maintain attribution to the author(s) and the title of the work, journal citation and DOI.
Published under licence by IOP Publishing Ltd

implemented with the oil palm FFB ripeness grading system to improve the quality and ensure a high accuracy fresh fruit bunch (FFB)ripeness grading. The proposed RBES algorithm uses the basic concept of rule base, called IF-THEN. The oil palm grading system ripeness grading is implemented with the features and techniques of FFB ripeness without further training and testing. These rules are applied to the different region of interests (ROIs)of the oil palm FFB. The system used the concept of a classifier, which is a function used to assign a class label for the calculation of the probability based on the attribute values that describe an instance [11].

2. Methods and techniques

The classification method RBTES is introduced based on three general steps. First, the experimental result of the FFB ripeness classification based on the supervised machine learning techniques is determined. Second, the rule-based expert system concept [12] is driven using specific ROIs of the oil palm FFB. Third, new rule-based trees are introduced based on the concept of DTs [13, 14] for higher accuracy of the oil palm FFB ripeness grading system, as shown in Figure 1.

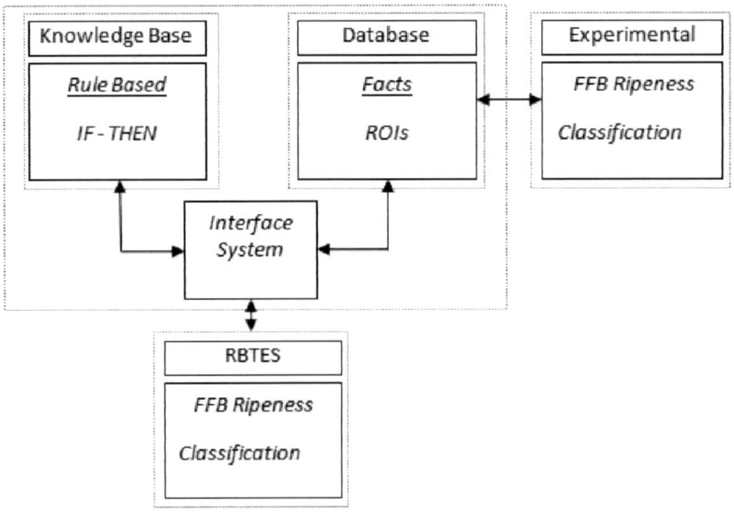

Figure 1. Diagram of the proposed RBES.

2.1. *Experimental FFB ripeness classification*

Several experiments were conducted on the different models (colour, texture, and thorns) of the oil palm FFB grading system. The ripeness of the three different oil palm FFBROIs illustrated in Figure 2 was tested with the different feature extraction techniques for each individual model. The three different supervised machine-learning techniques, namely, artificial neural network (ANN), k-nearest neighbours (KNN), and support vector machine (SVM), were implemented with each unique test to investigate and to recognize experimentally the proper methodology for the oil palm FFB ripeness grading system.

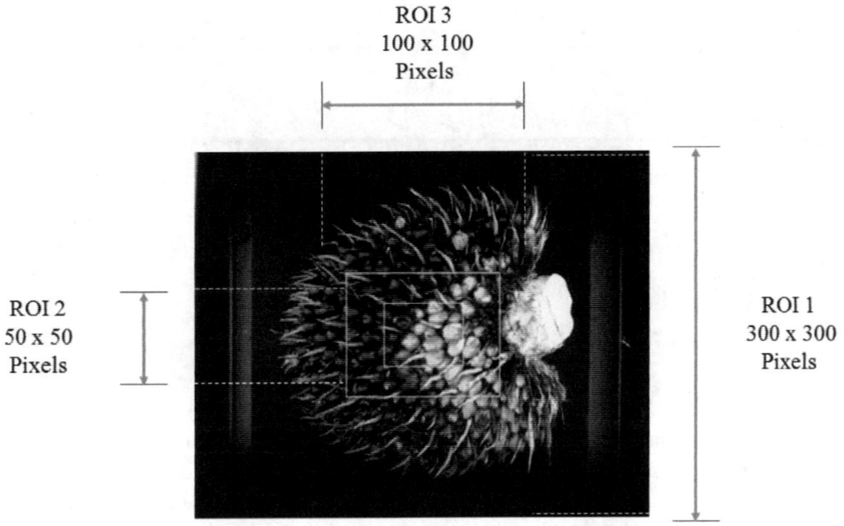

Figure 2. Oil palm FFB image ROIs.

2.2. Expert system

The oil palm expert system (knowledge-based system) is a common term used to describe a rule-based processing system, which consists of three elements, namely, database, knowledge base, and interface engine.

2.2.1. Database

A database or working memory is method to derive data regarding the oil palm FFB ripeness. The oil palm FFB database includes a set of facts used to match against the IF (condition) parts of rules stored in the knowledge base. The oil palm FFB ripeness data were established by the expert of the Malaysian Oil Palm Board (MPOB) based on the study visit done during the early stage of the research. Based on the basic elements illustrated above, the data of the ripeness of the oil palm FFB were experimentally examined, and considered as inputs to the rule-based expert system.

2.2.2. Knowledge base

The knowledge base is defined as the set of if-then-else rules and known facts, which contain the domain knowledge useful for problem solving [15]. In the rule-based expert system of the oil palm FFB, the knowledge is represented as a set of rules. Each rule specifies a relation, recommendation, directive, strategy, or heuristic of the FFB, and has the IF (condition) THEN (action) structure ELSE (action). When the condition part of a rule is satisfied, the rule is decided, and the action part is executed. Since the system has three ROI ripeness inputs, the three inputs of ROIs have three possibilities. Based on the three inputs and one output, there are three rule-based possibilities for each output (under ripe, ripe, and overripe). The ith $(i = 1...3)$ rule-based is defined as follows:

$$\textbf{IF}\ r1 \text{ is } R_1^i \text{ and } r2 \text{ is } R_2^i \text{ and } r3 \text{ is } R_3^i,\ \textbf{THEN}\ t_i \text{ is } T_i(i=1...3),\ \textbf{ELSE}\ t_i \text{ is } T_i(i=2) \quad (1)$$

where $r1$, $r2$, and $r3$ denote three input values for ROI1, ROI2, and ROI3, respectively; t_i denotes the output; R_1^i, R_2^i, and R_3^i $(i=1...3)$ denote the three input rule-based for the ith rule that are in (under ripe, ripe and overripe) the three different ROIs, and $T_i(i=1...3)$ denotes the output rule-based of the consequence of the ith rule, which are under ripe, ripe, and overripe.

2.2.3. *Interface system*

An inference system, which contains the reasoning logic, processes the rules and data. The system links the rules given in the knowledge base with the facts provided in the database. Classification and prediction use rule-based algorithms. The rule-based method is a data mining supervisor learning algorithm for performing the classification and the prediction by generating if-then rules that cover all possible cases in a dataset. Finally, the RBES was built based on the close relations of the three different FFB ripeness classes, namely under ripe, ripe, and overripe.

1. The target of the oil palm FFB harvester is the ripe class, which is normally used for palm oil industries due to its reach with oil, lesser water, and fatty acid. By contrast, the under ripe contains much water and the overripe has a high rate of fatty acid.
2. Based on the logic sequence of the ripeness stage, the harvester can harvest the under and overripe classes. However, it is not logical to harvest the under ripe class as overripe class as well as overripe class as under ripe class.
3. Based on the above facts, the expert system rule-based was set as:

Rule 1: **IF** *ROI1***AND***ROI2***AND***ROI3are***OVER RIPE**
OR *ROI1***AND***ROI2 are* **OVER RIPE**
OR *ROI1***AND***ROI3 are* **OVER RIPE**
OR *ROI2***AND***ROI3 are* **OVER RIPE**
THEN *FFB* is **OVER RIPE**
Rule 2: **ELSE IF** *ROI1***AND***ROI2***AND***ROI3are***UNDER RIPE**
OR *ROI1***AND***ROI2are* **UNDER RIPE**
OR *ROI1***AND***ROI3 are* **UNDER RIPE**
OR *ROI2***AND***ROI3 are* **UNDER RIPE**
THEN *FFB* is **UNDER RIPE**
Rule 3: **ELSE IF** *ROI1***AND***ROI2***AND***ROI3are***RIPE**
OR *ROI1***AND***ROI2 are* **RIPE**
OR *ROI1***AND***ROI3 are* **RIPE**
OR *ROI2***AND***ROI3 are* **RIPE**
THEN *FFB* is **RIPE**
Rule 4: **ELSE** *FFB* is **RIPE**

3. Results

The RBES is a prediction system proposed and implemented with the oil palm FFB ripeness grading system. The aim of the RBES is to optimize the quality of the oil palm grading system result, and to ensure the high accuracy of the FFB ripeness classification. The RBES algorithm uses a basic rule based on the concept called the IF-THEN-ELSE rule. This rule is based on the facts of oil palm FFB ripeness, and improves the oil palm grading system ripeness implementation with FFB ripeness features and techniques without further training and testing. FFB ripeness is classified in terms of ripeness classes by using a rule-based DT, which considered the differentiation between ROIs.

3.1. ROIs

The concept of the oil palm FFB expert system rule-based DT is based on the three ROIs of the FFB. The expert system was successfully implemented on the ROI ripeness with significant result accuracy. Figure 3 shows the optimization of the Olifera oil palm FFB ripeness grading system based on the ROIs of the colour model with different feature extraction and machine learning techniques.

The best results of the rule-based system were selected for further testing stages. The application of this system on ROIs for statistical colour feature extraction with the KNN classifier reached 93%. The statistical colour feature extraction with the SVM classifier was at 92%, and the colour histogram feature extraction with the SVM classifier was at 92% as well. These findings are shown in Figure3(a, b, and e). Moreover, the higher result of the ROIs was selected if their results were higher than the rule-based result, such as the ROI1 of statistical colour feature extraction with the ANN classifier at 93%, ROI1 of the colour histogram feature extraction with the KNN classifier at 93%, and ROI3 of the colour histogram feature extraction with the ANN classifier at 94%, as shown in Figure3(c, d, and f).

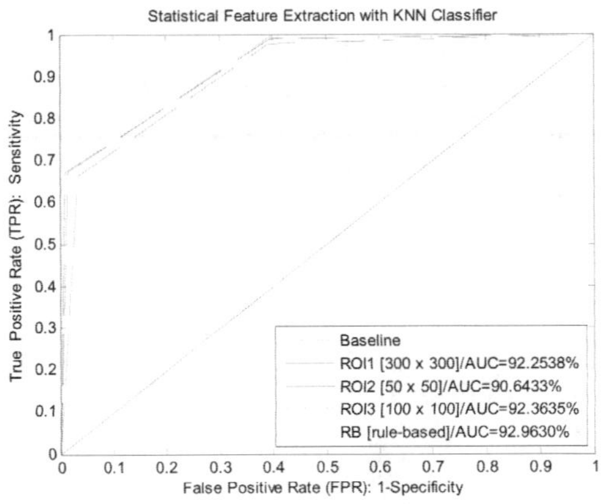

(a)

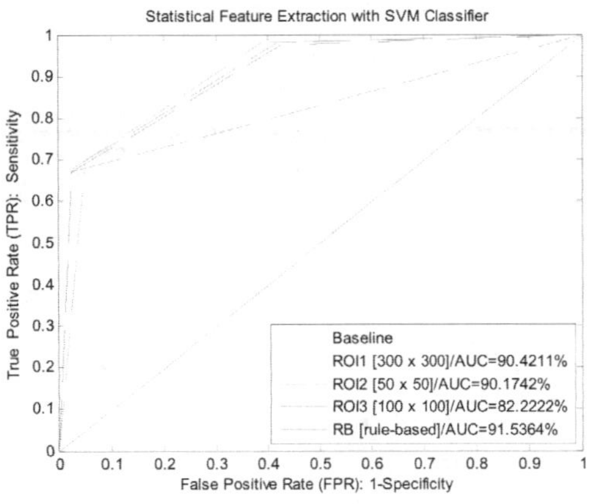

(b)

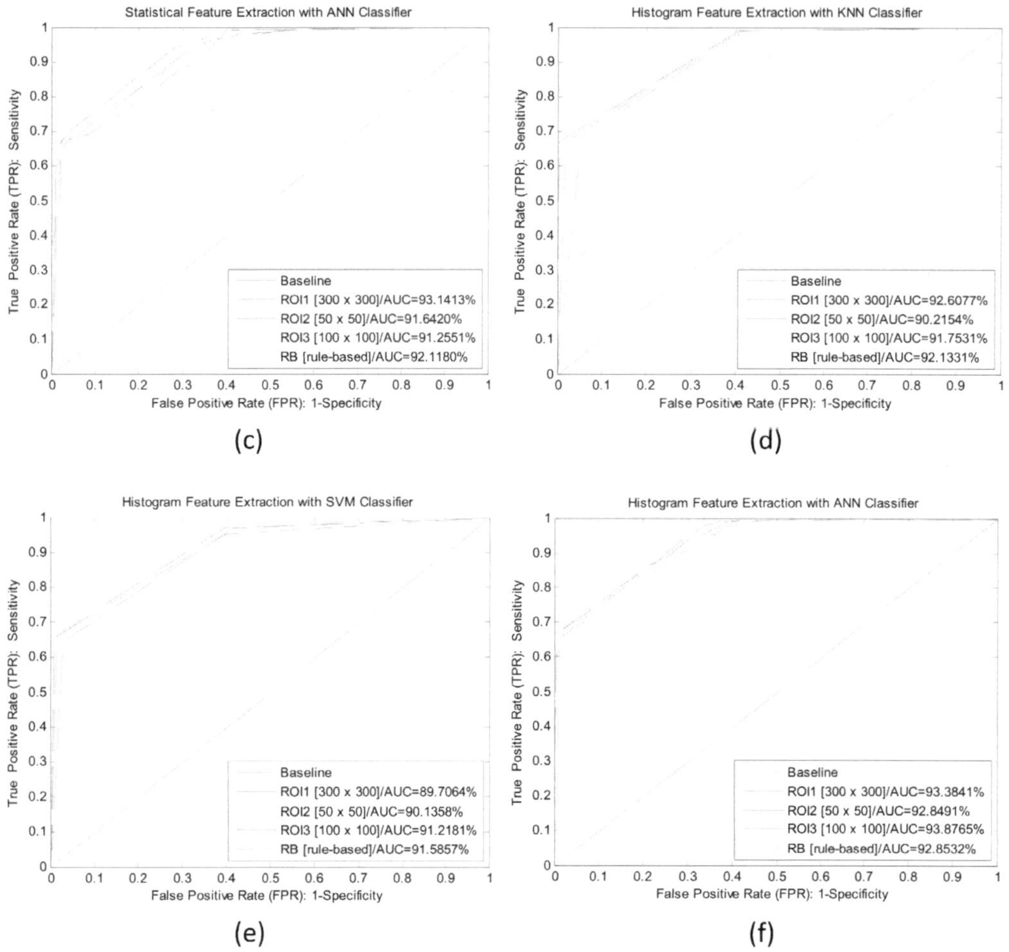

Figure 3. Rule-based of Nigrescens FFB for ROIs: (a) Statistical-KNN, (b) Statistical-SVM, (c) Statistical-ANN, (d) Histogram-KNN, (e) Histogram-SVM, and (f) Histogram-ANN.

4. Conclusion

A new approach was developed under the name of expert rules-based system based on the image processing techniques results of the three different oil palm FFB region of interest (ROIs), namely; ROI1 (300x300 pixels), ROI2 (50x50 pixels) and ROI3 (100x100 pixels). Generally, the results show that the best rule-based ROIs for statistical colour feature extraction with KNN classifier at 94% were chosen as well as the ROIs that indicated results higher than the rule-based outcome, such as the ROIs of statistical colour feature extraction with ANN classifier at 94%, were selected for further FFB ripeness inspection system. The efficiency of the grading rate performance based on rule-based ROIs were significantly higher on the KNN and SVM than the ANN supervised machine learning in all cases. However, the rule-based system was built based on the different cases of FFB ripeness and provided trusted results despite lesser accuracy than the ROIs.

Table 1. Rule-based results of Olifera FFB of the ROIs test computing based on feature extraction techniques and classifiers.

Techniques	FFB ripeness accuracy (%)											
	ANN				KNN				SVM			
	ROI 1	ROI 2	ROI 3	**RB**	ROI 1	ROI 2	ROI 3	**RB**	ROI 1	ROI 2	ROI 3	**RB**
Statistical colour	93	92	92	**92**	92	91	92	**93**	90	90	82	**92**
Colour histogram	93	93	94	**93**	93	90	92	**92**	90	90	91	**92**
GLCM	92	91	92	**92**	79	80	90	**91**	75	79	80	**79**
BGLAM	92	93	93	**92**	90	91	92	**92**	92	90	92	**92**
Gabor	83	79	83	**79**	78	90	79	**80**	76	88	89	**90**

Acknowledgement

This research work was supported by University Putra Malaysia (UPM), Geospatial Information Science Research Centre (GISRC), Malaysian Palm Oil Board (MPOB), Ministry of Science, Technology and Innovation (MOSTI) Malaysia.

References

[1] D.-W. Sun, Computer Vision Technology for Food Quality Evaluation. (2006). In *Proceedings of the International Symposium on Novel Technology in Food Processing and Engineering*, National Chung Hsing University, Taichung, Taiwan,

[2] E. N. Malamasa, E. G. M. Petrakisa, M. Zervakisa, L. Petitb and J.-D.Legat, (2003). Image and Vision Computing 21, 171

[3] V. Leemans and M.-F.Destain,(2004). Journal of Food Engineering 61, 83

[4] J. B. Njoroge, K. Ninomiya, N. Kondo and I. Toita, (2002).Automated Fruit Grading System using Image Processing. In *S1CE02-0822*, Auk 5-7.2MI2, osaks, pp 1346

[5] T. Morimoto, T. Takeuchi, H. Miyata and Y. (2000). Hashimoto, Computers and Electronics in Agriculture 26, 171

[6] M. Z. Abdullah, L. C. Guan, K. C. Lim and A. A. Karim, (2004).Journal of Food Engineering 61, 125

[7] M. Z. Abdullah, A. S. Fathinul-Syahir and B. M. N. Mohd-Azemi, (2005). Transactions of the Institute of Measurement and Control 27 (2), 65

[8] R. Marmoa, S. Amodiob, R. Tagliaferrid, V. Ferrerib and G. Longo, (2005). Computers & Geosciences 31, 649

[9] M. Rashidi and K. Seyfi, (2007).Wourld Journal of Agricultural Sciences 3 (6), 735

[10] K. H. Ghazali, M. M. Mustafa and A. Hussain, (2008).Machine Vision System for Automatic Weeding Strategy in Oil Palm Plantation using Image Filtering Technique. In *Proceedings of World Academy of Science, Engineering and Technology*, Vol. 31, pp 655

[11] C. A. Edeki, (2012). A Comparative Study of Data Mining and Statistical Learning Techniques for Prediction of Cancer Survivability. In *School of Business and Technology*, Capella University, Doctor of Philosophy, p 99

[12] B. G. Buchanan and R. O. Duda, (1982).Principles of Rule-Based Expert Systems Department. Stauford University, Stanford, pp 1

[13] A. Mustapha, (2012). Development of a Computer Assisted Vertebral Fracture Assessment System (Avfas) For Vertebrae Irregularity Classification. In *Faculty of Engineering and Built Environment*, University Kebangsaan Malaysia, Bangi, Doctor of Philosophy, p 293

[14] R. Fanguy and V. Raghavan, (2003).Generating Rule-Based Trees from Decision Trees for Concept-based Information Retrieval. In *IEEE/WIC International Conference on Web Intelligence*, Halifax, Canada, pp 119

[15] M. Jazzar and A. Jantan, (2008). IJCSNS International Journal of Computer Science and Network Security 8 (4), 1

IOP Conf. Series: Earth and Environmental Science **20** (2014) 012019 doi:10.1088/1755-1315/20/1/012019

Simulation of coastal wave spectra energy from ENVISAT satellite data

Maged Marghany
Institute of Geospatial Science and Technology (INSTeG),Universiti Teknologi Malaysia,81310 UTM, Skudai, Johor Bahru, Malaysia.

E-mail: maged@utm.my,magedupm@hotmail.com

Abstract. In the last two decades, scientists have developed several powerful techniques to retrieve energy from natural sources such as a sun radiations, oceans and winds. This study is aimed at stimulating wave energy from large scale synthetic aperture radar (SAR) during different monsoon periods. In doing so, the nonlinear velocity bunching algorithm is used to retrieve the information of ocean wave spectra parameters such as significant wave height, directions, and energy on offshore, midshore, and onshore. Therefore, the maximum peak of the wave energy spectra density of 1.4 m^2 s has occurred during northeast monsoon period. It is clear that the mid-shore and onshore has the highest peak of 0.8 and 1.37 m^2 s, respectively as compared to offshore. In conclusions, a nonlinear algorithm of velocity bunching can be used to retrieve the significant wave height from synthetic aperture radar (SAR). In addition, SAR can be used to map the distribution of ocean wave spectra energy and determined the potential energy zone in Malaysia coastal waters.

1. Introduction

Ocean wave plays significant role in renewable energy industries over the world. It is important to investigate the potential zone of wave energy in coastal zones. The coastal zone is closed area of land and it is easy to supply the generated energy from wave energy farms by cables to land zone [4,6,7,8]. Therefore, the implementation of wave energy farms in coastal zone requires accurate information about the potential wave energy. In this regard, using conventional techniques for wave energy investigation by being a time consuming as it is required to deploy several wave rider buoys over a large scale area of the coastal zone.

Synthetic Aperture Radar (SAR) has been recognized as a powerful tool for modeling ocean waves and forecasting over an area of 300 km x 300 km. Hence, the sediment transport could be modeled by the wave spectra information extracted from a SAR image[1]. Currently, a number of investigations have been carried out on the assimilation of SAR wave mode data into wave forecasting models. This is because the SAR image spectrum has turned out to be far removed from the actual wave spectrum and rather complicated post-processing is necessary for extracting quantitative wave information In this regards, previous studies were carried out by Hasselmann and Hasselmann, [2], and Li et al., [5] by developing an inversion algorithm to map SAR wave spectra into ocean wave spectra. Hasselmann and Hesselman [2] introduced a non-linear algorithm which was developed by Vachon et al.,[11] to model the significant wave height based on the azimuth cutoff. Vachon et al. [11] defined the azimuth cutoff as the degree to which the SAR image spectrum is constrained in the azimuth direction. The azimuth cutoff is affected by the wind and wave condition in a quasi-linear forward-mapping model [11]. Marghany [8] utilized the azimuth cutoff model which was developed by Vachon et al. [11] to

Content from this work may be used under the terms of the Creative Commons Attribution 3.0 licence. Any further distribution of this work must maintain attribution to the author(s) and the title of the work, journal citation and DOI.
Published under licence by IOP Publishing Ltd

estimate the significant wave height. This study is aimed to retrieve the wave energy spectra from ENVISAT ASAR data using the velocity bunching algorithm along the east coast of Malaysia.

2. Velocity bunching algorithm

According to Pierson and Moskowitz [10] the significant wave height spectrum H_s can be related to the one sided directional wave number spectra density $S(\vec{K},\phi)$ by the following formula

$$H_s = 4[\int\int S(\vec{k},\phi)\,dk_i\,dk_j]^{0.5} \quad (1)$$

where k_i and k_j are the wave numbers in the azimuth and range directions, respectively, and $\vec{K}$ is the wave-number magnitude of k_i and k_j in the azimuth and the range directions, respectively.. The ocean wave spectrum was obtained by input of in situ wave parameters (dominant wave number, wave propagation direction ϕ and significant wave height H_s. The actual ocean wave spectrum density $S(\vec{K},\phi)$ can be calculated from the following equation [6,10],

$$S(\vec{K},\phi) - \frac{\sqrt{S(k_i,k_j)d\vec{K}}\; e^{\left[-2(5.88\,U^{-2}/\vec{K}\cos\phi)^2\right]}}{\left[\vec{K}^2 - (0.5\,gU^{-2})\vec{K}\cos\phi + 2.4U^{-2}\right]^2} \quad (2)$$

where U is the wind speed and ϕ is the wave direction. The actual wave spectra contours have been determined by using Eq. 2. Furthermore, Eq. 2 has been used with quasi-linear model to map the ENVISAT ASAR wave spectra onto ocean wave spectra.

In this study, a single ENVISAT ASAR image (Table 1) frame comprising of 512 x 512 pixels and lines was extracted. The ENVISAT ASAR images are a two dimensional sampling of the ocean wave field and thus a two-dimensional (2-D) Fourier transfer has to be utilized [1,6].When the Fourier transfer was selected, the output domain is the two-dimensional frequency spectrum of the input image [6].

Table 1: Physical Parameters of ENVISAT ASAR.

Parameters	Values
Radar Wavelength (cm)	5.6
Ground Resolution (m)	30
Incident Angle (°)	25
Polarization	HH/VV

To map observed SAR spectra into the ocean wave spectra, a nonlinear model was applied. The nonlinear theory is explained below: According to the Gaussian linear theory, the relation between ocean wave spectra $S(\vec{K},\phi)$ and ENVISAT ASAR image spectra $S_Q(\vec{K})$ could be described by tilt and hydrodynamic modulation (real aperture radar (RAR) modulation) and velocity bunching modulation. The tilt modulation is linear to the local surface slope in the range direction i.e. In the plane of radar illumination. The tilt modulation in general is a function of wind stress and wind direction for ocean waves and ENVISAT ASAR. According to Vachon et al.,[11] the tilt modulation is the largest for HH polarization. The hydrodynamic interaction between the scattering waves (ripples) and longer gravity waves produced a concentration of the scatterer on the up wind face of the swell. Following Vachon et al., [11] ENVISAT ASAR image spectra can map into ocean wave spectra under the assumption of the quasi-linear modulation transfer function $S_Q(\vec{k})$ which is given by

$$S_Q(\vec{K})=R(K)H(k_i;K_c)\left[\frac{S(\vec{K},\phi)}{2}\left|T_{lin}(k_{i,j})\right|^2+\frac{S(-\vec{K},\phi)}{2}\left|T_{lin}(-k_{i,j})\right|^2\right] \quad (3)$$

where $H(k_i;K_c)$ is an azimuth cut-off function that depends upon azimuth wave number k_i and range wave number k_j, the cut-off azimuth wave number K_c and $R(\vec{K})$ is the ENVISAT ASAR point spread function. ENVISAT ASAR point spread function is a function of the azimuth and the range resolutions [7,11]. According to Vachon et al., [11] T_{lin} is a linear modulation transfer function which is composed of the RAR (the tilt modulation and hydrodynamic modulation), and the velocity bunching modulation. The RAR modulation transfer function (RAR MTF) is the coherent sum of the transfer function associated with each of these terms, i.e.

$$T_{lin}(\vec{K})=M_t(\vec{K})+M_d(\vec{K})+M_v(\vec{K}) \quad (4)$$

The tilt modulation $M_t(k)$ can be described by vachon et al., [11] as follow

$$M_t(\vec{K})=k_j\frac{4\cot\theta}{1\pm\sin^2\theta}e^{(\frac{i\pi}{2})} \quad (5)$$

where k_j is the range wave number and θ is the local incident angle of the radar beam. Following Vachon et al., [11], the hydrodynamic modulation transfer function can be given by

$$M_d(\vec{K})=4.5\omega K\frac{\omega-i\mu}{\omega+\mu^2}\sin^2\phi \quad (6)$$

where $\vec{K}$ is long wave number, i is $\sqrt{-1}$, ω is the angular frequency of the long waves, ϕ is the azimuth angle and μ is the relaxation rate of the Bragg waves which is 0.5/s. According to Vachon et al. [11] velocity bunching can contribute to equation 4 (linear MTF) based on the following equation which was given by Vachon et al. [11]

$$M_v(\vec{K})=\frac{R}{V}\omega\left[\frac{k_i}{\vec{K}}\sin\theta+i\cos\theta\right] \quad (7)$$

where *R/V* is the scene range to platform velocity ration. In order to estimate the significant wave height from the quasi-linear transforms, we adopted the algorithm that was given by Vachon et al., [11] and modified by Marghany (2003) to be appropriate for the geophysical conditions of tropical coastal waters:

$$\lambda_c=\beta\left(\int_{H_{s0}}^{H_{sn}}\sqrt{H_s}\,dH_s+\int_{U_0}^{U_n}\sqrt{U}\,dU\right) \quad (8)$$

where is λ_c cut-off, H_s and U are the in situ data of significant wave height and wind speed. The measured wind speed was estimated at 10 m height above the sea surface. β is an empirical value which results of *R/V* multiplied by the intercept of azimuth cut-off C when the significant wave height and the wind speed equal zero. A least squares fit was used to find the correlation coefficient between cut-off wavelength and the one calculated directly from the ENVISAT ASAR spectra image by equation (8). Then, the following equation was adopted by Marghany [7] to estimate the significant wave height (H_{sT}) from the ENVISAT ASAR images

$$H_{sT} = \beta^{-2} \int_{\lambda_{c0}}^{\lambda_{cn}} (\lambda_c)^2 \, d\lambda_c \qquad (9)$$

where β is the value of $\left(c\frac{R}{V}\right)^{-1}$ and H_{sT} is the significant wave height simulated from ENVISAT ASAR images. The significant wave height H_s is estimated by using the velocity-bunching model. Finally, the in situ wave spectra measurements have been collected using AWAC wave rider buoy during the ENVISAT ASAR overpass on 28 December 2010.

3. Results and discussions

Figure 1 shows the ENVISAT ASAR satellite data along the coastal waters of Johor. This data was acquired on 28 December 2010 (Figure 1). The study area is located along the coast of Johor in the southern, eastern part of Peninsular Malaysia. The area is approximately 20 km of Johor (Figure 1), located in the South China Sea between 1° 57′ N to 2° 15′ N and 103° 51′ E to 104° 15′ E. Further, the coastline is bordered with varying width of alluvial plains. Consistent with Mastura [9] and Zelina et al., [12], this area lies in an equatorial region dominated by two seasonal monsoons and two transition monsoon. The southwest monsoon lasts from May to September while the northeast monsoon lasts from November to March. The transition monsoon periods involve April which is between end of northeast monsoon and the beginning of the southwest monsoon periods. Further, the second transition monsoon is October which is between the end of the southwest monsoon and beginning of northeast monsoon periods [6,12].

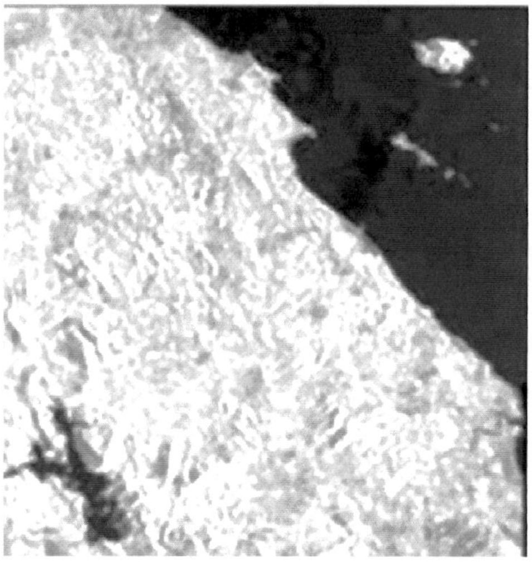

Figure 1. ENVISAT ASAR data were used in this study.

Figure 2 shows the simulated wave spectra that derived from AWAC wave rider buoy, ENVISAT ASAR data, and modeled by using the velocity bunching model. Clearly, the wavelength is varied from 100 m to 200 m. The main direction of the wave spectra is the northeast. The monsoon winds affect the direction and magnitude of the waves. Further, Marghany [8] stated that strong waves are prevalent during the northeast monsoon when the prevailing wave direction is from the north (November to March), while during the southwest monsoon (May to September), the wave directions are propagated from the south. According to Marghany et al., [6] the maximum wave height during the northeast monsoon season is 4 m. The minimum wave height is found during the southwest monsoon which is less than 1 m [6,8].

IOP Conf. Series: Earth and Environmental Science **20** (2014) 012019 doi:10.1088/1755-1315/20/1/012019

(a) (b) (c)

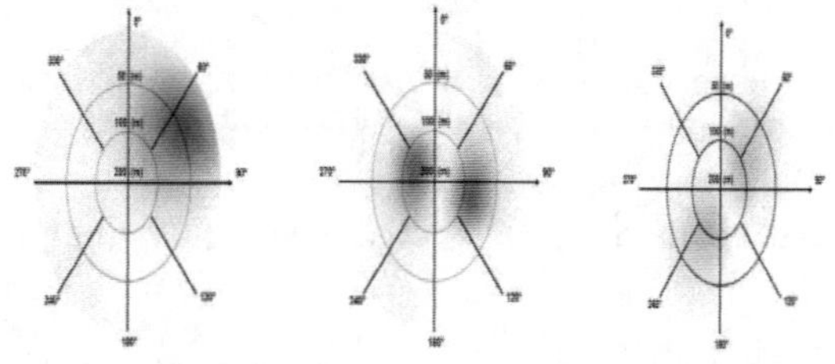

Figure 2. Ocean wave spectra derived from (a) AWAC, (b) ENVISAT ASAR and (c) velocity bunching model.

According to Marghany [7] velocity bunching model produces precisely the information of wave spectra. Indeed, it is able to solve the matter of non-linearity between SAR observed spectra and ocean wave spectra. In other words, it can be used to map SAR observed spectra into real ocean spectra. This result confirms that studies of Vachon et al., [11]; Marghany [8] and Marghany et al., [6]. Figure 3 shows the wave energy density spectra simulated from the velocity bunching model. It is interesting to find that the offshore and onshore wave spectra energy is 1.4 m^2s and 1.37 m^2s, respectively that are larger than midshore. According to Komar [4] offshore is the first region of ocean which received highest wind stress and that caused a high frequency of wave fluctuations. Therefore, the increment of wave spectra energy on shore is due to that the wave is experiencing the breaking status. It can be said that, the ENVISAT ASAR data can be used to simulate the pattern of the wave spectra energy.

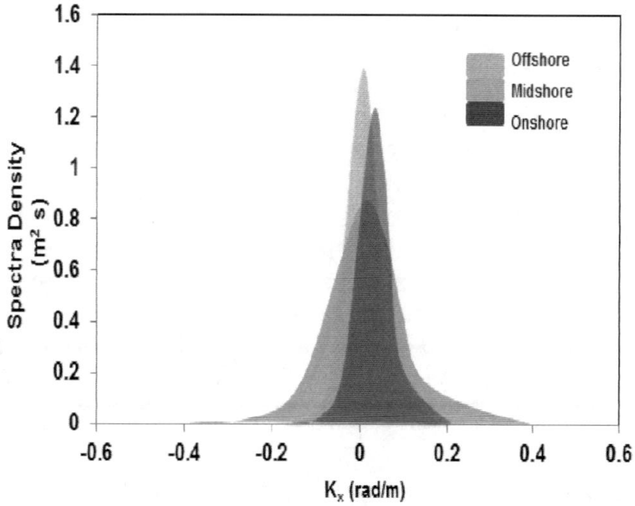

Figure 3. Wave spectra energy simulated from the velocity bunching algorithm.

4. Conclusions

The paper has demonstrated a method for retrieving wave spectra energy from ENVISAT ASAR satellite data. In doing so, nonlinear velocity bunching algorithm is implemented to retrieve significant wave height. The study shows that the maximum peak of the wave energy spectra density of 1.4 m^2 s has occurred during northeast monsoon period. In addition, wave spectra energy are varied along coastal water. Clearly, the mid-shore and onshore have the highest peak of 0.8 and 1.37 m^2 s, respectively. In conclusions, a nonlinear algorithm of velocity bunching can be used to retrieve the significant wave height from ENVISAT ASAR satellite data. In addition, ENVISAT ASAR data can be used to map the distribution of ocean wave spectra energy and determined the potential energy zone in Malaysia coastal waters.

5. References

[1] Cornet, Y., Anne C. L, Andre, O., 1993, Significantivity of wavelength computation using spatial auto-correlation theory on SAR Image of ERS-1. Proceedings of Second ERS-1 Symposium. Hamburg: Germany. 941-944.

[2] Hasselmann, K., and Hasselmann S.,1991, On the Nonlinear Mapping of an Ocean Spectrum and Its Inversion. *J. Geophys. Res.,* (96)10,713-10,799.

[3] Krogstad, H.E., and Schyberg, H.,1991, On Hasselman's nonlinear ocean –SAR transformation. Proceedings of IGARSS'91, Espo, Finland, June 3-6, 1991, 841-849.

[4] Komar, P.D ., 1976, Beach processes and sedimentation. Prentic-Hall, New Jersey.

[5] Li, X. M., Lehnera S., , He M. X., 2008, Ocean wave measurements based on satellite synthetic aperture radar (SAR)and numerical wave model (WAM) data–extreme sea state and Cross sea analysis. Int. J. Remote Sensing, (29) 6403–6416.

[6] Marghany,M., Hashim, M., and Cracknell A.P., 2011, Simulation of shoreline change using AIRSAR and POLSAR C-band data, *Environmental Earth Sciences*, (64)1177–1189.

[7] Marghany, M., 2003, ERS-1 modulation transfer function impact on shoreline change. International Journal of Applied Earth Observation and Geoinformation, (4) 279-294.

[8] Marghany, M., 2001, TOPSAR wave spectra model and coastal erosion detection. International Journal of Applied Earth Observation and Geoinformation, (3) 357-365.

[9] Mastura, S.,1987, *Coastal Geomorphology of Desaru and its Implication for Coastal Zone Management*. Monograph no.13. Bangi: Uinveristi Kebangsaan Malaysia.

[10] Person, W.J., and Moskowitz, L., 1964, A proposed spectral from fully developed wind seas based on the similarity theory of S.A. Kitaigorodskii. *Journal of Geophysical Research*, (69) 5181-5190.

[11] Populus J., Aristaghes, C., Jonsson,L., Augustin, J.M., Pouliquen, E.,1991, The use of SPOT data for wave analysis. *Remote Sensing of Environment*, (36) 55-65.

[12] Tukey, J.W., 1961, Discussion, emphasizing the connection between analysis of variance and spectrum analysis. *Technometrics*, (3) 191-219.

[13] Vachon, P.W., and Raney, R.K., 1991, Resolution of the ocean wave propagation direction in SAR imagery.IEEE Transactions on Geoscienceand Remote Sensing, (29)105-112.

[14] Vachon, P.W., Krogstad, H.E., and Paterson,J.S., 1994, Airborne and spaceborne synthetic aperture radar observations of ocean waves. Atmosphere-Ocean, (32) 83-112.

[15] Vachon P.W., Liu, A. K., and Jackson, F.C., 1995, Near-shore wave evolution observed by airborne SAR during SWADE. Atmosphere-Ocean, (2) 363-381.

[16] Zelina, Z.I., A., Arshad, S.C.Lee, S.Japar, A.T.Law,R.A. Nik Mustapha, M.M., Maged (2000). East Coast of Peninsular Malaysia. *Sea at The Millennium: An Environmental Evaluation. In (Ed) Charels Sheppard. Oxford.Vol II, 345-359.*

IOP Conf. Series: Earth and Environmental Science **20** (2014) 012020 doi:10.1088/1755-1315/20/1/012020

Assessing shoreline response to three submerged breakwaters at Kerteh Bay, Terengganu, Malaysia using Landsat imagery

Lawal Abdul Qayoom Tunji, Khamaruzaman Wan Yusof,
Ahmad Mustafa Hashim and Nasiman Sapari

Civil Engineering Department, Universiti Teknologi PETRONAS, 31750 Perak, Malaysia; Tel: + 601923629

E-mail: tunjilal@yahoo.com

Abstract. As part of a project to determine the exact structural and environmental parameters governing the mode and magnitude of salient formation behind a submerged breakwater, a remote sensing technique is being adopted to assess the extent of erosion/accretion at Kerteh Bay, Terrengganu, Malaysia. Multi-temporal Landsat satellite images of coarse resolution for the years of 1994, 2000, 2006, 2009 and 2012 were acquired for this purpose. The images were subsets divided into smaller areas of interest and classified using supervised classification of support vector machine. The classified image is then vectorized to extract shoreline based on waterline in each of the subset rasters images. Tidal correction were adopted to correct the waterline/shoreline to the mean sea level (MSL) datum. Comparison of corrected shorelines was carried to obtain the extent of erosion/accretion at the Kerteh Bay, Terrenganu, Malaysia. It was observed that substantial accretion was observed between the years 1994-2006 at the upper part of the study area, the part between northern part and the southern part also experienced accretion but not as much as compared to northern part for the same year. Erosion was noted between the years 2006-2012 for all of the areas of the study area but the rate slowed down between the years 2009-2012 for all the areas. Slope estimated from the imageries were compared with in situ slope of the same area, this served as a validation for the method used.

1. Introduction

1.1 Background

Emergent coastal structures, such as groynes, detached offshore breakwaters, and sea walls have been successfully adopted as coastal protection measures for many decades (3, 18). This type of breakwaters is common in the US and Europe (3), and even more so in Japan, where (17) reported the completion of over 4,000 emergent breakwaters by the mid-1980s. However, these types of highly intrusive and aesthetically unappealing engineering structures are becoming increasingly unpopular among the more environmentally aware modern communities. As a result, submerged breakwaters (SBWs), which do not impair amenity or aesthetics, are becoming a preferred option for coastal protection (1, 15 and 11).

Content from this work may be used under the terms of the Creative Commons Attribution 3.0 licence. Any further distribution of this work must maintain attribution to the author(s) and the title of the work, journal citation and DOI.
Published under licence by IOP Publishing Ltd

However, SBWs have rarely been adopted for coastal protection in the past and therefore, their efficacy remains largely unknown. Furthermore, enhanced shoreline erosion has been reported in the lee of the structures at several SBW projects. (15) carried out a comprehensive review on documented projects of submerged breakwater in different parts of the world, unfortunately not many of these type of project are available, from the ten projects reviewed erosion were noticed at the lee of seven (7) of the projects.

1.2 Study Area

The study area is located within a town called Kerteh in the district of Kemamam in Southern Terengganu, Malaysia, about 30 km or 20 minutes' drive north of Chukai. Kerteh is the base of operations for Petronas in Terengganu, overseeing the oil platform operations off the state's coast. Kemamam is a district of 2,536 km^2 area with a population of 174,876. It geographical location is 4° 31′ 38″ N and 103° 28′ 9″ E. The stretch of the beach protected is approximately 2100 m. The study area is characterized with much of its coast to be a series of large and small hook-shaped bays, fully exposed to direct wave attack (especially during the NE-monsoon) from the South China Sea. The geomorphologic feature of Kerteh bay is such that its development is controlled by protruding headlands. Most of the bays along this region are considered to be in dynamic equilibrium; this is when constant supply of material from upcoast or within its embayment is passing through the bay and beyond the downcoast headland. The littoral drift rate, associated with the dynamically stable configuration of Kerteh Bay, has been computed to be some 210,000 m^3/yr of which more than 80% is transported during the NE-monsoon period

The cause of the coastal erosion at the study area Kerteh bay, was studied by (20). Some major causes were highlighted by the researchers. The beach platform at Kerteh bay is such that there exists a continual longshore sediment transport from upcoast to downcoast, disruption of this dynamic stability may easily occur when upcoast sediment supply is (partly) cut off which can result into erosion of the coast leading to a larger indentation of the bay configuration. If the entire upcoast sediment supply is cut off, the bay would become even more indented until littoral drift ceases. Another factor identified is the cross-shore sediment transport, although it was reported that this factor doesn't have as much effect as the longshore sediment transport. The direction and intensity of this transport phenomenon are ruled by the wave steepness, geometry of the seabed slope and the size of the seabed particles. The beach will accrete at moderate wave conditions and recede under severe wave attack

The third factor is the supply of sand by S. Kerteh river discharge. Unlike the larger rivers in the region such as S. Terengganu, S. Dungun and S. Kemaman, it was found that S. Kerteh River only drains a very limited catchment and it is unlikely that its sediment yield will be of significance for beach stability. The last factor identified to be affecting the stability of the coastal area within Kerteh Bay is considered to be the human activity such as removal of natural dune systems and vegetation.

An interesting phenomenon that affects the coastal erosion at Kerteh Bay is the upcoast sediment supply from Paka Bay which is largely transferred into Kerteh Bay through offshore bar bypassing at the northern end of the bay. The "supply point" on to the coast of Kerteh Bay is located immediately updrift from Rantau PETRONAS Complex, which makes this coastal stretch particularly vulnerable to any disruption of the equilibrium situation. This is reflected in the shoreline mapping from 1966 to 1987; the observed erosion over this period would indicate an average deficit in the upcoast sediment supply of some 40,000 m^3/yr. The causes and persistency of this deficit in unknown, but quite likely originate from S. Kerteh influence (disruptive of bar bypassing) and from shore developments within the upcoast Paka Bay, undertaken since the late sixties.

From the findings of the cause of coastal erosion at Kerteh Bay, (20) proposed some mitigation measures. By careful examination of the geomorphologic situation of the Kerteh Bay and acknowledging the cause of erosion at the Rantau Petronas Complex, various defense schemes were proposed. An artificial supply of sand (beach nourishment, perched beaches), structures to prevent waves from reaching erodible materials such as bulkheads, seawalls, revetments and offshore breakwaters and the last being structures to slow down the rate of littoral transport such as groynes (trapping the sediment) or offshore breakwaters (reducing the wave energy in the coastal zone) were all among the methods proposed.

The beach nourishment and use of three offshore submerged breakwaters were adopted for the purpose of mitigating the coastal erosion at the affected area as it the time of the study carried out by (20). Subsequently, three submerged breakwaters with beach nourishment to mitigate the coastal problem at a specific time. The solution could be considered relatively effective considering the problem at that time but no monitory survey has been carried out since the installation to evaluate the performance of these structures but erosion has been noticed at the south and north of the protected area. The objective of this study is to assess the shoreline response towards submerged breakwaters that was constructed at Kerteh Bay, Malaysia using Landsat Imagery.

2. Material and Methods

2.1 Field profile survey

Total station method of survey has been in place for quite a while. It is method of survey that involves the use of total station equipment to perform survey. In this field work, it was used to carry out beach profile survey, among the parameters that total station measures are angles, both vertical and horizontal angle and also distances, both slopes distances and horizontal distances. These parameters were adopted to compute the elevation at each point on each profile line. Figure 1 show the survey path followed with respect the SBWs installed, total of 2.6 kilometres was survey.

The equipment's used for the survey are as follows;

1. Total station (1 unit)
2. Mini-prism (2 unit)
3. Tripod (2 units)
4. 100 meter rule (1 unit)
5. Tape (1 unit)
6. Walkie talkie (2 units)
7. Staffs (2 units)
8. Prism (1 unit)

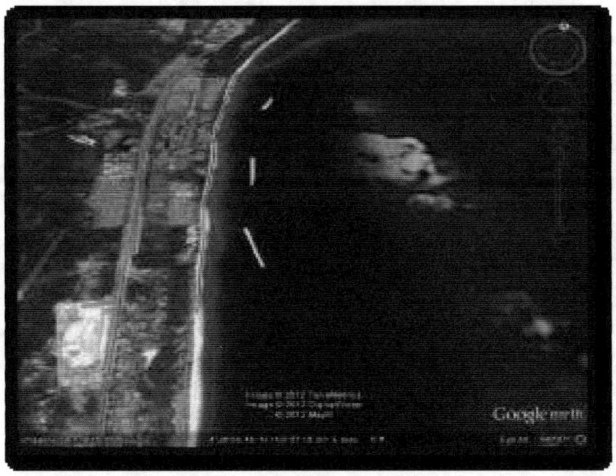

Figures 1: Survey path.

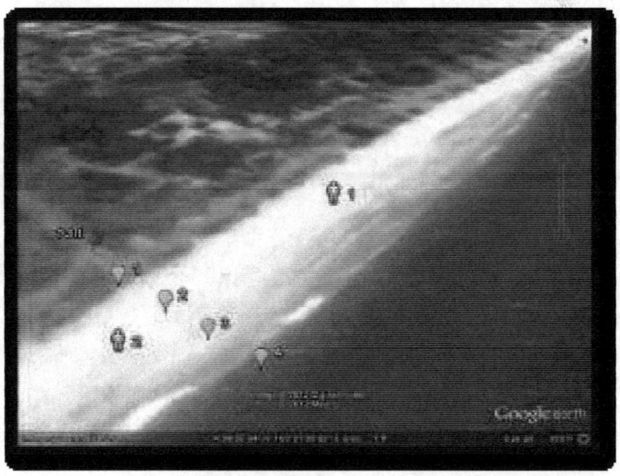

Figure 2: Survey description.

The steps of survey are as follows;

1. Identify/locate position of permanent bench marks, if none create temporary bench marks to identify each profile line.
2. Set out the equipment for survey.
3. Ensure to level the equipment properly.
4. Ensure to place the total station at a convenient position so as to capture as many profile lines as possible, if possible 3 profile lines.
5. The survey should involve at least two persons, one person operates the total station and the other moves the prisms from one spot to the other.
6. Determine the coordinate of the first bench mark, if not available use an arbitrary bench mark (NEZ: 1000, 1000, 100).
7. From the first bench mark capture the first Back sight, and record the Vertical angle, Horizontal angle and Slope distance, the equipment computes the appropriate NEZ from these values. This survey is only interested in the Z value which is the elevation.
8. Also from the first bench mark capture the first point on the first profile line Fore sight, Also record the vertical angle, horizontal angle and slope distance. This should be done for all of the point on the first profile line without change the position of the total station.
9. Repeat this for all of the other 25 profile lines.

2.2 Remote sensing

In recent years the numerous studies of quantitative aspects of coastline structure variously focused on a range of data extraction approaches that can be categorized as sensor, band and visual selection. In sensor selection, active and passive sensors are selected to delineate waterlines between water bodies and coastal zones. Although synthetic aperture radar (SAR) images have advantages over those based on other approaches, including near infrared (NIR) and shortwave infrared (SWIR), optical imagery is effective depending on broad ground coverage, cost-effectives and availability of data1996: (23). Thus, many studies have assessed the relative merits of various sensors (22) Landsat series are useful for deriving waterline information because they represent long-term historical data and provide repetitive, synoptic multispectral images with global coverage (23) Landsat series are useful for deriving waterline information because they represent long-term historical data and provide repetitive, synoptic multispectral images with global coverage (23). This study involves data collection over a global coverage and repetitiveness is of necessity. Such imagery characteristics were found by the adoption of Landsat imagery.

2.2.2 Waterline extraction

In the present study 15 cloudless satellite images from LANDSAT satellites were collected from the years 1996, 2000, 2006, 2009 and 2012, three images were collected for each of the year. The spatial resolution of the images used is 15 m, Table 1contains the shooting time of the image, image resolution, image precise date and day, tidal level corresponding to each of the scenes.

2.2.3 Determination of shoreline

To determine the shoreline a sophisticated method called one-line model or shoreline change model is adopted. The method of shifting waterlines to the tidal-datum-based shoreline position on mean sea level (MSL) is based on the concept of one-line model for shoreline evolution and called one-line shift method (OSM). It is assumed that beach moves offshore or onshore with one bottom profiles. Fig 3 shows three beach profiles at three different times: t_i $(i = 1,2,3)$. At time t_i, the waterline is located at x_i, away from the origin of the transformed coordinates, the corresponding water depth is h_i above or below MSL. When the sea surface is at MSL, the MSL-datum-based shoreline is located at z_i away from the origin. Figure 3 illustrates an example of a beach profile moving from the right to the left.

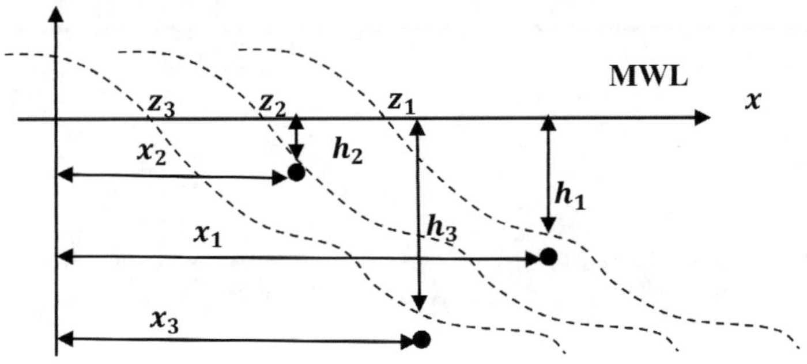

Figure 3: Beach profiles at three different times and symbolic notation.

If the extracted waterline from satellites images at time $t_2$and t_3 are located at x_2 and x_3, respectively, x_3>x_2. Extracted waterline without consideration of tidal effect imply that the beach moves from left to the right. This inference conflicts with assumption for Figure 3 shifting the extracted waterlines to the MSL-datum-based shoreline position is necessary to accurately estimate the beach movement.

Table 1: Satellite imageries descriptions with corresponding tidal levels.

Masters Image					Slave Image				
Date	Time	Sensor	Resolution (m)	Tidal level (cm)	Date	Time	Sensor	Resoltion (m)	Tidal level (cm)
4/11/1994	10:42:09 A.M	TM	15	-23	4/11/1994	10:42:09A.M	TM	15	-23
					3/7/1995	11:15:00 A.M	TM	15	-50
					9/2/1995	11:15:00 A.M	TM	15	70
2/7/2000	10:56:10A.M	TM	15	-21	2/7/2000	10:56:10 A.M	TM	15	-21
					4/3/2000	11:14:29 A.M	TM	15	-35
					5/8/2001	11:12:01 A.M	TM	15	85
9/11/2006	11:11:34 A.M	TM	15	43	9/11/2006	11:11:34 A.M	TM	15	43
					10/29/2006	11:11:43 A.M	TM	15	-23
					4/17/2005	11:10:27 A.M	TM	15	-10
2/7/2009	11:11:50 A.M	TM	15	-115	2/7/2009	11:11:50 A.M	TM	15	-115
					7/1/2009	11:12:24 A.M	TM	15	17
					1/22/2009	11:11:45 A.M	TM	15	-87
4/20/2012	11:16:09A.M	TM	15	27	4/20/2012	11:16:09 A.M	TM	15	27
					5/22/2012	11:16:39 A.M	TM	15	103
					7/23/2011	11:15:31 A.M	TM	15	37

IOP Conf. Series: Earth and Environmental Science **20** (2014) 012020 doi:10.1088/1755-1315/20/1/012020

Based on past research it is reasonable to assume that a beach face has an approximately uniform slope. The bottom slope, s, is defined by

$$\frac{h_i}{x_i - z_i} = s, (i = 1,2,3) \qquad (1)$$

In which the subscript i denotes the physical quantity at time t_i and both s and z_i are unknown at this point. It is assumed that the beach moves at a steady velocity during a period of t_1 to t_3. The moving speed of the beach profile can be expressed as;

$$v = \frac{z_j - z_i}{t_j - t_i} = \frac{\Delta z_{ji}}{\Delta t_{ji}} \quad (i, j = 1, 2, 3) \qquad (2)$$

Two different times, between $t_1$and t_2 and between t_2 and t_3, are chosen. Equating the moving speeds during two time differences and then inserting the result into eq (2) gives the bottom slope as

$$s = \frac{\Delta h_{32} - \frac{\Delta t_{32}}{\Delta t_{21}} \Delta h_{21}}{\Delta x_{32} - \frac{\Delta t_{32}}{\Delta t_{21}} \Delta x_{21}} \qquad (3)$$

Where the operator means the difference of a physical quantity from time t_i to t_j. Substituting eq. (3) into eq (2) yields the MSL-based shoreline position

$$Z_i = x_i - \frac{h_i\left(\Delta x_{32} - \frac{\Delta t_{32}}{\Delta t_{21}} \Delta x_{21}\right)}{\left(\Delta h_{32} - \frac{\Delta t_{32}}{\Delta t_{21}} \Delta h_{21}\right)} \qquad (4)$$

When three sequential waterlines are detected at t_i $(i = 1, 2, 3)$, the corresponding offshore distance xi is known and the water depth hi is obtained by eq. (1). The MSL-datum-based shoreline position is finally determined by eq. (4). A non-dimesionalized ratio of two time difference on Eqs. (3) And (4) shows that the unit of t is independent of estimating the shoreline position. Time unit in the calculation of time difference is hour. When the shoreline positions determined by the proposed OSM at times ti are obtained, the movement of the beach is then evaluated for further investigation. (2)

3. Results

As stated in Section 2, Landsat multi-temporal images of the study area with 15 m spatial resolution were acquired. These images are 15 m resolution of panchromatic form. Figure 4 shows the unprocessed image. Three images each were acquired for the years 1994, 2000, 2004, 2009 and 2012. The three images in a year were required for the purpose of correcting tides for each year. Using the OSM method of tidal correction it is required to use images of a particular area taken at most six months apart, this will ensure the images used for the tidal correction are taken at a time when there is no or negligible change in bottom slope.

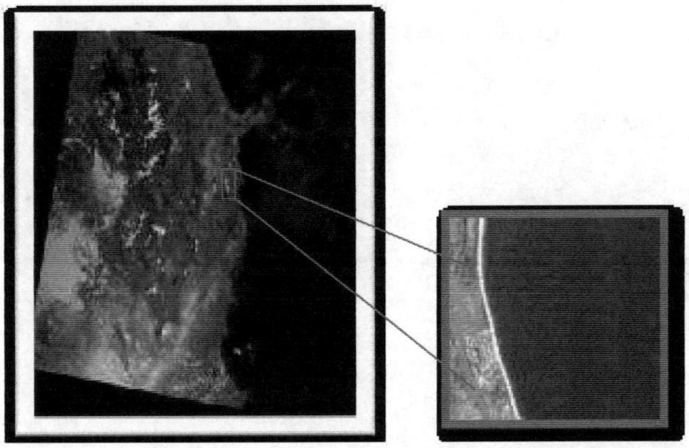

Figure 4: Showing sub-setting of unprocessed image.

After the application of tidal correction using the OSM method, changes were evaluated using the overlaying method. This was performed in the ENVI 4.9 software. The graphic representation of this process is as shown in Figure 5.

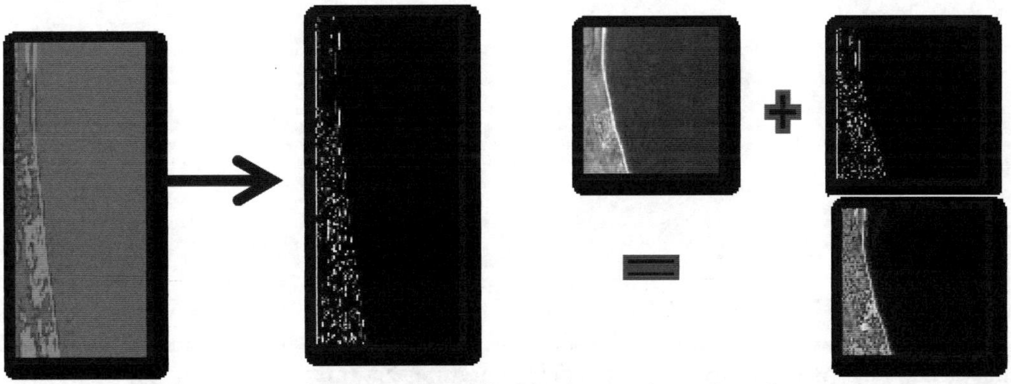

Figure 5: Vectorization of classified images and overlay of vector on raster image.

As part of the process to obtain the key parameters that govern the mode of shoreline change and the magnitude at the lee of a breakwater, beach profile surveys were also conducted. The study area was divided into three major parts for the sake of analysis of the beach survey. The divided study area consists of a staff quarters, a golf course and a school. The staff quarters is situated at the northern part of the study area, the golf course is in between the staff quarters and the school. The school can be regarded as situated at the southern part of the study area. Figures 5 through 7 shows graphical representation of analysis of erosion that occurred at the staff quarters, golf course and school respectively as obtained from the remote sensing analysis.

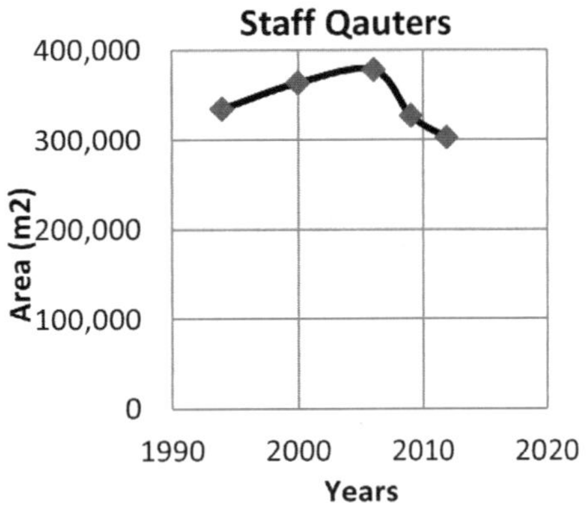

Figure 5: Showing erosion at the staff quarters.

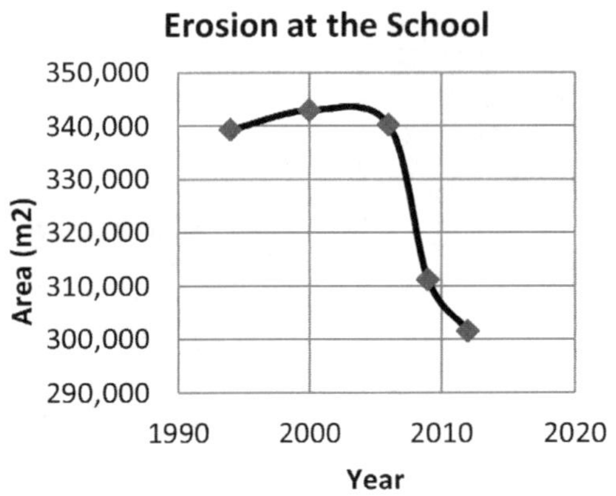

Figure 6: Showing erosion at the school.

Figure 7: Showing erosion at the Golf course.

Results from the beach profile survey conducted at the study area are presented in Figure 8 through 12. These plots show a comparison of survey results conducted at different times.

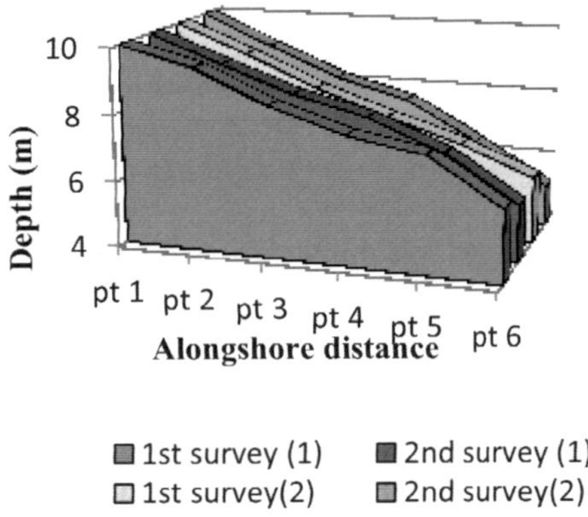

Figure 8: Plots of survey line 1 &2 for 1st and 2nd survey

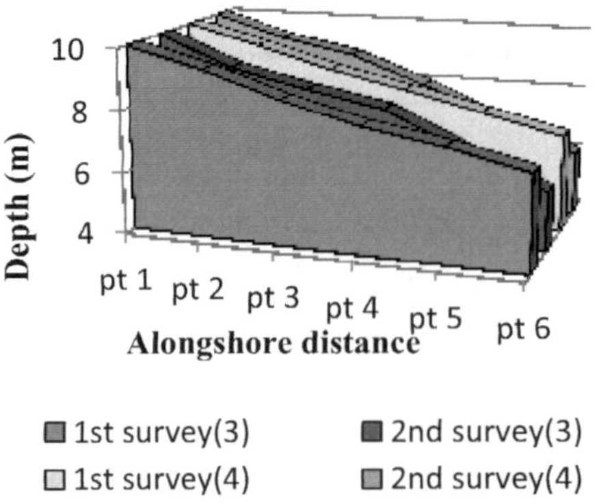

Figure 9: Plots of survey line 3 & 4 for 1st and 2nd survey

IOP Conf. Series: Earth and Environmental Science **20** (2014) 012020 doi:10.1088/1755-1315/20/1/012020

The study area was divided into 60 survey lines with survey lines 50 meters apart. Six points on each survey line were surveyed for the 1st and 2nd survey exercise. Then these points are compared in plots as shown in Figures 3.4 through 3.8. This comparison shows the extent of erosion or accretion between the 1st and 2nd survey exercise.

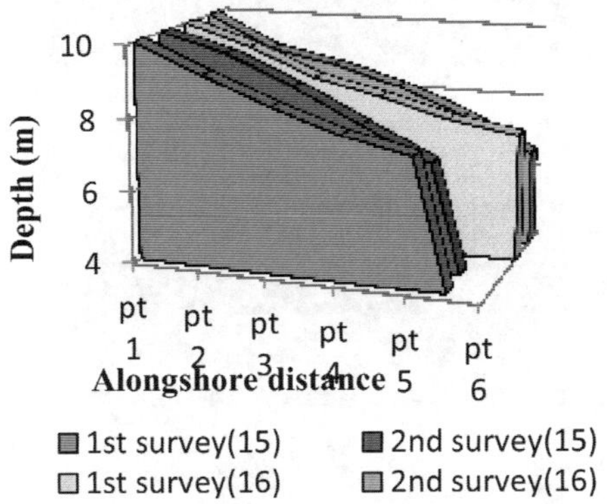

Figure 10: Plots of survey line 15 & 16 for 1st and 2nd survey

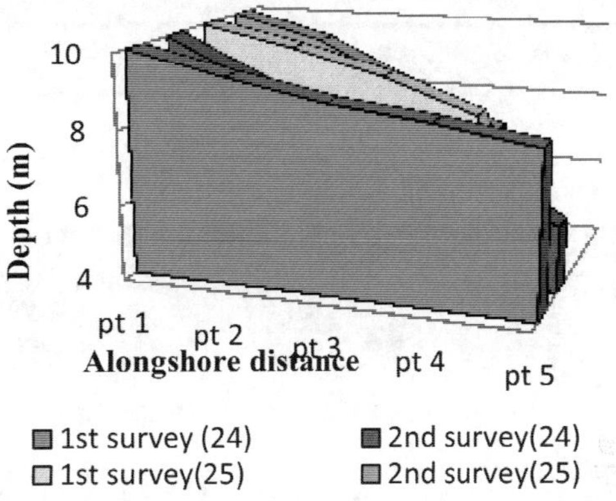

Figure 11: Plots of survey line 22 & 23 for 1st and 2nd survey

While Figures 8 and 9 shows plots of surveys line towards the southern part of the study area, Figure 10 shows plots of survey somewhere in between the southern part (school) and the northern part (staff quarters).

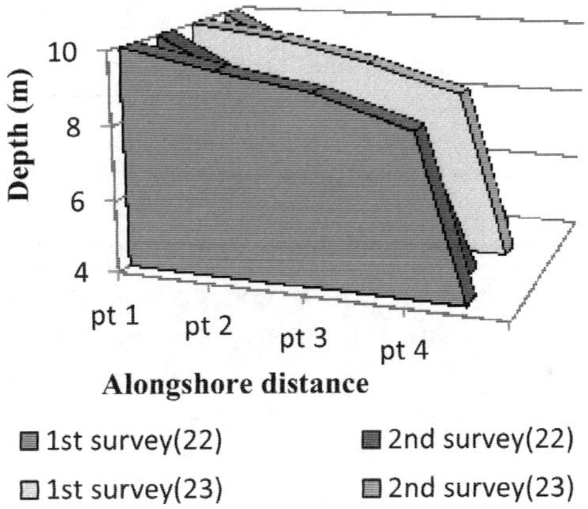

Figure 12: Plots of survey line 24 & 25 for 1st and 2nd survey

Figure 12 show plots representing line 22, 23, 24 and 25, these are the survey lines at the northern part of the study area, taken just in front of the staff quarters in the study area. To complement the graphical plots of the two surveys carried out on the study area, pictures were taken during the two surveys. A close look of the pictures amplifies the results obtainable from the plots of the surveys. Figures 13, shows pictures of the study area during the first survey and Figure 14 shows pictures of the study area during the second survey.

Figure13: Pictures of 1st survey

Figure 14: Pictures of 2nd survey

4. Discussion

4.1 Remote sensing

Landsat images acquired were processed as described in Section 2. The tidal correction was done using the OSM method as described in Section 2. The results obtained by processing these medium Landsat images indicate erosion and accretion were taken place at different years at the three different parts of the study area.

It was found that accretion occurred at the Staff Quarters, which is the northern part of the study area between the years 1994-2000, the area affected by accretion can is about 28,800 m^2 of area. Between 2000-2006, there was also accretion at the same area but lesser intense than that of 1994-2000. The area affected was about 14,400 m^2. In the year 2009-2012 erosion of magnitude 51,075 m^2 were observed.

For Golf course area, between 1994-2000, lesser accretion was noted as compared to the Staff quarters area. The accretion covered an area about 10,800 m^2. At the Golf course area, accretion occurred between the years 2000-2006, covering an area 25,200 m^2. Excessive erosion occurred between 2006-2009 in this area. The magnitude of the erosion is about 66375 m^2. In the subsequent years 2009-2012 erosion of about 8235m^2 occurred.

In the school area, accretion was also noted between the years 1994-2000 covering an area of 3600 22 m^2 and unlike the other two areas (staff quarters and golf course) less erosion occurred between 2000-2006 covering an area of 2700 m^2. However, erosion during the years 2006-2009 and 2009-2012 were at higher rate with values of 29025 m^2 and 9675 m^2 respectively.

4.2 Field survey

The second field trip revealed an alarming erosion rate that had occurred between the period of the first and second visit. Visual observation of the site reveals more erosion occurred at the northern part of the study area as compared to the southern part of the study area. Figures 11 and 12 indicate that more serious erosion occurred at the northern part as compared to the southern part. Figures 11 and 12 are plots of the profile lines 22, 23, 24 and 25, which represent survey lines at the northern part of the study area and Figure 8 and 9 shows plots of profiles 1, 2, 3 and 4, which represent profile lines at the southern part of the study area. The erosion rate in the area between the northern part and the southern parts can be considered as moderate. More erosion was found in the northern part as compared to the southern part as depicted in Figure 10.

5 Conclusion

It can be concluded that erosion is still occurring at the study area. Over a period between March, 2012 and July 2012 a reasonable amount of erosion had occurred at the protected area (study area).

The Landsats images indicate that substantial accretion occurs at the staff quarters and golf course between the years 1994-2006. Less accretion was found at the school area as compared to the staff quarters and golf course areas for the same period. Erosion was more intensive between the years 2006-2012 for all of the areas but the rate slowed down between the years 2009-2012 in all of the three areas. Landsat imagery was found to be useful in the assessment of the erosion rate the study area.

References

[1] Burcharth H., Hawkins S. Zanuttigh B. and Lamberti A.,(2007) Environmental Design Guidelines for Low Crested Structures, Elsevier, The Netherlands 400 pp.

[2] Wei-Wei Chen and Hsein-Kuo Chang, (2009), Estimation of shoreline position and change from satellite images considering tidal variation, Estuarine, Coastal and Shelf 84 (2009 54-60pp.

[3] Dean R. G and R. A Dalrymple, Coastal Processes with Engineering Applications, Cambridge University Press (2001) 488p.

[4] Dean, R. G., Browder, A., Goodrich, M.S., Donaldson, D.G., 1994. Model tests of the proposed P.E.P. reef installation at Vero Beach, Florida. Tech. Rep. UFL/COEL-94-012, Coastal and Oceanographic Engineering Department. University of Florida, Gainesville, FL. 28 pp.

[5] Deguchi, I., Sawarangi, T., 1986. Beach Fill at two coasts of different configuration. Proc. 20th International Conference on Coastal Engineering ASCE, Taipei, Taiwan, pp. 1032-1046.

[6] Douglas, L., Weggel, J.R., (1987) Performance of a perched Beach-Slaughter Beach, Delware. Proc. Coastal Sediments '87. ASCE, pp. 1385-1398.

[7] Funakoshi, H., Shiozawa, T., Tadokoro, T., Tsuda, S. 1994. Drifting characteristics of littoral sand around submerged breakwater. Proc. International Conference on Hydro-technical Engineering for Port and Harbor Construction, Yokosuka, Japan, pp. 1157-1178.

[8] Hanson H and Kraus N, Shoreline response to a single transmissive detached breakwater, Proceedings of 22nd International Conference on Coastal Engineering ASCE, Delft, The Netherlands (1990). Pp. 2034-2046

[9] Komar, P.D., 1998. Beach Processes and Sedimentation. Prentice Hall, Upper Saddle River, New Jersey, 544pp.

[10] Kraus, N. C., Gingerinch, K.J., Rosati, J.P., 1989, Duck85 surf zone sand transport experiment. U.S Army Engineer Waterways Experiment Station, Tech. Report CERC-89-5

[11] Lamberti, A., Mancinelli, A. Italian experience on submerged barriers as beach defence structures. Proc. 25th International Conference on Coastal Engineering. ASCE, Orlando, USA, pp. 2352-2365.

[12] Nir, Y., (1982). Offshore artificial structure and their influence on the Isreal and Sinai Mediterranean Beaches. Proceedings of the 18th International Conference on Coastal Engineering, 1857-1855 pp.

[13] Pope, J., & Dean, J. L. (1986). Development of design criteria for segmented breakwaters, Proceedings 20th international conference on coastal engineering. American Society of Civil engineers, Tapei. Pp. 2144-2158.

[14] Ranasinghe R., Hacking N., and Evans P., Multi-function artificial surf breaks: a review, Report No. CNR 2001.015, NSW Dept. of Land and Water Conservation, Parramatta, Australia (2001)

[15] Ranasinghe R., Turner I., and Symonds G., Shoreline Response to Submerged Strutures: A review Coastal Engineering. Vol. 53, 65-79

[16] Ranasinghe R. and S. Sato, Beach morphology behind impermeable submerged breakwater under obliquely incident waves. Coastal Engineering (2007). Pp. 1-24

[17] Seiji, W. N., Uda, T., Tanaka, S., Statistical study on the effect and stability of detached breakwaters. Coastal Engineering in Japan 30 (1), 121-131.

[18] Silvester R. and Hsu J., Coastal stabilization, World Scientific Publishing Co., Singapore (1997) 578 pp.

[19] Stauble, D. K., Tabar, J.R., Smith, J.B., (2000) Performance of a Submerged breakwater along a hardbottom influenced coast: Vero Beach, Florida. Proc. 13th National Conference on Beach Preservation Technology, Melbourne, Florida, USA, pp. 175-190.

[20] Tilmans W.M.K, W.H.G. Klomp & H.H. de Vroeg 1992. Coastal erosion – the Kerteh Case. Analysis of causative factors and mitigative measures using dedicated mathematical modeling tools, *International Colloquim on Computer Applications in Coastal and Offshore Engineering.*

[21] Tomassicchio U., Submerged breakwaters for the defense of the shoreline at Ostia: field experiences, comparison . Proc 25th International Conference on Coastal Engineering. ASCE Orlando, USA, pp. 2404 -2417.

[22] Yamano, H., Shimazaki, H.S., Matsunaga, T., Ishoda. A., McClennen, C., Yokoki, H., Fujita, K., Osawa, Y., Kayanne, H., (2006). Evaluation of various satellite sensors for waterline extraction in a coral reef environment: Majuro Atollm Marshall Islands. Geomorphology 82, 398-411

[23] Zhao, B., Guo, H., Yan, Y., Wang, Q., Li, B., (2008). A simple waterline approach for tidelands using multi-temporal satellite images: a case study in Yangtze delta. Estuarine, Coastal and Shelf Science 77, 134-142

A new method to determine eroded areas in arid environment using Landsat satellite imagery

Aydda A*, Algouti Ah, Algouti Ab, Essemani M, Taghya Y

Department of Geology, GEOBASSMA Laboratory, University of Cadi Ayyad, Faculty of Sciences Semlalia, BP2390, Marrakech, Morocco.

*E-mail Address: ayddaali@gmail.com

Abstract. Erosion (by water or wind) is an increasing problem for many local authorities and government agencies throughout the world. The identification of eroded areas in arid and humid regions can be very useful for environmental planning and can help reduce soil and sediment degradation in these regions. In this work we present a new method to determine eroded areas in arid environment. In this method were explored lithological data to determine eroded areas. These data were collected in the field using GPS (Global Positioning System) checkpoints and geological maps. For that, two lithological maps of the study areas were analysed to determine lithological data change. Those two maps were obtained from the classification algorithm by applying the maximum likelihood on two Landsat satellite images. After images classification and validation a change detection technique was adopted to determine eroded areas. This method was applied in northern part of Atlantic Sahara desert to confirm their potentiality.

1. Introduction

Erosion is any process, natural or man-made, by which material is removed from one location and deposited in another. In order for erosion to occur, an erodible material must be exposed to some form of energy or eroding force. Erosion is one of the principal mechanisms of desertification at national and regional levels. Is an increasing problem for many local authorities and government agencies throughout the world. The identification of eroded areas in national and regional levels can be very useful for environmental planning and can help reduce soil and sediment degradation in this region.
The use of remote sensing techniques has been shown to have potential for identification of eroded surfaces on regional scales [1, 2, 3, 4, 5, 6, 7]. Several methodologies applied to mapping of erosion area include spectral data [3, 5, 8], vegetation indices [2, 9, 10] and combinations of remote sensing and morphological data [11].

The Landsat Thematic Mapper (TM) and the Enhanced Thematic Mapper (ETM+) have been used to monitory changes in landforms and land degradation mechanisms [12], and forest cover change and deforestation mechanisms [13, 14]. Further, Many studies have shown the importance of Landsat imagery for lithological mapping in arid environment [15, 16]. No studies explored the lithoilgical data to define eroded areas.

The arid environment cover more than one fifth of the Earth's land, and they are found on every continent. These areas exist under a moisture deficit, which means they can frequently lose more

Content from this work may be used under the terms of the Creative Commons Attribution 3.0 licence. Any further distribution of this work must maintain attribution to the author(s) and the title of the work, journal citation and DOI.
Published under licence by IOP Publishing Ltd

moisture through evaporation than they receive from annual precipitation. When vegetation cover is scarce and soil and sediments are clear were explored lithological data to determine eroded areas in these zones.

2. Methodology

The methodology for this study is shown in Figure 1. In this method were explored lithological data to determine eroded areas. These data were collected in the field using GPS (Global Positioning System) checkpoints and geological maps. For that, two lithological maps of the study areas were analysed to determine lithological data change. Those two maps were obtained from the classification algorithm by applying the maximum likelihood on two Landsat satellite images. This classification was adopted for both sets of Landsat images by delimiting polygons around representative points collected in field. After images classification and validation a change detection technique was adopted to determine eroded areas.

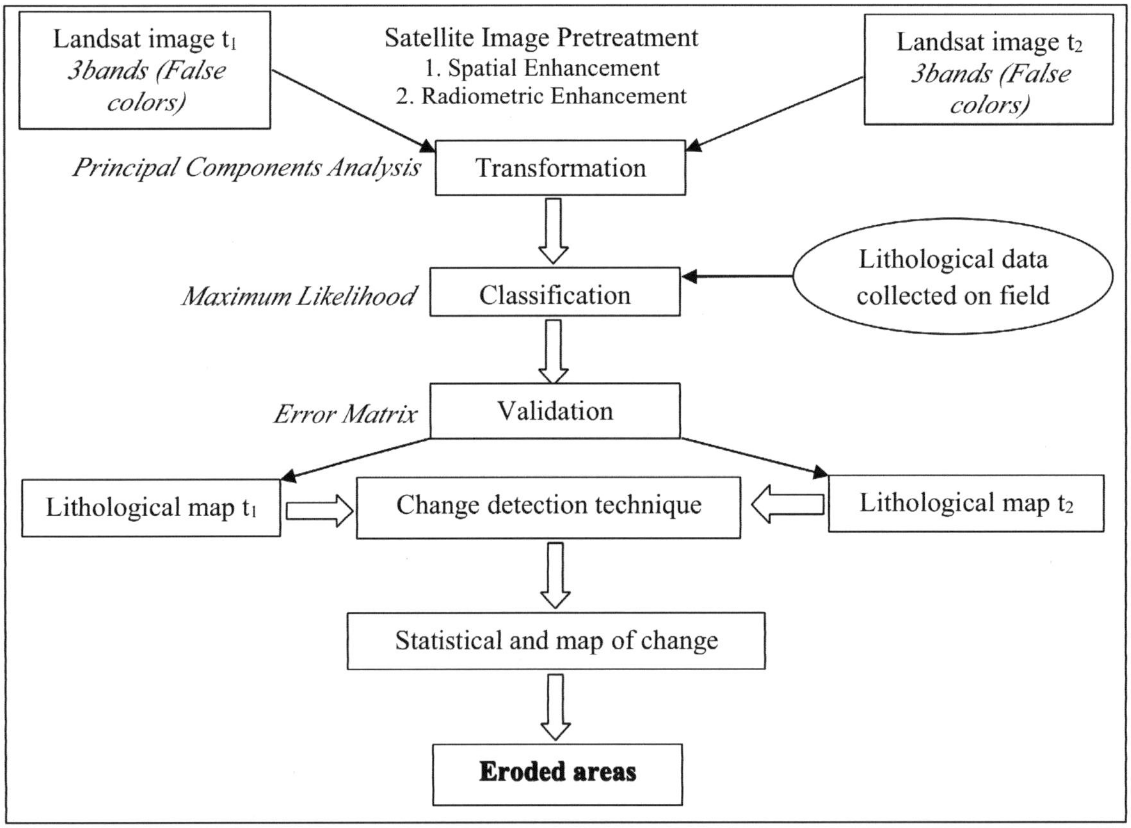

Figure 1. The methodology of this study.

3. Case of study: A northern part of Atlantic Sahara desert (SW of Morocco)

3.1. Study area

The study area is located in northern of Sahara Atlantic (SW of Morocco). It is a vast coastal platform, extending from 12°0′ 35″ to 12°23′ 55″W in longitude and 27°53′ 6″ to 28°8′ 27″N in latitude (figure 2).

Climatically, this region is a part of the Boreal domain of maritime trade winds, where precipitation is less than potential evapotranspiration. This wind is one of the most regular winds in the world [17]. Thus, the dry season extends for about 12 months. The amount of precipitation is about 33 mm per year. The yearly mean maximum air temperature is 26 °C and the yearly mean minimum air temperature is 19 °C. The extreme aridity of this area is primarily caused by large-scale atmospheric subsidence due to the Açores anticyclone.

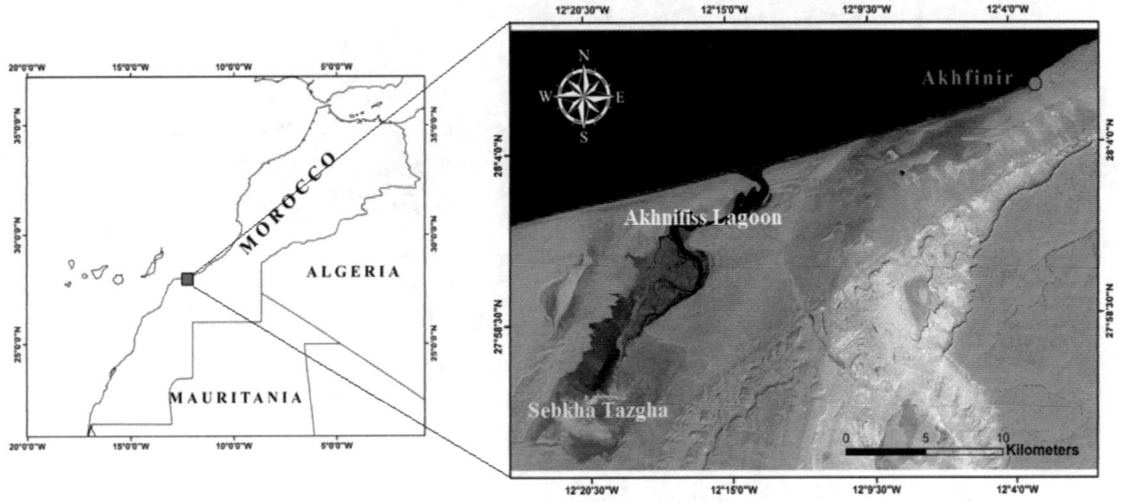

Figure 2. Location map of the study area.

3.2. Lithological data

The surface of the Akhfinir region is dominated by flat layering of hard rocks at the surface (Moghrebian Sandstone-Limestone Slab). This flagstone, is overlain by Pleistocene consolidated dunes, and broken, in another place, by sabkha depression where Miocene marls and silts were deposed. Ideally the surface lithology interpretation is validated in the field using GPS checkpoints and geological maps.

In this region eight facies were distinguished: sand, beach rocks, salt crust, sand-limestone slab, alluvium facies, marl and silt, limestone, and mud. Description of theses facies classes are presented in Table 1.

Table 1. Description of different facies classes of the study area.

Class	Description
Sand	is principal component of dunes
Salt crust	is sabkha components, is sand and clay cimented by the alkaline minerals
Beach rocks	is Pliocene terrigenous sandstones facies
Sand-limestone slab	is a hard rocks Moghrebian facies
Alluvium facies	is part of the recent Pleistocene sediments, are conglomeratics sand
Marl and silt	are Miocene transgressive deposit of sabkhas
Limestone	is Senonian (Later Cretaceous) transgressive deposit
mud	is a principal component, with sand, in mudflats and salt marshes

3.3. Results

The results of classification are shown in figure 3 and 4. The results obtained from the application of confusion matrix to the validation data show that all of the facies are better classified, with a high overall prediction accuracy, and low rates of both commission and omission errors.

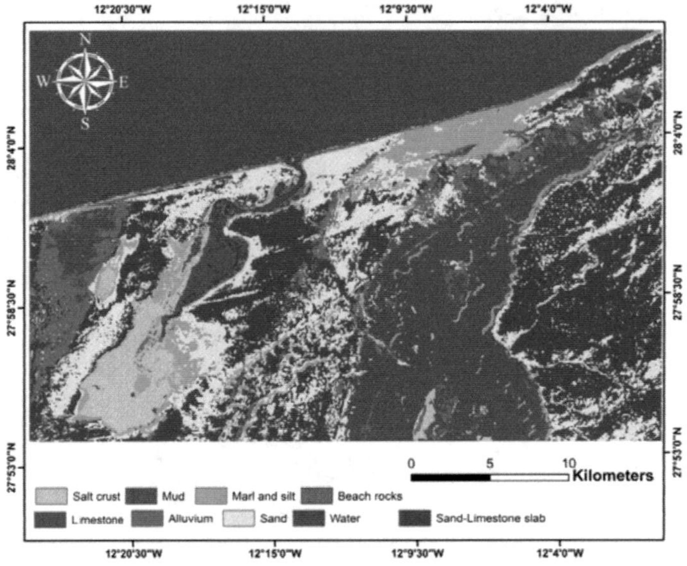

Figure 3. Lithological map of study area after classification using Landsat TM 1987.

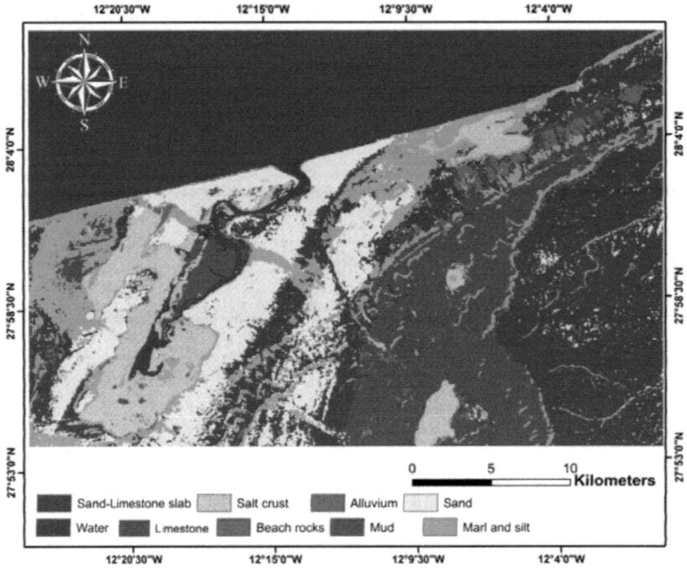

Figure 4. Lithological map of study area after classification using Landsat ETM 2011.

A change detection statistics and the difference map operations were carried out for the classification results of 1987 and 2011 images in order to produce the statistical data about the spatial distribution of different lithological classes (table 2) and a eroded areas map (figure 5). Examination of statistical change (table 2) is beneficial to ascertain the reasons behind the observed changes.

Table 2. Change in lithological classes between 1987 and 2011 (unit: Km^2). Classes signification 1: Water, 2: Limestone, 3: Salt crust, 4: Alluvium, 5: Mud, 6: Sand, 7: Marl and silt, 8: Beach rocks, 9: Sand-Limestone slab.

		1987									
		1	2	3	4	5	6	7	8	9	Sum
2011	1	279.11	0.92	0	4.72	0	1.18	0.41	9.03	0	295.37
	2	0	138.01	0.3	5.13	0	5.16	0.73	0.03	0.41	149.77
	3	0.03	4.39	29.07	3.09	0.56	8.19	18.77	1.24	8.74	74.08
	4	0	5.62	0	7.22	0	0.73	0.03	0	0.26	13.86
	5	0	0	1.72	0	10.5	0.02	0.22	0.79	0.03	13.28
	6	0.01	0.43	5.36	0.62	1.22	65.6	2.42	0.22	59.77	135.65
	7	0.02	13.74	2.8	22.76	0.06	15.68	29.19	0.35	18.37	102.97
	8	0.07	0.01	0.82	0.01	0.1	1.22	2.42	1.25	0.77	6.67
	9	0	6.15	0.21	16.69	0	26.81	1.25	0.01	174.24	225.36
	Sum	279.24	169.27	40.28	60.24	12.44	124.59	55.44	12.92	262.59	
Change		16.13	-19.5	33.8	-46.38	0.84	11.01	47.53	-6.25	-37.23	

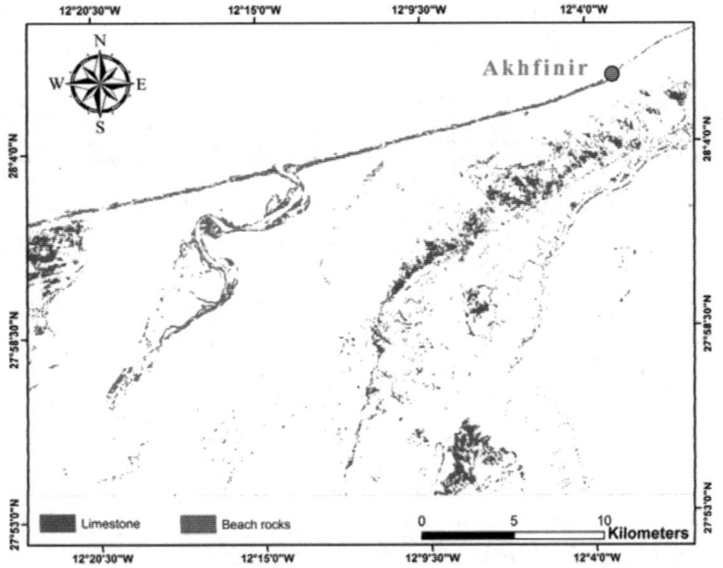

Figure 5. Eroded areas in region of study.

4. Conclusion

The objective of this study was to determine eroded areas in arid environment using new method based on lithological changing data. In this study a change detection technique was applied on two classified lithological map to provide eroded areas map. Further, the statistical change results were beneficial to ascertain the reasons behind the observed changes.

The results of application of this method in northern part of Atlantic Sahara desert confirm their potentiality. The obtained erosion map could be used as a decision tool for sediment erosion management.

References

[1] Lambin E.F 1996 Change detection at multiple temporal scales: seasonal and annual variations in landscape variables. *Photogrammetric Engineering and Remote Sensing* 62 (8), 931–938.

[2] Lambin E.F and Ehrlich D 1997 Land-cover changes in Sub-Saharan Africa (1982–1991): Application of a change index based on remotely sensed surface temperature and vegetation indices at a continental scale. *Remote Sensing Environment,* 61(2), 181–200.

[3] Haboudane D, Bonn F, Royer A, Sommer S and Mehl W 2002 Land degradation and erosion risk mapping by fusion of spectrally based information and digital geomorphometric attributes. *International Journal of Remote Sensing* 18, 3795–3820.

[4] King C, Baghdadi N, Lecomte V and Cerdan O 2005 The application of remote-sensing data to monitoring and modelling of soil erosion. *Catena* 62, 79–93.

[5] Lu D, Batistella M, Mausel P and Moran E 2007 Mapping and monitoring land degradation risks in the Western Brazilian Amazon using multitemporal Landsat TM/ETM+images. *Land Degradation and Development* 18, 41–54.

[6] Alatorre L.C and Beguería S 2009 Identification of eroded areas using remote sensing in a badlands landscape on marls in the central Spanish Pyrenees. *Catena* 76, 182-90.

[7] El Baroudy A.A 2011 Monitoring land degradation using remote sensing and GIS techniques in an area of the middle Nile Delta, Egypt. *Catena* 8, 201-208.

[8] Lenney M.P, Woodcock C.E, Collins J.B and Hamdi H 1996 The status of agricultural lands in Egypt: The use of multi temporal NDVI features derived from Landsat TM. *Remote Sensing Environment,* 56, 8-20.

[9] Tripathy G.K, Ghosh T.K and Shah S.D 1996 Monitoring of desertification process in Karanataka state of India using multi-temporal remote sensing and ancillary information using GIS. *International Journal of Remote Sensing* 17, 2243–2257.

[10] Prince S.D, Becker-Reshef I and Rishmawi K 2009 Detection and mapping of long-term land degradation using local net production scaling: Application to Zimbabwe. *Remote Sensing of Environment* 113, 1046-1057.

[11] Liberti M, Simoniello T, Carone M.T, Coppola R, D'Emilio M and Macchiato M 2009 Mapping badland areas using LANDSAT TM/ETM satellite imagery and morphological data. *Geomorphology* 106. 333-343.

[12] Yang X, Ding Z, Fan X, Zhou Z and Ma N 2007 Processes and mechanisms of desertification in northern China during the last 30 years, with a special reference to the Hunshandake Sandy Land, eastern Inner Mongolia. *Catena* 71, 2–12.

[13] Zhang Q, Devers D, Desch A, Justice C.O and Thownshend J 2005 Mapping tropical deforestation in Central Africa. *Environmental Monitoring and Assessment,* 101 (1-3), 69-83.

[14] Townshend J.R, Masek J.G, Huang C, Vermote E.F , Gao F , Channan S, Sexton J.O , Feng M, Narasimhan R, Kim D, Song K, Song D, Song X.P, Noojipady P, Tan B, Hansen M.C, Li M and Wolfe R.E 2012 Global characterization and monitoring of forest cover using Landsat data: opportunities and challenges, *International Journal of Digital Earth,* 5:5, 373-397,

[15] Sultan M, Arvidson R.E, Sturchio N.C and Guinnes, E.A 1987 Lithologic mapping in arid regions with Landsat TM data: Meatiq dome, Egypt. *Geological Society of America Bulletin* 99, 748–762.
[16] Gad S and Kusky T 2006 Lithological mapping in the Eastern Desert of Egypt, the Barramiya area, using Landsat thematic mapper (TM). *Journal of African Earth Sciences* 44, 196–202.
[17] Elbelrhiti H 2005 *Morphodynamique des barkhanes: Etude des dunes du Sud-Ouest Marocain.* Thèse Université Paris VII, France. 165p.

IOP Conf. Series: Earth and Environmental Science **20** (2014) 012022 doi:10.1088/1755-1315/20/1/012022

GIS based real time assessment of wildfire and other changes in a forest: A review

Dr Vishweshwar Kallimani, Dr Bala Chandra, Nihit Vyas, Jyoti Kallimani
Manipal International University
#1, Persiaran Manipal,71800, Putra Nilai,
Darul Khusus, Nilai,Negeri Sembilan
Malaysia

Email: vnihit@gmail.com

Abstract. Global warming has been one of the major concerns of the century. The causes of this problem are the green-house gases. Due to the manmade mistakes the carbon dioxide (CO_2) emission has reached highest of its level in recent years. Forest fires, haze and deforestation have contributed in increasing the CO_2 emissions to the atmosphere. Diagnosing these problems in the initial stage would definitely avert the major losses. This paper deals on detecting the changes in forest scenarios due to fire, felling of trees, etc. using a Geographical Information System (GIS) information analysis system using a pattern recognition concept, and detect the unusual changes and alert the system in early stage.

1. Introduction

South Asia pacific countries have been found suffering due to the haze conditions frequently. Haze level in Malaysia on 9th Feb 2014 reached the level from 29 to 167 (unhealthy level). Blazes are deliberately lit every year to clear land for palm oil and wood pulp plantations. Haze from fires on Sumatra is an annual problem in Southeast Asia [20]. Hundreds of the haze-causing agricultural fires burning in western Indonesia's Riau province are occurring on land managed by timber, palm oil and logging companies. Figure 1 briefs the data on the logging of trees in a rich forest from logging to plantations.

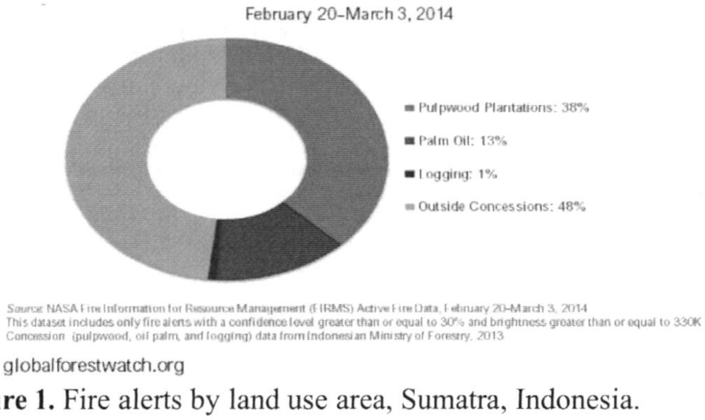

Figure 1. Fire alerts by land use area, Sumatra, Indonesia.

The major threat of the century is global warming. Annually the carbon dioxide (CO_2) emission has reached an alarming stage of spewing of 32 Billion Tons of CO_2 to the atmosphere [19]. The man made inventions and mistakes have contributed in adding a greenhouse gases to the atmosphere. Human activities contribute to increase the emission of greenhouse gases. This is due to the

Content from this work may be used under the terms of the Creative Commons Attribution 3.0 licence. Any further distribution of this work must maintain attribution to the author(s) and the title of the work, journal citation and DOI.
Published under licence by IOP Publishing Ltd

transportation, industrialization, energy sectors and deforestation carried out for farming, or due to forest fires and mining activities.

2. Deforestation: modern day plague

The remains of forests cover 30% of the land. Deforestation is the process whereby natural forests are cleared through logging and/or burning. Deforestation is clearing forests in large scale. At the current rate the world's rain forest would be vanished in hundred years. The cause is due to the factors like wildfires, overgrazing which may prevent the growth of young trees. Deforestation negatively effects the environment such as loss of habitats, trees. Trees also play a critical role in absorbing the greenhouse gases that fuel global warming. A larger green-house gas can be expected with fewer forests and thus inducing a climate change [21]. Earth's climate is changing. The results so far are as follows:

- Global temperature have risen 5° C
- Glaciers have melted and retreated dramatically
- Ecosystems around the world are being altered

2.1. Carbon dioxide emissions

Healthy forests help absorb greenhouse gasses and carbon emissions "Tropical forests hold more than 210 Giga tons of carbon, and deforestation represents around 15 percent of greenhouse gas emissions," according to the [22]. Major players for carbon emissions are Electricity and heat (24.9%), Transportation (14.3%), Industry (14.7%), deforestation (11.3) and afforestation is -0.4 %, as shown in the figure 2.

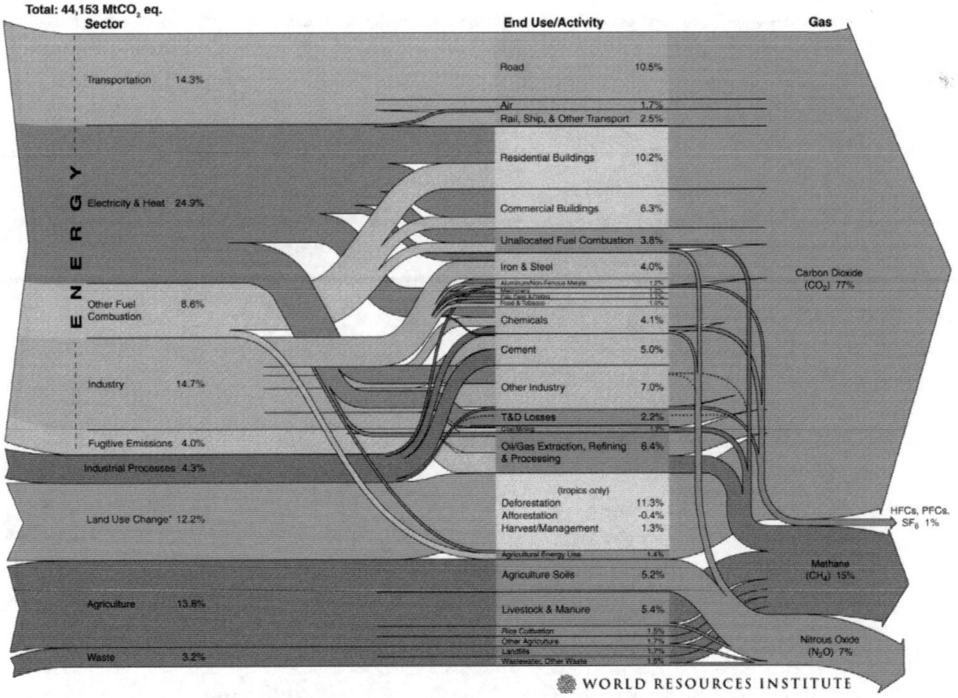

Figure 2. World greenhouse gas emissions in 2008.

Some other statistics:

- About half of the world's tropical forests have been cleared [23].
- Forests currently cover about 30 percent of the world's land mass [24].
- Forest loss contributes between 12 percent and 17 percent of annual global greenhouse gas emissions (Figure 2).

3. Tracking system for changes in forest

The menace of damages made to the forest by the humans is mainly due to felling of trees, fire, poaching, farming etc have some imbalancing effects on the eco system. There is a need of an accurate real time system which will detect the changes in forests at an early stage and alert the concerned authorities to enact. Equally educating people, governance, and legal actions are also necessary to avoid the forest accidents. Many government, state, and private forestry organizations and agencies today utilize geospatial technology such as Geographical Information System (GIS) and remote sensing for various applications supporting analysis, assessment, and management of our forests. The geospatial technologies help in the detection of forest fires, analysis of deforestation and felling of trees, prevention of poaching etc.

4. Geographical information systems (GIS) and wildfire prevention

The first known use of the term "Geographic Information System" was by Roger Tomlinson in the year 1968. A geographical information system integrates the use of hardware and software to acquire, store, analyze, manipulate, manage and represent geographic information. The basic applications of a GIS involve geographical mapping, meteorological survey, study of ecology and geology, natural resource management (oil, gas, forestry etc), disaster management, sociology, archeology, defense, intelligence, public safety (i.e emergency management and criminology), global positioning systems and regional planning. Earth's natural and man-made features are represented in spatial data form. A GIS references these real-world spatial data elements to a coordinate system. These features can be separated into different layers; a GIS system stores each category of information in a separate "layer" for ease of maintenance, analysis, and visualization [8]. A GIS can be classified into four categories which are not necessarily mutually exclusive.

- Four Dimensional GIS: Designed to analyse the geographical data with three dimensions of space and one dimension of time.
- Multimedia and Hypermedia GIS: Facilitates an access to a wide range of geo-referenced multimedia data by selecting resources from a geo-referenced image map base.
- Web GIS: A web based GIS which allows the user to access new forms of multimedia geo-representations on the web.
- Virtual Reality GIS: Utilises Virtual Reality Modelling Language (VRML) to create, manipulate and explore geo-referenced virtual environment.

When it comes to wildfire management, GIS provides accurate and comprehensive information to analyse, visualize, and prioritize values at risk, such as housing developments, utility infrastructure, wildlife, and natural or cultural resources. Within the scope of wildfire management, wildfire prevention comprises of all the activities that can be performed in order to reduce or prevent wildfire occurrences as well as to minimise wildfire potential impacts and damage. For the purpose of mitigation a GIS plays a vital role in mapping fire-prone areas, monitoring fuel load and risk modelling; for preparedness and response it helps in fire detection, predicting spread/direction of fire, early warning and coordinating fire-fighting efforts. For recovery a GIS can be used as a damage assessment tool useful for mapping the extent of burn, understanding biological responses due to differential surface heating (i.e fire severity) and quantifying extent and pattern of burned areas.[10]

This paper provides an overview of the real time system in identifying the fires and changes in the forest land. The intension is to avoid massive damages caused due to the bush fire or wild fire which is needed to be identified at the smaller scale and alert the concerned system. The system is intended to identify the felling of trees, illegal logging, and alert the system. Here an important thing to be noticed is fire to be pointed at the early stage and the location and alert the system, as soon the accident occurs. The watching of the forest area, capturing of images, comparison with test case image and the standard image is done iteratively for round the clock.

5. Image processing for wildfire detection

Image processing is the study of any algorithm that takes an image as input and returns an image as output. The image processing involves image display and printing, image editing and manipulation, image enhancement, feature detection and image compression. Latest advancements in digital camera technologies and content based video processing have led to the development of sophisticated vision based fire detection systems. The two major categories for wildfire detection are sensor based systems and image based systems.

5.1. Colour models for fire and smoke

Images with fire or smoke samples are analysed in order to create colour models. The rules defined for *RGB* colour space to detect possible fire pixel [1][2][5] are transformed into *YCbCr* for analysis. The implicit fuzziness in the rules obtained from repeated experiments and the impreciseness of the decision variable can be encoded in a fuzzy representation [3]. The output is expressed as a number between 0 and 1which will give the probability that a pixel is a fire-pixel or smoke-pixel.

Figure 3 explains different stages of the fire alerting system. *RGB* image is converted to *YCbCr* and the data is analysed by a predefined fuzzy logic rule set.

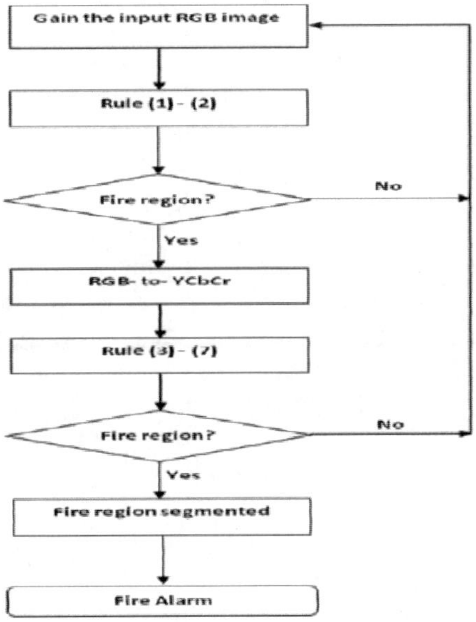

Figure 3. Flow chart of algorithm for forest fire detection [7].

For detection of fire pixel the greater the difference between *Y(xy)* and *Cb(xy)* components of pixel higher the probability that it is a fire pixel because the luminance information which is related to the intensity is naturally expected to be dominant for a fire pixel.
Hence for a fire pixel to exist-

$$Y(x, y) \geq Cr(x, y) \geq Cb(x,y) \quad (1)$$

Table 1 represents the fuzzy logic rule set used to encode impreciseness of the decision variable.

Table 1. Rule table for fuzzy interference system [3].

		$Cr(x,y) - Cb(x,y)$			
$P_f(x,y)$		**NS**	**PS**	**PM**	**PB**
$Y(x,y)$-$Cb(x,y)$	**NS**	LO	LO	LO	LO
	PS	LO	LO	HI	HI
	PM	LO	HI	HI	ME
	PB	LO	ME	ME	ME

Smoke pixels are formulated as follows,

$$
\begin{aligned}
&|R(x\ y) - G\ (x\ y)| \leq Th \\
&|G\ (x\ y) - B\ (x\ y)| \leq Th \qquad\qquad (2) \\
&R\ (x\ y) - B(xy) \leq Th
\end{aligned}
$$

where *Th is* a global threshold ranging from 15 to 25.

The smoke pixels however do not show chrominance characteristics like fire pixels. At the beginning, when the temperature of the smoke is low, it is expected that the smoke will show colour from the range of white-bluish to white. Toward the start of the fire, the smoke's temperature increases and it gets colour from the range of black-grayish to black. Most smoke samples have a grayish colour. Thus recognition of fire and smoke pixel can be carried out using colour model.

6. Satellite and aerial remote sensors

Satellite Imaging Corporation provides advanced remote sensing techniques, multi-spectral and panchromatic image data processing services (including orthorectification), haze reduction and removal, pan sharpening, image enhancements, geo-referencing and mosaicing, and color/grayscale balancing. Optional satellite imaging features may incorporate specialized processing procedures, which are used to analyze land cover and change detection, extraction of culture data, normalized difference vegetation index (NDVI) classification and mapping, lithological classification, environmental analysis, urban development, and monitoring emergencies[25].

6.1. Detection of abnormalities

The task is to develop a pattern recognition system for the detection of abnormalities in the images. The goal is to classify the images into number of categories or classes. The system will be presented with satellite images and it must detect the abnormalities in the image if found such as fire/smoke/felling of trees/new plane land. The proposed method of forest fire detection uses the statistical approach of pattern recognition that requires careful attention to the issues such as the image acquisition, definition of pattern classes, pattern representation, feature extraction and selection, clustering and segmentation, classifier design and learning, selection of training and test samples, and performance evaluation and analysis [13][2][3][12][16].

6.1.1. Statistical pattern recognition

Most of the commercial pattern recognition systems use statistical pattern recognition in which a pattern is represented by a set of n features, or attributes, viewed as a n-dimensional feature vector [11]. The effective way of establishing decision boundaries between the pattern classes is very important here. The proper use of statistical decision theory is expected to establish decision boundaries effectively. The recognition system is operated in either training (learning) or classification (testing) [4][11].

The input image is preprocessed for example, color domain transformation, filtering etc (Figure-4). The background removal requires certain initialization. The basic background modeling or statistical technique or soft-computing technique can be used to model the background in a known background environment. In order to address issues in foreground extraction, the background removal techniques are used. The extracted object is classified to a class (for example, a fire or smoke) using a classifier. A qualitative and quantitative evaluation of existing state of the art techniques and proposed technique is done. The qualitative parameters include noise, clouds, obstructions and camouflage. Quantitative parameters such as precision is considered [12].

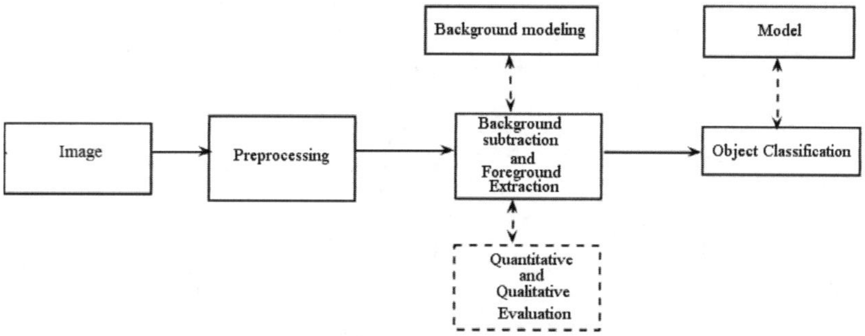

Figure 4. Feature extraction for classification of fire and smoke.

To determine the similarity between two entities (points, curves, or shapes) of the same type template matching is used. In template matching, a template (typically, a 2D image) of the test image or a prototype of the pattern to be recognized is available. The feature comparison between test image and standard image would be performed by classifiers such as Euclidean distance classifier, or KNN for different values of *k* [13].

Euclidean distance:

$$d(x,y) = |x-y| = \sqrt{\sum_{i=1}^{n} (xi - yi)^2} \quad (3)$$

The available training set is used to optimize the similarity measure. Because the template matching is computational intensive process, the high end processing systems are used for this. The success of the template matching depends upon the image quality, change of the viewpoint and intra class variations among the patterns. If the deformation is difficult to explain or to model, deformation template models or rubber sheet deformations can be used for pattern matching [14].

The benefit of considering the static state of the background image is that in the method of background subtraction, intact target region can be extracted, however if the background image is not updated in time, the extracted target may be vague and inaccurate. For forest environment, the whole scene does not keep still due to waving trees, changing weather, varying light, moving shadow, shaking camera, and so on. Therefore, compared with moving estimation, color based segmentation is more suitable for forest fire extraction [16]. Celik et al. [2] described the color features of fire in RGB color space, and decided whether one pixel belongs to the fire region.

6.1.2. 3D color model for segmentation
The 3D color model can be used in the process of segmentation. For segmentation of possible flame regions, color values of each pixel in an image are checked with a pre-determined color distribution, which represents the range of possible fire colors in a color model such as RGB space. As shown in Figure 5, the flame pixels are segmented manually from the fire regions of sample images.

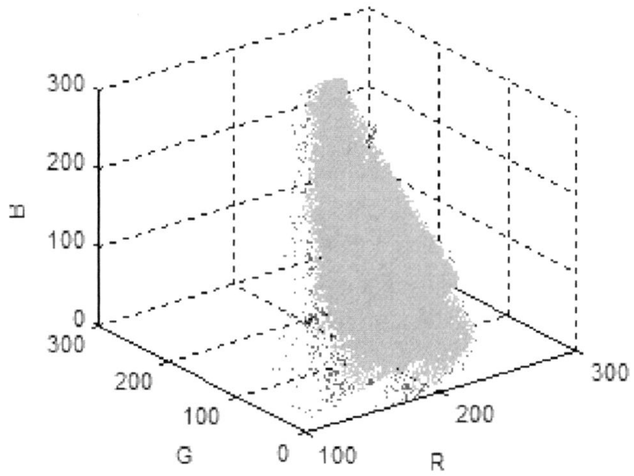

Figure 5.3D color model for image segmentation [2].

The threshold values along R, G and B axis can be used to define a rough space for fire color. To build a more precise color model, 3D shape of the point cloud is represented by Gaussian mixture model (GMM), and the pixel whose color within the range of the GMM distribution model can be taken as a candidate fire pixel [16].

7. Real time forest monitoring system (RFS)
The system involves pattern recognition, features extractions and pattern matching techniques (Figure 6). The satellite image is monitored for every n seconds depending on the season and features of the new case image are extracted. The sample test case (image) features are then compared with a standard image features, by the pattern matching techniques. The system with compare and analyse the proximity value of the test case and standard case and decides the situation is normal or abnormal. If abnormal, then the alerting system is activated and send an alerting message to the concerned people such as ministry/police/forest/disaster management departments [9][12][13][14].

The task is to develop a pattern recognition system for the detection of abnormalities in the images. Classification of the images in to number of categories or classes is the main goal in this task. The system will be presented with satellite images and it must detect the abnormalities in the image if found such as fire/smoke/felling of trees/new plane land.

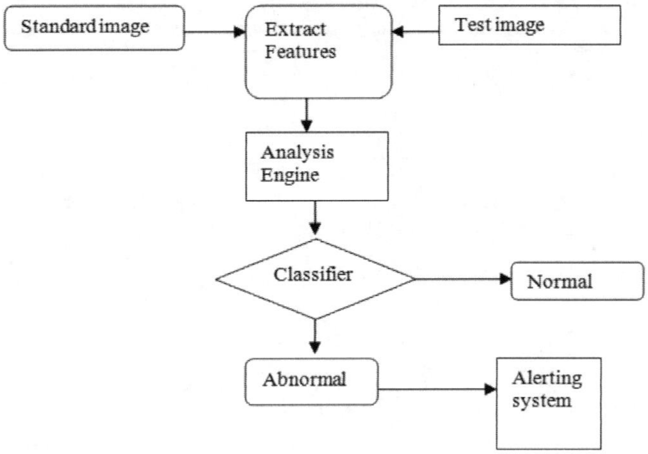

Figure 6.RFS flowchart.

The first stage is to capture the image, and the next stage involves the image processing steps such as image enhancement in order to assist the system by presenting it only useful information for further analysis. The feature selection and extraction is the next stage and the numerical values are used in comparing the features of the standard image (Figure 7). The distance between these image features would yield a decision of the conditions normal or abnormal. The similarity between two feature vectors and choosing an appropriate measure for it determines the result of the image.

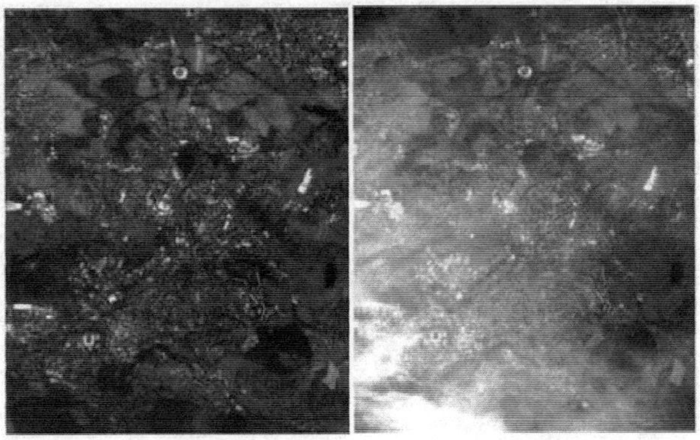

Figure 7.Standard image vs test image [26].

There are five possible cases. The pattern recognition system can be designed around a single classifier, which assigns unknown image directly to one of the five classes of the tree structure as shown in figure 8. A separate classifier was used for each node, and each of them perform a decision in two ways normal and abnormal cases. The second node indicts the image as a normal one, are tested and classified in either smoke, or fire or plane land (new portion in the image) and so on. The similarity in vectors would be measured by the Euclidean distance [13].

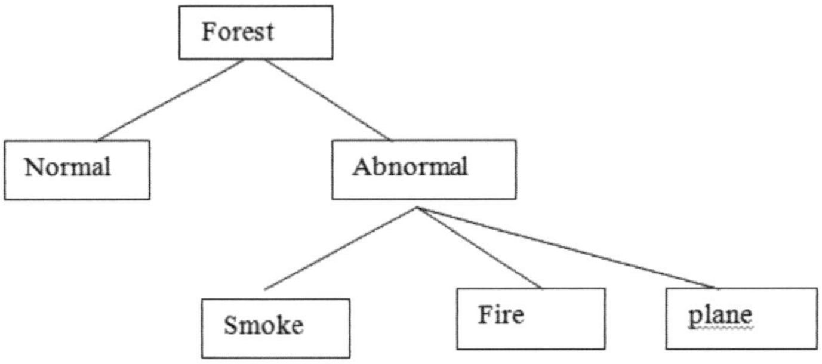

Figure 8. Pattern recognition classes.

7.1. Template matching

The matching of test image template with features matrix, with the standard image template features matrix with certain measure is determined the process of decision making. A first step is to define a measure or a cost measuring the distance or the similarity between the known reference patterns and the unknown (test pattern) which is a pattern matching process yielding % match. Each pattern is expressed in terms of vector matrix with selected features, the distance measure is here is the Euclidean distance.

A similarity measure between two vectors SM_{xy} can be defined as,

$$SM_{xy} = 1/(1+ \alpha d_{xy}) \qquad (4)$$

where α is a positive constant. The higher the value of d_{xy}, the lower the similarity between vector x and y [18].

The minimum distance between test vector and standard vector is considered as the most matching vector. Higher the distance indicates the abnormalities indication with reference to the new image and the system classifies it as an abnormal case. Further analysis was performed to determine the last set of classification in deciding whether smoke or fire or plane land. Based on the final decision an alerting system would be activated if the decision falls on the abnormal class. Once the abnormality is detected by the RFS system, an alerting mechanism is activated to alert the authorities. In crisp, if any untoward incident in forest, such as fire, smoke, felling of trees etc, RFS detects the changes observed in the test images received form satellite and analyse the condition whether normal or abnormal. This capturing of image and analysis would be carried out in n seconds depending on the season, moisture content, rain, etc.

8. Conclusion

In order to reduce the ill-effects of deforestation, certain measures such as early detection of abnormalities by using the detection and analysis models and focusing on indicating the abnormalities are considered. The models can be extended for applications such as obtaining census data for endangered animals like elephants, tigers etc.

References

[1] Celik, T., Demirel, H., Ozkaramanli, H., "Automatic Fire Detection in Video Sequences", *European Signal Processing Conference*, EUSIPCO-06, Sept. 2006.

[2] Celik T., Demirel H., Ozkaramanli H., Uyguroglu M.: Fire Detection in Video Sequences using Statistical Color Model. IEEE International Conference on Acoustics, Speech and Signal Processing, pp. II213-216. (2006)
[3] Celik, T., Demirel, H., Ozkaramanli, H., Fire And Smoke Detection Without Sensors: Image Processing Based Approach, 2007, 15th European Signal Processing Conference, Poznan, Poland.
[4] Celik, T., Demirel, H., Ozkaramanli, H., "Fire Pixel Classification Using Fuzzy Logic And Statistical Colour Model", *Icassp 2007.*
[5] Chen, T., Wu, P., Chiou, Y., "An early fire-detection method based on image processing", *Proc. IEEE Internat. Conf. on Image Processing*, ICIP'04, pp. 1707-1710, 2004).
[6] Klir, G. J., Yuan B., *Fuzzy Sets and Fuzzy Logic*, Prentice Hall, 1995.
[7] Vipin, V., Image Processing Based Forest Fire Detection, 2012, International Journal of Emerging Technology and Advanced Engineering.
[8] Prakash, A., Geographical Systems- An Overview, 2010, Indian Institute of Information Technology.
[9] López, A.,Ballester, M., Guisandez, M. Operational Remote Sensing in Fire Prevention: An Overview, 2009.
[10] South African National Space Agency, 2012, Remote Sensing in Disaster Management, African Leadership Conference.
[11] Statistical Pattern Recognition: A Review Anil K. Jain, Fellow, IEEE, Robert P.W. Duin, and Jianchang Mao, Senior Member, IEEE Transactions on patter analysis and machine intelligence, Vol. 22, No.1, January 2000
[12] D. Conte, P. Foggia, G. Percannella, F. Tufano, and M. Vento, "An experimental evaluation of foreground detection algorithms in real scenes," EURASIP J. Adv. Signal Process, vol. 2010, pp. 7:1–7:10, Feb. 2010. [Online]. Available: http://dx.doi.org/10.1155/2010/373941
[13] Sergios Theodordis, Konstantinos Koutroubas, 2006, Pattern Recognition, ed 3. Academic Press.
[14] U. Grenander, General Pattern Theory. Oxford Univ. Press, 1993.
[15] R. Bajcsy and S. Kovacic, ªMultiresolution Elastic Matching,º Computer Vision Graphics Image Processing, vol. 46, pp. 1-21, 1989
[16] Jianhui Zhao, Zhong Zhang, SVM Based Forest Fire Detection Using Static and Dynamic Features, ComSIS Vol. 8, No. 3, June 2011
[17] David Crisp, NASA Jet Propulsion Laboratory, Pasadena, CA, USA, et al., 2009, "The Need for Atmospheric Carbon Dioxide Measurements from Space:Contributions from a Rapid Reflight of the Orbiting Carbon Observatory", NASA
[18] Sankar Pal, Simon, C.K Shiu, 2004,Foundations of Soft Case-Based Reasoning, Wiley
[19] http://www.nasa.gov/pdf/363474main_OCO_Reflight.pdf
[20] Ministry of Natural resources and environment, Malaysia (www.nre.gov.my/english)
[21] World Resources Institute (www.nre.gov.my/english)
[22] World wild life federation (www.wwf.org)
[23] Food and Agriculture Organisation, (www.fao.org)
[24] National Geographic,(http://environment.nationalgeographic.com)
[25] http://www.satimagingcorp.com/svc/forestry.html
[26] http://www.satimagingcorp.com

IOP Conf. Series: Earth and Environmental Science **20** (2014) 012023 doi:10.1088/1755-1315/20/1/012023

Monitoring of shoreline changes using remote sensing (case study: coastal city of Bandar Abbas)

E Tamassoki[1], H Amiri[2]and Z Soleymani[3]
[1] M.S.C graduated of de-desertification, Hormozgan University, Bandar Abbas, Iran
[2]M.S.C student of remote sensing, Hormozgan University, Bandar Abbas, Iran
[3]M.S.C student of de-desertifucation, Hormozgan University, Bandar Abbas, Iran

Email: ehsantamasoki@yahoo.com

Abstract. Shoreline change is one of the most common natural processes that prevail upon coastal areas. The most important aspect of managing coastal areas is identifying the location and change over time of shoreline. This requires frequent monitoring of the shoreline using satellite imagery over time. We have used imagery from the Landsat TM-5 sensor from 1984,1998 and 2009 in order to monitor shoreline changes using the Max Likelihood Classification method (MLC) in Bandar Abbas city. Monitoring showed that during the period from 1984 to 1998 the area of coastline of Bandar Abbas increased 804.09 hectares. The increase over the next 11-year period was as less, at only 140.81 hectares. In 2009 there was a drastic decrease in shoreline, with the total length of shoreline decreasing from330 km to 271 km during the period from 1984 to 2009.Results showed that in each period in which the area of coastline advanced, changes in length of shoreline had been less prominent.

1.Introduction

The geographic interface of the shoreline, the boundary between land and sea, along with various land uses, and other effects of geological and hydrodynamic phenomena and their effects have made shoreline change one of the most common processes in coastal areas. These changes can have different manifestations over short and long time periods. The recognition of sediment erosion, transport, and depositions one of the most important and sensitive marine phenomena has a direct impaction the design of ports and marine structures, and in coastal management in general.

Changes in erosion and sedimentation due to increasing precipitation have been observed in coastal areas and estuaries, canals, and coastlines. Therefore, identifying the prevailing procedure to determine volatility and change in position of shoreline is the most important parameter required in managing coastal areas. Therefore it is important to determine measurements of the coastline during

Content from this work may be used under the terms of the Creative Commons Attribution 3.0 licence. Any further distribution of this work must maintain attribution to the author(s) and the title of the work, journal citation and DOI.
Published under licence by IOP Publishing Ltd

different times and to be able to compare and evaluate them for inclusion into the prognosis and decisions as to the future status and location. Among the studies that have been conducted in southern Iran are the monitoring of shoreline changes and geomorphology in Assaloye[6]. Results of this study showed that approximately 85% of landforms in these areas have changed due to the construction of industrial plants, anf that the estuaries east of the study area reminded unchanged because there is habitat of Mangrove forests which is protected by government. In addition to the importance shoreline, another important aspect is the sustainable development and protection of the coastal environment. Monitoring of coastline areas is needed over time, as shoreline are the most important and dynamic of natural phenomenon [8] and changes in one part affects the other parts. Failure of observe this fact so has caused many coastal projects to be unsuccessful, or to have a negative impact on the surrounding areas. Today local studies are an important priority when doing projects of management engineering. The absence of a comprehensive database of coastal areas in Iran created many problems in a number of our building projects, for example in the making of seaports and piers. Remote sensing technology is an effective method for obtaining the required data, and this method doesn't have the usual limitation of time and place [1]. So for optional management of the coast, the location in formation of past present and future coastline and how it changes is essential. And now, the newest and most economical method for this purpose is the use of satellite data. In this case, Landsat images which reflect both visible bands, as well as infrared bands, are widely used to monitor coastlines, for example the use of *TM* and ETM^{+} satellites, and Landsat [5]. Hence, in this paper, we used *TM-5* images from 1984, 1998 and 2009 to monitor the coastline city of Bandar Abbas. The importance of coast line to Hormozgan Province is demonstrated by some important ports, for example Jask, Kong, Sirik and Shahid Rajaei. The province because of the ports 9as the biggest commercial port in Iran, Shahid Bahonar, Lenge, and also Ghashm island which can manage their coastal areas by utilizing remote sensing techniques.

2.Study area

The area studied includes the coastal Bandar Abbas city which is located in the territory between *20′ 55* ° to *57°* east longitude to *40′ 26°* to *20′ 27°* north latitude. This area is a widespread area in the Persian Gulf which is adjacent to the Oman sea in the Hormoz canyon . The Persian Gulf is located between the Arabic block and the Zagrous mountains. Elongation of this basin is due to its geological structures. During the Jurassic and Cretaceous periods the srea was situated roughly where the current Zagros mountains are in Iran , but then it gradually moved toward the South West. Considering that the material at times is carried by wind in the Persian Gulf in a mass of dust and sand As a result this material together with a mixture of marine sediments creates sediment of sea - wind in this area. Limestone Aalyts is created due to wave action on the Persian Gulf coast which sorted out on their diameter. Aalyts are often found in shallow areas in the Strait of Hormuz [2]. These conditions together with and the influence of hydrodynamic processes lead to relatively rapid changes in the coastals rather than geomorphology of other systems (Figure 1).

IOP Conf. Series: Earth and Environmental Science **20** (2014) 012023 doi:10.1088/1755-1315/20/1/012023

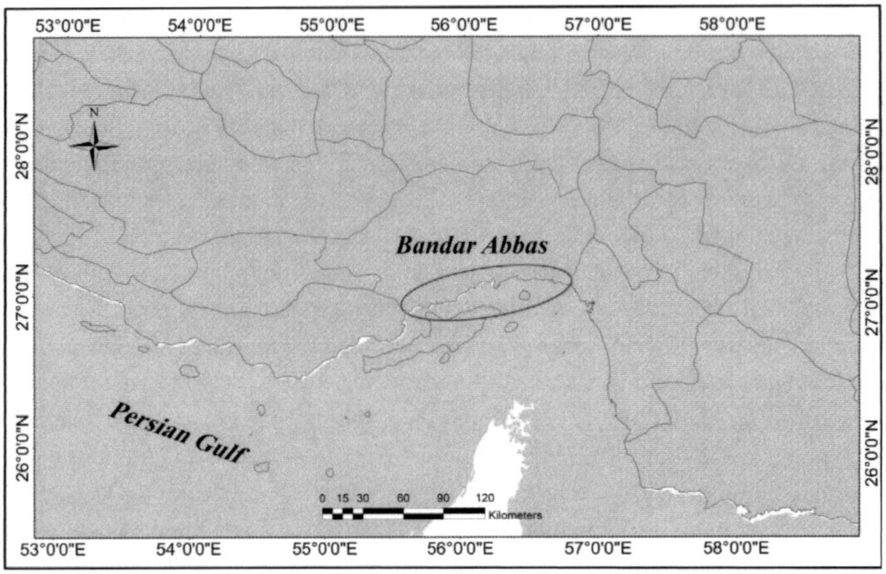

Figure 1: Geographical location of the study area.

3.Materials and methods

In this study we quantitatively interpreted the shoreline during three periods being 1984, 1998 and 2009 utilizing satellite imagery *TM-5* with a spatial resolution of 30 m . Each of these images were taken in a similar month, so they are comparable in terms of time. Images were analyzed after geometric and radiometric corrections were made to the images utilizing the software *ENVI4.8*. Since the purpose of this paper is to monitor changes to the shoreline, the best band combination to detect movement in the boundary between land and water, is the color *RGB 543*. In addition, the spectral bands have a low correlation and hence have more informative [5]. For images analysis we used the method to classify Maximum Likelihood and maps of the shoreline were arranged and presented in ArcGIS. The method of maximum likelihood classification (MLC) is the most common methods of classification of satellite imagery. The method of Parametric maximum likelihood classification acts on basic Basin rule. We found, due to use of the statistical parameters involving basic probability, variance, covariance and average classes that this method in the best method compared to algorithms [3]. This method is based on the probability that a pixel belongs to each classes of m and is checked and t the pixel is then assigned to a class which has the maximum probability.

$$\in W_i \text{ if } p(x \mid W_i) > p(x \mid w_i) \text{ for all } j \neq I \quad (1)$$

This means that the pixel with vector of spectral of x will belong to the class of spectral w. If the probability of belonging to pixels to class of p $(x \mid w_i)$ be greater than the probability of belonging to the other classes, the first stage of classification is the calculation of probability of p $(x \mid w_i)$. Because picture data for classification is provided in several bands, it im plies therefore that the following equation for each vector of spectral X for each class of data is:

$$p(X|w_i) = \frac{1}{(2\pi)^{\frac{f}{2}}|\Sigma_i|^{1/2}} \exp\left[-\frac{1}{2}(X - M_i)^T \Sigma_i^{-1}(X - M_i)\right] \tag{2}$$

Which in that w_i= *class of I ,exp[ ] = e* (the base of logarithms) Exponentiation of desired number , X is the value of a pixel of M_i and is the mean of all pixels values of training of classes and $|\Sigma_i|$ is the determinant of the matrix of covariance of class, Σ_i^{-1} matrix of variance-covariance and f is the number of bands . Therefore it is enough that the mean and variance of each of the classes save to compute the likelihood function [7]. Sometimes due to spectral overlap, the probability calculated for some classes is close together and therefore the greatest probability is very small. In such cases it cannot be confidently attributed to this of pixels and is used as a threshold for control. If we use the threshold as a base we may find or create some pixels that are unclassified which must be tagged in the process of post-processing [3] . For this purpose, the beach and the sea were used as two separate classes of classification. Finally, to determine the accuracy of the classification obtained, we utilized satellite images of the 30 ground control points which were randomly selected. Kappa index was calculated on the changes of shoreline was 0.88 and overall accuracy was 91%.

Images was used of three of images of *TM-5* Landsat satellite
These images show the shorelines of city of Bandar Abbas which had taken on the times of following:

- Image 1984
- Image 1998
- Image 2009

4.Discussion
In this study, we interpreted quantitative changes of shoreline in city of Bandar Abbas during a 25-year period. According to previous studies the original formation of these areas is composed predominantly of tertiary geological organizations and made of sandstone and marl. These formations have periodic stratification and are very sensitive to weathering and degradation processes. The climatic conditions commonly associated with seasonal and torrential rains cased these sediments to continuously deposit on the coastline resulting in the therefore coastline being completed. Due to the performance of marine processes especially waves and the gentle slope of the coastline and coastal fine sediment and silt, resulted in the coastline receding [9]. The shoreline was be exposed to erosion and sedimentation because of changes to the shoreline. This resulted follow spatial the dynamic processes of erosion and sedimentation including how erosion of waves or continuity of waves [4]. The first satellite images were taken in 15 May 1984 by the Landsat *TM-5* (Figure 2). The shoreline at this time has been certained by the maximum likelihood classification method and is clear and marked with a green color in Figure 3.

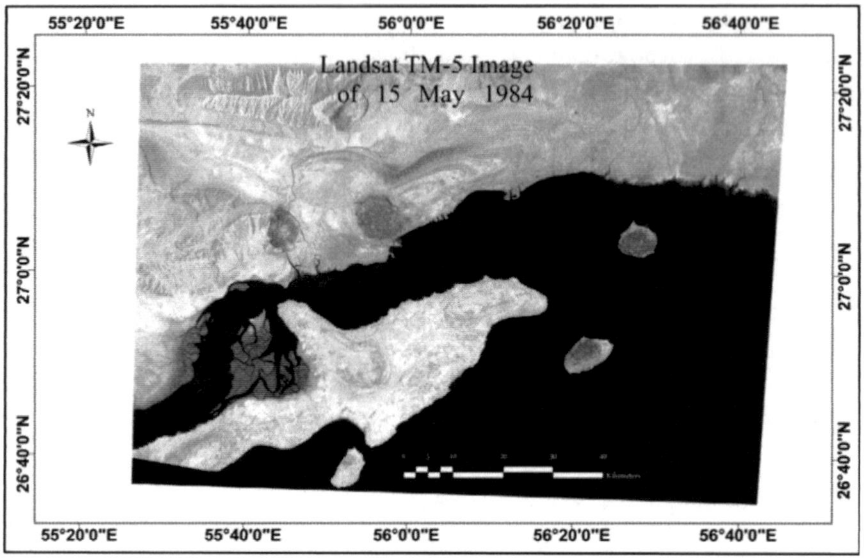

Figure 2: Image sensor TM-5 in 15 May 1984.

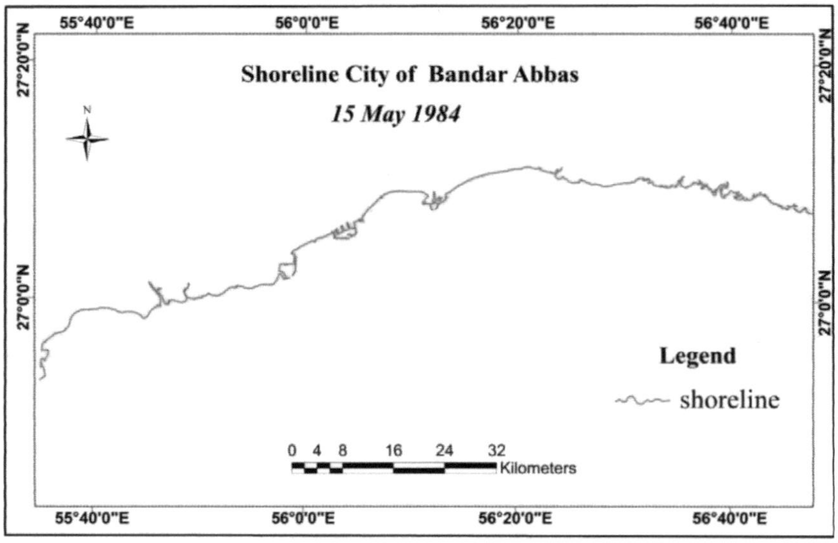

Figure 3: Map of Shoreline on 15 May 1984, was constructed on the basis of the Method of max likelihood classification.

A study of the changes to the shoreline using maximum likelihood classification showed that coastal city of Bandar Abbas has been completed and is 804/09 hectare in extent after 14 years. The Coastline of 1998 is illustrated in blue.(Figure 5).

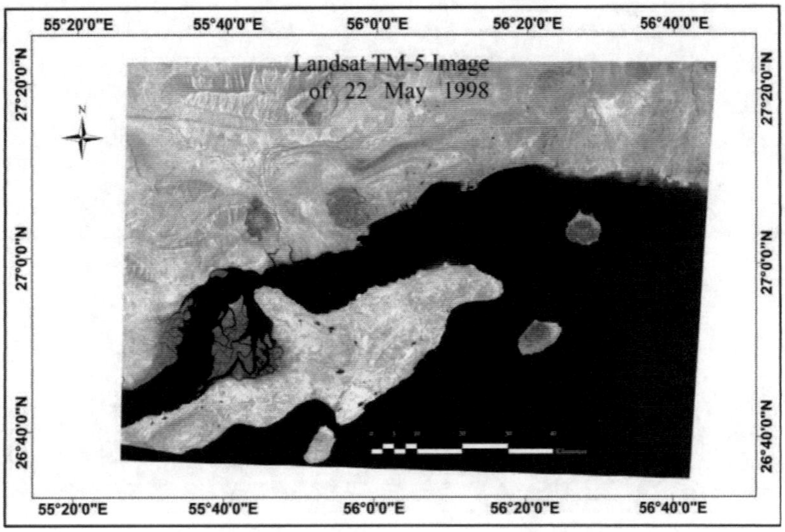

Figure 4: Image sensor TM-5 in 22 May 1998.

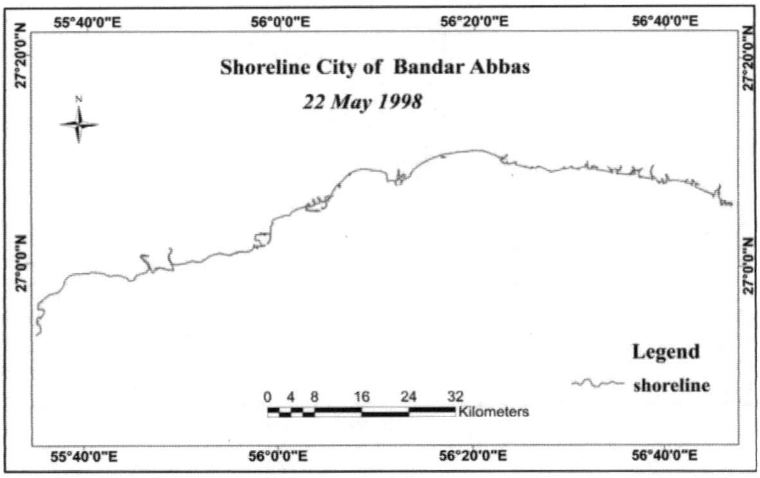

Figure 5: Map of the coastline of 22 May 1998 was constructed on the basis the Method of max likelihood classification.

The third satellite image take after 25 years from the first image and 11 years after the second image. This is satellite image Landsat *TM-5* of 20 May 2009 (Figure 6). The shoreline classification maximum likelihood method is shown in red (Figure 7). Monitoring of changes of coastline in the period 1998 to 2009 showed the coastline of city of Bandar Abbas has been completed and is 140 /81 hectares in extent which was much lower rather than the last period in 1984 to 1998 (Fig. 8).

Figure 6 : The image sensor TM-5 taken on 20 May 2009.

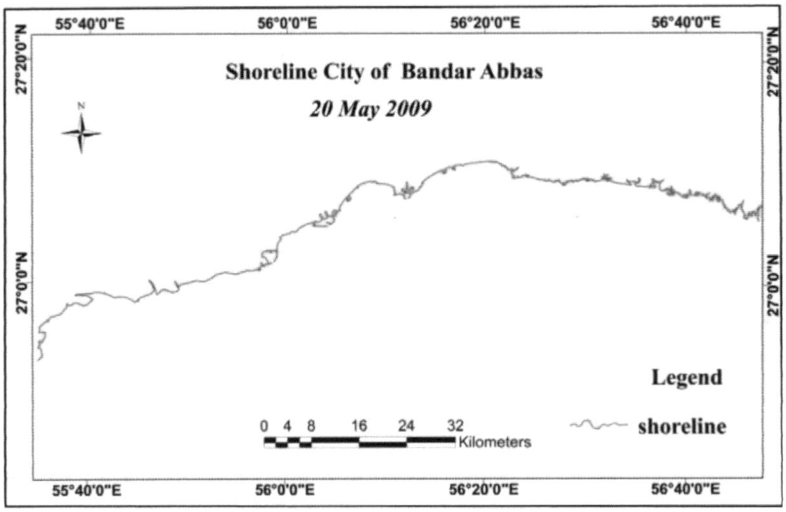

Figure 7: Map of the shorelines of 20 May 2009 constructed on basis the Method of max likelihood classification.

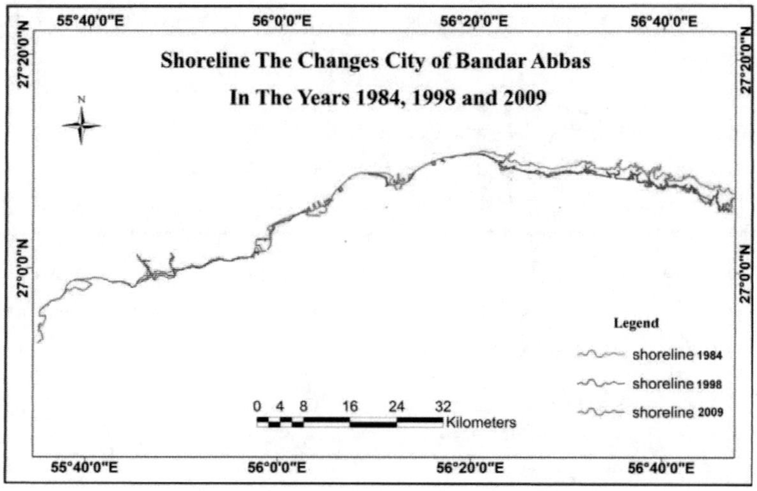

Figure 8: Map of changes of to the shoreline obtained from TM-5 images over a 25-year period based on the basic Method of max likelihood classification.

Shoreline changes in this study show, despite the fact that most progress of the shoreline was during 1984 to 1998, the biggest changes in shoreline occurred in 2009 when the length of shoreline increased by 272 km and in 1984 to 330 km in again in 2009 whilst in 1998 the length of the shoreline was very low (table 1). A comparison of satellite imagery indicated that the greatest changes occurred in the east coast city of Bandar Abbas. The shoreline, due mainly to the presence of numerous estuaries which appeared crinkled, was increased during 2009.

The shoreline length was greater than in 1998. While this changes in 1998 is not very sensible. Despite the instability of shoreline in east of the city of Bandar Abbas, there were no significant changes on the Western shoreline. The changes to the shorelines around the docks of Shahid Haghani and Shahid Rajai were minimal.

Table 1: Length of shoreline have been calculated for the different years.

Year	Length of shoreline (km)
1984	272
1998	275
2009	330

5.Results

One of the most common processes on prevailing coastal area results in shoreline changes. These shoreline changes occur by geological and hydrodynamic phenomena. Understanding changes to shorelines is the most important parameters needed for management of coastal areas. Continual monitoring and studying of sattelir\te images of the coastline is required to study coastline in different times. In this study we have used images from the sensor *TM-5* of satellite Landsat for the monitoring of changes to the shoreline of Bandar Abbas city in periods of 25 years using the Method of max likelihood classification (MLC).

Results showed the during the initial 14-year period from 1984 to 1998 shoreline city of Bandar Abbas 804/09 hectares has been completed. During the next 11-year period from 1998 to 2009, the amount of progress was less and was 140.81 hectares in extent. The most pronounced changes to the shoreline occurred in 2009 resulting in the Length of shoreline increasing from 272 km in 1984 to reach 330 km in 2009 Consider that the length of shoreline in 1998, was 275 km, which is a very small change. The results obtained in this study indicated that during each period in which the shoreline made more progress changes in the Length of shoreline is very low.

6.References

[1] Alesheikh A.A Sadeghi Naeeni F and Talebzadeh A 2003 "Improving classification Accuracy using External Knowledge" GIM International, Aug Vol 17 No 8 pp: 12-15.
[2] Darvish Zadeh A 2010 Geology of Iran. The Institute for Amir Kabir Publications Fourth Edition.
[3] Fatemi S and Rezaei B 2010 Principles of Remote Sensing Publications Azadeh.
[4] Khaledi Sh 2001 Natural Disasters Shahid Beheshti University Publishing.
[5] Moore, L.J.,2000 "Shoreline mapping techniques", Journal of Coastal Research, 16(1), 111-124. Royal Palm Beach (Florida), ISSN 0749-0208.
[6] Naimi Nezamabad A Ghahroudi Tali M 2010 Monitoring of Shoreline Changes and Geomorphologically Persian Guif Using Remote Sensing and Geographic Information System (Case Study: Coastal Zone of Assaloyeh) Scientific- Research Journal Geographical Space Summer Issue 30 pp: 45-61.
[7] Richards J.A. 1998. Remote sensing and Digital Image Analysis. An Introduction 2 nd ed. Springer, Berlin, Heidelbrg.
[8] Winarso, G. and S. Budhiman, 2001"The potential application of remote sensing data for coastal study", Proc. 22nd Asian Conference on Remote Sensing.
[9] Yamani M and Nohegar A 2006 Coastal Geomorphological in the East Strait of Hormoz, Hormozgan University Publishing.

IOP Conf. Series: Earth and Environmental Science **20** (2014) 012024 doi:10.1088/1755-1315/20/1/012024

GIS for public health : A study of Andhra Pradesh

B. Shrinagesh[1], Markandey Kalpana[2], Baktula Kiran[3]
[1]Associate Professor, Department of Geography, Nizam College, Hyderabad, India
[2]Professor, Department of Geography, Osmania University, Hyderabad ,India
[3]Research Scholar, Department of Geography, Osmania University – Hyderabad, India

Email: geosri74@gmail.com

Abstract. Geographic information systems and remote sensing have capabilities that are ideally suited for use in infectious disease surveillance and control, particularly for the many vector-borne neglected diseases that are often found in poor populations in remote rural areas. They are also highly relevant to meet the demands of outbreak investigation and response, where prompt location of cases, rapid communication of information, and quick mapping of the epidemic's dynamics are vital. The situation has changed dramatically over the past few years. GIS helps in determining geographic distribution of diseases, analysing spatial and temporal trends, Mapping populations at risk, Stratifying risk factors, Assessing resource allocation, Planning and targeting interventions, Monitoring diseases and interventions over time. There are vast disparities in people's health even among the different districts across the state of Andhra Pradesh largely attributed to the resource allocation by the state government. Despite having centers of excellence in healthcare delivery, these facilities are limited and are inadequate in meeting the current healthcare demands. The main objectives are to study the prevalent diseases in Andhra Pradesh, to study the infrastructural facilities available in A.P. The methodology includes the Spatial Database, which will be mostly in the form of digitized format. The Non-Spatial Database includes both secondary data as well as the primary data.

1. Introduction

Geoinformatics technology can aid in epidemiological investigation and outbreak response thereby reducing the health hazards in the communities before, during and after epidemic episodes. GIS enhances the quality of spatial and non spatial data for analysis and decision making by providing, integrated approach to disease control and surveillance at local, regional and national levels. Effective response requires considerable commitment to preparedness, risk reduction, prevention and control, and timely economical support. Advance planning in terms of health education, warning and evacuation, emergency health care provision and infrastructure protection can be effectively prepared and implemented and GIS can be a key tool in this regard. Health care is a large, growing, and complex sector of economies around the world. India, because of its diverse culture and practices, faces many complex challenges in providing the best health care for its people and at the lowest possible cost. Currently the use of health informatics is in its infancy and

Content from this work may be used under the terms of the Creative Commons Attribution 3.0 licence. Any further distribution of this work must maintain attribution to the author(s) and the title of the work, journal citation and DOI.
Published under licence by IOP Publishing Ltd

the data is manual and nonintegrated. Geographic information systems has the capabilities that are ideally suited for use in infectious disease surveillance and control, particularly for the many vector-borne neglected diseases that are often found in poor populations in remote rural areas.[1-3] They are also highly relevant to meet the demands of outbreak investigation and response, where prompt location of cases, rapid communication of information, and quick mapping of the epidemic's dynamics are vital. The situation has changed dramatically over the past few years. GIS helps in determining geographic distribution of diseases, analyzing spatial and temporal trends, Mapping populations at risk, Stratifying risk factors, Assessing resource allocation, Planning and targeting interventions, Monitoring diseases and interventions over time. The analysis functions use the spatial and non-spatial attributes in the database to answer questions about the real world. e.g., composite map could be created from exposure information, combined with the knowledge of health/ disease status (e.g. water borne diseases) by house location or water sources could be included in the study to investigate relation between polluted water and disease risk. Data on potential confounding or modifying factors (e.g. age, gender, sex, income, settlement type) could also be incorporated into data analyses. Health professionals have used maps when analyzing associations between location, environment and diseases.GIS in public health is used for basic mapping and involves spatial analysis of disease occurrences and contributing environmental factors.GIS helps us in visual analysis and exploratory analysis along with estimation of relationship between measures of disease incidence [5]. Health research using GIS includes demographic, political, environmental, ecological, topographical, hydrological, and climactic, land use, public infrastructure, transportation, health infrastructure and epidemiological data[6-7]. A GIS is an information system designed to work with data referenced by spatial / geographical coordinates [4].

There are vast disparities in people's health even among the different districts across the state of Andhra Pradesh largely attributed to the resource allocation by the state government. Despite having centers of excellence in healthcare delivery, these facilities are limited and are inadequate in meeting the current healthcare demands. Most public health facilities lack efficiency, are understaffed and have poorly maintained or outdated medical equipment. Andhra Pradesh faces a huge need gap in terms of availability of number of hospital beds per 1000 population. Moreover, Andhra Pradesh faces a shortage of doctors, nurses and paramedics that are needed to propel the growing healthcare industry.

1.1 Objectives

The objective of the paper is to carryout a health mapping of Andhra Pradesh through GIS that would bring in an interactive spatial analysis tool enabling the concerned departments to perform re-districting, re-locating health jurisdictions for effective utilization of health infrastructure. Such an automated facility to generate proposals at the district level and review at the State level were lacking as a system. In a state- wide implementation of Health Information System, sharing health related information with other districts is one of the primary goals, so that consistent and current data is available with the districts. The main objectives are to study the prevalent diseases in Andhra Pradesh, to study the infrastructural facilities available in A.P.

1.2 Methodology

The methodology includes the Spatial Database, which will be mostly in the form of digitized format. The non-spatial Database includes secondary data from various govt. sources.

2. Uses of GIS in Public Health Studies

This includes: 1. Identifying areas where a particular disease is prevalent; 2. Identifying vulnerable population; 3. Identify factors responsible for diseases; 4. Incidences of each disease; 5. Identify health care centres; 6. To determine morbidity and mortality rates; 7. Water polluting sources; 8. Diseases zonation mapping; 9. Target and plan the remedial measures.

IOP Conf. Series: Earth and Environmental Science **20** (2014) 012024 doi:10.1088/1755-1315/20/1/012024

Representation of disease incidences data can be displayed as dot maps to show health event, choropleth maps to display death and illness rates in geographical area, and diagram maps to present quantitative data within map.

2.1 Study Area

Andhra Pradesh is the fifth largest State with an area of 2,76,754 km^2, accounting for 8.4% of India's territory. The State has the longest coastline (972 km) among all the States in India. Andhra Pradesh is endowed with a variety of physiographic features ranging from high hills, undulating plains to a coastal deltaic environment. As per the graph, it shows the health indicators for the state of Andhra Pradesh since 1961. The total population of the state is 8,46,65,533 with 50.2% of male population and 49.8% of female population, with a density of 308 persons/ sq.km, The sex ratio for 2011 was 992 which is higher compared to the last census.

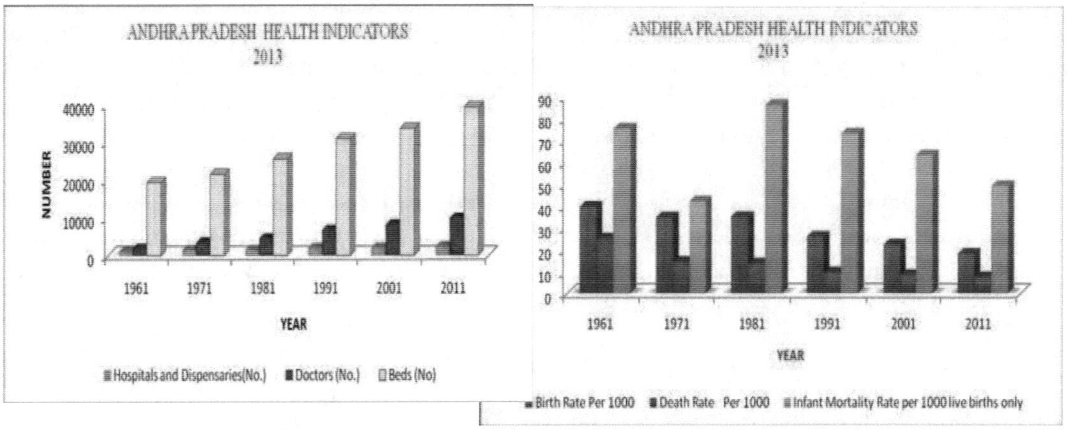

Figure 1. Health indicators of Andhra Pradesh.
Source: AP Statistical Data

The Crude Birth Rate per 1,000 Population is 18.4% for the year 2011 in the State. The Crude Death Rate per 1,000 Population is 7.5 for the year 2011. The Infant Mortality Rate is 43 for the year 2011 per 1,000 live births.

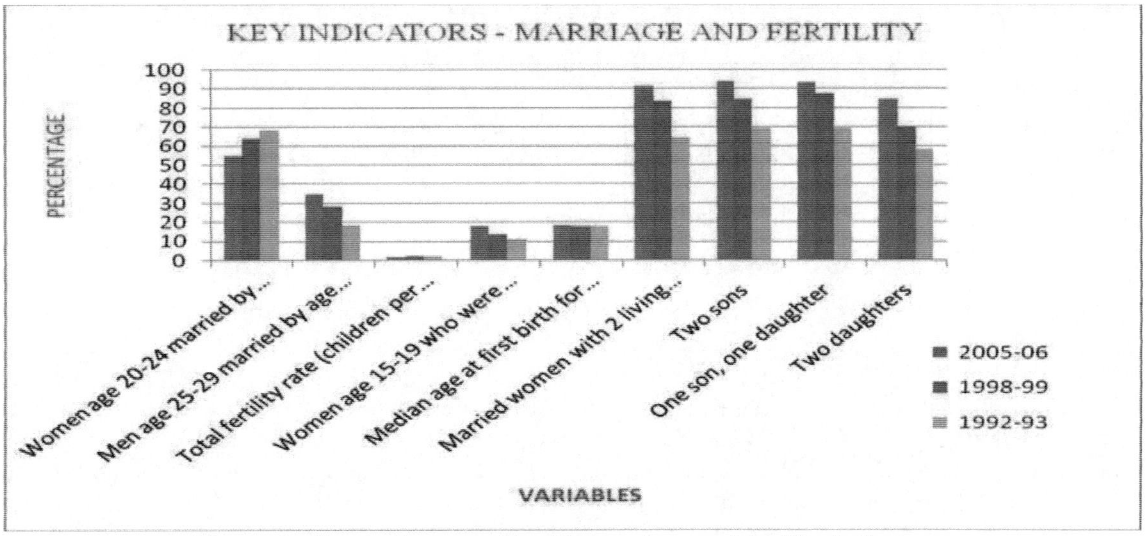

Figure 2. Key indicators of health, marriage and fertility indicators.
Source: AP Statistical Data

The graph above shows the key indicators of Andhra Pradesh under marrigage and fertility variables. 54.7 % of women get married by the age of 18years and 34.8 % of men get married by the age of 21 years. The total fertility rate of women has come to 1.79 % in 2006 from 2.25 % in the year 1999.

3. State of Health

Diseases caused by poor water supply and sanitation are the source of most of the illnesses attributed to poor environment. According to a study by the World Bank (2001), environmental factors account for about 20% of all the diseases in Andhra Pradesh. More than 90 % of this is associated with traditional environmental problems such as lack of access to protected water supply and sanitation, and indoor air pollution resulting from the use of biomass fuels. Environmental factors account for 22-23 % of diseases in rural areas, and for 18- 19 % of diseases in urban areas (World Bank, 2011). Children are most vulnerable to illnesses due to poor household environment and deaths of children under five accounts for nearly two-thirds of such cases. The next vulnerable group is rural women who are particularly affected by exposure to smoke from dirty cooking fuels, in addition to the impact of unsafe water and poor sanitation. Bacterial contamination presents the highest and most immediate health risk affecting 65 % of those depending on dangerously contaminated water. It accounts for 11.3 % of the total diseases in the state Within the state, there are wide variations in access to tap water across districts. In a few districts like Hyderabad, West Godavari, Chittoor, Anantapur and Ranga Reddy more than 60% of the total households have access to tap water. Overall, access to tap water is poorer in the districts of Coastal Andhra region than in districts in both Telangana and Rayalseema.

In Srikakulam and Vizianagaram only 10 and 18 % of the total households respectively have access to tap water. Hand pump is the second most important source of drinking water in almost all districts in Andhra Pradesh. There is a rural-urban gap in the coverage of households with access to tap water and the coverage in rural areas is much lower than the urban levels. While nearly 70 % of urban households (3.0 m households) have access to tap water, only 40 % in rural (5.1 m households) areas are covered. Only in West Godavari and Chittoor districts do more than 60 per cent of the rural households have access to tap water followed by Nizamabad, Ranga Reddy, Medak, Nalgonda and Anantapur with more than 50 % of the total households. The National Habitation Survey of drinking water status in rural habitations reports that only 4 3 % of the habitations are fully provided with drinking water. Except Chittoor, in almost all other districts the majority of habitations are only partially covered. In Vishakhapatnam, a considerable number of habitations do not have proper supply of drinking water. The National Habitation Survey 2003 also points out that the water quality in over 70 % of the total habitations in Nalgonda is affected by fluoride. The percentage of fluoride- affected habitations is also very high in Anantapur (23 %), Karimnagar (20 %), Prakasam (17 %) and Guntur (14 %) districts.

The below table shows the health infrastructure facilities in Andhra Pradesh. In Andhra Pradesh, the availability of Primary Health Centres,(PHC) Sub – Centres(SC) and Child Healthcare Centres (CHC) is relatively better than the all-India average and many states in India, except the south Indian states of Kerala and Tamilnadu.

There are 1,709, 12,476 and 292 PHC, SC and CHCs respectively serving around 60 million rural population in Andhra Pradesh . Since the inception of mandals, a sub-administrative division in the district, the State had made it mandatory that there should be one PHC in each mandal. The number of PHCs is more than the number of mandals (i.e.1,126). This seems to be inadequate in terms of population norm14 (1 PHC per 30,000 persons) according to which 60 million rural population should be served by 2,000 PHCs, which means that 291 more PHCs are required. Similarly, an additional 172 CHCs are required. But in terms of sub-centres (existing 12,476), AP seems to have 524 more SCs than the norm (12,000). However, people located in rural areas continue to be the most disadvantaged in terms of access to health care facilities. There are a number of villages

IOP Conf. Series: Earth and Environmental Science **20** (2014) 012024 doi:10.1088/1755-1315/20/1/012024

where there may not be any basic medical facility available. Even where facilities are available, their functioning's dismal in rural areas. Many villages do not have any qualified medical practitioners.

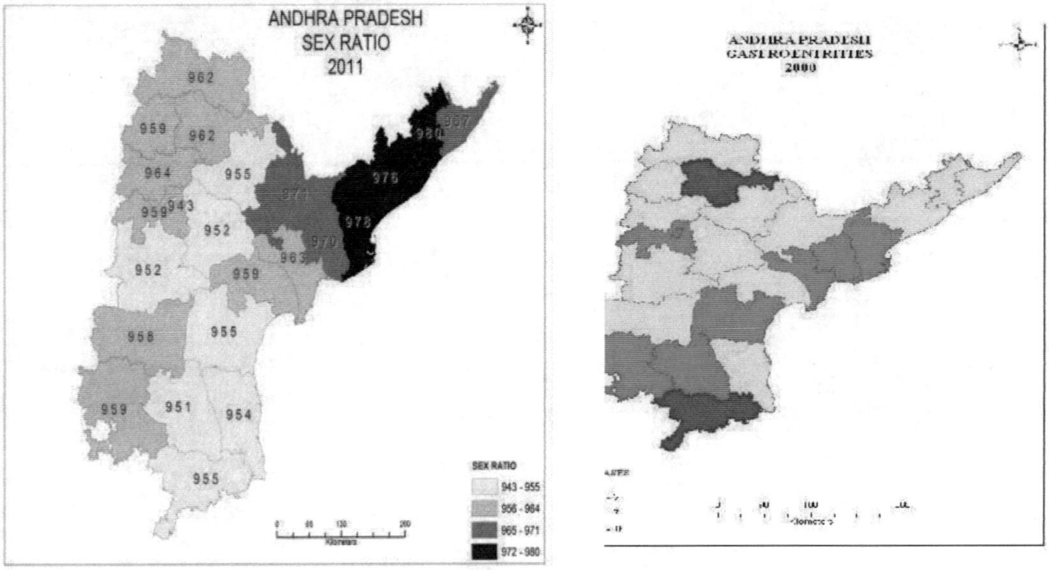

Figure 3. Map showing sex ratio and gastroenteritis.
Source: A.P Health department.

Table 1. Health infrastructure facilities.
Source: AP Statistical Data.

S N	District	No.of Sub-centres	No.of PHCs	No.of CHCs	No.of Area Hosp.	No.of Dist. Hosp.	No.of 24x7 PHCs	No.of 108 Ambul-ances	No.of MHUs
1	Srikakulam	478	76	14	2	--	36	26	20
2	Vizianagaram	442	66	10	2	1	33	26	17
3	Visakhapatnam	584	85	13	1	1	35	41	20
4	East Godavari	809	103	20	7	1	33	36	24
5	West Godavari	627	73	16	7	1	31	34	21
6	Krishna	620	78	9	5	1	28	35	20
7	Guntur	696	77	16	3	1	32	37	24
8	Prakasam	559	85	14	3	--	37	32	20
9	Nellore	483	74	15	2	1	28	26	20
10	Chittoor	651	92	9	9	1	37	40	24
11	Kadapa	462	68	12	1	--	34	28	22
12	Anantapur	609	80	11	6	1	44	39	23
13	Kurnool	555	83	18	1	1	45	29	20
14	Mahabubnagar	681	84	14	4	1	62	33	32
15	Rangareddy	397	48	9	4	1	28	41	14
16	Hyderabad	53	85	10	3	1	10	24	--
17	Medak	536	68	8	4	1	36	27	21
18	Nizamabad	416	42	14	2	1	29	27	14
19	Adilabad	469	68	13	6	--	39	31	24
20	Karimnagar	576	73	16	3	1	35	32	22
21	Warangal	608	64	14	4	1	36	40	20
22	Khammam	583	66	13	5	1	35	31	27
23	Nalgonda	582	71	4	7	1	37	37	26
	Total	12476	1709	292	91	19	800	752	475

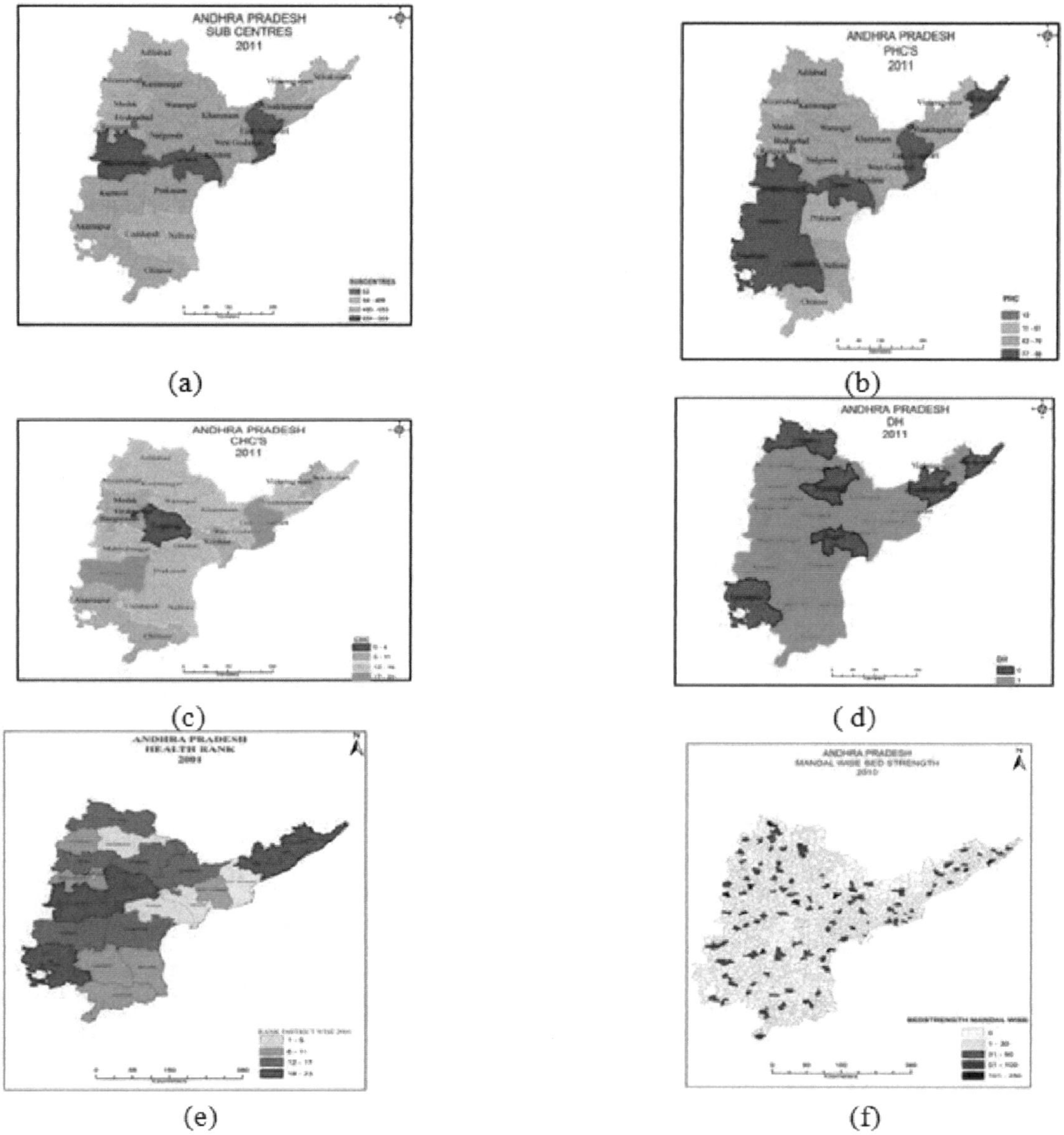

Figure 4. Maps showing infrastructuralfacilities: (a) Sub-centres (b)PHC (c)CHC (d)DH (e)Health Rank (f) Mandal-wise bed strength.

While the hospitals at secondary and tertiary level are over-utilized, the primary level services, especially PHCs and sub-centres, are underutilized mainly due to non-availability of doctors at these health centers, especially in rural areas. These primary health centers account for around 5 % of out- patients (OP) and in-patients (IP) in rural Andhra Pradesh.

4. Conclusion

The performance of Andhra Pradesh has been relatively better in the all-India context in terms of a few demographic indicators particularly with respect to demographic transition because of a dramatic reduction in total fertility rate (TFR) and decline in the growth of population. Primary health care services should be accountable to the local governments and ensure better services for the money spent. Due to lack of adequate resources, there was stagnation in the size and degeneration in the quality of care at the public hospitals. There is overcrowding and more than full capacity utilization of public hospitals, particularly at the secondary and the tertiary levels. There is still widespread unmet demand for medical care among the poor and the government should expand facilities in the public sector to meet this demand. The recent investments in the secondary level hospitals and perceptible improvement in the quality of care attracted more and more patients particularly from the poorer sections. Thus public hospitals at present are self-targeted, i.e., used mostly by the poor. The bed and doctor population ratios are abysmally low even after taking private facilities into account. The lack of control on the cost and quality of care is another major problem in introducing the social insurance. Encouragement should also be given to not-for-profit hospitals. There should be regulation of private medical care and no partnership with private hospitals. The government should promote institutions to regulate the private sector. Third, there is a need to promote community health insurance schemes (e.g. SEWA's scheme) in order to provide health services at low cost to the poor in rural areas. A.P. has introduced Arogyasri programme. The effectiveness of this scheme can be judged only in the future. Fourth, health sector development has to be integrated with the overall process of development. For example, there is no synergy between the processes directed at improving drinking water facilities, sanitation and public hygiene, access to elementary education, nutrition and poverty alleviation and the processes that improve access to public health and medical services. An integrated approach is important to improve the performance of the health sector. Therefore, it is suggested that the current public policy of encouraging corporate hospitals and public- private partnership in health care needs to be reviewed given their far-reaching implications and the public health care system needs to be further strengthened and extended.

References

[1] Balamurugan, G., (2005), Remote Sensing, GIS Modeling for Characterizing the Fractured Aquifer

[2] System in Hosur region, Tamil Nadu, India, Department of Earth Sciences, IIT, Mumbai, p 45-60.

[3] Kamini J, Satish J and Raghavswamy V (2006), Spatio- temporal Analysis of Land use in Urban

[4] Mumbai ,using Multi Environmental Health Perspectives, Vol.105(6), pp. 598-605.

[5] Kondo H, Yasuda T, Hasizume M, Koido Y and Yamamoto Y(2002), Post Flood Epidemics of

[6] Infectious Diseases in Mozambique, Pre Hospital Disaster Medicine, Vol.17(3), pp. 126-133

[7] Martin,D, 2002, GIS- Socioeconomic applications, pp1-8, 162-184.

[8] Rytkonen, M., (2004), Not all maps are equal: GIS and Spatial Analysis in Epidemiology, International Journal for Circumpolar Health, pp 9-24.

[9] Sensor Satellite Data and GIS techniques Journal of Indian Society for Remote Sensing, Vol.34(4), pp. 385-396

[10] Watson J, Gayer M and Connoly M (2007) Epidemics after Natural Disaster, Emerging Infectious diseases, Vol.13 (1), pp. 1-5

An effective road management system using web-based GIS software

Nik Mohd Ramli Nik Yusoff [1] *, Helmi Zulhaidi Mohd Shafri[1,2], and Ratnasamy Muniandy[1]

[1]Department of Civil Engineering, Faculty of Engineering Universiti Putra Malaysia (UPM), 43400 Serdang, Selangor, Malaysia
[2]Geospatial Information Science Research Centre (GISRC), Faculty of Engineering, Universiti Putra Malaysia (UPM), 43400 Serdang, Selangor, Malaysia

*E-mail: nmramli86@gmail.com

Abstract. Recently, there is a rapid development of road transportation network. This situation arises due to the number of vehicle on the road that keeps increasing year by year. Thus, this will increase the possibility of dangerous situations to the road users if roads are not being maintained appropriately. Therefore, in order to keep the road in safe condition, the road management activity should be improved. A complete system for road management has been applied in the developed countries for the past decades. However, the usage of application programming interface (API) from GIS software has limitation to make modification in terms of interactivity of the system. Open source software was proposed as a way to assist in reducing the development cost of developing road monitoring system. In this paper, open source software was proposed as a way to assist in reducing the development cost in developing road management system. OpenLayers framework was used as map rendering platform. The system can be used by administrator of road network to update the road information. At the same time, it can be used by the road user as well to view the information regarding road-related incidents.

1. Introduction

Recently, there has been a rapid development of road transportation network throughout the world due to increased human population and economic activities. The increase of vehicle on the road has demanded the commitment to improve the current road transportation systems. Current road transportation system helps the authority in managing road infrastructure effectively. It can support in terms of monitoring the road infrastructures extensively. However, to keep the road infrastructure in safe condition, a comprehensive monitoring and managing system should be developed. By providing a comprehensive road management system, the occurrence of road accidents also can be minimized [1]. Several countries have developed road management system for past decades for their countries. However, the existing system still can be improved for future work and wider scope [2].

In previous studies, road management system has proved to be able to manage the road network system properly. [3] implemented internet-based system for highway pavement management system by integrating spatial and attribute database in the system. Meanwhile, [2] has proposed road maintenance information portal system to facilitate the sharing and retrieving of information among project participants. These examples use web-based GIS concept in order to construct the systems.

Content from this work may be used under the terms of the Creative Commons Attribution 3.0 licence. Any further distribution of this work must maintain attribution to the author(s) and the title of the work, journal citation and DOI.
Published under licence by IOP Publishing Ltd

However, the usage of application programming interface (API) from GIS software has limitation to make modification in terms of interactivity of the system [4].

The usage of commercial software required the product license and need to be renewed every year. Consequently, the developer needs to spend some amount of money to maintain the software. In this paper, open source software was proposed as a way to assist in reducing the development cost in developing road management system. This paper proposed the development of road management system based on WebGIS.

Geographic information system (GIS) is a powerful tool and has capability to handle and process spatial data in a large volume. GIS is defined as a system that able to input, store, retrieve, update, display, and comprehensively analyses spatial data based on computer hardware and software [5]. It is also known as desktop GIS. GIS becomes a widespread technique that can create maps, integrate information, visualize and solve problems, and develop valuable solutions [6]. Applying GIS for road management system have been used before in developed countries [7]. It is an effective system to monitor all the activities that occur on the roads and store the records for future development. Furthermore, this system can be as decision support system (DSS) for road development and construction in the future. To make this system ease to access, it is developed as internet platform.

Web based GIS or WebGIS is an Internet-based platform that provides client-side applications using WWW protocols running on the worldwide web which can embed geographic information data as well as non-geographic data [3, 8]. It is a new type of GIS which is accompanied by the rapid development of Internet [9]. WebGIS comprises various advantages compared to desktop GIS [8].

Therefore, the objective of this study is to propose an effective road management system to improve the current system. Open source web-based GIS software is used in developing the system in order to minimise the maintenance cost. By implementing this system, it provides a comprehensive system that can monitor and manage road conditions which is also able to be accessed by users through the internet.

2. Workflow of methodology

The flowchart of this study was designed to guide the authors to achieve the objective. In achieving the objectives, six methodological steps were designed and shown in figure 1. Requirement analysis is a process that usually taken to identify problems exists in the study area. This step involves various procedures and materials. The preliminary study was conducted to identify the latest and previous systems that able to support spatial data for road management system. Interview session also was conducted with government agency in Malaysia such as Lembaga Lebuhraya Malaysia (LLM) and experts in road management system. Thus, the lacking of the current systems was identified.

From preliminary studies, the information gained was utilized to develop a comprehensive web-based GIS system that able to overcome the lacking in the current systems. The standard procedure in system development was applied in this study which started with system design, software and hardware survey, database design, and programming code. Then, the system was tested to identify the reliability and robustness in serving the functions for administrators and distributing information to users.

Hardware and software selection are vital components of GIS field. High performance of hardware is needed to support the speed of data and processing for a web. The next component that essential in web development is database platform. Database platform is used to store all information in one location for manipulating, processing, and lastly presenting to users through the web browser. In database design, it showed the overall database structure include the elements of the database. The elements of the database consist of entities, attributes, and the relationship between entities and attributes. System development is a phase to make configuration of software, data entry, and data editing of the attributes in the web-based GIS system. After the system has finished being developed, it needs to be tested or validated in term of the quality and competency in providing the function to users.

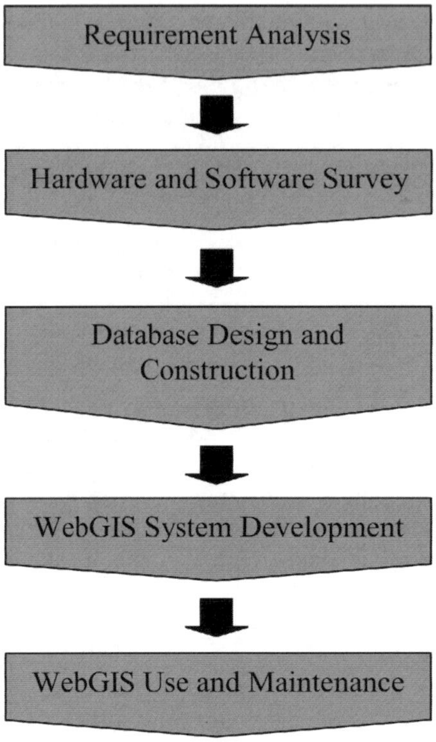

Figure 1. Methodology for development process of the system.

2.1. System design architecture

Web has capability to input, process, and lastly visualize the results on screen using internet platform. System design is the main procedure in web development. It shows the overall components in the system. Development of the web has three vital components to make it run efficiently. The architecture of system design is shown in figure 2.

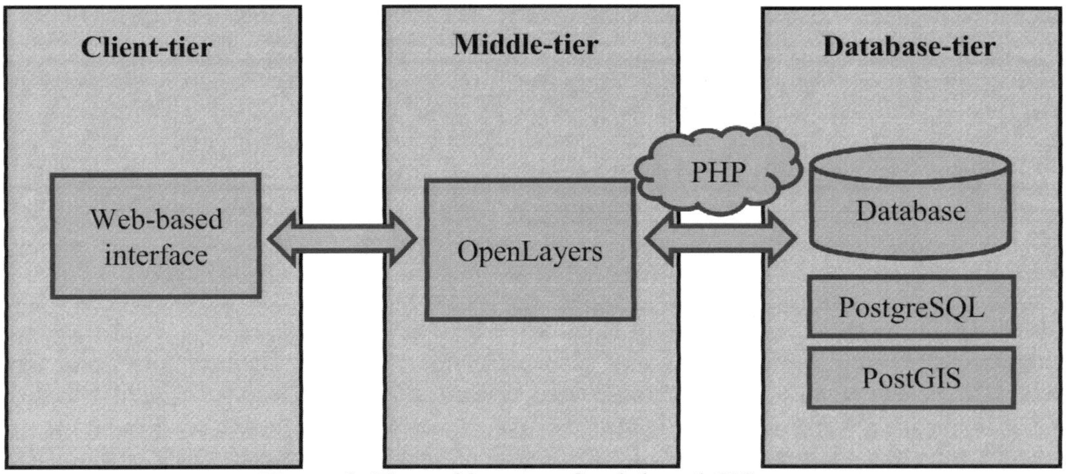

Figure 2. The architecture of web-based GIS system.

2.2. Client tier architecture

Client-tier consists of web browser to display information to users. This tier also called interface to receive user request for processing. Various web browsers can support the client tier such as Internet Explorer, Mozilla Firefox, Google Chrome and others. Web browser allows users to input their request. In this study, the interface was developed using hypertext markup language (HTML) as the main programming scripting of the interface. The contents of the web page were written in the form HTML element before publishing on the web browser. At the same time, JavaScript language was embedded in the HTML scripts to generate interactive and dynamic visualization.

Client tier is the main page of the system that deals with the users. The graphic user interface (GUI) receives user requests through the web browser. Then, the requests will be transferred to the database tier for processing. The data needed will be retrieved from the central database in the database tier. Subsequently, the result will be transferred again through web tier and visualize on the client tier for user.

2.3. Middle tier architecture

Middle tier is the most important tier of the whole system. The function in this tier is to control the whole process of the system. After client tier received requests from user, the results will be sent back to user via middleware. It serves as the intermediate medium between client tier and database tier for webGIS system. The connection required HTTP and SQL that can translate between them.

OpenLayers framework was attached in this tier for map visualization. OpenLayers is a powerful and flexible that functions for displaying maps in web browser using JavaScript library. It is foundation of all webs mapping framework such as Google Maps, OpenStreetMap, Google Satellite, Yahoo Maps, and Bing Maps [10]. This research used three web mapping applications which included Google Maps, Google Satellite, and OpenStreetMap to look in variety of interface display as shown in figure 3.

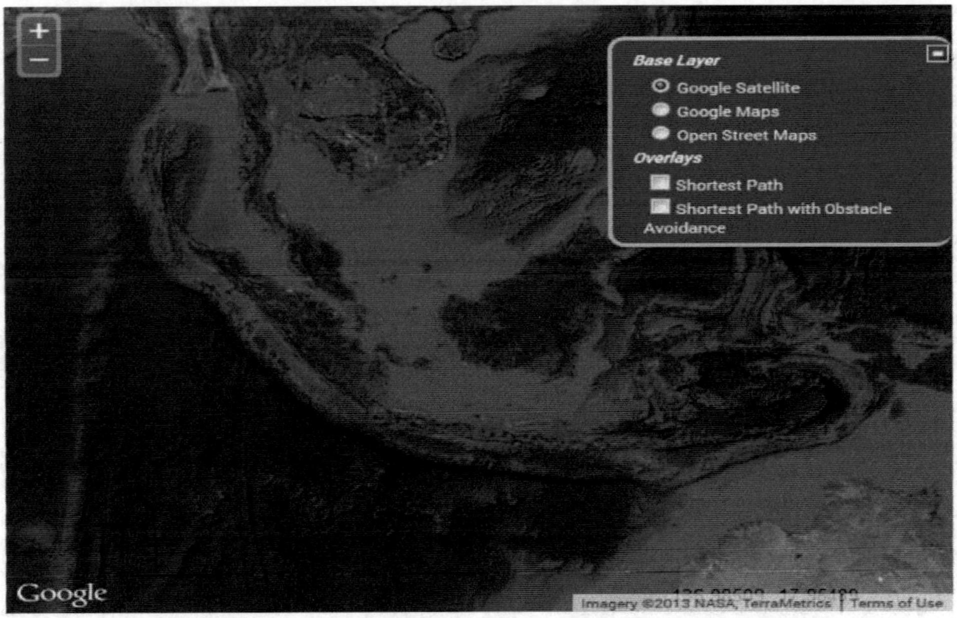

Figure 3. OpenLayers framework for web mapping rendering.

2.4. Database tier architecture

Database tier or server tier is a tier that store and retrieve information in this system. It is the database management system (DBMS) for this web. PostgreSQL/PostGIS database was placed in the database tier to manage and provide data when processing were run. Hypertext preprocessor (PHP) as server-side scripting language was used in the database tier to communicate with the client-tier. PHP code was embedded in HTML code. When clients invoke the HTML page, the Web server executes the scripts, which in turn access the data from PostgreSQL/PostGIS database. The Web server then organizes the data into HTML page and sends it back to the browser. Thus, users can view road information and shortest path route that retrieved from database on the web page.

2.5. Open source software and usage

The development of open source software for GIS usage has been implemented in the last 10 years [4]. Open source software is free license software that has capability like the license software in running and modifying program for any purpose [11]. Nowadays, it becomes popular and reliable to current system. For GIS usage, the software must have capability for creating, managing, analyzing, and visualizing geographic data [4]. To minimize the cost, this study utilized open source software for web development and attached with free database system software as well. Hence, this study can be categorized as low cost project and can be implemented in real situation to replace the current system.

One of the important components in web development is web server. This component serves and enables the information from database to be accessed using world wide web (WWW) protocol. Apache web server also known as apache HTTP server is the well-known open source web server platform. Apache is a high performance web server with full features functionalities. For database platform, this study applied the concept of integration of PostgreSQL and PostGIS as the database management system which transfers data through internet protocol. This database was used because it was most advance open source database that available today [12].

2.6. Development of web-based GIS

This system was developed on a workstation with window 7 operating system. The hardware specification of the server was a 2.4GHz Intel® core™ i5 processor with random access memory (RAM) of 4GB. The server runs WampServer to retrieve data from postgreSQL database using the PHP language. The database used was PostgreSQL 9.1 with PostGIS 1.5 spatial extension to support geographic data.

Three web mapping systems were rendered in OpenLayers framework which was OpenStreetMap, Google Satellite Maps, and Google Maps. By developing this system, it can be used for administrative usage to input all information of roadwork activities that occur on the road. Moreover, it can be a control station to monitor the construction works on the road. Symbol icon will appear on the map included with progress status of the roadwork. At the same time, it can be used by the road user as well to view the information regarding road activities happen on the road.

GIS consists of five elements and two of the elements are hardware and software. Selection of software should be done in proper way because some software has different advantages and disadvantages. The selected software must be evaluated in term of functionality so that it not affects the system when it is running. In order to make this system cost-effectively, free software was used. PHP 5.4 was used as server-side scripting language and PostgreSQL 9.2 as the database platform. JavaScript programming language was used to write coding for client-site to produce interactive and user-friendly web page. Then, web browser such as Mozilla or Internet Explorer will function as user input and displays the output after processing.

3. Results and discussions

The GUI was developed to integrate all functions proposed in the developed web-GIS with roads data. The interface was developed using HTML and combined with JavaScript code to make it more dynamic. In the main interface, there are five tabs on the header of the interface. All the available functions are shown on the interface for user consideration. Figure 4 shows the main interface for web-GIS for road management system.

Figure 4. Main interface for WebGIS for road management system.

The five menus that were created on the main interface were Home, Admin, User, Partners, and Contact Us. Each menu has specific function to serve users when interact with the system. Home menu provides homepage of the web-GIS interface. When users click on this tab, it will return users to the homepage. The second menu is Admin. This menu was designed special for administrative usage. To keep the data precisely, only registered users can edit and update information in the database. This is because some current systems allow public users to input the information freely. This situation can reduce the authorization of data in the database. Therefore, to make sure all the inserted data free from offense, admin function was enabled in this system.

The next menu is User menu. In this menu, it consist the main content of web-based GIS system that shows the base map using OpenLayers framework. Three base maps were rendered in the framework. Users can choose which base map they prefer by selecting on the right menu. This function was created in this system due to enhance the visualization of base map. The second function of User menu is generating shortest path analysis. Users need to input their origin and destination nodes in the provided columns. If they need the specific time for travelling, they can insert the date and time in the provided columns. The reason this system chooses to use node value as the input unit instead of street name because it can detect the exact location of the source and target on the map. Thus, it increased the accuracy of the system. But sometimes it difficult for users to get the correct location based on the node value.

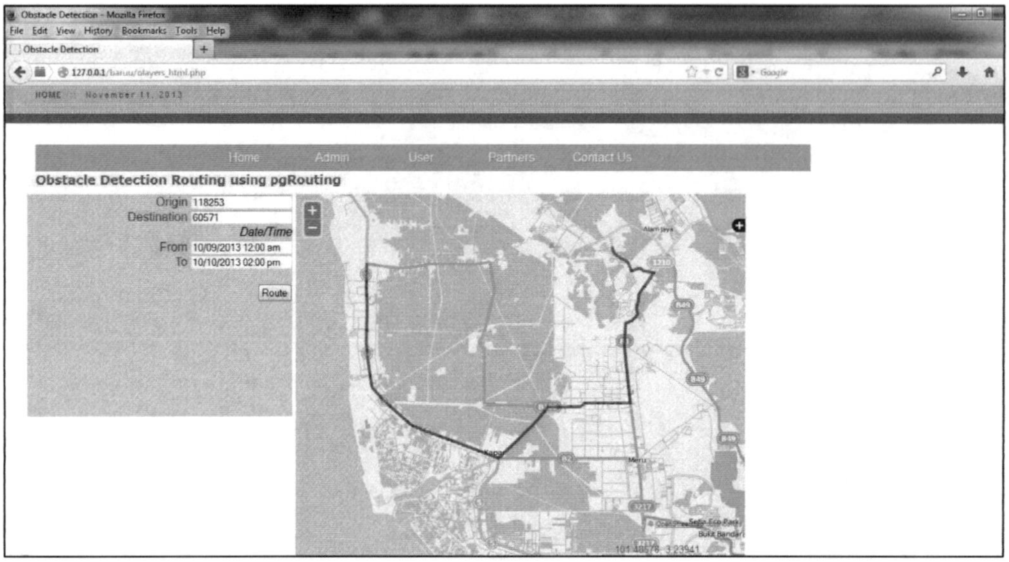

Figure 5. Shortest path between two points.

Partner is the fourth menu that exists in the interface of the system. However, this tab is not likes other tabs. This tab used dropdown menu that listed the potential partners from government and private departments that related to road management in Malaysia such as Lembaga Lebuhraya Malaysia (LLM), Jabatan Kerja Raya (JKR), and PLUS. This menu was created in this system to integrate some information within different departments. The last menu of this system is Contact Us. The purpose of this menu is to provide the information about the person who users can contact regarding to this webGIS system.

4. Conclusion

In this paper, a comprehensive system development for road management system was developed by using open source software. The free software was used because it becoming popular in nowadays. WebGIS become the core part of the system since it can support road data which consist of spatial and non-spatial data. Development of this system involves two parts which are system design and system development. This system will give benefits to road administrator to gather information in a single database and can work as decision support system for road maintenance. At the same time, it can provide visualization for public users to plan their journey. By accessing this system, the administrator from authority departments can update information in the database easily. Only registered users are allowed to update the information in the database in order to maintain the authenticity of information. However, the road users can visualize the base maps in the interface. They can interact with the system by flying through, pan, slide, and viewing road condition with roads information on screen. Shortest path analysis also was provided in the system to enhance the capability of the system. The future work and improvement of road management system, we suggest that the data are collected from related agencies to insert in the database. The database will become more updated and the quality of data is certified. Other features such as photos and videos can be uploaded in the database to improve map visualizations.

Acknowledgments

The authors would like to thank the Universiti Putra Malaysia (UPM) for providing us the research grant under the research university grant scheme (RUGS) and graduate scholarship in completing the study.

References

[1] Ma X 2008 Transportation Development Strategic Plan about 2010 Guangzhou Asian Games *Journal of Transportation Systems Engineering and Information Technology* **8(4)** 16-22

[2] Kubota S, Sugawara T, Kosawada T and Abe A 2011 Web GIS-based Information Portal System for Road Maintenance *Proceedings of the 19th International Conference of Geoinformatics* 1-5

[3] Feng X and Quanwen L 2010 Development of Highway Management System Based WebGIS *Proceedings of the WASE International Conference of Information Engineering (ICIE)* pp 136-138

[4] Steiniger S and Hunter AJ 2012 The 2012 Free and Open Source GIS Software Map–A Guide to Facilitate Research, Development, and Adoption *Computers, Environment and Urban Systems* 39(0) 136-150

[5] Zhang H, Feng Z and Gao R 2011 Research and Development of Intelligent Decision System for Highway Maintenance Based on GIS *Proceedings of the Electronic and Mechanical Engineering and Information Technology (EMEIT)* 1945-1948

[6] Sadoun B and Saleh B 2010 A Geographic Information System (GIS) to Define Indicators for Development and Planning in Jordan *Proceedings of the 2010 International Conference on e-Business (ICE-B)* 1-7

[7] Yusoff NMRN, Shafri HZM, Muniandy R, and Wali IM 2014 Development of Obstacle Avoidance Technique in Web-based Geographic Information System for Traffic Management using Open Source Software *Journal of computer science* **10(8)** 1259-1268

[8] Xie F 2010 Design and Implementation of Highway Management System Based WebGIS *Journal of Networks* **5(12)** 1389-1392

[9] Xinkai Y, Jie S, Fengtao Y, Yongfeng D and Zhenchao D 2009 Study of Intelligent Public Transport Monitoring System Based on WebGIS *Proceedings of the Second International Conference of Intelligent Networks and Intelligent Systems (ICINIS)* 78-81

[10] Perez AS 2012 *OpenLayers Cookbook* (Birmingham, UK: Packt Publishing)

[11] Moreno-Sanchez R, Anderson G, Cruz J and Hayden M 2007 The Potential for The Use of Open Source Software and Open Specifications in Creating Web-based Cross-border Health Spatial Information Systems *International Journal of Geographical Information Science* **21(10)** 1135-1163

[12] Anderson G and Moreno-Sanchez R 2003 Building Web-Based Spatial Information Solutions around Open Specifications and Open Source Software *Transactions in GIS* **7(4)** 447-466

Traffic Analysis and Road Accidents: A Case Study of Hyderabad using GIS

M. Bhagyaiah[1], B. Shrinagesh[2]

[1]Research Scholar, Department of Geography, Osmania University, Hyderabad, India.

[2]Associate Professor, Department of Geography, Osmania University, Hyderabad, India.

E-mail: josebhag@gmail.com

Abstract. Globalization has impacted many developing countries across the world. India is one such country, which benefited the most. Increased, economic activity raised the consumption levels of the people across the country. This created scope for increase in travel and transportation. The increase in the vehicles since last 10 years has put lot of pressure on the existing roads and ultimately resulting in road accidents. It is estimated that since 2001 there is an increase of 202 percent of two wheeler and 286 percent of four wheeler vehicles with no road expansion. Motor vehicle crashes are a common cause of death, disability and demand for emergency medical care. Globally, more than 1 million people die each year from traffic crashes and about 20–50 million are injured or permanently disabled. There has been increasing trend in road accidents in Hyderabad over a few years. GIS helps in locating the accident hotspots and also in analyzing the trend of road accidents in Hyderabad.

1. Introduction

The process of rapid and unplanned urbanization has resulted in an unprecedented revolution in the growth of motor vehicles world-wide. The alarming increase in morbidity and mortality owing to road traffic incidents (RTI) over the past few decades is a matter of great concern globally [1, 2, 4, 5]. Currently motor vehicle accidents rank ninth in order of disease burden and are projected to be ranked third in the year 2020[7-9, 11-13, 15-16]. In India, more than 70,000 people get killed due to RTI every year, and this needs to be recognized as an important public health issue[3,14,17]. Very few studies have attempted to understand the epidemiology of risk factors associated with RTI in Indian cities [10, 18]. The present study aims to examine the magnitude of this multifaceted problem in a rapidly developing Hyderabad metropolis and analyse the causative factors.

Hyderabad, a 400 year old city is the state capital of Andhra Pradesh. It lies on the Deccan Plateau, 541 meters (1776ft) above sea level, over an area of 625 sq.km. of municipal corporation and 7200sq.km.of Hyderabad Metropolitan area, is the fifth largest city in India, with a population of above 8 million. Unlike other Indian metros it continues to attract considerable migrant population due to its strategic geographical location, multilingual and cosmopolitan culture, tremendous growth potential and investment

Content from this work may be used under the terms of the Creative Commons Attribution 3.0 licence. Any further distribution of this work must maintain attribution to the author(s) and the title of the work, journal citation and DOI.
Published under licence by IOP Publishing Ltd

friendly economic policy has propelled the overall population growth of the city and the surrounding areas. The increasing pressure of the burgeoning population is putting Hyderabad's Transport system under constant pressure. Due to its prominence as a major high-tech center, it is one of the fastest growing with a population density of more than17,000 persons per sq.km. The rapid rate of urbanization with increased economic activity has encouraged migration to the twin cities, which led to an increase of personal, public, and para (3 and 6 seat autos) transit vehicles, industrial output, and increasing burden on the city's infrastructure.

Hyderabad along with the surrounding ten Municipalities constitutes the Hyderabad Urban Development Area (HUDA) and has been growing at an average rate of 9%.The need of the hour is a robust system that is dependable, comfortable, affordable and sustainable. Population with an estimation to touch 13.64 million by 2021, and the increase in vehicular number has definitely created crisis in Hyderabad. The increase in the vehicles since last 10 years has put lot of pressure on the existing roads and ultimately resulting in road accidents. It is estimated that since 2001 there is an increase of 202 percent of two wheeler and 286 percent of four wheeler vehicles with no road expansion. Motor vehicle crashes are a common cause of death, disability and demand for emergency medical care [6].

The main reason for road accidents in the city of Hyderabad is attributed to physical characteristics. The study tries to assess the increase in vehicular number and its effect on road accidents, also tries to locate the hotspots of road accidents and its causes.

The main objectives of the study are

1. To study the cause of road traffic accidents in Hyderabad city.
2. To identify and analyze the risk factors and accident prone sites (Hot spots), which form the crucial determinants of the RTI.
3. To understand the nature, type and mode of occurrence of accidents.
4. To study the nature and type of injuries.

2. Methodology

The methodology adopted in the study is based on secondary data from various govt. sources namely; Census of India, Ministry of Transport, Hyderabad Traffic Police, Road Transport Authority etc. The study is also based on primary survey where data was collected through observation and survey methods. The use of GIS was also done as it helps in locating the accident hotspots and also in analyzing the trend of road accidents in Hyderabad.

The classified traffic counts data collected at the various busy points and inner ring road points were Analyzed to arrive at hourly flows and daily flows as per direction and class of vehicles. Traffic volume analysis is essential to appreciate traffic characteristics along these lines, mid-block sections and at Inner corridor. Traffic analysis was done for all the above locations to get the information of hourly variations, compositions and peak hour flows. The Origin-Destination data was analyzed to get the information of occupancy, trip purpose, and trip length for passenger traffic.

The analysis regarding traffic at various locations in the city suggest that the traffic flow on NH-65 near Malakpet is highest with 2, 21,922 vehicles/day; while a minimum traffic is observed at Bollaram with only 8,293 vehicles/day. On the other side the maximum traffic flow is observed at Begumpet ROB with 2, 16,235 vehicles/day; along the inner corridor the maximum traffic flow is observed at Chaderghat bridge with 1,19,009 vehicles/day; It can be seen that the share of Private vehicles are in the range of 42% and 84%. Public Transport is observed to be around 10% Non-Motorised Transport was observed to be around 14% at Moula-Ali station which was highest among all the locations. Volume was also observed to be lowest among all the locations (3,569 vehicles).

Table 1. Average daily traffic at various locations

S.No	Location	Passenger Vehicles				Total vehicles
		Private	IPT	PT	NMTs	
1	Faluknuma ROB	39,061	14,834	3,643	1,020	58,558
2	Kandikal Gate Colony	8,708	5,168	113	1,317	15,306
3	UppuGuda	8,218	6,343	95	882	15,538
4	TalabKatta	23,438	11,015	139	2,779	37,371
5	Yakutpura	9,324	5,403	117	1,476	16,320
6	On NH-65 near Malakpet	1,68,573	33,139	9,073	1,645	43857
7	Nimboliadda	28,496	7,414	1,359	1,812	39,081
8	Tilak Nagar	61,154	12,116	974	1,533	75,777
9	APHB Colony	34,354	7,815	47	3,107	45,323
10	Vidyanagar	45,067	7,598	1,843	819	55,327
11	Adikmet Road	35,186	5,738	1,521	329	42,774
12	Arts College	9,993	2,182	11	578	12,764
13	Sitaphalmandi	26,891	4,978	307	646	32,822
14	AlugaddaBhavi	1,04,297	18,851	5,880	884	25615
15	Lallaguda Gate	12,147	2,802	63	667	15,679
16	Safilguda	33,666	5,277	330	608	39,881
17	Bolaram	7,231	692	79	291	8,293
18	Hi-tech City MMTS Station RUB	76,593	16,629	1,271	227	94,720
19	Sanath Nagar ROB	92,904	15,206	7,148	741	115,999
20	Fateh Nagar ROB	64,222	13,199	1,103	902	79,426
21	Begumpet ROB	2,16,325	34,691	6,788	903	42,382
22	Ministers Rd - Necklace Rd, James	41,152	5,140	37	508	46,837
23	Ranigunj	72,312	10,551	3,242	979	87,084
24	Rashtrapathi Road	70,132	15,683	2,864	1,765	90,444
25	Oliphant Bridge-Rathifile	60,091	27,878	8,168	1,336	97,473
26	Rail Nilayam RUB	1,04,329	13,496	1,953	641	16,090
27	Lalaguda ROB	38,107	7,520	2,170	917	48,714
28	Tippukhan Bridge	32,941	4,414	2,063	666	40,084
29	Attapur Bridge	85,472	18,719	1,943	1,437	107,571
30	PuranaPul Bridge	45,654	9,671	159	4,353	59,837
31	Muslim Jung	54,157	11,405	183	4,010	69,755
32	Nayapul Old Bridge	49,742	12,411	3,098	2,441	67,692
33	Nayapul New Bridge	73,987	15,953	2,379	2,648	94,967

34	Shivaji Bridge - Salar Jung	40,008	5,691	2,261	2,179	50,139
35	Chaderghat Causeway	58,817	14,526	2,796	962	77,101
36	Moosarambagh	68,068	13,064	1,153	1,597	83,882
37	Nagole Bridge	77,904	11,668	3,516	427	93,515

Source: Transport Department

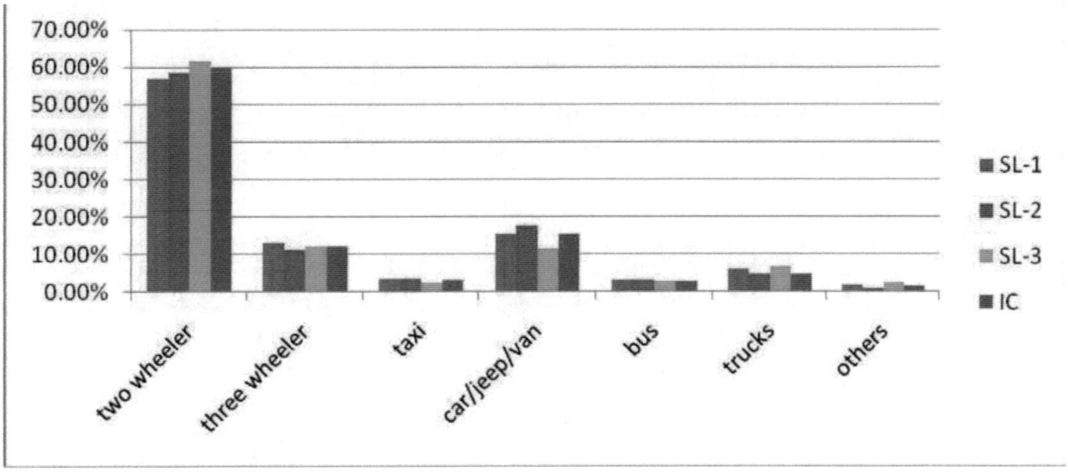

Figure 1. Vehicle movement in Hyderabad.
Source: Hyderabad Traffic Police.

3. Analysis

Nearly 3,400 people die on the world's roads every day. Tens of millions of people are injured or disabled every year. Children, pedestrians, cyclists and the elderly are among the most vulnerable of road users. In 2013, deaths due to road traffic crashes were the tenth leading cause of death in all age groups in India. According to the WHO,1,97,135 people were killed in 300 000 road traffic crashes in India in 2013 Although the leading causes of death in India in numerical terms are cardiovascular diseases, respiratory infections and diarrhoeal diseases, the ranking of causes of death changes when age-at-death is considered. Road traffic crashes are the third leading cause of death for people between 5 and 44 years of age. It is estimated that the morality rate in road traffic crashes has risen from less than 30 per 100 000 in the early 1970s to more than 50 per 100 000. Despite this, road safety has not received enough attention in India as well as in many other developing countries.

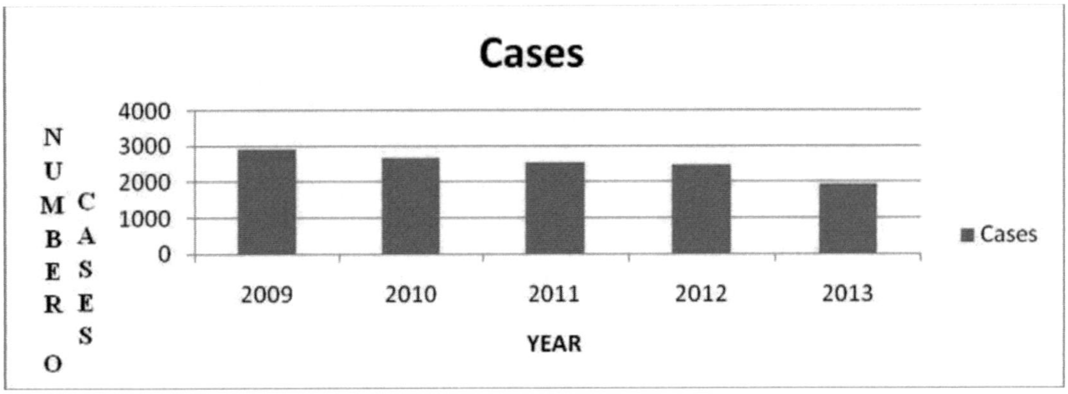

Figure 2. Injured cases for last five years.

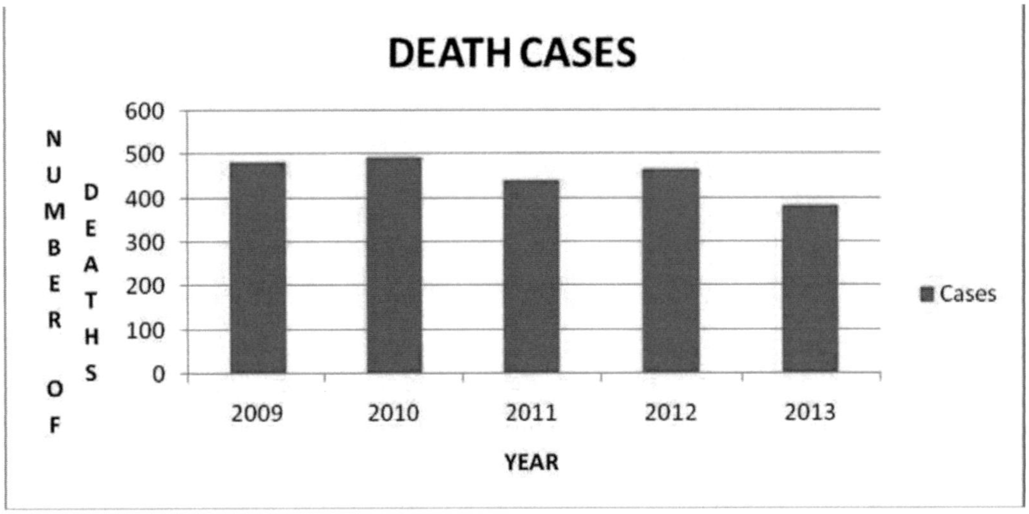

Figure 3. Death cases in Hyderabad.

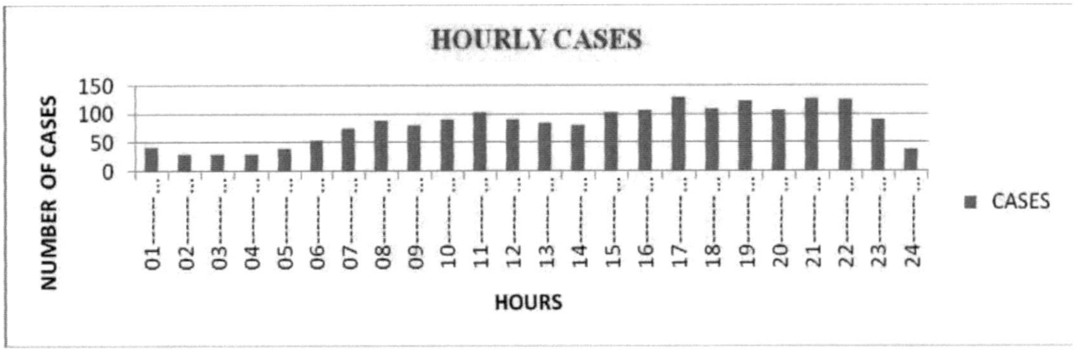

Figure 4. Hourly accident cases in Hyderabad.

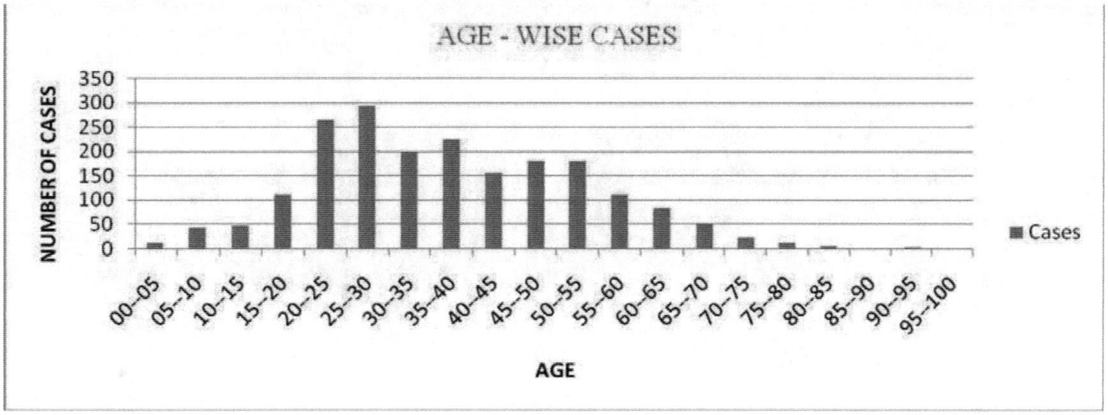

Figure 5. Accident victims age.

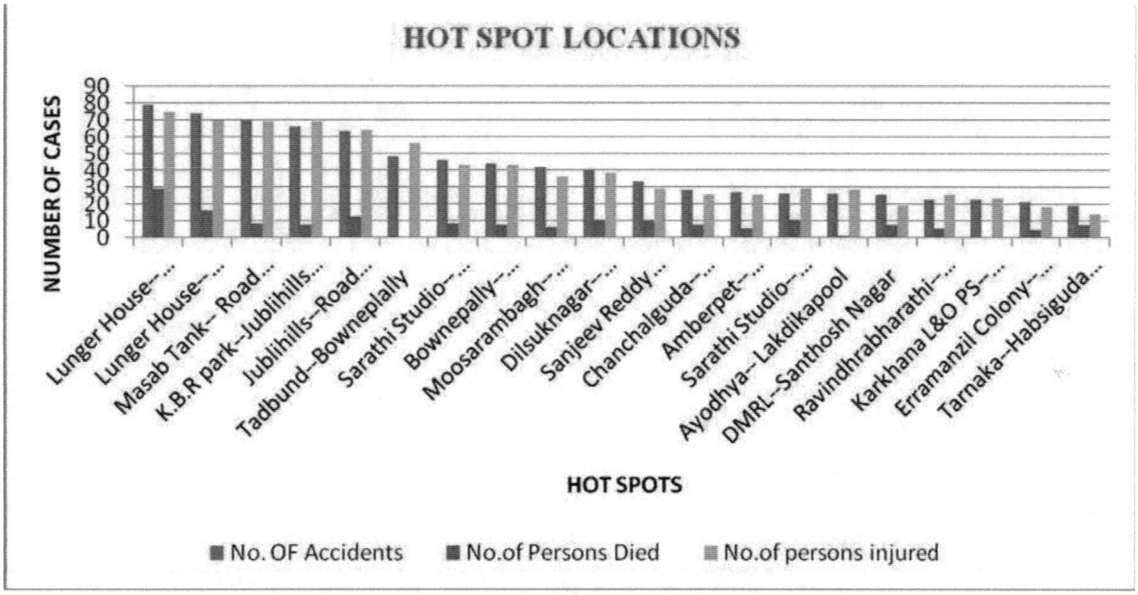

Figure 6. Accident hotspot locations.

4. Results & Discussions

A total of 2,990 cases of road traffic incidents were recorded in the police database for 2013, in which 411 (16.08%) people were killed. In the same year, 316 cases of road traffic crashes resulting in 353 deaths were reported in the newspaper. The majority of those who died due to these crashes were males. Seventy per cent of those killed were between 16 and 49 years of age. Pedestrians and riders of two- wheelers were the most vulnerable. Collision with a vehicle caused 86.4% of all crashes and 60% of the victims died before reaching a hospital. The maps reveal the nature of incidents and also the vulnerable spots in the city. The maps give us the various types of injuries and type of vehicle involved.

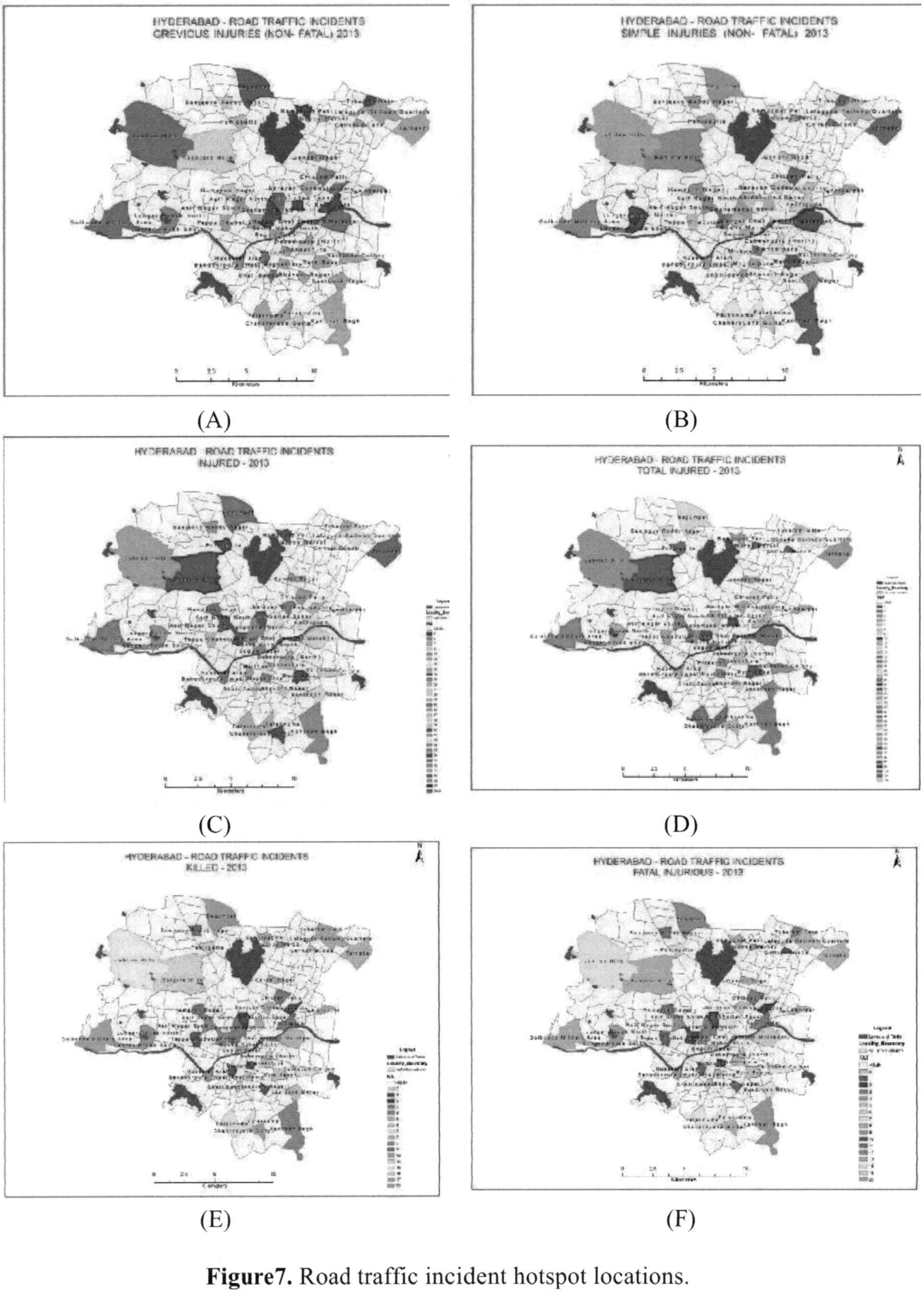

Figure7. Road traffic incident hotspot locations.
Source: Hyderabad Traffic Police, 2013.

The available data were not comprehensive enough to provide a thorough basis for planning intervention strategies to reduce fatalities due to road crashes. These data also highlight that there is a need to strengthen surveillance. The data currently available do not give the true magnitude of the burden due to road traffic crashes as there could be underreporting; reliability of these data needs strengthening as discrepancies were noticed between the two data sources; and these data are not comprehensive enough to plan proper intervention strategies to reduce fatalities and injuries due to road crashes. As it cannot be expected that the newspapers would cover all the above because it is not their prime responsibility, we focus on the police database for recommendations as it is the legal source of information on fatalities and injuries due to road crashes. It is important to know the correct number of people killed in road traffic crashes to understand the real burden of road crashes. It is possible that some cases of road traffic crash deaths are not reported to the police.

The possible reasons for this could be suboptimal coordination between the police and hospitals in the event of death occurring in hospitals due to a road crash; not all deaths are reported to police by the public; and because not all deaths related to road traffic crashes occur in hospitals. It is also possible that someone who was recorded as 'injured in road traffic crash' in the police database died at a later date but this is not reflected in the police database, which can result in underreporting of deaths due to road traffic crashes. A proper data collection mechanism with involvement of all the major stakeholders on whom this data collection depends needs to be put in place, including regular updating of the police database to understand the real burden of road crashes. There is also a need to extend the coverage of data on road traffic crashes collected by the police to include all the factors needed to assess the true nature of road crashes.

The reporting of road crashes is not adequate as the physical environment at the time of the crash, status of the people involved (alcohol or drug use and behavior), use of protective gear for two-wheeler riders (helmet), condition of the vehicles involved in the crash are not covered in the police database. All this information is important to have a comprehensive understanding of the nature of road crashes, and plan and evaluate proper intervention strategies. The reliability and accuracy of the data collected are also important. For example, in the police database the nature of crash was recorded as 'head-on collision', and the cause of crash as 'negligence of driver' for the entire road crashes that resulted in death. We explored the reasons for this documentation. Many of the forms received did not have 'nature of crash' filled, and it was assumed to be 'head-on collision'. Second, recording of the cause of crash as 'negligence of driver' in the police report did not require any proceedings in a court of law, and thereby served to reduce the administrative burden, as the resources available with the police are limited.

5. Conclusion

In conclusion, the police database and newspaper reports provide insights into the magnitude and nature of fatalities due to road traffic crashes. The limitations of the police database, which is the legal source of information on fatalities resulting from road traffic crashes, indicate a need for strengthening the road traffic crash surveillance system so that reliable, accurate and adequate data on road traffic crashes and the resulting fatalities and injuries can be collected. This could then form the basis for planning effective intervention strategies to improve road safety in the city. More effort is needed to have a comprehensive understanding of the various aspects of road traffic crashes, and the recommendations made for strengthening surveillance could serve as an initial step towards reducing fatalities and injuries due to road crashes in the long term.

References

[1] Afukaar FK, Antwi P, Ofosu-Amaah S. Pattern of road traffic injuries in Ghana: Implications for control. Inj Control Saf Promot 2003;**10**:69–76.

[2] Centers for Disease Control. CDC Surveillance Update. Atlanta:Centers for Disease Control and Prevention; 1988.

[3] Government of India. Code of Criminal Procedure—1973. New Delhi:Ministry of Home Affairs, Government of India; 1973.

[4] Holder Y, Peden M, Krug E, Lund J, Gururaj G, Kobusingye O (eds). Injury surveillance guidelines. Geneva:World Health Organization; 2001.

[5] Jacobs G, Aaron-Thomas A, Astrop A. Estimating global road fatalities. TRL Report 445. London:Transport Research Laboratory; 2000.

[6] Joint Transport Commissioner and Secretary. Vehicles registered in twin cities. Hyderabad:Government of Andhra Pradesh; 2003. http://www.aptransport.org (accessed October 2003). [7] Kapp C. WHO acts on road safety to reverse accident trends. Lancet 2003;**362**:1125.

[7] Krug E (ed). Injury: A leading cause of the global burden of disease. Geneva: World Health Organization; 1999.

[8] Krug EG, Sharma GK, Lozano R. The global burden of injuries. Am J Public Health 2000; **90**:523–6.

[9] Mohan D. Road traffic injuries—a neglected pandemic. Bull World Health Organ 2003;81:684–5.

[10] Murray CJL, Lopez AD, Mathers CD, Stein C. The Global Burden of Disease 2000 Project: Aims, methods, and data sources (revised). Global Program on Evidence for Health Policy Discussion Paper No. 36. Geneva:World Health Organization; 2001.

[11] Nantulya VM, Reich MR. The neglected epidemic: Road traffic injuries in developing countries. BMJ 2002;**324**:1139–41.

[12] Nantulya VM, Sleet DA, Reich MR, Rosenberg M, Peden M, Waxweiler R. Introduction: The global challenge of road traffic injuries: Can we achieve equity in safety? Inj Control Saf Promot 2003;**10**:3–7.

[13] National Crimes Records Bureau. Accidental deaths and suicides in India—1999. New Delhi:Ministry of Home Affairs, Government of India; 2001.

[14] Odero W, Garner P, Zwi A. Road traffic injuries in developing countries: A comprehensive review of epidemiological studies. Trop Med Int Health 1997;2: 445–60.

[15] Odero W, Khayesi M, Heda PM. Road traffic injuries in Kenya: Magnitude, cases and status of intervention. Inj Control Saf Promot 2003;**10**:53–61.

[16] Registrar General of India. Population totals: India, Census of India 2001. New Delhi: Ministry of Home Affairs, Government of India; 2001. http://www. censusindia.net (accessed October 2003).

[17] World Health Organization. The World Health Report 1999: Making a difference. Geneva:World Health Organization; 1999.

Integration of geospatial multi-mode transportation systems in Kuala Lumpur

[1]M A Ismail, [2]M N Said

Department of Geoinformation, Faculty of Geoinformation and Real Estate, Universiti Teknologi Malaysia, 81310 UTM Skudai, Johor, Malaysia

Email: [1]azizolismail@gmail.com, [2]m.nor@utm.my

Abstract. Public transportation serves people with mobility and accessibility to workplaces, health facilities, community resources, and recreational areas across the country. Development in the application of Geographical Information Systems (GIS) to transportation problems represents one of the most important areas of GIS-technology today. To show the importance of GIS network analysis, this paper highlights the determination of the optimal path between two or more destinations based on multi-mode concepts. The abstract connector is introduced in this research as an approach to integrate urban public transportation in Kuala Lumpur, Malaysia including facilities such as Light Rapid Transit (LRT), Keretapi Tanah Melayu (KTM) Komuter, Express Rail Link (ERL), KL Monorail, road driving as well as pedestrian modes into a single intelligent data model. To assist such analysis, ArcGIS's Network Analyst functions are used whereby the final output includes the total distance, total travelled time, directional maps produced to find the quickest, shortest paths, and closest facilities based on either time or distance impedance for multi-mode route analysis.

1. Introduction

An efficient transportation network can stimulate economic transformation, physical development and improve mobility. Urban areas are populated by over half the world's population and are anticipated to witness most of the population growth in the next forty years [1]. With an increasing population, the demand for transportation will also increase [2]. In certain location such as urban areas, human mobility usually happens over a multi-mode transportation network. People tend to use more than one mode of transportation to arrive at a particular destination due to many reasons. Accordingly, when analyzing transportation systems people should not simply consider each mode of transport separately but should look at it as a multi-mode transportation systems with relationships and dynamics between components. Moreover, each mode of transport has its weaknesses and strengths and by using a combination of modes potentially can cancel their negatives out, offering an advanced platform for more flexible, efficient, sustainable and reliable transportation [3].

There are works done beyond the urban territories such as reported by Frederic S *et al.,* [4] who highlighted the development of databases regarding territories where multi-mode transport systems are being established, all around the world. These databases are not only to be used for integrated mobility information services, but also for new analysis to improve the global multi-mode transportation over a territory. In fact, transportation planners and decision-makers are increasingly considering multi-mode transportation strategies to support sustainable transportation associated with urban development [5]

Content from this work may be used under the terms of the Creative Commons Attribution 3.0 licence. Any further distribution of this work must maintain attribution to the author(s) and the title of the work, journal citation and DOI.
Published under licence by IOP Publishing Ltd

[6]. This is of a particular importance when more people are getting mobile with the availability of new public transportation options that connects the routes.

In Malaysia, mainly in Kuala Lumpur, the network becomes more complex and complicated. Common transportation features that can be observed include multi-layer highways and major roads with many intersections. In fact, urban public transportation such as Light Rapid Transit (LRT), Kereta api Tanah Melayu (KTM) Komuter, Express Rail Link (ERL), and Monorail exist to boost economic development in Kuala Lumpur. Currently, public transportation that operates within Klang Valley area is managed by various separate companies. For instance, RapidKL operates on LRT which consist of two lines namely Kelana Jaya Line and Ampang Line. RapidKL is also in charge of the KL Monorail. Kereta api Tanah Melayu Berhad (KTMB) on the other hand operates on KTM Komuter, while Express Rail Link Sdn Bhd operates on ERL. However, none of these network lines lie on the same platform as each other, for example KL Monorail line is not connected physically with any other line.

Determination of an optimal path within a highly complex transportation network is not an easy task, especially for those who are unfamiliar with the local transportation system. Users might get confused when travelling from one place to another. Integration of multi-mode transportation systems would not only ease travellers' transfer from origin to destination, but also make their journey a more enjoyable [7] and less stressful experience. Complicated network of route systems requires critical analysis which can only be realized with the implementation of GIS to improve the movement of people. Moreover, they are not integrated, i.e. they exist as separated systems.

Even though some of well-known route planning systems (see Figure 1) are making efforts to integrate more transportation modes, it is unfortunately performed totally separated for each mode i.e. one mode at a time. In fact, several problems have been identified for route determination of Kuala Lumpur area using Google Maps such as lack of or missing pedestrian paths. The combination of modes is not logic, and the pinpoints of origin-destination are not snapped precisely. This can lead to misunderstanding of information generated and might get users into a troublesome situation.

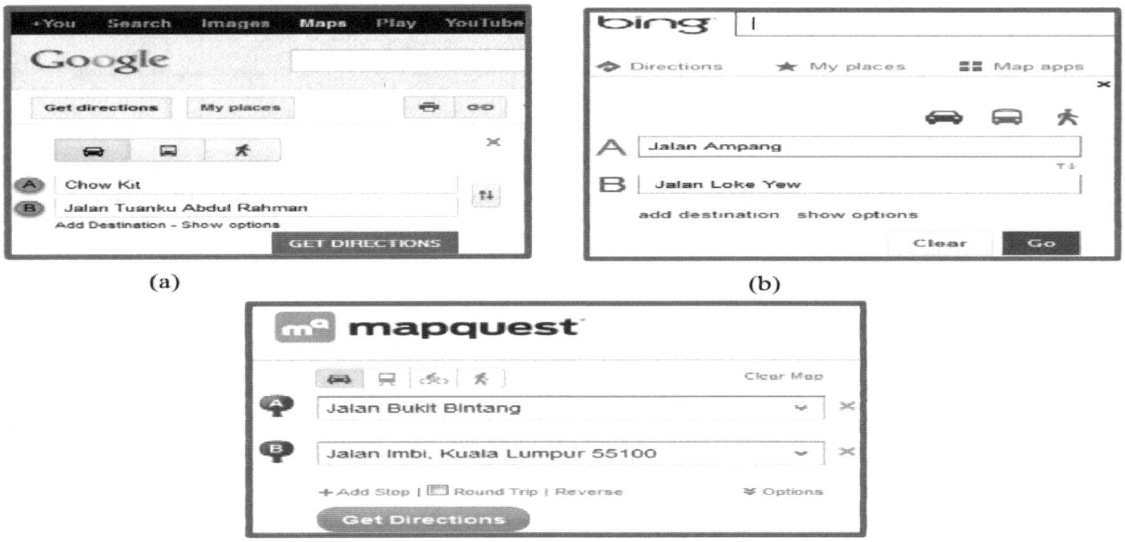

(c)

Figure 1. Separation of modes in (a) Google Maps (b) Bing Maps (c) MapQuest.

To date, there is no such work which involves the modelling and integrating of urban public transportation in Kuala Lumpur. So, the main aim of this study is to develop an intelligent and integrated data model of route network systems within a GIS environment.

2. Modelling multi-mode transportation framework

The metropolitan city of Kuala Lumpur, Malaysia has been selected as the study area. This area contains most of the transit systems (LRT, Monorail, ERL, KTM Komuter) in Malaysia. This Federal Territory covers an area of 243 square kilometres [8]. Under the 1984 Kuala Lumpur Structure Plan, the city centre was designed as the principal core to provide specialized metropolitan services, national and international commercial and business activities, central government activities, and much more.

2.1. Designing abstract connector

One of the crucial parts in designing a geospatial multi-mode transportation network is the integration of available networks (road, rails, pedestrian) into a single layer. In this research, the abstract connector (see Figure 2) is adopted which acts as a 'bridge' to connect between different modes of transportation. It also reflects the transfer process within the movement of people. The solid lines as shown in Figure 2 represent physical networks while the dashed lines illustrate the abstract connector links. The physical network edges contain required attributes such as travelled time, speed limit, length, etc. The physical nodes (shown with green circles in Figure 2) represent locations associated with X, Y coordinates as required attributes.

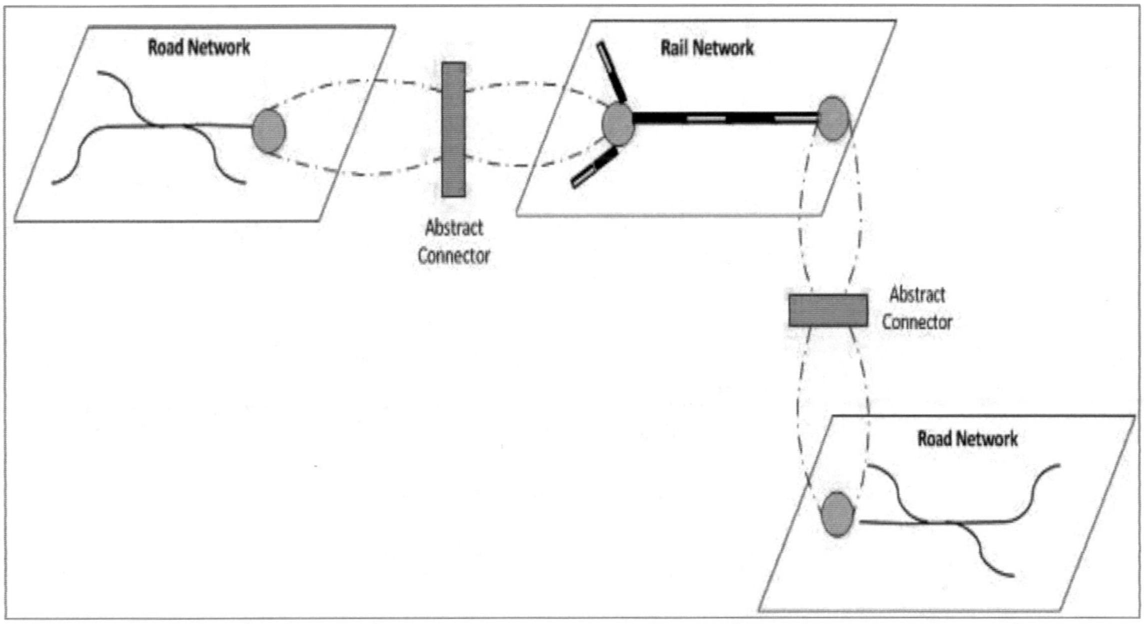

Figure 2. Adopting an abstract connector system in a multi-mode transport network.

In the Kuala Lumpur area, transit stations are accessible using the pedestrian network. The pedestrian network plays a key role in mode transfer. Thus, abstract connectors are properly situated at the intercept point between transfer link and the walkways that lead to the transit stations. It is to be noted that there are two types of transfer; one is transfer within the same mode (e.g. transfer from KTM Sentul Pelabuhan Klang Line to Seremban Rawang Line), the other is between different modes (e.g. transfer from KL Sentral's LRT to Tun Sambathan Monorail).

2.2 Building network dataset
Building network dataset is a compulsory procedure. It is to ensure the data is ready for network analysis to take place. There are 5 steps required in order to create a network dataset which are (i) naming the network dataset and choosing source feature classes (ii) assigning connectivity policy, connecting groups (iii) setting elevation policy (iv) specifying the attributes for the network dataset, and (v) configuring directions.

For the travel time calculation, the formula as shown in Equation 1 is used. It is very important to ensure the unit is correct and comply with the formula. The distance refers to the road length in m unit while the speed limit is converted from km/hr into m/minute.

$$\text{Travel\_time (minutes)} = \text{Distance (m)} / (\text{Speed limit} \times (1000/60)) \quad (1)$$

A measure of the amount of resistance, or cost, required to traverse a path in a network is known as impedance. If the impedance is the distance (m), the optimal route is the shortest route. On the other hand, if the time (minutes) is chosen as impedance, the optimal route is the quickest route. Figure 3 shows how a user can define the suitable impedance for a journey.

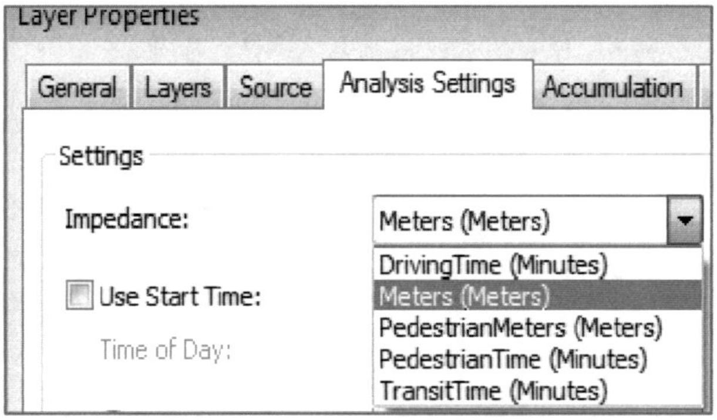

Figure 3. The choice of impedance.

2.3. Assumptions of the analysis
The analysis is based on two assumptions, i.e. (a) for the travel time, it is calculated in the condition that there is no traffic jam in the streets and (b) travel speed is considered as the maximum speed limit. The maximum speed limit is assigned based on the Malaysia's road class as presented in Table 1.

Table 1. Maximum speed limit based on the road classes.

Road Class	Speed (km/hr)
Highway	110
Federal	90
State	90

2.4 Geodatabase design and development
Geodatabase design is the process of producing a detailed data model of the database to meet end user requirements. The Entity Relationship Model (ERM) is most commonly used during conceptual design. The resulting Entity Relationship Diagrams (ERD) provides a data-centric and structural view of the database. A new option is to use a Unified Modelling Language (UML) which contains class diagrams that also incorporate structural views of the database. Figure 4 depicts the conceptual design adopted for the data model.

IOP Conf. Series: Earth and Environmental Science **20** (2014) 012027 doi:10.1088/1755-1315/20/1/012027

For this research, the major (required) feature classes (map layers) are road networks, and transit stations and lines (LRT, ERL, KTM Komuter, monorail) – with the correct network topology. Meanwhile, the supporting feature classes include Points of Interests (POIs), transfer points, connecting stations and interchange stations. As accuracy is not an issue for this research, the point and line data are extracted from Google Earth's images, Google Maps, websites and Garmin BaseCamp 4.2.5 map data. Figure 5 shows the geodatabase model developed for this application based on MS1759. The personal geodatabase is divided into 3 main feature datasets such as the built environment, demarcation, and transportation. Therefore, it enables the geodatabase to more accurately model the network dataset relationships, and maintains data integrity, resulting to more reliable output.

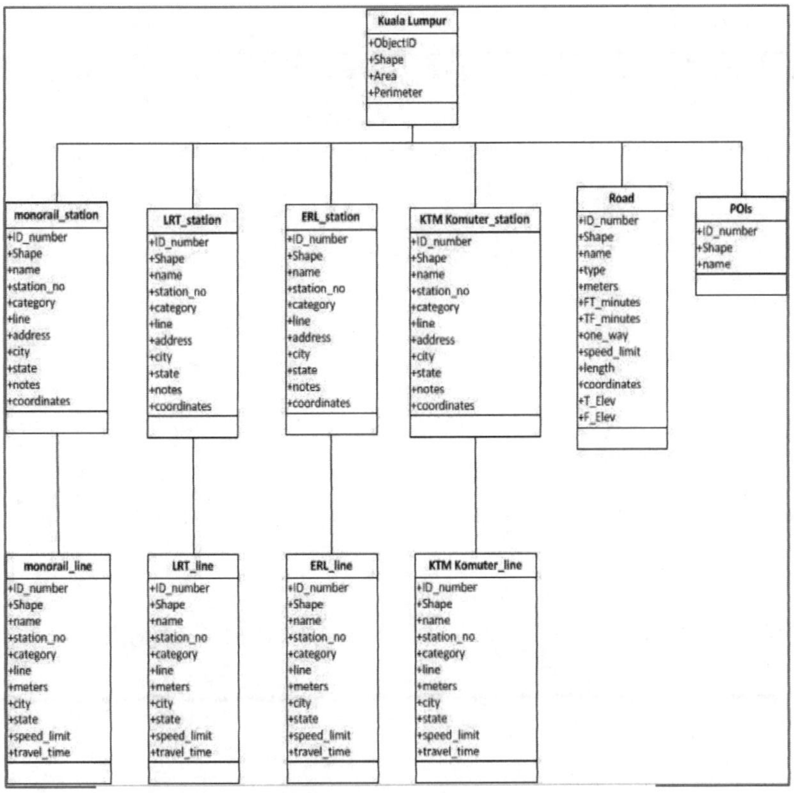

Figure 4. Database conceptual (UML) design.

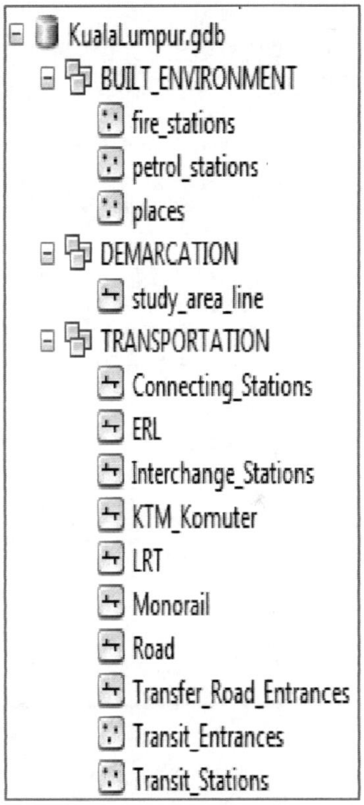

Figure 5. Geodatabase.

3. Application testing

The framework presented in the previous section is used to model and analyze the public transportation systems in Kuala Lumpur's metropolitan area. The user can visually select the origin-destination without having to input manually information regarding the facilities on the networks. Moreover, with the database associated, it will assist users to have a better interpretation of the network characteristics and minimize the potential for errors generating in the network data files. The well-known Dijikstra's algorithm embedded within Network Analyst extension makes it possible to perform analysis such as optimal path and closest-facility determination.

3.1. Route analysis in multi-mode transportation network

In the real world, commuters and travellers frequently use several modes of transportation such as walking, driving, and taking the trains. Figure 4 displays the transit systems available in Kuala Lumpur's city including the LRT, Monorail, KTM Komuter, ERL, roads (driving) and footpaths (walking). For LRT, it comprises of 3 lines, the Ampang Line, the Kelana Jaya Line and the Sri Petaling Line. Likewise, the KTM Komuter consists of 2 lines which are the Sentul-Pelabuhan Klang Line and the Seremban-Rawang Line.

To demonstrate a multi-model route analysis, the starting point or source is set to Chowkit Monorail Station and the destination is KL Sentral KLIA Transit ERL. The bold pink colour in Figure 6 highlights the best route to travel from the source to destination. Meanwhile, Figure 7 shows the direction, total time and total distance for multi-modal transportation. In this scenario, the individual travelling uses 5 modes of transportation, Monorail, LRT, KTM, walking and ERL.

Figure 6. Map of multi-mode transportation analysis.

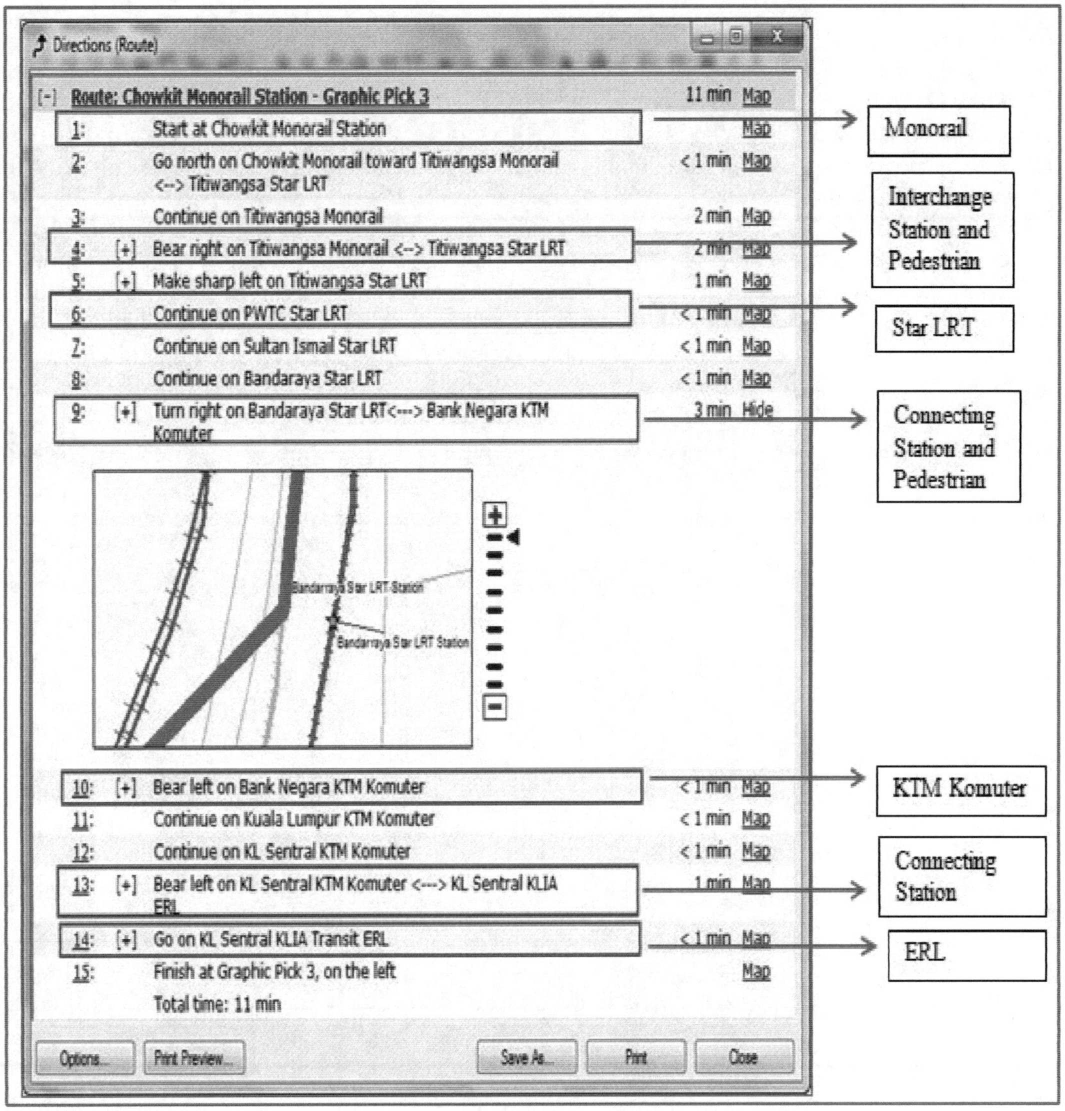

Figure 7. The directions, total time and total distance for multi-modal transportation.

It has to be noted that the definition of connecting stations and interchange stations adopted for this application is as follows:

- **Connecting station**: Is the station that connects two different train modes. For example, Putra Station connects PWTC Star LRT and KTM Komuter
- **Interchange station**: Interchanges between the stations requires an individual to exit the station and then proceed to cross a footbridge that crosses a major roadway. For example, Titiwangsa station is a rapid transit interchange station that allows for transfers between the Ampang Line rapid transit system (formerly known as Star, and the Ampang and Sri Petaling Lines) and the KL Monorail.

3.2. Closest facility analysis

Closest facility is a type of network analysis for finding the closest locations (facilities) from sites (incidents), based on the impedance chosen. For instance, finding transit stations near a current location. The users can specify a cutoff threshold beyond which the algorithm will not search for a facility. Figure 8 displays the results of finding the closest transit stations from a particular location (denoted as "A" in Figure 8) by using pedestrian mode. The green line is the monorail network whereas the blue and orange line represent the LRT network for the Sri Petaling Line and the Ampang Line respectively. As network analysis function is executed, the output shows there are 3 nearby transit stations; Titiwangsa Monorail Station, Titiwangsa Star LRT Station and Chowkit Monorail Station. The yellow line colour shows the direction to those stations. As depicted in Figure 9, the closest transit station is Chowkit Monorail Station which is about 8 minutes from the current position. The result does not only give information about total distance and total time, but also the direction and map to the location.

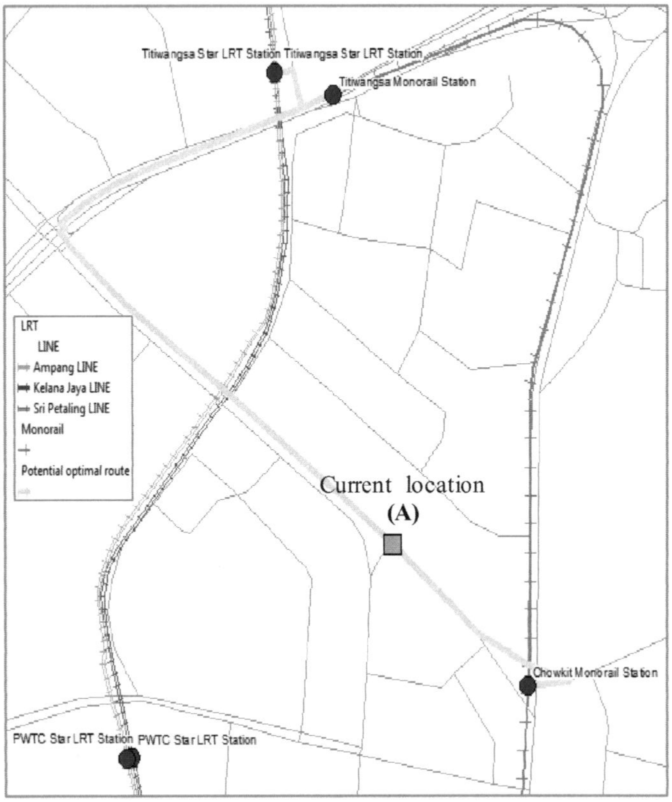

Figure 8. Closest facility analysis.

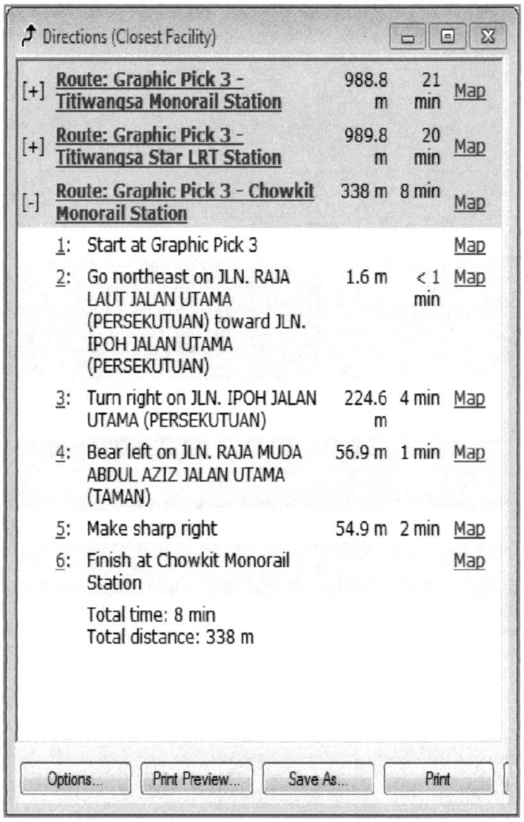

Figure 9. Least time, distance and direction of closest facilities.

4. Conclusion

Throughout this study, all of the urban transportation systems (LRT, Monorail, KTM Komuter, ERL, driving and walking) in the city of Kuala Lumpur are successfully integrated into a single database model by adopting an abstract connector approach. This is to support the real needs of people when moving from one place to another in a situation that has many modes of transport available. All of these facilities are modelled in such a way that they are interconnected and an individual can travel from the start point to the end point. With a proper design of the database, users can freely choose the mode of transportation by either selecting the single or multi-mode. This data model supports the display of the information regarding the total time, total distance, directions and a map to the selected destination. It is particularly useful to assist travellers planning their journey. This type of analysis can be used to find not only the shortest or quickest routes but also for other network applications such as modelling hydrologic flow, traffic flow, and service areas. Consequently, by using GIS, complicated network of routes can be visualized and worked out in a very precise manner.

Acknowledgements
This work is financed by Zamalah / Institutional Scholarship provided by Universiti Teknologi Malaysia and the Ministry of Higher Education of Malaysia.

References
[1] Spickermann A, Grienitz V, A.von der Gracht H 2013 *Heading towards a multimodal city of the future?: multi-stakeholder scenarios for urban mobility, technological forecasting and social change* ISSN 0040-1625
[2] Farahani R Z, Miandoabchi E, Szeto W.Y., Rashidi H 2013 A review of urban transportation network design problems *European J. of Operational Research* **229** 281-302
[3] Mahrous R W 2012 *Multimodal transportation system : modeling challenges* Master's Thesis University of Twente
[4] Frederic S, David J, Christophe D 2008 *Potimart : Open Source Transport GIS Software For Multimodal Network Analysis*. Potimart paper for ITS Europe in Geneva
[5] Krygsman S 2004 *Activity and travel choice(s) in multi-modal public transport systems* Ph.D's Thesis Utrecht University, The Netherlands
[6] Matthew J Eckelman 2013 Life cycle assessment in support of sustainable transportation *Environ. Res. Lett.* **8** 021004
[7] Luigi dell'Olio, Angel Ibeas, Patricia Cecín, Francesco dell' Olio 2011 Willingness to pay for improving service quality in a multimodal area *Transportation Research Part C: Emerging Technologies* **19** 1060-1070
[8] Mohamad J 2005 Urban transport and growth management strategies: a tale of two southeast asian cities at the dawn of the new millennium *Geografia: Malaysian J. of Society and Space 1* **1** 11-22

Regulation of land attitudes in Kazakhstan

Gulnara Kurmanova
PhD Doctor (Doctor of Economics), Professor
Head of the Cadastre Department, Kazakh Agrotechnical University
E-mail: kurmanova_gul@mail.ru

Abstract. Land relations play an important role in the life of any state. The article is devoted to the questions of the land relations development in Kazakhstan. Tasks, stages and results of land reform are considered in it. The analysis has shown that land reform has not affected on a condition of land fund in a good way: the huge areas have been transmitted into stock lands, arable land has sharply decreased, fallow lands that have negatively affected on a quality of the lands and economy of republic was formed. In addition, the problems and ways of land relations perfection in republic were considered.

1. Stages of land reform

Change of political building, finding by Kazakhstan an independent sovereign state status has caused the change of ground building in our republic. Realization of a strategic rate of economy modernization is closely connected with inclusion of ground resources in a market turn, by effective utilization and protection of the grounds. N.A.Nazarbaev repeatedly specified the major value of the ground for the development of a national economy. He marked, that « ... the ground, alongside with the natural riches and trial estate, is one of whales on which the economy keeps».

Any reform is a change, transformation, a reorganization in any sphere of a public life, but at the same time it should not destroy bases of existing economic-social structure. Therefore, spent land reform in republic is a process of purposeful stage-by-stage change of ground attitudes. These changes included:
- Transformation of institutes of the landed property, ground possession and using ;
- Formation of the economic mechanism of regulation of ground attitudes;
- Redistribution of ground fund;
- Perfection of the government management by ground resources.

Realization of land reform has led to the change of ground attitudes. Ground attitudes are closely connected with an agrarian as the further development of agriculture directly depends on character of development of ground attitudes, their conformity to the new economic conditions. Historical experience convincingly testifies that the question on the ground is a basis of an agrarian policy, and dependence of agrarian questions is capable to push out for long years from public attention all other problems connected with an agrarian reform. Land reform is consequence of a spent agrarian reform which with the development of productive forces level and production relations has changed a character of ground attitudes.

First steps in market attitudes in sphere of land tenure have been made in 1987 when after a long interdiction an opportunity of rent attitudes in an agriculture has been given. Call « To return the owner to the ground » has sounded these years which has been lost, not as a result of lead in postrevolutionary years of the ground nationalization and a private property denying, owing to hasty

Content from this work may be used under the terms of the Creative Commons Attribution 3.0 licence. Any further distribution of this work must maintain attribution to the author(s) and the title of the work, journal citation and DOI.
Published under licence by IOP Publishing Ltd

liquidation of new economic policy (SEP) and lead, as a rule by violent methods, associations of 25 million allotments of individual country facilities in 250 000 collective-farms and 5000 state-farm lands.

However, the development of commodity-money attitudes sphere went slowly, on what, certainly, the settled conservative methods of authority had affected. Radical crisis in the ground policy has come in the beginning of 1990th years, since time of finding republic sovereignty. In the first years of land reform the problem of transition to market attitudes, creations of the free ground market were put, but the most important was not considered - practically there was no state regulation. Therefore, subsequently many mistakes have been admitted at the creation of the developed market infrastructure that has proved the necessity of the state intervention in questions of ground attitudes regulation.

Land reform in our republic was spent stage by stage. Conditionally it can be divided into 6 stages:

1 stage (1990-1993) - during this period « The Ground code » (1991), « About land reform » (1991), « About a country facilities » (1990), « About the ground tax » (1991) have been accepted. Thus, the legal base for carrying out of land reform has been created, conditions for equal in rights functioning variety of forms of managing on the ground as bases of multistructure economy and at the same time preservation of property on the ground were created, transition to paid land tenure was carried out, the ground tax was entered, the basic directions of land reform and a way of its realization were defined. This stage has been legislatively fixed in the ground Code of republic put into operation in 1991.Many clauses of the all-union Law in view of social and economic and ecological features of different territories and regions of republic, in essence, have been detailed in it.

Accepted in 1991 « The Law on land reform » [2] has defined the problems and the basic directions of the land reform. One of substantive provisions of the law was the granting of grounds for lifelong inherited possession to citizens of republic for conducting a country and personal part-time farm, animal industries, gardening and country construction. Besides the law had been stipulated a creation of special ground fund of area with its subsequent redistribution with the purpose of the ground rational use. The part of special ground fund could be transferred in conducting city councils for conducting commodity manufacture. In 1992 «RK LAW « On property privatization features of the state agricultural enterprises » [3] was accepted. As a result of privatization of state farms and privatizations of collective farms new forms of managing (CFE, cooperative societies, joint-stock companies, etc.) have been formed. At formation of CFE the structure of family was considered, the size of the ground area was defined proceeding from an average ground share which was defined by the decision of labour collective. During this period the decision on an establishment of feature of settlement in the connection with allocation of agricultural lands for population cattle pasturage, truck farming, gardening, mowing was accepted.

Certainly, legislative and normative-legal documents accepted at this period have played the certain positive role at the creation of new forms of managing on the ground by means of reorganization of state farms and privatization of collective farms. However, with introduction of market attitudes in republic the exclusive state ownership on the ground became the deterrent factor of these processes. There were sharp discussions concerning an opportunity of granting of the ground to a private property during that period among scientific community and workers in the field of land management and ground cadastre, in mass media of republic, as well as in other former union republics. It was a question of legitimacy p.5 clauses 3 of the Ground Code of republic in which the following has been written down: « Sale and purchase, donation, the mortgage and autocratic exchange of the ground areas are forbidden ». Such decision accepted by Parliament of republic at that time, was explained by fear of the ground could be redeemed by businessmen of shadow economy, as a rule, not having experience in an agriculture, with a view of gamble and extraction of unearned incomes.

The national expert of the project on improvement of an estimation system , registration and the account of farmland of German society on technical cooperation (GTZ) Anotshenko has stated a following estimation of the given land reform stage. He considered, that the laws accepted during this period, being the initial stage of market attitudes, did not provide a full complex of the market mechanisms applied in a world practice in agriculture. So, normative-legal certificates of the given period contained regulations about the state pattern of ownership on the grounds of agricultural purpose, about the state control over quality and efficiency of ground using, defined an opportunity of withdrawal of the ground areas and their allocation by the state bodies to other land users, etc. As a result of reform the legal status of the grounds has not exchanged - the state remains the unique proprietor and the manager of the ground, the volume of its rights was kept former irrespective , in whose actual using the ground was. The basic purpose of agriculture reform beginning of that period was the possibility of the private rural manufacturer category occurrence, the businessman, the state ground received in using was the basic object of manufacture. Land reform as now and then is considered as the basic mechanism of an agriculture efficiency increasing [4].

2-nd stage (1994-1995) was the period of adaptation of the ground legislation to the market conditions. During this period Decrees of RK President concerning regulation, developments and perfection of ground attitudes have been accepted: « About some questions of ground attitudes regulation "," On the further perfection of ground attitudes ». All this has allowed to enter institute of sale and purchase of the right of land tenure, the right of lifelong inherited possession, the right of rent of the ground areas, i.e. the major attributes of a private property necessary for those who directly worked on the ground were practically realized to be insured from any arbitrariness and to have an opportunity to transfer it to descendants. And the term « constant possession » for all other economic structures, engaged an agricultural production, emphasized a priority of an agriculture, specificity of the ground which are being unusual real estate. Main, eternal and irreplaceable means of the agricultural production, demanding the extremely careful attitude to the ground and its protection from the negative natural phenomena and anthropogenous influence.

With this purpose base rates of these rights estimated cost have been approved. In this case estimated cost was a basis at definition of the uniform ground tax for country farms, i.e. uniform ground tax CFE was equal to 0.1 % from estimated cost of the ground area. Besides this stage also is characterized by the sanction of various operations with the ground, namely the mortgage of the rights to the ground and their donation. At the given stage privatization of the state agricultural enterprises and formation of new forms of managing on the ground was carried out therefore, owners of conditional shares have the right to concede these rights to other persons. But such transactions on the grounds of agricultural purpose almost did not take place anywhere for the different reasons. The formed country facilities having received property share, still possessed an opportunity to give a mortgaging guarantee and to receive credits in banks. However, the 2-nd stage of land reform has coincided with the main reform on privatization and groundless, sometimes hasty reorganization of state farms and transformation of collective farms in non state agro formations (Joint-Stock Company, Open Company), finally led to mass complication of a financial condition of facilities, encumbrance them by debts and leading many of them up to bankruptcy. Having lost support of large-scale enterprises, CFE, and more large-scale enterprises, not having other mortgaging guarantee, except for certificates on the right to landed property, without the document on the right to dispose of the ground as proprietors, could not receive the credit in bank. In the conditions of general lack of money, they could not sell also a part or completely their right to vitally inherited possession or simply the ground possession.
Thus, the initial stages of land reform had both positive, and the negative parties. Many farms were created formally with the purpose of reception of cheap credits and the grounds in lifelong inherited possession. Besides soil-protective system of agriculture developed during many years have been destroyed an absence of material base have been affected. All this has led to deducing of the significant areas from land revolution.

The 3-rd stage (1995-2000) is marked by acceptance of the important documents - Decrees of the President, having valid laws: « About the ground », « About the state registration of the rights to real estate and transactions with it », « About the mortgage of real estate », The Law « About a country facilities ». At this stage according to RK Constitution the exclusive state ownership on the ground is cancelled. From the moment of acceptance of the Decree « About the ground » for its realization by RK Government 25 decisions have been developed and approved. The decree « About the ground » has solved the main question of principle about the status of the ground, having recognized it an object of the real estate. The given stage was characterized by the following innovations:
- The bases, limits and conditions of transferring the ground areas in a private property to physical and legal persons were established;
- The state registration of the rights to the ground, real estate and transactions with it was entered;
- Preconditions for a civil turn of the grounds were created;
- The order of the mortgage of the ground areas and the rights of land tenure was defined;
- Norms of free-of-charge transfer and the order of granting of the grounds in a private property were established;
- The legal regime of the organization and functioning of country (farmer) facilities was defined [5].

All we remember wide discussion expediency of a private property on the ground at discussion of the new project of RK Constitution. As a result by the decision of a national referendum 30.08.95 y RK Constitution had been accepted where in clause 6 legitimacy of a private property on the ground on the bases had been recognized, in the conditions and limits established by law. The private property has been entered in the limited quantity: for the development of a personal part-time farm, gardening and country construction due to the grounds, settlements, for individual housing construction, due to other categories, except for the grounds of agricultural purpose. As a result in a private property there were 1,3 % of ground fund. However, the grounds of agricultural purpose already as CFE were given to all non state legal persons constant using.

In January, 1999 the project of "Law" About additional changes in President Decree has been published, the valid law « About the ground » from 22.12.95. Amendments to item 44 were made in it and two new were added. It has made the bill inaccessible to discussion to the broad audience of readers. Developers, having reacted to remarks of readers, have replaced the bill with the full text of the Law « About the ground ». Almost half a year went active discussion of the new bill in working group of Mazhilis into which the scientists-landowners, lawyers, experts of the different structures concerning the ground have entered. Then the Mazhilis after hot discussion, has approved the bill in the first reading. It has been published in 23.06.99y, but after the arisen protests on places has been withdrawn by the Government. In the bill which has been born for national discussion, distribution of a private property on the grounds which was being used by CFE, and also on the grounds carried in reserve special fund, all approximately on 40 million in hectares of agricultural purpose grounds was provided. Such offer as a whole did not resist to item 6 of RK Constitution, however, under the insisting of developers, the bill provided the repayment of the ground given in a private property. And up to the repayment it was given to peasants on the right time rent.

Meanwhile, it is served as the main incentive motive of expansion of the right of a private property on the grounds of agricultural purpose: to enable peasants to receive credits by granting to banks a right forming document of the ground proprietor. This opportunity was promised three times by the President of Republic to peasants in the meetings. In the conditions of a crisis situation in the village, full lack of money, granting of the private property right under condition of the repayment was useless for the majority of peasants. Contrary to agrarian objections, in every possible way emphasized loss of an overall objective - opening of a way for the reception of credit, monetarist approach has prevailed - aspiration to receive means for updating a profitable part of the budget more quickly. Soon on places

there were the people who have understood, that the reception of the right of a private property under condition of the preliminary repayment will not make peasants happy.
It certificates the history of clearing of peasants from a serfdom in 1861 with the right of the ground repayment, hung a lot of years as burden on peasants, stolypin reform which has caused the formation of reserve army work and, for this reason, protested not only against expansion of a private property on the grounds of agricultural purpose.

A.A.Satybaldin marked the following lacks of Kazakhstan's land reform carrying out, influencing on functioning of new market ground attitudes:
- Instability of the market economy, the purchased disproportions, crisis processes and tendencies;
- Incompleteness of legal base and discrepancy of the normative-legal documents acting on that moment adjusting market ground attitudes;
- Slow movement to the multistructure land property;
- Default of taxation stimulating function, a weak level of investment policy in the field of effective using of agricultural lands, high risk of mortgaging, hypothecary operations;
- Impreparation of a great mass of the population to the conditions of independent development of an agricultural production on the received ground [6].

The 4-th stage (2001-2003) is characterized by acceptance on January, 24th, 2001 the Law « About the ground ». The ground legislation specifies the competence of the state bodies in the field of regulation of ground attitudes, powers of local representative bodies and institutions of local government were certain. The right of constant land tenure was stipulated to give only to the state legal persons. Rent was defined as the basic form of using the grounds of agricultural purpose by non state land users. Term of time using was reduced from 99 years till 49 years, secondary land tenure was forbidden, the legal regime of conditional ground shares, and also target dates of the order was specified by it.

In this plan the state put the requirement to owners of conditional ground shares to accelerate their realization (within a year) and in case of default - to deprive of their right to a ground share. Such requirement was explained by an observance of stability of wildlife management, however, first, the principle of stability of wildlife management does not mean the necessity of its perfection from time to time, secondly, instead of threat of the termination of this right not issued the ground share, many scientists offered to keep at that time the right of an output to reception of the share by way of carrying out of land management from the grounds of special ground fund in view of a cost estimation and quality received by it in a nature site of the ground.

Actually, from 2.3 million citizens of republic received certificates on conditional ground shares, 2.2 million were already defined. According to the data RK Agency published at that time on management of ground resources, dynamic process of formation CFE proceeds. However from 91,5 thousand CFE majority of owners of conditional ground shares have brought them in authorised capitals of companies and cooperative societies. Division of land tenure reorganized by state and collective farms on conditional ground shares was planned with intention by their further personification and naturalization, first of all, in case of CFE formation. However, the citizens who have received the certificates, were not ready to it neither psychologically, nor morally. Besides in the conditions of the created crisis situation, from a hopelessness, not wishing to remain in private with the arisen difficulties, the overwhelming majority of peasants has gone on a line of the least resistance and have handed over property share and ground shares in different companies and cooperative societies. Many of them have not made the wrong choice, but other part quite could leave their structure and dispose their share in another way - to form their own CFE or be busy with average or small business.

IOP Conf. Series: Earth and Environmental Science **20** (2014) 012028 doi:10.1088/1755-1315/20/1/012028

Carrying out of land reform in Kazakhstan at the given stage has been carried out in the following basic directions:
- Transition to new ground system- creation and formation of qualitatively new system and structure of land tenure, formation of the ground proprietors class ;
- Registration of right attested documents on the ground, formation of land-cadastral proceedings on the ground areas;
- Realization of the state control over use and protection of the grounds;
- Protection the citizens' rights on the ground , legal persons and the state;
- Creation and conducting a ground cadastre and monitoring of the grounds;
- Perfection of ground market attitudes;
- Development of the ground market [7].

At the given stage process of agricultural enterprises reforming was practically completed. During transformation of collective farms, denationalization and privatization of the agricultural organizations on January, 1st, 2003 nearby 138 thousand CFE were organized to which 31.5 million in hectares of the grounds were fixed, 3.5 thousand economic companies of different kinds were created. However, practice has shown, that the lead privatization of state farms has given also negative result in the conditions of other factors negative actions. In particular, in facilities with excessive greater areas of the grounds at the developed crisis situation it was not necessary to think of their rational use. As a result elemental, spontaneous, grounds throwing in a deposit, and also full refusal from part of the grounds with the transition in the grounds stock were carried out. So, on 1/1/2003y the ground stock included 124696,5 thousand in hectares.

The 5-th stage of land reform which has legislatively fixed the right of a private property to the grounds of agricultural purpose began in June, 20th, 2003 with the acceptance of RK Ground code [8] and had established norms of regulation ground legal relationship in agrarian sector in the conditions of various patterns of ownership on the ground. Thus, using of the grounds on the right of land tenure was provided also. It means, that introduction of a private property on the grounds of agricultural purpose does not provide their obligatory repayment in the property. The private property and the right of time land tenure in parallel can operate.
The repayment of the grounds of agricultural purpose for conducting a country farm and a commodity agricultural production in a private property is carried out on a paid basis. The ground legislation establishes the flexible system of the ground areas transfer conditions in the property:
- For full cadastral cost;
- For a reduced price, accepted on all administrative and territorial units at a rate of 75 % from cadastral (estimated) cost of the ground area;
- By credit - till 10 years.

With the purpose of effective utilization of agricultural purpose grounds, preservation of soil fertility the Ground Code from 2003 cancels the right of secondary land tenure - subrent. On other categories of the grounds transfer to secondary land tenure by non-state land user, and also the fulfilment of other transactions by the right time long-term onerous land tenure it is authorized under the condition of the repayment of the right to these grounds at the state.

Conditionally it is possible to consider 2011y as the 6-th stage of land reform in connection with an acceptance of RK Law № 464-IV 3 RK « About modification and additions in some acts of RK concerning regulation of ground attitudes » 7/20/2011y [10]. In particular, it regulates the attitudes connected:
- With the grounds reservation for maintenance of settlements development, accommodation of planned objects engineering, transport and social infrastructure, strategic objects, objects of defense and safety, creation of especially protected natural territories, development of minerals deposits;

IOP Conf. Series: Earth and Environmental Science **20** (2014) 012028 doi:10.1088/1755-1315/20/1/012028

- With regulation of price zoning questions territories of settlements in a view of market mechanisms;
- With the settlement of the ground areas registration order in a private property;
- With the definition of ground area compulsory withdrawal norms for no purpose use or with infringement of the legislation.

2. Results of land reform

The analysis of the ground use for years of reform in our republic has shown, that land reform not in the best image has affected on condition of ground potential. So, at the beginning of land reform the relative density of the grounds of agricultural purpose made 80, 1 %, and in 2012 there was a sharp recession (table 1).

Table 1. Dynamics of ground fund on categories of the grounds, th.hectars.

The name of the grounds categories	1991 y.	2012 y.	Difference (-) - reduction (+) - increase
1. Land for agricultural purpose	218 375,8	93 428,2	-124 947,6
2. Land of settlements	3 747,2	23 789,8	+20 042,6
3.Industry land	18 796,8	2 620,8	-16 176,0
4. Land of especially protected natural territories	775,1	5 776,5	+5 001,4
5 The land of wood fund.	10 179,2	23 059,9	+12 880,7
6. The land of water fund	819,9	4 113,2	+3 293,3
7. The land of stock	18 952,3	108 385,4	+89 433,1
The grounds total	**271 646,3**	**261 173,8**	**-10 472,5**
Including the grounds used in territory of other states	149,8	0,9	-148,9
The Earth, used by other states	993,7	1s1 317,3	+10 323,6
Territory of republic	**272 490,2**	**272 490,2**	**-**
Note: Given Agencies of RK on management of ground resources on 1/1/2013.			

Apparently from table 1, the most valuable grounds were sharply reduced to 125 million hectare due to transition of their greater part in the grounds of a stock.
Agricultural lands are prevailing kind of holdings in all categories of the grounds. The most valuable agricultural lands are, mainly, in the structure of the grounds of agricultural purpose. The structure of agricultural lands within republic is presented in figure 1.

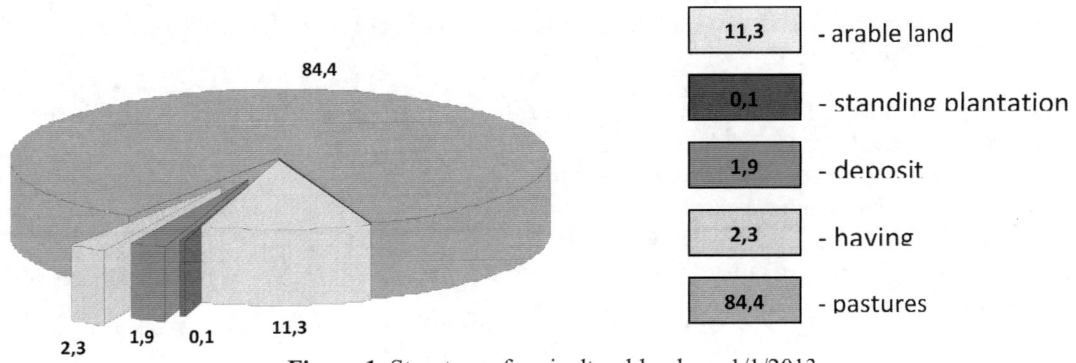

Figure 1. Structure of agricultural lands on 1/1/2013.

Apparently from figure 1, the greatest relative density make pastures - 84,4 %. The reduction of an arable land and increase in laylands is the negative part of the land reform (figure 2).

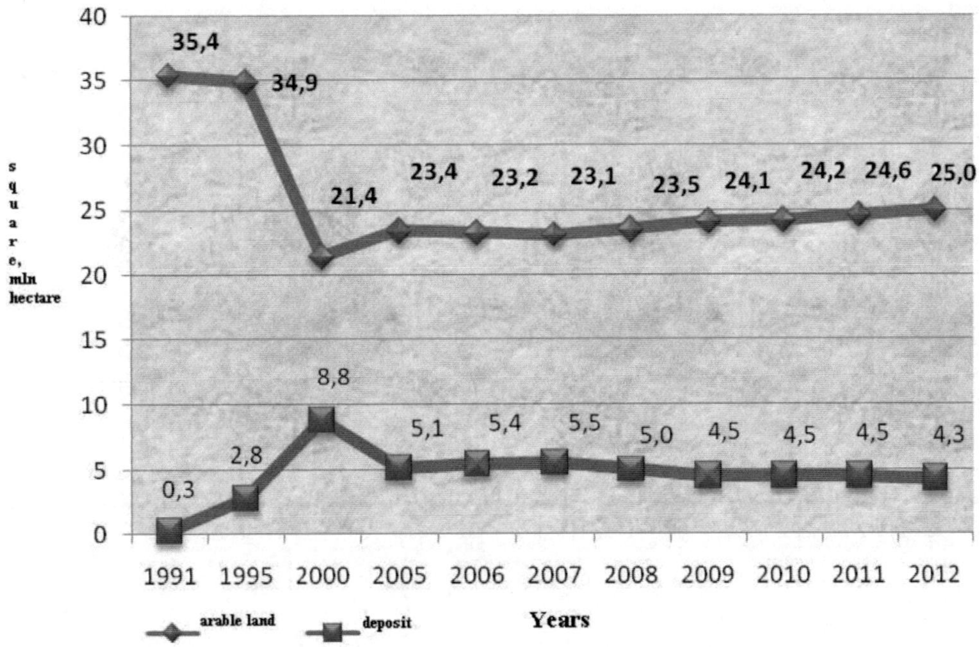

Figure 2. Dynamics of an arable land and deposit area, one million in hectares
Note: Given Agencies of RK on management of the ground resources

From figure 2 it is visible, that for the period from 1991 to 2000 from structure of an arable land 14 million hectares of low productive grounds has been deduced that has made 39.6 % of its area, and for all analyzed period - 10,4 mln of hectares.

Carrying out of land reform in republic has caused change of managing forms on the ground. Realization of this direction was carried out in 2 directions:

- By formation of small-scale enterprises - country (farmer) facilities - due to conditional ground shares;

- By formation of the large agricultural enterprises (joint-stock company, Open Company) on the basis of former state farms and collective farms.

As a result of formation of various forms of managing on the ground the variety of subjects of managing can be divided into the following categories in Kazakhstan:
- State agricultural enterprises. It includes scientific research institutes, organizations, educational institutions, subsidiary agricultural enterprises etc;
- Large non state agricultural enterprises Joint-Stock Company, Open Company);
- Personal subsidiary farms (population facilities), are the basic manufacturers of animal industries production, therefore in a greater degree cover needs in it;
- The agroindustrial companies.

The greatest relative density is made by the country facilities. According to RK Agency on management of ground resources [9] on January, 1st, 2013 205 thousand facilities are registered in republic to which 54,6 % of the grounds for agricultural purpose are fixed (figure 3).

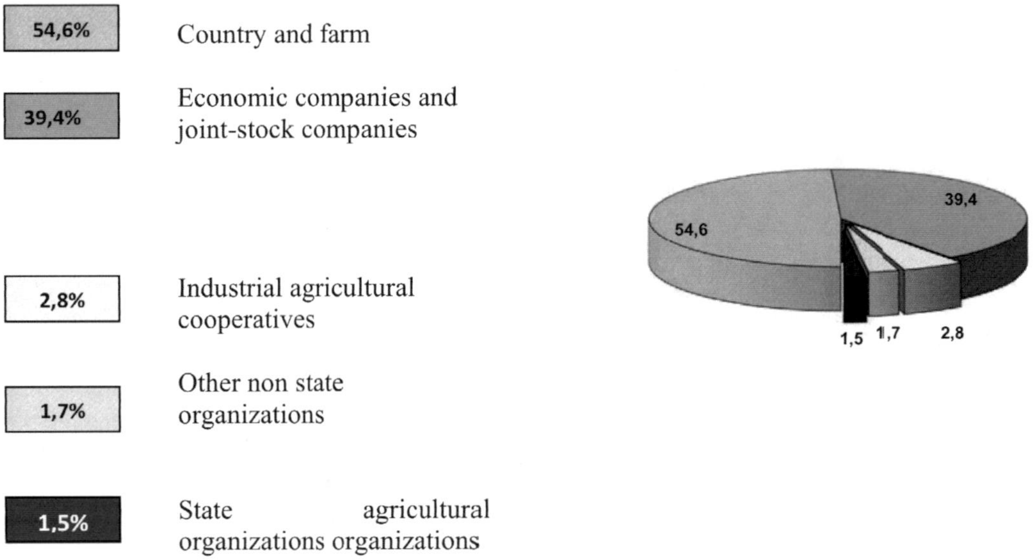

Figure 3. Distribution of the grounds on the forms of managing

Introduction of a private property on the ground is the result of the land reform. However, under the prospective forecasts the mass repayment of the grounds has not happened. So, on 1/1/2013y 0,7 % from a total area of ground fund of republic is in a private property. From a total area of the grounds which are being in a private property, on the grounds of agricultural purpose 59,9 %, the grounds of settlements - 31,8 %, the grounds of the industry - 8,3 % are necessary.

3. Problems and ways of ground attitudes perfection in Kazakhstan

Realization of RK Ground Code positions, and also the performance of the state and branch programs actions directed on the perfection of management by the ground resources, regulations of ground attitudes, on maintenance of the ground rational use, preservations of the soil fertility and ground resources protection is the basic direction in realization of ground policy at the present stage.

Now actions on rational use and protection of ground resources are carried out, works on creation and conducting the state ground cadastre and automated information systems are performed, on registration of the ground areas by citizens and legal persons for the various purposes, on perfection of paid land tenure development and the market of the ground, studying of a soil condition and vegetative cover, the ground mapping and on maintenance with necessary geodetic production, conducting of the grounds monitoring, realization of the state control over observance of the ground legislation, carrying out of land management and by many other kinds of the works which are necessary in the modern conditions.

Inventories of the grounds, calculated for 2012-2014 with a view of not used grounds revealing definition is conducted at present. Offers on improvement of a qualitative condition ground holdings will be given by the results of inventory. So, on preliminary results of inventory discrepancies of registration data with actual parameters, absence of monitoring behind occurring changes of the ground conditions are revealed.

An involving in an agricultural turn of the stock grounds is still a serious problem So, on 1/1/3013,108.4 million of hectares in republic is in this category. In this connection, it is necessary to develop the program on involving and land development of a stock for agricultural and other purposes, and also the mechanism of granting of these grounds with agricultural commodity producers.
Work on optimization of an arable land area, elimination of low productive turn parts is carried out. It is necessary to develop actions on further fallow lands using.

Introduction of innovative projects of the territory organization, improvement of soil-protective system of agriculture, realization of anti erosion and other nature protection actions have paramount value in improvement of the grounds using of agricultural purpose. In this connection, we consider the performance of the following kinds of works is necessary:

- To develop mechanisms of economic incentives provision, rational use of the grounds for agricultural purpose, preservation and restoration of soil fertility;
- To develop programs and projects on land development with investment appeal;
- To make projects of the inter farm organization agro formed territory on the basis of the landscape-ecological approach, introduction and development of zone moisture- resource saving systems of agriculture and scientifically-proved crop rotations;
- To increase development of contract designs on improvement of the irrigated grounds use, struggle with the soil salinization, wind and water erosion.

Works on creation of a multipurpose ground cadastre model with the new approach to the ground using in the conditions of market economy, to its transition into the automated technologies and creation of the ground-information system which are being simultaneously a component of nation-wide information systems, entering in « The Electronic government » will be continued in republic. In this plan it is necessary to lead works on modernization of program and technical maintenance, development of new systems and subsystems of separate components and elements, actualizations of a database about the ground areas, to an end of all system delivery AIS GZK to commercial operation.
Studying of the ground qualitative conditions, carrying out of soil, geobotanical researches, soil appraisal with application of geoinformation technologies and on this basis a creation of national database about soil and vegetation, first of all for the grounds of agricultural purpose is required. Such database allows to predict the development of negative processes and to develop actions on prevention of agricultural grounds leaving.

References

[1] K.A. Sagadiev Reforms: an analytical sight. The first book (articles, letters, speeches, interview). - Almaty: SP PPC «GAUKHAR», 2006. - p 352.
[2] K.A. Sagadiev Reforms: an analytical sight. The second book (articles, speeches, interview). - Almaty: SP PPC «GAUKHAR», 2006. – 286 p.
[3] The Comming of Post - Industrial Society: A Venture in Social Forecasting. - New York: Basic Books, 1973. - P. 123 - 129.
[4] Bohman M, Cooper. J at al The Use and Abuse of Multifunctionality. Economic Research Service. - November, 1999. - Pp. 45-48.
[5] B.Ospanov, Z. Dyusenbekov Land resources and land reform in Republic of Kazakhstan. - Astana: Agency of RK on the management of ground resources, 2001. – p 352 .
[6] Cagdas V., Demir H and Gur M. Land Consolidation, Valuation and Cadastre/FIG XXII International Congress. - Washington, D.C.U.S.A, 2002, april 19-26. - P. 24.
[7] P.F. Loyko Problems of ground policy and the government ground resources in Russian Federation // Land management, a cadastre and monitoring of the grounds. - 2008. - № 2. - p. 5-19.
[8] N.V Komov. The role of the ground factor in the creation of Russia socially focused market economy // Land management, a cadastre and monitoring of the grounds. - 2008. - № 4. - p. 5-11.
[9] L.I. Abalkin Russia agrarian tragedy // Questions of economy.-, 2009. - № 9. - p. 4-14.
[10] M.A. Gendelman and others Land management designing. The textbook. - Astana, 1999. – p.583
[11] M.A. Gendelman, G.K. Kurmanova, T.E. Karbozov, K.N. Zhaksybaeva, G.K. Zhagiparova Land reform and land management in RK // Scientifically-practical magazine « Land management, cadastre and monitoring of the grounds ».-, 2006. - № 1. - p. 40-47.
[12] A.A. Varlamov , N.V.Komov ,V.S. Shamanaev, V.N. Khlystun State regulation of ground attitudes. - M.: Kolos, 1998. – p.264
[13] V.V Kosynski Acceleration of ground transformations in the country // Land management, cadastre and monitoring of the grounds. Scientifically-practical magazine.-, 2008. - №7. – P.3-4
[14] Z.D. Dyusenbekov The problems of ground attitudes transformation at transition to the market economy // The collection of proceedings of the international conference on a theme: « The role of the state and the right in the market economy of a transition period » (on November, 6-7th, 2003). - Almaty, 2003. – P. 110-118.
[15] T.I.Espolov ., R.Y.Kuvatov , U.K. Kerimova. Increase of RK agriculture efficiency in the conditions of integration with the foreign markets. - Almaty, 2004. – p.598.
[16] The summary analytical report on a condition and use of RK grounds for 2012 - Astana, 2013. – p.250.
[17] B.Ospanov, Z. Dyusenbekov Land resources and land reform in Republic of Kazakhstan. - Astana: Agency of RK on the management of ground resources, 2001. – p 352 .
[18] Z.D. Dyusenbekov Problems and methods of transition to the paid land tenure // Asia - economy and life. - 1997. - № 27. - P.4.
[19] Z.D Dyusenbekov. The problems of ground attitudes and mechanisms of transition to the paid land tenure // International agricultural magazine. Astana, 1998. - № 1. - P. 11-14.
[20] Z.T. Seifullin The regional principles and methodical aspects of the taxation in Kazakhstan // Researches, results. - Almaty: KazNAU, 2007. - №1.-P. 18.
[21] V.A Pavlova. The formation of the ground area cost in a control system of ground resources // Land management, cadastre and monitoring of the grounds.-, 2008. - № 8. - P. 32-35.
[22] Z.D. Dyusenbekov Land reform in RK //International agricultural magazine. - 1999. - № 2. - P. 32-38.
[23] V.V. Grigoruk The formation of the competitive environment in the market of the ground areas // Works of International scientifically-practical conference. - Kokshetau 2005 - T. 1. - P. 24-27.

IOP Conf. Series: Earth and Environmental Science **20** (2014) 012028 doi:10.1088/1755-1315/20/1/012028

[24] V.V. Miloserdov Russian model and foreign experience of land reform // International agricultural magazine. - 1995. - № 2. - P.18.

IOP Conf. Series: Earth and Environmental Science **20** (2014) 012029 doi:10.1088/1755-1315/20/1/012029

Spatial Analytic Hierarchy Process Model for Flood Forecasting: An Integrated Approach

Abd Nasir Matori[1], Dano Umar Lawal[2], Khamaruzaman Wan Yusof[1], Mustafa Ahmad Hashim[1], Abdul-Lateef Balogun[2]

[1,] Assoc. Professor, Civil Engineering Department, Universiti Teknologi PETRONAS, Malaysia

[2] PhD student, Civil Engineering Department, Universiti Teknologi PETRONAS

E-mail: nasrat@petronas.com.my

Abstract. Various flood influencing factors such as rainfall, geology, slope gradient, land use, soil type, drainage density, temperature etc. are generally considered for flood hazard assessment. However, lack of appropriate handling/integration of data from different sources is a challenge that can make any spatial forecasting difficult and inaccurate. Availability of accurate flood maps and thorough understanding of the subsurface conditions can adequately enhance flood disasters management. This study presents an approach that attempts to provide a solution to this drawback by combining Geographic Information System (GIS)-based Analytic Hierarchy Process (AHP) model as spatial forecasting tools. In achieving the set objectives, spatial forecasting of flood susceptible zones in the study area was made. A total number of five set of criteria/factors believed to be influencing flood generation in the study area were selected. Priority weights were assigned to each criterion/factor based on Saaty's nine point scale of preference and weights were further normalized through the AHP. The model was integrated into a GIS system in order to produce a flood forecasting map.

1. Introduction

In reality, flood control decision making tries to lessen the flood damages; to reduce the depth of floodwater in floodplain zones; and also to reduce the flooding period to support victims. These issues are usually too sophisticated and ill-structured to be considered from the study of just one criterion, characteristic, or perspective hoping that it will result in the most desirable decision [1]. In fact, this kind of single dimensional strategy is only over-simplicification of the real flood challenges at hand, which can result in unrealistic decisions. A much more desirable approach is the simultaneous consideration of all suitable criteria which are associated with the specific floodplain management problems. This important issue should be putting into cognizance in the development of decision making models. Dealing with such issues comprises the key point of interest in Multi-Criteria Decision Analysis (MCDA).

MCDA comprises of a complex aspect of operational research, which is specialized in the development and execution of decision support tools and methods to address sophisticated decision challenges involving numerous criteria, goals, or objectives of contradicting flood challenges. The various tools and methods available from MCDA are not just some mathematical models aggregating criteria, opinions, or attributes, but in addition are decision support oriented. Support serves as a major concept in MCDA, meaning that the models are certainly not designed by way of a straightforward sequential way in which the decision maker's role is passive. Rather, a repetitive method is used to analyse the priorities of the decision maker and represent them as regularly as it can be in a suitable decision model. This repetitive and interactive preference modelling process comprises the main foundation of the decision support orientation of MCDA. It is among the fundamental unique features of the MCDA in contrast to traditional stochastic and optimization decision making strategies [2]. In this study, the principles of the Multi-Criteria Decision Making (MCDM) in the context of Analytic Hierarchy Process (AHP) have been integrated into GIS to simulate and forecast the flood susceptible zones. AHP is a mathematical modeling technique for multi-criteria decision making ([3-5]. It was

Content from this work may be used under the terms of the Creative Commons Attribution 3.0 licence. Any further distribution of this work must maintain attribution to the author(s) and the title of the work, journal citation and DOI.
Published under licence by IOP Publishing Ltd

developed by Saaty, a mathematician in the early 70s. The AHP method helps to specify numerical weights representing the relative importance of factors, elements, criteria for flood susceptibility models [5-8]. AHP allows both qualitative and quantitative approaches to solve complex decision problems. The technique structures/decomposes problems into a hierarchy of elements or factors influencing a system by incorporating levels. In quantitative aspects, the AHP can prioritize a set of attributes and distinguish in general the more important factors from the less important factors. The pair-wise comparison judgments were made with respect to the attributes of one level of hierarchy given the attribute of the next higher level of hierarchy (from the main criteria to the sub-criteria).

The majority of the work and literature available on the application of AHP in geo-engineering is dedicated to the generation of relative weights of influential factors and integration of the weights into a GIS environment. AHP's comprehensive literature collections can be found at http://www.expertchoice.com. For instance, Dai et al[9] generated relative weights of influential factors affecting urban (Lanzhou city, China) geo-environment using AHP, where the suitability potentials of urban land use were evaluated using GIS. Recently, Dano et al[10] used AHP in finding the most suitable sites to propose a new recreational park for Universiti Teknologi Malaysia. Similarly Ayalew et al[11] employed pairwise comparisons to come up with the relative weights of landslide controlling factors, where he further used GIS to develop a landslide risk assessment map for Tsugawa area of Agano River, Niigata Prefecture, Japan. Moreover, Komac[12] adopted the multivariate statistical analysis where he came to the conclusion that the use of the AHP method provides a way of defining the factor's weights in the linear landslide susceptibility model. Its vast application is based on its simplicity, ease of use, and profound flexibility inherent in the approach [13]. The objectives are to develop a flood susceptibility map based on the views of two category of experts (hydrologists and geologists) using some hydrogeological indexes. The data preparation and integration were carried out in a GIS environment.

2. Study area

The State of Perlis is bounded by Thailand in the north, and Kedah in the south, while its western coastline borders the Straits of Malacca as illustrated in Fig. 1. It is bounded by longitudes 100 ° 07' 02"E/ 100 ° 22' 33"E and latitudes 6 ° 43' 19"N/6 ° 15' 13"N respectively. Perlis is chosen due to her prone to flooding as what had happened in 2010.

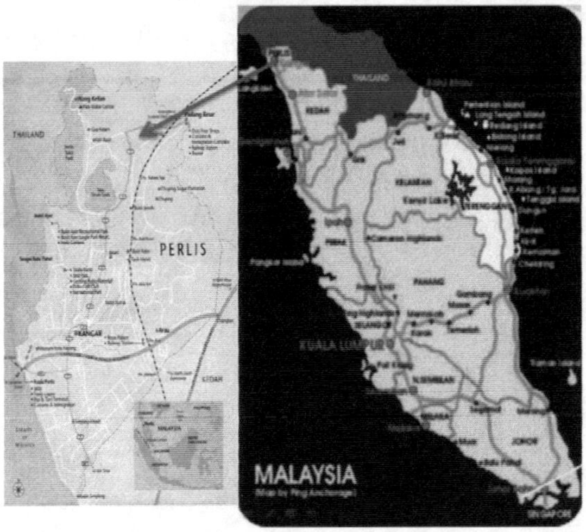

Figure 1: Map of Peninsular Malaysia showing the study area Perlis.

3. Methodology

3.1. Weighting the flood-Influencing criteria/factors

Based on the method of the MCDM, each factor's weight was computed and assigned to the factors discussed previously in GIS environment. [14-16] described MCDM as a method that permits each criterion to be weighted with respect to its relative importance/influence. Experts' judgments were used in assigning the weight of each criterion. The experts' ratings were aggregated using a Geometric Mean method as follows: Geometric Means = $((X_1)(X_2)(X_3)....(X_N))^{1/N}$; where X = Individual score and N = Sample size (Number of scores). Therefore, the experts' preferences helped in carrying out the pairwise comparison matrix. As mentioned earlier, AHP technique developed by [17] was used in producing the relative weights based on the Saaty's scale of influence 1 to 9 as shown in Table 1. The weights of the factors were generated based on the steps described by [18].

3.2 AHP integration into GIS system

In this stage, the AHP results were integrated into a GIS system to simulate the flood vulnerable areas. The combination of all the thematic layers (the factors) using the Weighted Linear Combination (WLC) method in accordance to the computed weights was carried out in a GIS. The Weighted Linear Combination, or simple additive weighting, relies on the concept of a weighted average where continuous criterions are standardized to a collective numeric range, and then combined by means of a weighted average [19]. The overlay of the thematic layers was to obtain the final simulation of the flood forecasted zones. The results of the final analysis managed to indicate the potential flood areas in the study area.

4. Results & Discussion

4.1 Pairwise comparisons of the factors

It is normally presumed that MCDA started at the beginning of 60s. A good number of experts of MCDA considered that their discipline stems mainly from the early work on goal programming as well as study of Drobne and Lisec [19] and Simon [20]. Simon[20] proposes a framework for analysing human decision-making processes by differentiating between the intelligence, design, along with choice phases MCDA techniques are numerical algorithms that define suitability of a particular solution on the basis of the input criteria and weight together with some mathematical or logical means of determining trade-offs when conflict arise. By this technique, a weight value ranges from 1 to 9 was assigned to each factor by the experts to reflect their relative significance. Using the WLC method, all the map layers (the factors) were overlaid in the final GIS spatial analysis for flood susceptible zones simulation.

The WLC technique can be carried out using any type of GIS system possessing the overlay. This permits the evaluation criterion map layers to be overlaid in order to obtain the composite map layer which is output. Therefore, the output of this WLC method gave a map which simulated the most potential flood susceptible zones of Perlis. Figure 2 shows the flood forecasting model. The responses of the experts were computed in Table 1 and 2 below: Where, RF = Rainfall; GL = Geology; ST = Soil type; S = Slope; LU = Land use.

Table 1: A matrix of pair-wise comparisons of five criteria for the AHP process.

	RF	GL	ST	S	LU
RF	1.00	4.92	3.41	2.30	3.61
GL	0.20	1.00	2.05	0.39	0.48
ST	0.29	0.49	1.00	0.70	0.46
S	0.44	2.58	1.42	1.00	0.60
LU	0.28	2.08	2.17	1.66	1.00
SUM	2.21	11.07	10.05	6.05	6.15

Table 2: Determined relative criterion weights.

	RF	GL	ST	S	LU	Weights
RF	0.45	0.44	0.34	0.38	0.59	0.44
GL	0.09	0.09	0.20	0.06	0.08	0.11
ST	0.13	0.04	0.10	0.12	0.07	0.09
S	0.20	0.23	0.14	0.17	0.10	0.17
LU	0.13	0.19	0.22	0.27	0.16	0.19
SUM	1.00	1.00	1.00	1.00	1.00	1.00

Table 3: Determined Consistency Ratio (CR).

		STEP I :									
RF	1.00	4.92	3.41	2.30	3.61						
GL	0.20	1.00	2.05	0.39	0.48						
ST	0.29	0.49	1.00	0.70	0.46						
S	0.44	2.58	1.42	1.00	0.60						
LU	0.28	2.08	2.17	1.66	1.00						
		*					STEP II : SUM1/WEIGTH				
	0.44	0.11	0.09	0.17	0.19						
		=				SUM1					
RF	0.44	0.52	0.32	0.38	0.70	2.36		0.44			5.36
GL	0.09	0.11	0.19	0.07	0.09	0.54	/	0.11	=		5.14
ST	0.13	0.05	0.09	0.12	0.09	0.48		0.09			5.14
S	0.19	0.27	0.13	0.17	0.12	0.88		0.17			5.27
LU	0.12	0.22	0.20	0.28	0.19	1.02		0.19			5.25
									SUM2		26.16

Table 4: Random Inconsistency Indices.

n		3.00	4.00	5.00	6.00		7.00	8.00	9.00	10.00
RI		0.58	0.90	1.12	1.24		1.32	1.41	1.45	1.49

The Consistency Index (CI), which is a measure of departure from consistency, was calculated using the formula:

$$CI=\frac{\lambda - n}{n - 1} \qquad (1)$$

Where, n = number of factors (i.e. 5) and λ = average value of the consistency vector determined in step ii above.

$\lambda = 5.36 + 5.14 + 5.14 + 5.27 + 5.25 = 26.16/5 = 5.23$

Based on equation (3), CI = 5.23-5/5-1 = 0.06

IOP Conf. Series: Earth and Environmental Science **20** (2014) 012029 doi:10.1088/1755-1315/20/1/012029

In order to assess the robustness of the expert view the Consistency Ratio (CR) was calculated using equation (2). The result is shown in Table 3.

$$CR=\frac{CI}{RI} \quad (2)$$

Where, RI is the random inconsistency index whose value depends on the number (n) of factors being compared; for n = 5, RI = 1.12 as illustrated in Table 4 [17].
Using Equation (4),

$$CR = \frac{0.06}{1.12} = 0.05$$

Therefore, since 0.05 < 0.1, it indicates that there is a realistic degree of consistency in the pairwise comparison and as a result, the weights 0.44, 0.11, 0.09, 0.17 and 0.19 can be assigned to rainfall, geology, soil type, slope and land use respectively.

4.2 AHP spatial flood forecasting model

The combination of all parameters was carried out in GIS using the WLC method. This is based on the weights generated from the AHP. The Weighted Linear Combination (WLC) formula is shown below:

$$FSZ = \sum_{wixi} \quad (3)$$

Where, *FSZ* is the Flood Susceptible Zones; w_i is the factors weights which must sum up to 1, and x_i is the criterion score i. All these relevant data were uploaded into the GIS system before analysis was carried out using a total of five thematic maps: land use, geology, rainfall distribution, soil type and slope gradient. Figure 2 presents the result of the GIS spatial simulation model for flood susceptible zones. In Figure 2, the light blue, yellow, red, white and pink colors represent very highly, high, moderate, little as well as hardly susceptible zones to flooding. These are basically based on the ratings of the flood influencing factors given by the experts.

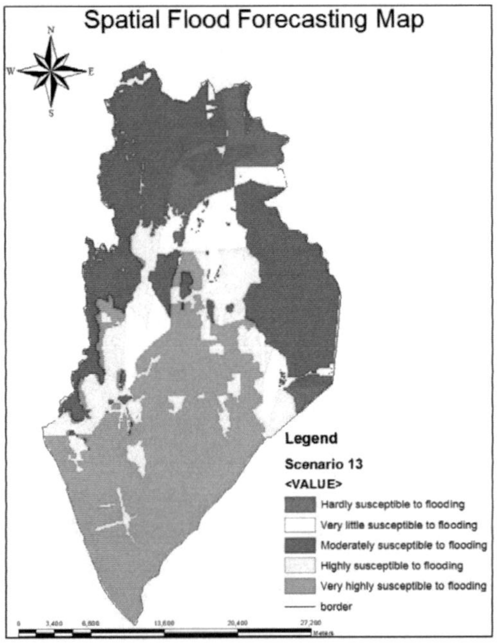

Figure 2: Map of forecasted flood susceptible zones of Perlis.

5. Conclusion

In this study, a GIS-based Analytic Hierarchy Process was successfully applied in simulating the flood susceptible zones for Perlis State. In recent years, GIS proved profound potentials in combating flood catastrophes. The social and economic losses as a result of flood disasters, technologic crises, as well as global epidemics are growing; needing more effective group decision making under uncertainty. In this study, the efficiency of GIS-based AHP as spatial multi-criteria forecasting tool was discovered. Hence, this system will help civil authorities such as planners and emergency agency in taking positive and in-time steps during the pre-disaster situations. It will similarly help them during post-disaster activities to assess damages and losses caused as a result of flood.

6. References

[1] Komac, M. (2006). A Landslide Susceptibility Model Using the Analytical Hierarchy Process Method and Multivariate Statistics in Perialpine Slovenia. Geomorphology .74(1–4), 17-28.

[2] Gale T. (2006). Encyclopedia of Management. ''S. Neil, Ed., International Directory of Business Biographies-F-L.

[3] Saaty, T. L., (1980). The Analytic Hierarchy Process: Planning, Priority Setting, Resource Allocation. McGraw-Hill, New York.

[4] Cheng, E. W. L., and Li, H. (2001). Analytic Hierarchy Process: An Approach to Determine Measures for Business Performance. Measuring Business Excellence, 5(3), 30-37.

[5] Chen, D. C., Chen, P. C., Lee, Y. C., You, S. U. and Jao, H. C. (2011). Using an Analytic Hierarchy Process to Develop Competencies on Mould Product Creativity for Vocational College Students. World Transactions on Engineering and Technology Education (WIETE) .9(1), 54-59.

[6] Emmanouloudis D, Myronidis D, and Ioannou K. (2008). Assessment of Flood Risk in Thasos Island with the Combined Use of Multi-criteria Analysis AHP and Geographical Information System. Innovative Applications of Informatics in Agriculture and Environment .2, 103-115 ESRI.

[7] Sinha R, Bapalu G, Singh L, and Rath B. (2008). Flood risk Analysis in the Kosi River Basin, North Bihar Using Multi-Parametric Approach of Analytical Hierarchy Process (AHP). Journal of Indian Society of Remote Sensing.36, 335-349.

[8] Meyer V, Scheuer S, and Haase D. (2009). A Multi-criteria Approach for Flood Risk Mapping Exemplified at the Mulde River, Germany. Natural Hazards .48, 17-39.

[9] Dai, F.C., Lee, C.F., and Zhang, X.H. (2001b). Assessment of Landslide Susceptibility on the Natural Terrain of Lantau Island, Hong Kong. Environmental Geology .61, 257-271.

[10] Dano, U.L., Matori, A. N., and Balogun, A. (2011). A Geographic Information System and Multi-Criteria Decision Analysis in Proposing New Recreational Park Sites in Universiti Teknologi Malaysia. Modern Applied Science, 5(3), p39.

[11] Ayalew, L., Yamagishi, H., and Ugawa, N. (2004). Landslide Susceptibility Mapping Using GIS-based Weighted Linear Combination, The Case in Tsugawa Area of Agano River, Niigata Prefecture, Japan. Landslides.1, 73-81.

[12] Komac, M. (2006). A Landslide Susceptibility Model Using the Analytical Hierarchy Process Method and Multivariate Statistics in Perialpine Slovenia. Geomorphology .74(1–4), 17-28.

[13] Laemlaksakul, V., and Bangsarantrip, S. (2008). Analytic Hierarchy Process for Design Selection of Laminated Bamboo Chair. Proceedings of the International Multi-Conference of Engineers and Computer Scientists, , 2

[14] Pradhan,B. "Flood susceptible mapping and risk area delineation using logistic regression, GIS and remote sensing," Journal of Spatial Hydrology, vol. 9, 2010.

[15] Yahaya,S., Ahmad, N. and Abdalla, R. "Multicriteria analysis for flood vulnerable areas in Hadejia-Jama'are River basin, Nigeria," European Journal of Scientific Research, vol. 42, pp. 71-83, 2010.

[16] Eastman, J.R. "Multi-criteria Evaluation and Geographical Information Systems Eds. Longley PA, Goodchild MF, Magurie DJ and Rhind DW, Second ed. John Wiley and sons, New York," vol. 1, pp. 493-502., 1996.

[17] Saaty,T.L. "The Analytic Hierarchy Process: Planning, Priority Setting, Resource Allocation. McGraw-Hill, New York." 1980.

[18] Malczewski ,J. " GIS and Multicriteria decision analysis. John Wiley and Sons, Inc. United States of America, ," pp. pp. 177-192., 1999.

[19] Drobne,S. and Lisec, A "Multi-attribute decision analysis in GIS: weighted linear combination and ordered weighted averaging," Nature, vol. 4, pp. 28, 2009.

[20] Simon, H. "The new science of management decisionPrentice-Hall," Englewood Cliffs, NJ, 1977.

IOP Conf. Series: Earth and Environmental Science **20** (2014) 012030 doi:10.1088/1755-1315/20/1/012030

Investigation of flood pattern using ANOVA statistic and remote sensing in Malaysia

Norsuzila Ya'acob [1, 2,*], Nor Syazwani Ismail[1], Norfazira Mustafa[1] and Azita Laily Yusof[1, 2]

[1]Faculty of Electrical Engineering, Universiti Teknologi MARA, 40450 Shah Alam Selangor, Malaysia
[2]Wireless Communication Technology (WiCoT), Faculty of Electrical Engineering, Universiti Teknologi MARA, 40450 Shah Alam, Selangor, Malaysia*Institute of Space Science, Universiti Kebangsaan Malaysia, 43600 Bangi, Selangor, Malaysia.

norsuzilayaacob@yahoo.com,wani_nureen@yahoo.com,faziramustafa@yahoo.com, azita968@salam.uitm.edu.my

Abstract. Flood is an overflow or inundation that comes from river or other body of water and causes or threatens damages. In Malaysia, there are no formal categorization of flood but often broadly categorized as monsoonal, flash or tidal floods. This project will be focus on flood causes by monsoon. For the last few years, the number of extreme flood was occurred and brings great economic impact. The extreme weather pattern is the main sector contributes for this phenomenon. In 2010, several districts in the states of Kedah neighbour-hoods state have been hit by floods and it is caused by tremendous weather pattern. During this tragedy, the ratio of the rainfalls volume was not fixed for every region, and the flood happened when the amount of water increase rapidly and start to overflow. This is the main objective why this project has been carried out, and the analysis data has been done from August until October in 2010. The investigation was done to find the possibility correlation pattern parameters related to the flood. ANOVA statistic was used to calculate the percentage of parameters was involved and Regression and correlation calculate the strength of coefficient among parameters related to the flood while remote sensing image was used for validation between the calculation accuracy. According to the results, the prediction is successful as the coefficient of relation in flood event is 0.912 and proved by Terra-SAR image on 4th November 2010. The rates of change in weather pattern give the impact to the flood.

1. Introduction

Nowadays, at the rural area or urban area were face with the natural flood and flash flood in the certain periods. The flood happened without any sign to show that it will happen. The residents basically were not alert and beware about the flood and impact to them. Normally, they alert about flood disaster. Moreover, the increasing of development and human activities can have significant influence on the process that contributes to disaster of flood. To avoid this problem, the suitable technologies can be developed to control and prevent the flood risk and impacts for short or long terms [1]-[3]. In the flood prediction, remote sensing is been reliable tools and can be used in prediction process. It provided the researcher with a sort of multi-date satellite images for monitoring and recording the changes of process of past flood events.

Content from this work may be used under the terms of the Creative Commons Attribution 3.0 licence. Any further distribution of this work must maintain attribution to the author(s) and the title of the work, journal citation and DOI.
Published under licence by IOP Publishing Ltd

Remote sensing can be defined as combination of information from the places such as distance which occur with the camera device based on the ground or sensors or cameras based on ships, aircraft, satellites, or other spacecraft. In archaeology, it generally refers to the use of aerial or satellite imagery to study cultural patterning on the ground. Aerial photography has been long used by archaeologists, but recent development of new sensing and data processing technologies has led to new possibilities for archaeological studies.

In this project, Terra-SAR image was used to compare the prediction. TerraSAR-X is a high resolution X-band radar sensor with three operational models such as Spotlight, Strip Map and Scans AR. Spotlight is the most sophisticated radar imagery available which has 1 m spatial resolution. Its steerable radar beam in flight direction (like a spotlight) illuminates the ground for the longest time possible to ensure this high resolution. For large area coverage the Scans AR mode (16 m resolution) allows areas up to 100 000 km^2 to be acquired within a week owing to its large scene size [4]-[7]. The main focus of this project are to identify the parameters involve in the prediction of flood followed by investigate the pattern and analyses or compare them with the previous flood.

2. Methodology

The main parameter for this research will focused on temperature, humidity, rainfalls and wind speed value were collect at station involved. The image was captured by Terra-SAR satellite for validation.

2.2 Data Collection

Data includes the amount off rainfalls, wind speed, temperature and humidity in Titi Gajah area as shown in Figure 1. The data of amount of rainfalls are covered from three stations such as Bukit Kayu Hitam, Kepala Batas and Alor Setar stations that nearest to the study area. For the humidity, wind speed and temperature data were taken the mean relatives for daily reading and covered for period between Augusts until November 2010 in the same station. The information the flood occurred were collected from the local newspapers and also the images from TERA-SAR at the Malaysian Remote Sensing Agency (ARSM)

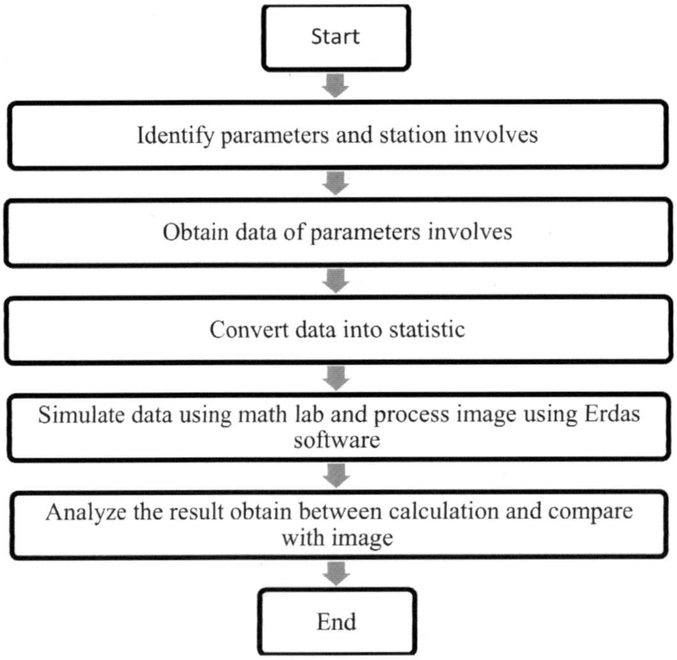

Figure 1: Flowchart of Reseach Methodology.

2.3 Analytical Techniques

In terms of the analysis part, the statistic method was used to preliminary treatment of data. The ANOVA statistic was chosen as a method to calculate the analysis for statistic. ANOVA is the techniques that used the two hypotheses to determine the results and namely as null hypothesis *(H_0)* which the assumption under investigation being tested. The H_0 is statement that "there is no effect" or "there is no change". The possible outcomes in testing a H_0 are "reject" or "fail to reject". Reject the H_0 means that there is significant evidence in the data and in the test to justify reject H_0. When H_o was rejected the data is said to be statically significant. Fail to reject H_0 means that there is NOT enough evidence in the data and the test to justify reject H_0. The alternate hypothesis *(H_1)* state that it adopt if there is strong evidence (sample data) against the null hypothesis *(H_0)*[8]. There are equations involved in ANOVA and showed in Table 1.

Table 1. ANOVA Statistic Table.

Source	**Degree of freedom**	**Means square**	**F statistic**
A	a-1	SSa/dfA	MSA/MSB
C	c-1	SSb/dfB	MSC/MSB
AC	(a-1)(c-1)	SSac/dfAC	MSAC/MSB
B	N-ac	SSb/dfB	

Where:
a= Number of levels of independent variable A.
c= Number of levels of independent variable C.
ac= Number of cell in the experiment.
N= Total number of observations in experiment
n_1= Number of observations in cell 1, etc.

$$SS_{total} = \sum X^2 - \frac{(\sum X)2}{N} \tag{1}$$

$$SS_A = \sum \frac{\sum(for\ each\ row)^2}{n\ for\ each\ row} - \frac{(\sum X)^2}{N} \tag{2}$$

$$SS_C = \sum \frac{\sum(for\ each\ column)^2}{n\ for\ each\ column} - \frac{(\sum X)^2}{N} \tag{3}$$

$$SS_{between} = \frac{(\sum X_1)^2}{n_1} + \cdots \frac{(\sum X_{ac})^2}{n_{ac}} - \frac{(\sum X)^2}{N} \tag{4}$$

$$SS_{AC} = SS_B - SS_A - SS_C \tag{5}$$

$$SS_B = SS_{total} - SS_{between} \tag{6}$$

It is very difficult to predict flood level since it fluctuate in high non-linear. The precision or complexity of traditional measures such as the hydrological statistical method and the multi-dimension mixed regression algorithm is not always satisfactory.

3. Result
This part will discuss the results and analyses from the parameters involved and their relation with the flood in November 2010 at Titi Gajah area.

3.1Parameters Analysis
There are parameters were identified related to the floods which are:

3.1.1 Temperature
Actually, the change of one weather elements can produce the changes in regional climate [6]. The increasing of temperature will be affected the amount of clouds and amount of rainfalls. Table 2 showed that the value of $F_{statistic}$ calculation 0.99 is greater than the $F_{probability}$ value 0.4887. It can be proved that the temperature also apart of parameter involved in the flood.

Table 2. ANOVA statistic of temperature.

Source	Degree of freedom	Means square	F statistic	F probability
Months	2	5592.86	16.46	0
Days	29	418.69	0.99	0.4887
Errors	87	427.88		
Total	119			

In Malaysia, the average of temperature is between 23°C until 32°C. If the value is more than that it will consider as extreme temperature. Figure 2 will showed that the range of temperature in August until November 2010.

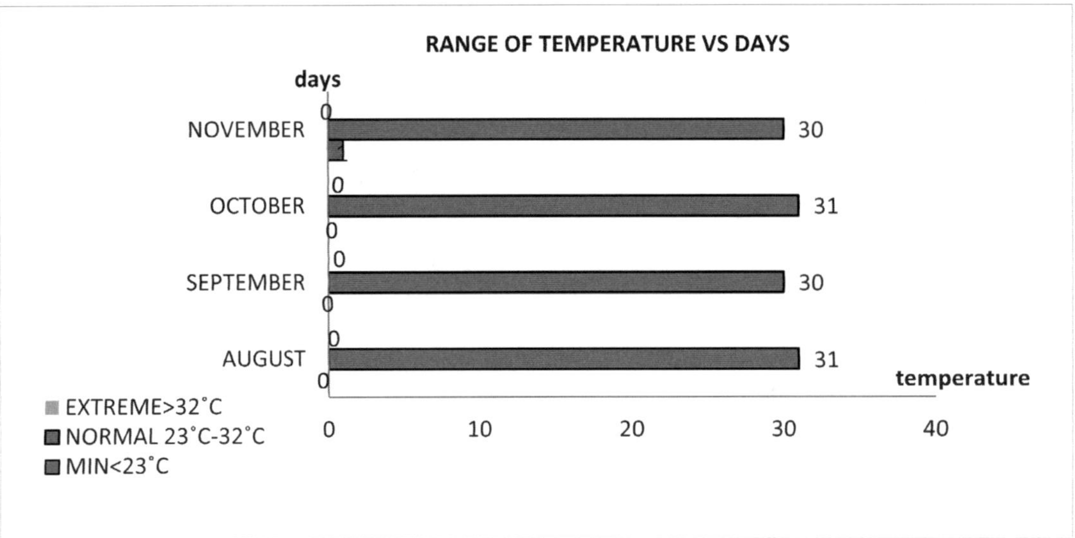

Figure 2. Range of temperature.

The extreme changes of temperature give the effect to the surrounding such as vaporization, wind speed, amount of rainfalls and also the humidity. When the temperature decrease from the normal temperature, La Nina phenomena will be happened and cause the heavy rainfall to the area involved. Based on the La-Nina Phenomena, it can be related that between August to October, Malaysia in wet and cold condition period. Addition to the monsoon season, the changing temperature gives the effect to pattern of rainfall in Titi Gajah Area. It can be concluded here that the small changes in temperature either decrease or increase were effect to the surrounding. Since ANOVA statistic showed 0.99, it can be concluded that the temperature reading has relation to the flood.

3.1.2 Humidity

Air's water vapours carrying capacity increases with increasing temperature, and decreases with decreasing temperature. Referring to Table 3, Statistic Prediction showed the probability the humidity was involved which $F_{statistic}$ value is 0.98 compared to the $F_{probability}$ 0.5082.

Table 3. ANOVA Statistic of humidity.

Source	Degree of freedom	Means square	F statistic	F probability
Months	2	559.86	13.07	0
Days	29	418.69	0.98	0.5082
Errors	87	427.88		
Total	119			

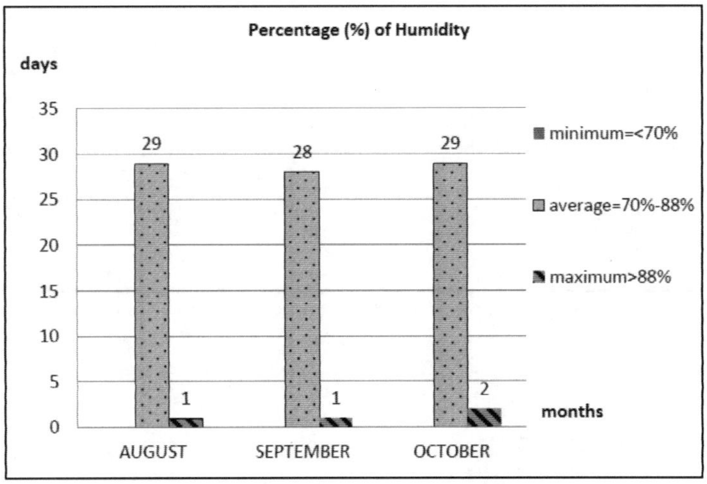

Figure 3. Percentage of humidity versus days.

According to Figure 3, the maximum readings of humidity were records every month. The maximum reading showed that heavy rains were happened on the days and surrounding becomes wet. When humidity becomes higher, process of vaporization will be increase rapidly and make the cloud becomes heavy. Then extreme rainfalls will be happened quickly. Titi Gajah area was surrounded by paddy field area. The maximum range was found in the north-western peninsular (Alor Star) where the average relative humidity varies from 72% to 87% in February.

It noted that in Peninsular Malaysia, the minimum relative humidity is usually available in January and February, but for the east coast states, namely Kelantan and Terengganu, where the relative humidity is a minimum in March. The maximum relative humidity was typically is in November. As in the case of temperature, diurnal variation of relative humidity is greater than the annual change.

The average daily minimum can reach as low as 42% during the dry months and as high as 70% during humid months. However, the mean daily maximum did not change much from one place to another, which is more than 94% and may reach as high as 100%. States in the northwest, the northern Kedah and Perlis and Sabah have biggest daily humidity changes. It can be conclude that the increase percentage of humidity proportional due to the temperature reading. Both of elements were given the impact to the flood. The increases of humidity also effect to the rainfalls pattern in Titi Gajah.

3.1.3 Rainfalls

When the humidity was increase, quantities of rainfalls also increase because the vaporization process. The probability it was involved can be shown in Table 4 analysis, system showed that the value of $F_{statistic}$ calculation is 0.92 which is greater than the $F_{probability}$ value. The null hypothesis, *Ho* state that rainfalls are not involved in flood pattern is rejected. It showed that the data is the concrete evident for the flood pattern.

Table 4. ANOVA Statistic of rainfalls.

Source	Degree of freedom	Means square	F statistic	F probability
Months	2	1332.45	3.14	0.264
Days	65	391.01	0.92	0.6425
Errors	195	424.23		
Total	263			

Heavy rains induce flooding when they occur about three or more times in a month during the period [1]-[3]. Based on Figure 4, Bukit Kayu Hitam station showed that the highest reading of rainfalls 263 mm on the October and 393 mm on November followed by Alor Setar station 344 mm on October.

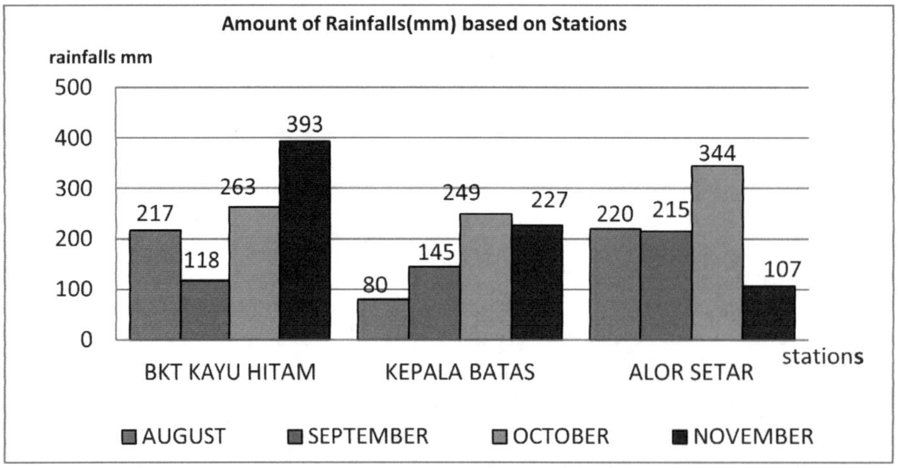

Figure 4. Amount of rainfalls based on station.

The extreme rainfalls were record in Bukit Kayu Hitam and made overflow from river nearest to Bukit Kayu Hitam and at the same time Titi Gajah also sustain heavy rain and caused the flood. The rainfall pattern in the area other than the west coast of Peninsular Malaysia showed two maximum rain period separated by two periods of minimum rainfall. Main maximum usually occurs in October and November, while the secondary maximum occurring in April to May. In the northwest, the main minimum occurs in January and February, while the secondary minimum occurs in June and July. In other places, the primary minimum occurs in June and July, while the secondary minimum occurs in February. It can be conclude the heavy and very heavy rain can be happen when the increases rapidly of the temperature readings and percentage of humidity in the air.

3.1.4 Wind Speed

Lastly, the element was identified and give the impact to the flood is wind speed. In Northeast Monsoon occurs in November until March of the year accompanied with rumbustious and waterspout. Table 5 below showed the statistic of wind speed involved. Based on Table 5, the prediction also give the result that wind speed is a part of parameter involved in the flood. Statistic results showed that the value of $F_{statistic}$ calculation 1.33 is greater than the $F_{probability}$ value of 0.1544.

Table 5. ANOVA Statistic of wind speed.

Source	Degree of freedom	Means square	F statistic	F probability
Months	2	6.19697	14.51	0
Days	29	0.56951	1.33	0.1544
Errors	87	0.42697		
Total	119			

The northeast monsoon season usually commences in early November and ends in March. During this season, steady easterly or north-easterly winds of 10 to 20 knots prevail. The winds over the east coast states of Peninsular Malaysia may reach 30 knots or more during periods of strong surges of cold air from the north (cold surges). In Malaysia, Wind movement will change the moving ocean. Low air pressure occurs due to an increase in the surface temperature of the sea water. This causes the wind patterns change from the sea shore and air moisture accumulated and subsequently causes the phenomenon of heavy rain.

As Malaysia is mainly a maritime country, the effect of land and sea breezes on the general wind flow pattern is very marked especially during days with clear skies. On bright sunny afternoons, sea breezes of 10 to 15 knots very often develop and reach up to several tens of kilometres inland. On clear nights, the reverse process takes place and land breezes of weaker strength can also develop over the coastal areas.

3.2 Total Prediction Pattern

To prove that all the parameters have relationship due to flood, all the parameters data were combined. Table 6 showed the result of combination of all parameters in flood prediction. According to Figure 5, the temperature was donating the higher percentage relation to the flood followed by Humidity then Rainfalls and lastly winds speed. The heavy cloud will be move according to wind speed from high pressure to low pressure then make the extreme rainfalls to the area involve. If the extreme rainfalls were happened in several days, chances flood might be happened also increase.

Table 6. Combination results of all parameters.

Source	Degree of freedom	Means square	F statistic	F probability
Months	2	127523	445.45	0
Days	119	443.4	1.55	0.0012
Errors	357	286.3		
Total	479			

According to Figure 5, the temperature was donating the higher percentage relation to the flood followed by humidity then Rainfalls and lastly winds speed. The climate changing gives the impact to the flood for the example when the temperature increase in 1 degree Celsius, the humidity in the surrounding also increases proportionally. It happened because the process of vaporization will happened rapidly compared to usually. The heavy cloud will be move according to wind speed from high pressure to low pressure then make the extreme rainfalls to the area involve. If the extreme rainfalls were happened in several days, chances flood might be happened also increase.

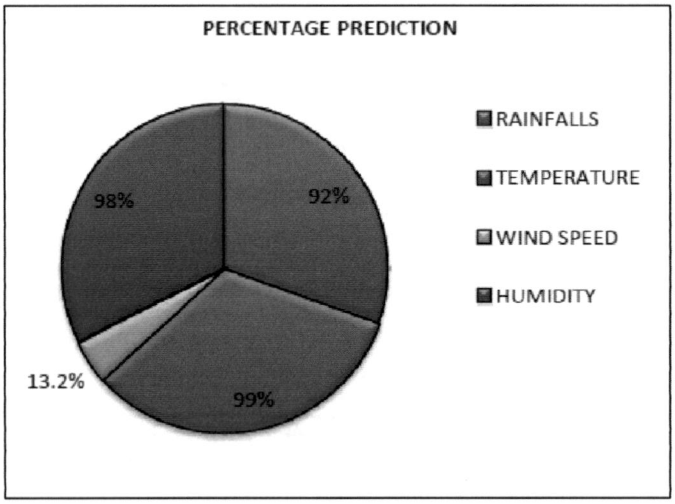

Figure 5. Percentage of parameters in flood pattern.

3.3 Regression and Correlation Test

After applying the implicitly gain in the prediction, the values of R^2 and r^2 are equal as shown in Figure 6. So the data model and the correlation have the same values of 83.21. To get the coefficient of the prediction, the values of correlation must be square root and the actual value of the coefficient is 0.912. According to coefficient table, the strength of coefficient can be classified as very high coefficient.

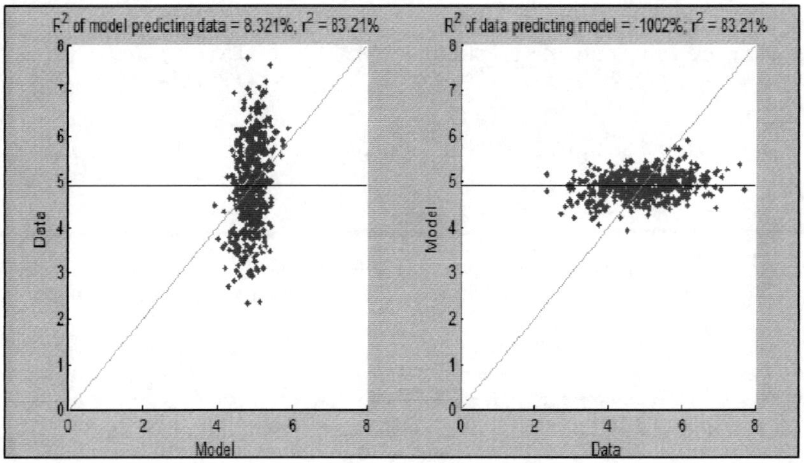

Figure 6. Values of R^2 and r^2 after implicitly gain.

3.4 Validation Image to Calculation
Image from Terra-SAR satellite showed that the Flood happen on November 2010 in Kampung Titi Gajah as shown in Figure 7. Based on the all analysis on the correlation of the parameters involved in the flood on 4th November 2010, it can be conclude that all the parameters show strong correlation. Since the value of coefficient is 0.912, all the parameters give the impact to the flood. The correlation pattern also gives impact if there is a change in one of the elements.

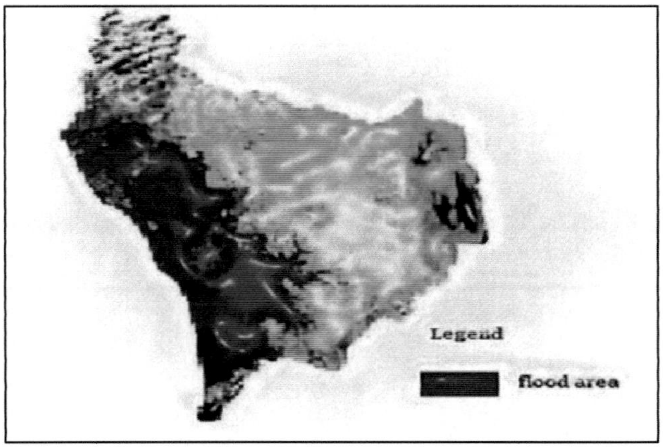

Figure 7. Terra-SAR satellite image on 4th November 2010.

4. Conclusion
It can be conclude that the rainfalls, temperature, wind speed and the humidity are the parameters involved in the process of flood. Each the element in space weather have the relation to each other. The statistic in show that each parameter satisfied to the target result less than 1. The changes in temperature will give effect to the humidity reading then followed by amount of rainfalls and lastly the wind speed at the surrounding was involved. Since the coefficient of pattern is 0.912, the correlation is very high strength and it is successful. Then the result from image of Terra-SAR satellite is a concrete evident showing that in 4th November 2010, there was flood happened at that time.

Acknowledgment
This research is sponsored by Universiti Teknologi MARA (UiTM) through the university research grant 600 RMI/DANA/5/3/RIF (194/2012). The appreciation also goes to Malaysian Remote Sensing Agency (ARSM), Department of Irrigation and Drainage Malaysia and Malaysia Meteorology Department for provided the discharge history data to complete this project.

References

[1] Akintola, F O (1986). Rainfall Distribution in Nigeria (1892 – 1983). Ibadan: Impact Publishers Nig.Ltd.
[2] Ayoade, J O (1976). A Preliminary Study of the Magnitude, Frequency and Distribution of Intense Rainfall in Nigeria. Hydrological Science Bulletin
[3] Babatola, J S (1996). Recent Changes in Rainfall Patterns and its Implication for Flood Occurrence in Ondo, Nigeria. Ondo Journal of Arts and Social Sciences, 1 (1), pp. 125 – 136.
[4] J. Mittermayer, M. Younis, B. Bräutigam, T. Fritz, R. Kahle, R. Metzig, Verification of TerraSAR-X System. IEEE Geoscience and Remote Sensing Symposium (IGARSS), Barcelona, Spain, 2007 .
[5] R. Metzig, B. Bräutigam, D. Polimeni, J. Böer, M. Bachmann, J.Mittermayer. TerraSAR-X Instrument Characterization / Verification. Accepted at: 7th European Conference on Synthetic Aperture Radar, Friedrichshafen, Germany, 2008.
[6] J. Mittermayer, U. Steinbrecher, A. Meta, N. Tous-Ramon, S. Wollstadt,M. Younis, J. Marquez, D. Schulze, C. Ortega, TerraSAR-X System Performance and Command Generation. Accepted at: 7th European Conference on Synthetic Aperture Radar, Friedrichshafen, Germany, 2008.
[7] J. Marquez-Martinez, C. Gonzalez, M. Younis, S. Wollstadt, R. Metzig,U. Steinbrecher, N. Tous-Ramon, A. Meta, J. Mittermayer, In-Orbit SAR Performance of TerraSAR-X. In: IEEE Geoscience and Remote SensingSymposium (IGARSS), Barcelona, Spain, 2007.
[8] Campolo, M., Soldati, A., and Andreussi, P. 2003. Artificial Neural Network Approach to flood Forecasting in the River Arno. Hydrological Sciences Journal/ Journal des Sciences Hydrologiques, Vol. 48(3), pp. 381-398
[9] T. Back, Evolutionary Algorithms in Theory and Practice, Oxford University Press, 1998.
[10] Campolo, M., Soldati, A., and Andreussi, P. 2003. Artificial Neural Network Approach to flood forecasting in the River Arno, Hydrological Science Journal/ Journal des Sciences Hdrologiques,Vol.48(3),pp.381-398

IOP Conf. Series: Earth and Environmental Science **20** (2014) 012031 doi:10.1088/1755-1315/20/1/012031

Landslide susceptibility mapping along PLUS expressways in Malaysia using probabilistic based model in GIS

Norbazlan M.Yusof [1,3], Biswajeet Pradhan[2]

[1]PLUS Berhad, Persada PLUS, Subang Interchange,KM15, New Klang Valley Expressway, 47301 Petaling Jaya, Selangor, Malaysia
[2]Department of Civil Engineering, Faculty of Engineering, University Putra Malaysia, 43400 UPM, Serdang, Selangor Darul Ehsan, Malaysia

Email : bazlan@plus.uemnet.com

Abstract
PLUS Berhad holds the concession for a total of 987 km of toll expressways in Malaysia, the longest of which is the North-South Expressway or NSE. Acting as the 'backbone' of the west coast of the peninsula, the NSE stretches from the Malaysian-Thai border in the north to the border with neighbouring Singapore in the south, linking several major cities and towns along the way. North-South Expressway in Malaysia contributes to the country economic development through trade, social and tourism sector. Presently, the highway is good in terms of its condition and connection to every state but some locations need urgent attention. Stability of slopes at these locations is of most concern as any instability can cause danger to the motorist. In this paper, two study locations have been analysed; they are Gua Tempurung (soil slope) and Jelapang (rock slope) which are obviously having two different characteristics. These locations passed through undulating terrain with steep slopes where landslides are common and the probability of slope instability due to human activities in surrounding areas is high. A combination of twelve (12) landslide conditioning factors database on slope stability such as slope degree and slope aspect were extracted from IFSAR (interoferometric synthetic aperture radar) while landuse, lithology and structural geology were constructed from interpretation of high resolution satellite data from World View II, Quickbird and Ikonos. All this information was analysed in geographic information system (GIS) environment for landslide susceptibility mapping using probabilistic based frequency ratio model. Consequently, information on the slopes such as inventories, condition assessments and maintenance records were assessed through total expressway maintenance management system or better known as TEMAN. The above mentioned system is used by PLUS as an asset management and decision support tools for maintenance activities along the highways as well as for data quality checking and integrity. In this study, TEMAN data were further analysed and subsequently integrated with landslide susceptible map for Gua Tempurung and Jelapang area in Perak.

1. Introduction

Landslides carry high economic and social loses to many organizations in Malaysia and not exception to highway industry like PLUS Berhad. Occurrences of landslides are prevalent in hill complexes both in the highlands and lowlands including along the highways. These landslides have caused loss of lives and properties in recent years. While agriculture in landslide occurring areas has caused severe soil erosion downstream, hill construction projects for infrastructure and residential purposes were the main triggering factors of landslides.

[3] To whom any correspondence should be addressed

Content from this work may be used under the terms of the Creative Commons Attribution 3.0 licence. Any further distribution of this work must maintain attribution to the author(s) and the title of the work, journal citation and DOI.
Published under licence by IOP Publishing Ltd

Planners have taken cognizance that developable areas in the gentle terrain have become scarce and encroachments into the more sensitive hilly areas particularly in the highlands are inevitable. Development planning in the hills based on landslide considerations is of paramount importance. In this context a national landslide hazard zoning map is a prerequisite to assist in decision making for approving developments in the prone areas [54]. To further strengthen this initiative, landslide detection and monitoring system should also be developed to for timely mitigation measures.

Landslides have always posed serious threats to settlements and structures in Malaysia that support transportation, natural resources and tourism. They cause considerable damage to highways, waterways, properties, livestock and pipelines. Though most of these landslides occurred on cut slopes or embankments alongside roads and highways in mountainous areas still there are records of landslides in other areas. Few landslides occurred near high-rise apartments and in residential areas, causing death to human being.

In the recent years, remote sensing and GIS have played major roles in carrying out analytical analysis in natural hazard particularly in development of landslide susceptibility map (LSM) for authorities [1, 33, 34]. The advancement in GIS technology provides a significant contribution in analyzing and producing LSM [50, 52, 53, 55, 59, 63]. They are studies done by experts and scientists around the world in Frequency Ratio (FR) and probabilistic analysis to produce LSM [1, 11, 17, 64, 65].

Recently, a large number of landslides have triggered along east coast highways and other highways in peninsular Malaysia. The extent of damages can be reduced or minimized if a long term early warning system predicting the landslide prone areas would have been in place. The landslides that occurred in the New Klang Valley Express Highways (NKVE) region in the year 2003 have woken up the highway authorities and other organizations towards the seriousness of landslide management and prevention. The October 2002 landslide in Kuala Lumpur is fresh in the memory of the people as completely destroyed a few houses and killed six members of a family. Landslides in Malaysia are mainly triggered due to the tropical rainfall and flash floods causing failure of the rock surface along the fracture, joint and cleavage planes. The geology of the country is quite stable but continuous development and urbanization leads to deforestation and weathering, erosion of the covered soil masses causing serious threat to slopes.

In 2006 and 2010, there were studies done in landslide hazard and risk analysis for Penang Island [34, 54] using a frequency ratio and logistic regression model. These models did not provide the weightage for landslide causative parameters. Weight determination can only be done through the ANN (Artificial Neural Networks) and fuzzy logic models amongst others [7, 18, 27, 53, 54, 55, 56, 64]. There was a study [77] done regarding soil erosion to see the impact of development affecting Ringlet reservoir in Cameron Highland. However their study is very limited for the soil erosion along the reservoir area only. EWarns, a real-time GPS based transmitter has been developed [72] to monitor the rainfall information for some stretches of highways and tourism locations in Cameron Highland. In last few years, landslide hazard evaluation using GIS and data mining such as fuzzy logic, and artificial neural network methods have been applied by researchers in different countries [16, 22, 48, 36, 37]. However their result output cannot be directly used in the Malaysian landslide hazard analysis. This is due to the changes in the geographical environment set up, litho types and different climatic condition etc. The local geographical settings cause different landslide types according to completely different mechanisms and are absolutely incomparable.

There are two main issues highlighted in this paper. Firstly is regarding the combination of twelve (12) parameters extracted from highly optical and non-optical based satellite images by using remote sensing techniques to form a highly reliable landslides susceptibility map at two different sites. e.g. soil and rock slopes. Secondly is how landslides susceptibility databases been integrated with TEMAN databases which are comprised of inventory, condition assessment and maintenance records to further improve the current slopes ranking in PLUS.

2. Study area and data

Gua Tempurung and Jelapang in Perak are the two locations which were selected for this analysis. These locations are shown in Figure 1. The selections of the study area are based on the following criteria:-

a. There are scars of slope failure at the higher part of the slopes in the west outside of the right of ways (ROW). Although they do not pose any threats to the highway but geomorphological study of this area should be conducted to determine the slopes stability at the east side area which may have a direct impact on the highway.
b. There are lineaments across in the Jelapang area which may have higher influence to the stability of rock slopes of the area.
c. There is debris flow event (debris flow) in the Gua Tempurung area in 2004 that was originated from outside of the ROW. Further studies should be undertaken to prevent the event from recurring.
c. Criteria of water catchments for both study area are different.

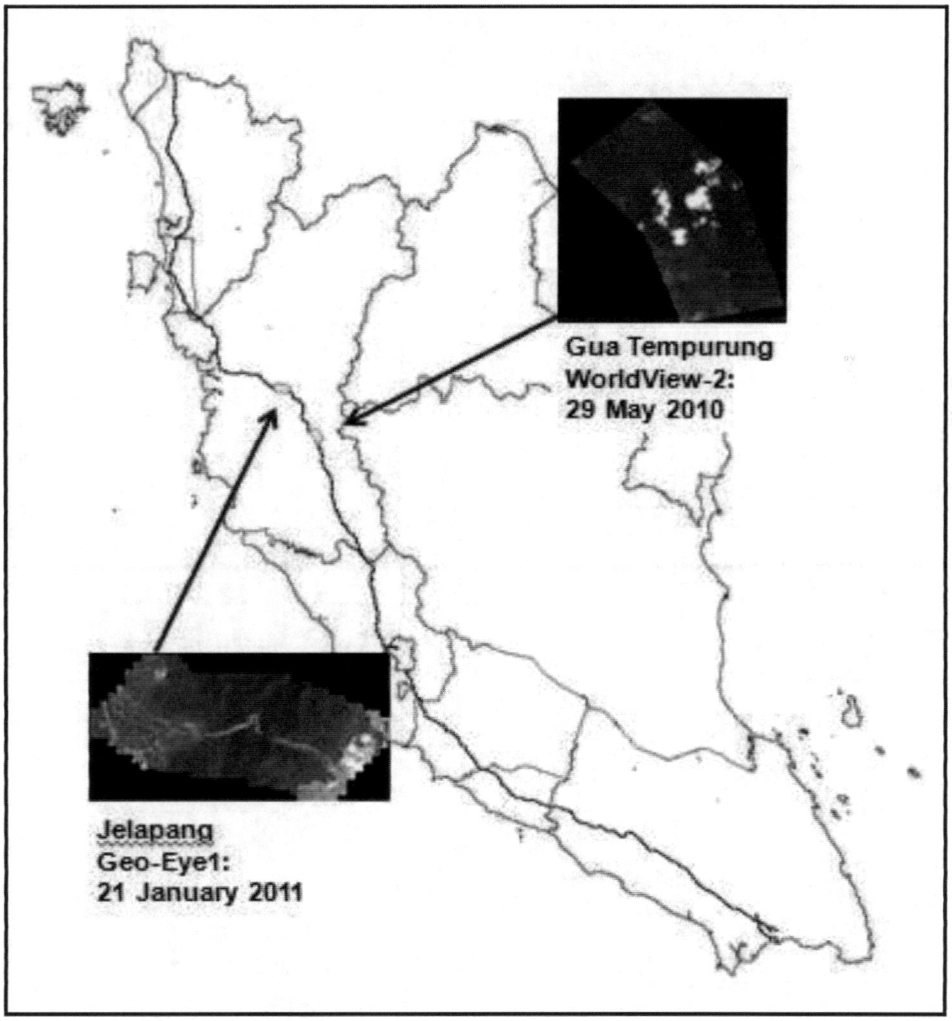

Figure 1. Location map of the study area.

3. Methodology

The research methodology is implemented into five phases: data preparation, modelling, GIS works (e.g. preparation of LSM), the accuracy assessment of the model, the data integration and preparation of landslide hazard maps. Illustration of the overall methodology is shown in Figure 2.

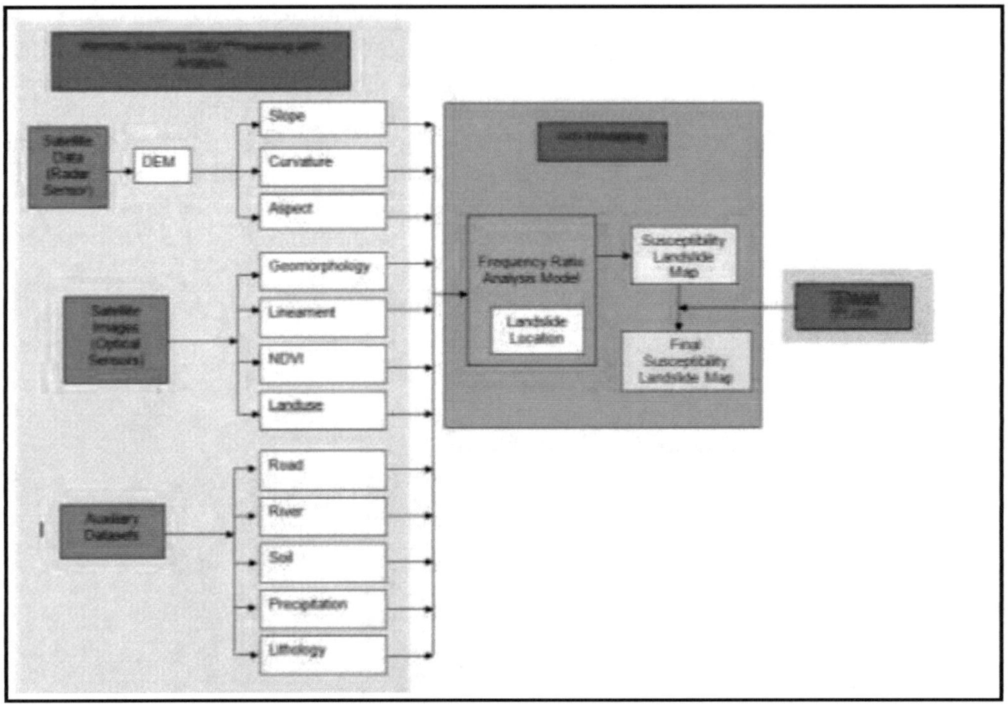

Figure 2. Overall methodology adopted in this study.

3.1 *Data preparation*

Several types of data have been used to extract the important parameters for GIS modelling. The data sources and their parameters are shown in Table 1.

Table 1. Data source.

Bil.	Data Sources	Extracted Parameters
1.	Geology Map	Lithology
2.	Geology Map/ Image Satellite Landsat TM	Geomorphology
3.	Land use map	Soil type
4.	Rainfall Data	Total of rainfall
5.	Topography Map	Drainage Road
6.	IFSAR Data	Lineament Cut slope Aspect angle Curvature
7.	High resolution satellite image (WorldView-2/GeoEye)	Normalized Difference Vegetation Index (NDVI) Land use

3.2 *Cut slope*
Cut slope is correlated with gravitational force. Hence, a steep cut slope is more likely to collapse rather than gentle cut slope. The classification of cut slopes in this study is shown in Table 2.

Table 2 : Classification of slope angle.

Bil.	Slope Angle
1.	0-15°
2.	16-25°
3.	26-35°
4.	>35°

3.3 *Curvature*
Curvature is one of the factors that contribute to landslide and divided into few types such as concave, convex or flat. Concave is capable of storing more water rather than a flat surface and a convex surface is proven otherwise. Hence, the concave surface is highly potential that could trigger landslide compare to the other surface types. The classification of curvature is summarized in Table 3.

Table 3 : Curvature classification.

Bil.	Curvature
1.	Convex
2.	Flat
3.	Concave

3.4 *Aspect angle*
Aspect angle means orientation of slope surface in 360° winds direction. Aspects angle are classified as flat angle (-1 °), North (337.5 ° -360 °, 0 ° - 22.5 °), the Northeast (22.5 ° - 67.5 °), East (67.5 ° - 112.5 °), Southeast (112.5 ° - 157.5 °), South (157.5 ° - 202.5 °), West (202.5 ° - 292.5 °) and Southwest (292.5 ° - 337.5 °). The aspect angle influences landslides thru certain factors such as exposure to the sun, wind (air factor) and rain. The aspect angles' classification is shown in Table 4.

Table 4 : Aspect angle classification.

Bil.	**Aspect Angle**
1.	Flat (-1)
2.	N (0°-22.5°)
3.	NE (22.5°-67.5°)
4.	E (67.5°-112.5°)
5.	SE (112.5°-157.5°)
6.	S (157.5°-202.5°)
7.	SW (202.5°-247.5°)
8.	W (247.5°-292.5°)
9.	NW (292.5°-337.5°)

3.5 *Lineament*
Lineament is known as a straight feature such as faults, sinkholes lines and lines of volcanoes. In the study of landslides, lineament is associated with the presence of cracks in the rock that affect slope stability. The lineaments classification is shown in Table 5.

Table 5 : Lineament classification.

Bil.	Distance from lineament (m)
1.	100
2.	200
3.	300
4.	400
5.	>400

3.6 *Normalized difference vegetation index (NDVI)*
Normalized difference vegetation index (NDVI) is used to indicate a plant cover that influences the presence of water in an area. In this research, satellite images (high spectral resolution) are used to extract this information. Its value is between -1 and +1 with positive values indicate areas with vegetation cover, while a negative value means otherwise. The NDVI classification is shown in Table 6.

Table 6 : NDVI classification.

Bil.	NDVI Value	NDVI Classification
1.	< -0.5	Unclassified
2.	(-0.5) - 0.1	Cloud
3.	0.1 – 0.2	Water
4.	0.2 – 0.4	Fairly Vegetated, bare land
5.	> 0.4	Dense Vegetated

3.7 *Geomorphology*

Geomorphology describes the landform of an area. Generally it is divided into five main groups, alluvial, denudational, marine, karst and others. Each group has its own criteria, which representing its topography. For example, alluvial is located at flat area compared to denudational which is located at high topographic area. The detailed of geomorphological classification is shown in Table 7.

Table 7 : Geomorphology classification.

Bil.	Group of Geomorphology	Geomorphology Classification
1.	Alluvial	Active Floodplain Floodplain Infilled Valley Waterbody Paneplain Alluvial Plain
2.	Denudational	Denudational Hill Residual Hill Piedmont Scarp Structure Hill Pediment Isolated Hill Structure Denudational Hill Structure Denudational Hill (With Folding)
3.	Karst	Limestone Cuesta Isolated Limestone Hill
4.	Marine	Sand Bar/Sand Beach Coastal Ridges And Swales Mud Flat
5.	Others	Land Outside Of Coastal Zone Sea Water

3.8 *Land use*

Land use shows land cover of an area that is due to human activities. In general, low vegetated area is more likely to have landslides against the highly vegetated area. In this research, the classification of land use is shown in Table 8.

Table 8 : Land use classification.

Bil.	Land Use Classification
1.	Agricultural Land
2.	Barren Land
3.	Moderately Vegetated Area
4.	Sparsely Vegetated Area with Less Ground Cover
5.	Urban and Associated Area
6.	Water Body
7.	Thickly Vegetated Area

3.9 *Road*

Slope that is nearer to the road can cause instability to the area. Hence, the distance from the constructed road to the slope is one of the factors that is considered for this research. Technical review reveals that the closer distance from road to slope, the higher possibility of landslide. The distance from road to the slope can be classified in Table 9: -

Table 9 : Road classification.

Bil.	Distance from Road (m)
1.	40
2.	80
3.	120
4.	160
5.	200
6.	>200

3.10 *River*

The slope area that is closer to the river is more likely to have landslide due to erosion. The closer distance from slope to river is the higher possibility of landslide occurrences. The distance from the road to the river can be classified in Table 10.

Table 10 : River classification.

Bil.	Distance from river (m)
1.	50
2.	100
3.	150
4.	>150

3.11 *Soil type*

Soil type is one of the Quasi-static factors that need to be considered in landslide analysis. Usually landslide occurred on slopes that are characterized as loam and sand. This is because these two types of soil are less elastic than the clay and peat. The soil classification is shown in Table 11.

Table 11 : Land use classification.

Bil.	Land Use Classification
1.	Organic
2.	Sand
3.	Loamy
4.	Sandy Loam
5.	Sandy Clay Loam
6.	Clay Loam
7.	Silty Clay Loam
8.	Sandy Clay
9.	Silty Clay
10.	Clay
11.	Peat

3.12 *Total rainfall*

Total rainfall is one of dynamic variables factor that is closely related to landslide. This is because water from rainfall can be easily seeping into the ground and subsequently increased ground water level. Eventually this situation will weaken the soil structure. However, based on this fact the landslide is actually proportional directly to the amount of rainfall in the area. The rainfall classification is shown in Table 12.

Table 12: Rainfall classification.

Bil.	Total Rainfall Annually (mm)
1.	1-1000
2.	1001-1200
3.	1201-1500
4.	1501-1750
5.	1751-2000
6.	2001-2250
7.	2251-2500
8.	2501-2750
9.	2751-3000
10.	3001-3250
11.	3251-3500
12.	3501-3750
13.	3751-4000
14.	>4001

3.13 *Lithology*

Lithology or rock type is one of the key factors that contribute to landslides. Generally there are three types of rocks and they are igneous, sedimentary and metamorphic. The classification of lithology is shown in Table 13.

IOP Conf. Series: Earth and Environmental Science **20** (2014) 012031 doi:10.1088/1755-1315/20/1/012031

Table 13: Lithology classification.

Bil.	Lithology Classification
1.	Sand (mainly marine)
2.	Clay and silt (marine)
3.	Peat, humic clay and silt
4.	Clay, silt, sand and gravel – undifferentiated (continental)
5.	Shale, mudstone, siltstone, hyalite, slate and hornfels
6.	Sandstone/metasandstone
7.	Conglomerate
8.	Limestone / marble
9.	Schist
10.	Ignimbrite
11.	Acid to Intermediate volcanic : mainly pyroclastic, rhyolites to dacitic composition
12.	Intermediate to basic volcanism : mainly pyroclastic
13.	Acid Intrusive (undifferentiated)
14.	Intermediate Intrusive (undifferentiated)
15.	Basic intrusive, mainly gabbro
16.	Ultrabasic intrusive, commonly altered to serpentine
17.	Vein quartz
18.	Clay, silt, sand, peat and minor gravel
19.	Cross-bedded sandstone with subordinate conglomerate and shale/mudstone.Volcanics are locally present.
20.	Interbedded sandstone, siltstone and shale; widespread volcanics, mainly rhyolitic to dacitic tuffs. ConglomeSert local.
21.	Phyllite, slate and shale with subordinate sandstone and schist. Prominent development of limestone throughout the succession.
22.	Phyllite, slate, shale and sandstone; argillaceous rocks are commonly carbonaceous. Limestone and acid to intermediate volcanics.
23.	Phyllite, schist and slate; limestone and sandstone locally prominent. Some interbeds of conglomerate and chert and rare volcanics.
24.	Schist, phyllite, slate and limestone. Minor intercalations of sandstone and volcanics
25.	Sandstone/metasandstone with subordinate siltstone, shale and minor conglomerate.

3.14 *GIS modelling*

GIS modelling that is used to produce LSM for this research is Frequency Ratio Model (FR) [1, 33, 35, 54, 56, 59]. FR is the ratio of the total pixel in landslide area against total pixel of research area. FR is calculated to obtain the probability of landslides occurrence in each class. Probability values for each class are then will be added together to get landslide susceptibility index (LSI) [54].

$$LSI = Fr_1 + Fr_2 + Fr_3 + \ldots + Fr_n \quad (1)$$

Subsequently, LSI is used to generate a LSM. The values of this index are classified into four classes, namely very high, high, medium, and low which indicates the landslides probability.

IOP Conf. Series: Earth and Environmental Science **20** (2014) 012031 doi:10.1088/1755-1315/20/1/012031

3.15 *Accuracy assessment model*

Area under the Curve (AUC) is used to determine the accuracy of the model for this study [34]. For this purpose, a graph of cumulative percentage of landslides ratio is compared with cumulative percentage of landslide pixel for Jelapang and Gua Tempurung area and they are plotted separately. The data for the Jelapang and Gua Tempurung is in Figure 6 and Figure 94, respectively. AUC values obtained from each of the plotted graph determined the accuracy of analysis results for this area.

4. Results and discussion

4.1 The following are some of the samples for parameters extraction from FR technique for optical and non-optical based data :-

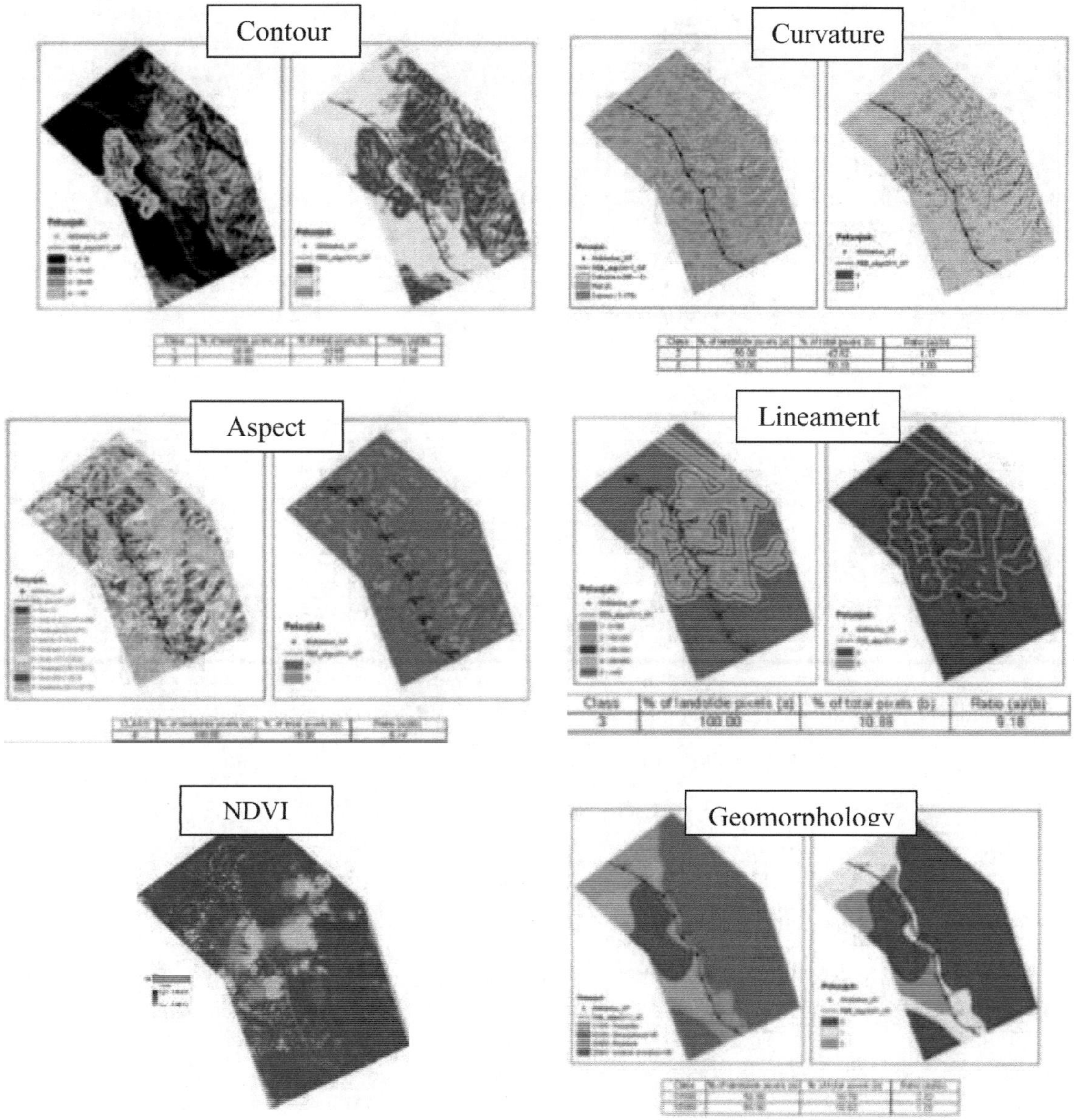

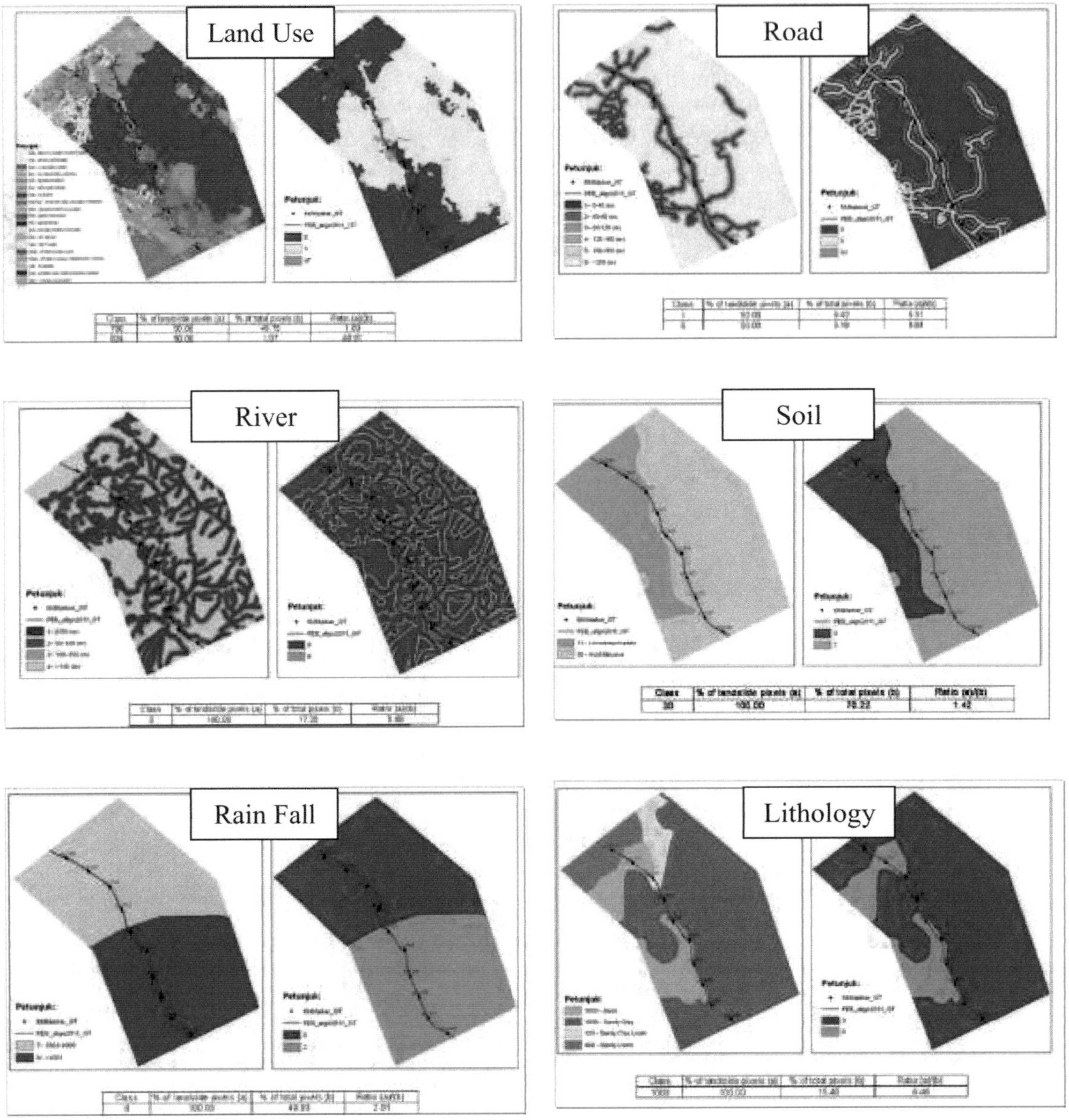

Figure 3. Samples of parameters extraction from satellite data.

4.2 *Landslide susceptibility map of Gua Tempurung*

They are two types of LSM for Gua Tempurung area produced in this study, namely: -

a) LSM for the entire study area (Figure 4) and
b) LSM for engineered slopes at Gua Tempurung (Figure 5).

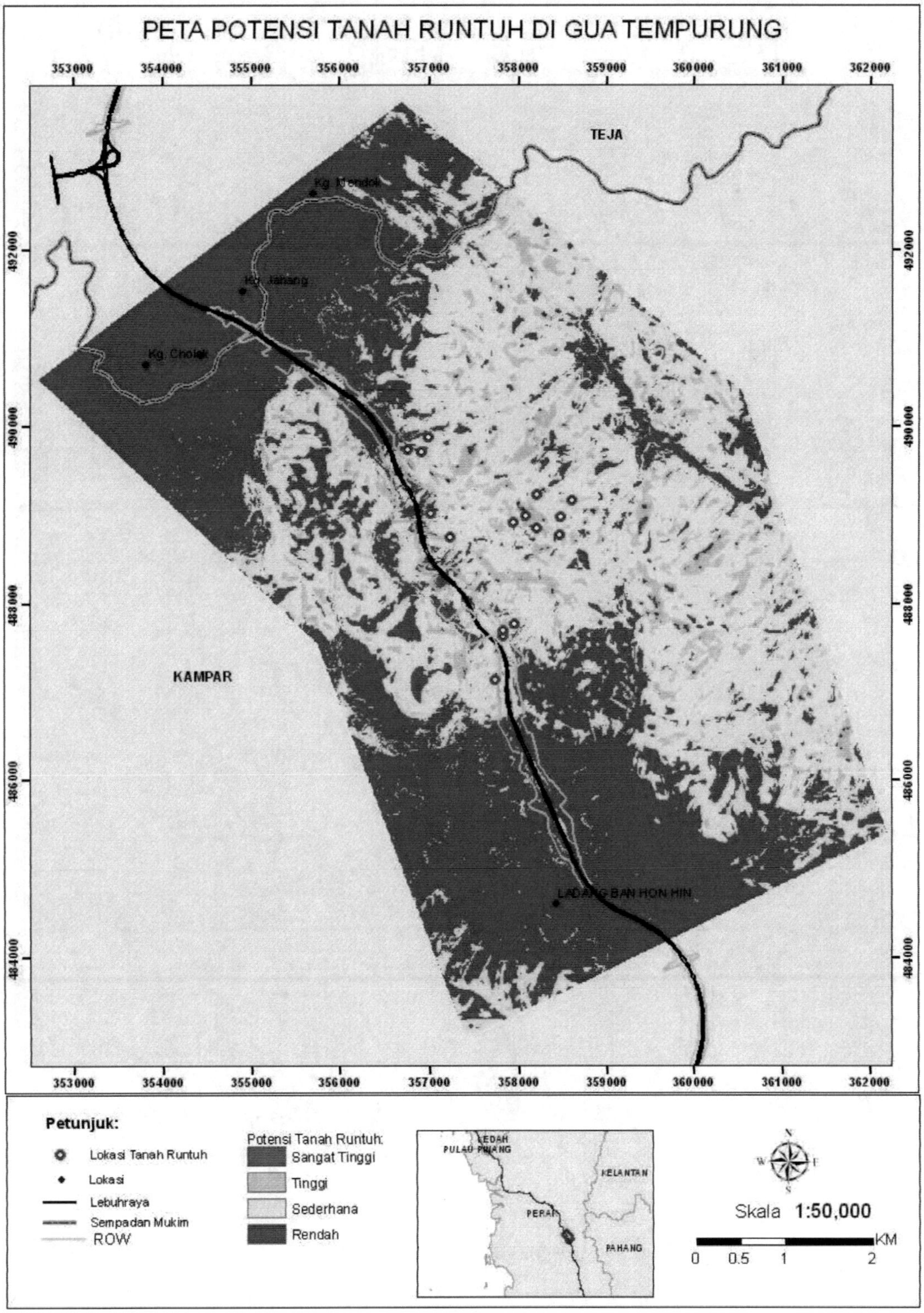

Figure 4. LSM for Gua Tempurung.

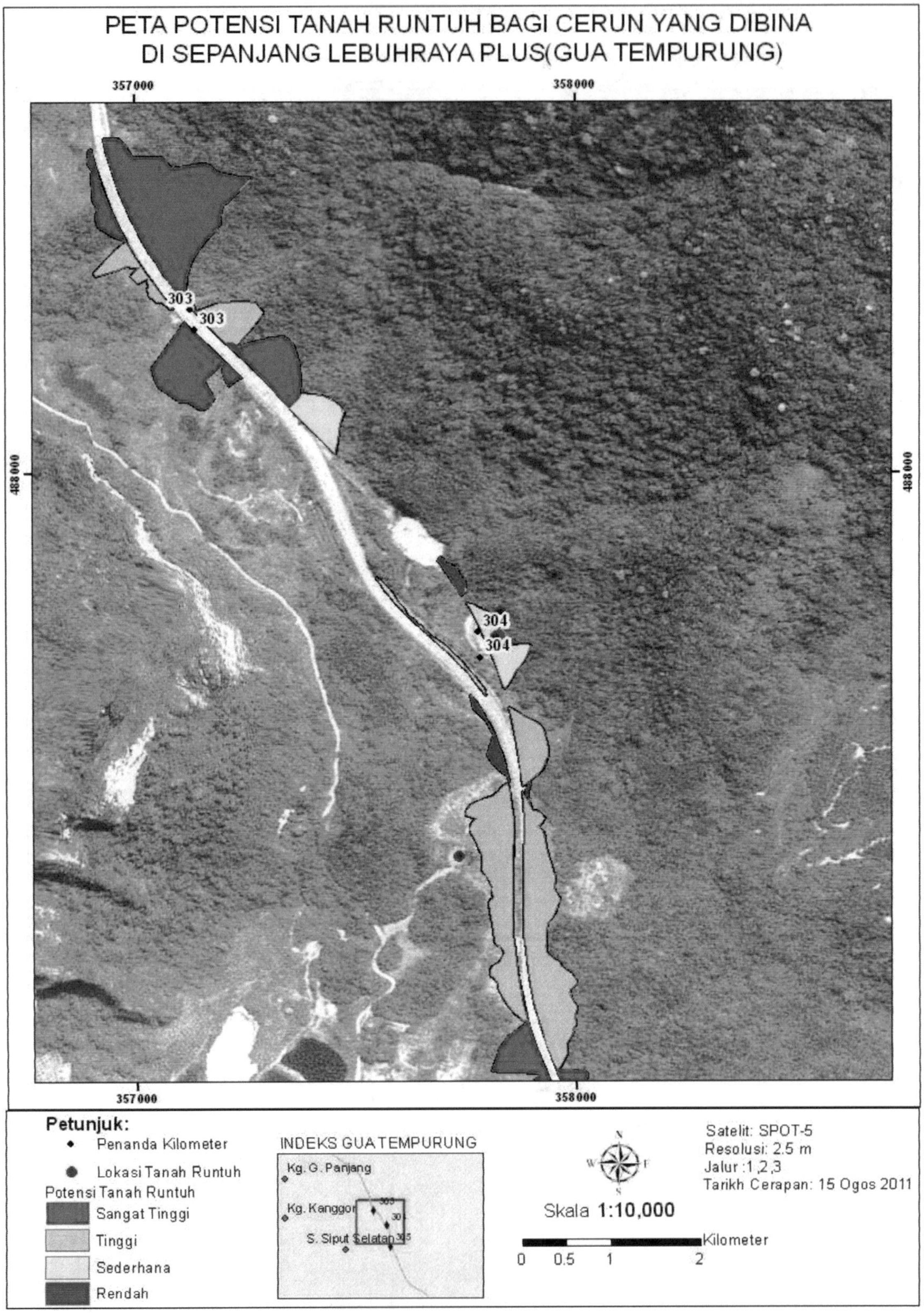

Figure 5. LSM for engineered slopes at Gua Tempurung.

4.3 *Accuracy assessment model*

Based from the following graph in Figure 6, the landslide cumulative ratio versus landslide cumulative pixel shows the area under the curve (AUC) is at 0.7406 which gives an accuracy of 74% for the model of Gua Tempurung. From this graph, it is shown that 51.55% of landslides at Gua Tempurung is located in lower class, 39.5% is in the medium class while 8.7% and 0.3% are in high and very high class, respectively.

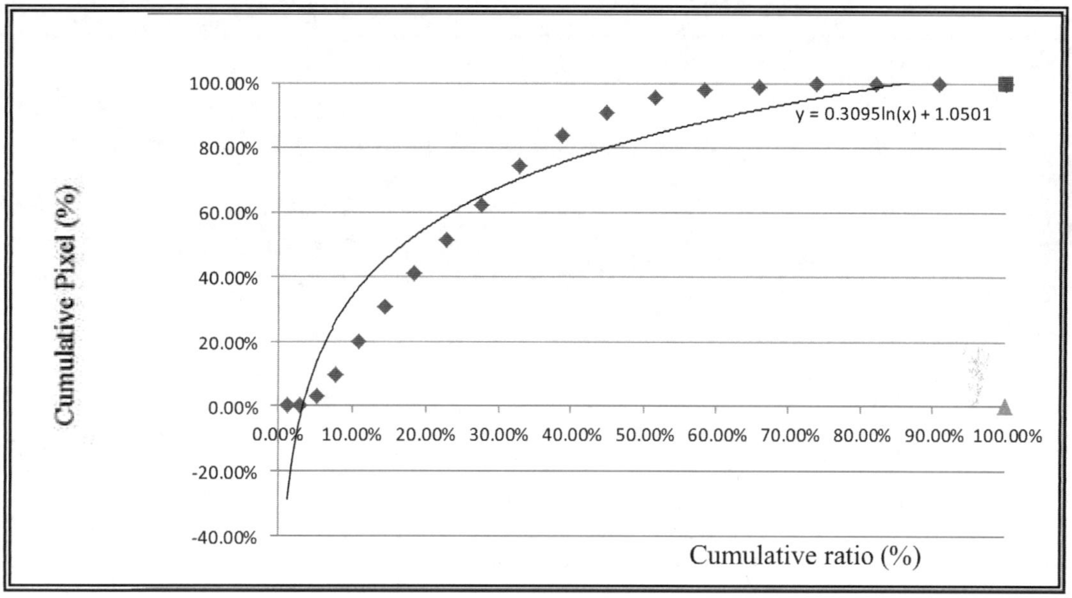

Figure 6: Cumulative ratio against pixel.

4.4 *Landslide susceptibility map for Jelapang Area*

LSM for the entire area of research in Jelapang is shown in Figure 7 while Figure 8 shown LSM for engineered slopes in Jelapang area.

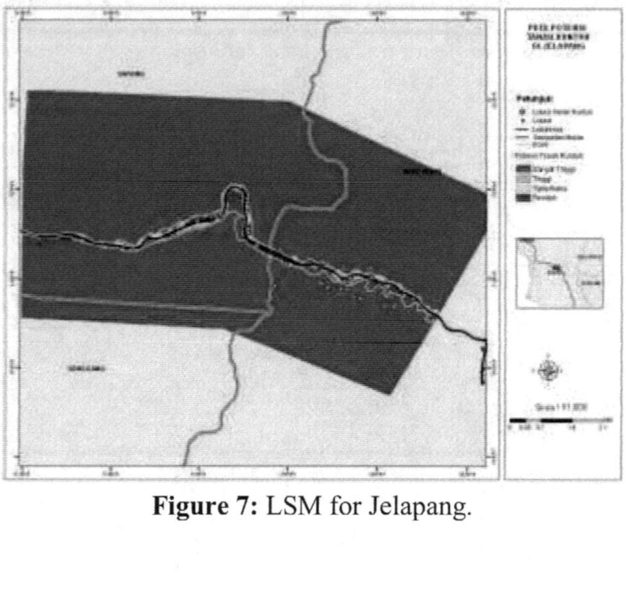

Figure 7: LSM for Jelapang.

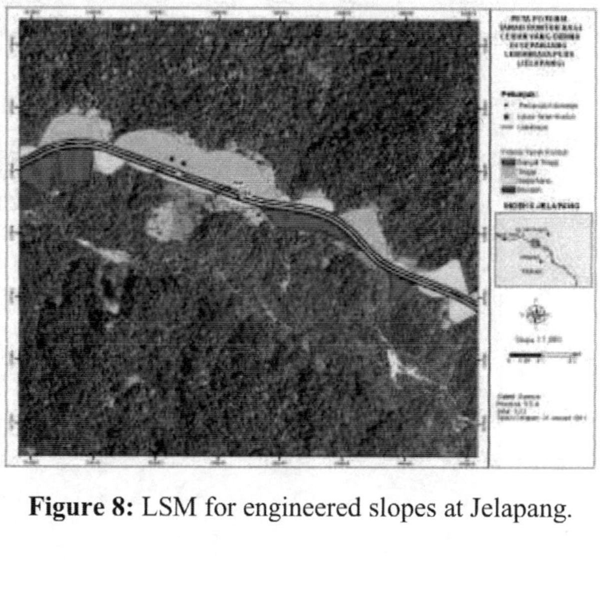

Figure 8: LSM for engineered slopes at Jelapang.

4.5 *Accuracy assessment model*

From the following graph in Figure 9, cumulative ratio versus cumulative pixel shows that the 98% of landslides at Jelapang are located in Lower class, 1.11% in medium class while 0.28% and 0.02% are in high and very high class, respectively. Area under curve (AUC) value obtained is 0.98 which gives an accuracy of 98% for this model.

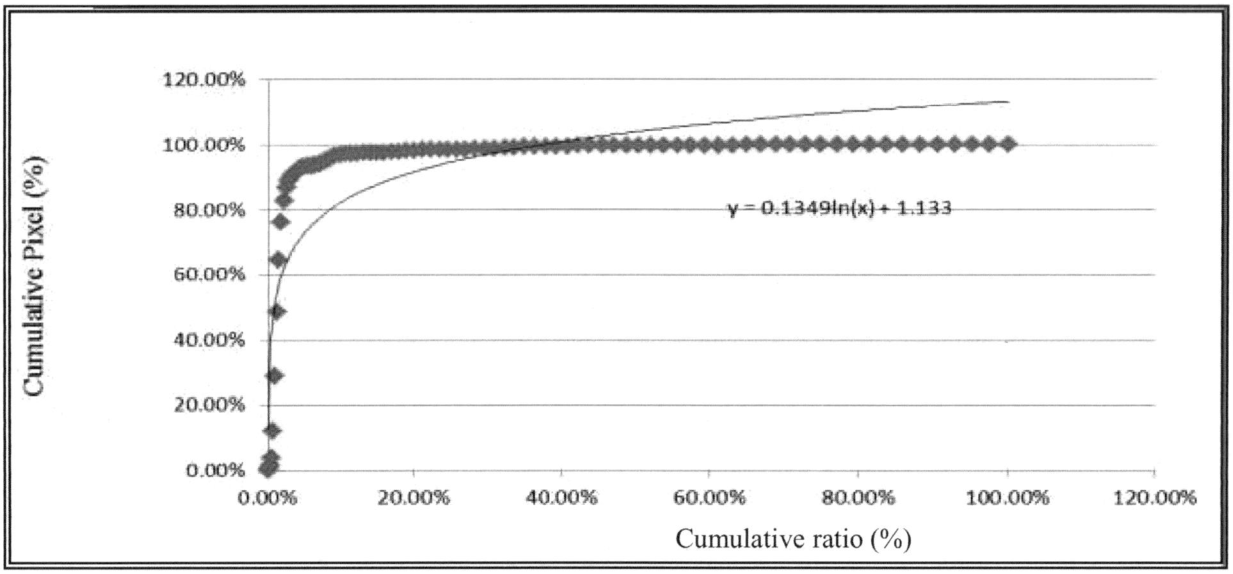

Figure 9: Cumulative ratio against pixel for Jelapang.

4.6 *Observation and analysis*

The followings are the observation on LSM (e.g. Figure 3, 4, 6 and 7). They are four (4) classes of susceptible landslides for both locations. The area is shown in percentage as follows:

Table 14: Percentage of landslides susceptible at Gua Tempurung and Jelapang.

Bil.	Landslides Susceptible Classification	Percentage at Gua Tempurung (%)	Percentage at Jelapang (%)
1	Very High	0.3	0.1
2	High	8.7	0.3
3	Medium	39.5	1.1
4	Low	51.5	98.5
Total		100	100

They are 43 landslides susceptible locations identified at both research locations. The breakdown of the landslides susceptibility locations following its classes are as follows:

Table 15: Total landslides susceptible locations at Gua Tempurung and Jelapang.

Bil	Landslides Susceptible Classification	Number of landslides at Gua Tempurung	Number of landslides at Jelapang	Total
1.	Very High	0	2	2
2.	High	2	5	7
3.	Medium	7	5	12
4.	Low	8	14	22
Total		17	26	43

The following are the main parameters that are dominant in which could trigger landslides in both research areas.

Table 16: Main criteria of landslides at Gua Tempurung dan Jelapang.

Bil	Parameters	Criteria at Gua Tempurung	Criteria at Jelapang
1.	Slope Angle	35° – 90°	35° – 90°
2.	Geomorphology	Denudational Hill	Denudational Hill
3.	Land use	Sandy Clay	Sandy Clay
4.	Precipitation	2251 – 2500mm	1751-2000 mm

There are 152 numbers of engineered slopes which are maintained by PLUS along the research area. The breakdown of the landslides susceptibility locations after integrated with TEMAN data is shown in table 17 as follows :-

Table 17: Numbers of slope based on its classification for Gua Tempurung and Jelapang.

Bil	Landslides Susceptible Classification	Number of Slope at Gua Tempurung	Number of slope at Jelapang	Total
1	Very High	9	1	10
2	High	4	5	9
3	Medium	15	60	75
4	Low	22	36	58
Total		50	102	152

5. Conclusion

Frequency ratio based mapping of landslide for Gua Tempurung and Jelapang area showed that these locations even located at hilly area but they are very much stable, there are only 7 out of 26 locations which are shown in Gua Tempurung and 2 out of 17 locations at Jelapang area are categorised as high and very high for landslides. The parameters that influenced slope failure for both locations were identified namely slope angle, geomorphology, soil type and precipitation. The quality of this model are then proven via calculation of AUC which shows that the landslide susceptibility map produced in this research paper can be used for planning and preventive measures. The integration with TEMAN data has further improved the LSM for the areas.

In conclusion, frequency ratio is one of the useful statistical based methods for LSM. LSM is a fundamental towards understanding better of hazard and disaster risk maps. There are many useful GIS based techniques either qualitative or quantitative to analyse the correlation between landslides and its influence parameters.

References

[1] Akgun, A.; Dag, S. & Bulut, F. (2008). Landslide susceptibility mapping for a landslide prone area (Findikli, NE of Turkey) by likelihood-frequency ratio and weighted linear combination models. Environmental Geology, Vol. 54, 1127-1143

[2] Atkinson, P.M. & Tatnall, A.R.L. (1997). Neural networks in remote sensing. International Journal of Remote Sensing, Vol. 18, 699-709

[3] Atkinson, P.M. & Massari, R. (1998). Generalized linear modeling of susceptibility to landsliding in the central Apennines, Italy. Computer & Geosciences, Vol. 24, 373-385

[4] Ayalew, L.; Yamagishi, H.; Marui, H. & Kanno, T. (2005). Landslides in Sado Island of Japan Part II. GIS-based susceptibility mapping with comparisions of results from two methods and verifications. Engineering Geology, Vol. 81, 432-445

[5] Baeza, C. & Corominas, J. (2001). Assessment of shallow landslide susceptibility by means of multivariate statistical techniques. Earth Surface Processes & Landforms, Vol. 26, 251-1263

[6] Basheer, I.A. & Hajmeer, M. (2000). Artificial neural networks: fundamentals, computing, design, and application. Journal of Microbiological Methods, Vol. 43, 3-31

[7] Caniani, D.; Pascale, S.; Sdao, F. & Sole, A. (2008). Neural networks and landslide susceptibility: a case study of the urban area of Potenza. Natural Hazards, Vol. 45, 55-72

[8] Catani, F.; Casagli, N.; Ermini, L.; Righini, G. & Menduni, G. (2005). Landslide hazard and risk mapping at catchment scale in the Arno River Basin. Landslides, Vol. 2, No. 4, 329-343

[9] Chang, T.C.; Chao, R.J. (2006). Application of back-propagation networks in debris flow prediction. Engineering Geology, Vol. 85, 270-280

[10] Chau, K.T. & Chan, J.E. (2005). Regional bias of landslide data in generating susceptibility maps using logistic regression: Case of Hong Kong Island. Landslides, Vol. 2, 280- 290

[11] Clerici, A., Perego, S., Tellini, C. & Vescovi, P. (2002). A procedure for landslide susceptibility zonation by the conditional analysis method. Geomorphology, Vol. 48, 349-364

[12] Cruden, D.M. & Varnes, D.J. (1996). Landslide types and processes, In: A. K. Turner and R. L. Schuster (eds), Landslides: Investigation and Mitigation, TRB Special Report, 247, National Academy Press, Washington, pp. 36–75

[13] Dahal, R.K.; Hasegawa, S.; Nonomura, S.; Yamanaka, M.; Masuda, T. & Nishino, K. (2008). GIS-based weights-of-evidence modelling of rainfall-induced landslides in small catchments for landslide susceptibility mapping. Environmental Geology, Vol. 54, 311-324

[14] Dai, F.C. & Lee, C.F. (2003). Landslide characteristics and slope instability modeling using GIS, Lantau Island. Hong Kong. Geomorphology, Vol. 42, 213-228

[15] Dai, F.C.; Lee, C.F.; Li, J. & Xu, Z.W. (2001). Assessment of landslide susceptibility on the natural terrain of Lantau Island, Hong Kong. Environmental Geology, Vol. 40, 381- 391

[16] Ercanoglu, M & Gokceoglu, C. (2002). Assessment of landslide susceptibility for a landslideprone area (north of Yenice, NW Turkye) by fuzzy approach. Environmental Geology, Vol. 41, 720-730
[17] Ercanoglu, M.; Gokceoglu, C. & Van Asch, T.W.J. (2004). Landslide susceptibility zoning north of Yenice (NW Turkey) by multivariate statistical techniques. Natural Hazards, Vol. 32, 1–23
[18] Ermini, L.; Catani, F. & Casagli, N. (2005). Artificial neural networks applied to landslide susceptibility assessment. Geomorphology, Vol. 66 No. 1-4, 327-343
[19] Gokceoglu, C.; Sonmez, H. & Ercanoglu, M. (2000). Discontinuity controlled probabilistic slope failure risk maps of the Altindag (settlement) region in Turkey. Engineering Geology, Vol. 55, 277–296
[20] Gomez, H. & Kavzoglu, T. (2005). Assessment of shallow landslide susceptibility using artificial neural networks in Jabonosa River Basin, Venezuela. Engineering Geology, Vol. 78, No. 1-2, 11-27
[21] Hines, J.W. (1997). Fuzzy and neural approaches in engineering. Wiley, New York.
[22] Lee, S. (2004). Application of likelihood ratio and logistic regression models to landslide susceptibility mapping using GIS. Environmental Management, Vol. 34, No. 2, 223-232
[23] Lee, S. (2005). Application of logistic regression model and its validation for landslide susceptibility mapping using GIS and remote sensing data. International Journal of Remote Sensing, Vol. 26, No. 7, 477-1491
[24] Lee, S. (2007a). Comparison of landslide susceptibility maps generated through multiple logistic regression for three test areas in Korea. Earth Surface Processes and Landforms, Vol. 32, 2133-2148
[25] Lee, S. (2007b). Application and verification of fuzzy algebraic operators to landslide susceptibility mapping. Environmental Geology, Vol. 52, 615-623
[26] Lee, S. & Dan, N.T. (2005). Probabilistic landslide susceptibility mapping in the Lai Chau province of Vietnam: focus on the relationship between tectonic fractures and landslides. Environmental Geology, Vol. 48, 778–787
[27] Lee, S. & Evangelista, D.G. (2006). Earthquake induced landslide susceptibility mapping using an artificial neural network. Natural Hazards & Earth System Sciences, Vol. 6, 687-695
[28] Lee, S. & Lee, M.J. (2006). Detecting landslide location using KOMPSAT 1 and its application to landslide-susceptibility mapping at the Gangneung area, Korea. Advances in Space Research, Vol. 38, 2261–2271
[29] Lee, S. & Min, K. (2001). Statistical analysis of landslide susceptibility at Yongin, Korea, Environmental Geology, Vol. 40, 1095-1113
[30] Lee, S.; Chwae, U. & Min, K. (2002a). Landslide susceptibility mapping by correlation between topography and geological structure: the Janghung area, Korea. Geomorphology, Vol. 46, 49-162.
[31] Lee, S.; Choi, J. & Min, K. (2002b). Landslide susceptibility analysis and verification using the Bayesian probability model. Environmental Geology, Vol. 43, 120-131
[32] Lee, S. & Choi, U. (2003). Development of GIS-based geological hazard information system and its application for landslide analysis in Korea. Geosciences Journal, Vol. 7, 243- 252
[33] Lee, S. & Pradhan, B. (2007). Landslide hazard mapping at Selangor, Malaysia using frequency ratio and logistic regression models. Landslides, Vol. 4, 33-41
[34] Lee, S. & Pradhan, B. (2006). Probabilistic landslide risk mapping at Penang Island, Malaysia. Journal of Earth System Sciences, Vol. 115, No. 6, 1-12
[35] Lee, S. & Sambath, T. (2006). Landslide susceptibility mapping in the Damrei Romel area, Cambodia using frequency ratio and logistic regression models. Environmental Geology, Vol. 50, 847–855
[36] Lee, S.; Ryu, J.H.; Min, K. & Won, J.S. (2003a). Landslide susceptibility analysis using GIS and artificial neural network. Earth Surface Processes & Landforms, Vol. 27, 1361- 1376
[37] Lee, S.; Ryu, J.H.; Min, K. & Won, J.S. (2003b). Landslide susceptibility analysis using artificial neural network at Boun, Korea. Environmental Geology, Vol. 44, 820-833
[38] Lee, S.; Ryu, J.H.; Min, K. & Won, J.S. (2006). The application of neural networks to landslide susceptibility mapping at Janghung, Korea. Mathematical Geology, Vol. 38, No. 2, 199-220

[39] Lee, S.; Choi, J. & Min, K. (2004a). Probabilistic landslide hazard mapping using GIS and remote sensing data at Boun, Korea. International Journal of Remote Sensing, Vol. 25, 2037-2052

[40] Lee, S.; Ryu, J.H.; Won, J.S. & Park, H.J. (2004b). Determination and application of the weights for landslide susceptibility mapping using an artificial neural network. Engineering Geology, Vol. 71, 289-302

[41] Lee, S.; Ryu, J.H. & Kim, I.S. (2007). Landslide susceptibility analysis and its verification using likelihood ratio, logistic regression, and artificial neural network models: case study of Youngin, Korea. Landslides, Vol. 4, No. 4, 327-338

[42] Moody, A.; & Katz, D.B. (2003). Artificial intelligence in the study of mountain landscapes. In Bishop, M. P. and Shorder, J. F. (Editors), Geographic Information Science and Mountain Geomorphology: Springer, Berlin, pp. 219–249

[43] Neaupane, K.M. & Achet, S.H. (2004). Use of backpropagation neural network for landslide monitoring: a case study in the higher Himalaya. Engineering Geology, Vol. 74, No. 3-4, 213-226

[44] Nefeslioglu, H.A.; Gokceoglu, C. & Sonmez, H. (2008). An assessment on the use of logistic regression and artificial neural networks with different sampling strategies for the preparation of landslide susceptibility maps. Engineering Geology, Vol. 97, 171-171

[45] Santacana, N.; Baeza, B.; Corominas, J.; Paz, A.D. Marturiá, J. (2003). A GIS-based multivariate statistical analysis for shallow landslide susceptibility mapping in La Pobla de Lillet Area (Eastern Pyrenees, Spain). Natural Hazards, Vol. 30, 281–295

[46] Ohlmacher, G.C. &, Davis, J.C. (2003). Using multiple logistic regression and GIS technology to predict landslide hazard in northeast Kansa, USA. Engineering Geology, Vol. 69, 331-343

[47] Paola, J.D. & Schowengerdt, R.A. (1995). A review and analysis of back propagation neural networks for classification of remotely sensed multi-spectral imagery. International Journal of Remote Sensing, Vol. 16, No. 16, 3033-3058

[48] Pistocchi, A.; Luzi, L. & Napolitano, P. (2002). The use of predictive modeling techniques for optimal exploitation of spatial databases: a case study in landslide hazard mapping with expert system-like methods. Environmental Geology, Vol. 41, 765–775

[49] Pradhan, B. (2010a). Remote sensing and GIS-based landslide hazard analysis and cross validation using multivariate logistic regression model on three test areas in Malaysia. Advances in Space Research, Vol. 45, No. 10, 1244-1256

[50] Pradhan, B. (2010b). Use of GIS-based fuzzy logic relations and its cross application to produce landslide susceptibility maps in three test areas in Malaysia. Environmental Earth Sciences (article on-line first available) DOI 10.1007/ s12665-010-0705-1

[51] Pradhan, B. (2010c). Manifestation of an advanced fuzzy logic model coupled with Geoinformation techniques to landslide susceptibility mapping and their comparison with logistic regression modelling. Environmental & Ecological Statistics (article on-line first available) DOI 10.1007/ s10651-010-0147-7

[52] Pradhan, B. (2010d). Application of an advanced fuzzy logic model for landslide susceptibility analysis. International Journal of Computational Intelligence Systems, Vol.3, No. 3 (September, 2010), 370-381

[53] Pradhan, B. & Buchroithner, M.F. (2010). Comparison and validation of landslide susceptibility maps using an artificial neural network model for three test areas in Malaysia. Environmental & Engineering Geoscience, Vol. 16, No. 2, 107-126

[54] Pradhan, B. & Lee, S. (2010a). Delineation of landslide hazard areas using frequency ratio, logistic regression and artificial neural network model at Penang Island, Malaysia. Environmental Earth Sciences, Vol. 60, 1037 – 1054

[55] Pradhan, B. & Lee, S. (2010b). Regional landslide susceptibility analysis using backpropagation neural network model at Cameron Highland, Malaysia. Landslides, Vol. 7, 13-30

[56] Pradhan, B. & Lee, S. (2010c). Landslide susceptibility assessment and factor effect analysis: back propagation artificial neural networks and their comparison with frequency ratio and bivariate logistic regression modeling. Environmental Modelling & Software, Vol. 25, 747-759

[57] Pradhan, B. & Lee, S. (2009). Landslide risk analysis using artificial neural network model focusing on different training sites. International Journal of Physical Science, Vol. 3. No. 11, 1-15

[58] Pradhan, B. & Lee, S. (2007). Utilization of optical remote sensing data and GIS tools for regional landslide hazard analysis by using an artificial neural network model. Earth Science Frontier, Vol. 14, No. 6, 143-152

[59] Pradhan, B.; Lee, S. & Buchroithner, M.F. (2010a). Remote sensing and GIS-based landslide susceptibility analysis and its cross-validation in three test areas using a frequency ratio model. Photogrammetrie, Fernerkundung, Geoinformation, Vol. l, No. 1, 17-32

[60] Pradhan, B.; Lee, S. & Buchroithner, M.F. (2010b). A GIS-based back-propagation neural network model and its cross application and validation for landslide susceptibility analyses. Computer Environment & Urban Systems, Vol. 34, No. 216-235

[61] Pradhan, B.; Lee, S. & Buchroithner, M.F. (2009). Use of geospatial data for the development of fuzzy algebraic operators to landslide hazard mapping: a case study in Malaysia. Applied Geomatics, Vol. 1, 3–15

[62] Pradhan, B.; Lee. S.; Mansor, S.; Buchroithner, M.F.; Jallaluddin, N. & Khujaimah, Z. (2008) Utilization of optical remote sensing data and geographic information system tools for regional landslide hazard analysis by using binomial logistic regression model. Journal of Applied Remote Sensing, Vol. 2, 1–11 DOI:10.1117/ 12.821511

[63] Pradhan, B.; Oh, J.J. & Buchroithner, M.F. (2010c). Weight-of-evidence model applied to landslide susceptibility mapping in a tropical hilly area. Geomatics, Natural Hazards & Risk 1(3):199-223 doi:10.1080/ 19475705.2010.498151

[64] Pradhan, B. & Pirasteh, S. (2010). Comparison between prediction capabilities of neural network and fuzzy logic techniques for landslide susceptibility mapping. Disaster Advances, Vol. 3, No. 3, 26-34

[65] Pradhan, B.; Sezer, E.; Gokceoglu, C. & Buchroithner, M.F. (2010d). Landslide susceptibility mapping by neuro-fuzzy approach in a landslide prone area (Cameron Highland, Malaysia). IEEE Transactions on Geoscience & Remote Sensing, Vol. 48, No. 10 (article on-line first available) doi:10.1109/ TGRS.2010.2050328

[66] Pradhan, B. & Youssef, A.M. (2010). Manifestation of remote sensing data and GIS on landslide hazard analysis using spatial-based statistical models. Arabian Journal of Geosciences, Vol. 3, No. 3, 319-326

[67] Pradhan, B.; Singh, R.P. & Buchroithner, M.F. (2006). Estimation of stress and its use in evaluation of landslide prone regions using remote sensing data. Advances in Space Research, Vol. 37, 698 -709

[68] Schalkoff, R.J. (1997) Artificial neural networks, New York, NY: Wiley Suzen, M.L. & Doyuran, V. (2004a). Data driven bivariate landslide susceptibility assessment using geographical information systems: a method and application to Asarsuyu catchment, Turkey. Engineering Geology, Vol. 71, 303–321

[69] Suzen, M.L. & Doyuran, V. (2004b). A comparison of the GIS based landslide susceptibility assessment methods: multivariate versus bivariate. Environmental Geology, Vol. 45, 665–679

[70] Swingler, K. (1996). Applying Neural Networks: A practical guide. Academic Press, New York.

[71] Tangestani, M.H. (2004. Landslide susceptibility mapping using the fuzzy gamma approach in a GIS, Kakan catchment area, southwest Iran. Austrian Journal of Earth Sciences, Vol. 51, 439–450

[72] T.k.Hui, F.H. Ali (2006). A Near Real-time Early Warning System on Erosion Hazards. American Journal of Environmental Sciences 2 (4): 146-153, 2006

[73] Varnes, D.J. IAEG Commission on Landslides (1984) Landslide hazard zonation – a review of principles and practice. UNESCO, Paris 63

[74] Yesilncar, E. & Topal, T. (2005) Landslide susceptibility mapping: A Comparison of logistic regression and artificial neural networks methods in a medium scale study, Hendek region (Turkey). Engineering Geology, Vol. 79, 251-266

[75] Youssef, A.M.; Pradhan, B.; Gaber, A.F.D. & Buchroithner, M.F. (2009) Geomorphological hazard analysis along the Egyptian red sea coast between Safaga and Quseir. Natural Hazards & Earth System Sciences, 751-766

[76] Zhou, W. (1999). Verification of nonparametric characteristics of back propagation neural networks for image classification. IEEE Transactions on Geosciences & Remote Sensing, Vol. 37, 771-779

[77] Z.A. Roslan, K.H. Tew (1997). Remote sensing application in determination of land use management factors of the USLE, Human impact on erosions and sedimentations (proceeding of Rabat symposium 6, April 1997) IAHS publ. no. 245, 1997

IOP Conf. Series: Earth and Environmental Science **20** (2014) 012032 doi:10.1088/1755-1315/20/1/012032

Landslide susceptibility mapping using decision-tree based CHi-squared automatic interaction detection (CHAID) and Logistic regression (LR) integration

Omar F. Althuwaynee, Biswajeet Pradhan*, Noordin Ahmad

Department of Civil Engineering, Faculty of Engineering,
University Putra Malaysia, 43400, Serdang, Selangor DarulEhsan, Malaysia

*Email.biswajeet24@gmail.com or biswajeet@lycos.com (Corresponding author)

Abstract. This article uses methodology based on chi-squared automatic interaction detection (CHAID), as a multivariate method that has an automatic classification capacity to analyse large numbers of landslide conditioning factors. This new algorithm was developed to overcome the subjectivity of the manual categorization of scale data of landslide conditioning factors, and to predict rainfall-induced susceptibility map in Kuala Lumpur city and surrounding areas using geographic information system (GIS). The main objective of this article is to use CHi-squared automatic interaction detection (CHAID) method to perform the best classification fit for each conditioning factor, then, combining it with logistic regression (LR). LR model was used to find the corresponding coefficients of best fitting function that assess the optimal terminal nodes. A cluster pattern of landslide locations was extracted in previous study using nearest neighbor index (NNI), which were then used to identify the clustered landslide locations range. Clustered locations were used as model training data with 14 landslide conditioning factors such as; topographic derived parameters, lithology, NDVI, land use and land cover maps. Pearson chi–squared value was used to find the best classification fit between the dependent variable and conditioning factors. Finally the relationship between conditioning factors were assessed and the landslide susceptibility map (LSM) was produced. An area under the curve (AUC) was used to test the model reliability and prediction capability with the training and validation landslide locations respectively. This study proved the efficiency and reliability of decision tree (DT) model in landslide susceptibility mapping. Also it provided a valuable scientific basis for spatial decision making in planning and urban management studies.

1. Introduction

Landslide hazards have a large risk impact, particularly on main cities, especially in the tropical developing countries, mainly because of rapid urbanization trends. Man-made activities have a vital role in disturbing the soil of hill slopes and thus increase infrastructure vulnerability index and fatality ratio[1].

The population of Malaysian increased from 6 million in 1960 to more than 28 million in 2012 (Figure 1). The above statistics shows that the population had changed 256% during the last 50 years as reported by the Department of Statistics, Malaysia. This consistently allied with excessive exploitation of lands, especially in Kuala Lumpur and commercial regions in Selangor estate, by

Content from this work may be used under the terms of the Creative Commons Attribution 3.0 licence. Any further distribution of this work must maintain attribution to the author(s) and the title of the work, journal citation and DOI.
Published under licence by IOP Publishing Ltd

establishing new settlements, roads, and other infrastructure projects. Despite of the fact that abundance of relevant studies have been carried out in this study area, landslide is still a recurring hazard.

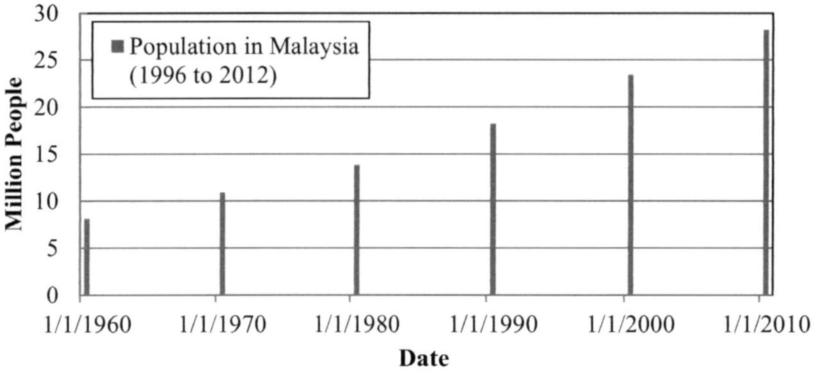

Figure 1. Variation of the population in Malaysia between 1960 to 2012

DT techniques, is popularly known as a multivariate automatic classification approach in landslide literature [2]. Some of the recently developed algorithms of decision tree pattern recognition are; Classification and Regression Tree (CART) [3], Improved algorithm ID3 "Iterative Dichotomizer" C4.5[4], CHAID[5] are used for prediction and classification purposes, [6]. A final report published in B.C. Ministry of forest, [7] have considered the road design parameters such as; width, angle, slope, cut, and fill of the road, that analysed using CHAID, to study their effects on road landslide vulnerability.

Multivariate statistical CHAID method produces an estimated significant chi-square level of interaction between conditioning factors and landslides. Pre-processed arbitrary classification of conditioning factors and tree pruning task are not required. It is worth to mention that, CHAID application considered a novel technique in landslide hazard mapping.

Multivariate statistical approaches, such as discriminant analyses and LR models have been successfully employed in landslide susceptibility mapping [8, 9]. The LR method is able to perform variable regression even if the variable has an abnormal distribution [10]. On the contrary, it requires few data pre-processing on theoretical assumptions than discriminant analysis like a bivariate classification [11]. This will raise the uncertainty due to classification mechanism. Therefore, an auto classification technique is preferable.

The current research aims to perform landslide susceptibility mapping for the part of the in Kuala Lumpur city and surrounding (Malaysia) using ensemble CHAID and LR model. The possibility of adopting this theory must be appreciated in a medium-scale landslide prediction assessment.

2. Study area

Kuala Lumpur and vicinity areas, plays a major role in economic and social development in Malaysia. During the monsoon, the area receive high amount of precipitation that cause instability in slopes [12]. The study area is enclosed between 2°56'N to 3°20'N latitude and 101°29'E to 101°54'0E longitude, with approximate area of 1975 km^2 (Figure 2).

The major types of land cover in the study area comprises of settlement, peat swamp forest, and abandoned mining, grassland and few shrub areas. The overall temperature of the area ranges between 29 to 32° C. The average precipitation vary from 58 to 240 (mm/month), which trap large amount of water periodically, leading to a high pore-water pressure that decreases the shear stress of slopes (Malaysian Meteorological Services Department). More about geological and geomorphological characteristics of the study area can be seen in Althuwaynee et al. [13].

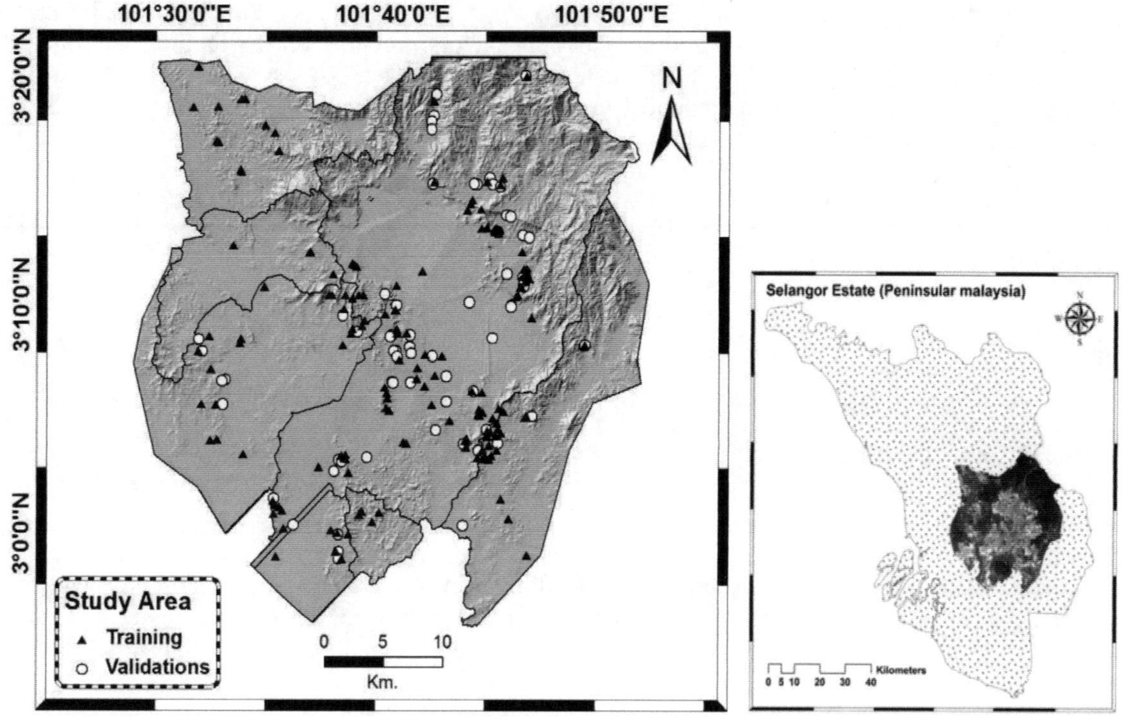

Figure 2. Study area.

3. Methodology

Two multivariate approaches (CHi-squared Automatic Interaction Detection CHAID and LR) were used in ensemble model, The methodology employed here was applied in three main stages;

1. The NNI method tested the spatial pattern of landslides, and the result showed a clustered pattern tendency. Hence, the output will be used as a training dataset.
2. Running the CHAID model, then observing the terminal nodes which carries the lowest chi-square values among the classified layers.
3. Terminal nodes were used as input dataset in LR model, the resultant probability coefficients of LR analysis, helped to find the best fitting function.
4. Finally, all the LR probability coefficients were combined with their corresponding factors, then the landslide probability map LSM was produced.

3.1. Data

Since this article follow-up article to [13, 14], the basic data and other characteristics of the study area can be referred to these articles. The landslide inventory consists of 219 landslides, were collected over the past 25 years, mainly by using remote sensing sources like aerial photos and SPOT 5 images. Landslide conditioning factors used in the modelling were derived from topographic, soil and geology maps.

3.2. CHi-squared Automatic Interaction Detection (CHAID)

CHAID model is considered as one of the main DT techniques, has been popularly used in many regression. Some tangible characteristics and applications of CHAID method can be found in [15]. The major enhanced characteristics of CHAID are; categorical or ordinal data which can be modelled. Both dependent variable and conditioning factors were presented as nominal, ordinal or continuous data. Althuwaynee et al.[9] discussed the CHAID approach in more detail.

The performance of classification iteration stops whenever there is no significant Chi-Square value between the dependent variable and conditioning factors. For that reason, highest Chi-Square value nodes comes first in the resultant tree, whereas, the terminal nodes carry the lowest Chi-Square values. CHAID method considers a statistical test that copes with data type and target nature. Like categorical data, Equation 1 of Pearson chi–squared will be used:

$$X^2 = \sum_{j=1}^{J} \sum_{i=1}^{I} \frac{(n_{ij}-m_{ij})^2}{m_{ij}} \tag{1}$$

where,

$$n_{ij} = \sum_{n \in D} fnI(x_n = i \cap y_n = j)$$

n_{ij} ,is the observed cell frequency and
m_{ij}, is the estimated expected cell frequency for $(x_n = i, y_n = j)$ following the independence model. The corresponding p-value given by p= $Pr(x_d^e > x^2)$[16].
The output of the CHAID method has the tree structure which consists of branches that represent automated classification of each significant conditioning factor, with a specific percentage of landslides (1) and non-landslides (0), in each node. Top of the tree is composed by a unique cell carry node (# 0) which represents the dependent variable information, such as the total number of training sites (landslides and non-landslides). The rest of the tree is formed by nodes representing the conditioning factors classification, which carries certain information such as node number, class interval limit, counts of landslides and non-landslides.
Pre-setting of the model's criteria is essential, because that will positively or negatively affect the tree size and processing time. Here, the main criteria components are; growth limit, merging value, and parent and child nodes[17]. The minimum value of each criterion represents the limit of stopping the tree growth process.

3.3 Logistic regression (LR)

Logistic regression (LR) relies on measuring the results with dichotomous variables such as 1 and 0 or true and false. Moreover, LR builds a statistical model to predict the logit transformation of dependent variable (landslide) occurrence probability [8, 18].

LR compute the changes in the likelihood that falls in each category to find the best fitting model to describe the relationship between landslide conditioning factors, then ranking the factors according to the highest numerical code among dependents [19]. Equation 2 was used to fit the dependent variables:

$$z = b_o + b_1 x_1 + \ldots\ldots + b_p x_p, \tag{2}$$

where, *Z* represents the linear combination of independent variables, absence or presence of a landslide, and variables value from $-\infty$ to $+\infty$. By contrast, fewer parameters with fewer cells are advisable for short and reliable operation process and reasonable equation limits.
The probability (*p*), estimate the probability of occurrence, denotes any pixel that is susceptible to slope failure, and it can be represented as the conditional probability in the LR model by the following expression:

$$\text{Logit}(p) = \ln\left(\frac{p}{1-p}\right) = \frac{1}{1+e^{-z}}, \tag{3}$$

where, p is the number of independent variables, b_0 is the constant or intercept of the equation, and $b_1, b_2,\ldots, b_p$ are the coefficients of the independent variables $x_1, x_2,\ldots, x_p$.

The independent binary variable with values of one or zero indicates the presence or absence of a category [20]. A forward stepwise logistic regression approach was used. Finally, the regression indices of the predictors were imported to ArcGIS software, to perform the final probability map using Equations 2 and 3.

4. Results and discussion

CHAID was successfully applied using the 130 cluster pattern landslides representing 2500 cells tested by NNI. A total of 14 causative factors (such as slope, aspect, land cover, soil type, lithology, altitude, NDVI, curvature, surface roughness, SPI (Stream Power Index), distance from road, distance from faults, distance from drains, and precipitation) represented the independent variables. Model criteria setting was carried out for optimizing the processing time and achieving the desirable results. Then Pearson Chi-square statistic test was used to control the classification significant level of CHAID result [9]. Most of the conditioning factors that showed a strong logical relationship with slope failure were classified by CHAID. The terminal nodes that represents the optimum decision consist of SPI, soil type, slope, aspect, roughness, distance from faults, distance from road, distance from drainage, land cover, elevation.

In the next step, terminal nodes were nominated according to the susceptibility level against landslide using LR method.

Multi-co linearity test had to be performed (i.e., ANOVA of inflation factor, tolerance, and co-linearity index). Test shows that all terminal spatial factors are not interdependent (Tol > 0.3) [21]. The maximum likelihood estimation method was used for coefficient extraction by an iterative process. The analysis was carried and the result showed a highest significant relationship was found between landslide occurrence and the following conditioning factors; SPI, soil type, land cover, distance from road, distance from fault, elevation, and distance from drainage.

The difference in the −2 log likelihood (−2LL) is considered as an indicator of model improvement over the null model. The lowest value of the −2LL represents the best step fit of the model to the data and explains the value decrement until the final iteration step. Cox/Snell's and Nagelkerke's R-square was used to measure the usefulness of the model. A higher R-square value of Cox/Snell and Nagelkerke corresponds to a better model (Table 1).

Table 1. Statistics summary of LR model.

Model Summary				Hosmer and Lemeshow Test		
Step	-2 Log likelihood	Cox & Snell R Square	Nagelkerke R Square	Chi-square	df	Sig.
12	3573.843^b	.048	.064	21.973	8	.005

a. Estimation terminated at iteration number 3 because parameter estimates changed by less than .001.

b. Estimation terminated at iteration number 4 because parameter estimates changed by less than .001.

Table 2 shows the Wald statistics for each model, the highest Wald value, implies the most significant value for the occurrence of landslides. The P-value represents the anticipated probability of rejecting the null hypothesis, and defined by statistical relationship between variables at the 95% confidence level. Hence if the P-value is less than 0.05 then the null hypothesis is likely not true.

Table 2. LR model statistics of the regression.

Variables in the Equation									
		$\mathcal{B}$	S.E.	Wald	df	Sig.	Exp(B)	95% C.I.for EXP(B)	
								Lower	Upper
Step 12[1]	spi	.009	.002	27.682	1	.000	1.009	1.006	1.012
	soil	.005	.002	9.020	1	.003	1.005	1.002	1.008
	road4	.014	.003	19.267	1	.000	1.014	1.008	1.021
	lcover2	.011	.003	11.000	1	.001	1.011	1.004	1.018
	fault6	-.009	.002	19.437	1	.000	.991	.987	.995
	drain1u	.004	.001	8.261	1	.004	1.004	1.001	1.006
	elev3	.007	.002	8.819	1	.003	1.007	1.002	1.011
	fault2	-.010	.004	6.908	1	.009	.990	.983	.997
	road2	.006	.002	9.341	1	.002	1.006	1.002	1.010
	lcover	-.005	.002	5.685	1	.017	.995	.990	.999
	soil3	.004	.002	5.342	1	.021	1.004	1.001	1.007
	spi1u	-.004	.002	4.714	1	.030	.996	.993	1.000
	Constant	-1.505	.508	8.786	1	.003	.222		

Finally, the susceptibility map prepared by multiplying the $\mathcal{B}$ (probability weight) values, which represent the positive/ negative weight of factors in LR . The prediction rate curve test was carried out on the integrated landslide susceptible map LSM (Figure 3a) using the validation data. LSM accuracy showed AUC=70% (Figure 3b). The prediction value of CHAID classification showed 73.4% of landslides locations and 73.3% for non-landslide locations (Table 3). The reason behind that low prediction accuracy of LSM, is the large study area compared by number/size of evidence (training data). Large difference of study area to landslide inventory (landslide locations area to study area equal to 1:6,512), the classification tree show 151 of terminal nodes. Thus we recommend of using a reasonable ratio between size and number of landslide and scale of study area, good example of the reasonable ratio was found in previous research done by the same authors [9].

Table 3. CHAID prediction value.

Risk		**Classification**			
		Observed	Predicted		
Estimate	Std. Error		0	1	Percent Correct
.267	.006	0	1843	673	73.3%
		1	666	1836	73.4%
		Overall Percentage	50.0%	50.0%	73.3%

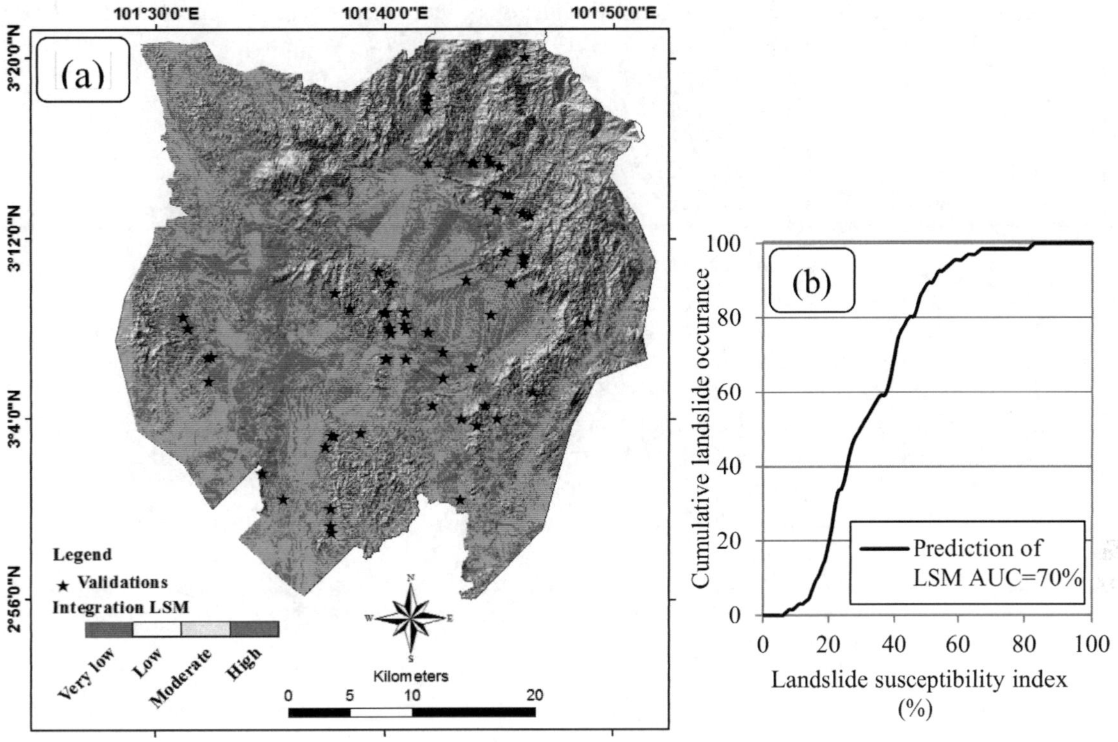

Figure 3. (a) Landslide susceptibility map (b) Prediction rate curve.

5. Conclusion

Highly hazardous nature of landslides and disastrous consequences especially on the socio-economic sectors, has become an indisputable fact worldwide. This has led for the planners to develop solutions on a short-and long-term basis. In recent years, landslide prediction mapping s has moved by bounds and leaps by utilizing the computation power of GIS. The main objective of this research was to use the DT-based CHAID method to perform the best classification fit for each landslide conditioning factor, then terminal nodes of the CHAID tree were integrated with LR model to find the corresponding coefficients of best fitting function, and also to assess the optimal terminal node. The classification prediction of CHAID showed 73% of the total landslide location successfully predicted. Then LSM was produced by integrating the terminal nodes in the LR model. Prediction rate curves test was used to evaluate the prediction performance of LSM, and it achieved AUC=0.70. As a conclusion, the model showed a good fit for logical automatic classification for landslide conditioning factors, without any interference from experts. Moreover, to facilitates the design productivity for roads embankment, slope rehabilitations and design, which may be useful in decision making.

Acknowledgement

The authors gratefully acknowledge the financial support from the UPM-RUGS project grant number: 6300114.

References

[1] P. Aleotti, "A warning system for rainfall-induced shallow failures," *Engineering Geology,* vol. 73, pp. 247-265, 2004.

[2] D. T. Bui, *et al.*, "Regional prediction of landslide hazard using probability analysis of intense rainfall in the Hoa Binh province, Vietnam," *Natural Hazards,* vol. 66, pp. 707-730, 2013.
[3] Y.-K. Yeon, *et al.*, "Landslide susceptibility mapping in Injae, Korea, using a decision tree," *Engineering Geology,* vol. 116, pp. 274-283, 2010.
[4] J. R. Quinlan, *C4. 5: programs for machine learning* vol. 1: Morgan kaufmann, 1993.
[5] J. Magidson, *SPSS for Windows CHAID reléase 6.0*: SPSS Incorporated, 1993.
[6] A. S. Bozkir and E. A. Sezer, "Predicting food demand in food courts by decision tree approaches," *Procedia Computer Science,* vol. 3, pp. 759-763, 2011.
[7] Terratech, "Inventory of forest landslide occurrence in the kamloops forest region," Terratech Western profile consultants LTD1994.
[8] B. Pradhan, "Landslide susceptibility mapping of a catchment area using frequency ratio, fuzzy logic and multivariate logistic regression approaches," *Journal of the Indian Society of Remote Sensing,* vol. 38, pp. 301-320, 2010.
[9] O. F. Althuwaynee, *et al.*, "A novel ensemble bivariate statistical evidential belief function with knowledge-based analytical hierarchy process and multivariate statistical logistic regression for landslide susceptibility mapping," *Catena,* vol. 114, pp. 21-36, 2014.
[10] S. Menard, "Coefficients of determination for multiple logistic regression analysis," *American Statistician,* vol. 54, pp. 17-24, 2000.
[11] L. Ayalew and H. Yamagishi, "The application of GIS-based logistic regression for landslide susceptibility mapping in the Kakuda-Yahiko Mountains, Central Japan," *Geomorphology,* vol. 65, pp. 15-31, February 1, 2005 2005.
[12] B. Pradhan, "Manifestation of an advanced fuzzy logic model coupled with Geo-information techniques to landslide susceptibility mapping and their comparison with logistic regression modelling," *Environmental and Ecological Statistics,* vol. 18, pp. 471-493, 2011.
[13] O. F. Althuwaynee, *et al.*, "Application of an evidential belief function model in landslide susceptibility mapping," *Computers & Geosciences,* vol. 44, pp. 120-135, 2012.
[14] O. F. Althuwaynee, *et al.*, "Prediction of slope failures using bivariate statistical based index of entropy model," in *Humanities, Science and Engineering (CHUSER), 2012 IEEE Colloquium on*, 2012, pp. 362-367.
[15] StatisticsSolutions. (2012). *CHAID.* Available: http://www.statisticssolutions.com/academic-solutions/resources/directory-of-statistical-analyses/chaid/
[16] S. Baker and R. D. Cousins, "Clarification of the use of CHI-square and likelihood functions in fits to histograms," *Nuclear Instruments and Methods in Physics Research,* vol. 221, pp. 437-442, 4/1/ 1984.
[17] I. S. Naftulin and O. Y. Rebrova, "Application of C&RT, CHAID, C4. 5 and WizWhy algorithms for stroke type diagnosis," in *Artificial Intelligence and Soft Computing*, 2010, pp. 651-656.
[18] K. Devkota, *et al.*, "Landslide susceptibility mapping using certainty factor, index of entropy and logistic regression models in GIS and their comparison at Mugling–Narayanghat road section in Nepal Himalaya," *Natural Hazards,* vol. 65, pp. 135 -165, 2013.
[19] S. Bai, *et al.*, "Combined landslide susceptibility mapping after Wenchuan earthquake at the Zhouqu segment in the Bailongjiang Basin, China," *Catena,* vol. 99, pp. 18-25, 2012.
[20] G. C. Ohlmacher and J. C. Davis, "Using multiple logistic regression and GIS technology to predict landslide hazard in northeast Kansas, USA," *Engineering Geology,* vol. 69, pp. 331-343, 2003.
[21] M. Van Den Eeckhaut, *et al.*, "Statistical modelling of Europe-wide landslide susceptibility using limited landslide inventory data," *Landslides,* vol. 9, pp. 357-369, 2012.

IOP Conf. Series: Earth and Environmental Science **20** (2014) 012033 doi:10.1088/1755-1315/20/1/012033

A survey of drought and variation of vegetation by statistical indexes and remote sensing (Case study: Jahad forest in Bandar Abbas)

E Tamassoki[1], Z Soleymani[2], F Bahrami[3] and H Abbasgharemani[4]
[1] M.S.C graduated of de-desertification, Hormozgan University, Bandar Abbas, Iran
[2]M.S.C student of de-desertification, Hormozgan University, Bandar Abbas, Iran
[3]M.S.C student of watershed management, Hormozgan University, Bandar Abbas, Iran
[4]M.S.C student of remote sensing & geographic system, Islamic Azad University, Yazd, Iran

Email: Zahra66.soleymani@gmail.com

Abstract. The damages of drought as a climatic and creeping phenomenon are very enormous specially in deserts. Necessity of management and conflict with it is clear. In this case vegetation are damaged too, and even are changed faster. This paper describes the process of vegetation changes and surveys it with drought indexes such as statistical and remote sensing indexes and correlation between temperature and relative humidity by Geographical Information System (GIS) and Remote Sensing (RS) in forest park of Bandar Abbas in successive years. At the end the regression and determination-coefficient for showing the importance of drought's survey are computed. Results revealed that the correlation between vegetation and indexes was 0.5. The humidity had maximum correlation and when we close to 2009 the period of droughts increase and time intervals decrease that influence vegetation enormously and cause the more area lost its vegetation.

1.Introduction

Drought is a climatic phenomenon that is known as a creeping ones because its effects are revealed slow. Iran are influenced by drought in recent decades, alternatively and its effects on human life are obvious [11] as its occurrence frequency is noticeable especially in arid and semi- arid regions. One of these region is BandarAbbas that has most occurrence (50%) [9]. Asiyayi [2] points out that each index is valid just for specific operation in specific regions because of variable meteorological conditions are responsible for creating drought, so we can use same drought indexes in same climatic regions.
Presenting appropriate and long- term data of climatic and hydrological parameters is really necessary. It is impossible to study drought without these data. Although precipitation is main factor in control,

Content from this work may be used under the terms of the Creative Commons Attribution 3.0 licence. Any further distribution of this work must maintain attribution to the author(s) and the title of the work, journal citation and DOI.
Published under licence by IOP Publishing Ltd

creation, extension and continuation of drought, evaporation and transpiration are most important factors that can explain drought behavior. In spite of above, precipitation is known as a most appropriate and quite accessible climatic parameter for creating and computing drought indexes because of calculating problems of evaporation and transpiration. In fact, indexes that are based on precipitation data are more suitable in comparison with complicated hydrological indexes. Standardized Precipitation Index (SPI) is accuracy and validity enough that is used for Iran, Khorasan province[2] that is accepted by scientific societies and operators so SPI is incorporated in this study. Because of drought and its environmental damages are not restricted to a region and they can risk the whole environment in long and medium terms, finding scientific solutions for reducing drought impacts is essential. In regard to vegetation significance for governmental and nongovernmental organization in the world and critical problem of removing green spaces due to drought and other disasters such as flood and fire that are followed by environmental damages and desertification, providing more information about vegetation conditions such as its amount and dispersal is vitally important. To use satellite data, extensive study on vegetation is possible. Vegetation indexes can be considered for declining radiation effects of other ground phenomena and major concern in vegetation information. In fact, vegetation indexes are demanded for distinguishing vegetation from other phenomena that it is not possible by single- bands [10]. Vegetation indexes make it possible to apply mathematical analysis on data. To operate Geographical Information System (GIS) and Remote Sensing (RS), we can study these regions. Capability of GIS in integrating layers that are made in RS is more effective than traditional procedures and results are more clear and more accurate [4] and to use of them we can assess vegetation changes in successive years. Normalized Difference Vegetation Index (NDVI) is sought to survey and assess vegetation [12]. Drought intensity is determined in a 30 years statistical period for 20 stations in 3 temporal scales, 6, 12 and 24 months that results indicated Fars province had trouble with drought in many years and its intensity in recent years is much more than prior years and other results reveal, short- term droughts have much fluctuations and variation and they are more sensitive to humidity variations and severe droughts have long continuation periods in long term temporal scales. Tipaniat and Nitin [12] surveyed the area variation of Mangro Forests by ETM^+ of Landsat satellite RGB- NDVI (Red-Blue-Green Normalized Difference Vegetation Index) classification technique, in Kraby, Thailand.

Feyzizadeh and Hajimirrahimi [6] applied TM images of Landsat and HDR (High Dynamic range) of Spot satellites and studied green spaces variation of Tabriz and stated that more than 46% of green spaces of Tabriz had destroyed in a years of interested time period. Some researchers [14-13- 8- 3- 15] surveyed the reply of plants to weather fluctuations by NDVI (Normalized Difference Vegetation Index).The correlation between NDVI and precipitation is estimated almost 200-500 mm in growth season [7]. Yang et al [15] studied correlation between temperature and NDVI (Normalized Difference Vegetation Index) of Nebraska in U.S.A and found out a high correlation between NDVI (Normalized Difference Vegetation Index) and soil temperature so it can be confirmed that temperature has a direct impact on plant growth. Sensors of satellites receive different information from vegetation that different vegetation to some extent can be distinguished by analyzing these information.

One of the most important factors that influence vegetation conditions are climatic factors. One of the most significant applications of remote sensing is assessing quantitative correlation between vegetation and climatic factors in regional and world scales. Precipitation and temperature influence on water equation directly and affect soil moisture and plant growth [1].

Assessment of space correlation between NDVI (Normalized Difference Vegetation Index) and climatic factors is employed severally and there is a high correlation between NDVI and rainfall extent in dry lands. The correlation between NDVI and temperature is less but considerable. Regarding such as these researches, correlation between precipitation and NDVI is discovered and basing on them, we can acknowledge that weather variations causes vegetation variations. The main purpose of our research is drought survey with vegetation variations of Jahad Bandar Abbas Forest park, using satellite images and correlation between drought indexes and climatic factors and vegetation variations in 25 years time period.

2.Material and Methods

Case study: The case study selected for studying is Jahad forest park, a 2190 ha area located in Bandar Abbas city, Hormozgan Province. It is constructed in 1993 and its central portion is implemented at the distance of 8 kilometers from east of city. This park is situated between 560 24 23 and 560 24 33 of east length and 27 16 37 and 27 16 54 north width (Figure 1). This park ends to the old path of Bandar Abbas in north, national properties in south and Hormozgan university in east. A very good cover with *Prosopis juliflora* is observed in This area. The north of region were planted few years ago. *Hammada saliconiaca* can be mentioned that it grows in dry lands and deserts. The natural and native species such as *Prosopis Mimosaceae*, *Acacia ehrenbergiana*, *Ziziphus spina-christi*, *Acacia nilotica* and *Prosopis juliflora* are covered this region. Besides ecological and conservative impacts, this cover makes a beautiful view for region.

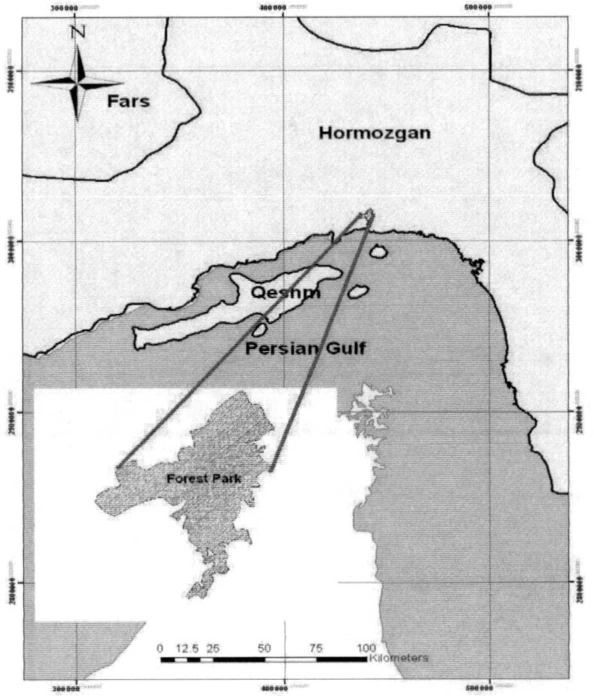

Figure 1: Geographical position of Jahad Forest park of Bandar Abbas.

The same and long term time base of statistical periods are used in order to survey and comparison of drought indexes.

3.Standardized Precipitation Index (SPI)

This index is presented by Mackee et al in 1995. Many researchers point out to its flexibility of drought-based. SPI index (Standardized Precipitation Index) is computed in base of long-term precipitation registeration for each region. For determining SPI index, first step is suiting a αdistribution to precipitation data for each station, next step is converting accumulation probability from acquired αdistribution to accumulation standard normal distribution with zero average and one variation [10]. Positive amounts of SPI (Standardized Precipitation Index), show more than precipitation medium and negative amounts, indicate less than precipitation medium. Whenever SPI amounts are negative and show -1 or less, drought is occurred and its positive ones reveal the end of drought (Table 1) [2].

$$g(x) = \frac{1}{\beta^x \Gamma(\alpha)} x^{a-1} \varepsilon^{-x/\beta} \qquad x > : \alpha \qquad (1)$$

$\alpha = Shape\ parameter$

$\beta = scale\ parameter$

$x =$ precipitation amount

$\Gamma(\alpha) = \alpha$ function

Table1: Classification of drought based on SPI index.

Classification of SPI index	
Quite severe wet year	2 <
Very moist	1.55to 1.99
Medium wet year	1 to 1.49
Almost normal	0.99 to -0.99
Medium drought	-1to -1.49
Severe drought	-1.99 to -1.5
Quite severe drought	-2>

4.Normalized Difference Vegetation Index (NDVI)

NDVI index sometimes is called NVI [2]. In this research ENVI 4.7 software of remote sensing was operated for extracting vegetation index from Tm-5 sensor images with 30 m resolution that was downloaded from http://earthexplorer. Usgs.gov/. All of 3 downloaded images belonged to May, in turn there was possibility of comparison and monitoring of changes. The temporal point of images were 15 May 1984, before establishing park, 22 May 1998, 5 years after establishing park and 20 May 2009, 16 years after establishing park. During this temporal extent, region was faced with several drought periods. The NDVI of three images above is more than 0.5. In fact, NDVI reveals density of vegetation. More NDVI demonstrates denser vegetation. Whereas vegetation indexes are designed based on spectrum features of vegetation, amounts resulted from them reflects vegetation richness within aimed pixel [2] and NDVI (Normalized Difference Vegetation Index) is calculated as:

$$NDVI = NIR - R/NIR + R \qquad (2)$$

Then, we took regression between images by EDRISI KLIMANJARO software and results were analyzed. At the end, SPSS statistics 17 was applied for statistical analysis of data of climatic parameters, statistical indexes and normalized vegetation index.

5.Results

As results indicates, during 25 years period, 1984, 2003, 2004, 2007 had mean drought periods and 1985, 1994, 2001 had severe drought. Closing to 2009, drought periods have increased and interval between them have decreased (table 2 and figure2).

Table 2: Data of Standarded Precipitation index.

Year	Intensity of drought	SPI index
1980	almost normal	-0.707637316
1981	almost normal	-0.032190447
1982	almost normal	0.975114317
1983	almost normal	-0.32118949
1984	mean drought	-1.01158475
1985	sever drought	-1.739238441
1986	almost normal	-0.754647306
1987	almost normal	-0.416620251
1988	almost normal	-0.058180841
1989	almost normal	0.258038969
1990	almost normal	-0.610751044
1991	almost normal	0.164224222
1992	almost normal	0.782053717
1993	almost normal	0.175548352
1994	sever drought	-1.981818637
1995	almost normal	0.228822869
1996	almost normal	0.66937414
1997	almost normal	0.708254908
1998	almost normal	-0.010440846
1999	almost normal	-0.866809398
2000	almost normal	-0.021277067
2001	sever drought	-1.68472519
2002	almost normal	-0.854047686
2003	mean drought	-1.178597318
2004	mean drought	-1.323182471
2005	almost normal	-0.462500596
2006	almost normal	-0.425006386
2007	mean drought	-1.3438112
2008	almost normal	-0.960339275
2009	almost normal	-0.105438931

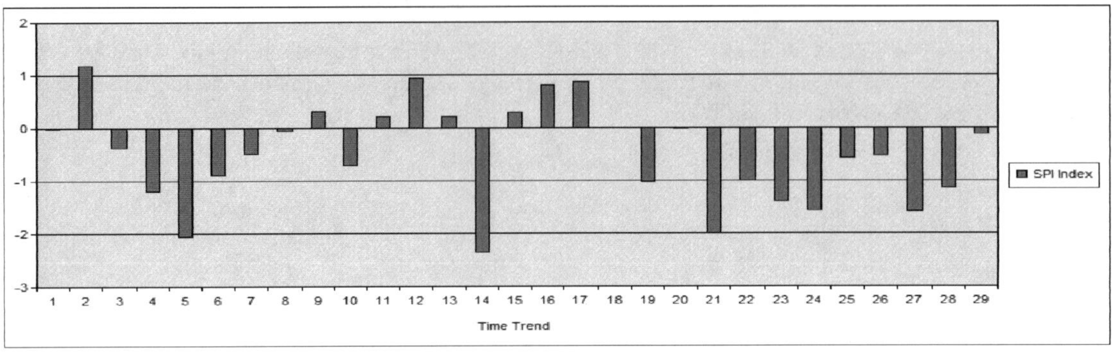

Figure2: Column Diagram of Standardized Precipitation Index during 1980-2009 in Hamdidi Bandar Abbas Station.

6.Normalized Difference Vegetation Index

Different amounts of NDVI (Normalized Difference Vegetation Index) display different Vegetation, for instance NDVI of dispersed Vegetation start from 0.05 [5]. Considering Jahad Bandar Abbas Forest park is placed within warm coastal dry lands, this research assessed Vegetation with NDVI more than 0.05. In image of 15 May 1984, before park established, most of Vegetation is wild. Drought indexes reveal that there is not any drought in region between 1980-1984. In this time NDVI more than 0.05 cover almost 6.5 percent of area that is equal with 146.4859 ha (Figure 4). In second image of 22 May 1998, 5 year after park established, planting of trees was started by natural resources office. Drought indexes show there is no continues drought between 1984-1998 in region except 1985 and 1995 that took place quite severe drought. In this time up to 1998 the Vegetation with NDVI more than 0.05 establish in 44.25 percent of area that is equal with 969.3614 ha, it means the Vegetation is increased equivalent to 82.8755 ha in this interval time (Figure 5).

In third image of 20 May 2009, 16 years after park established, drought indexes indicate drought since 1998 to 2009. In this temporal extent, 16.16 percent of area have vegetation with NDVI more than 0.05 that is equal with 354.1106 ha (Figure 6). Extracting of NDVI (Normalized Difference Vegetation Index) found vegetation increasing between 1984 to 1998, during this time there were no continuous drought periods but region was faced with continuous drought periods after 1998, in turn vegetation reduced (Figure1).

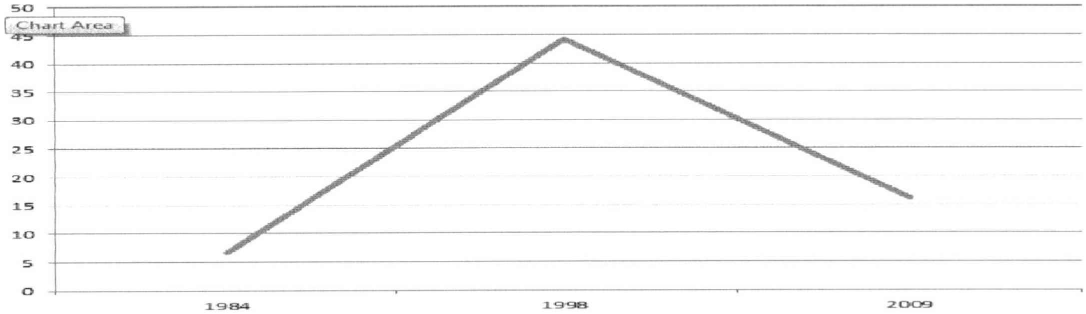

Figure3: Vegetation percentage changes during 25 years as ha.

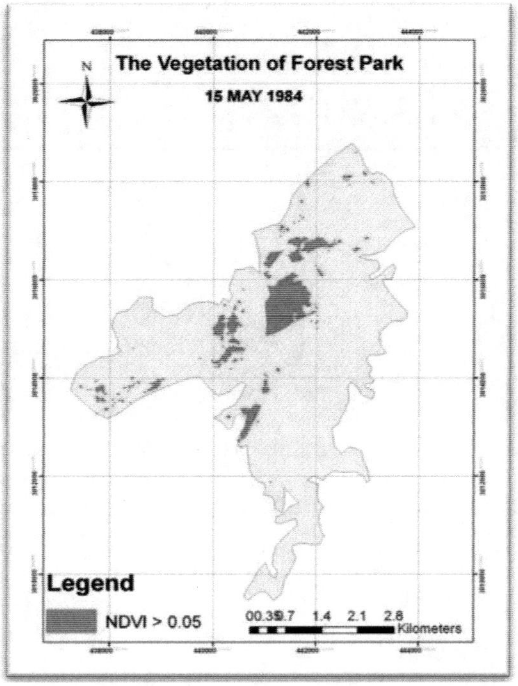

Figure 4: Vegetation index of 15 May 1984.

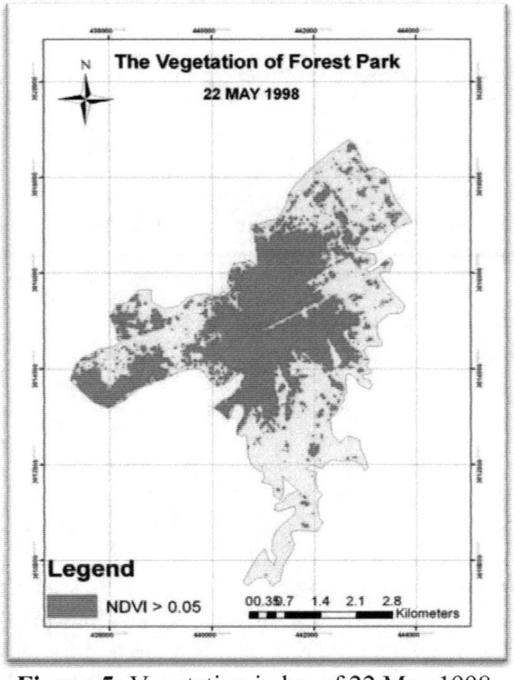

Figure 5: Vegetation index of 22 May 1998.

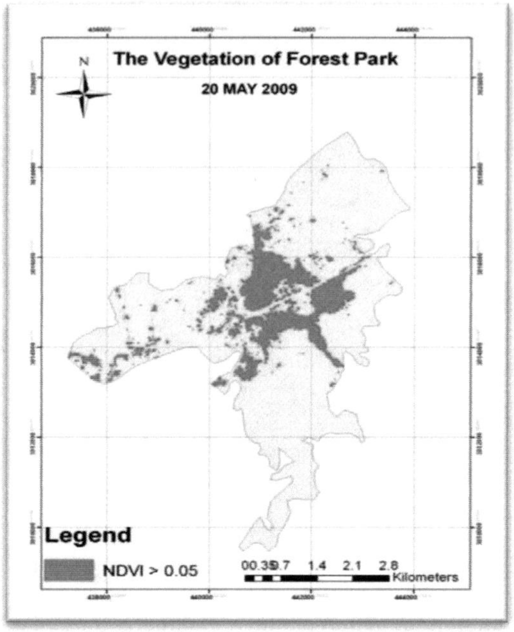

Figure 6: Vegetation index of 20 May 2009.

7.Discussion and conclusion

Statistical analysis of area vegetation, climatic parameters such as temperature and relative humidity and drought indexes of objected region exhibit most correlation between vegetation area and relative humidity as 0.94 and after that most correlation is between vegetation area and temperature, there is also fairly high correlation with precipitation and drought indexes (table3).

Table 3: Correlation matrix between vegetation area, climatical parameters and indexes.

	Relative humidity	temperature	Precipitation	**SPI**
correlation matrix	0.94	0.91	0.77	0.75

After estimating vegetation in 1984, 1998 and 2009 and extracting NDVI, each image entered in Idrisi Kilimanjaro software for statistical calculations.

It has been found that correlation between NDVI of 1984 and 1998 as 0.53, whereas determination coefficient between these two years is 0.28, this points out that 0.28 percent of NDVI of 1998 is same with NDVI of 1984. Also correlation examination between NDVI of 1998 and 2009 ones points to correlation as 0.49 with determination coefficient as 0.24, this indicates a same vegetation between 1998 and 2009 as 0.72 with determination coefficient as 0.51 that denotes a common negotiation as 0.51 between 1998 and 2009 (table 4).

Table4: Correlation coefficient matrices (r) and determination coefficient (R^2) between vegetation.

Year	**R**	**R^2**	**R**	**R^2**	**R**	**R^2**
1984	1	1	0.53	0.28	0.49	**0.24**
1998	0.53	0.28	1	1	0.72	**0.51**
2009	0.49	0.24	0.72	0.51	1	**1**

As results are obvious, after each period, vegetation changes and its area percentage reduced considerably. It is likely to continue vegetation reduction process that it demonstrate necessity of monitoring and drought management because not only this issue cause environmental problems but also is followed by social- economic problems that their effects are uncompensated or it is compensated, in long way. For drought effects, main components of adjustment and preparing, should contain, drought monitoring, primary awareness, damage estimation and response to it.

Although a disaster like drought does not display its effects suddenly and endangers human's life and properties in long time, with appropriate information about drought monitoring and its beginning time we can assess its intensity and extension and restrict its calamity and reduce environmental and economical damages [2].

8.Reference

[1] Akhtari R, Mahdiyan H and Morid Sh 2006 Analysis spatial drought indexes SPI and EDI in Tehran province, Iran water resource resreches vol 3 pp:27-38.
[2] Asiyayi M 2006 Drought indexes, Sokhan Gostar press..
[3] Di L Rundquist, D C and Han L 1994,Modelling Relationships Between NDVI and Precipitation During Vegetation Growth Cycles, International Journal of Remote sensing Vol 15 pp:1221-2136.
[4] Farzaneh a 1992 GIS application in natural resources Forest and range magazine 14 pp12-15.
[5] Fatemi s and Rezayi Y 2006 remote sensing basics Azadeh Press.
[6] Feyzizadeh B and Hajimirrahimi 2007 Revelation green spaces of Tabriz using objectivity showing procedures of urbon GIS, Collection papers of first conference on urban GIS
[7] Milich, L., and Weiss, E.,2000, GAC NDVI Images: Relationship to Rainfall and Potential Evaporation in the Grazing Lands of The Gourma (Northern Sahel) and in the Croplands of the NIGER-Nigeria Border(Southern Sahel) International Journal of Remote Sensing, 2 pp:261-280.
[8] Mola A R and Nicholson S E 1990 A Study of Rinfall and Vegetation Dynamics in the African Sahel Using Normalized Difference Vegetation Index Journal of Arid Environments 19, pp: 1-24.
[9] Nazemossadat S 2001 Does it rains? Extra drought and precipitation in Iran and theire relation with elnino- south fluctuations Shiraz University Press P120.
[10] Salajegheh A Fateh Abbadi A and Najafi Haji Vorm A 2008 compersion of numerial nervous and time series in drought predication (Case Study: Khorasan Razavi province), Iran Seience and watershed management engineering magazine.
[11] Shamsipoor A Alavi Panahe S Mohammadi H 2010 Assessment of vegetation analysis in Kashan and thermo indexes of Noaa- AVHRIR in drought effectiveness.

[12] Tipaniat,V and KTNitin.2003 Asatellite based Monitoring of Changes in Mangroves in Krabi, Thailand,Proceedings of Map Asia Conference,Malaysia.

[13] Tucker C J Vanpraet C L Sharman M. J and Van Ittersum, G 1985 Satellite Remote Sensing of Total Herbaceous Biomass Production in the Senegalese Sahel 1980-1984 Remote Sensing of Environment 17 pp: 233-249.

[14] Yang L Wylie B Tieszen, L.l and Time-integrated NDVI of Grassland over the U.S. Northern and Central Great Plains Remote Scencenig Environ 65 pp:25-37.

[15] Yang Y Yang L and Merchant J W 1997 AN Assessment of AVHRR/NDVI Ecoclimatological Relation in Nebraska USA International Journal of Remote Sensing 18 pp:2161-2180.

IOP Conf. Series: Earth and Environmental Science **20** (2014) 012034 doi:10.1088/1755-1315/20/1/012034

Building EOS capability for Malaysia – the options

M D Subari[1] and A Hassan[1]
[1]Associate Professors, Perdana School for Policy Study, Universiti Teknologi Malaysia
E-mail: m.subari@utm.my

Abstract: Earth observation satellite (EOS) is currently a major tool to monitor earth dynamics and increase human understanding of earth surface process. Since the early 80s, Malaysia has been using EOS images for various applications, such as weather forecasting, land use mapping, agriculture, environment monitoring and others. Until now, all EOS images were obtained from foreign satellite systems. Realising on the strategic need of having its own capability, Malaysia embarked into EOS development programs in the early 90s. Starting with TiungSAT-1, a micro-satellite carrying small camera, then followed by RazakSAT, a small satellite carrying 2.5 m panchromatic (PAN) medium-aperture-camera, the current satellite program development, the RazakSAT-2, designed to carry a 1.0m high resolution PAN and 4.0m multi-spectral camera, would become a strategic initiative of the government in developing and accelerating the nation's capability in the area of satellite technology and its application. Would this effort continue until all needs of the remote sensing community being fulfilled by its own EOS? This paper will analyze the intention of the Malaysian government through its National Space Policy and other related policy documents, and proposes some policy options on this. Key factors to be considered are specific data need of the EOS community, data availability and the more subjective political motivations such as national pride.

1. Introduction – the EOS applications in Malaysia

Earth observation satellite (EOS) is currently a major tool to monitor earth dynamic and increases human understanding of the earth surface process as well as the atmospheric dynamics. The early applications of remote sensing (RS) sciences and applications in Malaysia were initially for meteorological applications using TIROS/NOAA in the 60s. RS applications for land resources and environmental studies started in 70s with the availability of images from the LANDSAT series of satellites. For an equatorial region such as Malaysia, the main limitation of optical-base images was due to continuous cloud cover over the region and severe raining seasons especially at the beginning and end of year. In the 90s, the use of radar EOS images mainly from the RADARSAT satellites manages to overcome some of the problem. Currently both the optical and radar images are widely used, however, repeat capabilities of each satellite is still not sufficient for operational RS purposes [4].

For some time, satellite images from the LANDSAT and SPOT satellite series, are the most widely used EOS images for various land and marine applications. The SPOT satellites with

Content from this work may be used under the terms of the Creative Commons Attribution 3.0 licence. Any further distribution of this work must maintain attribution to the author(s) and the title of the work, journal citation and DOI.
Published under licence by IOP Publishing Ltd

better spatial resolution and stereo capability have enhanced applications in areas such as topographical mapping. Current QuickBird (0.6m), WorldView-2 (0.5m) and GeoEye-1 (0.5m) sub-meter resolution images has enabled details mapping such as utilities to be carried out. Most of these applications are done post-processed.

For many operational RS applications, of which mostly decision making-related such as disaster-related monitoring and real-time surveillance, where real time or near-real time data supply are needed, these data are not sufficient. Apart from its insufficiency in repeat imaging capabilities, these satellites are mostly polar-orbiting or in Sun Synchronous Orbiting (SSO), hence with longer revisit time, further restricting the use of EOS images for operational purposes. Other shortcomings on the current supply of EOS images includes data availability, normally long-period of delivery, data suitability and data price.

2. **The EOS program – the capacity building way**

Demand of EOS images for various RS applications is ever increasing in both public and commercial sectors, with dominance in the public sector. EOS images and RS applications has become an important administrative tool. Ironically, as highlighted earlier, until now, all EOS images were obtained from foreign satellite systems.

Realising on the strategic need of having its own capability, Malaysia has embarked into the EOS development program since early 90s. Starting with TiungSAT-1, a micro-satellite carrying small camera, then followed by RazakSAT, a small satellite carrying 2.5 m panchromatic (PAN) medium-aperture-camera (MAC), the current satellite program development, the RazakSAT-2, designed to carry a 1.0m high resolution PAN and 4.0m multi-spectral camera, would become a strategic initiative of the government in developing and accelerating the nation's capability in the area of satellite technology and its application.

The question to ask (1) would be; what is the direction of the country on EOS? Would this capability building effort continue until all needs of the RS community being fulfilled by its own EOS?

EOS is normally a government-led program [6], as part of a technology capability building effort, with key objective of this initiative is to fulfill part-of the domestic RS user community requirement on specific spectral frequency and spatial resolution need. It may be mixed also with the more subjective, political motivations such as national pride.

There are debates on whether government in developing countries should just rely on EOS images globally available or the need to own and operate their own satellites [7]. The general conclusion would be to adapt the earlier option, but there are subtle realities, however, that challenge this conclusion. It is true that EOS data are available from the international community which can be shared with countries that do not own satellites, including also from commercial providers that sell high resolution satellite imagery.

The problem remains, however, that developing countries cannot always get the data they need when they need it [7]. This may be because the data collected by other countries does not account for the technical requirements of a particular user in a developing country, such as temporal frequency, spatial resolution, spectral frequency or geographic coverage. Also, organizations from developing countries that wish to share international data may be obliged to establish bilateral agreements with each data producer and keep those agreements up to date in order to ensure access (through licensing agreement), which generally an expensive exercise. Meanwhile, high resolution optical data, which is particularly useful for projects in urban planning, is mainly available from commercial providers and could be very expensive.

Long term solution for the country in fulfilling the need of local RS community is much needed. In the absent of any official policy stand of the government, this could be difficult to answer. This paper will analyze the intention of the Malaysian government in this issue, through its National Space Policy and other related policy documents, and proposes some policy options on this. Key factors to be considered are specific data need of the RS community, data availability and the more subjective political motivations such as national pride.

3. The EOS program roadmap - the basis

The satellite development program

Malaysia has ventured into satellite making since mid-1990 through the purchase of its first two communication satellites, MEASAT-1 and MEASAT-2 from Boeing Ltd. The MEASAT satellites are owned by Binariang Sdn. Bhd. (now Measat Satellite System Sdn. Bhd.), a private entity providing satellite communication-based services in the country as well as globally. Although it is a commercial purchase, some local engineers were sent to Boeing during the development period, to acquire some level of knowledge and know-how in satellite making.

The real transfer-of-technology (ToT) program in satellite technology was done with Surrey Satellite Technology Ltd. (SSTL), United Kingdom in the 90s. 10 engineers from Malaysia were attached with SSTL throughout the development of TiungSAT-1, a micro-satellite which was launched in 2000 and operated for more than 3 years. This project was undertaken by the Space Science Studies Division (BAKSA), a division then under the Ministry of Science, Technology and Environment (MOSTE) Malaysia. Subsequently, a government-linked company (GLC), Astronautic Technology Sdn. Bhd. (ATSB) was formed as a special vehicle to continue undertaking the task of capacity building in satellite technology for the nation.

The National Space Agency (ANGKASA) of Malaysia (incorporating BAKSA) was formed in 2003 to spearhead the development of the space sector for the country. Utilizing grant from ANGKASA, ATSB developed RazakSAT (2004-2009), an R&D remote sensing satellite jointly with a South Korean company, SatRecl. RazakSAT, carrying a MAC was successfully developed, launched, going through its Launch and Early Operation Phase (LEOP) and operated for about eight months. Unfortunately it ceased operation in mid-2010 due to some components malfunctioned, believed to have been caused by the excessive exposes to the solar radiation particles in the region of the South Atlantic Anomaly. During the course of the development and operations of RazakSAT, about 100 engineers and scientists were involved, plus few tenths more of expertise in related areas such as satellite operations and image processing.

The current EOS development program, the RazakSAT-2 satellite, is a continuation of the nation's strategic capacity building initiatives in satellite technology. A development project under ANGKASA (2013-2015), with ATSB as the main contractor, the intended outcome of this project is not only an operational high resolution remote sensing satellite, but achieving the total spectrum of the capacity building aspects in satellite technology for the country.

The National Space Policy

Although work on the development of such policy has started way back in 2005, until now (2013) the National Space Policy (NSP) for Malaysia has not been officially endorsed. It is supposedly the strategic document that charted the way-forward and providing the framework of development in the space sector for the country [6].

The document envision on building the nation's capability to embrace space as a strategic sector for the national wellbeing towards achieving Vision 2020 and beyond. This could be achieved by developing the country's potential in the space sector to support the development of the new economy, and strengthening the national security. The listed objectives of the policy are;

1. To build up space infrastructure and industry for economic benefit and safeguarding the nation's sovereignty.
2. To empower the civil society in enriching their quality of life through information from innovative application of space technology.
3. To capitalize/harness on space as the frontier for new knowledge generation towards contributing to scientific and technological advancement.
4. To ensure critical mass of talent in space related sector to support the realization of the space policy.

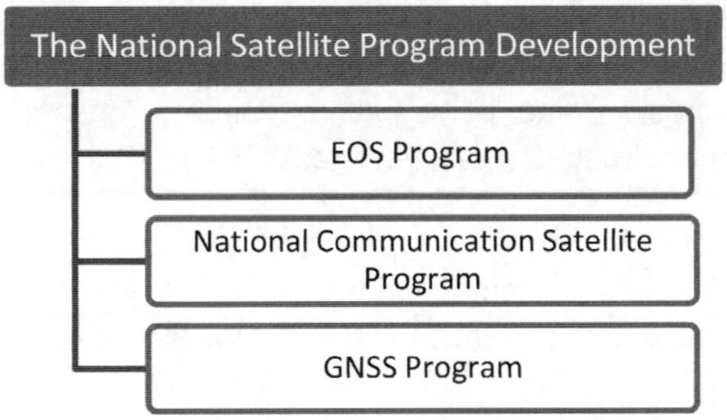

Fig.1: The National Satellite Program Development.

The National EOS program

The NSP draft outlined the direction of the satellite program development for the country until 2020. These are under Program 3: Accelerating Space Technology and Infrastructure Development, with related plan of action in the National Satellite Program Development that includes the EOS Program, the National Communication Satellite Program and the Global Navigation Satellite System (GNSS) Program.

Under the EOS program, it is outlined that;

> *'The Malaysia's EO satellite series (RazakSAT) should be continued because it will play a very important role in resources investigation and the monitoring of environment disasters. EO satellite data can be applied to numerous fields as agriculture, forestry, water resources, land use, cartography, mineral exploration and ocean investigations. The technology and expertise acquired in developing the EO satellite can then be applied to the development of the more sophisticated communication satellite'.*

4. The EOS capacity building program - to build or to acquire?

The current EOS development program, the RazakSAT-2 satellite, is a continuation of the nation's strategic capacity building initiatives in satellite technology. A development project under ANGKASA (2013-2015), with ATSB as the main contractor, the intended outcome of this project is not only an operational high resolution remote sensing satellite, but achieving the total spectrum of the capacity building aspects in satellite technology for the country.

The question to ask (2) would be; would this effort of capacity building continue until all needs of the local RS community being fulfilled by our own EOS?

The ability of a nation to drive innovation and economic progress depends on its science and technology capacity. This was the thought process for the development of the National Science and Technology Policy (NSTP), the second NSTP (NSTP II) and the National Action Plan for Industrial Technology Development (Fadzilah and Krishna, 2007). The ability shall include all factors including the manpower and personnel to conduct research and to carry out the technological development, infrastructure, investment, institutional arrangement and regulatory framework.

Satellite technology development has long been recognized worldwide as an indicator of science and technology ability. It has been the domain of technologically advanced countries and regions such as the United States, Europe, Japan and recently China. However an increasing number of developing countries, Malaysia included, have started to enter this strategic sector especially through joint development of small satellites. It is known that engineering knowledge and work experience gained in these satellite programs is often directly transferable to other technology fields.

Several issues of pertinent that could lead to certain policy options are highlighted as follows.

i. Digital Earth and EOS

As the country moves towards a fully developed nation by 2020, digital-economy adoption requires the setting up of digital infrastructures such as the Digital Earth (DE). DE, or technically known as the Spatial Data Infrastructure (SDI) would enabled the full capitalization of geospatial data among users and data providers in the country [1].

An EOS, as providers of the basic data to an SDI, in the form of various satellite images, would be an important component in the realization and maintenance of the country's SDI. China for example has placed great importance on the development of Earth observing satellites [3]. The four satellite series in China, which include resource satellites, environment satellites, meteorological satellites and ocean satellites, have been developed within the past few decades. They have also planned to have more high-resolution remote sensing satellites for building their next generation DE.

Ii. Building the EOS program – who's job is it?

The development and operation of an EOS system requires a significant commitment. Without a clear return on the investment, and an expensive venture as such, it is the government financial mechanism that would be the best one to support such implementation. The private sector on the other hand, is expected only to involve in the downstream applications of it, in further enhancing the justification of such program. In fact, most EOS program implementation is a not-for-profit making venture, but more of a national need, i.e. more on the justification of strategic need, rather than economy. Again, China has clearly shown that it is the government's job to do it [3].

iii. Fulfilling the domestic RS community need and sustaining of the EOS program

As the case elsewhere, local RS community certainly has a wide range of EOS image needs. A recently conducted user requirement survey [5] has indicated that slightly more than 50% of demand are in utility, infrastructures and engineering works. This clearly indicates that images of higher resolution (1m or better) are much needed by the local RS users. On the signal side, optical images are still the preferred choice for such applications, for ease of use and cost. For this need, the optical payload was chosen in the current government EOS program development, the RazakSAT satellite series. As such, it is important for the EOS program to be continued and sustained by the government. The survey has also indicated that the next EOS program development for the country would be satellites carrying radar payload.

iv. The operational needs of EOS

It has been mentioned earlier that the business of getting data from foreign service providers are faced with issues such as data unavailability, or data available does not account for the technical requirements needed such as temporal frequency, spatial resolution, spectral frequency or geographic coverage. Another major shortcoming of current foreign EOS services is the ability to provide the needed image at the needed time for operational purposes. This is understood simply due to the nature of the EOS operational perimeter. With most EOS flying at SSO, the repeat visit could well be more than a week, hence for an operational purposes, such as monitoring a forest fire, the gap in imaging opportunity, is just too long. Even with multiple of EOS available, this gap is still a major cause. In addition to that, for most operational applications, where optical images are preferred, continuous cloud cover would add on to the problem.

v. To sit on the upper bench – the ability to contribute

EOS images have been integrated in disaster-management initiatives such as the Sentinel Asia, and UN-SPIDER, where Malaysia is a member. These initiatives works on a mutual cooperation, with members of the consortia contributes in assisting other members stricken by disasters, through data obtained via their infrastructures, for example, supplying images from their EOS. Within such an arrangement, the ability to contribute would indirectly give the member-country the credits. The more significant the contributions are, the more credit it will get. It is time for Malaysia, a leading member country of ASEAN, to have our own EOS – else the country is merely recipient of these initiatives or end users!

5. The options

To ensure full capitalization of the sector, it seems that the way forward is to build the country's capability in EOS, mainly in the small satellite missions. So far, this seems to be the current strategy of the Malaysian government. With full awareness that these will not targeting to make the country independent from using foreign satellite images (which even space faring nations is not doing), but having the capability to develop, launch and operate critical and the much needed EOS mission. In implementing this strategy, we believe both options of acquiring some and built some should be taken.

Under the NSP draft, it has been outlined that the direction of the EOS program development falls clearly under Program 3: Accelerating Space Technology and Infrastructure Development, where the EOS Program development is one of the three satellite development program to be pursued. Another important program in the NSP is one under Program 1: Exploring Space for New Knowledge & Scientific Discovery, where it is indicated that a specific orbital designation is to be favored, i.e. on enhancing the Near Equatorial Orbit (NEqO).

With these, we like to propose options for the development of the EOS program in the country, such as follows.

i. To continue with the EOS capacity development program

Currently we have the capability and capacity to design and develop (assemble, integrate, test), small (<1 ton) satellite bus and related sub-systems. We are also capable to launch (utilizing launch service provider) and operate such system. As for RazakSAT, the optical payload, an MAC was jointly developed with foreign technology partner, SatRecl a South Korean company. For RazakSAT2, it is still to be jointly developed with a foreign technology partner. We need to have R&D on payload development, either optical, radar, hyper-spectral or others. These would then complete the capacity building process.

ii. To continue with the RazakSAT satellite series

The optical payload would still be the preferred payload for most applications. For this, the RazakSAT satellite series (optical payloads) should be continued. Once the RS community built up applications using the RazakSAT2 imageries, the dependency on the continuous supply of the image will be established, and then it will be the responsibility of the government to ensure continuity of such service. Planned spatial resolution of 1.0m panchromatic and 5.0 m multi-spectral of RazakSAT2 would seem to satisfy majority domestic RS users. Future camera resolution should be determine prior to new mission development.

iii. To have radar satellite program

Being at the equatorial region, having a continuous cloud cover is a challenge. A radar payload satellite would greatly enhance the imaging needs of the local RS community. Limiting the satellite mass to be less than 1 ton, which we already have capability of then the development would be focus only to the payload of the mission. That would probably a mini-version radar. This could be targeted as the next development after the RazakSAT2 program.

iv. To continue pursuing the utilization of the NEqO

Malaysia has pioneered the NEqO with RazakSAT. The NSP draft has also recommended the pursuance on the capitalization of NEqO. NEqO simply gives the opportunity in high imaging repeatability for Malaysia, hence giving a huge advantage for operational RS applications such as forest fire monitoring. With a repeat of about 14 times vicinity imaging per day (including night time of course!), this could really enhance the monitoring operations.

The coming RazakSAT2 satellite unfortunately will fly the SSO. These decision is temporarily taken in consideration of several reasons, amongst them could be due to;

- The need to optimize ROI of RazakSAT2. With SSO orbit, capability to image the entire globe, hence increasing the potential of economic earning to the country.
- That the local RS communities are customs to commercial image processing software which is mainly based on the SSO images. Hence, non-SSO images (such as NEqO) posed difficulties for them.
- Local researchers are still not ready to capitalize NEqO images that have variation in sun-angle, shadows and sun-illuminations.

These are but challenges that have to be overcome by local RS experts and users before the next missions NEqO EOS could be realized.

6. Conclusions

The country has made the right policy in developing its local capability in the EOS program. Although it does not aim to make the country independent from using foreign satellite images but rather having the capability to develop, launch and operate critical and the much needed selected EOS mission. It is still the role of the government to ensure the successful implementation of this strategic initiative. Full capability is almost achieved except for some areas such as payload. With the acceptance and implementation of the National Space Policy, these could well be achieved by 2020.

References

[1] Abu Hanifah F, and Majeed Z A (2007) Implementing national spatial data infrastructure (NSDI) in Malaysia. Paper presented at the *Joint International Symposium and Exhibition on Geoinformation, ISG/GNSS,* Johor Bahru, Malaysia, 5-7th November.

[2] Fadzilah A D and Krishna V V (2007) *Science and Technology Policy and the Dynamics Underlying the Malaysian Innovation System*, in Tim Turpin and V V Krishna (eds) *Science and Technology Policy and Diffusion of Knowledge in Asia-Pacific* Edward Elgar Publishing UK ISBN 9781 84376 7800.

[3] Huadong G 2012. China's EOS for building a Digital Earth *International Journal of DE*, Vol.5, Issue.3, pages 185-188.

[4] Mohamad I S M, Hashim M and Ahmad S 2004 Research Strategies for RS Development in Malaysia *Paper presented at the National Workshop on Research Strategy for Space Development* the National Space Agency, MOSTI unpublished report.

[5] The Malaysian Remote Sensing Agency (ARSM) 2012 User Requirement Survey of EOS Image in Malaysia unpublished report.

[6] The National Space Agency 2012 Draft version of the National Space Policy unpublished report.

[7] Wood D and Weigel A 2011 Building technological capability within satellite programs in developing countries *Acta Astronautica* 69 (2011) 1110–1122

magHD: a new approach to multi-dimensional data storage, analysis, display and exploitation

Christophe Angleraud
Chief Technology and R&D Officer, Magellium, Toulouse, France

E-mail: christophe.angleraud@magellium.fr

Abstract. The ever increasing amount of data and processing capabilities – following the well-known Moore's law – is challenging the way scientists and engineers are currently exploiting large datasets. The scientific visualization tools, although quite powerful, are often too generic and provide abstract views of phenomena, thus preventing cross disciplines fertilization. On the other end, Geographic information Systems allow nice and visually appealing maps to be built but they often get very confused as more layers are added. Moreover, the introduction of time as a fourth analysis dimension to allow analysis of time dependent phenomena such as meteorological or climate models, is encouraging real-time data exploration techniques that allow spatial-temporal points of interests to be detected by integration of moving images by the human brain. Magellium is involved in high performance image processing chains for satellite image processing as well as scientific signal analysis and geographic information management since its creation (2003). We believe that recent work on big data, GPU and peer-to-peer collaborative processing can open a new breakthrough in data analysis and display that will serve many new applications in collaborative scientific computing, environment mapping and understanding. The magHD (for Magellium Hyper-Dimension) project aims at developing software solutions that will bring highly interactive tools for complex datasets analysis and exploration commodity hardware, targeting small to medium scale clusters with expansion capabilities to large cloud based clusters..

1. Introduction

As the capacity of data storage & processing increases, the spatial and spectral resolution of raster or vector data also increases, as well as the complexity of the phenomena that the scientists or the engineers attempt to explain. Strangely enough, the media used to explore and display these ever more complex data landscapes did not evolve that much in the last decades: keyboard, mice and colour screens.

Although screens are bigger, flatter and more colourful, mice are thinner and more reactive; we must admit that since Google™ Earth has brought a brilliant dynamic GIS visualization environment on everyone's desktop, no important new technique have come around to help understanding easier and deeper the vast amounts of data available.

Recently though, so called Cloud Computing has quickly turned computer farms into large arrays of hard disks, allowing peta-bytes of data to be persisted and processed. However, low cost and long lasting persistence is now a reality but, to our knowledge, no software is yet taking advantage of these

Content from this work may be used under the terms of the Creative Commons Attribution 3.0 licence. Any further distribution of this work must maintain attribution to the author(s) and the title of the work, journal citation and DOI.
Published under licence by IOP Publishing Ltd

new environments for providing efficient data access, visualization and analysis on commodity hardware, as far as geospatial information is concerned.

magHD is attempting to make use of high performance computing frameworks for hyper-dimensional datasets, using the best of the breed distributed storage, distributed processing as well as volume data exploration strategies used in medical imaging in order to create a highly interactive visualization environment that will facilitate understanding of complex phenomena.

We also believe that real-time interaction with end-users is one of the keys to achieve this goal by using eye persistence unconscious abstraction performed the brain. magHD will also provide new ways of interacting with the data, for example by using smartphones and tablets to provide simultaneous collaborative displays, yet preserving individual explorations and analysis.

The applications of magHD could be potentially quite large both in the civil and military domains: scientific visualization, climate analysis, GIS science, smart augmented reality GPS, complex environmental analysis, assessment and picturing (Recognized Environmental Picture & Rapid Environment Assessment), command & control...

1.1 Goals and organization of this document

This document has been written to provide a common view of the magHD project to its participants as well as to seek for potential early adopters and technical or financing partners.
It is organized in 4 sections:

1. Data storage and indexing
2. Data access and processing
3. Interactive data visualization
4. Current status & foreseen use cases

1.2 Previous work

Our work is mainly built upon existing technologies and similar approaches used to explore complex datasets used mainly in medical and scientific imaging. Therefore we do not claim a highly innovative scientific work but rather try to transpose these existing techniques into a new area of use: GIS and earth sciences where vast amount of geospatial data have to be explored and combined together into a consistent exploitation and visualization environments

In the field of computer science, many new techniques have emerged these last 10 years with regards to "big data" and "cloud" solutions. Among those, we would like to cite here the efforts of the Apache Foundation with its Hadoop [1] framework. This framework is quite comprehensive and offers a large collection of technologies that can be used in the building of our multi-dimensional data analysis, exploitation and visualization environment.

Despite the existence of such large framework, we consider that it generally lacks some native capacity to properly index and distribute the kind of data that environmental analysis requires. Namely, we believe that an essential part still missing is a proper multi-dimensional indexing and distributed storage solution. Some technologies are providing entry level spatial indexing solutions, such as mongoDB [2], Lucene Spatial API [3], or Spatial4J [4] for Solr [5], but these entry level are focused solely on the 2D indexing (longitude, latitude) case for mapping applications. Some work has also been done on using Hadoop and/mapreduce style of parallel processing in order to perform 2D spatial operations: [6], SMJR [7] and Pigeon [8].

In order to build a more generic solution, that could be used with general purpose multi-dimensional datasets (such as meteorological forecast fields for temperature, wind speed that would require simultaneous access to and query of X, Y, Z, T and vector value fields), one would need an embedded distributed multi-dimensional indexing scheme. Some interesting extensions to the previous distributed 2D spatial indexing work have been published: [9] and [10].

Therefore, we can see that "big data" or "cloud" infrastructures are on the verge to provide efficient scalable solutions for multi-dimensional data. The next question is raised by the amount of data and the large differences of data types that scientists and end users will have to cope with on a single (or dual) screen environment. We believe that to provide efficient analysis and display environments it is necessary to provide highly interactive processing capabilities to provide new ways of combining complex data. These kinds of approaches have been used in the medical imaging area, in order to ease the exploration of volumetric datasets. We can cite here the techniques called "importance driven visualization" [11].

We also would like to cite existing software packages that offer interesting features that show partial feasibility of our approach: Paraview [12] and Makai Voyager [13].

2. Distributed multi -dimensional data store

In order to provide a highly interactive environment able to cope with very large multi-dimensional datasets, the first challenge to tackle is proper indexing and storage.

Recent progresses have been made lately on big datastores through distributed file systems (DFS) as well as noSQL (Not Only SQL) databases. These new technologies are now reaching a first degree of maturity, proving sustainable performances and reliability but the applications are dedicated to textual processing (web pages and social networking indexing).

magHD will make use of the best available software techniques such as distributed RAM cache, distributed databases and distributed disk storage using multi-dimensional on-the-fly indexing to provide efficient data search and retrieval. We do not intent to re-invent the wheel but, to our knowledge, no cloud environment provides yet efficient techniques for building distributed multi-dimensional indexes.

We aim at using the most efficient spatial indexing techniques, such as R-tree, kd-tree, Gist, in a fully distributed cluster to allow for parallel search and extraction in a peer-to-peer kind of communication scheme.

We also believe that the rather old approach to data modelling used in GIS and relational databases is a strong limit to data re-use and sharing. Therefore, our approach will be to favor tag like attributes and to favor multiple overlapping models to hold the information. This dynamic non-restrictive approach will make use of dynamic filtering and combination (using the map-reduce paradigm) in order to extract semi-structured information in destination to the end user applications.

This approach should also be quite appropriate for handling different representations of common objects. For example, the set of attributes for a given road could show common values (name, id, number of lanes…) as well as distinct ones according to the end users application (traffic information, pavement type, administrative authority…). This arrangement bed on tag sets belonging to different namespaces will also allow multiple views matching different normative requirements to be extracted from the same data set. In a similar way, the geometry of the road itself could be multiple: central axis for GIS mapping applications and separate lanes and kerb for road construction/maintenance applications. The geometries could also handle different level of details to be hold together.

Given this requirement of flexibility, we believe that the distributed key-value stores or the document oriented noSQL databases are a good fit, provided, multi-dimensional indexing is added (some solutions such as mongoDB or memBase are already providing simple 2-dimensions indexes, we want to broaden this to any number of dimensions).

This native multi-dimensional indexing will allow magHD to cope with a wide range of applications requiring different types of data, such as geosciences, meteorology/climatology, oceanography, hydrography, geography... The proposed approach will mix Big Data storage issues with native indexing and distribution to provide the best performing solutions for fast data search and access.

3. Data Access and processing

The coupling of data storage issues with native indexing and distribution over machine clusters represent the main building block of efficient access and processing.

Our idea is to promote distributed processing strategies on machine clusters that favour the execution of processing on the very same node where storage happens. This strategy is not new; it has been vastly used on large clusters for web 2.0 applications as well as through the Map/Reduce paradigm promoted first by Google and then by the Hadoop open source cloud framework.

The extension of this paradigm to data processing dedicated to multi-dimensional data sets that we want to achieve within magHD is based on the direct use of the native multi-dimensional index to identify the storage nodes holding data intersecting the area of interest of the processing request.

The processing requests will therefore be split and distributed on all the nodes holding intersecting data. Thus we intend to extend our Ingrid data processing framework to make its workflow engine able to run into a fully distributed environment.

Specific attention will also be given to data blocks storage and transfer. It appears to us that JPEG2000 is a very versatile and efficient storage format and could be very efficiently used in such distributed storage environments (for earth observation images but as well for a lot of earth science gridded data sets). Our intent regarding JPEG2000 is to forget the file storage structure and to make direct use of the wavelet packets entities to spread them across the distributed data store.

One of the main asset of our processing framework will be its ability to discretize the processing requests into gridded processing chunks (the Map phase of Map/Reduce) so that the results of overlapping processing requests could be stored into an interim distributed cache. For example, a processing request that requires belongs to a real-time display cession, will always be split with exactly the same discretized bounding boxes and resolutions to maximize the chances that the processing results will be reused by the next partially overlapping visualization request. Therefore, the definition of these discretization grids need to be shared among all the processing nodes and exploitation work-stations, this will be part of the processing chains metadata.

Our main focus being interactive visualization, we want to take advantage of the pauses occurring during the human manipulation of the software, when the end user is performing analysis tasks. This means that during an interactive cession, we have bursts of data I/Os and CPU during view motion but then the CPU or GPU resources are much less solicited and could therefore be lent to other users.

Consequently, we envision a collaborative environment where each node could, in turn, play the role of a server or a client of other nodes, at any time. Thus, the available resources would therefore be used at their maximum potential at the global cluster level.

IOP Conf. Series: Earth and Environmental Science **20** (2014) 012035 doi:10.1088/1755-1315/20/1/012035

As a wrap up of the previous section, we provide in Figure 1 an overview of the different components that magHD is being built upon.

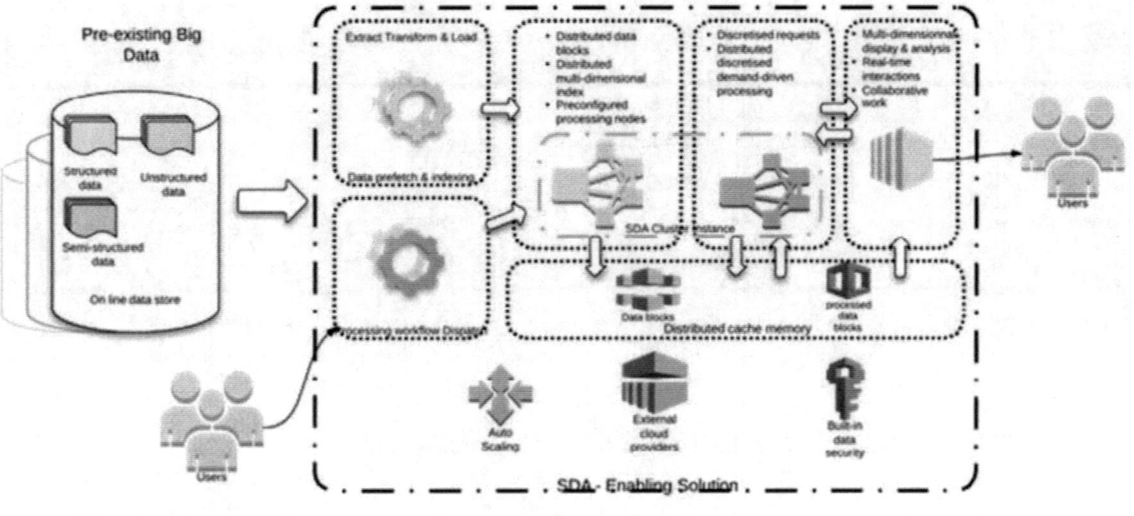

Figure 1. magHD Architecture overview.

4. Inter active Data Visualization

More data makes users more confused. The vast majority of current data exploitation and exploration software propose layered data organizations, each layer being independently rendered through style sheets on the screen. The GIS software proposes to filter out some data based on the displayed scale in order to reduce screen clutter.

We believe that it is possible to make better use of the current graphics engine by using much more interactive techniques.

4.1 Real time interaction

First, adopting fast rendering GPU-aided techniques can open new fields of interaction with the end-user, thus, allowing psycho-visual perception to increase the amount of information exposed to the user's brain.

For example, using rapid cursor manipulation to dynamically explore time or to change particles density or contrast of images can help to understand phenomena if, and only if, the users gestures are immediately changing the displayed data, in real-time. The unconscious associations of the gesture with what the eyes are perceiving allow temporal integration of complex changes by the brain.

Particle animation techniques could also be used to show data cube contents without occluding the background information. Again, to allow for psycho-visual integration, the particles have to be displayed very smoothly. This is possible thanks to GPU programming techniques mainly used in video-games called shaders (sort of GPU hosted highly parallelized micro-software).

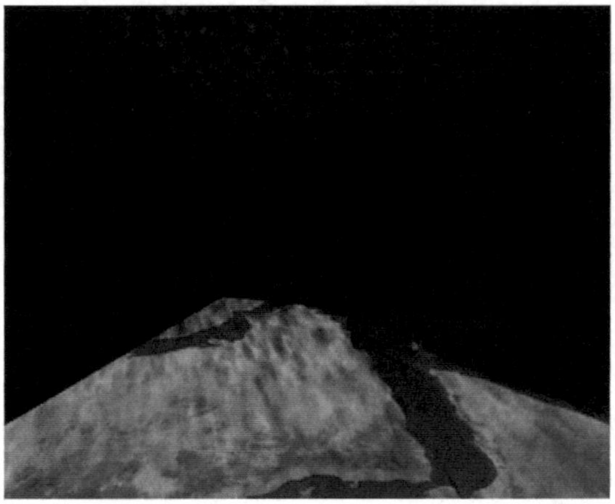

Figure 2. Particules used to show a cubic wind field while keeping the earth background visible.

Figure 3. Dynamic interactive plane clipping attached to camera motion.

Second, we believe that allowing real-time navigation through complex datasets is a way to allow intuitive understanding of the displayed contents. For example, the forward-backward motion of the camera can be used to move a temperature clipping plane in order to show local variations that can reveal specific phenomena to take into account.

4.2 Importance driven appearance

In order to reduce the clutter on screen, we believe that the data appearance should be able to change by itself, according to the other kind of information displayed at the same screen location. Our goal is to make data more intelligent so that its appearance could change in presence of other overlapping data.

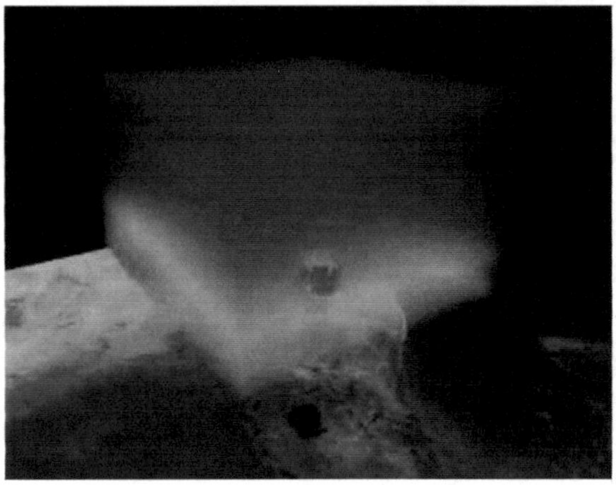

Figure 4. Importance driven building drills a transparency hole through a temperature cube.

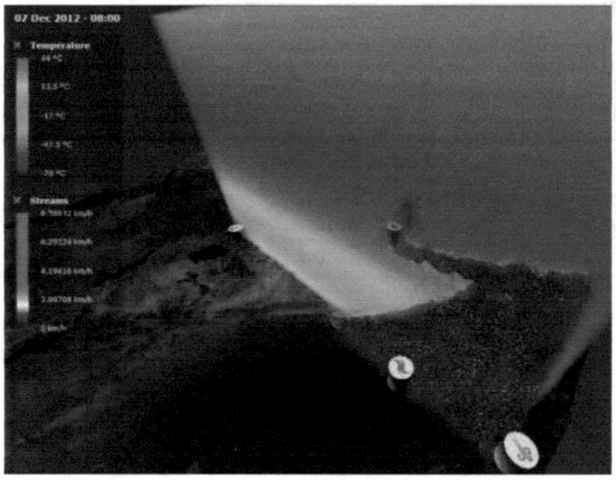

Figure 5. In this case we requested that the current field was more important than the temperature field.

This concept is borrowed from medical volume visualization techniques, called "importance-driven visualization". This technique aims at automatically changing the transparency of voxels occluding more important voxels that could be hidden. We propose to adopt the same principles and to extend it to other properties than transparency.

Namely, we envision a new way of specifying data rendering: each data record displayed on the screen could embed invisible importance information (that would be derived from the requirements of the end users). This importance information would then be interpreted by the graphic engine during the rendering passes in order to adapt the appearance of less important objects. For example, it's the transparency, blurriness, color saturation, filling / wireframe, line width, etc. could be modified dynamically in order to reveal the important features but keeping less important information slightly visible to deliver the overall contextual (more global) information.

For example, as shown on figure 3&4 controlling transparency can help to keep the most valuable information in front of the eye even though it is located behind other, less important information that would normally be occluding the more important one. We could make use of this feature for new generation GPS devices that would make use of collaborative distributed vehicle locations (like AIS for ships) in order to show potentially colliding vehicles occulted by buildings or any large obstacle on the screen of a smart augmented reality3D GPS.

5. Current status & foreseen use cases

At this stage, we have developed a prototype implementation which shows the feasibility of the magHD concepts. Currently, the developments are mainly dedicated to the visualization engine. We have a first version of the magHD SDK running and aim at being able to release a first operational version of this SDK by the end of 2013, as some illustrations of this paper show.

Meanwhile, we have also started some experimentation with distributed storage and processing, buts this area will still need some work, especially concerning multi-dimensional distributed indexing and proper use of distributed caches.

We envision multiple applications for this technology: scientific visualization & general public showcases, oil & gas sub-sea and sun-surface visualization of complex data sets, climate change monitoring, defense REP (Recognized Environmental Picture) & REA (Rapid Environment Assessment) applications, defense and civil command & control, lightweight embedded 3D guiding (augmented reality GPS) and we are currently seeking partnerships or early adopters in order to further investigate advanced visualization and analysis techniques and to build actual applications.

6. Conclusion

magHD has now proven the feasibility of most of its challenging features and is now ready to enter into a new R&D phase dedicated to large distributed storage & processing and collaborative work. Version 1 of its Hyper-Dimension data visualization engine will be ready by the end of this year. It will be package as an SDK ready to be used for developing new innovative smart data-centric applications.

IOP Conf. Series: Earth and Environmental Science **20** (2014) 012035 doi:10.1088/1755-1315/20/1/012035

7. References

[1] Apache, "Hadoop," [Online]. Available: http://hadoop.apache.org.
[2] MongoDB, Inc., "mongoDB," mongoDB, [Online]. Available: http://www.mongodb.org
[3] Apache Foundation, "Lucene Spatial API," Apache, [Online]. Available: http://lucene.apache.org/core/4_0_0/spatial/.
[4] Apache Foundation, "Spatial4J," Apache, [Online]. Available: https://github.com/spatial4j/spatial4j.
[5] Apache Foundation, "Solr," Apache, [Online]. Available: http://lucene.apache.org/solr/.
[6] E. Tanin, A. Harwood and H. Samet, "Using a distributed quadtree index in peer-to-peer networks.," VLDB J., vol. 16, no. 2, p. 165–178, 2007.
[7] S. Zhang, J. Han, Z. Liu and K. Wang, "SJMR: Parallelizing spatial join with MapReduce on clusters," in CLUSTER'09. IEEE, 2009.
[8] A. Eldawy and M. F. Mokbel, "Pigeon: A Spatial MapReduce Language," in ICDE14, 2014.
[9] C. Du Mouza, W. Litwin et P. Rigaux, «A Framework for Distributed Spatial Indexing in Shared-Nothing Architectures, » BDA, 2007.
[10] C. Zhang, F. Li and J. Jestes, "Efficient Processing of k Nearest Neighbor Joins using MapReduce," in EDBT, 2012.
[11] I. Viola, A. Kanitsar and M. E. Groller, "Importance-driven feature enhancement in volume visualization.," Visualization and Computer Graphics, IEEE Transactions, vol. 11, no. 4, pp. 408-418, 2005.
[12] Kitware, "Paraview," [Online]. Available: http://www.paraview.org/.
[13] Makai Ocean Engineering, "Makai Voyager," [Online].Available: http://voyager.makai.com

Fuzzy B-spline optimization for urban slum three-dimensional reconstruction using ENVISAT satellite data

Maged Marghany
Institute of Geospatial Science and Technology (INSTeG),Universiti Teknologi Malaysia,81310 UTM, Skudai, Johor Bahru, Malaysia.

E-mail:maged@utm.my,magedupm@hotmail.com

Abstract. A critical challenges in urban aeras is slums. In fact, they are considered a source of crime and disease due to poor-quality housing, unsanitary conditions, poor infrastructures and occupancy security. The poor in the dense urban slums are the most vulnerable to infection due to (i) inadequate and restricted access to safety, drinking water and sufficient quantities of water for personal hygiene; (ii) the lack of removal and treatment of excreta; and (iii) the lack of removal of solid waste. This study aims to investigate the capability of ENVISAT ASAR satellite and Google Earth data for three-dimensional (3-D) slum urban reconstruction in developed countries such as Egypt. The main objective of this work is to utilize some 3-D automatic detection algorithm for urban slum in ENVISAT ASAR and Google Erath images were acquired in Cairo, Egypt using Fuzzy B-spline algorithm. The results show that the fuzzy algorithm is the best indicator for chaotic urban slum as it can discriminate between them from its surrounding environment. The combination of Fuzzy and B-spline then used to reconstruct 3-D of urban slum. The results show that urban slums, road network, and infrastructures are perfectly discriminated. It can therefore be concluded that the fuzzy algorithm is an appropriate algorithm for chaotic urban slum automatic detection in ENVSIAT ASAR and Google Earth data.

1. Introduction

The United Nations termed an urban slum as a tumble-down area of the city which portrayed by poor quality housing, dirtiness, and lacking in tenancy security. In this regard, urban slums refer to housing areas that were once relatively affluent but which deteriorated as the original dwellers moved on to newer and better parts of the city, but has come to include the vast informal settlements originated in cities in the developing world. In addition, slum buildings vary from simple shacks to permanent and well-maintained structures. Most slums lack clean water, electricity, sanitation and other basic services [20]. Therefore, Ismail et al., [21] argued that urban slum is considered as cancer cells or pockets within the healthy urban area that is required critical surgical intervention. They proposed a new method to identify urban slum which is based on Coexistent Urbanism approach. They proposed approach that is considered three groups of factors to differentiate between slums. These involve basic and socioeconomic factors, urban analysis factors and fabric morphological factors. Finally they applied space Syntax to slums network, then slums can be classified By a devised normalized index called Coexistence Potential index which measures effectiveness and Neediness for intervention.

Content from this work may be used under the terms of the Creative Commons Attribution 3.0 licence. Any further distribution of this work must maintain attribution to the author(s) and the title of the work, journal citation and DOI.
Published under licence by IOP Publishing Ltd

The accurate technology is required to identify and monitor the change and growth of the urban slums. Recent advances in space technology might show excellent promises for land use and change monitoring [9,16]. In this context, remote sensing and Google Earth plays a tremendous role in monitoring land use spatial variations. Mahogany [9] has proposed that a should utilize Google Earth for urban slum investigation. There are many other studies have utilized the optical remote sensing satellite for urban slum monitoring. High resolution imagery has been used for many years [16]. Some of these studies have established the new rule which is based on infrastructures and urban planning morphology to identify the urban slum. Kit et al. [5] used very high resolution satellite imagery from the Quick Bird satellite with resolutions of 2.44-2.88m and 0.61-0.72 m, respectively to identify the urban slums. Further, they concluded that remote sensing technologies provide urban managers with the to increase the efficiency of their often very limited resources and to target support and improvement measures at those particularly in need. The relatively low hardware requirements and absence of software licensing costs to also make the technique feasible for use in developing countries. According to Holden [4], the poor in the dense urban slums are the most vulnerable to infection due to (i) inadequate and restricted access to safety, drinking water and sufficient quantities of water for personal hygiene; (ii) a lack of removal and treatment of excreta; and (iii) a lack of removal of solid waste, particularly the organic fraction, which attracts vermin. Therefore, Holden [4] stated that modern and reliable technologies is required to solve and improve the problem of urban slums. According to Kohli et al. [14] the absence of a unique and consistent definition of a slum hinders the development of objective, universally applicable slum detection methods using remote sensing and OOA, which require a clear conceptualization of the object of interest. According to Dekker [15], SAR data are suitable for monitoring urban land use. In fact, various bands of L, C, or X bands; with a wide range of incidence angles between 35° and 50°; and HH and HV polarization (HV interesting to investigate urban land use or quad pol monitoring urban environments). He stated that ENVISAT ASAR satellite data with 30 m resolution cannot detect small objects. On the contrary, the same system can be used in combination with higher resolution satellites such as RADARSAT-2 SAR, TerraSAR X and airborne SAR systems for monitoring land use. Amarsaikhan et al. [17] and Brook et al. [18] have also shown how the combined use of optical and SAR data sets can enhance urban feature detection and enhancement.

Therefore, Google Earth is often considered as a rapid source for information. In fact, it includes all the imagery for each country of the world. In other words, Google Earth is a database viewer that combines satellite data to reconstruct a three-dimensional model (3-D) of the ocean and Earth's surface. The overall objective of this work is to investigate spaceborne SAR data and Google Earth for automatic detection of urban slum in Cairo, Egypt. The main hypotheses is to reconstruct three-dimensional of urban slum in SAR satellite data and Google Earth uses fuzzy B-spline algorithm.

2. Data acquisition

This multi-colored composite ENVISAT Advanced Synthetic Aperture Radar (ASAR) wide swath mode image over Egypt's capital city of Cairo. Clearly ENVISAT ASAR image shows the Nile River is running from south to north. Radar images measure surface roughness rather than reflected light. The color in the image was produced by a combination of the three acquisitions of ENVISAT ASAR. A color is assigned to each date of acquisition: red for 7 June 2004, green for 20 November 2003 and blue for 8 April 2004, with a spatial resolution of 12.5 meters (Figure 1).

Figure 1. Composite colour image of ENVISAT ASAR over Cairo, Egypt.

3. Fuzzy B-spline algorithm

Following Marghany et al., [9] a fuzzy number is defined whose range is given by the minimum and maximum values of DTM along each kernel window size. Furthermore, the identification of a fuzzy number is acquired to summarize the estimated DTM data in a cell and it is characterized by a suitable membership function. The choice of the most appropriate membership is based on triangular numbers which are identified by minimum, maximum, and mean values of DTM estimated. Furthermore, the membership support is the range of digital elevation data in the cell and whose vertex is the median value of DTM data [1].

There are two basic notions that we combine together [1,3]: (i) confidence interval and (ii) presumption level. A confidence interval is a real value interval which provides the sharpest enclosing range of urban 's elevation spatial variation values [9,10]. Based on assumption issued by Anile [1], that μ -level is an estimated truth value in the [0, 1] interval on our knowledge level of the digital terrain model (DTM). The 0 value corresponds to minimum knowledge of urban 's topography, elevation gradients, and 1 to the maximum urban 's topography elevation gradients. A fuzzy number is then prearranged in the confidence interval set, each one related to an assumed level $\mu \in [0, 1]$. Moreover, the following must hold for each pair of confidence intervals which define a number: $\mu \succ \mu' \Rightarrow \xi \succ \xi'$ [9].

Following Marghany et al. [8], let us consider a function $f : \xi \rightarrow \xi'$, of N fuzzy variables $\xi_1, \xi_2, ..., \xi_n$. Where ξ_n are the global minimum and maximum values of urban 's topography, elevation gradients along the space. Based on the spatial variation of the urban 's elevation gradients, the fuzzy B-spline algorithm is used to compute the function f [7]. Following Marghany et al., [9] ξ *(i,j)* is the urban elevation value at location *i,j* in the region *R* where *i* is the horizontal and *j* is the vertical coordinates of a grid of m times n rectangular cells. Let *N* be the set of eight neighbouring cells. The input variables of the fuzzy are the amplitude differences of urban 'elevation ξ defined by [1]:

$$\Delta\xi_N = \xi_i - \xi_0, N = 1, ..., 4 \quad (1)$$

where the ξ_i, *N*=1, 4 values are the neighbouring cells of the actually processed cell ξ_0 along the horizontal coordinate *i*. To estimate the fuzzy number of urban's elevation ξ_i which is located along the vertical coordinate *j*, we estimated the membership function values μ and μ' of the fuzzy variables ξ_i and ξ_j, respectively by the following equations were described by RÖvid et al. [11]

$$\mu = \max\left\{\min\left\{m_{pl}(\Delta\xi_i) : \xi_i \in N_i\right\}; N = 1....,4\right\} \quad (2)$$

$$\mu' = \max\left\{\min\left\{m_{LNl}(\Delta\xi_i) : \xi_i \in N_i\right\}; N = 1....,4\right\} \quad (3)$$

Equations 2 and 3 represent the urban elevation in 2-D, in order to reconstruct fuzzy values of the urban elevation in 3-D, then the fuzzy number of the DTM elevation in z coordinate is estimated by the following equation proposed by Russo [11]

$$\xi_z = \Delta\mu MAX\{m_{LA}\left|\xi_{i-1,j} - \xi_{i,j}\right|, m_{LA}\left|\xi_{i,j-1} - \xi_{i,j}\right|\} \quad (4)$$

where ξ fuzzy set of urban 's elevation values in z coordinate which is function of i and j coordinates i.e. $\xi_z = F(\xi_i, \xi_j)$. Fuzzy number F_O for urban 's elevation in i,j and z coordinates then can be given by

$$F_O = \{\min(\xi_{z_0}, ..., \xi_{z_\Omega}), \max(\xi_{z_0}, ..., \xi_{z_\Omega})\} \quad (5)$$

where Ω =1, 2, 3, 4,
The fuzzy B-spline algorithm has been implemented with ENVISAT ASAR data which acquired over Cairo, Egypt.

4. Results and discussion

Figure 2 shows the selected study area from the Google Earth tool that suggested the location of urban slum. This area is located away from Misr Al Qadima between 29 58 12 N at 30 00 36 N and 31 12 36 E to 31 15 00 E. This was selected based on the criteria of low quality building, poor infrastructures, overlapping between high quality and low quality buildings, high building density, hazardous location, poor environmental condition, durability. Figure 3 shows that urban slum are dominated by poor roof material, absence of roads, irregular roads, lack of vegetation, irregular shape of settlement association with neighboring areas, texture and locality. This confirms the study of Kohli et al. [14].

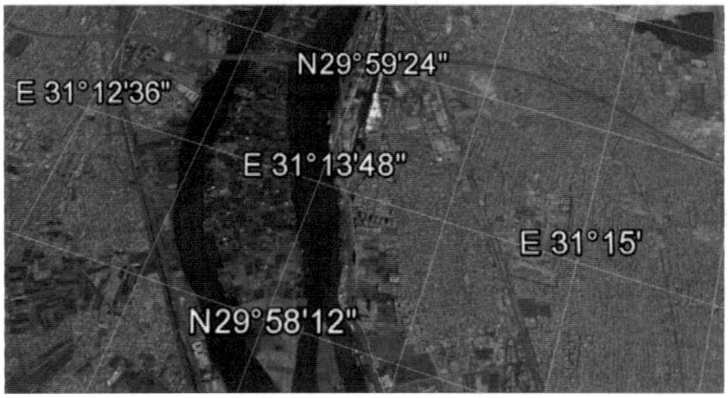

Figure 2. Selected location of urban slum.

Figure 3. Physical indicators for urban slum.

Clearly, Figure 4 contains uncertainties due to speckle. Speckle is considered a major obstacle for SAR data interpretation.[19]. Thus, adaptive filters such as Lee, Kuan, Frost and the co-occurrence matrix are often used to enhance the edges of urban features.[9,15]. However, these filters are not totally able to reduce speckle noise in SAR data. In addition, the ENVISAT ASAR resolution of 30 m does not allow the detection of small objects such as slum dwellings [15].

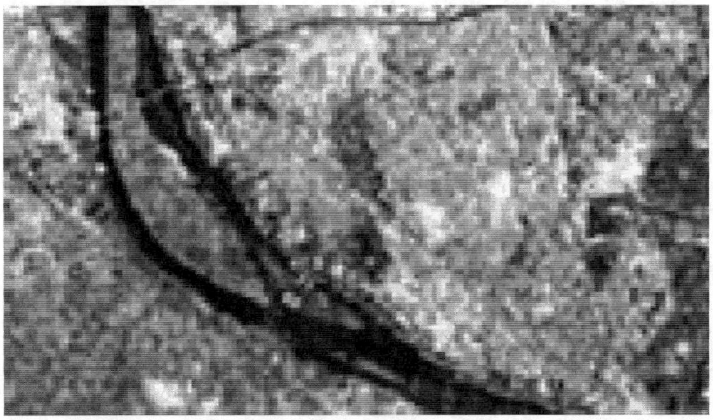

Figure 4. Urban slum selected in ENVISAT ASAR stellite data.

Figure 5 shows the result obtained by using the fuzzy B-spline algorithm. It is clear that the 3-D visualization discriminates between infrastructures and buildings. Roads, buildings and infrastructures are clearly displayed in both ENVISAT ASAR and Google Earth (Figure 5). This is due to the fact that the fuzzy B-spline, is considered as a deterministic algorithm, which is described here to optimize a triangulation only locally between two different points [1][3][6][9]. This corresponds to the feature of deterministic strategies of finding only sub-optimal solutions usually. The visualization of infrastructures of urban slums is sharp with the ENVISAT ASAR and Google Earth because each operation on a fuzzy number becomes a sequence of corresponding operations on the respective μ-levels and the multiple occurrences of the same fuzzy parameters evaluated as a result of the function on fuzzy variables [1];[7], [9].

(a) **(b)**

Figure 5. 3-D view using fuzzy B-spline algorithm (a) ENVISAT ASAR and (b) Google Earth.

5. Conclusion

This study has demonstrated a method to visualize 3-D of urban slums using ENVISAT ASAR and Google Earth imagery. Some urban slums in Cairo city were selected as test area. Then, the fuzzy B-spline was implemented to reconstruct a 3-D visualization of the urban slums. The results show that the high density of housing spatial distributions, inherent in many dense slum areas, together with the high level of infrastructure, can be clearly observed.. In conclusions, the ENVISAT ASAR and Google Earth could be used as tool for urban slum monitoring. Further, fuzzy B-spline shows excellent tool for 3-D visualization and discrimination between infrastructures and housing features in remotely sensed data specially with SAR satellite data. With the current operational status of many high resolution SAR satellites, the proposed method described in this paper becomes very useful for urban slum studies.

6. References

[1] Anile A. 1997 *Report on the activity of the fuzzy soft computing group*, Technical Report of the Dept. of Mathematics, University of Catania, March 10 p.

[2] Boss S.K. 2012 http://www.franklin.edu/file_source/franklin/files/slc/GEinIntroCourses.pdf.

[3] Fuchs, H. Z.M. Kedem, and Uselton, S.P.,1977, Optimal Surface Reconstruction from Planar Contours. *Communications of the ACM*, 20, 693-702.

[4] Holden, R., 2008, Urban Sanitation Technologies: The Challenges of Reaching the Urban Poor. Paper presented at IRC Symposium: Sanitation for the urban poor partnerships and governance. Delft, The Netherlands, 19-21 November 2008.

[5] Kit O., Lüdeke,M.,and Reckien D., 2012, Texture-based identification of urban slums in Hyderabad, India using remote sensing data. Applied Geography. 32, 660-667

[6] Marghany,M., Hashim M., and Cracknell A., 2009, 3D Reconstruction of Coastal Bathymetry from AIRSAR/POLSAR data. Chinese Journal of Oceanology and Limnology. 27(1), 117-123.

[7] Marghany, M. and Hashim, M., 2010, Lineament mapping using multispectral remote sensing satellite data. International Journal of the Physical Sciences. 5(10), 1501-1507.

[8] Marghany, M., Hashim M., and Cracknell A., 2010, 3-D visualizations of coastal bathymetry by utilization of airborne TOPSAR polarized data. *International Journal of Digital Earth*, 3(2),187 – 206.

[9] Marghany M., 2012, Three-dimensional lineament visualization using fuzzy B-spline algorithm from multispectral satellite data. In Escalante-Ramirez B. (ed.) "Remote Sensing - Advanced Techniques and Platforms". InTech - Open Access Publisher, University Campus STeP Ri, Croatia. pp.213-232.

[10] Russo, F.,1998,Recent advances in fuzzy techniques for image enhancement. IEEE Transactions on Instrumentation and measurement. 47, pp: 1428-1434.

[11] Rövid A, Várkonyi A.R. and Várlaki P., 2004, 3D Model estimation from multiple images IEEE Int. Conf. on Fuz Sys FUZZ-IEEE 1661.

[12] Urbanpeek (2012). http://urbanpeek.com/2011/06/13/a-wake-up-call-cairo-slums. [Access 3rd October 2012].

[14] Kohli D, Richard S, Norman K, Stein A, 2012, An ontology of slums for image-based classification. Computers, Environment and Urban Systems. 36 (2), 154–163.

[15] Dekker R 2006, Monitoring urban development using ENVISAT ASAR *1st EARSeL Workshop of the SIG Urban Remote Sensing Humboldt-Universität zu Berlin, 2-3 March 2006, http://www.earsel.org/workshops/SIG-URS-2006/PDF/Session3_Dekker.pdf.*

[16] Genderen, J.L. van, 1989, High resolution satellite data for urban monitoring. Int. Jnl. of Remote Sensing, 10, 257-258.

[17] Amarsaikhan, D., Blotevogel, H.H., Genderen, J.L. van, Ganzorig,M., Nergui,B., 2010, Fusing high-resolution SAR and optical imagery for improved urban land cover study and classification. International Journal of Image and Data Fusion, 1, 83-97.

[18] Brook, A., Ben-Dor, E. And Richter, R., 2013, Modelling and monitoring urban built environment via multi-source integrated and fused remote sensing data. International Journal of Image and Data Fusion, 4, 2-32.

[19] Huang, Y., Genderen, J.L. van, and Veen van, B.F., 1997, The ITC filter: a new adaptive filter for SAR speckle reduction. In proceedings of the second international airborne remote sensing conference and exhibition.

[20] Ward, W. P., 2008, The Life and Times of Cardinal Wiseman, 1. Biblio Bazaar. p. 568.

[21] Ismail,A., ,Bakr, H., and Anas, S., 2013, A hybrid GIS space syntax methodology for prioritizing slums. Coordinates, IX, 44-50.

Using game engine for 3D terrain visualisation of GIS data: A review

Ruzinoor Che Mat[1], Abdul Rashid Mohammed Shariff[2], Abdul Nasir Zulkifli[1], Mohd Shafry Mohd Rahim[3] and Mohd Hafiz Mahayudin[1]

[1]School of Multimedia Technology and Communication, Universiti Utara Malaysia, 06010 UUM Sintok, Kedah

[2]Geospatial Information Science Research Centre (GIS RC),Universiti Putra Malaysia, 43400 Serdang, Selangor

[3]Faculty of Computing, Universiti Teknologi Malaysia, 81310 UTM Skudai, Johor

E-mail: ruzinoor@gmail.com

Abstract. This paper reviews on the 3D terrain visualisation of GIS data using game engines that are available in the market as well as open source. 3D terrain visualisation is a technique used to visualise terrain information from GIS data such as a digital elevation model (DEM), triangular irregular network (TIN) and contour. Much research has been conducted to transform the 2D view of map to 3D. There are several terrain visualisation softwares that are available for free, which include Cesium, Hftool and Landserf. This review paper will help interested users to better understand the current state of art in 3D terrain visualisation of GIS data using game engines.

1. Introduction

In recent years, a lot of research has been conducted with regard to 3D terrain construction. Most of the research has developed new applications for viewing and creating in 3D. Eventually this helps in promoting tourism and future planning for disaster and security [1]. It has been proven that the 3D terrain can be perceived a lot better compared to normal 2D map which requires certain level of comprehension in map reading. Currently, there are few categories of 3D terrain visualisation engines that are available on the internet using the game engine and open source. Each of these categories has advantages and disadvantages. Therefore, the aim of this paper is to provide a comprehensive review of 3D terrain visualisation of GIS data generated from the game engine. This paper begins with a brief history of game engine and continues with the detail of game engines such as Torque 3D, Unreal Development Kit (UDK), Unity3D and CryEngine. The second section covers the available 3D terrain visualisation software. The third section explains terrains visualisation using game engine and shows the terrain created using the game engine. The fourth section discusses on 3D terrain visualisation and 3D terrain visualisation using game engine and finally the fifth section concludes this paper.

[1] To whom any correspondence should be addressed.

Content from this work may be used under the terms of the Creative Commons Attribution 3.0 licence. Any further distribution of this work must maintain attribution to the author(s) and the title of the work, journal citation and DOI.
Published under licence by IOP Publishing Ltd

2. Game engine

Game engine refers to a set of tools that help the developers to create games a lot faster and easier using commands that used to require a lot of coding through plug-in or asset provider of the game engine [2]. In the earlier days, game was limited to hardware capabilities. The earliest development of arcade games started in 1966 [2], and one of the most well known arcade game known as Pong was introduced [3]. This game utilised the advancement of technologies of arcade games which ported into the consoles. As the technologies grew, the game was then ported into the PC platform around 1990. During this time, the term "game engine" was introduced. Since then, many different types of game engine have been introduced including Torque 3D, UDK, Unity3D and CryEngine. Figure 1 shows the architecture of game engine that was introduced by Lewis and Jacobson [4].

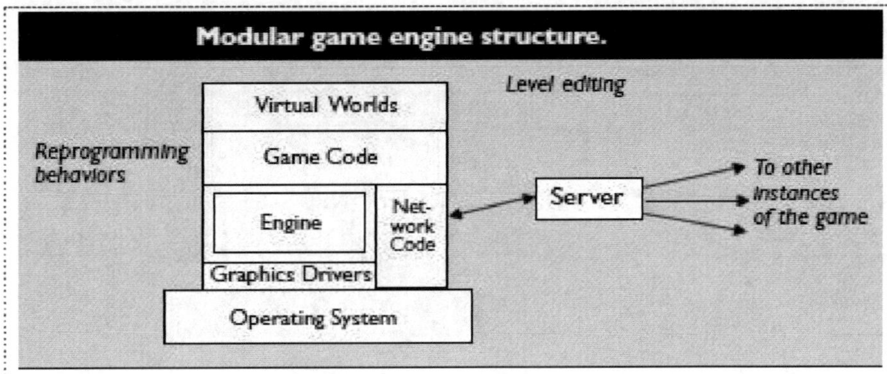

Figure 1. The architecture of game engine (Source: Lewis and Jacobson [4]).

2.1 The Torque Game Engine (TGE)

TGE is a game engine published by GarageGames [5]. Torque started around as a company by the name of Dynamix, then later after acquiring the tribes 2 source codes from Source, it have been changed to Torque 3D. The engine uses C++ as its main script which allows the user to have a better editing of the engines. Torque 3D supports PhysX, modern shader features, and advanced deferred lighting mode. TGE runs on most windows platforms, Mac and Linux. TGE also provides documentation for the users in creating and designing their own games. The database of this game engine supports MySQL and uses Torque script.

Shiratuddin and Thabet [6] discussed details on how TGE could be utilised to develop Design Review System from manual to 3D complete visualisation system. Besides that they also utilized this engine for application such as a Southern Miss's Innovation and Commercialization Park [7]. Figure 2 shows the architecture of Torque 3D game engine as explained by Nilson and Söderberg [8].

2.2 Unreal Development Kit (UDK)

UDK was released by Epic Games Inc. and it is a free version of Unreal Engine 3 [9]. It has been used by professional studios and indie developers to create well known games such as Gears of Wars, X-men Origins and many others. UDK provides the developer with same functionality as paid version. UDK allows for visual scripting using Unreal kismet and its main script is called UnrealScript. Navarro et al. [10] explained that UDK has a lot of functionalities including skeleton animations. It uses NVidia's PhsyX and UDK behaviour editor. Shiratuddin and Thabet [11] explained the architecture of UDK in their research on "Virtual Office Walk-through Using a 3D Game Engine" as shown in figure 3.

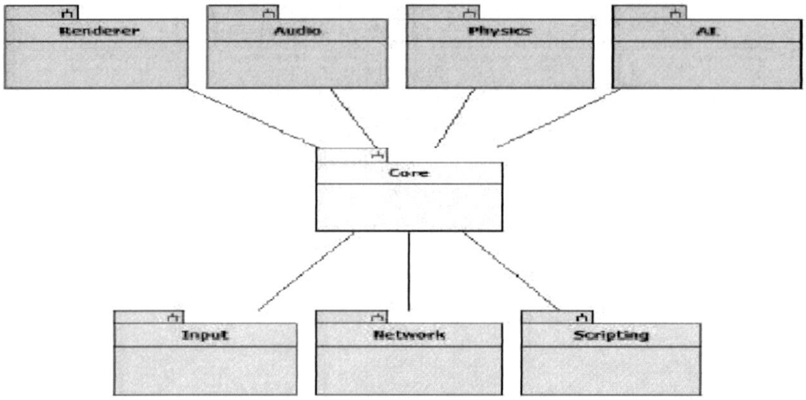

Figure 2. Torque game engine architecture (Source: Nilson and Söderberg [8]).

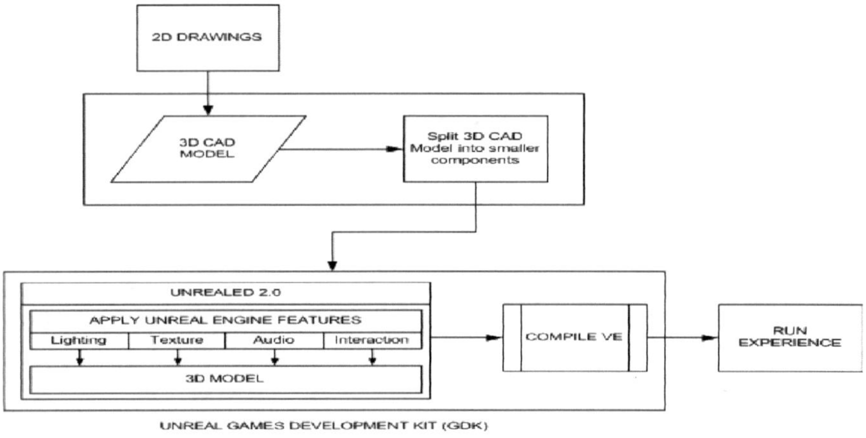

Figure 3. The UDK game engine architecture (Source: Shiratuddin and Thabet [11]).

2.3 Unity3D

Unity3D is a game engine created by Unity Technologies [12]. It can be used to create online and offline games. Unity3D provides a wide range of assets that help developers to create games as well as documentations and tutorials. It utilises the concept of assets which work like a plug-in where the user can import or just copy the asset file into a specific folder. Unity3D uses JavaScript, C# and Boo as the main programming languages. The main advantage of Unity3D is that it can run on multi-platform OS including Windows, Mac, Linux and also mobile. It is capable of importing other 3D models in fbx, sbx and obj formats. Wang et. al. [13] presented the architecture of Unity3D as shown in figure 4.

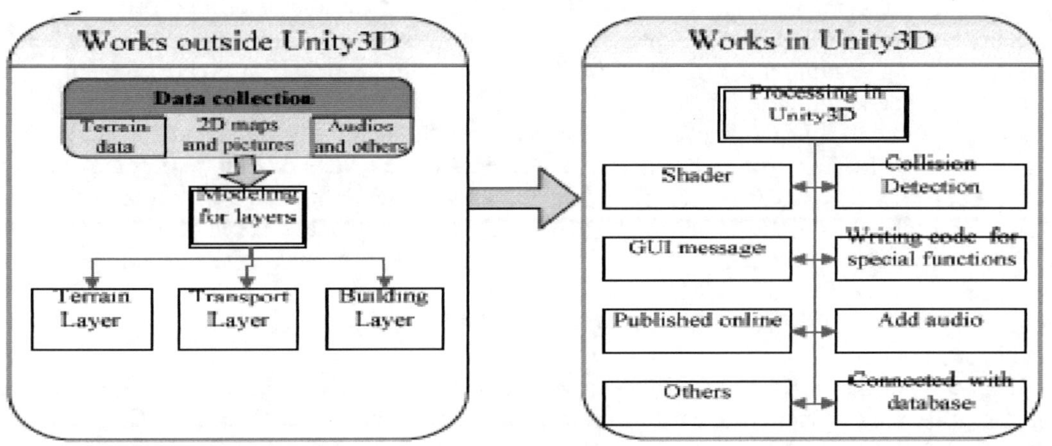

Figure 4. Unity3D game engine architecture (Source: Wang et. al. [13]).

2.4 CryEngine

CryEngine is a game engine created by Crytek, a developer based in Germany [14]. CryEngine provides a much rich environment to create games in a sandbox using "WYSIWYG" method. There have been four releases for CryEngine and each release is an improvement to the others. CryEngine is free for non-commercial development which allows developers to have first hand on the game engine. CryEngine itself uses sandbox editor to add objects into the world terrains. Trenholme and Smith [15] explained the detail functionalities of the CryEngine. Meanwhile, Herrlich [16] explained the features of CryEngine. Figure 5.0 shows the sandbox editor view of CryEngine.

Figure 5. The sandbox editor view of CryEngine (Source: Herrlich [16]).

3. 3D terrain visualisation software

In recent years, a lot of 3D terrain modelling software as well as terminologies have emerged. The terminologies which include digital elevation model (DEM), triangular irregular network (TIN) and digital terrain model (DTM) are normally used for terrain data type and they are frequently used in software that have the capability to read and display 3D terrains. Ruzinoor et al. [17] reviewed the techniques and software for 3D terrain visualisation of GIS data and explored different types of techniques for visualising 3D terrain [18]. The VRML technology has been used by Ruzinoor et al. [19] for visualising terrain online. Horne [20] the original author of 3DEM created the software for commercial market and after a few years of development, he halted the project but still share the full version of the software for public used. The software has the capability of importing and generating 3D terrain from DEM data. This software also provides a surface view of Mars terrains. Zuse Institut Berlin & Lenné3D [21] created the BioSphere3D. This software has the capability to display the landscape of globe terrain by using data such as DEM, satellite and aerial images, 3D models (Collada), 3D plant models (Lenné3D Flora3D) and shapefiles. Cesium is a 3D terrain visualiser that uses globe view concept which has the capability to zoom and explore the whole world terrain. It uses WebGL libraries to load the shapefiles in producing the terrain. It is based on entire cloud and fully dependent on scripting [22].

Besides that, Earth3D was developed by Dominique Andre Gunia [23] using Java and C++ as its core programming codes. The data for this software is generated from NASA, USGS, CIA and the city of Osnabrück. GenesisIV is a landscape software created by Geomantics [24]. The advantages of this software is the capability in generating the photorealistic terrains. G3DGMV (3D Graphical Map Viewer) was developed by Yost [25] for terrain visualisation which uses DEM and DLG as its main data source. It also provides the functions for saving as image and creating guide based on user preferences. Hftool was introduced by Stock [26] which is a Unix based terrain viewer and the output from this software can be converted into different types of file formats. Integrated data viewer (IDV) was created by Unidata [27] and is based on java programming. The software is able to create an animation of satellite images, viewing a grid view of terrain plot and also displaying a data based on globe.

LandSerf a terrain visualisation software was created by Jo Wood [28] and it is also a Java based terrain visualisation. It can handle multiple files of GIS format and also provides high interactivity visualisation. This software also able to import and export raster format and able to integrate with Garmin GPS receiver. Guth [29] has introduced MicroDEM which is a portable terrain viewer where it can view DEM data and also able to merge the data. Virtual Globe was created by Sintef, a research organization in Scandinavia that cover a lot of research area until it was taken over by Norkart for further development. This software uses globe as its main view and data such as SRTM, GLOBE and GTOPO30 to display terrain model. This software was also developed using Java [30]. OpenEV was originaly developed by Atlantis Scientific Inc. [31], which is currently known as Vexcel. As of 2006, it was acquired by Microsoft and has ceased involvement in the project. OpenEV is a terrain viewer that uses OpenGL for rendering and it supports various types of raster and vector formats to create elevation.

4. 3D terrain visualisation software using game engine

As game engine development progresses, several new add-ons have been included such as terrain editor. Most game engines have their own terrain editor which helps in creating realistic terrains. In Unity3D, the developer is able to design a realistic terrain using terrain editor which can be viewed using the first person view. The terrain editor in Visual 3D allows the user to import DEM data to generate the terrain [32]. Torque game engine also provides the terrain editor as explained by Yang et. al. [33] on how the terrain can be created using TGE. Figure 6.0 shows an example of terrain generated using TGE. Besides that, UDK also provides a terrain editor that supports DEM data that can be manipulated using Unreal engine [34]. The DEM of TGE allows GIS data to be loaded but the data must be in a compatible format in order to generate the terrain. While Unity3D also has the capability to load GIS data and generating the terrain from the data and uses script to control the data to be more interactive. CryEngine requires a bit of works to get the GIS data into the game engine. Figure 6-9 show the images of terrain generated from different types of game engines which include Torque, UDK, Unity3D and CryEngine.

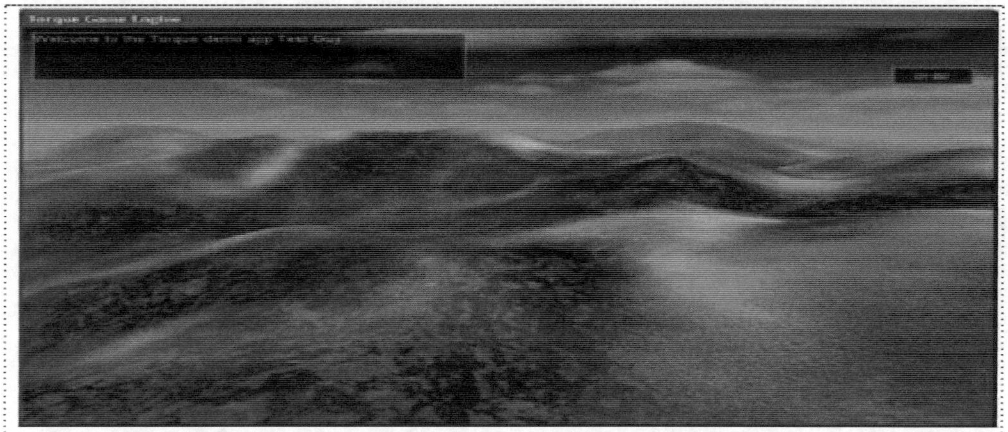

Figure 6. Images of terrain created by Torque engine. (Source: Yang et. al. [33]).

Figure 7. Image of terrain created by the UDK game engine. (Source: Epic Games Inc. [9]).

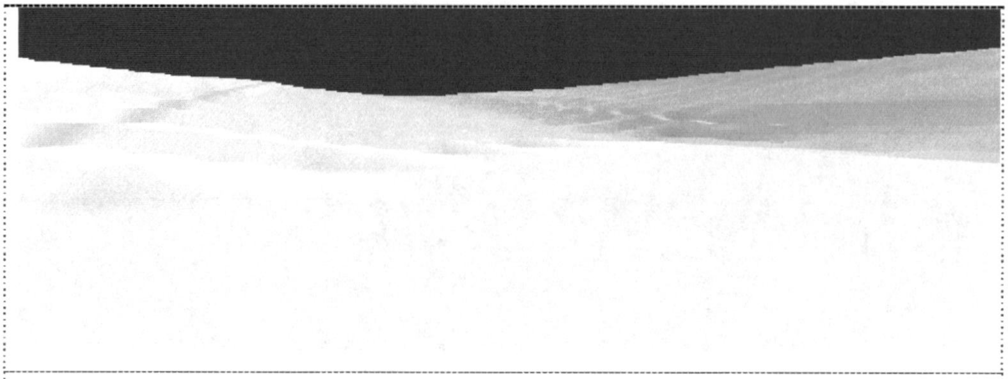

Figure 8. Image of terrain created by Unity3D game engine.

Figure 9. Image of terrain created by CryEngine (Tracy & Tracy [35]).

5. Discussion

Terrain visualisation software was first introduced in the early 1990's whereby during this time the computing capability of the hardware was still limited. Table 1 shows a list of terrain visualisation software where most of them use C++ and Java as their core programming tools. Most of these software (Biosphere3D, GenesisIV, IDV, Landserf, MicroDEM, Norkart Virtual Globe and OpenEV) has their own graphical user interface (GUI) tools. Besides that, G3GDMV and Hftool use other tools such as GTK, GNOME and SGI with improved GUIs. All of these software was developed to load data locally. Some of the software does not read the data, but uses OpenGL technology for their rendering. Figure 10 shows the evolution of terrain visualisation software from 1995 till now.

IOP Conf. Series: Earth and Environmental Science **20** (2014) 012037 doi:10.1088/1755-1315/20/1/012037

Table 1. Terrain visualisation software.

Terrain software	Year	Language used
IDV	1997	JAVA
Hftool	1998	Linux/SGI
G3GDMV	1999	GTK/ Gnome/Linux
Norkart Virtual Globe	2001	JAVA
OpenEV	2001	OpenGL, C ,GTK
Landserf	2004	JAVA
GenesisIV	2005	C++
MicroDEM	2007	C++
3DEM	2009	SGI/Microsoft OpenGL libraries
Biosphere3D	2007	C++, OpenGL 2.0
Cesium	2012	JavaScript/WebGL

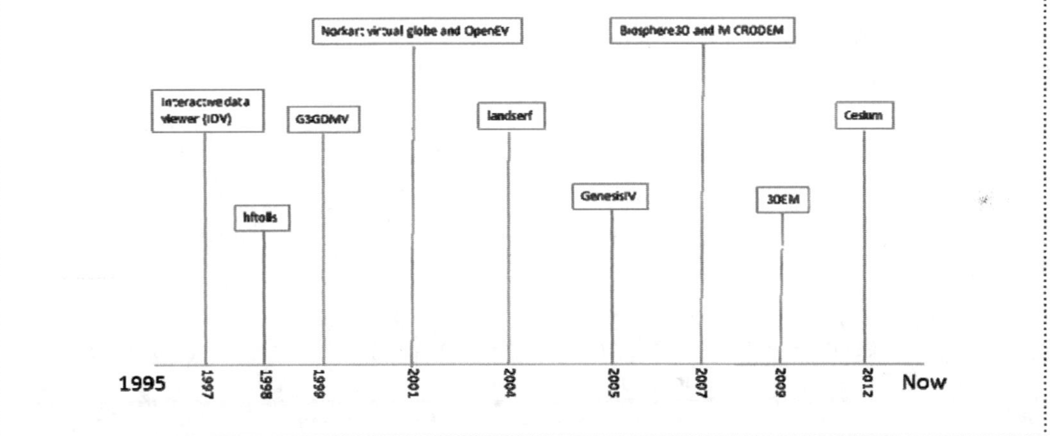

Figure 10. The evolution of terrain visualisation software.

Based on the game engines that have been discussed in this paper, most of them have built-in terrain editor. It is easier for the user to generate the terrain without intensive coding, although using GIS data in game engine would require a few works on coding itself and some of the game engines provide plug-in which can be used by the user to create the terrain. An article by Green [34] looked at how DEM can be used inside UDK to view its data, Unity3D also allows GIS data to be viewed in specific format. The script written by Jeff Craighead allows Unity3D to be used with GIS data [12]. TGE also allows loading of GIS data but it requires the data to be loaded into a compatible format. Cryengine also requires a bit of works to get the GIS data into the game engine. This shows that most of the game engines have the capability in generating the terrain from GIS data. Table 2 compares the different types of game engines. It shows that most of the game engines were developed a few years ago and until now they can still be used for developing new applications.

Table 2. Comparison on different types of game engine.

Game Engine	Year created	Level Editor	GIS support	Source code and SDK	Documentation
Torque 3D	2001 Originate from Tribes 2 game	World Editor Material Editor Terrain Editor Shape Editor Forest Editor River Editor Decal Editor Decal Road Editor Mesh Road Editor	Custom imports	https://github.com /GarageGames/Torque3D Available version 3.5	Available at http://docs.garagegames.com/tge/official/
CryEngine	2004	Sandbox Terrain editor Sound Editor	Custom imports	Available as a package	Available at http://freesdk.crydev.net/dashboard.action
Unity 3D	2005	Game object Mesh Effect Physics Terrain Asset can be imported into a single scene, also uses plugin with editor	Custom imports	SDK available for free as Unity3D source codes is available for purchase	Available at http://wiki.unity3d.com/index.php/Main_Page
Unreal Development Kit	2009 originate from Unreal 3 game engine	Geometry Mode Terrain Mode Mesh Paint Mode Static Mesh Mode Landscape Mode Foliage Mode	Custom imports	http://wiki.beyondunreal.com /UnrealScript_source_code Available in multiple versions.	Available at http://www.unrealengine.com/en/udk/documentation/

6. Conclusion

As the technologies march forward into the future, several new technologies and new software to visualise terrain emerge and that include game engines. These software allows editing and creating of 3D terrains which can be a reliable source of reference to all. Game engines provide different approaches to the GIS data. Game engines also provide vast potential for customization in making GIS data useful in various applications such as tourism, future disaster planning, security and many others. With the help of game engines, GIS data can be viewed from different types of perspectives and angles.

7. References

[1] Wyld DC. 2010 The virtual tourist: using the virtual world to promote the real one *Advances in Competitiveness Research*.**18**.

[2] Wikipedia. Game Engine. 2014 [cited 2014 12 January]; Available from: http://en.wikipedia.org/wiki/Game_engine.

[3] Wikipedia. Arcade. 2014 [cited 2014 12 January]; Available from: http://en.wikipedia.org/wiki/Arcade_game.

[4] Lewis M, Jacobson J. 2002 Game engines in scientific research: Introduction. *Communications of the ACM*.**45**(1):27-31.

[5] GarageGames.com. Torque game engine. 2014 [cited 2014 12 January]; Available from: http://www.garagegames.com/products/torque-3d/overview.

[6] Shiratuddin MF, Thabet W. 2011 Using a 3D Game Engine to Develop a Virtual Design Review System. *Journal of Information Technology in Construction*.**16**:39-68.
[7] Shiratuddin MF, Fletcher D, 2006 Development of Southern Miss's Innovation and Commercialization Park Virtual Reality Environment. *Proc. 6th Int Conf on Construction Applications of Virtual Reality*; Orlando, Florida.
[8] Nilson B, & Söderberg, M. 2007 Game Engine Architecture.
[9] Epic Games Inc. Unreal Development Kit. 2014 [cited 2014 10, January]; Available from: http://www.unrealengine.com/udk/downloads/.
[10] Navarro A, Pradilla JV, Rios O. 2012 Open Source 3D Game Engines for Serious Games Modeling. *Modeling and Simulation in Engineering*.
[11] Shiratuddin MF, Thabet W. 2002 Virtual office walkthrough using a 3D Game Engine. *International Journal of Design Computing*.**4**.
[12] Unity3D. Unity3D. 2014 [cited 2014 10 January]; Available from: http://unity3d.com/.
[13] Wang S, Mao Z, Zeng C, Gong H, Li S, Chen B, 2010 A new method of virtual reality based on Unity 3D. *Proc. 18th International Conference Geoinformatics 2010* Beijing University, Beijing, China: IEEE.
[14] CryTEK. Cry Engine 3. 2014 [cited 2014 12 January]; Available from: http://www.crytek.com/cryengine/cryengine3/.
[15] Trenholme D, Smith SP. 2009 Computer game engines for developing first-person virtual environments. *Virtual reality*.**12** (3):181-7.
[16] Herrlich M, 2007 A Tool for Landscape Architecture Based on Computer Game Technology. *Proc. International Conference on Artificial Reality and Telexistence*; Esbjerg, Denmark: IEEE.
[17] Ruzinoor CM, Shariff ARM, Pradhan B, Rodzi Ahmad M, Rahim MSM. 2012 2012/06/01 A review on 3D terrain visualization of GIS data: techniques and software. *Geo-spatial Information Science*.**15**(2):105-15.
[18] Ruzinoor CM, Shariff ARM, Mahmud AR, Pradhan B. 2012 *3D Terrain Visualisation for GIS: A Comparison of Different Techniques*. In: Buchroithner M, editor. True-3D in Cartography: Springer Berlin Heidelberg. p. 265-77.
[19] Ruzinoor CM, Sharif ARM, Mahmud AR, Pradhan B, 2008 Online 3D Real Time Terrain Rendering Algorithm for GIS Data: A Conceptual Idea. *Proc. 3rd International Workshop on 3D Geoinformation 2008*; Seoul, South Korea
[20] Horne R. 3DEM. 2014 [cited 2014 10 January]; Available from: http://freegeographytools.com/2009/3dem-website-is-gone-but-3dem-still-available-here.
[21] Zuse Institute Berlin, lenne3D Gmbh. Biosphere3D: Real-time open source Landscape Visualization. 2007 [cited 2014 10 January]; Available from: http://www.biosphere3d.org/.
[22] Analytical Graphics Inc. Cesium. 2014 [cited 2014 2 January]; Available from: http://cesiumjs.org/index.html.
[23] Gunia DA. Earth 3D. 2014 [cited 2014 10 January]; Available from: http://www.earth3d.org/index.html.
[24] Geomantics. Genesis IV. 2014 [cited 2014 10 January]; Available from: http://www.geomantics.com/genesis4.htm.
[25] Yost N. Introduction to g3DGMV (3D Graphical Map Viewer). 2014 [cited 2014 10 January]; Available from: http://g3dgmv.sourceforge.net/index.html.
[26] Stock MJ. Hftools. 1998 [updated 1998-09-24; cited 2014 15 January]; Available from: http://markjstock.org/pages/hftools.html.
[27] Unidata. Integrated Data Viewer (IDV). 2014 [cited 2014 12 January]; Available from: http://www.unidata.ucar.edu/software/idv/.
[28] Wood J. LandSerf. 2014 [cited 2014 20 January]; Available from: http://www.landserf.org/.
[29] Guth PP. MicroDEM. 2014 [cited 2014 20 January]; Available from: http://www.usna.edu/Users/oceano/pguth/website/microdem/microdem.htm.

[30] Norkart Company. Norkart Virtual Globe. 2014 [cited 2014 20 January]; Available from: www.virtual-globe.info.
[31] OpenEV. OpenEV. 2014 [cited 2014 20 January]; Available from: http://openev.sourceforge.net/index.php?page=features.
[32] Realmware Corporation. Visual 3D. 2014 [cited 2014 10 January]; Available from: http://www.visual3d.net/game-engine/features.
[33] Yang T, C. WB, Lobb RJ. 2004. Game Engine Support for Terrain Rendering in Architectural Design [Doctoral dissertation]. Private Bag 92019, Auckland, New Zealand. University of Auckland
[34] Green D. Terrains from DEMs: Using Digital Elevation Models 2012 [cited 2014 4 January]; Available from: http://udn.epicgames.com/Three/TerrainsFromDEMs.html.
[35] Tracy D, Tracy S. 2011. CryEngine 3 Cookbook: Over 90 Recipes Written by Crytek Developers for Creating Third-generation Real-time Games: (Packt Publishing Ltd)

Acknowledgments
The authors would like to express their sincere appreciation to Ministry of Education, Malaysia for funding this project under the Prototype Reseach Grant Scheme (PRGS). Without their support and helps, this project project could not be success.

Application of support vector machine for classification of multispectral data

Nurul Iman Saiful Bahari, Asmala Ahmad and Burhanuddin Mohd Aboobaider
Department of Industrial Computing
Faculty of Information and Communication Technology
Universiti Teknikal Malaysia Melaka
Hang Tuah Jaya, 76100 Durian Tunggal, Melaka, Malaysia

E-mail:nuruliman.sf@gmail.com, asmala@utem.edu.my and burhanuddin@utem.edu.my

Abstract. In this paper, support vector machine (SVM) is used to classify satellite remotely sensed multispectral data. The data are recorded from a Landsat-5 TM satellite with resolution of 30x30m. SVM finds the optimal separating hyperplane between classes by focusing on the training cases. The study area of Klang Valley has more than 10 land covers and classification using SVM has been done successfully without any pixel being unclassified. The training area is determined carefully by visual interpretation and with the aid of the reference map of the study area. The result obtained is then analysed for the accuracy and visual performance. Accuracy assessment is done by determination and discussion of Kappa coefficient value, overall and producer accuracy for each class (in pixels and percentage). While, visual analysis is done by comparing the classification data with the reference map. Overall the study shows that SVM is able to classify the land covers within the study area with a high accuracy.

1. Introduction

Land cover classification is one of the most important remote sensing applications. It has been widely used in many fields such as town planning, studies of environmental change, land resource planning, and geological mapping. Generally, a good classifier should be able to classify pixels into desirable land covers. The factors taken into consideration when selecting a classification method include accuracy, speed and practicality. Among the frequently used methods in classification are maximum likelihood classification (MLC) and artificial neural network (ANN). However, there are drawbacks to these classifications; ANN has been associated with over fitting and local minima problems [5], while MLC needs large training area and assumption that the data are normally distributed. In recent years, there have been an effort to develop better reliable classification methods; support vector machine (SVM) is one of them [10]. SVM is characterised by an efficient hyperplane searching technique that uses minimal training area and therefore consumes less processing time. The method is able to avoid over fitting problem and requires no assumption on data type. Although non-parametric, the method is capable of developing efficient decision boundaries and therefore can minimise misclassification. This is done through finding of optimal separating hyperplanes between classes by focusing on the training cases (support vectors) that lie at the edge of the class distributions, with the other training cases being excluded [13]. This study aims to carry out SVM classification on land covers over Selangor, Malaysia.

Originally, SVM is a binary classifier that works by identifying the optimal hyperplane and correctly divides the data points into two classes. There will be an infinite number of hypeplanes and SVM will select the hyperplane with maximum margin. The margin indicates the distance between the classifier and the training points (support vector). Figure 1 illustrates the basic idea of support vector machine [13]. A number of techniques can be used to expand the classifier from binary to multiclass i.e. one against all and

Content from this work may be used under the terms of the Creative Commons Attribution 3.0 licence. Any further distribution of this work must maintain attribution to the author(s) and the title of the work, journal citation and DOI.
Published under licence by IOP Publishing Ltd

IOP Conf. Series: Earth and Environmental Science **20** (2014) 012038 doi:10.1088/1755-1315/20/1/012038

one against one [9]. SVM can classify the data linearly or nonlinearly. Kernel function is used for nonlinear data. By comparison from previous studies, SVM produces classifications with a relatively high accuracy [5], [7], [11].

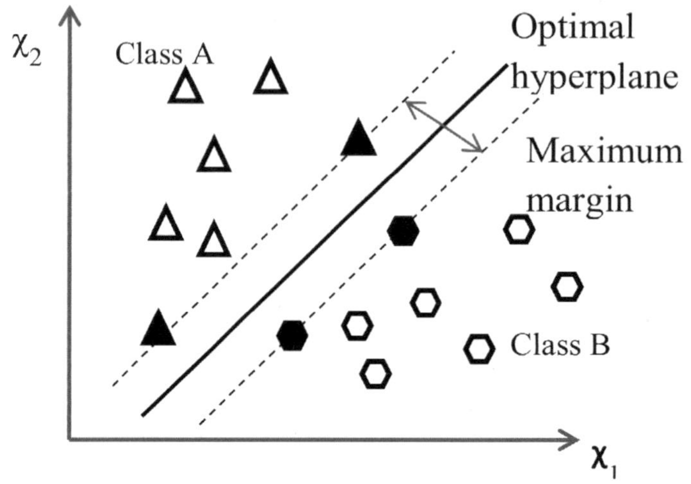

Figure 1. Basic idea of SVM.

Foody and Mathur [7] demonstrated that a single multiclass SVM classification can be applied and derive accurate classifications. They compared different classification methods, viz. discriminant analysis, decision trees, feed forward neural network, and SVM classification, and found that SVM yielded the highest accuracy. The outcome is similar to a study carried out by Shi and Yang [11] that showed SVM surpass the MLC in terms of quantitative accuracies. Candade and Dixon [5] also showed that SVM perform better than other classification. They made use of three different types of SVM kernels such as Polynomial, RBF and linear kernel and compared the results with ANN. They stated that the difference between the ANN and SVM maybe because ANN have the problem of over-fitting, local minima and sensitive to the dimensionality of the data while, SVMs show better accuracy even with a small number of training samples. In this paper, we present the application of SVM in classifying land covers recorded from the Landsat-5 TM satellite over the area of Port Kelang, located in Selangor, Malaysia.

2. Methodology

The Landsat-5 TM dataset, dated 11 Feb 1999 with a spatial resolution of 30×30m, was obtained from the Malaysia Remote Sensing Agency (ARSM). The dataset was subsetted spatially and spectrally. The area is about $840m^2$ within longitude 101°10'E to 101°30'E and latitude 2° 99' N to 3°15'N. The Landsat-5 TM is equipped with seven bands ranging from 0.45 to 2.35 μm, however, only 6 bands were used in this study (band 1, 2,3,4,5 and 7) since band 6 is a thermal band and therefore, is not relevant to this study. Initially, the land covers were visually analysed with the aid of the reference map produced in October 1991 by the Malaysian Surveying Department and ARSM. A total of 10 land cover was identified, i.e. industrial, oil palm, rubber, coastal swamp forest, coconut, dryland forest, cleared land, bare land, and sediment plumes. Pixels that representing water, cloud and cloud shadow were masked out because this study only focuses on land area [2].

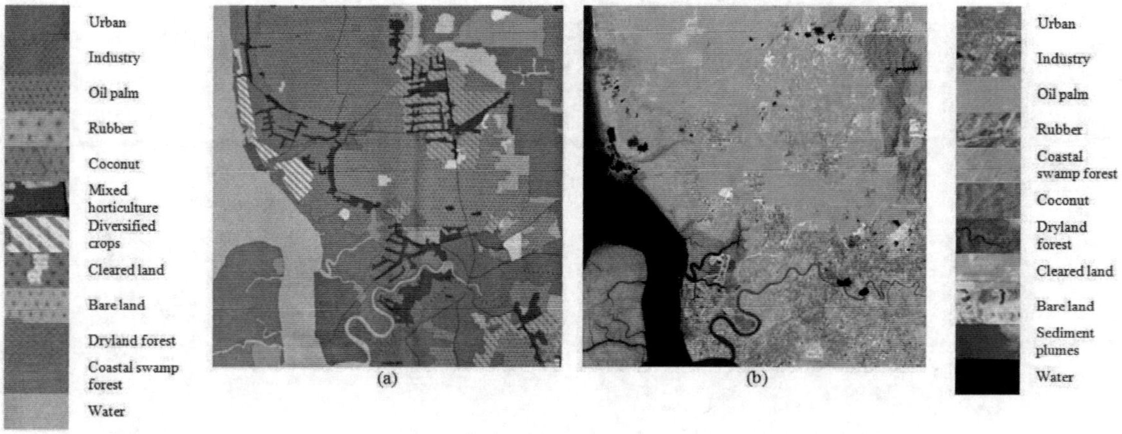

Figure 2. (a) The land cover map (b) The Landsat-5 TM with band 3, 2 and 1 in RGB layer.

Figure 2 above shows the study area at Port Klang. There are three types of plantation with palm oil as the major plantation followed by coconut and rubber. The palm oil plantation can easily recognize by the light green patches spread toward the north area. The coconut and rubber plantation are scattered at northwest and southeast part respectively. Moreover, there are two types of natural forest; in the northeast is dryland forest and coastal swamp forest in the southwest part of the study area. The industrial area (brighter pixel) covers the south part and the urban placement (pixel in red colour) is more concentrated toward southeast along the Klang River. The water bodies (pixel in blue colour) are mainly the Malacca Straits and Klang River. Sediment plume (brighter blue colour pixel) is located at the shore of the water body towards northwest part. The bare land and cleared land are scattered all over the place with the bare land more concentrated in the industrial and urban areas, while the cleared land is more concentrated at the oil palm plantation area.

Training areas were created by choosing polygons that contain training pixels representing the land covers. This step is the most crucial part, since inaccurate training pixels can lead to serious misclassification. Although SVM can classify with only small training areas, in this study, medium to large training areas are used. This is due to the fact that such training areas tend to produce classification with a high accuracy.

3. Results & discussion

Figure 3 below shows the classification performed by SVM with its legend and percentage area for each class. The 10 classes are coastal swamp forest (green), dryland forest (blue), oil palm (yellow), coconut (maroon), rubber (cyan), urban (red), industry (thistle), sediment plume (sea green), bare land (coral), cleared land (purple). Water, cloud and cloud shadow were masked white and excluded from further processing stages since this study focuses on land area only. Coastal swamp forest, dryland, oil palm, urban, industrial and sediment plume classes were easily recognised by their well-grouped pixels. While, rubber, cleared land, coconut and bare land classes are less grouped and more scattered over the places. The largest class is oil palm (34.9%), followed by urban (14.6%) and cleared land (11.9%). While two of the smallest areas classified are sediment plume (3.2%) and industrial (3.5%). Visually, overall performance of SVM in land cover classification is good as it can classify all pixels effectively.

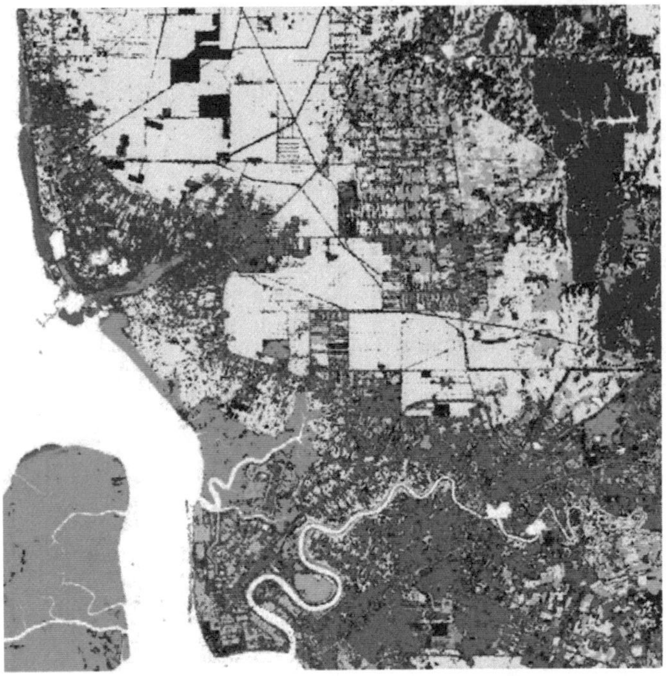

Class	Colour	Area (%)
Coastal swamp forest		8.7
Dryland forest		7.2
Oil palm		34.9
Cleared land		11.9
Sediment plume		3.2
Coconut		10.6
Bare land		1.0
Urban		14.6
Industrial		3.5
Rubber		4.4

Figure 3. SVM classification of Landsat-5 TM data and the class area in percentage. (Water, cloud and cloud shadow are masked white).

For accuracy assessment purposes, selection of ground truth pixels was done by random sampling. Accuracy analysis was carried by comparing the classified pixels with ground truth pixels using a confusion matrix [3]. Table 1 shows confusion matrices for SVM classification in terms of (a) pixels and (b) percentage. The results were presented in terms of producer accuracy and overall accuracy. Producer accuracy is the probability that a pixel in the classification falls into class x given the ground truth class is x and can be calculated using [8]:

$$\text{Producer accuracy} = \frac{C_{aa}}{C_{\bullet a}} \times 100\% \tag{1}$$

where,

C_{aa}= element at position a^{th} row and a^{th} column

$C_{\cdot a}$= column sums

Table 1 (c) below shows the producer accuracy for each class. Overall accuracy is the total percentage of pixels correctly classified and can be computed as:

$$\text{Overall accuracy} = \frac{\sum_{a=1}^{U} c_{aa}}{Q} \times 100\% \tag{2}$$

where,
Q= total number of pixels
U= total number of classes

The overall accuracy of 97.1% indicates that the SVM classification is good at separating pixel into their classes with very few of the pixels were not classified into its classes.

The Kappa coefficient is a measure of the agreement between variables and can be calculated using:

$$\kappa = \frac{\sum_{a=1}^{U} \frac{c_{aa}}{Q} - \sum_{a=1}^{U} \frac{c_{a\bullet} c_{\bullet a}}{Q^2}}{1 - \sum_{a=1}^{U} \frac{c_{a\bullet} c_{\bullet a}}{Q^2}} \tag{3}$$

where,
$c_{a\bullet}$ = row sums

Kappa coefficient value is always less than or equivalent to 1. A value of 1 indicates that the variables are in perfect agreement. The value of 0.96 shows that the SVM classification and the ground truth data are at a very good agreement [1]

Table 1. Confusion matrix for SVM classification.

		Ground Truth (Pixel)										
	Class	CSF	DF	OP	CL	SP	C	BL	U	I	R	Total classified pixels
	CSF	3621	0	0	0	0	0	0	0	0	0	3621
	DF	0	1536	4	0	0	1	0	0	0	7	1548
	OP	0		2504	12	0	77	0	0	0	0	2593
	CL	0	0	1	209	0	2	0	26	3	2	243
Classification Data (Pixels)	SP	0	0	0	2	445	18	0	0	5	0	470
	C	0	0	54	15	20	305	0	0	0	0	394
	BL	0	0	0	1	0	0	78	0	0	0	79
	U	0	0	1	12	0	6	0	537	0	0	556
	I	0	0	0	3	0	0	0	5	162	0	170
	R	0	0	2	4	0	0	0	0	0	165	171
	Total ground truth pixels	3621	1536	2566	258	465	409	78	568	170	174	9845

(a)

Ground Truth (Percent)

	Class	CSF	DF	OP	CL	SP	C	BL	U	I	R	Total classified (percent)
Classification Data (Percents)	CSF	100	0	0	0	0	0	0	0	0	0	36.78
	DF	0	100	0.16	0	0	0.24	0	0	0	4.02	15.69
	OP	0	0	97.58	4.65	0	18.83	0	0	0	0	26.37
	CL	0	0	0.04	81.01	0	0.49	0	4.58	1.76	1.15	2.47
	SP	0	0	0	0.78	95.7	4.4	0	0	2.94	0	4.77
	C	0	0	2.1	5.81	4.3	74.57	0	0	0	0	4
	BL	0	0	0	0.39	0	0	100	0	0	0	0.8
	U	0	0	0.04	4.65	0	1.47	0	94.54	0	0	5.65
	I	0	0	0	1.16	0	0	0	0.88	95.29	0	1.73
	R	0	0	0.08	1.55	0	0	0	0	0	94.83	1.74
	Total ground truth (percent)	100	100	100	100	100	100	100	100	100	100	100

(b)

Class	Producer Accuracy	
	(Pixel)	(Percent)
Coastal swamp forest (CSF)	3621/3621	100
Dryland forest (DF)	1536/1536	100
Oil palm (OP)	2504/2566	97.58
Cleared land (CL)	209/258	81.01
Sediment plume (SP)	445/465	95.7
Coconut (C)	305/409	74.57
Bare land (BL)	78/78	100
Urban (U)	537/568	94.54
Industry (I)	162/170	95.29
Rubber (R)	165/174	94.83

(c)

Almost all classes possess producer accuracy more than 90%, except for coconut and cleared land. The producer accuracy of 74.57% of coconut class is mainly because 18.83% of its pixels were classified as oil palm. This is due to the fact that coconut and oil palm has similar physical structure, producing similar spectral behaviour and misclassified as each other [1]. While for cleared land, producer accuracy of 81.01% is mainly because most of the pixels are being classified as all other classes except for coastal swamp forest and dryland forest. Cleared land representing the land without any vegetation or building like untarred road, tarred road, alley, orchard path, cleared land for next vegetation, and house divider.

However, for 30x30m resolution satellite data, all these features tend to be mixed with features from other classes like oil palm, coconut palm, residential building, industrial building, and rubber tree resulting in misclassification of the pixels. On the other hand, coastal swamp forest, dryland forest and bare land class shows 100% producer accuracy. This is because these three classes have very distinct features resulting in different spectral characteristics and not easily misclassified. In general, producer accuracy for each class shows that SVM can classify very well, except for bare land and coconut class. Such problem may be solved if higher resolution satellite data are used.

4. Conclusions

In this paper, we have studied the SVM performance for tropical land cover classification on Landsat TM data. The study area of Klang Valley has more than 10 land covers and classification using SVM has been done successfully without any pixel being unclassified. The overall accuracy and kappa coefficient of 97.1% and 0.96 respectively shows that this classifier can yield high classification accuracy and has high agreement between ground truth and classified data. It indicates that this classifier is a promising and reliable for remote sensing classification tool although not that well known compared to other methods such as ANN and MLC.

Acknowledgement

The authors thank Universiti Teknikal Malaysia Melaka (UTeM) for funding this study under FRGS grant (FRGS/2012/FTMK/TK06/03/02 F00143) and also the Malaysian Remote Sensing Agency (ARSM), Ministry of Science, Technology and Innovations (MOSTI) for providing the data.

References

[1] Ahmad A and Quegan S 2012 Analysis of maximum likelihood classification on multispectral data Applied Mathematical Sciences 6 6425 – 6436
[2] Ahmad A and Quegan S 2012 Cloud masking for remotely sensed data using spectral and principal components analysis Engineering, Technology & Applied Science Research (ETASR) 2 221 – 225
[3] Ahmad A and Quegan S 2013 Comparative analysis of supervised and unsupervised classification on multispectral data Applied Mathematical Sciences 7 74 3681 – 3694
[4] Altman DG 1991 *Practical Statistics for Medical Research* Chapman and Hall London 404
[5] Candade N and Dixon B 2004 Multispectral classification of Landsat images: A comparison of support vector machine and neural network classifiers. *ASPRS Annual Meeting Proc. Denver CO*
[6] Congalton R G and K Green 1999 *Assessing the Accuracy of Remotely Sensed Data: Principles and Practices* Lewis Publishers Boca Raton FL
[7] Foody G M and Mathur A 2004 A relative evaluation of multiclass image classification by support vector machines *IEEE Transactions on Geoscience and Remote Sensing 42* **6** 1335–1343
[8] Jensen J R 1986 *Introductory Digital Image Processing* Prentice-Hall Englewood Cliffs New Jersey 379

[9] Md Sap M N Kohram M 2008 Support Vector Classification of Remote Sensing Images Using Improved Spectral Kernels *J. Teknologi Maklumat 20*

[10] Mountrakis G, Im J and Ogole C 2011 Support vector machines in remote sensing: A review *ISPRS J. of Photogrammetry and Remote Sensing* 66 **3** 247-259

[11] Shi D and Yang X 2012 Support Vector Machine for Landscap e Mapping from Remote Sensor Imagery *AutoCarto 2012*

[12] Yao W and Han M 2011 Remote sensing image classification with parameter optimized Support Vector Machine based on evolutionary computation *Int. Workshop on Computational Intelligence (IWACI) 2011* 290–294

[13] Vapnik V 1995 *The Nature of Statistical Learning Theory* New York Springer Verlag

[14] Zhu H 2009 Classification of Urban Remote Sensing Image Based on Support Vector Machines *2009 17th Int. Conf. on Geoinformatics*

IOP Conf. Series: Earth and Environmental Science **20** (2014) 012039 doi:10.1088/1755-1315/20/1/012039

Gamification and geospatial health management

David Wortley
CEO and Founder GAITSS, Gamification and Immersive Technologies Strategic Solutions, 6E Coriander Road, Leicester, UK

E-mail: david@gaitss.net

Abstract. Sensor and Measurement technologies are rapidly developing for many consumer applications which have the potential to make a major impact on business and society. One of the most important areas for building a sustainable future is in health management. This opportunity arises because of the growing popularity of lifestyle monitoring devices such as the Jawbone UP bracelet, Nike Fuelband and Samsung Galaxy GEAR. These devices measure physical activity and calorie consumption and, when visualised on mobile and portable devices, enable users to take more responsibility for their personal health. This presentation looks at how the process of gamification can be applied to develop important geospatial health management applications that could not only improve the health of nations but also significantly address some of the issues in global health such as the ageing society and obesity.

1. Introduction

Gamification can be defined as a process which shapes the world (achieves goals/objectives) by influencing the actions, behaviours, characteristics and state of entities within the world (through the use of games strategies and enabling technologies). Gamification is related to but different from serious games and immersive technologies. Gamification is based on the assumption that the way society operates and functions is essentially equivalent to a game in which there are rules, prizes and penalties, winners and losers, competition and collaboration, individual players and teams. Success in society can be a mixture of merit (ability) and chance (fate/luck). Gamification uses a strategic approach to influencing and motivating human behaviours, development and performance by focusing on those factors which have the biggest impact on our actions and attitudes.

This paper concentrates on one of the most important global societal challenges today – public health and explores how enabling technologies and gamification strategies have the potential to address demographic lifestyle related health timebombs such as obesity, dementia, diabetes and cardiovascular diseases.

Content from this work may be used under the terms of the Creative Commons Attribution 3.0 licence. Any further distribution of this work must maintain attribution to the author(s) and the title of the work, journal citation and DOI.
Published under licence by IOP Publishing Ltd

2. Global health challenges

Obesity is a global problem, especially in the developed world (see Figure 1). Obesity is one example of a lifestyle related physical condition that is related to a variety of health problems, especially in later life. There is significant evidence that cardiovascular morbidity is strongly related to the following factors:

- Smoking
- Physical activity
- Nutrition
- Overweight and obesity
- Family history
- High blood cholesterol and other lipids
- High blood pressure
- Metabolic syndrome.

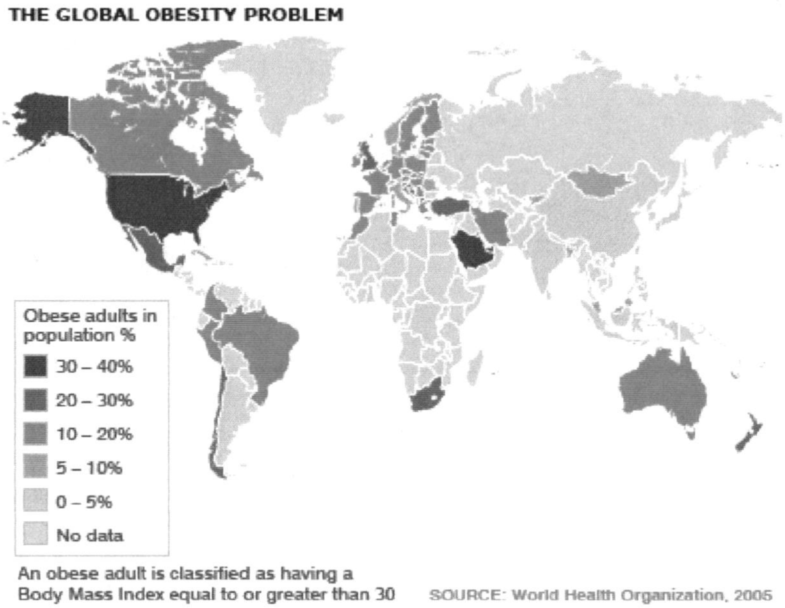

Figure 1. Incidence of obesity in adult population.

In the developed world, whether through National Health Services and Private Medical Insurance, the model for health service provision is based on a universal payment system in which everyone pays to fund the possibility of needing a medical intervention whether that medical intervention arises from unforeseen accidents or self-inflicted health neglect. As the world moves towards an ageing society with increasing proportions of the population in retirement and at greater risk of health problems, there will be insufficient resources to provide health care on demand for all conditions and all people.

Since many of the problems could be addressed by improved lifestyles in the general population, any strategies which can motivate and empower large sections of the population to adopt healthier lifestyles is going to become increasing important. The challenge is how to reduce the demands on the health services that are directly related to poor lifestyles. Governments tend to adopt policies designed to:

- Educate and inform
- Tax "unhealthy" products
- Legislate on food nutritional content
- Offer incentives to provide health beneficial services.

Whilst all these policies make some contribution to reducing unhealthy practices, they do not place any responsibility for lifestyle management on the individual citizen.

3. Gamification and enabling technologies

One of the most significant problems in empowering citizens [1] to alter lifestyles is the lack of knowledge and understanding of how to self-manage health. For those who can afford it, there has been an increasing demand for personal coaches and self-help organisations such as Weight Watchers, both of which provide feedback, education and motivation at a personal level. Today however there are an increasing number of consumer devices (see Figure 2) that track the key performance indicators that reflect health and provide tools on mobile devices such as smartphones and tablets to track and manage lifestyle related activities such as exercise, nutrition, sleep, weight, blood pressure etc.

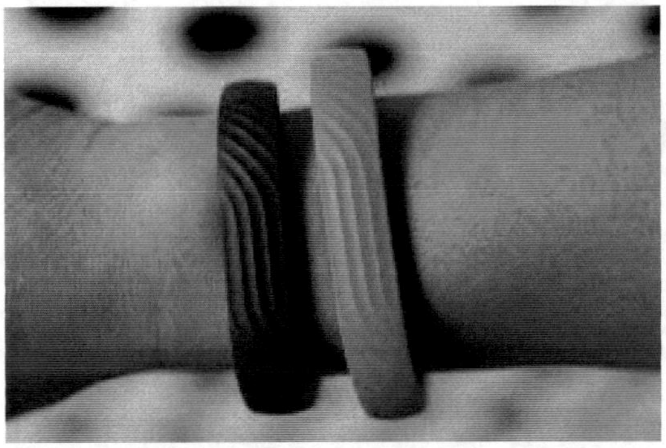

Figure 2. Jawbone UP lifestyle monitoring bracelets.

These technologies enable gamification strategies that can give individuals greater control and understanding of their personal lifestyle factors and thus reduce the potential burden on society the arises from unhealthy lifestyle practices. The results of added control and understanding, linked to positive motivation can be impressive (see Figures 3 & 4).

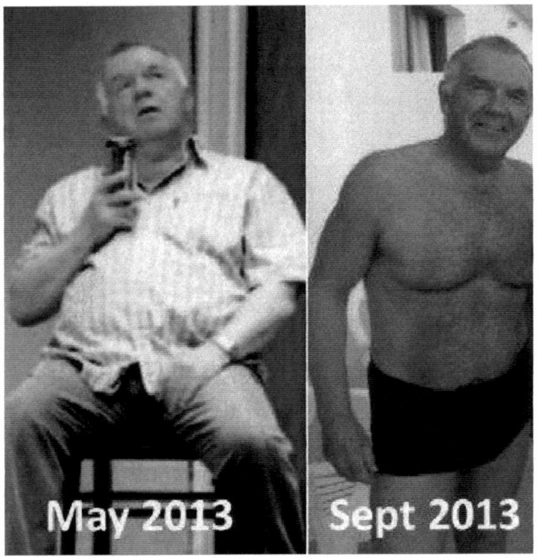

Figure 3 Impact of lifestyle monitoring technologies on weight.

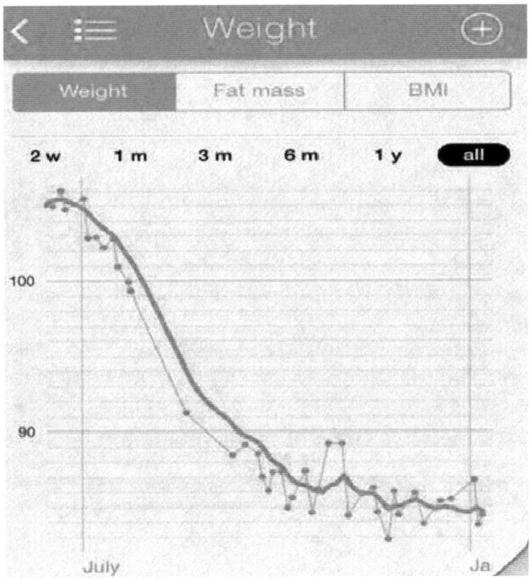

Figure 4 Withings Wifi weight chart.

4. Geospatial health management

The use of lifestyle monitoring devices together with a series of "win-win" relationships that can be brokered by the concept of co-production in an ecosystem or health promoting network, including the users, health and fitness professionals, insurance and nutrition industries. By adopting gamification practices that incentivise and reward not only healthy lifestyles but also the collection of key lifestyle data by each individual citizen, it becomes possible to generate "cloud-based" big data that can greatly enhance our ability to manage today and plan for tomorrow. As such devices and/or the mobile applications they use invariably have embedded GPS tracking, it becomes possible to not only analyse geospatial related health and lifestyle features to inform policy decisions, but also to use social networking techniques to motivate citizens in a geolocation to act collectively as a team to improve the health of their community.

5. Conclusion

Society is on the threshold of major challenges to the way we manage public health because the existing funding practices do not provide incentives or penalties for self-management of lifestyle. The use of consumer wearable lifestyle monitoring technologies with geospatial enabled data capture can not only provide better data for public health management but also motivation for better community health practices.

6. References

[1] Wortley D J 2012 *Gadgets to God* (Leicester, UK: Matador Publishing)

IOP Conf. Series: Earth and Environmental Science **20** (2014) 012040 doi:10.1088/1755-1315/20/1/012040

Some legal concerns with the use of crowd-sourced Geospatial Information

George Cho

Faculty of Education, Science, Technology & Mathematics,
University of Canberra, University Drive,
Bruce, ACT 2617, Australia

E-mail: George [dot] cho [at] canberra [dot] edu [dot] au

Abstract. Volunteered geographic information (VGI), citizens as sensors, crowd-sourcing and 'Wikipedia' of maps have been used to describe activity facilitated by the Internet and the dynamic Web 2.0 environment to collect geographic information (GI). Legal concerns raised in the creation, assembly and dissemination of GI by *produsers* include: quality, ownership and liability. In detail, accuracy and authoritativeness of the crowd-sourced GI; the ownership and moral rights to the information, and contractual and tort liability are key concerns. A legal framework and governance structure may be necessary whereby technology, networked governance and provision of legal protections may be combined to mitigate geo-liability as a 'chilling' factor in VGI development.

1. Introduction

Spatial law and policy occupies a fascinating phase in technological development as new legal relationships and operational imperatives arise. The 'grey' areas of interpretation and respective responsibilities are set to become more prominent. Geospatial information has been transformed by 'disruptive' technologies ('innovative') that have dictated a new *modus operandi*, from the collection of spatial data, storage, analysis to its use in information systems. The previous use of the floppy disk and compact disc (CD) to exchange data and information has given way to the universal service bus (USB), flash drives and cloud computing (CC). The displacement of CDs and digital video discs (DVD) as a storage media by CC has potentially made these physical media redundant. The information technology evolution has also witnessed innovative uses of the Internet from Web 1.0 to Web 2.0. In its latest guise, many new possibilities have emerged including the sharing of maps and spatial information in one common storage medium – the 'cloud'.

This paper outlines some preliminary findings and concerns that are caused by disruptive technologies. The literature has prominently focused on technology and apart from Scassa (2013), little attention has been paid in previous studies to legal concerns. This paper posits the view that there are many uncertainties that may need to be resolved before the operational and technological innovations may be deployed on a large scale. In part 2 of this paper the Web 2.0, crowd-sourced maps and *produser* environment is described. This discussion leads to parts 3, 4, & 5 that raise

Content from this work may be used under the terms of the Creative Commons Attribution 3.0 licence. Any further distribution of this work must maintain attribution to the author(s) and the title of the work, journal citation and DOI.
Published under licence by IOP Publishing Ltd

concerns with the quality of data and information, ownership especially of intellectual property rights, and legal liability in contract and tort respectively. The potential risks for each of these concerns have to be appropriately managed. Part 6 examines current legal frameworks and the legal governance of the collection and dissemination of geospatial data. The final part 7 summary and conclusion identifies five observations and suggested actions that may be taken to resolve the practical problems for *produsers*. Many of the findings reported and discussed here are preliminary ones but nevertheless provide a basis for proposing a governance framework for all parties.

2. Web 2.0, Crowd-sourced Maps, VGI and *Produsers*

This section lays the groundwork for a common understanding of the Internet environments beginning with Web 1.0 through to the collaborative milieu of Web 2.0 with innovative ways of harnessing the myriad applications. In particular, the spatial dimension and its mapping and recording is of particular interest especially with volunteered geographic information (VGI) and the different producers and users as participants in crises, emergencies and disaster management.

In the Web 1.0 environment, the World Wide Web (WWW) known synonymously as the Internet was envisioned by Sir Tim Berners-Lee as "a collaborative medium, a place where we [could] all meet and read and write".[1] Recent developments on the Internet and especially within Web 2.0, have extended the range of use – from the desk-bound to the mobile device. However, the more ubiquitous 'application' (apps) of mobile electronic devices has produced new legal concerns in general and liability in particular. More specifically, net-citizens (netizens) who contribute to the collection, use and dissemination of geospatial information have triggered a re-evaluation of traditional legal frameworks. The legal and policy framework is continually being cajoled to adapt the law to provide novel solutions for the digital age.

In the Web 2.0 environment the static pages of websites are used and updated by collaborative and cooperative means by users for dialogue, creation of content, and the online meeting of virtual communities. The new environment and other mobile technologies have opened up the landscape for nearly everyone to access the Internet, to create and share content. There is a democratisation in the production of information.

In this environment communication takes place anytime and anywhere. However, the management of such activities will invariably be a place-based where 'jurisdiction' becomes pre-eminent and the law of the land prevails. In the absence of specific laws to govern electronic-based activities in a jurisdiction, existing law and regulations may have to be adapted to cater for and address new challenges of technology. The adaptation of the law could be by way of analogising case by case or re-writing rules and regulations for 'special or exceptional cases'.

Furthermore, innovative ways of collaborating and disseminating information have been created to harness the dynamic power of Web 2.0. Paper maps have made way to electronic maps together with the so-called 'crowd-sourced' maps. *Ushahidi* ('witness' in Kiswahili) is the interactive map that was developed by concerned citizens during the disputed Kenyan elections of 2007. The interactive maps provided eyewitness accounts of violence across Kenya. The maps were accompanied by written text in *Ushahidi* and images were displayed on Google Maps. This example demonstrates the potential applications that could assist first responders and humanitarian organisations in the wake of natural disasters, crises and violent conflicts.

Volunteered geographic information (VGI) is "the widespread engagement of large numbers of private citizens, often with little in the way of formal qualifications, in the creation of geographic information".[2] VGI is part of a broader trend of "user-generated content" (UGC) that is becoming commonplace in Web 2.0-based applications. The VGI community is set to play an influential role in modern map making and they have sometimes been described as *produsers* defined below.[3]

Produser is a portmanteau of two words combining 'production' and 'usage' – a word popularised by Axel Bruns (2008).[4] The word refers to the type of user-led content creation that takes place in a variety of online environments such as Wikipedia. There is the blurring of boundaries between producing and using, hence the *produser*.

Examples of other VGI include need for information and maps in crisis situations, for example, during the Haitian earthquakes of 2010.[5] Volunteer mapping efforts that assists in relief efforts have since been used in the context of forest fires, floods, hurricanes and other disasters. Volunteers capture and record information using mobile devices and submit these to a website or central agency that analyses, reviews and then publishes the data on base maps, for example, those provided by OpenStreetMap or GoogleMaps.[6] Most of these contributors are novices and non-professional map creators. While reliability and accuracy of VGI might be questioned, the timely production of near-instantaneous maps during times of crisis has proven invaluable. Further examples of VGI include the Libyan crisis map in 2011; the United Nations Organisation for the Coordination of Humanitarian Affairs (UNOCHA) maps; Transparency Morocco; Mamdawrinch; and Syrian Tracker 2013.[7]

Scassa (2013) identified a series of potential risks from the perspective of the operator of a VGI site, the contributor of VGI and user of the product, service or application that is created using VGI. The article gave an overview of emerging issues of which developers, contributors and users should be aware when engaging in VGI-related activities. The lessons for all are timely.

The digital environment has made nearly everyone with a mobile device a map maker and a digital recorder. Nearly every phenomenon that is observable may be captured and mapped. Using cell phones, global positioning system (GPS) devices and placing information on the Internet using social media has elevated the importance of the spatial context. The newly discovered cartographer may be untrained in geographic information systems (GIS) technology but are now enabled to contribute to location-based services and bringing GIS to the masses.

While expectations are on the rise, so are the legal risks. Legal liability questions have been raised where the subscribed data might be biased and wrong leading to an endangerment to life. VGI service providers on their part have to ensure that they do not engage in conduct that is deemed "negligent" and to verify volunteer contributions.

The ubiquity of location-aware devices that automatically record personal movements, habits and other information linked to one's location raise other concerns such as privacy. Equally important are those relating to the ownership of geospatial information and the kinds of tacit agreements between producers and users.

Many of these questions have remained largely in the background, lightly discussed and seldom highlighted. In the following three sections, issues of quality of data and information, of ownership and of legal liability are reviewed to help prepare a proposed legal framework.

3. Concern # 1 Quality of data and information

Amongst the range of potential risks is the concern with the quality of the data contributed by the 'crowd' relative to those produced by professional mapping agencies. Data authoritativeness is closely linked to the background and training of the contributors. Traditional government sources of GI have always been considered more 'authoritative' because of the professional training and other quality checks in place. What is considered authoritative geospatial data may be debatable and appropriate criteria may be necessary.

However, what is not in dispute is that the production of authoritative geospatial data is costly, time consuming and usually undertaken by the public sector. Evidence suggest that many government mapping agencies might be struggling to obtain continuing funding to maintain geospatial data or the mandate to keep geospatial databases current. As a result it it seems that there may be a growing reliance on VGI data. While on the one hand, VGI data sources could be highly accurate, complete, in time and at a practical useable scale,[8] the potential risks are still largely unknown and little understood.

The life cycle of GI from the collection of data, cleansing, and assembly of results are prone errors and inaccuracies. Without careful editing and ground-truthing inaccuracies might creep in through misinterpretation of observations, digitising errors, resampling and the use of incorrect map projections as well as the inexpert final assembly of results.[9] For VGI data to be readily acceptable the issues of quality control and monitoring of VGI contributions are a prerequisite as the risks are neither hypothetical nor inconsequential. Tragic consequences could ensue from using 'bad' data.

The Canadian the Centre for Topographic Information, a division of Natural Resources Canada, has begun assessing the potential of a collaborative mapping model that combines contributions from provincial and municipal organisations as well as those from its citizens.[10] In Australia, the state of Victoria's Department of Sustainability and Environment has developed a Notification and Editing Service that encourages crowd-sourcing of GI. The OpenStreetMap Collaborative Prototype (OSMCP) project of the U.S. Geological Survey (USGS) and swisstopo [*sic*] Revision Service of the Swiss Federal Office of Topography are experimenting with similar ideas.

In the U.K. OpenStreetMap (OSM) is one of the better known examples of collecting and leveraging VGI. Initiated in 2004, the project was a response to the strict copyright laws that apply to Ordnance Survey maps and geospatial data. With OSM any interested party may view and edit geographic data in a collaborative way. Registered contributors may make additions, modify inaccurate features, delete stale or invalid data and generally edit features to improve the quality of the geospatial database. In addition to individual contributions, organisations have donated complete data sets to OSM. Ironically OSM does not use quality control experts to vet contributions. The quality of the data is refined over time through iterative corrections of the submitted data by subsequent contributions.[11]

The concerns about the quality of data also raises issues about who owns the contributed geospatial data in the mix. The rights to ownership of the intellectual property can be a 'real' issue especially when the end use is transformed from the altruistic ideals to the for-profit motives.

4. Concern #2 Ownership: Intellectual property and moral rights

Intellectual property rights (IPR) including the ownership of copyright is a pressing concern with VGI data. In most common law jurisdictions copyright is vested in the author of an original work that is the expression of an idea that also involves a degree of skill and judgement in the work.

In the U.S. 'originality' focuses on a creative spark, rather than on the 'sweat of the brow', whereas, in Canada 'originality' requires skill and judgement and not copied from a previous idea.

With electronic databases, concerns have been raised about whether the 'arrangement' of data qualify as sufficiently original. The views here may be contrasted to database protection laws in the EU especially the *sui generis* regime where protection is afforded not just to an original selection or arrangement of the data (similar to copyright law) but to the underlying data itself. In VGI applications and its products there may be independent rights in the underlying base map. Given the nature of any map is made up of layers of information there can conceivably be multiple copyright holders in any given work.

With many contributors adding to the volunteered geospatial database, there can be difficulties in tracing the heritage of the data and hence the ownership of the copyright. The 'confected' or synthetic data may 'lose' copyright status because of the difficulties associated with tracing the current and original owners of the data. It has been suggested that Creative Commons licences could assist in alleviating such a problem[12] and to avoid what Scassa (2013) has described as the 'Wiki' effect – where data are combined from thousands of disparate sources to form one or more coherent works. The perverse outcome may be that the confected data may become more inaccurate, corrupted with further errors and there may be no one to blame because there is no copyright holder or there may be too many owners to be held accountable.

The Australian Copyright Council has proposed several steps to manage and minimise potential liability risks when posting data on websites.[13] These include terms and conditions that state that the host service provider will not infringe copyright. In addition to indemnities, service providers must ensure that the agreements are legally binding. Service providers have also to educate users not to post material without the relevant copyright owner's consent, to retain the right to remove material if copyright is infringed, to monitor sites for potentially infringing material, and to moderate the process of accepting contributions with a clear statement as to how to complain and how to assert one's rights.

In cases where the VGI contributed data contain material that infringe copyright whether advertent or otherwise, efforts should be made to remove such material and steps put in place to prevent further infringement. Also the use of disclaimers to alert users to the limitations of the data can help mitigate and prevent litigation.

Other service provider-contributor agreements include the terms of service to be provided by the website, disclosure of information to third parties and data protection protocols including indemnities and warranties regarding the data in storage.

Traditionally geospatial data produced by government agencies have copyright vested in the Crown even in cases where the agency is the first publisher of VGI contributions. In these instances it is important to communicate the intention of the Crown to contributors and to emphasize the terms and conditions of use. In turn, VGI contributors may surrender ownership and use of their data and to limit the extent of the permitted uses.

There may be a binding agreement between the contributors and the websites using the contributed data to ensure that they have not infringed the rights of third parties and that they will indemnify site operators for any damages arising from any such infringements. Such an agreement could include a statement that unequivocally states that the contributors themselves have the rights to the contributed data. On the other hand, contributors may request attribution and acknowledgement of their materials. In most jurisdictions moral rights clauses in the legislation enable the assertion of such rights.

Moral rights, recognised by Australia, Europe and the Berne Convention – *droit moral* -- protects the integrity of a creator's works in terms of the right to attribution (claim to ownership of a work), right not to have authorship of a work falsely attributed and a right to integrity (a right to object to distortion, mutilation or other modification which might be prejudicial to the creator's honour or reputation).[14]

Ownership issues in terms of the consent to release the data containing private, confidential and sensitive information are also pertinent here. The website may have implicit access to all data found on its servers. However, there is the danger that the data could be disclosed to third parties innocently, advertently, inadvertently by accident, and deliberately by unauthorised persons. The key to such security risks hinge on the contractual relationship between the user and the service provider and how these are managed.

In some cases, there may be jurisdictional prohibitions on the transfer of data across international borders. The EU Data Protection Directive, for example, regulates the processing of personal data within the European Union.[15] Important components of this Directive include privacy and human rights.[16] On 25 January 2012 the European Commission unveiled a draft of the EU General Data Protection Regulation that will supersede the Data Protection Directive which may have direct implications on geospatial cloud computing usage and deployment of crowd-sourced geospatial information on websites.[17]

The potential for the deployment of poor/bad data on VGI sites, misuse of the data, and innocently 'illicit' data can lead to a number of potentially unpredictable consequences. Examples of such use of innocent illicit data include the unauthorised use of copyrighted data contributed by a third party data supplier, and the inappropriate use of personal information by VGI user organisations that may contravene privacy and data protection legislation.

There may also be ethical considerations in the release of private information on public documents such as published maps. Scassa (2013) gave the example of *Proposition 8 Maps* – a mash-up of Google maps and Prop 8 donors to a campaign in support of a ban on gay marriages in California.[18] The map revealed names, addresses and donation amounts of donors. Such revelations were unforseen by the donors and for some who may not have consented to the use of the information in this manner. The potential for mischief, subsequent harm and tort liability as a result of posting of the information could affect the VGI host, contributor and innocent bystander.

A breakdown in contractual relationships and the likely impact of tortious damage from using inaccurate and/or sensitive data point to a discussion of legal liabilities and personal responsibilities that follows next.

5. Concern #3 Liability: contractual and tortious

While appearing to be a semantic one, there may be difficulties in determining whether a VGI website is the source of a product, service or mere provision of information. In general, legal liability in common law attaches to negligent behaviour where a duty of care was owed and was breached and where there was a causal link between the negligent act and the loss suffered.

In the case of contributors in the 'crowd' who may not have any specialised training or expertise in mapping or geography, mistakes and errors may occur the data collection phase itself. A threshold test for liability is whether there was an exercise of due diligence of a reasonable person engaged in that activity. Liability may be unlikely for an innocent mistake.

Tortious liability may arise in a number of other ways. As a shield against any liability VGI site operators could insist that users accept the terms of use of their data.

VGI data supplier organisations may be liable for negligent contributions of erroneous data that result in injury or damage to persons who have relied on that data. Data accuracy is a critical element that VGI organisations should guard jealously. The extent of the responsibility for user generated content could hinge on the degree of quality control in the data it receives, the editing and filtering of the data, the nature and subject-matter of the data and the degree of control asserted by the data supplier organisation in the data. Under common law tort, the greater the control the greater the responsibility and hence exposure to liability should anything go wrong.

Professional geospatial data producers assume a higher standard of care. They may be sued for providing incorrect data, misleading or negligent information as well as defamatory information.[19] However, it would be difficult to hold amateur contributors to the same standard of care as the collection of their data may not have undergone as rigorous a process as the professionals. Their data may contain minor errors, and may be a synthesis of data from numerous sources.

Another concern with VGI activity is that of the exposure to liability of the volunteer contributor. These volunteers may be described as 'Good Samaritans'. In many jurisdictions, Good Samaritan law is not found on statute books but is a concept adopted by the courts and applied as public policy. However, practices vary in different jurisdictions.

In Australia, the various states have enacted Good Samaritan laws that protect professionals providing assistance at the scene of an accident or any emergency. Most of these provisions define a Good Samaritan (rescuer) as a person acting without expectation of payment or other consideration who comes to the aid of another. Hence, a Good Samaritan should not be liable for assisting in an emergency if all reasonable care and skill has been taken.[20]

However, the laws in much of Europe criminalize a failure to help people in an emergency. Indeed in Germany, the failure to provide first aid to a person in need of help is punishable under § 323c of the German Criminal Penal Code.[21]

In the U.S., courts have taken a view that there is no duty to rescue.[22] Robson's (2011) discussion of the so-called Good Samaritan laws in the U.S. unequivocally demonstrates that a person is not required to rescue another, even if the person could do so safely.[23] The discussion suggests that:

- A volunteer group that undertakes to rescue others exposes itself to liability.
- A duty to rescue arises only when a person puts another in danger but has a responsibility to mitigate the harm and danger.
- A duty to rescue is required where there exists a special relationship between the parties.

Emergency services using crowd-sourced geospatial information are in an invidious position. The 'Good Samaritan' duty of emergency services could be mitigated by not undertaking rescue services and in the case of collecting geospatial data in situ of the emergency, by not providing the crowd-sourced data and maps. It would be ironic to note that, the digital volunteers could be considered to assume the role of a passive observer and not communicate with the victims. Or, the digital volunteers could discourage reliance on the use of information with disclaimers and other notices with statements to the effect that the information are not to be relied upon in life threatening situations.

Other worries about legal liability, however, remain in the sense that crowd-sourced data might be misused; volunteers could get hurt in collecting the data; the legal responsibility for the data are unascertained and unascertainable; and the legal status of the volunteers themselves, as a loose collective of like-minded people who wish to help, is uncertain.[24] The U.S. Federal statute *Volunteer Protection Act of 1997 (VPA)* offers some protection to a broader range of volunteers such as operating under organisational structures of a particular kind and imposing limits on volunteer compensation.

In the case of contractual liability – relationship between the VGI website service provider and users – the 'duty' is expressed in the "Terms of Use" posted on the web page. Thus, for contributors, particular care must be exercised by not giving or promising any type of warranty or guarantee for the data submitted. To do so might expose and heighten the risk of legal responsibility for any errors. Contributors could refrain from providing any indemnification for the data and information submitted to the website.

From the preliminary findings of the above review, a legal framework may be needed to manage the viewpoints of the three major actors in the piece – the service provider as host, the 'crowd' as volunteer contributors, and the disaster manager both as contributor and user.

6. Current Legal Frameworks and Governance

A preliminary conclusion is that the VGI community in general, and crisis mapping community in particular, may need to develop general standards that address issues of quality, ownership and liability. Each of these issues come about where crowd-sourced maps are deployed whether in a disaster situation or in the calm surrounds of a planning studio. It seems that the consensus is that laws that regulate the use of technology, the governance of volunteered information and protection for all is undoubtedly a looming challenge for policy makers and governments. One view is that if the legal questions are not properly managed, tort liability might have "the potential to destroy the model before it realises its potential".[25]

Given the need for large-scale data and the ubiquity of geospatial data, technology has been developed to make these available anywhere, anytime and in any way. The Internet and sharing on the cloud and its use for geospatial applications promise a lot but many unresolved legal issues remain.

Since 2010, the Ordnance Survey (OS) GB, the National Mapping Agency in the U.K., has made available geospatial data to the public to encourage innovation and government transparency.[26] OS is by far the largest public sector agency that makes use of cloud computing as part of its web mapping service that deploys geospatial data to customer websites and enterprise systems, services that are hosted on the Amazon Web Services (AWS) platform.[27]

Increasingly one platform that is being used to host VGI and crowd-sourced maps is the 'cloud'. Cloud computing provide a means to undertake computing tasks, software, data access, data storage resources. Users of cloud computing (CC) take advantage of computing infrastructure but are not involved in the detailed technical setup.

The advantages, offered by CC generally and to geospatial cloud computing (GCC) in particular, to analysts, users, data providers and others, are unlimited. The benefits include lower costs, provide enlarged access to products and services and be able to do so online theoretically from anywhere in the world. The need for geospatial software may be leased or paid for on the basis of use and data used acquired on a needs basis through web services. This is indeed neo-geography *par excellence* and the 'cloud' is poised to radically change the way geospatial analytics are performed.

However, geospatial cloud computing raises other concerns such as the security of information and data in terms of unauthorised use of the services, the protection of personal, confidential and sensitive data, intellectual property rights especially of copyright and licensing the use of geospatial data and legal liability.

The Natural Resources Canada's *Primer on Cloud Computing* discusses a survey that lists ten different security risks. Among these risks include abuse and malicious use of cloud computing by spammers and hackers, weak application programming interfaces (API) that expose cloud computing users to confidentiality and integrity risks, technological vulnerabilities arising from shared technologies, data leakage, service and traffic hijacking, delegations of authority to the vendors and encryption issues.[28]

While the risk concerns appear credible especially with reports of data security lapses, helpful advice is available, for instance, from the European Network and Information Security Agency (ENISA).[29] In addressing some of these concerns an understanding of the roles and responsibilities of both the vendors and users are as important as assessing the threats and mitigating losses. Where these rights and duties have been established, the sharing of liability as well as the burden of the liability may be apportioned by mutual consent through arbitrators and mediators.

The International Bar Association (IBA) has been in the forefront of thinking in attempting to meet the challenges of technological developments. The IBA's proposed *International Geoinformation Convention* focuses on the reliability of information for the particular application and the limits the law should seek to impose on the undoubted power of the information. By addressing the many conflicting and overlapping existing rules and regulations associated with the collection and processing of GI, a new industry may develop with commercial certainty about rights and obligations of its many participants and also the confidence in moving in conformity with public opinion.[30] It may be posited that such an industry will not spontaneously blossom without the trellis provided by a robust international legal framework.

7. Summary and Conclusions

This paper has reviewed several concerns with the use of VGI-generated geospatial data. Five conclusions may be made in summary.

First, the Web 2.0 environment has witnessed a new wave of data collection methods, data use and data analytics. The challenge is how to manage this new world of geospatial data collection and dissemination. Cloud computing, the Internet of Things (IOT), open standards and open source all challenge traditional means of 'professional' data creation, collection and storage. With public sector funding of the collection of geospatial data becoming more restricted in straitened economic times, the public and private geospatial industry may have to turn to use shared crowd-sourced data. Protection by way of licencing and new methods of data ownership will have to be developed. In turn standards and new usage policies may also have to evolve. Yet, there remains tort liabilities related to questions of data assurance, integrity and ownership.

Second, there seems to be an acceptance of the terms of use of VGI data as well as the consequence of the negligent contributions of erroneous data. Also with amateur collectors and contributors of the data, there is an apparent lowering of the standards of care. The use of volunteered and crowd-sourced geospatial databases has highlighted the tensions with Good Samaritan laws especially for rescuers who may not be shielded legal liability. But the laws are mixed for different jurisdictions from the hard line U.S. position to the more benevolent Australian states. The challenge here for policy makers is that if there were to be no legal protection for volunteers, the ensuing litigation may destroy the VGI model before it reaches its full potential.

Third, cloud computing is the next 'storage' media that appeals to the needs of the neo-geography *par excellence*. Cloud computing is changing the way geospatial information collection and geospatial analytics are undertaken. However, the issues of security of information, the protection of personal, confidential and sensitive data, intellectual property and aspects of tort liability hover in the background.

Fourth, the record and reception of crowd-sourced geospatial data among different jurisdictions is mixed. For example, newly developed countries like China and India, and developing countries like Indonesia are very 'security conscious' and sensitive in so far as geospatial information and data are concerned. The fear of threats of invasion and border incursions persist.

Indonesia provides an extreme example where all geospatial data are collected by a central agency and no one else may collect such information. The Indonesian *Geospatial Information Act No. 4, 2011* provides the legal framework for acquiring accurate geospatial data and creates a regulatory framework for the administration of national geospatial information. Criminal sanctions are attached to deter the collection of geospatial data by private persons.[31]

Fifth, radical changes are needed to reduce *geo-liability*. This is because legal liability theories governing geospatial information at least in the common law world are largely under-developed and there is a tendency for the law to lag behind technology relative to resolving the growing problems and issues presented by technology and neo-geography. The disparities between legal and policy frameworks are evident.

Technological development in relative terms is without boundaries. Legal and policy frameworks tend to 'lag' and are not developed in a consistent manner. Such under-development leads to a degree of uncertainty and perhaps a 'chilling' effect on venture capitalists who might refrain from new and innovative investments for 'fear' of unknown legal repercussions.

On the other hand, addressing the legal and policy challenges head-on may lead to greater developments in geospatial information production and use. A legal framework that provides 'security' in all ways could also lead to long-term benefits to those concerned.

Given the three-way interlocking network of interests, the governance of VGI activity can be fiendishly difficult because of the various competing interests and tensions between each of the actors. This preliminary work has only focussed on the legal issues. The difficulties of legal analyses become more apparent in the absence of litigation and/or specific legislation in the field. Yet, the distinctiveness of the problem-set may warrant particular policies to address competing rights and interests and to emerge with solutions to practical problems in the use of geospatial VGI.

The final analysis suggests some kind of a 'relational' law is required within a framework where technology, networked governance and legal protection of geospatial information are woven into place. In such a scenario, it would be possible for volunteers to craft a code of practice that can be universally practised. Such a softly-softly self-regulated environment could prove fruitful and more easily acceptable and accepted by like-minded geospatial professionals and by the map using first responder community – the *produsers.*

References and End Notes

Ather, A (2009) 'A quality analysis of OpenStreetMap data' Department of Civil, Environmental & Geomatic Engineering, University College London at http://homepages.ge.ucl.ac.uk/~mhaklay/pdf/Dissertation-OpenStreetMap-Quality-Aather-2009.pdf accessed 08/08/2013.

Australian Copyright Council (2009) *Websites: User-generated Content and Web 2.0* at http://www.copyright.org.au/find-an-answer/browse-by-keywords/ accessed 08/08/2013.

Ball, M (2010) 'What's the distinction between crowdsourcing, volunteered geographic information, and authoritative data?' in *V1 Magazine* at http://www.vector1media.com/dialog/perspectives/16068-whats-the-distinction-between-crowdsourcing-vol accessed 08/08/2013.

Bruns, A (2008) *Blogs, Wikipedia, Second Life and Beyond: From Production to Produsage*, New York: Peter Lang.

Cho, G (2005) *Geographic Information Science: Mastering the Legal Issues*, Chichester: John Wiley & Sons Ltd.

Coleman, D, Georgiadou, Y & Labonte, J (2009) 'Volunteered Geographic Information: The nature and motivation of produsers', *International Journal of Spatial Data Infrastructures Research,* vol. 4.

Cloud Security Alliance (2010) *Top Threats to Cloud Computing V.1.0* available at https://cloudsecurityalliance.org/topthreats/csathreats.v1.0.pdf accessed 08/08/2013.

DiNucci, Darcy (1999)' Fragmented Future' *Print* 53 (4): 32 at http://darchd.com/fragmented-future.pdf accessed 06/07/2013.

European Network and Information Security Agency (2009) *Cloud Computing Information Assurance Framework* at http://www.enisa.europa.eu/act/rm/files/deliverables/cloud-computing-information-assurance-framework accessed 08/08/2013.

European Union (2012) General Data Protection Regulation, EC 25 January 2012 http://ec.europa.eu/justice/data-protection/document/review2012/com_2012_11_en.pdf.

European Union *Data Protection Directive 95/46/EC* on the Protection of Individuals with Regard to the Processing of Personal Data and on the Free Movement of such Data.

Goodchild, MF (2007) 'Citizens as sensors: The world of volunteered geography' at http://www.ncgia.ucsb.edu/projects/vgi/docs/position/Goodchild_VGI2007.pdf accessed 08/08/2013.

Hunter, G *et al.* (2009) 'Spatial data quality: Problems and prospects' in *Research Trends in Geographic Information Science*. Lecture Notes in Geoinformation and Cartography, DOI 10.1007/978-3-540-88244-2_8, Springer-Verlag Berlin Heidelberg, pp. 101-121.

Jackson, JZ & Vaurio, AM (1999) 'Good Samaritans in the Hospital', 158 *New Jersey Law Journal* 833.

Mantelero, A (2012) 'Cloud computing, trans-border data flows and the European Directive 95/46/EC: applicable law and task distribution', *European Journal of Law and Technology,* Vol. 3 Issue 2, 2012. Available at http://ejlt.org//article/view/96 accessed 08/08/2013.

Natural Resources Canada (2012) *Primer on Policy Implications of Cloud Computing*, Canadian Geospatial Data Infrastructure Information Product 20e Ottawa: Natural Resources Canada available at ftp://ftp2.cits.rncan.gc.ca/pub/geott/ess_pubs/291/291945/cgdi_ip_20e.pdf accessed 08/08/2013

Natural Resources Canada (2012) *Volunteered Geographic Information (VGI) Primer,* Canadian Geospatial Data Infrastructure Information Product 21e, Ottawa: Natural Resources Canada at http://geoconnections.nrcan.gc.ca/18.

Robson, ES (2011) 'Potential liability for crowd-sourced disaster response groups' at http://stipacommunia.wordpress.com/2011/09/26/potential-liability-for-crowdsourced-disaster-response-groups/ accessed 29/09/2011.

Robson, ES (2012) *Responding to Liability: Evaluating and Reducing Tort Liability for Digital Volunteers,* Washington, D.C.: Commons Lab, Woodrow Wilson International Centre for Scholars, September, 57 pp.

Scassa, T (2013) 'Legal issues with volunteered geographic information', *The Canadian Geographer,* v. 57(1), pp. 1-10. DOI:10.1111/j.1541-0064.2012.00444.x also at http://onlinelibrary.wiley.com/doi/10.1111/j.1541-0064.2012.00444.x/pdf accessed on 15/03/2014.

Zeiss, G (2011) 'Indonesia's Geospatial Information Act No. 4, 2011' found at http://geospatial.blogs.com/geospatial/2011/10/Indonesias-geospatial-information-act-no-4-2011-.html accessed 08/08/2013.

Zook, M *et al.* (2010) 'Volunteered geographic information and crowd sourcing disaster relief: A case study of the Haitian earthquake' (2010) 2 *World Health & Medical Policy* 7.

End Notes

[1] See 'Berners-Lee on the read/write web', BBC News 9th August 2005 at http://news.bbc.co.uk/2/hi/technology/4132752.stm accessed 07/07/2013. Web 2.0 was coined by Darcy DiNucci in 1999 and popularised at a Web 2.0 O'Reilly Media Conference in 2004. See DiNucci, Darcy (1999)' Fragmented Future' *Print* 53 (4): 32 at http://darchd.com/fragmented-future.pdf accessed 06/07/2013.

[2] Goodchild, MF (2007) 'Citizens as sensors: The world of volunteered geography' at http://www.ncgia.ucsb.edu/projects/vgi/docs/position/Goodchild_VGI2007.pdf accessed 08/08/2013.

---

[3] Coleman, D, Georgiadou, Y & Labonte, J (2009) 'Volunteered Geographic Information: The nature and motivation of produsers', *International Journal of Spatial Data Infrastructures Research,* vol. 4. Also see 'Produsage: A working definition' at http://produsage.org/node/9 accessed 08/08/2013.

[4] Bruns, A (2008) *Blogs, Wikipedia, Second Life and Beyond: From Production to Produsage*, New York: Peter Lang.

[5] See Zook, M *et al.* (2010) 'Volunteered geographic information and crowd sourcing disaster relief: A case study of the Haitian earthquake' (2010) 2 *World Health & Medical Policy* 7.

[6] See OpenStreetMap at http://www.openstreetmap.org and Google Maps at https://maps.google.com accessed 08/07/2013.

[7] The url hyperlinks respectively are: http://blog.standbytaskforce.com/2011/09/01/libya-crisis-map-report/; http://blog.ushahidi.com/2011/03/06/using-new-ushahidi-map-libya/ http://www.unocha.org/ochain/2012-13/; http://www.transparencymaroc.ma/TM/fr; http://www.mamdawrinch.com/?l=en_US; https://syriatracker.crowdmap.com/; http://irevolution.net/2012/03/25/crisis-mapping-syria/ accessed 14/03/2014.

[8] See Ball, M (2010) 'What's the distinction between crowdsourcing, volunteered geographic information, and authoritative data?' in *V1 Magazine* at http://www.vector1media.com/dialog/perspectives/16068-whats-the-distinction-between-crowdsourcing-vol accessed 08/08/2013.

[9] Hunter, G (2009) 'Spatial data quality: Problems and prospects' in *Research Trends in Geographic Information Science*. Springer Verlag, pp. 101-122.

[10] See Natural Resources Canada (2012) *Volunteered Geographic Information (VGI) Primer,* Canadian Geospatial Data Infrastructure Information Product 21e, Ottawa: Natural Resources Canada at http://geoconnections.nrcan.gc.ca/18.

[11] Ather, A (2009) 'A quality analysis of OpenStreetMap data' Department of Civil, Environmental & Geomatic Engineering, University College London at http://homepages.ge.ucl.ac.uk/~mhaklay/pdf/Dissertation-OpenStreetMap-Quality-Aather-2009.pdf accessed 08/08/2013.

[12] See http://www.creativecommons.org/licences. Creative commons licences give everyone from individual creators to large companies and institutions a simple, standardized way to grant copyright permissions to their creative work.

[13] Australian Copyright Council (2009) *Websites: User-generated Content and Web 2.0* at http://www.copyright.org.au/find-an-answer/browse-by-keywords/ accessed 08/08/2013.

[14] See *Berne Convention for the Protection of Literary and Artistic Works,* 1988.

[15] *Data Protection Directive 95/46/EC* on the Protection of Individuals with Regard to the Processing of Personal Data and on the Free Movement of such Data.

[16] In particular see Mantelero, A (2012) 'Cloud computing, trans-border data flows and the European Directive 95/46/EC: applicable law and task distribution', *European Journal of Law and Technology,* Vol. 3 Issue 2, 2012. Available at http://ejlt.org//article/view/96 accessed 08/08/2013.

[17] EU (2012) General Data Protection Regulation, EC 25 January 2012 http://ec.europa.eu/justice/data-protection/document/review2012/com_2012_11_en.pdf.

[18] See Prop 8 Maps (2012) 'Prop 8 Maps: A mash-up of Google Maps and Prop 8 Donors' at http://sfist.com/2009/01/09/mash-up_map_of_google_maps_and_prop.php; and http://www.huffingtonpost.com/2009/02/02/prop-8-donors-find-out-wh_n_163234.html accessed 15/03/2014.

[19] See the English case of *Bolam v Friern Hospital Management Committee* [1957] 1 WLR 582 and the discussion on the higher standard of care required for professionals practicing in their field of expertise in Cho, G (2005) *Geographic Information Science: Mastering the Legal Issues*, Chichester: John Wiley & Sons Ltd., pp. 370 ff.

[20] See for instance Australian Capital Territory (ACT) *Civil Law (Wrongs) Act 2002;* New South Wales (NSW) *Civil Liability Act 2002.*

[21] See 'Online NewsHour: French Legal System', September 3, 1997 at http://pbs.org/newshour/bb/law/july-dec97/french_9-3.html accessed 08/08/2013.

[22] Jackson, JZ & Vaurio, AM (1999) 'Good Samaritans in the Hospital', 158 *New Jersey Law Journal* 833.

[23] Robson, ES (2011) 'Potential liability for crowd-sourced disaster response groups' at http://stipacommunia.wordpress.com/2011/09/26/potential-liability-for-crowdsourced-disaster-response-groups/ accessed 29/09/2011. The landmark common law case is that of *M'Alister (or Donoghue) v Stevenson, HL* [1932] A.C. 562 and Lord Atkin's neighbour principle and in *Chapman v Herse* HCA (1961) 106 C.L.R. 112 duties to rescuers.

[24] Robson, ES (2012) *Responding to Liability: Evaluating and Reducing Tort Liability for Digital Volunteers,* Washington, D.C.: Commons Lab, Woodrow Wilson International Centre for Scholars, September, 57 pp.

[25] Robson (2011) *op. cit.* above.

[26] See OS OnDemand Web Map Service (WMS) at http://www.ordnancesurvey.co.uk/business-and-government/products/os-ondemand/index.html accessed 08/08/2013.

[27] See Amazon Web Services (AWS) at http://aws.amazon.com/ accessed 08/08/2013.

[28] See Natural Resources Canada (2012) *Primer on Policy Implications of Cloud Computing*, Canadian Geospatial Data Infrastructure Information Product 20e Ottawa: Natural Resources Canada available at ftp://ftp2.cits.rncan.gc.ca/pub/geott/ess_pubs/291/291945/cgdi_ip_20e.pdf accessed 08/08/2013 and the survey results reported in Cloud Security Alliance (2010) *Top Threats to Cloud Computing V.1.0* available at https://cloudsecurityalliance.org/topthreats/csathreats.v1.0.pdf accessed 08/08/2013.

IOP Conf. Series: Earth and Environmental Science **20** (2014) 012040 doi:10.1088/1755-1315/20/1/012040

[29] European Network and Information Security Agency (2009) *Cloud Computing Information Assurance Framework* at http://www.enisa.europa.eu/act/rm/files/deliverables/cloud-computing-information-assurance-framework accessed 08/08/2013.

[30] See International Bar Association at http://www.ibanet.org/Article/Detail.aspx?ArticleUid=6309c6e9-d561-4876-839a-5ecaacba361e accessed 22/08/2013.

[31] Zeiss, G (2011) 'Indonesia's Geospatial Information Act No. 4, 2011' found at http://geospatial.blogs.com/geospatial/2011/10/Indonesias-geospatial-information-act-no-4-2011-.html accessed 08/08/2013.

About the Author

George Cho PhD is a Professor in Geoinformatics and the Law at the University of Canberra, Australia and undertakes research on the legal aspects of geospatial information, and generally publishes articles on this subject and on aspects of cyber law. Direct comments to George [dot] cho [at] Canberra [dot] edu [dot] au.

Exploring the potentials of volunteered geographic information as a source for spatial data acquisition

Izyana Ariffin, Badariah Solemon, Rina Md. Anwar, Marina Md. Din, Nor Nashrah Azmi

College of Information Technology, Universiti Tenaga Nasional, Kajang, MALAYSIA

E-mail: izyana@uniten.edu.my, badariah@uniten.edu.my, mrina@uniten.edu.my, marina@uniten.edu.my, nashrah@uniten.edu.my

Abstract. The advancement of technologies nowadays enables participation by non-professionals, known as volunteers to participate in producing, sharing and consuming geographic information. Such information, termed as *volunteered geographic information* (VGI) has created a new approach of gathering geographic information. This paper discusses the traditional way of acquiring geographic information and potentials of VGI as an information source in GIS applications. We also review four commonly cited applications which rely on volunteers for their geographic information based on five criteria; the geometry type available in the applications, availability of user profile, average number of attributes used in the applications, data type of the information (raster or vector) and the domain the application belongs to. This review serves as a preliminary study in designing a GIS application used for asset management which aims at exploiting volunteers to produce geographic information related to assets.

1. Introduction

Technological revolution in the field of Information Technology nowadays has facilitated the rapid development of Geographic Information System (GIS). GIS can be defined as one type of information system that stores, displays, manipulates and analyzes spatial data particularly map-based information [1, 2]. The usage of GIS spreads across many application domains, namely transportation, travel, health surveillance, inventory management, military and many others [3]. However, despite its usefulness, GIS is very difficult to use; "there is little hope that community residents with no computing experience will be able to use the software", even with a proper training being provided [4, 5]. As technology advances, many improvements have been done by the software providers especially on the graphical user interface [3]. These improvements, with the help of embedded global positioning system (GPS) in mobile devices, have inadvertently enabled GIS to be accessible and consumed by a much wider range of users, especially the Internet users.

In the early days, mapping information in GIS were done only with military interest. In the 20th century, mapping information started to be used by civilian, mostly by National Mapping and Cadaster Agencies (NMCAs) [6]. However, the updating is seriously lagging and very little efforts were taken by the authorities to improve the level of details of existing maps [7]. Only in recent years that public has started to engage with the production as well as consumption of mapping information, as the result of Web 2.0 technologies. Successful implementation of OpenStreetMap (OSM), Google Earth and

Content from this work may be used under the terms of the Creative Commons Attribution 3.0 licence. Any further distribution of this work must maintain attribution to the author(s) and the title of the work, journal citation and DOI.
Published under licence by IOP Publishing Ltd

HDTraffic proves that public can be a reliable source for geographic information. In this paper, we aim at reviewing features available in commonly cited GIS applications that have successfully gained active participation by volunteers. This review serves as a preliminary study to a research that investigates the effectiveness of using VGI in distributed asset management. We discuss the common terminologies used to describe VGI, and explain how VGI can be harnessed as an alternative source of geographic information in GIS. We then explain the results of our review on the common features that are available in those successful applications. The review is used as a basis for designing an application which targets at utilizing volunteers to produce geographic information related to assets. We conclude our paper by discussing some issues in VGI and the future works that we find useful for our current research project.

2. Volunteered geographic information

The emergence of Web 2.0 technologies, miniaturization of GPS devices and advancement of broad band communication links have empowered the Internet users to be involved in producing geographic information. These Internet users, referred to many as *citizens* or *sensor* [7], who voluntarily produce, share and use geographic information, were termed by Goodchild as VGI [7]. Previously, Turner [8] described similar act of sharing location information with friends and visitors as *neogrography* while some others refer to similar nature of producing information as *crowdsourcing* geo-information. *Crowdsourcing* was first coined by Howe [9], in which he was inspired from the fact that "under the right circumstances, groups are remarkably intelligent, and often smarter than the smartest people in them". This was referred as "wisdom of crowds" by Surowiecki [5]. Apart from these, *participatory GIS* (PGIS), *public participants GIS* (PPGIS) and *community-integrated GIS* are among other terms that are commonly used to illustrate the approaches that are "context-and issue-driven rather than technology-led and seek to emphasize community development involvement in the production and/or use of geographical information" [10].

3. VGI as an alternative spatial data collection method

The data available in GIS is categorized into spatial and non-spatial information. Spatial information can be grouped into two (2) categories; raster and vector while non-spatial information, commonly known as attributes, are the physical information associated with the spatial information. The data sources for raster often come from digital remote sensing images and aerial photographs, while the data sources for vector come from global positioning system (GPS) and survey measurements. One of the commonly used method for collecting spatial data for GIS is known as field data collection [11]. In the past decades, this method was solely authoritarian; professional experts produce, dissemination is radial, and amateurs consume [12]. The method has been widely known to be very costly [4], in turn causes many organizations to avoid adopting GIS in their daily operational activities despite its capability in enhancing decision making process through the provision of basic data and complex analyses [1]. It relied fully on professionals due to the belief that the acquisition of geographic information requires extensive training, thus to be beyond the abilities of amateurs [12]. Thus, the accuracy of the data is considered of high quality due to the positional accuracy of the GPS device and techniques used by the professional land surveyor in capturing the geographic references. Additionally, due to the limitation of technology in the past, public only had local reference for geographic information, i.e. the local governments and agencies, who are able to provide detailed geographic information that can be verified and integrated at the national level [12].

The concept of utilizing public as source of geographic information was far from reality until the availability of Web 2.0 technologies. The advancement of spatial technologies, fostered by the proliferation of information supplied by the Internet users has resulted significant changes in the way organizations acquire geographic information to be used in GIS. VGI is now globally accepted as a reasonable means to collect and update geographic knowledge of local experts [4]. With VGI, not only the production and update of geographic information can be improved [13], organizations can also exploit volunteers, resulting in reduced operating cost [2]. Although the quality of VGI is still a major concern, volunteers are often more interested to map their immediate surroundings, which may contribute to surprisingly good quality of information [6].

Our main research objective is to look at the effectiveness of using volunteers to collect information related to assets which are geographically scattered. This research was originally driven by Coleman's questions; if an organization were to employ VGI for its source of information, how it is going to attract new volunteers, and how does it keep existing volunteers engaged [14]? Therefore, to design an application which targets volunteers to actively produce the geographic information, it is critical that we review the elements that exists in successful VGI projects, and consider the elements for our to-be-developed application. The following section discusses successful VGI projects that are commonly cited in literatures.

3.1 Successful existing VGI projects

The plethora of GPS-embedded mobile phone devices available in the market nowadays catalyzes many VGI-applications to come into usage. OpenStreetMap (OSM) (http://www.openstreetmap.org) project for example, has been very successful in collecting geographic information in most part of the world. Volunteers actively provide and consume geographic information in OSM, resulting the project to become the backbone for free and commercial printed or digital maps, navigation devices, location-based services or research initiatives [4]. Additionally, Waze mobile application (www.waze.com), has been widely used by drivers to get to destination by using least congested routes. Waze uses drivers' nearest telecommunication towers and GPS location to detect traffic jams and provide alternative routes for drivers to avoid heavy traffic roads. Drivers can also share information with nearby drivers if there are accidents or road maintenance at nearby location. In disaster and crisis management, Scipionus (http://gregstoll.dyndns.org/scipionus) and Ushahidi (http://www.ushahidi.com) become the information platform for disaster relief. The Internet users are able to follow the progress of crises, in real time through the eyes of those directly involved in the disaster [15, 16]. Wikimapia (http://www.wikimapia.com) is also one example of successful applications which relies on volunteers for its information source. Users can mark rectangles and polygons on a map to indicate places of interest, as well as update those information entered by other users. In addition to these, there have been numbers of VGI-applications actively used by the Internet users which include finding new bike routes (http://www.mapmyride.com), tracking infectious disease (http://healthmap.org), monitoring forest fires [17], locating cell phone coverage and Wi-Fi hotspot (http://opensignal.com), tracking local weather (http://weathersignal.com) and many others.

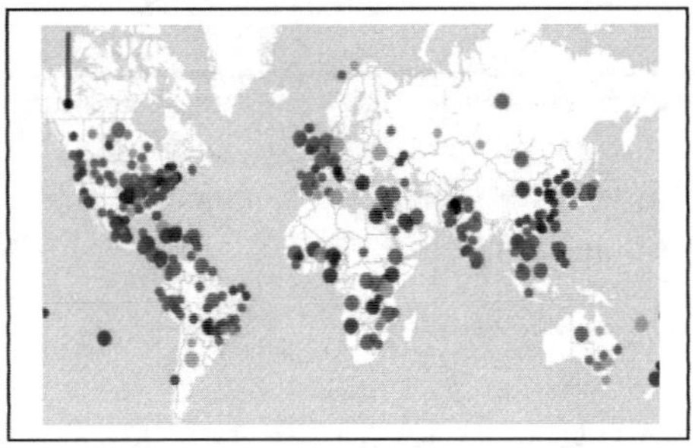

Figure 1. A snapshot of HealthMap application.
Source: http://healthmap.org

3.2 Reviewed projects

As mentioned earlier, this paper serves as a preliminary study for creating a VGI application for asset management. The application is targeting volunteers to supply geographic information associated to assets of an organization. To ensure the application is able to attract huge number of volunteers to participate, we selected four commonly cited projects which rely on public for the production of geographic information. We intended to evaluate applications used for asset management but very limited literatures were found. Many asset management applications only use crowdsourced information which is not associated with geographic references. As the application we are developing uses geographic location to map the asset location, it is essential to review similar nature of projects although the domain is different. The projects we selected are OSM, Waze, HealthMap and Wikimapia. The evaluation criteria used in the review are shown in table 1.

Table 1. Evaluation criteria.

Evaluation Criteria	Description
Geometry types	The geometry types of the information; point, polyline or polygon
User profile	The availability of login or user profiling function in the application
Average number of attributes	Associated physical attributes of the data for example name, size, lot number etc.
Type of data	Type of data produced by volunteers; raster or vector
Domain	The domain of the application for example health, navigation, utility, transportation, maping etc.

3.3 Evaluation criteria

The following sections explain the details of each criteria and its relevance to our study.

3.3.1 Geometry types

GIS has long been recognized for its difficulty to use [3-5], and volunteers are normally lacking of technical knowledge on the tools to use as well as the domain knowledge required in order to produce geographic information. Hence, creating a GIS application which is easier to be used by volunteers is always the highest priority. Nielsen suggested that in order to increase active participation, users should find the application "easy to contribute, without technical, logistic, legal or intellectual barriers" [18]. In GIS, point geometry is always easier to digitize as compared to polyline or polygon. Additionally, different users have different views of the information they want to see in the application, and different views would require different level of details, of geometry, imagery, attributes (semantics) etc. [19].

3.3.2 User profile

Many studies venture into finding what motivates volunteers to contribute information and what contributes to these volunteers' stickiness to the application. Although they do no gain financial reward, volunteers do provide information to enhance their personal reputation and own sense of worth in the online community [13]. Thus, volunteer's profile is required in order for him to be recognized among the online community members. Additionally, it has been suggested that volunteers need to be rewarded in order to obtain active participation [18]. To reward, being able to track who is the volunteer and assess the quality level of the information he contributes are mandatory. It is indeed one of the features that an application designer needs to look at when designing an interface for crowdsourcing application [20].

3.3.3 Number of attributes

Number of physical data and the complexity level of the attributes can deter volunteers from entering information. Complex attributes which require advanced technical knowledge may lead to users' frustration when describing the geographic information. The more associated information is required, the longer the time it takes for volunteers to key in information. With the rapid development of the interface used in crowdsourcing applications, it is expected that the data models will become more sophisticated [6]. Users normally prefer a simple and less complex application to use [20]; therefore we need to see how simple should a GIS application be in order to obtain huge participation by the volunteers.

3.3.4 Type of data

Using volunteers to map raster information requires a considerable skill in the use of photogrammetric techniques, expensive equipment for observation and analysis and substantial investment in large-format printing [5]. However, certain applications have made the interface simple that enables volunteers to consume information related to raster. We intend to review the type of data entered by volunteers in each of the application in order to measure the minimum level of expertise required by such volunteers.

3.3.5 Domain

Most of VGI applications that exists are designed for mapping the world. However, the information that comes from volunteers might be influenced by their interests or hobbies; volunteers tend to contribute information to community which they have something in common, and inadvertently form many social linked groups. These groups have totally different needs and different ways to keep them to be always motivated to volunteer [6]. It is good to gain an insight on the groups that exists to identify the most dominant domain that volunteers are in.

4. Review analysis

We visited all four projects mentioned earlier and analyzed the criteria we depicted in table 1. Table 2 summarizes the findings.

Table 2. Summary of review findings.

Project	*Geometry types*	*User profile*	*Average number of attributes*	*Type of data*	*Domain*
OpenStreetMap	*Point* Polyline Polygon	*Required*	*5-8*	*Vector*	*Mapping*
Waze	*Point* Polyline Polygon	*Optional*	*6*	*Vector*	*Navigation*
Wikimapia	*Polygon*	*Optional*	*7*	*Vector*	*Mapping*
HealthMap	*None*	*Required*	*7*	*None*	*Health*

4.1 Geometry types

OSM supports many types of geometry for mapping real world objects (RWOs) such as supermarket, kiosk, hospital, vending machine, post office, hotel and many others. To make it easier for users to find the objects to map, categories are used i.e. amenities, shopping and transport. In Waze, users are allowed to update information about traffic by adding point geometry to indicate hazard, accident or police appearance on a map. It also allows user to add newly constructed road by using '*Pave road*' function, which will track the starting point, the route and the end point of the new road (polylines). Besides updating through the mobile application, users are also encouraged to update the map used in Waze. Similar to OSM, Waze has many RWOs which are grouped into categories for easy finding. These include hospital, clinic, police station, university and many others. However, the list of RWO in Waze is more extensive when compared to OSM's. Unlike Waze, as Wikimapia was meant to encourage people to describe the world, the information provided by volunteers is linked to polygon geometry only. However, experienced users are allowed to add or update information related to polyline objects such as river, railroad and roads. The experience is measured based on the ratings on user profiles. On the other hand, HealthMap works totally different from OSM, Waze and Wikimapia. Since the application targets at getting volunteers to supply information regarding outbreaks, users describe the place by typing the place name to mark the location of the disease. Again, since the category of information is already specific (disease outbreak), volunteers do not have to select the category for the RWO, in this case, the disease location.

4.2 User profile

Before a volunteer can map inside OSM, login is required. However, only minimum number of information is required from a '*mapper*' when creating a login account, perhaps to increase the convenience of using such application. Unlike OSM, if one intends to Waze as a source for traffic information, login is not required. Only if a *'wazer'* plans to update the map used in Waze that login is made mandatory. This is perhaps due to the need for quality checking of the information provided by the '*wazer*'. Wikimapia operates just like Waze. The application allows users to provide attribute information related to the place i.e. name, description and categories without logging in. However, if a user wishes to provide information with related to polyline, login and experience are required as new users or low rating users are not allowed to contribute information related to polyline objects. Like in OSM, login is required if a volunteer wants to provide information regarding disease outbreak in HealthMap. Having a login function for the application that spreads outbreak information is seen very relevant, as any false information appears in the application may create panic situation among the citizens.

4.3 Average number of attributes

In OSM, the number of attributes available for users to key in is not consistent; it varies from one object to another, but on average, each object requires between 5-8 attributes associated with the object. However, in contrast, number of attributes required for updating map information in Waze is consistent for all objects. This definitely improves the usability of the application, as consistency is crucial in creating an effective user interface [21]. In Wikimpia and HealthMap, users are allowed to key in 7 associated attributes to the object they created. However, in HealthMap, only 1 object is available for user to map, which is the disease. Therefore the number of attributes can be considered consistent as in Waze.

4.4 Type of data

As expected, the data type produced by volunteers in OSM, Waze and Wikimapia is vector based. However, OSM does provide a function for volunteers to get the information available in OSM in raster format, and this has been made simple so that volunteers do not need to possess deep knowledge in photometry or remote sensing image processing. In HealthMap, as user does not draw any geometry in the application, hence, there is neither vector nor raster data being generated in the application.

4.5 Domain

OSM is clearly being exploited for collecting geographic information used for mapping activities. The information available in OSM is used to help map agencies fill in the information gaps in their existing maps. However, Waze is different. Although Waze also provides a platform for volunteers to update its maps, the main function of the application focuses at updating drivers with the latest nearby traffic information. The information is consumed for navigation purposes. Like OSM, Wikimapia is used mainly for mapping as the information supplied by volunteers is used to add more information to the existing maps. However, Wikimapia focuses more on describing the place rather than providing very accurate mapping information especially for streets. HealthMap, as the name suggests, is used for collecting information related to health domain.

4 Discussion & future works

The result of the review suggests that volunteers prefer to contribute geographic information pertaining to mapping. All applications use vector data by the volunteers, except for HealthMap which uses description to describe geographic location as opposed to geometry. To increase data quality, all applications require users to login before they can produce geographic information. However, this review is merely performed on only four commonly cited applications in literatures, and there are many other applications which rely on volunteers that were not included. In addition, as the review was performed on the breadth of the application rather than depth, further studies need to be conducted to evaluate each criteria in detail. This includes finding number of RWOs created by volunteers based on geometry and its type. A similar study has been done, but it focused on heavily edited objects only at few selected areas [22]. By knowing such details, we can identify popular geometry produced by the volunteers and perhaps further investigation can research into finding why such geometry is in favour. Although geometry should represents RWO actual characteristics - for example land parcel is in polygon and road is in plotline, this knowledge may not be possessed by the volunteers. In GoogleMaps, many tourist attraction places and shops are drawn as point instead of polygon, indicating that volunteers are clueless about RWO's suitable geometry. However, as Nielsen suggested, to increase participation, volunteers are encouraged to edit, rather than create, therefore the application owner or organization should provide templates for jobs to be done [18], in this context, the geometry of the object should be pre-defined.

Besides, finding details about the attributes can be further explored by looking at the complexity of the attributes. We can examine whether complexity of attributes plays an important role in increasing participation from volunteers. In the reviewed applications, the associated information to the geographic information is considered minimum and can be easily understood by non-professionals. However, in the future, mapping may evolve into three-dimensional, and the information produced by volunteers might have to be spatially analyzed, based on organizations' specific needs. Thus, there might be time when volunteers also need to contribute information that requires technical background. Additionally, as organizations start relying heavily on volunteers to produce information, the demand for level of information completeness may increase-organizations may start requesting volunteers to produce information in a way that it can be easily digested for decision making, without any further information processing done in between the information producers and consumers. Limitation of using VGI as a source for data acquisition also remains arguable by many. To what extent can we rely on these volunteers? How reliable is the information produced and why do they want to produce? This is what our research focuses to study. We intend to examine whether VGI can be applied in asset management domain, the type of geometry volunteers prefer to use, the motivation, and the elements required to sustain such volunteers. Although quality of the information produced by these unprofessional volunteers also is still a major concern, under certain circumstances, having poor quality data is always better than none. Until today, researchers pushed for more studies to be done in the area of VGI before it can be used as the primary source of information in any GIS application [23]. Majority still regard VGI as a complementary way of acquiring data instead of a substitution.

5. Conclusion

It has been observed that the number of projects that utilizes VGI as a new method for collecting data is rapidly growing. In this paper, we performed review on four commonly cited applications that make use of VGI and explored into five criteria that we considered important in designing our future application. It can be concluded that in order to attract volunteers to keep producing information, an application should consider deploying simple geometry to describe the RWOs, using minimum and preferably consistent number of attributes across all RWOs, implementing user profile functionality besides encouraging use of vector data as opposed to raster. However, as the extensiveness of this review is on the breadth rather than depth, more studies are needed to really identify the criteria of an application which can lead to increased volunteers participation.

References

1. Campbell, H. and I. Masser, *GIS In Organizations: How Effective Are GIS In Practice?* 1995: CRC Press.
2. Longley, P., *Geographic information systems and science*. 2005: John Wiley & Sons.
3. Goodchild, M.F., *Geographic information systems and science: today and tomorrow.* Annals of GIS, 2009. **15**(1): p. 3-9.
4. Goodchild, M.F., *Citizens as Voluntary Sensors: Spatial Data Infrastructure in the World of Web 2.0.* International Journal of Spatial Data Infrastructures Research, 2007. **2**: p. 24-32.

5. Goodchild, M., *NeoGeography and the nature of geographic expertise.* Journal of Location Based Services, 2009. **3**(2): p. 82-96.
6. Heipke, C., *Crowdsourcing geospatial data.* ISPRS Journal of Photogrammetry and Remote Sensing, 2010. **65**(6): p. 550-557.
7. Goodchild, M.F., *Citizens as sensors: the world of volunteered geography.* GeoJournal, 2007. **69**(4): p. 211-221.
8. Turner, A., *Introduction to neography*. 2006, O'Reilly Media Inc, Sebastopol CA.
9. Howe, J. (2006) *The Rise of Crowdsourcing*. **14**, 1-4.
10. Dunn, C.E., *Participatory GIS—a people's GIS?* Progress in human geography, 2007. **31**(5): p. 616-637.
11. Lwin, K.K. and Y. Murayama, *Web-Based GIS System for Real-Time Field Data Collection Using a Personal Mobile Phone.* Journal of Geographic Information System, 2011. **3**(4).
12. Goodchild, M.F., *in the World of Web 2.0.* International Journal, 2007. **2**: p. 24-32.
13. Coleman, D.J., Y. Georgiadou, and J. Labonte, *Volunteered geographic information: The nature and motivation of produsers.* International Journal of Spatial Data Infrastructures Research, 2009. **4**(1): p. 332-358.
14. Coleman, D.J., *Potential contributions and challenges of VGI for conventional topographic base-mapping programs*, in *Crowdsourcing Geographic Knowledge*. 2013, Springer. p. 245-263.
15. Roche, S., E. Propeck-Zimmermann, and B. Mericskay, *GeoWeb and crisis management: issues and perspectives of volunteered geographic information.* GeoJournal, 2013. **78**(1): p. 21-40.
16. Zook, M., et al., *Volunteered geographic information and crowdsourcing disaster relief: a case study of the Haitian earthquake.* World Medical & Health Policy, 2010. **2**(2): p. 7-33.
17. De Longueville, B., R.S. Smith, and G. Luraschi. *OMG, from here, I can see the flames!: a use case of mining location based social networks to acquire spatio-temporal data on forest fires*. in *Proceedings of the 2009 International Workshop on Location Based Social Networks*. 2009: ACM.
18. Nielsen, J., *Participation inequality: Encouraging more users to contribute.* Jakob Nielsen's alertbox, 2006. **9**: p. 2006.
19. Craglia, M., et al., *Next-Generation Digital Earth.* International Journal, 2008. **3**: p. 146-167.
20. Nakatsu, R. and E. Grossman, *Designing effective user interfaces for crowdsourcing: an exploratory study*, in *Human Interface and the Management of Information. Information and Interaction Design*. 2013, Springer. p. 221-229.
21. Johnson, J., *Designing with the mind in mind: Simple guide to understanding user interface design rules*. 2010: Morgan Kaufmann.
22. Mooney, P. and P. Corcoran, *Characteristics of heavily edited objects in OpenStreetMap.* Future Internet, 2012. **4**(1): p. 285-305.
23. Wilson, N., et al., *Interpreting Google flu trends data for pandemic H1N1 influenza: the New Zealand experience.* Euro surveillance: bulletin européen sur les maladies transmissibles= European communicable disease bulletin, 2009. **14**(44).

IOP Conf. Series: Earth and Environmental Science **20** (2014) 012042 doi:10.1088/1755-1315/20/1/012042

Comparative Analysis Study of Open Source GIS in Malaysia

Muhammad Zamir Abdul Rasid, Naddia Kamis, Mohd Khuizham Abd Halim
Faculty of Engineering, Universiti Putra Malaysia, Serdang, Malaysia

E-mail: naddiakamis@yahoo.com, zamir.rasid@yahoo.com, khuizham@gmail.com

Abstract. Open source origin might appear like a major prospective change which is qualified to deliver in various industries and also competing means in developing countries. The leading purpose of this research study is to basically discover the degree of adopting Open Source Software (OSS) that is connected with Geographic Information System (GIS) application within Malaysia. It was derived based on inadequate awareness with regards to the origin ideas or even on account of techie deficiencies in the open origin instruments. This particular research has been carried out based on two significant stages; the first stage involved a survey questionnaire: to evaluate the awareness and acceptance level based on the comparison feedback regarding OSS and commercial GIS. This particular survey was conducted among three groups of candidates: government servant, university students and lecturers, as well as individual. The approaches of measuring awareness in this research were based on a comprehending signal plus a notion signal for each survey questions. These kinds of signs had been designed throughout the analysis in order to supply a measurable and also a descriptive signal to produce the final result. The second stage involved an interview session with a major organization that carries out available origin internet GIS; the Federal Department of Town and Country Planning Peninsular Malaysia (JPBD). The impact of this preliminary study was to understand the particular viewpoint of different groups of people on the available origin, and also their insufficient awareness with regards to origin ideas as well as likelihood may be significant root of adopting level connected with available origin options.

1. Introduction

Open Source Software (OSS) is now a significant option means of arranging this output involving software program and possesses accumulated with reputation and make use of for the rewards and costs relative to this prominent private software program design [1]. One of several initial recognized advantages of OSS would be the price tag reduction, not just licensing income nevertheless for the particular progressively more readily available skills regarding OSS techies in the employees [2]. In Malaysia, a report by the OSCC mentioned that the overall cost of savings for open source software products was recorded to about RM 46.9 million in 2011 [3].

Although the main advantages of including Open Source Software program (OSS) to organizational devices are already widely suggested, OSS should be implemented in addition to use by prospects before these positive aspects can be recognized [4]. The objective of this research study is to find out the reason of poor adoption of Open Source GIS tools by Malaysian community. Meanwhile, this paper was also established to get information about the views and concerns from developers or organizations whom opt GIS in open-source platform rather than proprietary platform, and also to discover the technical tools deficiencies that discourage users from widely adopting them.

Content from this work may be used under the terms of the Creative Commons Attribution 3.0 licence. Any further distribution of this work must maintain attribution to the author(s) and the title of the work, journal citation and DOI.
Published under licence by IOP Publishing Ltd

Initial understanding was that all agencies including the Government sectors and Universities have been adopting OSS in their work. Report showed that the improvement was substantially high from 25 companies to 707 companies, which is 97% of final amount of agencies recorded [5]. Throughout this research, two methodologies consisting of interview session and also survey questionnaire were investigated in order to get a closer look at the current situation of open source GIS acknowledgment in Malaysia.

2. Literature Review

Technology changes since the 1980's derived the development of data storage as part of database management system. Result visualization and quantitative analysis are the most important elements in any GIS software programs. As mentioned by Filippo Randelli [6], the open source software (OSS) is basically similar to other database programs, it analyzes and store data; with unique characteristics of location based function where the information has a link with a unique point on the earth's surface.

Referring to the open source movement written research by Deek and McHugh [7], an international seek to advertise a great available model of computer software growth much more aligned with the accepted cerebral model of scientific discipline rather than the amazing modes associated with new technology, which were trait associated with contemporary business. The theory – as well as eyesight – is always to maintain controlled advances put together by computer software growth freely available for everyone to broaden and enhance after. So, this gives neighborhood associated with programmer-developers to create a collaborative sociable composition where by members change program code to add changes and deal with parasites. Computer programmers create these types of additions of the very own free will with regard to many different reasons, for example: increasing recognition from the FOSS group, improving it because of their very own requirements, improving his or her technological knowledge, improving it with regard to the use of people, receiving taken care of producing OSS, in addition to advertising OSS while movements [7]. Because of the actual openness with the course of action as well as the pool connected with competent, eager contributing factors, open origin computer software may create and also enhance quickly [8] [9].

In order to apply the OSS concepts, many authors have come out with different studies and findings on the subject matter. According to Anderson and Sanchez [10], Open Source GIS (OSGS) is a program for users to run for any purpose, modify and redistribute freely without any limitations or royalty payments. Open software allows for code modification and tools invention without any changes implied to the existing data functions and data processing. In addition, OSS is free software that available and accessible in the market such as GRASS, QGIS, SWAT, MapWindow and Diva GIS.

As the availability and accessibility are the advantages of Open Source Software, the capability of OSS in providing varieties of modules is also in line with proprietary or commercial software. It is a huge impact to the overall management as the modeling products can be referred to support the decision making in business strategy. For instances; GRASS provides GIS standards to support modeling and programming without the needs to develop the algorithms required for basic spatial processing. Scott et al. [11] explored the GRASS-GIS based system in spatial partitioning as to handle spatial heterogeneity issue of environmental modeling. Filippo Randelli [6] discussed the use of OSGS application from spatial analysis to business strategy on new logistic infrastructure as location based functions to track production chain.

Apart from modelling, the distribution and application of OSS may be large depending on the sectors applied. Nevertheless, many argue on the cost and benefits offered by the advent of this technology relative to the dominant proprietary. Different aspect influences the lower adoption of open source software in industry or government agencies. According to the survey and semi structured interview by Dhanaraj Thakur [12] in Canada and United States firms, many factors contributes to the level of adoption and acceptance; among the factors are industry structure, skills requirement, laws and policy. In developing countries, the open source approaches give benefits for socio economic development. There is a study done by Ademola A. Adenle, et al. [13] in area for biotechnology innovation as to deal with the intellectual property right issues. Nevertheless, the adoption of OSS can be applied in developing countries by the considerable amount of training and adequate resources as well as equip infrastructures.

In the modernisation era of IT technology, the piracy problem arises due to the abundance of software that affects the industry itself. OSS can be adopted as the strategic mean to prevent the problem without confronting the war but the free access dominate the piracy war. According to T. Pykäläinen [14], the OSS strategy is dependent on how and what condition OSS strategy can be adapted to prevent privacy and the effectiveness can be indicated.

The insight knowledge of OSS can be build up by understanding the fundamental or variable of the factors that favour OSS activities. According to Sebastian and Andreas (2013), [15], a culture characterized by interpersonal trust and self-determination values influenced the OSS activities and number of developers. Hence, the personnel of OSS users and workplace environments play a major role in OSS adoption and acceptance level.

3. Methodology

Two investigations were performed in order to fulfill the objective of this research study. The first investigation was primarily conducted through surveying form that is related to the general information regarding detail of users and their experience in Open Source GIS. This particular survey had been likewise conducted to consider the particular problematic elements connected with impacting GIS in open source platform compared to exclusive proprietary software. The second analysis was accompanied through an interview by site visiting the real implementation scenario of Open Source GIS in Malaysian public sector.

4. Data Collection and Analysis of Findings

A total of 27 volunteers from local Government Officers, student of Masters and PhD, as well as Lecturer from Local Universities have participated in the questionnaire form. Earlier in the survey, we found out that among the candidates, there were a very huge gap in terms of their level of experience in any GIS tools. These ranged from first time users until up to 10 years of experience. Considering experience as a major factor, only candidates dealing with GIS of 2 years or more were selected for the surveying results and analysis.

4.1 Awareness of open source concepts

In Part A, the level of awareness of open source GIS platform between the Government sectors and the University sectors we studied. The results from given questions were pointing out whether the candidates know about any open source GIS software, and about the acceptance level of using it in their work or research in the future. Results are shown on Figure 1.

In Government sector, 6 candidates have been known to be using GIS in open source platform and all of them prefer using it in their work. Only 3 candidates have not encountered with open source GIS yet, but 2 of them are willing to try. Whereas, in the University sector, only 4 candidates know about open source GIS platform and 3 of them are willing to use it in their work and research. Another 3 candidates have not encountered open source GIS yet, and only 1 of them are willing to try.

In Part A, we are able to get two distinctive results between the Governments sector against the Universities, which shows that the awareness and acceptance level of open source GIS platform is more dominant and higher among Government agencies rather than users from the Universities. Our view is that this occurrence may be due to regular training and campaign from the Open Source Competency Center (OSCC) agency under MAMPU, which has been more focus within the Government agencies only, while the Universities lack the attention from the agency itself, or most probably due to campaign have not yet reached the university intensively.

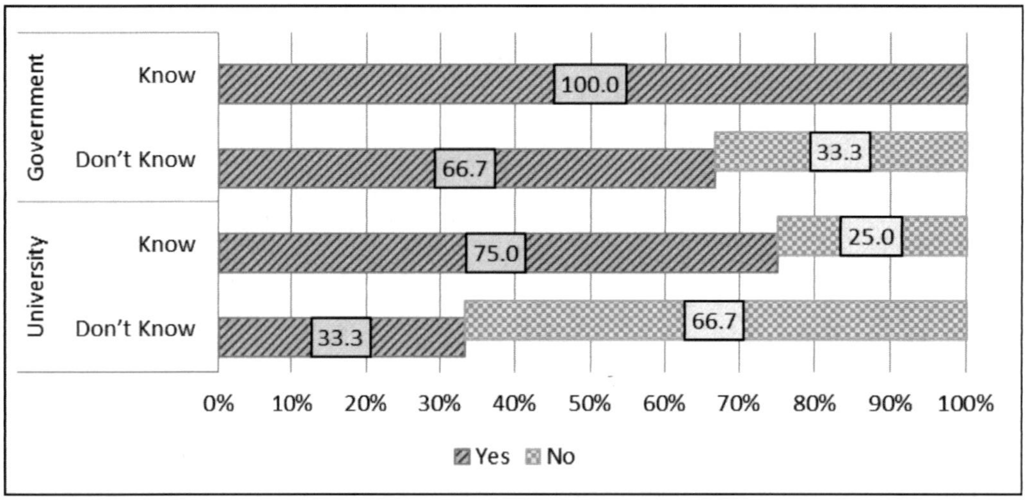

Figure 1: Acceptance of Open Source GIS Platform.

4.2. Comparative analysis between Open Source GIS and commercial GIS.

In Part B, a total of 8 questions were asked to the candidates about the implementation of open source GIS in Malaysia particularly. These questions were mainly to get an overlook about the few limitation factors which contributes to the acceptance of open source GIS in Malaysia, such as software design, security issues, system documentation, technical knowledge, software installation, user friendliness, bugs, and licensing factors.

In the results as shown on Figure 2, we can see that there is a wide range of understanding of Open Source GIS amongst the frequent users in Malaysia. Notably, most users are very well informed that there is no licensing limitation of using open source software. This also can be applied to reduce the cost of operating such software's in commercial GIS which require user to have subscription and licensing fees.

In the meantime, respondent main concerns were due to problem arising caused from installation with other extensions of the software that has not been bundled together as compared with most mainstream GIS. This was supported with results on open source GIS as not being user friendly especially for beginners, and also respondent agree that most open source GIS may require user to have additional technical knowledge especially in computer language and programming.

None the less, users perception are well balance in terms of few criteria's, such as difficulties in designing the output of open source GIS, security issues that increases threat from viruses, system documentations or system testing, and the presence of bugs that increases snags. These results can be viewed in such that users are neutral and have their own argument in their choices.

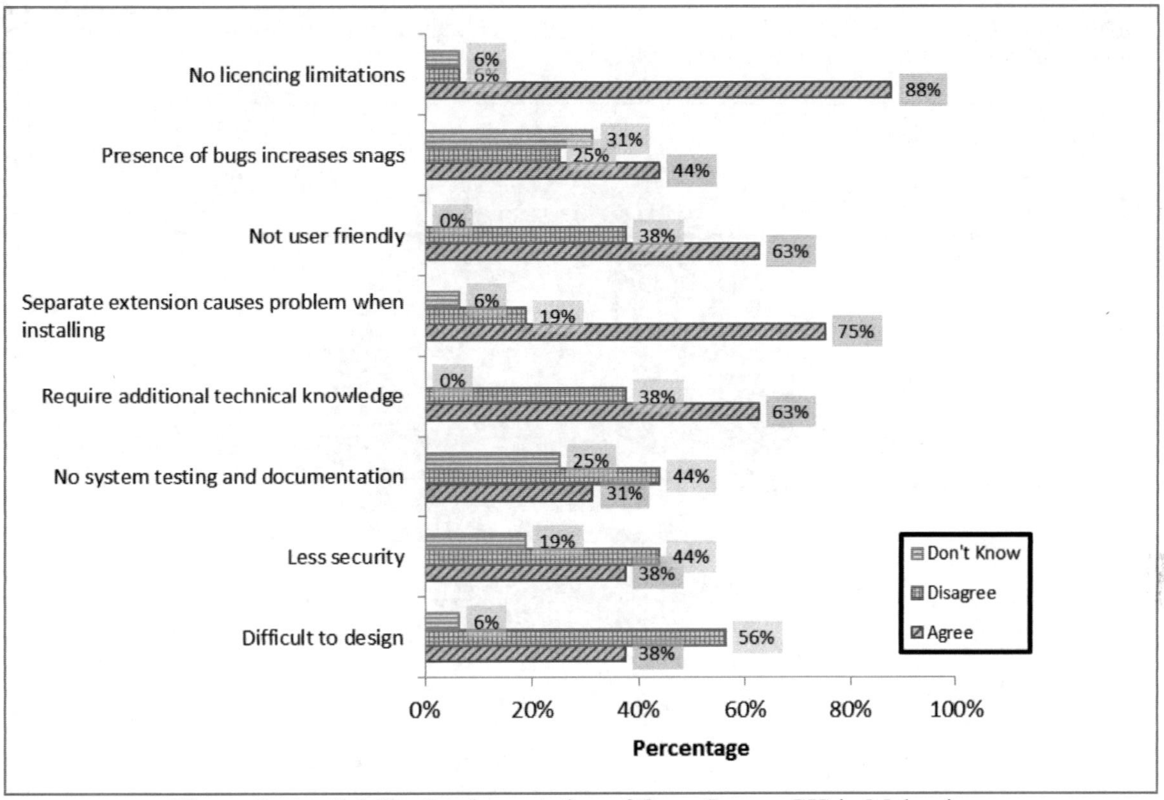

Figure 2: Availability Implementation of Open Source GIS in Malaysia.

5. Moving Forward of Open Source GIS in Malaysia

Open source software is an alternative geospatial technology despite commercial software that available in the current market nowadays. The interoperability offered by OSGS made it reliable in terms of sharing information and modification that useful for future research investigation.

In Malaysia, the establishment of a long-term Malaysian Public Sector Open Source Software (OSS) program is executed to steer the aim of being a develop nation. The OSS helps to pave the aim road and to strengthen the solid ICT plan as to build up the capacity among the people as not to dependent on 100% foreign built proprietary software. Malaysian Administration Modernisation and Management Planning Unit (MAMPU) is the main agency that coordinates the roadmaps for open source in Malaysia. Malaysia Government supports the use of Open Source Software through the established mechanism of OSS framework and Strategic Thrusts that was launched on 16th July 2004. This agenda have been to reduce the overall price tag associated with property, boost the freedom of choice intended for computer software use, hence increasing the growth of the local ICT as well as knowledge based society.

MAMPU through its Open Source Competency Centre (OSCC) has led the OSS pathway as the Phase III: Self Reliance that consists two term; short and medium term. According to MAMPU, the three strategic themes – Sustainment, Enablement and Empowerment are the framework in order to achieve the overall target objectives. The Sustainment component of Self Reliance is a continuous enhancement of OSS solution includes the capability of agencies' adoption of standards, policies and guidelines that established by MAMPU. The Enablement component enhances the IT personnel through OSS training and deployment of technology environment as to support the localized OSS operations. The Empowerment component accelerates the content growth of OSCC Knowledge Bank and e-market by distributed sourcing from agencies.

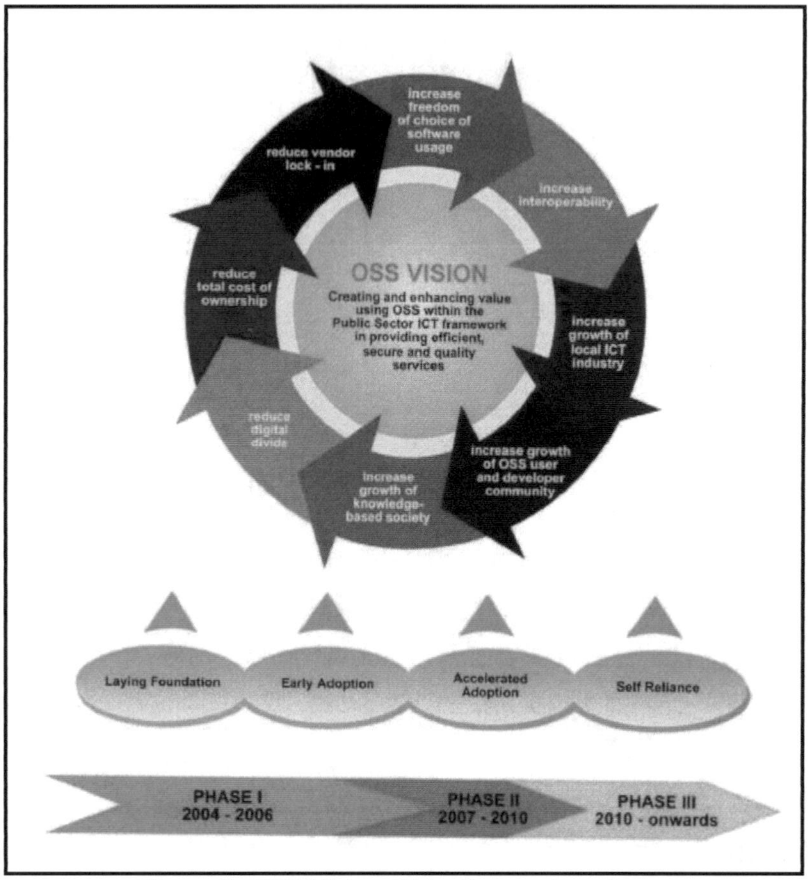

Figure 3: OSS Vision, Objectives and roadmap.

As to date, the OSS implementation in Malaysia is in progress as the set up Medium Term Roadmap for Phase III. Medium Term plan commences from 2013 to 2015 with the hopes that OSS innovation can be accelerated by the establishment of agencies' Smart Partnership. In addition to that, the actual creation associated with firm distinct boosted OSS program answers apart from targeting fresh OSS goods pertaining to home and also worldwide employment.

The key success of OSS implementation plan in Malaysia involves the respective agencies and the Malaysian Government's agenda for IT in the public sector. Nevertheless, the fundamental are being set up as early as the education approaches by the formulation of Learning Management System (LMS). The LMS Roll Out to Schools under Ministry of Education fostering the promotion of OS LMS and create awareness among 120 schools enrolled in Short Term plan.

Saipunidzam et al. [16] discussed this M-Learning as a brand new studying paradigm about the advancement regarding cell and cellular technologies. They had proved that Mobile Math is the developed model that suits to the mathematic lesson by mobile technology. According to Andrea Barker et al. [17] also found out in their study of which establishing nations can catch up with this specific brand new understanding paradigm throughout the educative venture on cell understanding.

Recent research study was done by Norazah [18], the learning management system (LMS) emphasized on the designing and developing the system prototype called E-Headship. Apart from that, the evalution results from 60 participants from Institution level known as Institut Aminuddin Baki (IAB) envisaged the succeeded in promoting the OS learning strategies to a higher degree.

Apart from education line, the OSS products has potential in Malaysia market but still not yet favorable in public sector agencies. According to Khadijah et al. [19], the selection criterion comprises the system quality, information quality and service quality that influences the OSS adoption. The characteristics defined each dimension that favors the factors of OSS implementation.

The Government of Malaysia supports the opportunities embedded in Open Source software through MAMPU's roadmap plan but there are still limitations on real practices and acceptance level among users. There are many aspects to be considered in implementing the OSS approaches either involves the target groups, the procedure and policy wise matters. Based on Nor Zairah et al. [20], the malfunction regarding OSS setup can be significantly influenced by means of organizational variables. Hence, appropriate guidelines and corrective actions should be taken into account to realize the OSS implementation.

5. Conclusion

This paper has a complete evaluation review of Open Source GIS platform amongst Malaysian individuals by Government servant and also University students and lecturers. This paper also acknowledges the current situation of adoption level from those frequent users of GIS.

The result from the first survey investigation has shown that the Government sectors are more mindful about the presence of open source GIS platform as compared to the University sector. This may be due to the campaigning work from respective agencies may lack promotions towards the Universities.

The second investigation through the interview session was mainly highlighting the current situation of open source GIS platform from an agency conducting such work. With the known advantages of operating open source GIS as well as the benefit of reducing the cost of operations, it clearly shows that open source GIS is the way to the future for Malaysian to implement and adapt.

None the less, this study is only a preliminary investigation thus requires more data and analysis for a comprehensive conclusion. Therefore, further research should be conducted with adequate numbers of participants and different field of users as to examine and validate the study analysis. The acceptance level of Open Source GIS also needs to be carried out in other variables as to enhance the research study.

7. Acknowledgement

The authors would like to thank the volunteers for helping out with the data collection for the surveying questionnaire. Out upmost and special thanks to En. Abbas Abdul Wahab of the Federal Department of Town and Country Planning Peninsular Malaysia for allowing us to have an interview session. Lastly we would like to thank our mentor, Assoc. Prof Dr Abdul Rashid bin Mohamed Sharif and the Universiti Putra Malaysia (UPM) for the completion of this research.

6. References

[1] Thakur, D. (2011). A limited revolution - The distribution consequences of Open Source Software in North America. *Technological Forecasting & Social Shange*, 244-251.

[2] Gallego, M. D., Luna, P., & Bueno, S. (2008). User acceptance model of open source software. *Computers in Human Behavior*, 2199-2216.

[3] OSCC, O. S. (2011). *Adoption Chart Report.* Malaysia: Malaysian Administrative Modernisation and Management Planning Unit.

[4] Gwebu, K. L., & Wang, J. (2010). Adoption of Open Source Software: The role of social identification. *Decision support systems*, 220-229.

[5] OSCC, N. (2011). *Government Agencies MyGOSS Award.* Malaysia: MAMPU.

[6] Randelli, F. (2012). Journal of GIS Trends 2012. *Open Source GIS Based Strategies for Firms: A Spatial Analysis Application To The Inland Terminal of Livorno*.

[7] Deek, F. a. (2008). Cambridge University Press. Open Source: Technology and Policy.

[8] Weber, S. (2004). *The Success of Open Source.* Cambridge, MA: Harvard University Press.

[9] Raymond, E. (2001). O'Reilly Media. In Raymond, *The Cathedral and The Bazaar: Musings on Linux and Open Source by An Accidental Revolutionary.* Sebastopol.

[10] Anderson, G., Sanchez, R. M., (2003). Building Web-Based Spatial Information Solutions around Open Specifications and Open Source Software. *Transactions in GIS* 7(4), 447– 466.

[11] Scott Mitchell, F. C. (2002). Proceedings of the Open source GIS - GRASS users conference 2002. *Advantages of Open Source GIS to Improve Spatial Environment Modelling*.

[12] Thakur, D. (2012). Technological Forecasting & Social Change 2012. *The Distributional Consequences of Open Source Software in North America*.

[13] Ademola A. Adenle, S. K. (2012). Journal of Technology in Society. *Analysis of Open Source Biotechnology in Developing Countries: An Emerging Framework for Sustainable Agriculture*.

[14] T. Pykalainen, D. Y. (2009). Journal of Strategic Information Systems. *Alleviating Piracy Through Open Source Strategy: An Exploratory Study of Business Software Firms in China*.

[15] Sebastian v.Engelhardt Andreas Freytag, (2013). Institutions, Culture and Open Source. *Journal of Economic Behavior & Organization, 2013.*

[16] Saipudnizam Mohamad, M. N. (August 2010). M-Learning: A New Paradigm of Learning Mathematics in Malaysia. *International Journal of Computer Science & Information Technology*, No.4.

[17] Andrea Barker, G. K. (2005). 4th World Conference of M-Learning 2005. *A Proposed Theoretical Model for M-Learning Adoption in Developing Countries*.

[18] Nordin, N. (July 2012). The Turkish Online Journal of Educational Technology. *Leveraging Open Source Software in Education Management and Leadership Training*, Volume 11 Issue 3.

[19] Khadijah Chamili, Y. Y. (2012). Volume 02 Issue 02. *Selection Criteria for Open Source Software Adoption in Malaysia Asian Transactions on Basic and Applied Sciences (ATBAS)*, 2221-4291.

[20] Nor Zairah Ab.Rahim, R. A. (2010). PACIS 2010 Proceedings. *Multiple Perspectives Technology Appropriation Analysis of Open Source Software Implementation Failure*, 110.

The convenience of temporary housing complexes in Iwate Prefecture constructed after the 2011 Tohoku Earthquake

Risa Shibutani and Noriaki Endo

Iwate University, Sch. of Humanities and Social Sciences, Morioka City, Iwate Prefecture, Tohoku District, Japan

Abstract. This study compares the convenience of temporary housing complexes in the Iwate Prefecture following the 2011 Tohoku Earthquake. The study was targeted at three major cities in the southern coastal area of Iwate Prefecture, namely, Kamaishi, Ofunato, and Rikuzen-Takata, that were most heavily struck by the earthquake-triggered tsunami.

We conducted a network analysis in geographical information system software using the coordinate data of several daily infrastructures. Temporary housing complexes within the defined service area of each infrastructure were assigned a score of 1.00. The main findings are summarized below:

1) The temporary housing complexes in Rikuzen-Takata City were less accessible to infrastructures (as evidenced by the low coverage area of 1.00 scores) than the other investigated cities.

2) The scores of Kamaishi City and Ofunato City were statistically similar, but complexes in Ofunato City were surrounded by slightly more infrastructures (greater coverage area of 1.00 scores) than Kamaishi City.

3) We identified more than the predicted number of blank areas in the targeted areas. Thus, we consider that support services for people living in such areas are urgently required, especially in the realms of daily shopping, banking, and healthcare.

1. Introduction

Following the 2011 Tohoku Earthquake [1] , many people were relocated to temporary housings [2] . These housings were assigned to persons who had lost their homes in the disaster and lacked the financial means to rebuild. In the Iwate Prefecture, which was seriously damaged by the earthquake-triggered tsunami, 94.6% of the temporary dwellings were occupied in December 2011.

However, temporary housing complexes are not always conveniently located for everyday living. For example, daily shopping and banking becomes problematic for residents located beyond a certain distance of the precincts.

Therefore, this study compared the convenience of temporary housing complexes in the southern coastal area of the Iwate Prefecture, constructed after the 2011 Tohoku Earthquake. The three target cities, Kamaishi, Ofunato and Rikuzen-Takata, had been most severely stricken by the tsunami that followed the earthquake.

The convenience of living environments is frequently evaluated by the distance between the dwelling and surrounding infrastructures [3] . Similarly, we found that the usability score for a

Content from this work may be used under the terms of the Creative Commons Attribution 3.0 licence. Any further distribution of this work must maintain attribution to the author(s) and the title of the work, journal citation and DOI.
Published under licence by IOP Publishing Ltd

IOP Conf. Series: Earth and Environmental Science **20** (2014) 012043 doi:10.1088/1755-1315/20/1/012043

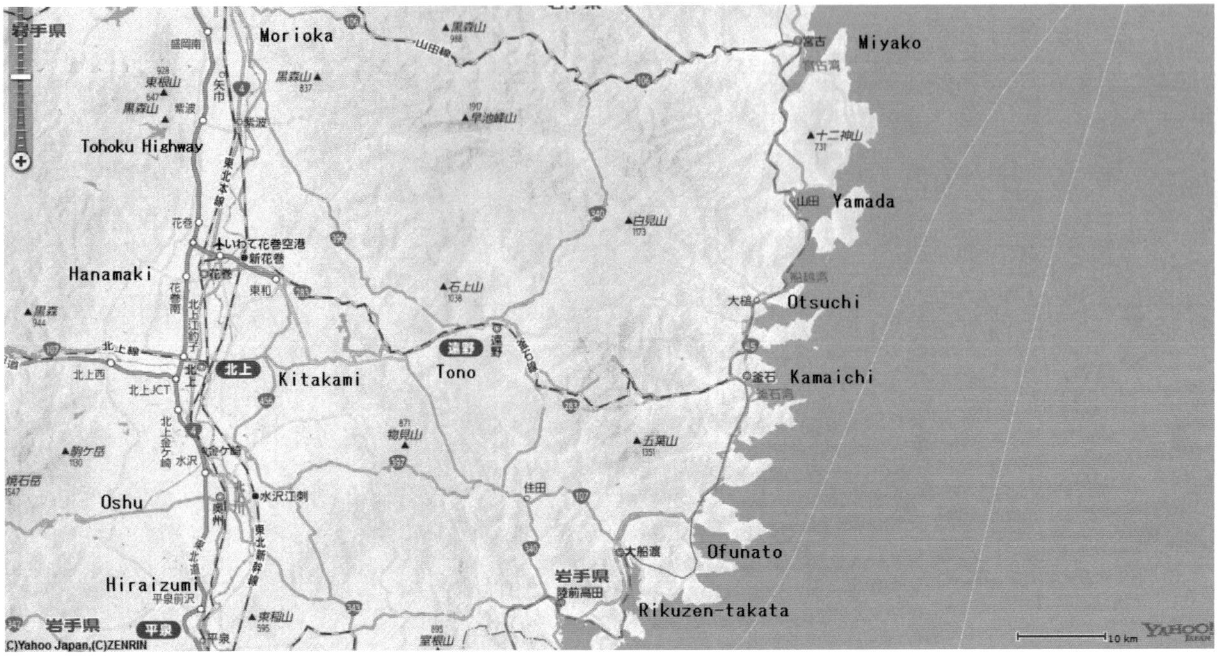

Figure 1. Southern coastal area of Iwate Prefecture.

Table 1. Populaions and Occupied Ratio of Temporary Housings.

	Populaions	Temporary housings	Occupied	Occupied Ratio(%)
Kamaishi City	39,574	3,164	2,865	90.5
Ofunato City	40,737	1,811	1,784	98.5
Rikuzen-Takata City	23,300	2,168	2,141	98.8
Iwate Prefecture	1,330,147	13,984	13,233	94.6

complex increases with decreasing distance from the infrastructures.

2. Methodology

2.1. Research area

The research focuses on three cities in the southern coastal area of the Iwate Prefecture [Fig. 1], [Table. 1], namely, Kamaishi City, Ofunato City and Rikuzen-Takata City, all of which were inundated by the tsunami that struck after the Tohoku earthquake in 2011.

2.2. Research materials

The data used in the study were the coordinate data of the temporary housing complexes [Fig. 2] and the infrastructures of daily life in December 2011.

2.3. Daily life infrastructures

The daily life infrastructures evaluated in this study are listed below. The assumed service areas are given in parentheses.

IOP Conf. Series: Earth and Environmental Science **20** (2014) 012043 doi:10.1088/1755-1315/20/1/012043

1)Commerce field
  a. Large supermarkets (3 km)
  b. Middle-sized supermarkets (1 km)
2)Service field
  a. post offices and banks (1 km)
3)Medical field
  a. large hospitals (3 km)
  b. private clinics (1 km)
4)Transport field
  a. bus stops (0.5 km)

2.4. Obtaining the research data
The temporary housing data were retrieved from the Iwate Prefecture website [4] .

Data on the daily life infrastructures were obtained from NTT i Town Pages (a telephone number search service available on the Internet) [5] .

Bus stop coordinates were available from digital maps compiled by the National Land and Traffic Agency of Japan.

The real infrastructure coordinates and states of the services were confirmed from Internet sites and maps.

2.5. Analysis
After plotting the infrastructures on the map, we conducted a network analysis using the ArcGIS Network Analyst function. This software enables the measurements of actual road distances in addition to linear distances between two features.

In this study, we created network datasets from simple road distances, neglecting road bends and speed limit data. From the network data sets, we determined the service areas of the infrastructures, representing individual features by polygons.

Based on the distance from the infrastructures and the service area covered by the infrastructure, we then assigned a usability score to each temporary housing complex.

3. Results 1: Comparizon of the characteristics of three cities
The infrastructures of the three target cities are compared in Figure 3.

Temporary housing complexes located within the defined service area of each infrastructure are assigned a usability score of 1.00. Complexes located within twice and thrice the defined service area are assigned scores of 0.66 and 0.33, respectively. Complexes located beyond three times the infrastructure service coverage score 0.00.

3.1. Kamaishi City
In Kamaishi City (as in the other target cities), the 1.00-score area coverage for convenience stores was small, and was also relatively low for small supermarkets. In contrast, the 1.00-score area for large supermarkets and clinics was relatively wide.

3.2. Ofunato City
The 1.00-score area coverage for almost all infrastructures was larger in Ofunato City than in Kamaishi and Rikuzen-Takata. Especially, buses, clinics, and large supermarkets were accessible to many temporary housing dwellers, while hospitals covered a relatively small accessible area.

3.3. Rikuzen-Takata City
The 1.00-score area coverage for almost all infrastructures was smaller in Rikuzen-Takata City than in Kamaishi and Ofunato. There is no large supermarket in this city, and no temporary housing complexes exist within the service area of convenience stores. The 1.00-score area coverage for small supermarkets, clinics, and buses are also comparatively small.

4. Results 2: Convenience of each infrastructure in the three cities
In each of the evaluated cities, infrastructures were generally located along main roads. Since the tsunami stricken areas were largely coastal, the temporary housing complexes tended to be located outside of the original downtown areas.

Of particular importance in this section is the blank area (0.00-scoring area) of each service. The coverage of the infrastructures in each city are described below.

4.1. Large supermarket (3 km) [Fig.4]
1) Kamaishi City: Large supermarkets are located from the coastal area to the internal area nearby national route 283 and the JR Kamaishi railway line. A temporary shop was constructed near Hirata following the disaster in December 07, 2011.

2) Ofunato City: Ofunato City hosts a shop in Takkon and a temporary shop in Ofunato Town.

3) Rikuzen-Takata City: No large supermarket currently exists in this city.

4.2. Middle-sized supermarket (1 km) [Fig.5]
1) Kamaishi City: Several service areas are located in the internal area close to national route 283. Two areas reside in Unosumai and Hirata.

2) Ofunato City: Most of the service areas are located near the Ofunato railway station, which is close to national route 45. Other service areas are near the Sanriku, Ryori, and Hosoura railway stations.

3) Rikuzen-Takata City: Several service areas reside in the downtown, Takekoma, and Otomo areas.

4.3. Convenience stores (0.5 km) [Fig.6]
1) Kamaishi City: Several service areas are located in the internal area close to national route 283. Two areas reside in Unosumai and Hirata.

2) Ofunato City: Most of the service areas are located near the Ofunato railway station nearby national route 45. Some service areas reside near the Sanriku, Ryori, and Hosoura railway stations.

3) Rikuzen-Takata City: Several service areas are located at Takekoma and Yahagi in the downtown area.

4.4. Post offices and banks (1 km) [Fig.7]
In Japan, post offices offer standard banking services. Hence, we treated both services as a single group.

1) Kamaishi City: This city hosts 7 banks and 15 post offices.

2) Ofunato City: This city hosts 4 banks and 8 post offices.

3) Rikuzen-Takata City: This city hosts 2 banks and 7 post offices.

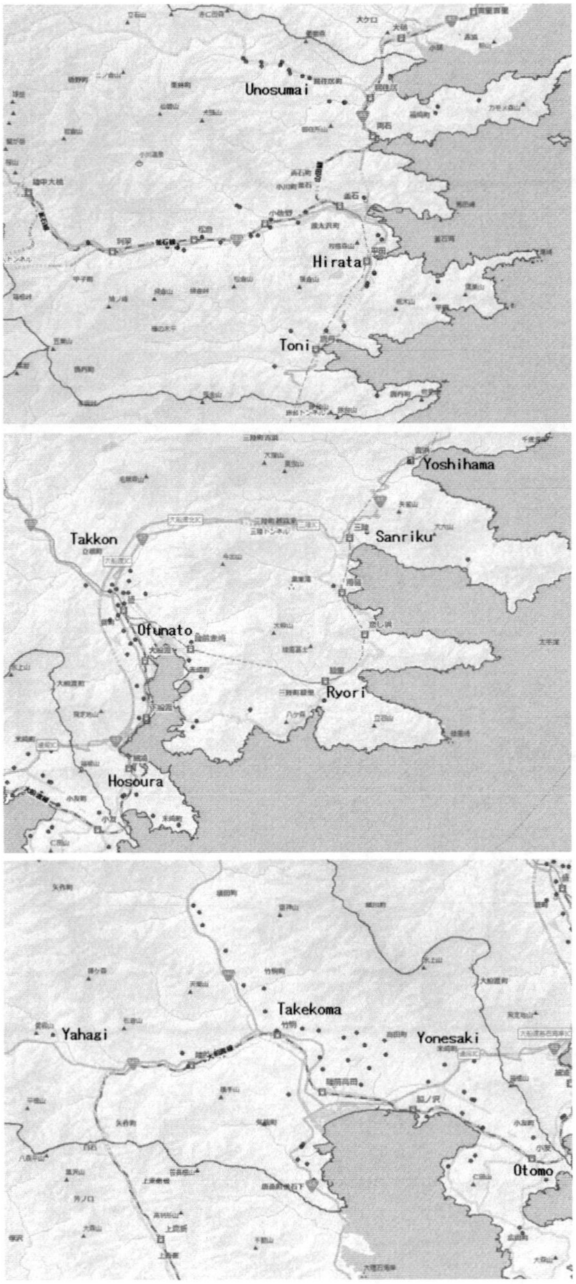

Figure 2. Locations of temporary housing complexes (Upper: Kamaishi City, Center: Ofunato City, Lower: Rikuzen-Takata City).

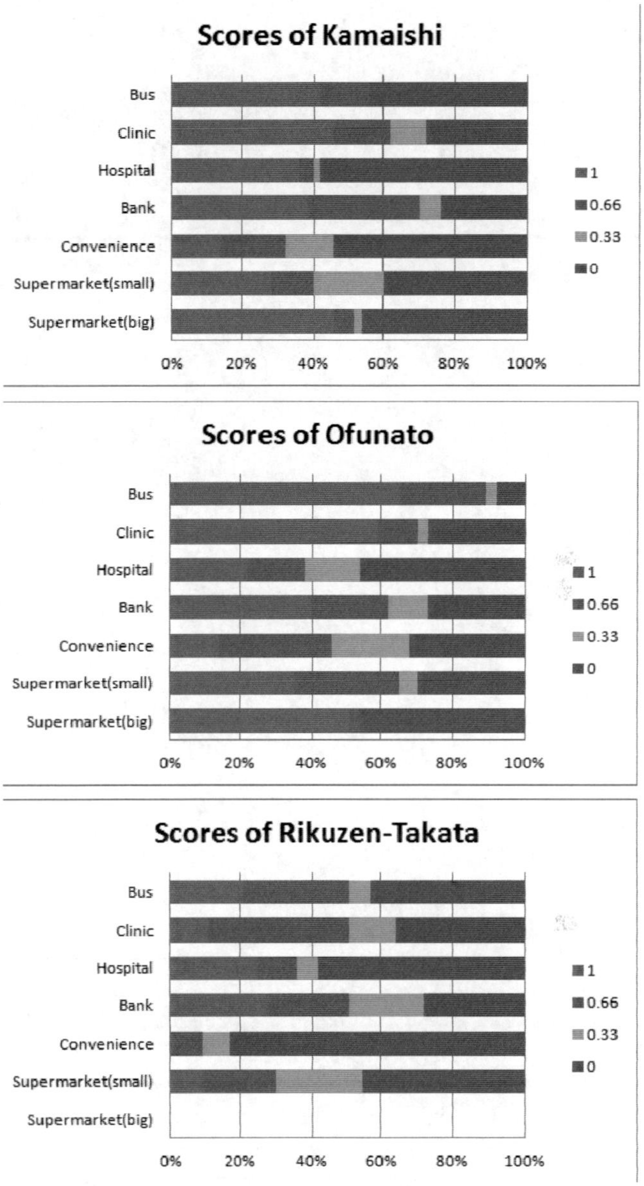

Figure 3. Comparisons of infrastructures of three targeted cities. Blue regions indicate the percentage of temporary housing complexes located WITHIN the defined service area of each infrastructure. Red and green regions denote the percentage of temporary dwellings within TWO and THREE times the defined service area, respectively. Violet regions indicate the percentage of temporary dwellings outside three times of the defined service area.

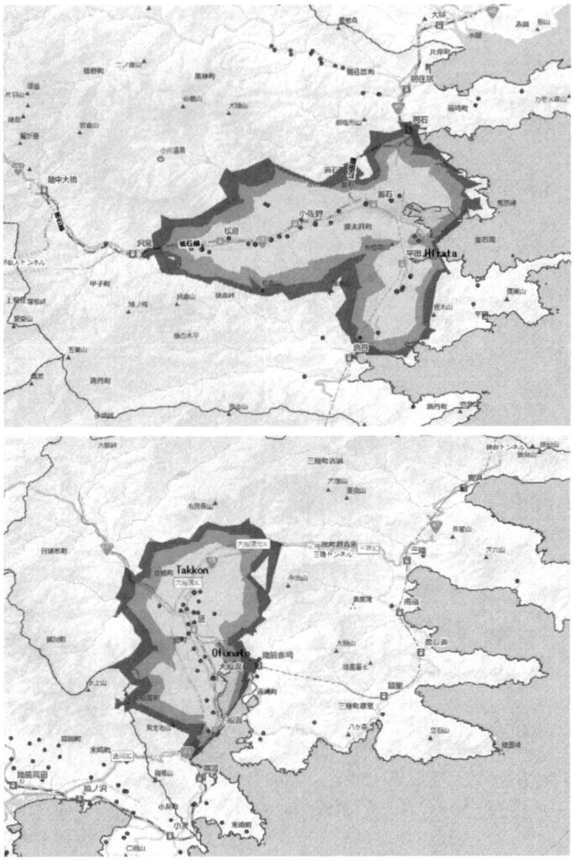

Figure 4. Locations of large supermarkets (Upper: Kamaishi City, Lower: Ofunato City). Green and red points indicate the infrastructures and the temporary housing complexes, respectively.

4.5. Large hospitals (3km) [Fig.8]

1) Kamaishi City: Four large hospitals lie along national route 283.

2) Ofunato City: A single large hospital exists in Ofunato Town.

3) Rikuzen-Takata City: There is a single large hospital in Yonesaki Town. This temporary hospital replaces the former hospital that was damaged by the tsunami.

4.6. Private clinics (1 km) [Fig.9]

1) Kamaishi City: Kamaishi City contains 29 private clinics. Several private clinics are located along national route 283; others are located in Kuribayashi, Unosumai, and the Hirata area.

2) Ofunato City: This city accommodates 32 private clinics. Most of these are located near the Ofunato railway station, which is close to national route 45. Some service areas exist near the Yoshihama, Ryori, and Sanriku railway stations.

3) Rikuzen-Takata City: The 12 private clinics in Rikuzen-Takata City are more separated than in the other investigated cities.

IOP Conf. Series: Earth and Environmental Science **20** (2014) 012043 doi:10.1088/1755-1315/20/1/012043

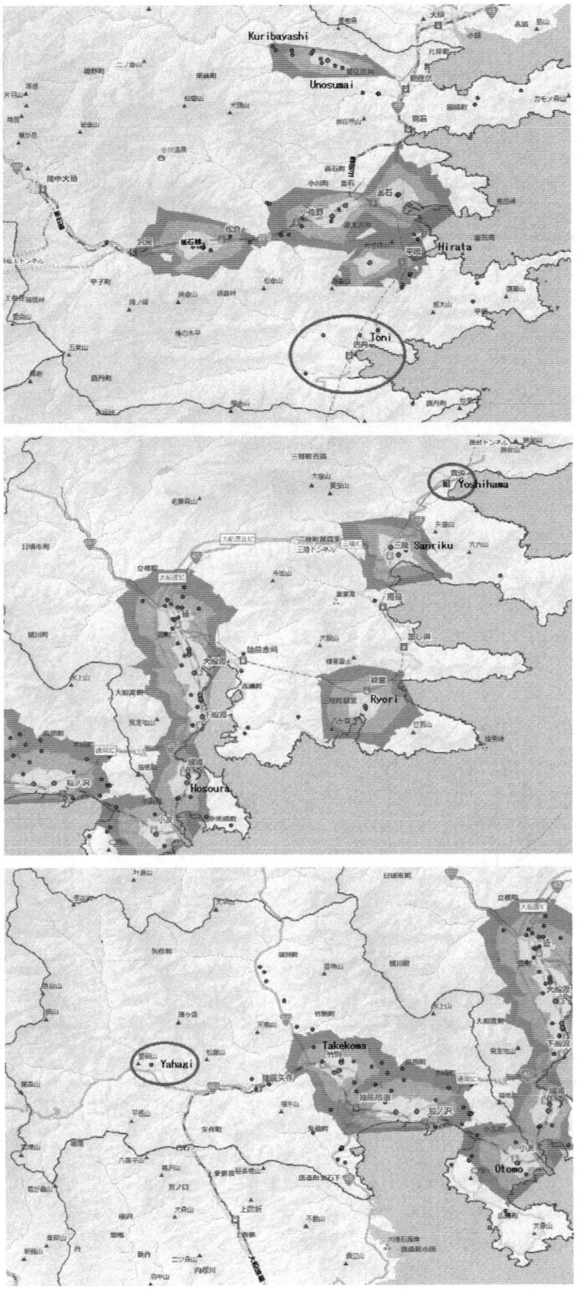

Figure 5. Location of middle-sized supermarkets (Upper: Kamaishi City, Center: Ofunato City, Lower: Rikuzen-Takata City). Green and red points indicate the infrastructures and the temporary housing complexes, respectively. Blank areas (where the score for each infrastructure = 0.00) are delineated by red ovals.

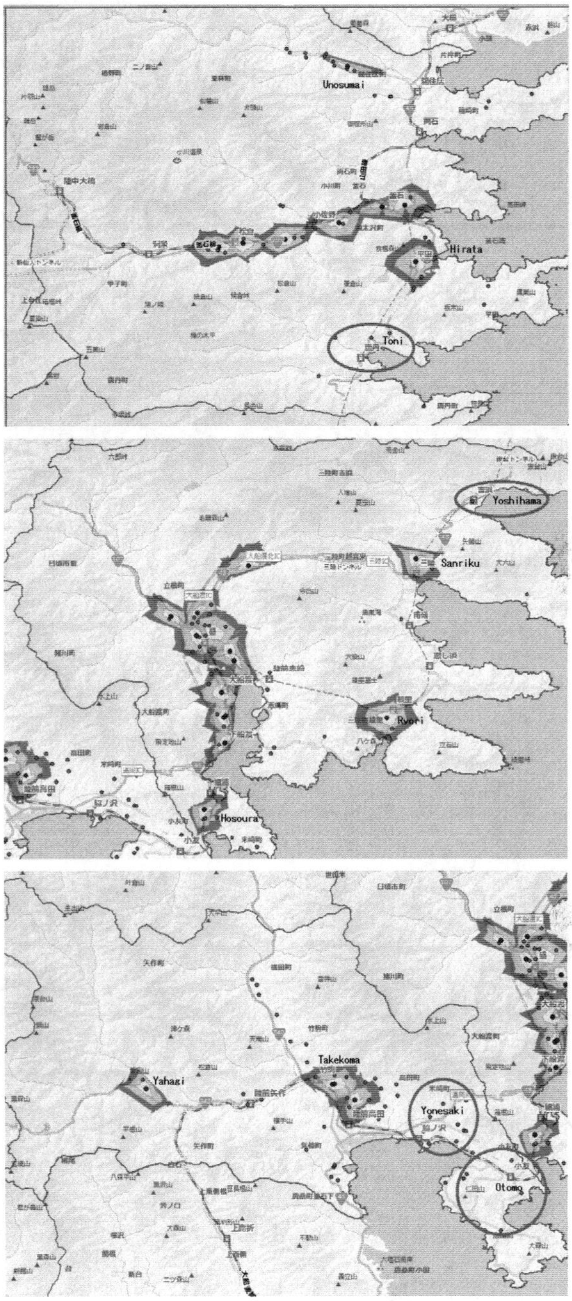

Figure 6. Location of convenience stores (Upper: Kamaishi City, Center: Ofunato City, Lower: Rikuzen-Takata City). Green and red points indicate the infrastructures and the temporary housing complexes, respectively. Blank areas are delineated by red ovals.

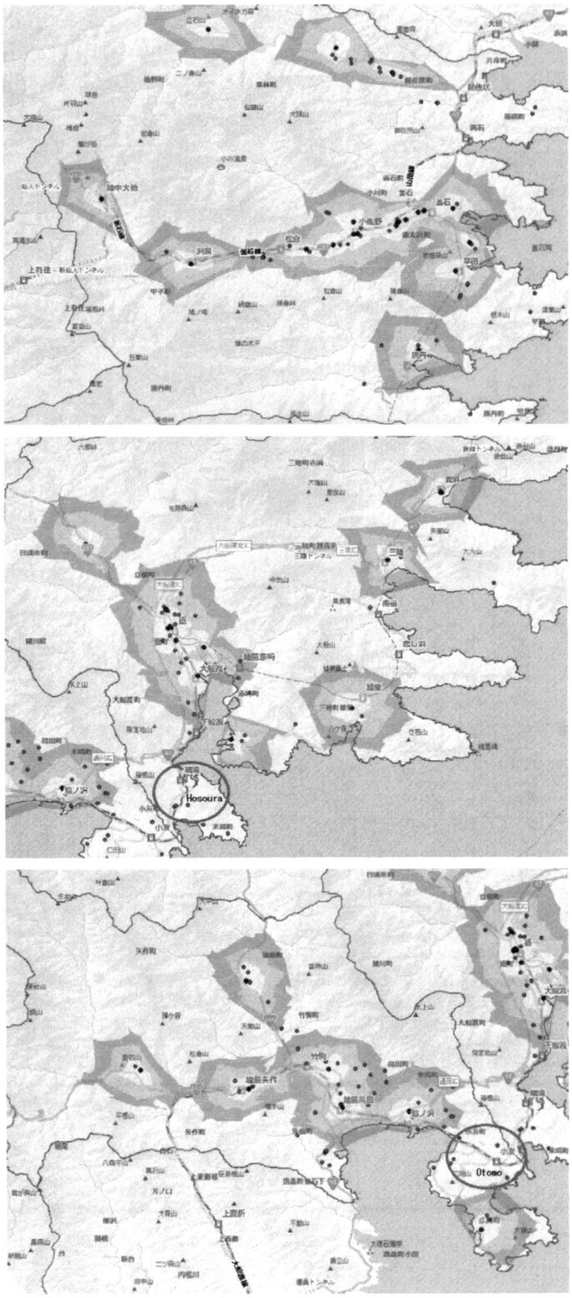

Figure 7. Location of post offices and banks (Upper: Kamaishi City, Center: Ofunato City, Lower: Rikuzen-Takata City). Green and red points indicate the infrastructures and the temporary housing complexes, respectively. Blank areas are delineated by red ovals.

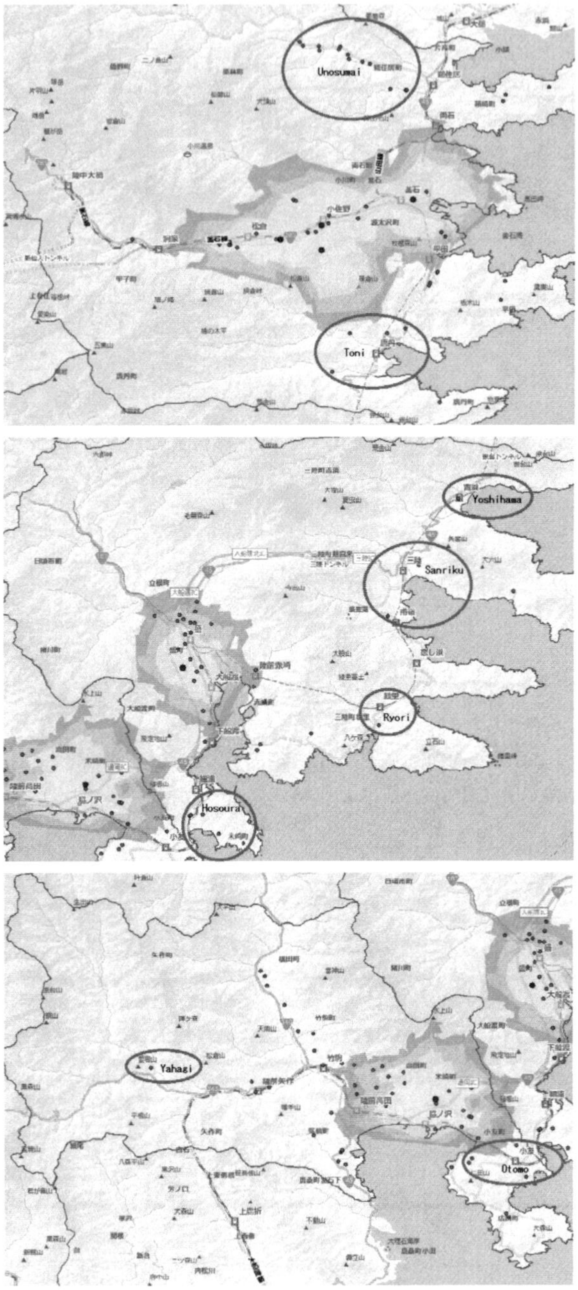

Figure 8. Location of large hospitals (Upper: Kamaishi City, Center: Ofunato City, Lower: Rikuzen-Takata City). Green and red points indicate the infrastructures and the temporary housing complexes, respectively. Blank areas are delineated by red ovals.

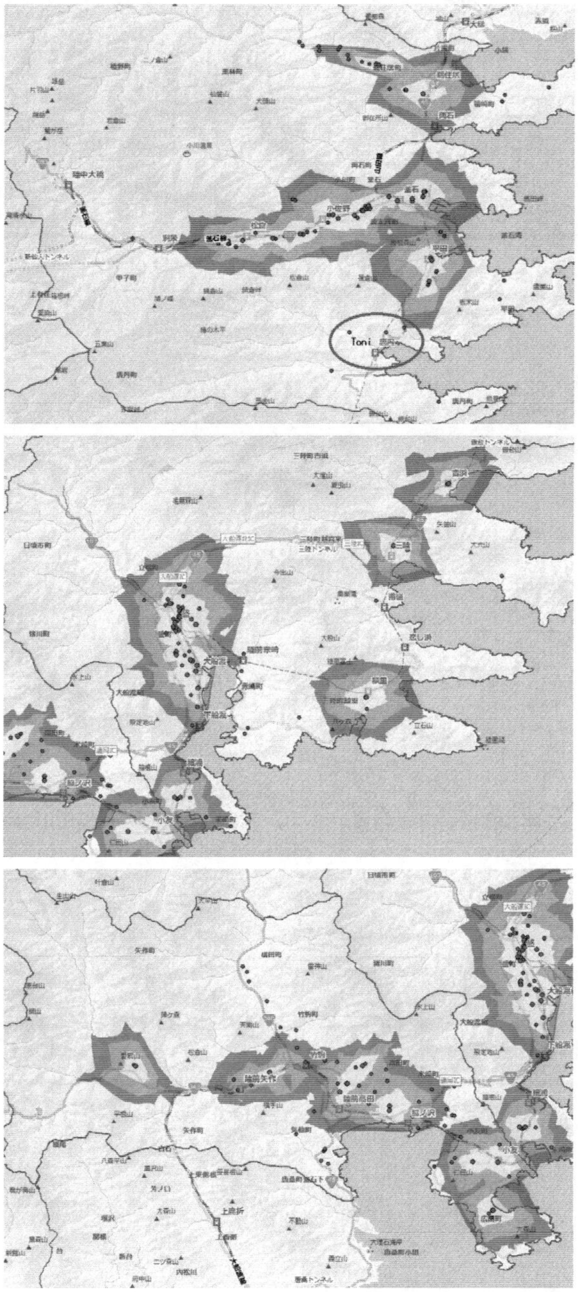

Figure 9. Location of private clinics (Upper: Kamaishi City, Center: Ofunato City, Lower: Rikuzen-Takata City). Green and red points indicate the infrastructures and the temporary housing complexes, respectively. Blank areas are delineated by red ovals.

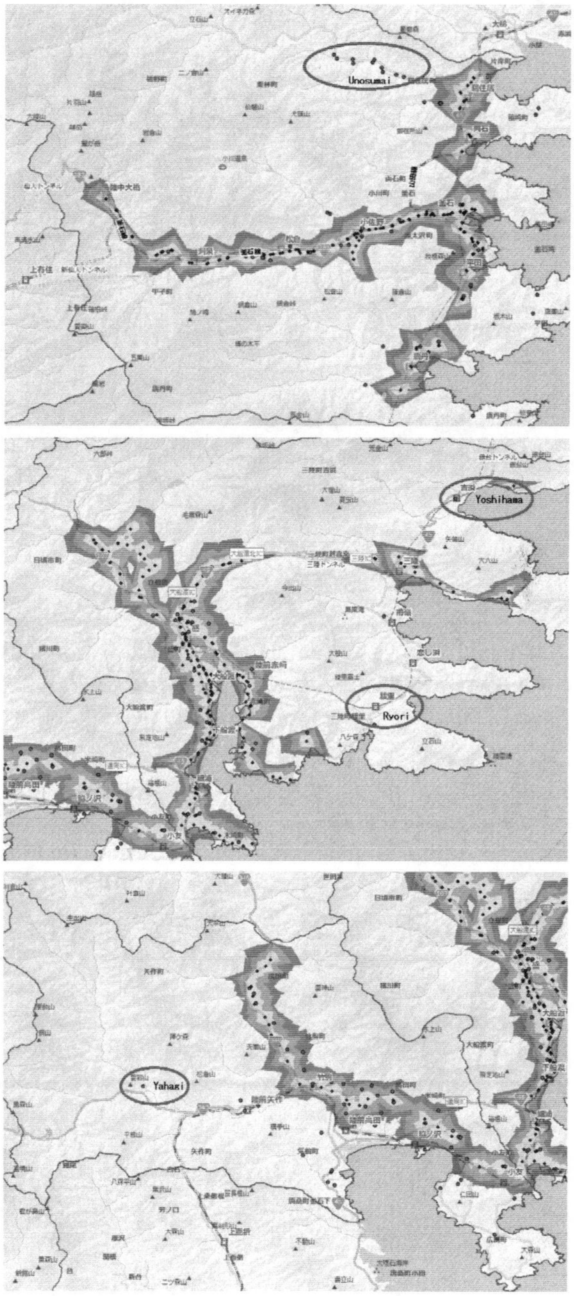

Figure 10. Bus stop locations (Upper: Kamaishi City, Middle: Ofunato City, Lower: Rikuzen-Takata City). Green and red points indicate the infrastructures and the temporary housing complexes, respectively. Blank areas are delineated by red ovals.

4.7. Bus stops (0.5Km) [Fig.10]

1) Kamaishi City: Most of the bus stops are located along national routes 45 and 283 near the residential areas.

2) Ofunato City: Most of the bus stops are located along national routes 9, 45, and 107 near the residential areas.

3) Rikuzen-Takata City: Most of the bus stops have been shifted from the tsunami-stricken national routes to the internal prefectural and city roads near the residential areas.

5. Discussion

5.1. City population and proportion of stricken urban area

Of the three investigated cities, Rikuzen-Takata City hosts the smallest population [Table. 1], and the proportion of its urban area struck by the tsunami was much larger. We link these factors to the low 1.00-scoring areas of the housing complexes in this city, that is, infrastructures are far less accessible to temporary housing dwellers in Rikuzen-Takata City than to the temporary housing occupants in the other cities.

5.2. Shape of the main city area

Although the populations of Kamaishi City and Ofunato City are statistically the same [Table. 1], the main urban areas are very differently shaped. Kamaishi City is narrower and longer than Ofunato City, and its urban area is much less intensive. This different topography probably explains the higher 1.00-scoring area of the temporary housing complexes in Ofunato City than in Kamaishi City, for almost all infrastructures.

5.3. Blank areas of the daily infrastructures

We identified more than the predicted number of blank areas (where temporary dwellings were not covered by daily infrastructures) in the targeted areas. Thus, we consider that support services for people living in such areas are urgently required, especially in the realms of daily shopping, banking, and healthcare.

Acknowledgements

We thank the people of the whole world who assisted the 3.11 affected people of Japan.

In addition, we thank the people who created and maintained the useful systems, which assisted the 3.11 affected people of Japan.

The authors would like to thank Enago for the English language review.

References

[1] Kazuro HIRAHARA and Takuo SHIBUTANI: The 2011 Tohoku Giant Earthquake Mw 9.0: An Overview (In Japanese), *Journal of Japan Society for Natural Disaster Science*, Vol.31(1), 3-22, 2012. Available at http://www.jsnds.org/contents/shizen_saigai_back_number/ssk_31_1_3.pdf

[2] Yosuke HIRAYAMA, Hiroshi MANO, Saki KASUYA, et al. : Housing Situationss after the Great East Japan Earthquake: The Case of Kamaishi City (In Japanese), *Journal of Architecture and Planning*, 77(679), 2157-2164, 2012.
Available at http://www.lib.kobe-u.ac.jp/repository/90001883.pdf

[3] Tomoko SEKINE: New Progress in Living Environment Analvsis using GIS (In Japanese), *Annals of the Japan Association of Economic Geographers*, Vol.47(4), 15-25, 2001.

[4] The Addresses of Temporary Housing Complexes in Iwate Prefecture (In Japanese):
http://www.pref.iwate.jp/kenchiku/saigai/kasetsu/009714.html
The file is available at
http://www.pref.iwate.jp/dbps_data/_material_/_files/000/000/009/714/kasetsu.pdf

[5] i Town Page of NTT Corporation (In Japanese):
http://itp.ne.jp/?rf=1

IOP Conf. Series: Earth and Environmental Science **20** (2014) 012044 doi:10.1088/1755-1315/20/1/012044

Computation of spatial significance of mountain objects extracted from multiscale digital elevation models

Dinesh Sathyamoorthy
Science & Technology Research Institute for Defence (STRIDE), Ministry of Defence, Malaysia

E-mail: dinesh.sathyamoorthy@stride.gov.my

Abstract. The derivation of spatial significance is an important aspect of geospatial analysis and hence, various methods have been proposed to compute the spatial significance of entities based on spatial distances with other entities within the cluster. This paper is aimed at studying the spatial significance of mountain objects extracted from multiscale digital elevation models (DEMs). At each scale, the value of spatial significance index *SSI* of a mountain object is the minimum number of morphological dilation iterations required to occupy all the other mountain objects in the terrain. The mountain object with the lowest value of *SSI* is the spatially most significant mountain object, indicating that it has the shortest distance to the other mountain objects. It is observed that as the area of the mountain objects reduce with increasing scale, the distances between the mountain objects increase, resulting in increasing values of *SSI*. The results obtained indicate that the strategic location of a mountain object at the centre of the terrain is more important than its size in determining its reach to other mountain objects and thus, its spatial significance.

1. Introduction

A spatially significant entity can be defined as an entity from which it is easy to reach all of its neighbouring entities (e.g., objects, hotspots, zones). A spatially significant entity has greater geometric proximity to the other entities, and should be at a strategic location and possess a relatively larger size [1-4]. The derivation of spatial significance is an important aspect of geospatial analysis and hence, various methods have been proposed to compute the spatial significance of entities based on spatial distances with other entities within the cluster, including kernel density estimation [5-7], spatial proximity matrix [8, 9], local indicators of spatial association [10], vulnerable localities index [11] and local clustering index [12]. These methods only consider the proximities of the entities, but not their geometric properties.

To this end, Sagar *et al.* [1] proposed a mathematical morphological based algorithm to compute the spatial significance of entities. The algorithm employed iterative dilations between origin and destination entities to determine the dilation distance. The computed dilation distances are then used to determine the spatial significance index *SSI* of each entity. As the effects of the morphological operations have intuitive interpretations using geometric terms of shape, size and location, the computed values of *SSI* are affected by the entities' boundary points that are furthest out with respect to the other entities. The algorithm was demonstrated using a cluster of water bodies and states in India.

Content from this work may be used under the terms of the Creative Commons Attribution 3.0 licence. Any further distribution of this work must maintain attribution to the author(s) and the title of the work, journal citation and DOI.
Published under licence by IOP Publishing Ltd

In this paper, the algorithm proposed in Sagar *et al.* [1] is employed to study the spatial significance of mountain objects extracted from multiscale digital elevation models (DEMs). Scale variations can constrain the detail with which information can be observed, represented and analysed. The term scale refers to a combination of both spatial extent, and spatial detail or resolution [13-16]. Changing the scale without first understanding the effects of such an action can result in the representation of patterns or processes that are different from those intended due to loss of detail, and variations in terrain parameters and landforms [15-18]. Hence, feature detection and characterisation often need to be performed at different scales of measurement. Wu *et al.* [19], Wood [20], Drăguţ *et al.* [21] and Poulos *et al.* [22] demonstrated that analysis of a location at multiple scales allows for a greater amount of information to be extracted from a digital elevation model (DEM) about the spatial characteristics of a feature.

2. Computation of spatial significance

Mathematical morphology is a branch of image processing that deals with the extraction of image components that are useful for representational and descriptional purposes. It has a well developed mathematical structure that is based on set theoretic concepts. The fundamental morphological operators are discussed in Matheron [23], Serra [24], Soille [25], and Najman and Talbot [26]. Morphological operators generally require two inputs; the input image A, which can be in binary or grayscale form, and the kernel B, which is used to determine the precise effect of the operator. Each cell in A is compared with B by moving B so that its centre hits the cell. Depending on the type of morphological operator employed, the cell value is reset to the value or average value of one or more of its neighbours.

Dilation (Equation 1) sets the cell values within the kernel to the maximum value of the cell neighbourhood. Binary dilation gradually enlarges the boundaries of regions of foreground cells, resulting in areas of foreground cells growing in size, and holes within those regions becoming smaller. Dilation can be performed iteratively for n times as in Equation 2.

$$A \oplus B = \{a + b : a \in A, b \in B\} \tag{1}$$

$$A \oplus nB = (A \oplus B) \oplus B \ldots n \text{ iterations} \tag{2}$$

Let a cluster of entities A be consisted of N number of entities denoted as A_1, A_2, A_3, … , A_N. An entity is designated as spatially most significant if its distance to the other entities in the cluster is the shortest. The dilation distance d from an origin entity A_i and a destination entity A_j is minimum number of dilation iterations required for A_i to occupy A_j:

$$d(A_{ij}) = \min(n : A_j \subseteq (A_i \oplus nB)) \tag{3}$$

Similarly, the dilation distance from A_j to A_i is computed as follows:

$$d(A_{ji}) = \min(n : A_i \subseteq (A_j \oplus nB)) \tag{4}$$

If both A_i and A_j have the same size, shape and orientation, $d(A_{ij}) = d(A_{ji})$. If there is a shape-size dissimilarity between the two entities, $d(A_{ij}) \neq d(A_{ji})$. A smaller entity would require a greater number of dilation iterations to occupy a larger entity and hence, the larger entity would have a smaller dilation distance.

SSI determines the spatial significance on an entity based on the distance from the other entities in the cluster. The lower the value of *SSI* of an entity, the higher is its significance, indicating that it has shorter distance to the other entities. The entity with the lowest value of *SSI* is the spatially most significant entity. The value of *SSI* for A_i is the minimum number of dilation iterations required for A_i to occupy all the other entities in the cluster, which would be the largest value of $d(A_{ij})$:

$$SSI(A_i) = \max_{\forall j}\left(\min\left(n : \left(A_j \subseteq \left(A_i \oplus nB\right)\right)\right)\right) \tag{5}$$

3. Methodology

3.1. Data set

The DEM in figure 1 shows the area of Great Basin, Nevada, USA, which is bounded by latitude 38° 15' to 42° N and longitude 118° 30' to 115° 30'W. The DEM was resampled to 925 m in both *x* and *y* directions. It is a Global Digital Elevation Model (GTOPO30) and was downloaded from the USGS GTOPO30 website [27]. GTOPO30 DEMs are available at a global scale, providing a digital representation of the Earth's surface at a 30 arc-seconds sampling interval. The land data used to derive GTOPO30 DEMs are obtained from digital terrain elevation data (DTED), 1-degree DEMs for USA and the digital chart of the world (DCW). The accuracy of GTOPO30 DEMs varies by location according to the source data. The DTED and the 1-degree dataset have a vertical accuracy of ±30 m while the absolute accuracy of the DCW vector dataset is ±2,000 m horizontal error and ±650 m vertical error [28]. Tensional forces on the terrain's crust and thins by normal faulting cause the formation an array of tipped mountain blocks that are separated from broad plain basins, producing a basin-and-range physiography [29-33].

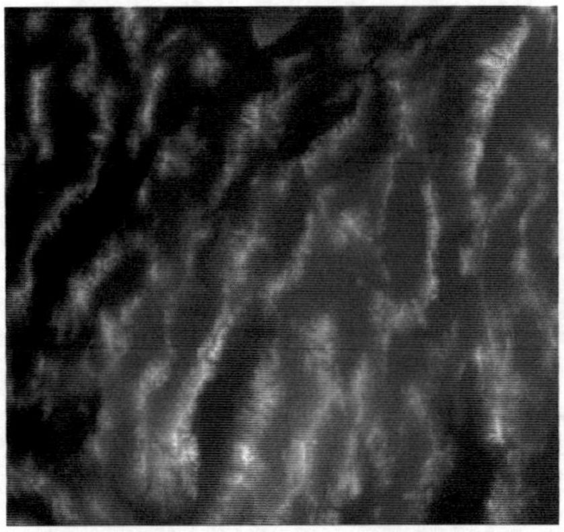

Figure 1. The GTOPO30 DEM of Great Basin. The elevation values of the terrain (1,005 to 3,651 m) are rescaled to the interval of 0 to 255 (the brightest cell has the highest elevation). The scale is approximately 1:3,900,000.

3.2 Generation of multiscale DEMs

In this paper, multiscaling is performed using the lifting scheme [34-35], which has proven to be a powerful multiscale analysis tool in image and signal processing [36-39], and has received recent attention in geospatial analysis [40-45]. This is due to its ability to preserve accurate surface profiles, in terms of waveform, shape and amplitude, without causing boundary destruction [46-48]. It is used to decompose an original dataset into low and high frequency subsets using the following three steps:

Step 1: Split

The original dataset $x[n]$ is divided into two disjoint subsets, even $x_e[n]$ and odd $x_o[n]$ indexed points.

Step 2: Predict

The odd and even subsets are often highly correlated. As this correlation structure is typically local, one subset can be used to predict the other subset. In this case, the even indexed subset is used to compute a course approximation of the odd indexed subset using the prediction operator P (Equation 6). The predicted odd indexed subset is smoother than the original odd indexed subset, with the high frequency components removed. Hence, the difference between the two subsets gives the high frequency subset $d[n]$ (Equation 7). The even indexed subset is left unchanged to become the input for the next step in the transform.

$$P(x_o[n]) = \frac{1}{2}(x_e[n] + x_e[n+1]) \quad (6)$$

$$d[n] = x_o[n] - P(x_o[n]) \quad (7)$$

Step 3: Update

The update operator U (Equation 8) uses $d[n]$ to compute the difference between the original dataset and its coarse approximation, which consists of the low frequency components of the original dataset. This is then applied to the even indexed subset to obtain the low frequency subset $c[n]$ (Equation 9).

$$U(d[n]) = \frac{1}{4}(d[n-1] + d[n+1]) \quad (8)$$

$$c[n] = x_e[n] + U(d[n]) \quad (9)$$

The above three steps form a lifting stage. Using a DEM as the input, an iteration of the lifting stage generates the complete set of multiscale DEMs $c_s[n]$ and the elevation loss caused by the change of scale $d_s[n]$. At each iteration, $c_s[n]$ only contains half of the points of the input for the iteration, and hence, the resolution of the generated multiscale DEM is reduced by half. At each iteration, the cells of the DEM that are modified are curvatures regions, while the unmodified cells are planar regions. The iterations are repeated until all curvatures in the DEM are removed, leaving only planar regions. For varying DEMs, the number of iterations required would be dependent on the surface profile; a rougher surface profile would require more iterations, while a smoother surface profile would require fewer iterations.

Multiscale DEMs of the Great Basin terrain are generated by implementing the lifting scheme for scales s of 1 to 20. As shown in figure 2, as the scale increases, the merging of small regions into the surrounding grey level regions increases, causing removal of fine detail in the DEM. As a result, the generated multiscale DEMs possess lower resolutions at higher degrees of scaling.

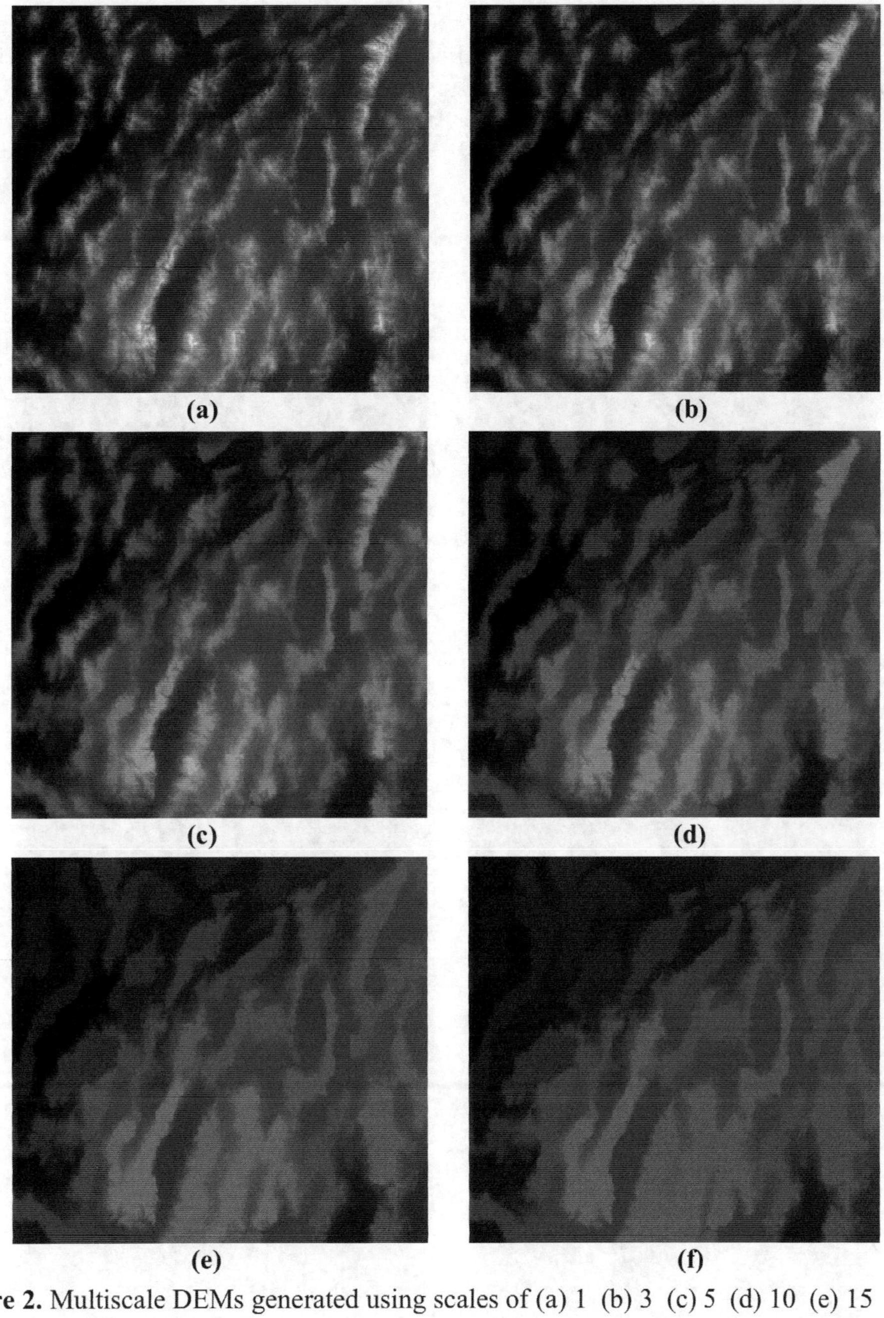

Figure 2. Multiscale DEMs generated using scales of (a) 1 (b) 3 (c) 5 (d) 10 (e) 15 (f) 20.

3.3 Mountain extraction

Mountains are portions of a terrain that are sufficiently elevated above the surrounding land (greater than 300 up to 600 m) and have comparatively steep sides. In a mountain, two parts are distinctive [49]:

1) The summit, the highest point (the peak) or the highest ridges
2) The mountainside, the part of a mountain between the summit and the foot.

IOP Conf. Series: Earth and Environmental Science **20** (2014) 012044 doi:10.1088/1755-1315/20/1/012044

The mountains of the generated multiscale DEMs are extracted using the mathematical morphological based algorithm proposed in Dinesh [50]. First, ultimate erosion is performed on the DEM to extract the peaks of the DEM. Conditional dilation is performed on the extracted peaks to obtain the mountains of the DEM. As shown in figure 3, the merge of small regions into the surrounding grey level regions and removal of fine detail in the DEM cause a reduction in the area of the extracted mountains.

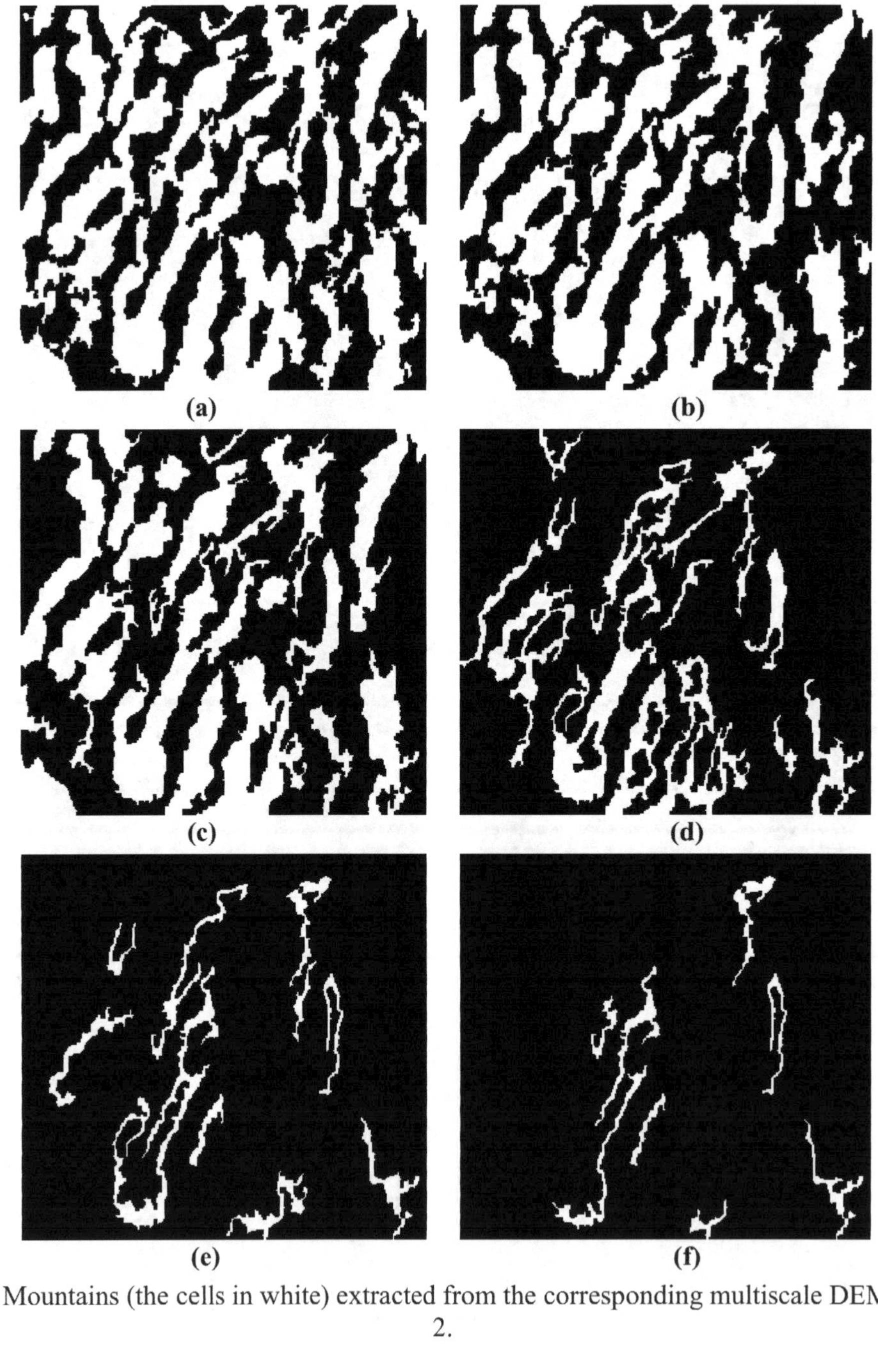

Figure 3. Mountains (the cells in white) extracted from the corresponding multiscale DEMs in figure 2.

4. Results & Discussion

The computed *SSI* values of the mountain objects extracted from the original DEM (figure 3(a)) are shown in table 1. In general, larger mountain objects, which have better reach to the other mountains objects, have lower value of *SSI*, indicating higher significance. The spatially most significant mountain object, with the lowest value of *SSI*, is mountain object 5 (figure 4(a)), which is the largest mountain object and is located at the centre of the terrain.

Table 1. Computed *SSI* values of the mountain objects extracted from the original DEM.

Label	Area (cells)	*SSI*
1	1227	281
2	10422	195
3	1353	273
4	298	284
5	14232	106
6	432	246
7	6444	172
8	311	267
9	1119	221
10	219	233
11	3574	247
12	3058	252
13	494	282
14	261	289

The *SSI* values for the spatially most significant mountain objects of the multiscale DEMs (figure 4) are shown in table 2. As the area of the mountain objects reduce with increasing scale, the distances between the mountain objects increase, resulting in increasing values of *SSI*. At scales 2 to 7, 9 to 11 and 13 to 20, the corresponding spatially most significant mountain objects have the same values (130, 151 and 154 respectively). In these cases, the spatially most significant mountain objects are objects with corresponding locations, requiring the same minimum number of iterative dilations to reach the boundary points that are furthest out with respect to the objects. This highlights the advantages of iterative dilations in computing *SSI* values as compared to proximity based algorithms that do not consider the geometric properties of the objects.

For scales 1 to 7 and 13 to 20, larger mountain objects located at the centre of the terrain are classified as spatially most significant. However, for scales 8 to 12, significantly smaller mountain objects located at the centre of the terrain are classified as spatially most significant. This indicates that the strategic location of a mountain object at the centre of the terrain is more important than its size in determining its reach to other mountain objects and hence, its spatial significance. For example, for scale 10 (figure 4(d)), the strategic location of mountain object 10 allows it better reach to the other mountain objects as compared to the larger mountain objects (Table 3).

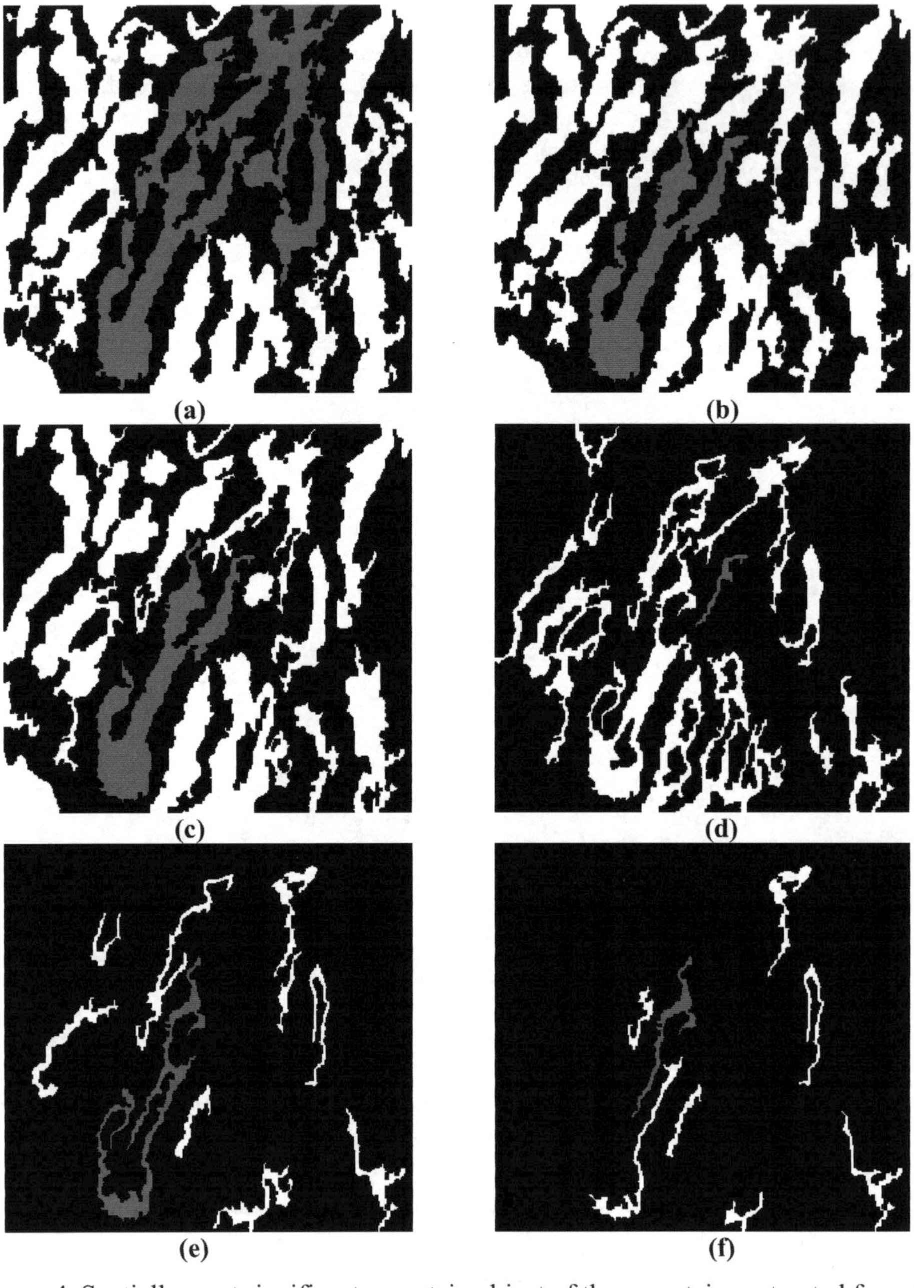

Figure 4. Spatially most significant mountain object of the mountains extracted from the corresponding multiscale DEMs in figure 2.

Table 2. Spatially most significant mountain object of each scale.

Scale	Num. of objects	Spatially most significant object	Area (cells)	*SSI*
1	14	5	14232	106
2	20	8	9581	130
3	21	8	5518	130
4	19	7	5350	130
5	19	7	5282	130
6	18	7	5280	130
7	18	7	5040	130
8	19	12	288	145
9	16	12	263	151
10	14	10	263	151
11	14	10	263	151
12	15	11	263	140
13	12	5	2670	154
14	10	3	2599	154
15	10	3	2183	154
16	10	2	2025	154
17	9	2	1946	154
18	8	2	1912	154
19	7	1	1402	154
20	8	3	629	154

Table 3. Computed SSI values of the mountain objects extracted from the DEM of scale 10.

Label	Area (cells)	*SSI*
1	571	242
2	1967	218
3	266	253
4	255	217
5	3330	154
6	1298	173
7	1420	182
8	2176	172
9	1152	187
10	263	151
11	221	207
12	803	214
13	229	236
14	1048	260

The results obtained thus far are based on analysis of a mountainous terrain with rough surface profile. The analysis is further extended for the GTOPO30 DEM of Great Plains, Nebraska (bounded by latitude 39° to 43° N and longitude 98° to 101° W) with a smoother surface profile (figure 5). The smoother surface profile of the terrain results in fewer multiscaling iterations (11) required to remove all the curvature regions. It is observed in figure 6 and table 4 that for scales 1 to 7, smaller mountain objects located at the centre of the terrain are classified as spatially most significant. For scales 8 to 11, with with no strategically located mountain objects at the centre of the terrain, larger mountain objects located at the corner of the terrain are classified as spatially most significant. For both cases, the corresponding spatially most significant mountain objects have the same values (165 for scales 1 to 7, and 169 for scales 8 to 11), as the objects require the same minimum dilation distances to the boundary points that are furthest out with respect to the objects.

Figure 5. The GTOPO30 DEM Great Plains. The scale is approximately 1:3,900,000.

5. Conclusion

In this paper, iterative morphological dilations were used to study the spatial significance of mountain objects extracted from multiscale DEMs. It was observed that as the area of the mountain objects reduce with increasing scale, the distances between the mountain objects increase, resulting in increasing values of SSI. The results obtained indicate that the strategic location of a mountain object at the centre of the terrain is more important than its size in determining its reach to other mountain objects and thus, its spatial significance.

7. Acknowledgement

The author is grateful to Assoc. Prof. Dr. Behara Seshadri Daya Sagar, Head of the Systems Science and Informatics Unit (SSIU), Indian Statistical Institute, for his suggestions that have helped strengthen this manuscript.

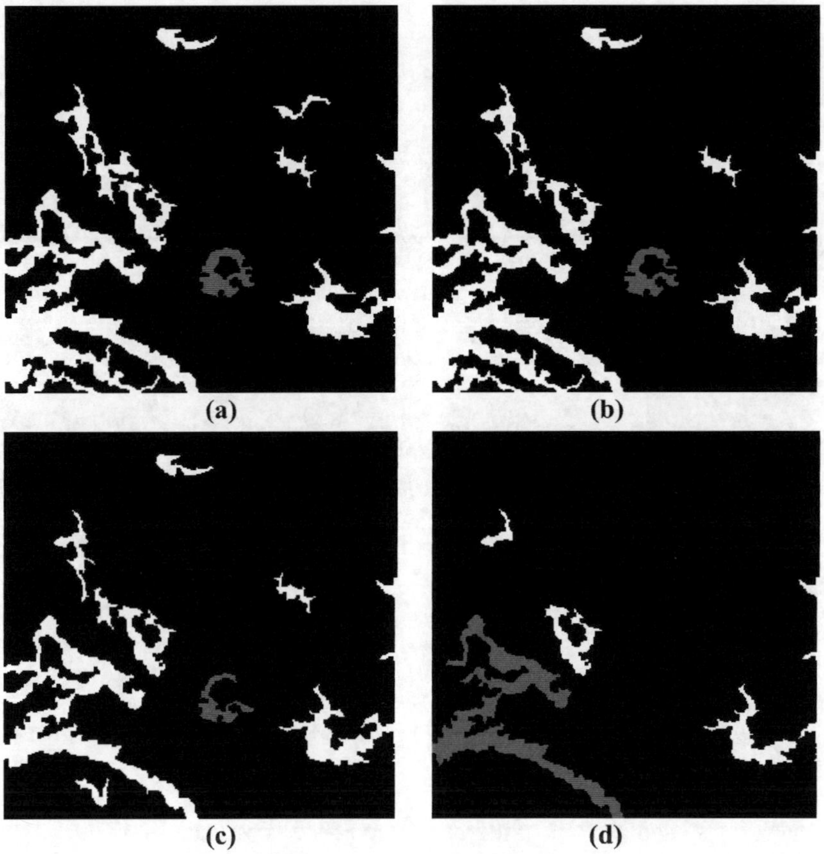

Figure 6. Spatially most significant mountain objects for the multiscale DEMs of Great Plains at scales of: (a) 1 (b) 3 (c) 5 (d) 11.

Table 4: Spatially most significant mountain objects for the multiscale DEMs of Great Plains.

Scale	Num. of objects	Spatially most significant object	Area (cells)	*SSI*
1	8	5	743	165
2	8	5	743	165
3	9	7	701	165
4	8	6	587	165
5	9	7	474	165
6	7	5	397	165
7	6	5	333	165
8	5	1	4269	169
9	6	1	4244	169
10	6	1	4170	169
11	5	1	3951	169

6. References

[1] Sagar B S D, Rajesh N, Ashok Vardhan S and Vardhan P 2013 Metric based on morphological dilation for detection of spatially significant zones *IEEE Geosci. Remote Sens. Lett.* **10** 500-504

[2] Anselin L 1992 *Spatial Data Analysis with GIS: An Introduction to Application in the Social Sciences* (Santa Barbara, California: National Center for Geographic Information and Analysis, University of California)

[3] Câmara G and Carvalho M S 2002 *A Tutorial on Spatial Analysis of Areas* (São José dos Campos, Brazil: National Institute for Space Research (INPE))

[4] Chainey SP 2010 *Spatial Significance Hotspot Mapping Using the NNI and Gi* Statistic* (London: UCL Jill Dando Institute of Crime Science)

[5] Silverman B W 1986 *Density Estimation for Statistics and Data Analysis* (London: Chapman & Hall)

[6] Schölkopf B and Smola A J 2002 *Learning with Kernels: Support Vector Machines, Regularization, Optimization and Beyond* (Cambridge, Massachusetts: MIT Press)

[7] Feitosa F, Câmara G, Monteiro A M V, Koschitzki T & Silva M P S 2007. Global and local spatial indices of urban segregation *Int. J. Geogr. Inf. Sci.* **21** 299-323

[8] Ding Y and Fotheringham A S 1992 The integration of spatial analysis and GIS *Comput. Environ. Urban* **16** 3-19

[9] de Aguiar A P D, Câmara G., Monteiro A M V, Cartaxo R and de Souza M 2003 Modelling spatial relations by generalized proximity matrices *Geoinfo 2003* (São José dos Campos, Brazil: 3-5 November 2003)

[10] Anselin L 1995 Local indicators of spatial association (LISA) *Geogr. Anal.* **27** 93-115

[11] Chainey S P 2008 Identifying priority neighbourhoods using the Vulnerable Localities Index. *Policing J. Policy Prac.* **2** 196-209

[12] Baohua L, Madden L V and Xu X 2012 Spatial analysis by distance indices: an alternative local clustering index for studying spatial patterns *Method. Ecol. Evol.* **3** 368-377

[13] Tate N and Wood J 2001 Fractals and scale dependencies in topography *Modelling Scale in Geographical Information Science* ed N Tate and P Atkinson (Chicester: Wiley) pp 35-51

[14] Li Z 2008 Multi-scale digital terrain modelling and analysis *Advances in Digital Terrain Analysis* ed Q Zhou, B Lees and G Tang (Berlin: Springer) pp 59-83

[15] Drăguţ L and Eisank C 2011 Object representations at multiple scales from digital elevation models *Geomorphology* **129** 183-189

[16] Goodchild M F 2011 Scale in GIS: An overview *Geomorphology*, **130**: 5-9

[17] Lam N, Catts D, Quattrochi D, Brown D and McMaster R 2004 Scale *A Research Agenda for Geographic Information Science* ed *M* Rechcígl, R B McMaster and E L Usery (New York: CRC Press) pp 93-128

[18] Summerfield M 2005 A tale of two scales, or two geomorphologies *T. I. Brit. Geogr.* **30** 402-415

[19] Wu S, Li J and Huang G H 2008 A study on DEM-derived primary topographic attributes for hydrologic applications: Sensitivity to elevation data resolution *Appl. Geogr.* **28** 210-223

[20] Wood J 2009. Geomorphometry in LandSerf *Developments in Soil Science, vol. 33: Geomorphometry: Concepts, Software, Applications* ed T Hengl and H I Reuter (Armsterdam: Elsevier), pp 333-349

[21] Drăguţ L, Eisank C and Strasser T 2011 Local variance for multi-scale analysis in geomorphometry *Geomorphology* **130** 162-172

[22] Poulos, M J, Pierce J L, Flores A N and Benner S G 2012 Hillslope asymmetry maps reveal widespread, multi-scale organization *Geophys. Res. Lett.* **39** L06406

[23] Matheron G 1975 *Random Sets and Integral Geometry* (Wiley: New York)

[24] Serra J 1982 *Image Analysis and Mathematical Morphology* (London: Academic Press) 1982.

[25] Soille P 2003 *Morphological Image Analysis: Principles and Applications* (Berlin: Springer Verlag)

[26] Najman L and Talbot H ed 2010 *Mathematical Morphology: From Theory to Applications* (New York: John Wiley & Sons)

[27] GTOPO30 1996 *GTOPO30: Global 30 Arc Second Elevation Data Set* Available online at: http://edcwww.cr.usgs.gov/landdaac/gtopo30/gtopo30.html Last access date: 1 June 2009

[28] Miliaresis G and Argialas D P 2002 Quantitative representation of mountain objects extracted from the Global Digital Elevation Model (GTOPO30) *Int. J. Remote Sens.* **23** 949-964.

[29] Howell D 1995 *Principles of Terrain Analysis: New Applications for Global Tectonics* (London: Chapman and Hall)

[30] Summerfield M 1996 *Global Geomorphology* (Longman: Essex)

[31] Summerfield M ed 2000 *Geomorphology and Global Tectonics* (New York: John Wiley & Sons)

[32] Miliaresis G 2008 Quantification of terrain processes *Lect. Notes Geoinform. Cartogr.* **XIV**: 13-28.

[33] Miliaresis G and Argialas D P 1999 Segmentation of physiographic features from Global Digital Elevation Model/GTOPO30 *Comput. Geosci.* **25** 715-728.

[34] Sweldens W 1996 The lifting scheme: A custom-design construction of biothorgonal wavelets *Appl. Comput. Harmonics Anal.* **3** 186-200.

[35] Sweldens 1997 The lifting scheme: A construction of second generation wavelets *J.Math. Anal.* **29** 511-546

[36] Claypoole R L and Baraniuk R G 2000 A multiresolution wedgelet transform for image processing *Wavelet Applications in Signal and Image Processing VIII, Volume 4119 of SPIE Proc.* ed M A Unser, A Aldroubi and A F Laine (Bellingham, Washington: SPIE) pp 253-262

[37] Starck J L 2002 Non-linear multiscale transforms *Multiscale and Multiresolution Methods: Theory and Applications* ed T J Barth, T Chan & R Haimes (Heidelberg, Germany: Springer) pp 239-266

[38] Guo S M, Chang W H, Tsai J S H, Zhuang B L and Chen L C 2008 JPEG 2000 wavelet next term filter design framework with chaos evolutionary programming *Signal Process.* **88** 2542-2553

[39] Abdul-Rahman, H S, Jiang, X J and Scott, P J 2013 Freeform surface filtering using the lifting wavelet transform *Precis. Eng.* **37** 187-202

[40] Hayat K, Puech W and Gesquère G 2008 Scalable 3-D terrain visualization through reversible JPEG2000-based blind data hiding *IEEE T. Multimedia* **10** 1261-1276

[41] Bernardin T, Cowgill E S, Gold R D, Hamann B, Kreylos, O and Schmitt, A. 2008 Real-time terrain mapping *Scientific Visualization: Challenges for the Future* ed H Hagen (Los Alamitos, California: IEEE Computer Society Press) pp 275-288

[42] Yang Z, Liu H, Yang N and Xiao X 2009 Study on multiscale generalization of DEM based on lifting scheme *Proceedings of the SPIE, Volume 7492* ed Y Liu and X Tang (Bellingham, Washington: SPIE) pp 74922E-74922E-6.

[43] Dinesh S, Ahmad Fadzil M H and Vijanth Sagayan A 2011 Computation of uncertainty of physiographic features extracted from multiscale digital elevation models using fuzzy classification *Image Processing, Image Analysis and Real-Time Imaging (IPIARTI) Symposium 2011* (Multimedia University (MMU), Cyberjaya, 6 October 2013)

[44] Ahmad Fadzil M H Dinesh S and Vijanth Sagayan A 2011 A method for computation of surface roughness of digital elevation model terrains via multiscale analysis *Comput. Geosci.* **37** 177-192.

[45] Chiao, L Y, Hsieh, C and Chiu, T S 2013 Exploring spatiotemporal ecological variations by the multiscale interpolation. *Ecol. Model.* **246** 26-33

[46] Jiang X Q, Blunt L and Stout K J 2001 Application of the lifting scheme to rough surfaces *Precis. Eng.* **25** 83-89.

[47] Jiang X Q, Blunt L and Stout K J 2001 Lifting wavelet for three-dimensional surface analysis *Int. J. Mach. Tool. Manu.* **41** 2163-2169

[48] Nonomura K, Ono M, Zhou L B, Shimizu J and Ojima H 2010 Design of digital filters for Si wafer surface profile measurement : Noise reduction by lifting scheme wavelet transform *Adv. Mater. Res.* **126-128** 732-737

[49] Neuendorf, K K E, Mehl, J P and Jackson, J A Eds. 2011 Glossary of Geology (Alexandria, Virginia: American Geological Institute)
[50] Dinesh S 2006 Extraction of mountains from digital elevation models using mathematical morphology *GIS Malaysia* **1** 16-19

A geospatial approach to evaluation of accessibility to secondary educational institution in Ogun State, Nigeria.

S A Ogunyemi[1], K H Muibi[1], O E Eguaroje[1], O O Fabiyi[2], A S Halilu[1]
[1]Cooperative Information Network, National Space Research and Development Agency, Obafemi Awolowo University Ile-Ife, Osun State, Nigeria.
[2]Regional Centre for Training in Aerospace Surveying, Ile Ife, Osun State, Nigeria

Email: samsonogunyemi@yahoo.com

Abstract. The study therefore examined the spatial distribution of secondary school and the accessibility levels to students in Yewa South local Government area of Nigeria and the factors that give rise to them with a view to providing a framework for effective placement of students. Primary and secondary data were used for the study. Primary data was collected using questionnaire and a hand-held GPS receiver used to capture the coordinate points of schools and other relevant data. Secondary data include administrative map, population figures of both students and Teachers, Names and addresses of the secondary schools in the study area, the school placement sheet and list of schools. The data analysis was carried out using *network analysis*. School location, number of secondary schools in each ward and the total area were used to determine the pattern of distribution of secondary schools in the study area. Settlements, roads, schools location, number of secondary schools, ward and school enrolments were used to generate both school accessibility and effective placement measure using network analysis. Network analyses were also performed to evaluate nearest school to student and a set of origin-destination (OD) matrix. However, the whole of 25 secondary schools in Yewa south were considered for network analysis ward by ward, and the results of the distance students travel from their settlements to their various schools were shown for all the secondary schools in each ward. Out of 415 Students, 210 students travel below 2km to their schools which constitute 50.60% total number of students in the local government area while 205 students travel above 2km to their schools which constitute 49.40%.

1. Introduction

Accessibility is a term often used in transport and land-use planning, and is generally understood to mean approximate ease of reaching [14,15]. It is believed that accessibility is a concept that has taken on a variety of meaning, including the amount of effort required by a person to reach a destination or the number of activities which can be reached from a certain location [4]. Accessibility is also defined as the extent to which the transport system enables individuals or goods to reach destination by means of transport mode(s) [5]. In other word, accessibility concerns both the pattern of activities and the links between the activities. The concept of accessibility is used in many contexts and in different ways, for example, as a goal in transportation policy, as a means in rural development policy, as an indicator of rural deprivation and as a variable in location analysis [9]. A location is defined to be accessible if the effort it takes to get there is acceptable to the target group; so the concept of accessibility incorporates not only the transport link between the origin and destination and the ability for travelling by the target group,

Content from this work may be used under the terms of the Creative Commons Attribution 3.0 licence. Any further distribution of this work must maintain attribution to the author(s) and the title of the work, journal citation and DOI.
Published under licence by IOP Publishing Ltd

but also the characteristics of the destination and the objective of the trip [7]. Accessibility is travel impedance (distance or time) between students' location and service points [1].

The public facility's accessibility is defined as the distance to which people must travel to receive the service, or from which a service is provided to the whole community of interest in administration, economy, education, health etc. (China standard GB/T.50280.98, 1998) cited by [9]. Public facilities are important because they provide both desirable services to and impose undesirable impacts on those who do not have access to its use in the areas. Education is therefore an important part of these public facilities.

Education plays an important role in our society. United Nations Educational, Scientific, and cultural Organization indicates that the development of education is important to the development of economy all over the world. World Bank points out that the GDP (Gross Domestic Product) of a country will increase 3% every year with increasing level of education. Education is a basic human right, and is indispensable for the realization of other human rights as a means for accessing broader social, economic, political and cultural benefits [17]. Because it is transformative and empowering, education contributes to building more just societies through reducing poverty and inequalities, enhancing acceptance of diversity, and promoting respect for the rule of law.

Secondary education serves as a link between the primary education and higher education, and plays a very important role in this respect. A child's future can depend a lot on the type of education he/she receives at the secondary level. Apart from establishing the roots of education of a child, secondary education can be instrumental in shaping and directing the child to a bright future. This stage of education serves as means to move on to a higher stage as well as to provide generic competencies that cut across various domains of knowledge as well as skills. Providing secondary education to all, both boys and girls, with a focus on quality education assumes greater meaning today in our society.

Ogun State is one of the three States created from the defunct Western State of Nigeria in 1976 [16]. At that time, 19 States were created by the Federal Military Government with the State Capital at Abeokuta [6]. Ogun State at present has 20 Local Government Councils with almost three hundred and ninety six secondary schools [10]. However, this is evident that Ogun State is the first among others to have the influence of Western Education in Nigeria. The provision of educational infrastructure and accessibility has always remained a problem because the demand is always on the increase as a result of the growing population and continuous agitation for standard education in Ogun State.

Presently, the clamour for more standard classrooms, standard library and library equipment, books and science laboratories among others in the face of very limited funds is a serious challenge to the government [10]. Under a Federal System of government, it is an acceptable norm that every state in a federating unit shall derive its policy from the National Policy. Ogun State to a large extent is therefore not an exemption. Consequently, the Ogun State Education Policy is derived from: (i) National Policy on Education; (ii) Ogun State Educational Law; and (iii) Ogun State Education Policy. With 100% transition being expected from primary schools into Secondary Schools and neighborhood schools as advocated under the Universal Basic Education (UBE), the State Government between 2003 and 2007 increased the number of secondary schools from 265 to over 396. This is also to provide access for all primary school graduates as provided for in the UBE Act to Junior Secondary Schools with separate management, administration and infrastructure [13]. Though, this is a welcome development, yet, an optimum access to secondary school education was not provided in Ogun State.

In Ogun State, accessibility to secondary education has persistently remained low. At the inception of secondary education in the country, accessibility to this level of Education was limited only to sons and daughters of the chiefs. However, there are high disparities in the rates of accessibility in the regions; some are closer to their schools while some are far away from their school location [11,12]. In Yewa South, in term of accessibility to secondary education, urban areas are more favoured than rural areas. In rural area, secondary schools operate below their optimum enrolment with girls more favoured than boys in the lower classes and vice versa in the upper classes.

IOP Conf. Series: Earth and Environmental Science **20** (2014) 012045 doi:10.1088/1755-1315/20/1/012045

In the use of GIS based accessibility analysis for education planning, where expected marked area size is a key factor in the location decision, there is a need for simple and effective indicators of school accessibility that can be easily visualized. Analysis based on the concept of accessibility is therefore ideally suitable to be integrated with GIS [7,8]. In this study, accessibility will be focused to analyze demand as the most important geographical factor in the secondary school planning.

1.1 Statement of research problem

The role of education cannot be overemphasized in the development of any nation [3]. Accessibility to education at all level helps to equip people with the potentials to overcoming the constraints to development [2]. In Yewa south local government, despite the quantitative expansion in secondary education, accessibility to this education level is still low. Disparities in attendance between urban and rural children are significant and schools are few in areas of real demand for secondary education. These disparities are perceived to be a result of a combination of factors which include accessibility, parents' attitude, social and economic factors. This study intends to investigate the spatial distribution and accessibility to secondary education in Yewa south Local Government in Ogun State.

1.2 Aim of the study

The aim of the study is to examine the spatial distribution of secondary school, identify the accessibility level to residents in the study area and the factors that give rise to them.

1.3 Objectives

- To identify spatial pattern of secondary schools in Yewa South local government.
- To examine the level of accessibility to secondary schools in Yewa south.
- To propose a framework for effective placement of students to secondary schools in the Local government.

1.4 Study area

Yewa South Local Government in Ogun State lies between Longitudes 2°47′24″E and 3°6′48″E of the Greenwich Meridian and Latitudes 6°37′46″N and 6°55′42″N of the Equator. The area is bounded in the East by Ifo and Ado – Odo/ Ota local Government and in the West by Ipokia Local Government and north by Yewa North. The Local government area is inhabited predominantly by the Yoruba speaking people of South Western Nigeria. It has a total land area of 629.38 square kilometers, with population of 150,850 (NPC, 2006). There are (15) fifteen public secondary schools and (10) ten registered private secondary schools in the study area (Ministry of Education Science and Technology). The Local Government is divided into ten (10) wards (Ilaro I, Ilaro II, Ilaro III, Iwoye, Idogo, Owode I, Owode II, Ilobi/Erinja, Oke – Odan and Ajilete).

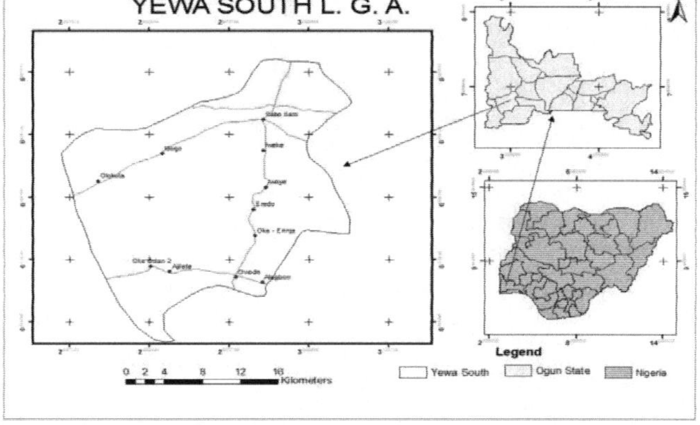

Figure 1. Yewa south local government.

2. Materials and method

2.1. Data sources and collection

The data used in this research include primary and secondary data. Primary data was collected using questionnaire and a hand-held GPS receiver. The GPS receiver is used to capture the geographic coordinate points of schools and other relevant data. Secondary data include administrative map, population figures of both students and teachers, names and addresses of the secondary schools in the study area, the school placement sheet and list of schools.

The administrative map was scanned and geo referenced to WGS 1984 UTM Zone 31N.

Personal geo database, feature dataset and feature classes for existing settlements, roads and wards were created in GIS environment. The existing roads, settlements and wards were therefore digitized from the georeferenced administrative map using the above stated feature classes. The digitized roads were converted to network dataset for distance analysis between the schools and the settlements. The data obtained from the administered questionnaire, School data and GPS point coordinates of secondary schools were typed into excel spread sheet for easy import into GIS environment. The coordinates of secondary schools typed in Microsoft excel were imported into GIS environment for conversion into point map.

The data analysis was carried out using *network analysis* in GIS environment. School location GPS coordinates, number of secondary schools in each ward and the total area were used to determine the pattern of distribution of secondary schools in the study area. Settlements, road network, schools location GPS coordinates, number of secondary schools, ward and school enrolments were used to generate both school accessibility and effective placement measure to schools using network analysis. Network analyses were also performed to evaluate nearest school to student and a set of origin-destination (OD) matrix. Neighbourhood Centroid -to -Schools were used for both nearest school and origin – destination analysis in order to find the shortest distance.

3. Results and discussion

3.1. Spatial pattern of secondary schools

The field survey and data collected reveal that there are 25 secondary schools in Yewa South local government area out of which fifteen (15) are public secondary schools and ten (10) registered private secondary schools. The local government is divided into ten (10) wards (Ilaro 1, Ilaro 2, Ilaro 3, Iwoye, Idogo/Ipaja, Owode 1, Owode 2, Ilobi/Erinja, Ajilete and Oke – odan). The below figures 2 and 3 show the distribution pattern of the secondary schools and settlement pattern in Yewa South Local Government.

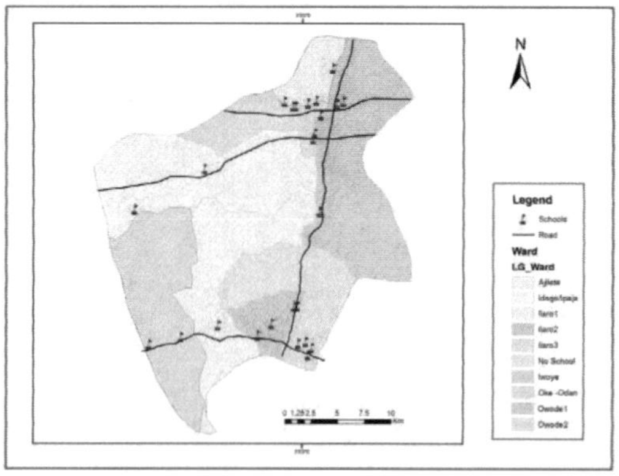

Figure 2. Distribution of schools

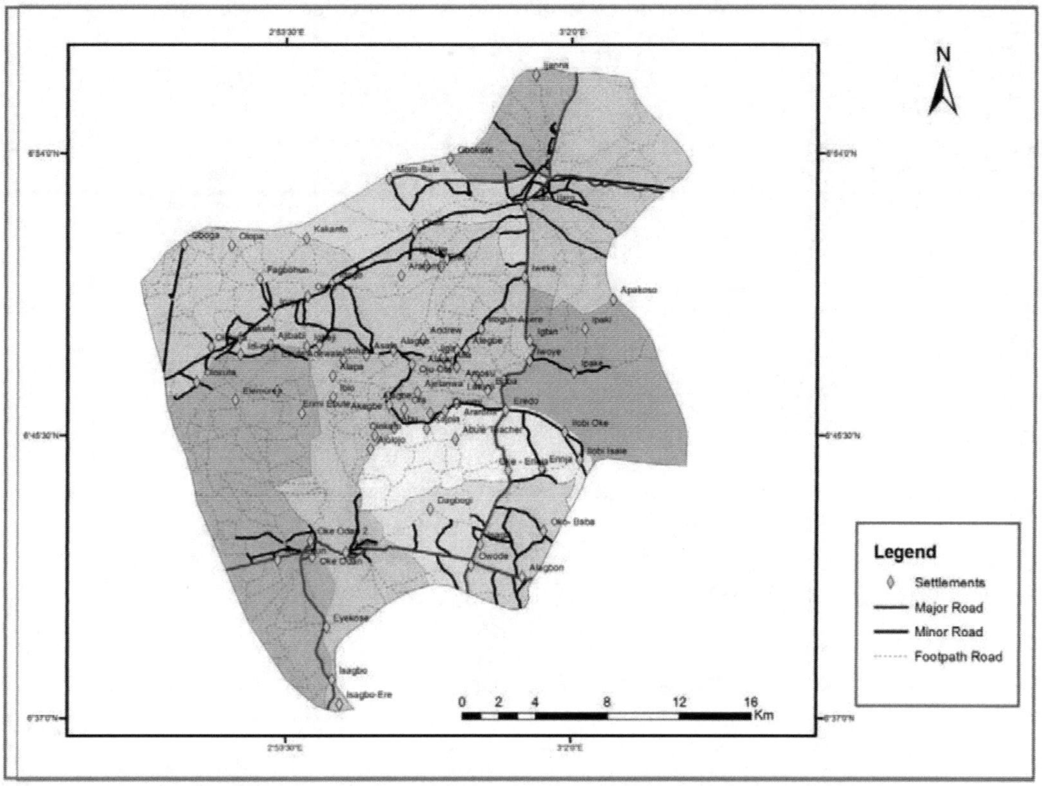

Figure 3. Distribution of settlements in Yewa south local government.

3.2. Spatial accessibility to secondary schools and travel distance analysis

Network analysis was performed in GIS environment to evaluate the accessibility pattern, nearest school to students and a set of origin to destination (OD) matrix was also carried out. Neighbourhood Centroid - to -Schools was used for both nearest school and origin – destination analysis in order to find the shortest distance. The road network analysis carried out between the students' residents and school locations show different categories of distances students need to travel to get to their various schools.

3.2.1. Travel distance analysis. The travel distance analysis between settlements and school locations for all the secondary schools in each ward reveal the mean distance of 0.88km for Yewa College High School, 0.97km Auntie Kemi Model School, 0.25km Poly Staff college, 1.39km Idogo/ Ipaja High School, 0.97km Area Community Secondary School, 0.94km Baptist High School, 0.63km Itolu High School, 0.27km Anglican high School, 0.25km Emmanuel high School, 0.42km Deuteronomy Secondary school, 1.05km Owode Secondary School, 0.60km Area Community High School, 0.62km Able God Secondary School, 2.51km Holy Child Secondary School, 1.15km Glorious Foundation School, 0.84km Army Day Secondary School, 1.77km Kings College, 2.18km Queen Unique Secondary School, 2.87km Oke Odan Grammar School, 0.13km Muslim Progressive High School, 1.67km Optimum Model School, 1.38km Ipade Ola School, 0.35km Oronna High School, 1.23km Iwoye High School and 1.19km Ajilete High School. In all, Muslim Progressive High School has the least mean distance with 0.13km and Oke Odan Grammar School has the highest mean distance with 2.87km.

However, effort was made to show the results of the whole of 25 secondary schools for network analysis ward by ward in the local government to see the travel pattern of students from their settlements to their various schools, and the result of the distance students travel from their settlements to their various schools was shown for all the secondary schools in each ward. See figure 4 to 12.

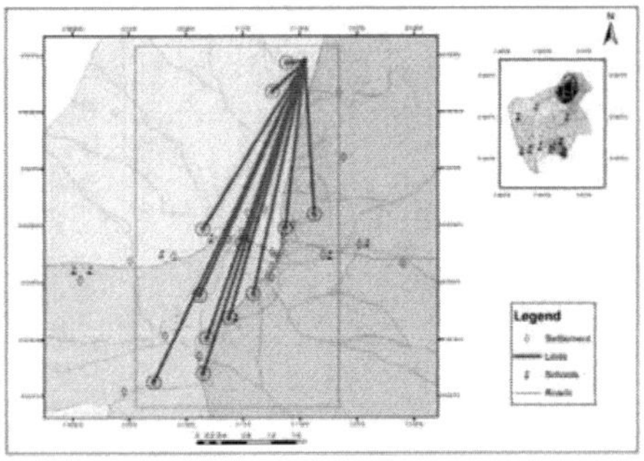

Figure 4a.Yewa college

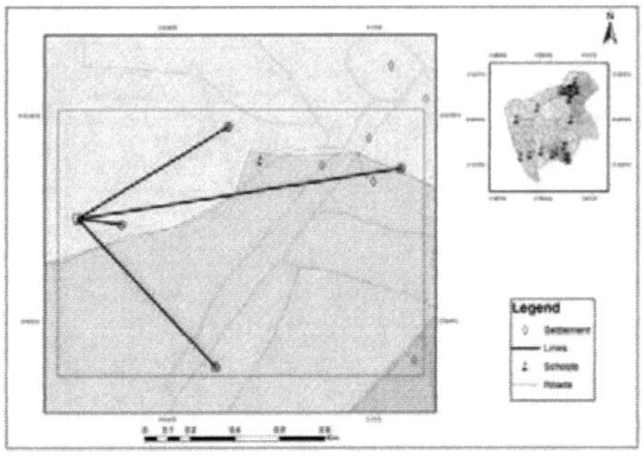

Figure 4b. Auntie kemi

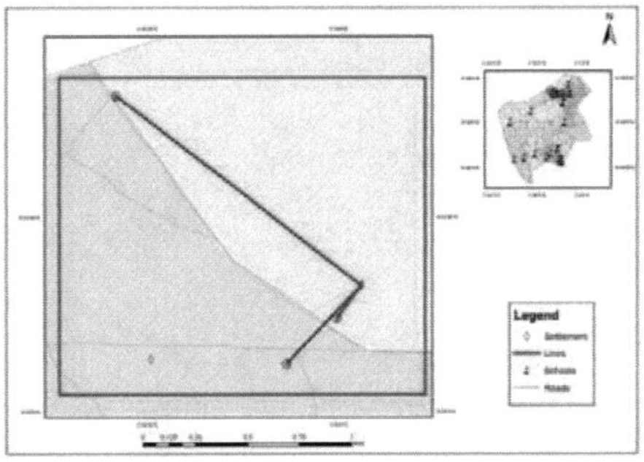

Figure 4c. Poly staff college

Figure 4. Result of distance travelled by students to school in Ilaro ward 1.

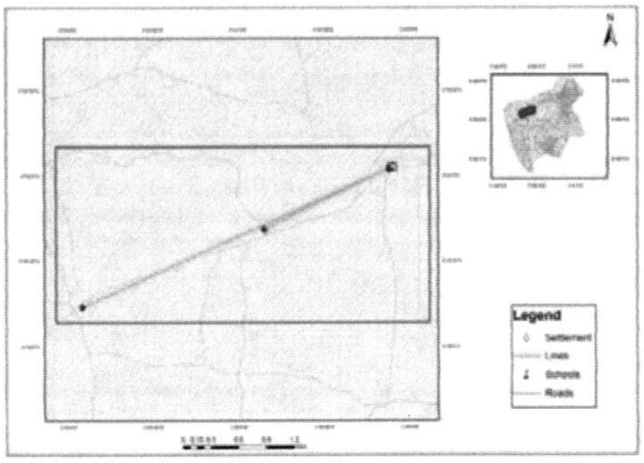

Figure 5a. Idogo/Ipaja high school

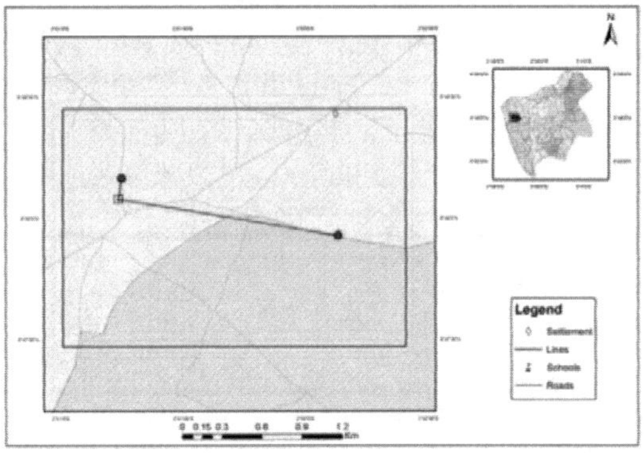

Figure 5b. Area high school

Figure 5. Result of distance travelled by students to schools in Idogo/ Ipaja ward.

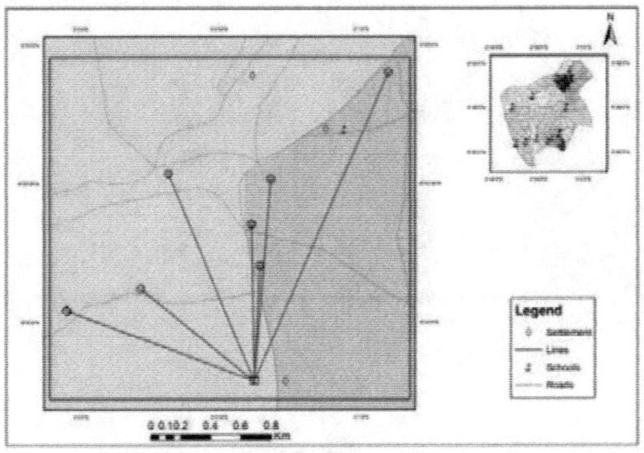

Figure 6a. Baptist high school

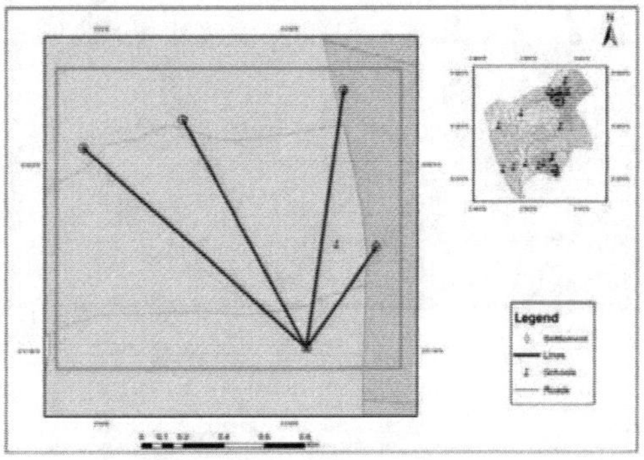

Figure 6b. Emmanuel high school

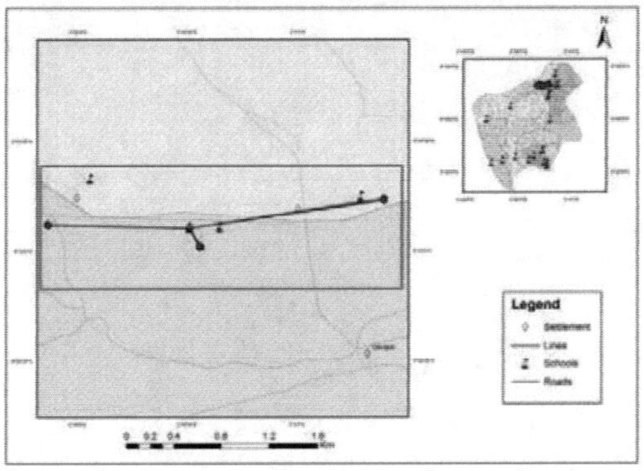

Figure 6c. Deuteronomy secondary school

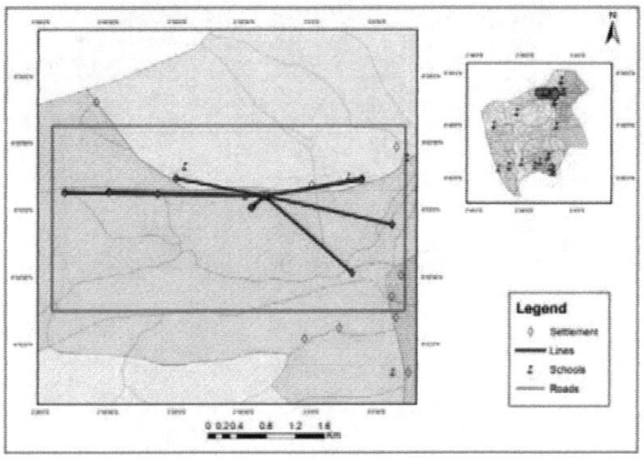

Figure 6d. Itolu high school

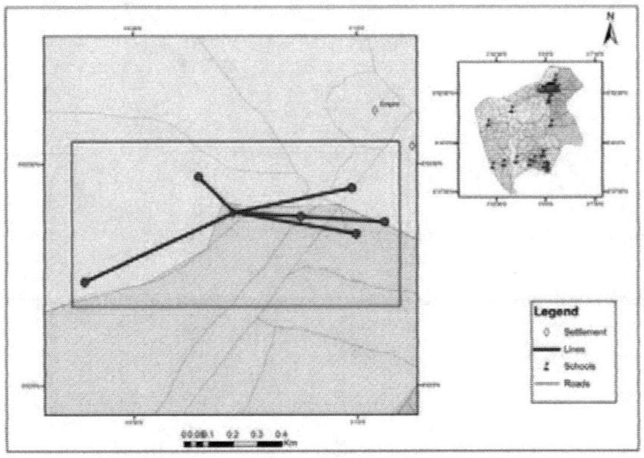

Figure 6e. Anglican high school

Figure 6. Result of distance travelled by students to schools in Ilaro ward 3.

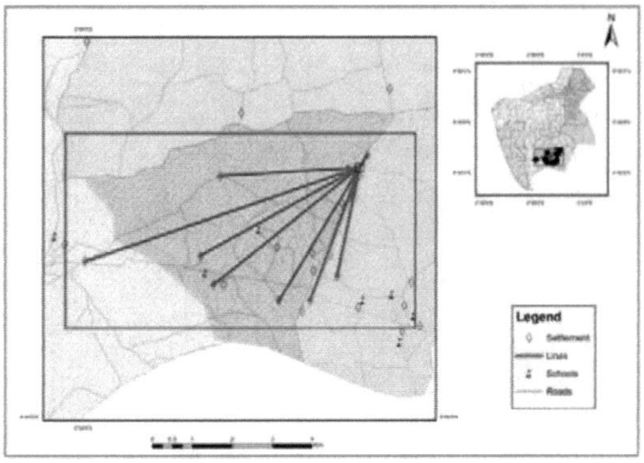

Figure 7a. Owode Secondary School

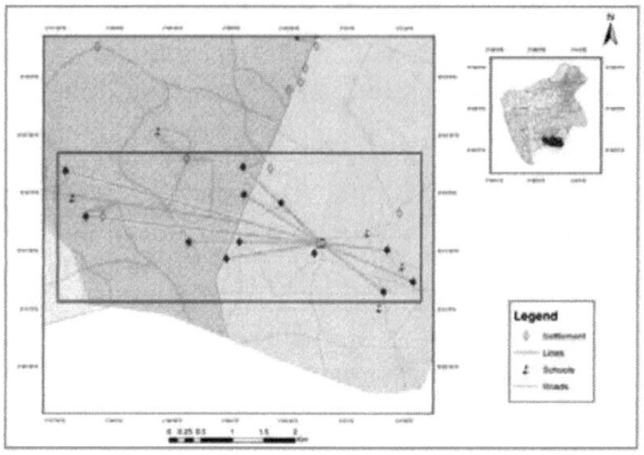

Figure 7b. Area Comm. High School

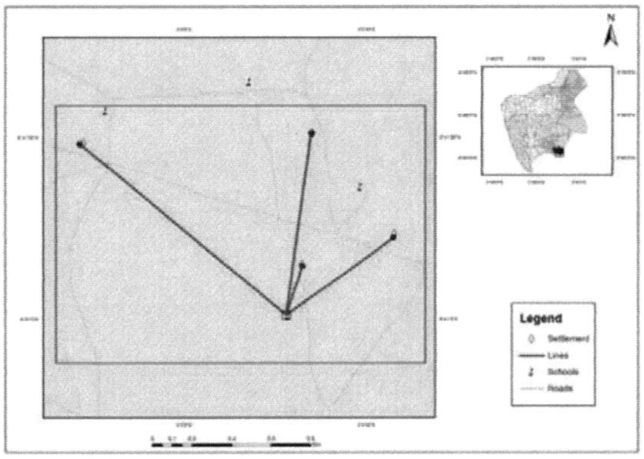

Figure 7c. Able God school

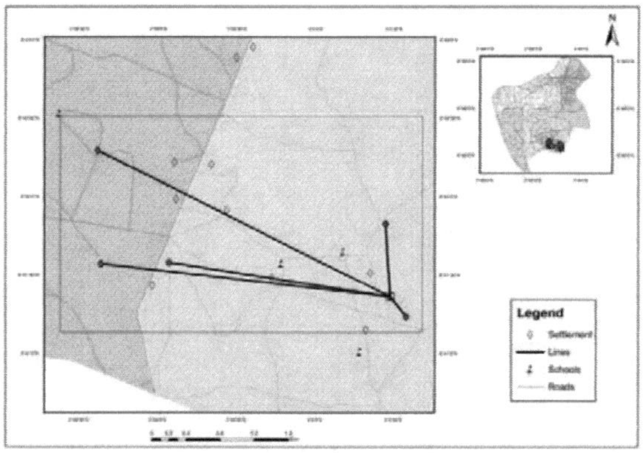

Figure 7d. Holy child school

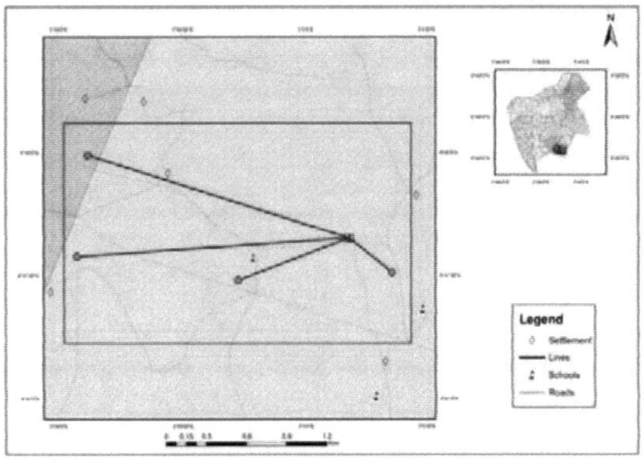

Figure 7e. Glorious school

Figure 7. Result of distance travelled by students to schools in Owode ward 2.

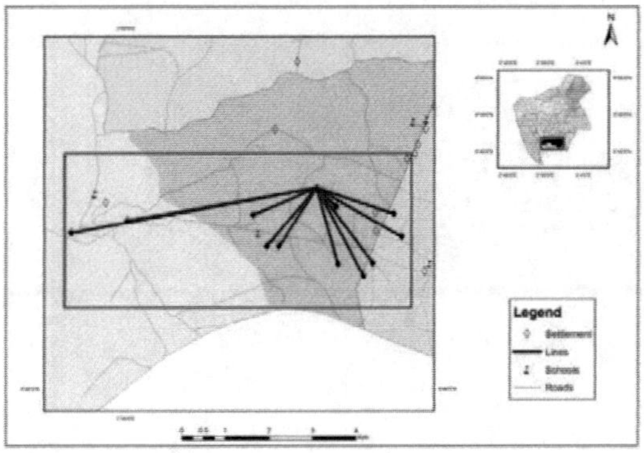

Figure 8a. Army day secondary school

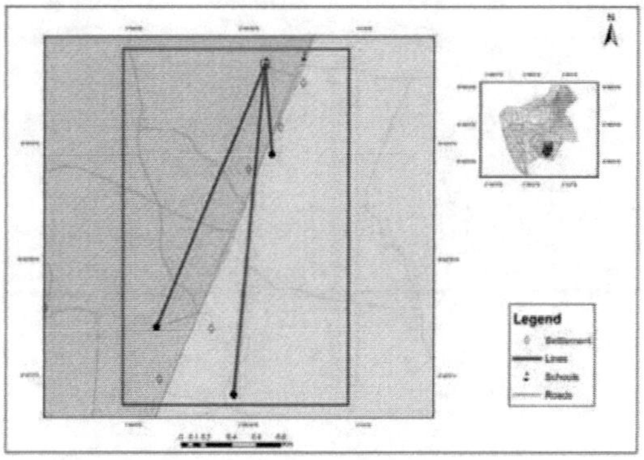

Figure 8b. Kings college

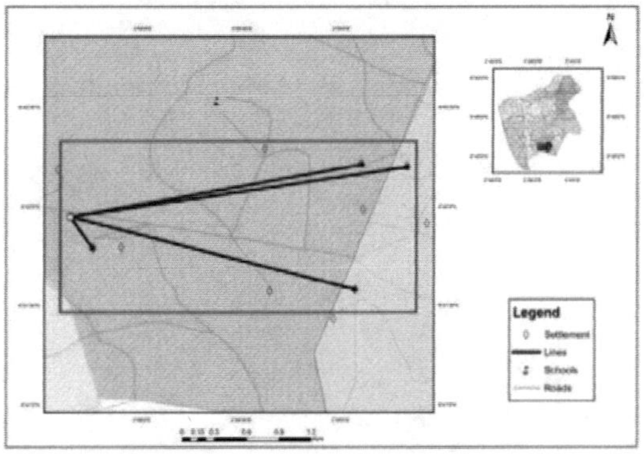

Figure 8c. Queen unique school

Figure 8. Result of distance travelled by students to schools in Owode ward 1.

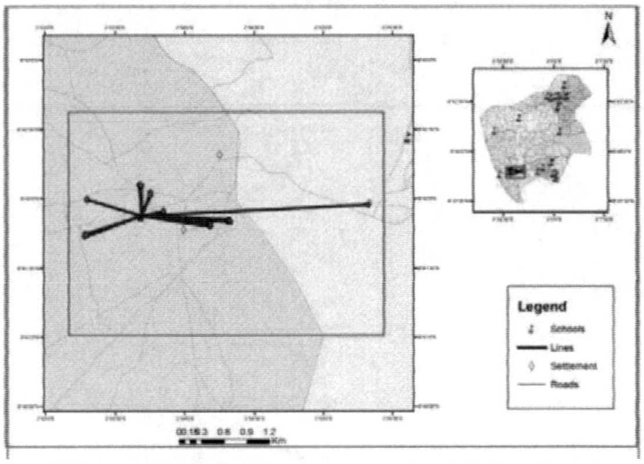

Figure 9a. Muslim progressive high school

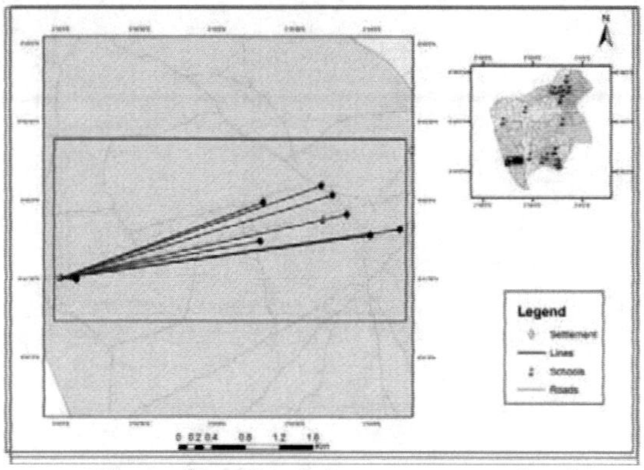

Figure 9b. Oke Odan grammar school

Figure 9. Result of distance travelled by students to schools in Oke Odan ward.

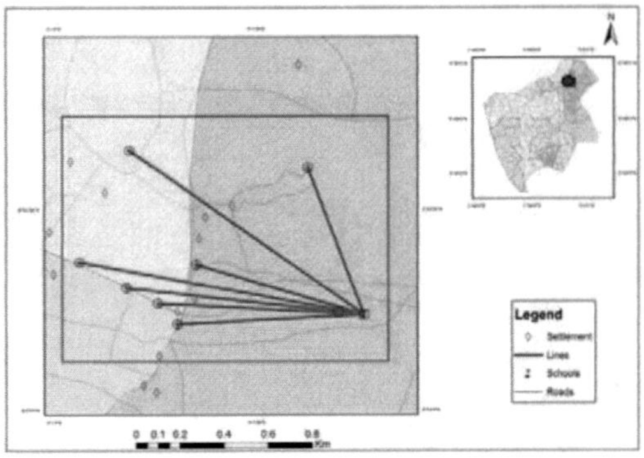

Figure 10a. Oronna high school

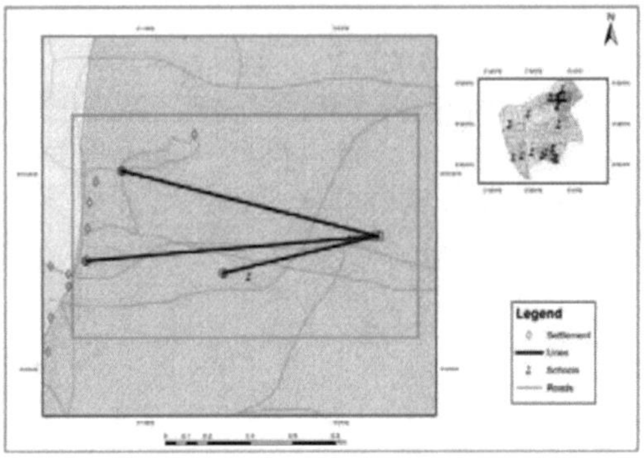

Figure 10b.Ipade Ola model school

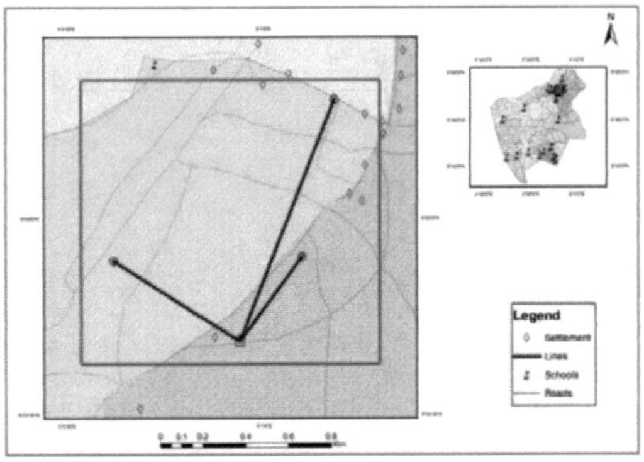

Figure 10c. Optimum model school

Figure 10. Result of distance travelled by students to schools in Ilaro ward 2.

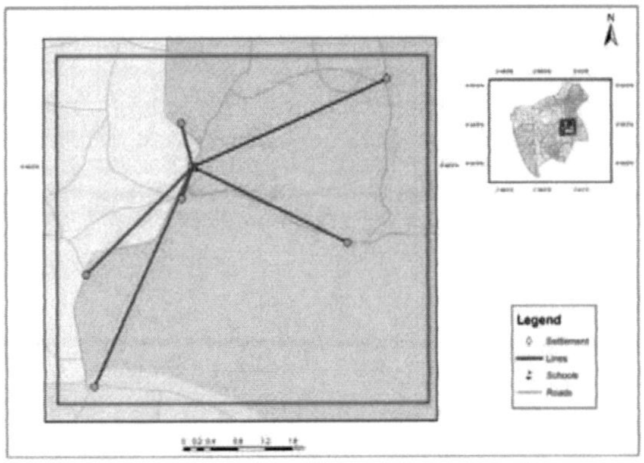

Iwoye high school

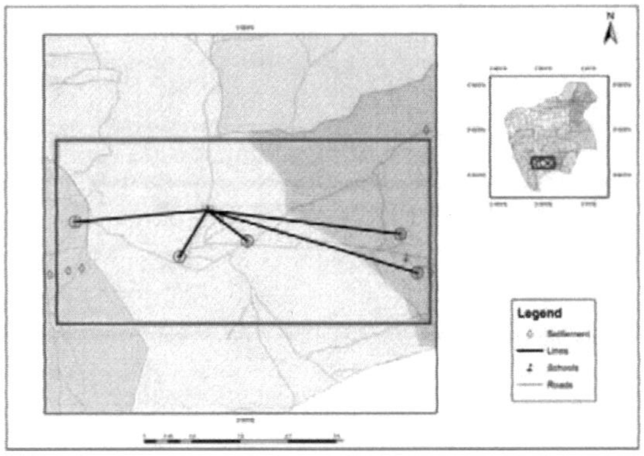

Ajilete high school.

Figure 11. Result of distance travelled by students to schools in Iwoye ward.

Figure 12. Result of distance travelled by students to schools in Ajilete ward.

IOP Conf. Series: Earth and Environmental Science **20** (2014) 012045 doi:10.1088/1755-1315/20/1/012045

3.3. Accessibility level assessment using UNESCO standard

The total sample in the Local Government area is four hundred and fifteen (415) students using total population of students in each school as a frame. To validate the result from network analysis using GIS, the accessibility level was also examined from the information obtained from the respondents. The responses of students with respect to distance they travel from their residents to their various schools were categorized based on the UNESCO standard of 2km (walking distance to school) to determine number of students that are disadvantaged in terms of distance to schools.

The first categories are the percentage of those that travel below 2km, these include; Oronna high School 86.6%, Optimum success College 66.67%, Ipade Ola Model School 100%, Muslim Progressive High School 94.44%, Deuteronomy Model School 100%, Anglican High school 100%, Emmanuel High School 76.92%, Baptist High School 58.33%, Area Community high school 100%, Auntie Kemi Model School 83.33%, Poly Staff College 100%, Ajilete High School 57.14%, Able God School 100% and Glorious Foundation School 57.14% while the second categories are the percentage of those that travel above 2km, these include; Oke Odan Grammer school 77.78%, Itolu High School 53.57%, Yewa College High school 81.88%, Queen Unique Secondary School 85.71%, Army Day Secondary school 73.08%, Area Community High School 57.14%, Owode Secondary school 65.85% and Holy Child School 66.67%.

Out of 415 respondents, 210 students travel below 2km to their schools which constitute 50.60% total number of students in the local government area while 205 students travel above 2km to their schools which constitute 49.40%. This means, students that travel below 2km distance to secondary schools constitutes the highest percentage which is informed by nearness of secondary schools to their residents while students that travelled above 2km distance to secondary school constitutes the lowest percentage which is informed by school facilities, peer group, absence of secondary school, availability of teachers and parental influence.

3.4. A model of spatial redistribution of secondary schools in the study area

3.4.1. Classroom capacity. The third objective of this study is to propose a model for redistribution of secondary schools in Yewa South. To achieve this objective, (UNESCO standard 2002) model was used (classroom capacity 1 teacher: 30 Students) with the total population of students in each classroom to get the number of classrooms needed.

Table 1. Senior school classroom capacity.

S/n	Name of School	Total Enrolment	No of Classroom	Classroom Capacity	Excess	No of classroom needed
1.	Yewa College	1434	14	420	1014	48
2.	Emmanuel High. Sch.	208	3	90	118	7
3.	Muslim High Sch.	504	8	240	264	17
4.	Area Comm. Sch.	996	9	270	726	33
5.	Anglican High Sch.	190	3	90	100	6
6.	Iyewa High Sch.	322	6	180	142	11
7.	Itolu High Sch.	1131	9	270	861	38
8.	Iwoye High school	142	3	90	52	5
9.	Oronna High sch.	511	5	150	361	17
10.	Owode Sec. Sch.	1190	9	270	920	40
11.	Army Day Sch.	1121	10	300	821	37
12.	Baptist High Sch.	80	3	90	-	3
13.	Idogo Ipaja High Sch.	53	3	90	-	3
14.	Area High Sch.	69	3	90	-	3
15.	Oke – Odan Gram Sch.	58	3	90	-	3
	Total	8009	91	2730	5379	271

Total population/ 30 = Number of classrooms

1434/30 = 47.8 ≈ 48 (forty eight classrooms) are needed in Yewa College senior secondary school.

Table 2. Junior school classroom capacity.

S/n	Name of School	Total Enrolment	No of Classroom	Classroom Capacity	Excess	No of classroom needed
1.	Yewa College	1747	15	450	1297	58
2.	Emmanuel High. Sch.	281	3	90	191	9
3.	Muslim High Sch.	398	7	210	188	13
4.	Area Comm. Sch.	2161	10	300	1861	72
5.	Anglican High Sch.	361	3	90	271	12
6.	Iyewa High Sch.	393	8	240	153	13
7.	Itolu High Sch.	1338	12	360	978	47
8.	Iwoye High school	474	3	90	384	16
9.	Oronna High sch.	789	7	210	579	26
10.	Owode Sec. Sch.	1253	10	300	953	42
11.	Army Day Sch.	1806	9	270	1536	60
12.	Baptist High Sch.	149	3	90	59	5
13.	Idogo Ipaja High Sch.	207	3	90	117	7
14.	Area High Sch.	233	3	90	143	8
15.	Oke – Odan Gram Sch.	380	3	90	290	13
	Total	11970	99	2970	9000	401

Total population/ 30 = Number of classrooms

$1747/30 = 58.2 \approx 58$ (fifty eight classrooms) are needed in Yewa College Junior secondary school.

The UNESCO sets 1:30 as a model for classroom capacity. This simply means one teacher per thirty students in a class. None of the secondary schools in the study area meets this standard. Therefore, effort was made to calculate the needed number of classroom as against the observed number of classrooms in each school. The calculation for the needed number of classrooms was done by considering the total population of students in each school. The total population of students was divided by thirty (30) to ascertain the number of classrooms needed and this was documented in tables 1 and 2.

4. Conclusions

This study investigated the spatial distribution and accessibility to secondary educational Institutions in Yewa South Local Government, Ogun State and the factors responsible. It also proposes framework for effective placement of students to secondary schools in the study areas, using UNESCO Standard (walking to school 2km and classroom capacity 1:30). Other analyses to aid decision for future planning were carried out using Network analysis within a GIS environment. This research has been able to demonstrate the dynamic capabilities of Geographic Information System application in spatial distribution and accessibility to secondary schools analysis in Yewa South Local Government Area.

This study will help zonal educational authorities and ministry of education to visualize the location of secondary schools on the map and also to consider the nature of accessibility when placing students to secondary school.

IOP Conf. Series: Earth and Environmental Science **20** (2014) 012045 doi:10.1088/1755-1315/20/1/012045

References

[1] Biba, S, K M Curtin, and G Manca. (2010) "A new method for determining the population with walking access to transit." International journal of geographical information Science: 347-364.

[2] Denga. D.I. (2000) Evolving a new education culture: The universal basic education focus international journal of research in basic and life-long education, VOL. 1 (1&2), pp. 1-6

[3] Eamon, M. K. (2005). Social-demographic, school, neighborhood, and parenting influences on academic achievement of Latino young adolescents. journal of youth and adolescence, 34(2), 163-175.

[4] Geurs K. T. and Ritsema J.R. (2001). Accessibility measures, review and applications. Bilthoven, Rijksinstituut voor Volksgezondheid en milieu (RIVM).

[5] Handy, S. L. (1997) Measuring accessibility: an exploration of issues and alternatives "Environment and planning 29: 1175 – 1194.

[6] Ijaduola K A (1998). Education in Nigeria: An historical perspective. Ijebu-Ode: published by Lucky Odoni (Nig.) Enterprises, pp 68 – 90.

[7] Jong T. and Ritsema J.R (1999), "Accessibility analysis and spatial competition effects in the context of GIS – supported service location planning." compute, environ. and urban system.

[8] Jong T., Ritsema J. and Toppen F. J. (2001). "GIS as a tool for location service center." proceedings of the second European GIS conference (EGIS): 509 – 517.

[9] Lu, Y. (2004). Evaluation of accessibility to primary schools, pp 26 – 31.

[10] Ogun State Education Hand Book (2007). Trends in educational development in Ogun State 229 - 233.

[11] Oni, A. O. (2007a). Graph-theoretic analysis of intra-community road network: Case study of Covenant University, Nigeria. Paper presented at the college seminar/workshop held at college of science & technology conference room, Covenant University, Ota, Nigeria.

[12] Oni, A. O. (2007b). Analysis of accessibility and connectivity of Ikeja arterial roads. Paper presented at the 1st national conference organized by department of estate management, Yaba College of Technology, Lagos.

[13] Owolabi, J. (2006). Quantitative methods of educational planning. Moboroda: Published by Lucky Odoni Enterprises, Nigeria.

[14] Sikdar P.K., Kanagadurai, B, Mahesh Chand and Singh A.K. (2002), " Rural road network planning based on functional accessibility", proceedings of the international conference on advances in civil engineering, january 3-5, IIT Kharagpur.

[15] Sikdar, P.K, Kanagadurai, B, Prasada Rao, I., and Jain, P.K., (2002) "GIS based functional accessibility approach for rural road network planning – a case study", ROTRAN 2002, IIT Kharagpur, Vol-II, pp 57-69.

[16] Taiwo C O (1980). The Nigerian education system: Past present and future lagos: Thomas Nelson (Nigeria) Ltd., pp 66 – 90.

[17] World Bank (2002). Human development sector for African region, Uganda post primary education.

IOP Conf. Series: Earth and Environmental Science **20** (2014) 012046 doi:10.1088/1755-1315/20/1/012046

Overview of Space Applications for Environment (SAFE) initiative

Ko Hamamoto*1, Toru Fukuda*1, Yoshimitsu Tajima*2, Wataru Takeuchi*3, Shinichi Sobue*4 and Tomoyuki Nukui*1

*1 Earth Observation Research Center, Japan Aerospace Exploration Agency, 2-1-1 Sengen, Tsukuba, Ibaraki 305-8505, Japan,
*2 School of Engineering, The University of Tokyo, Bunkyo-ku, Tokyo 113-8656, Japan
*3 IIS, The University of Tokyo, 6-1, Komaba 4-chome, Meguro 153-8505, Japan
*4 Remote Sensing Technology Center, 3-17-1 Toranomon, Minato, Tokyo 105-0001, Japan

E-mail: hamamoto.ko@jaxa.jp

Abstract. Climate change and human activities have a direc or indirect influence on the acceleration of environmental problems and natural hazards such as forest fires, draughts and floods in the Asia-Pacific countries. Satellite technology has become one of the key information sources in assessment, monitoring and mitigation of these disasters and related phenomenon. However, there are still gaps between science and application of satellite technology in real-world usage. Asia-Pacific Regional Space Agency Forum (APRSAF) recommended to initiate the Space Applications for Environment (SAFE) proposal providing opportunity to potential user agencies in the Asia Pacific region to develop prototype applications of satellite technology for number of key issues including forest resources management, coastal monitoring and management, agriculture and food security, water resource management and development user-friendly tools for application of satellite technology. This paper describes the overview of SAFE initiative and outcomes of two selected prototypes; agricultural drought monitoring in Indonesia and coastal management in Sri Lanka, as well as the current status of on-going prototypes.

1. Introduction

Space Application for Environment (SAFE) is a voluntary initiative aims to encourage environmental monitoring using space based technologies, specifically satellite remote sensing to identify and characterize changes in the environment through long-term observations that may be useful to reduce possible risks. SAFE also attempts to contribute to implementing adaptation programs related to disasters risk activities and further to other climate change associated problems using satellite remote sensing technology. Satellite remote sensing is identified as one of important source of information to identify some of the critical environment parameters such as perennial water resources, land cover, forests, estuaries and wetlands, mangroves areas with respect to spatial distribution and temporal changes.

SAFE considers continuous data provision with sensor capable of acquiring data in different wavelengths and varying spatial resolution could provide environment sensitive information in order

Content from this work may be used under the terms of the Creative Commons Attribution 3.0 licence. Any further distribution of this work must maintain attribution to the author(s) and the title of the work, journal citation and DOI.
Published under licence by IOP Publishing Ltd

IOP Conf. Series: Earth and Environmental Science **20** (2014) 012046 doi:10.1088/1755-1315/20/1/012046

to better understand the status of the environment and its changes. Initiated in the Asian-Pacific Regional Space Agency Forum (APRSAF), SAFE is a robust approach for long-term monitoring of the natural environment by using satellite remote sensing to evaluate current status, understand ongoing changes, monitor the nature of changes and, where possible, use as baseline information for hazard assessment, risk management and helping in developing mitigation and adaptation strategies. Furthermore, end users of SAFE activities are the agencies and experts working in various agencies that are responsible for the natural environment of their own country as well as responsibilities over regional and international directions. These agencies have their own mandate, functionalities and responsibilities to sustainable management of the environment.

2. SAFE overview

2.1. Background

Satellite data are widely used to augment the need of past data for continuous study of climate change and related phenomenon of the planet. The usage are varies and monitoring long-term phenomenon is said to be one of valuable contribution by space borne satellite data. Intergovernmental Panel on Climate Change (IPCC) use of earth surface temperature estimation is one of the key applications of space borne satellite data and it is reported that the use of satellite data will be the one of major information sources that could come handy in validating certain climate change indicators. On the other hand, observations of Earth from space have been done for well over 40 years and have contributed to many application areas including climate science. However, attempts to exploit this wealth of data are often hampered by a lack of homogeneity and continuity of satellite data and also by the insufficient understanding of the products and their uncertainties. This in fact has hampered the effective use of satellite data in Asia and the Pacific, specifically in developing countries.

The APRSAF was established in 1993, to increase the awareness on space based programs including earth observation and application of regional countries and to exchange views toward future cooperation (For details: http://www.aprsaf.org/). The forum provides a stage for space agencies in the Asia Pacific region as well as potential user agencies to exchange their ideas, share information and grow together as a region contributing to the advancement of space science. Among many positive directions under APRSAF, Japan Aerospace Exploration Agency (JAXA) proposed initiated SAFE as a scheme to collaborate with APRSAF member countries to promote the use of satellite data in understanding natural environment. At the APRSAF-15 held in Hanoi, Vietnam, members recommended and approved the official establishment of SAFE as a new initiative.

2.2. Objectives

SAFE was launched to facilitate the gap between science and application of satellite technology in real-world usage and providing opportunity to potential user agencies in the region to collaborate to develop prototype applications. The SAFE program and its objectives are dynamic in nature in order to address and find solutions to ever changing climate and environment related problems. The overall concept of SAFE is given in figure 1. The current emphasize of SAFE is to develop solution oriented applications using space based technologies in following areas but members are always encourage to communicate for other potential applications;

1. Forest resources management including mangrove forest; mapping, monitoring, non-forested vegetation and integration in carbon estimations,
2. Coastal monitoring and management; erosion and accretion of coasts, modelling sediment transportation, and relate to climate change,
3. Agriculture and food security; drought assessment, monitoring and effect on paddy areas, paddy area mapping and yield estimation,
4. Water resource management; rainfall estimation, anomaly assessment, identify water scarcities and extreme events such as drought and floods,

5. Development user-friendly tools and share modules and information on the use of satellite data primarily for above fields of applications.

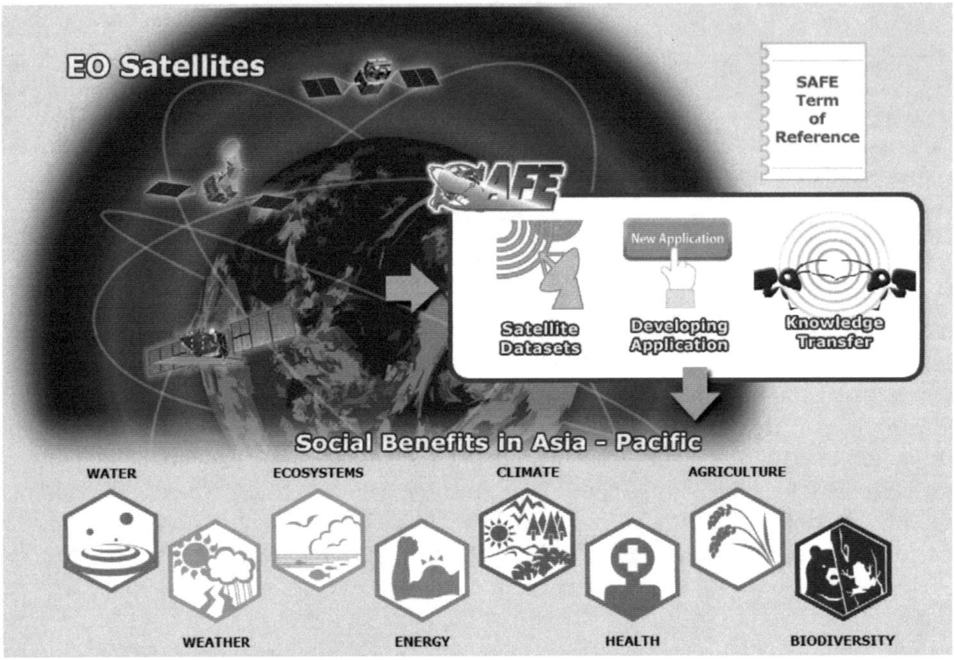

Figure 1. Concept of SAFE activity.

2.3. SAFE Implementation scheme

Principally, SAFE adopts a prototyping approach to demonstrate the value of space-based technologies to tackle issues related to environment and climate change. Typically, SAFE prototype is implemented by three groups; a prototyping executor, a technical supporter, and data & application creator. The team can collaborate other SAFE members to support their activities, and it is expected SAFE members response to the team's request. To ease the barriers of the start-up process, a given prototype activity is limited to two years.

SAFE secretariat announces the APRSAF community to submit requests for SAFE prototypes through various communication modes. These call for a proposal (CFP) is announced a couple of months in advance before a SAFE workshop, with necessary documentation including format template to submit new proposals to SAFE Secretariat. Since the activities are voluntary based, each proposed members have to provide necessary resources to analyze earth observation data provided by APRSAF member agencies and applied analyzed results to implement the problem addressed in the proposal in cooperation with operational governmental bodies, specifically with the end users. The project team member work together under the rules and guidance set forth by APRSAF.

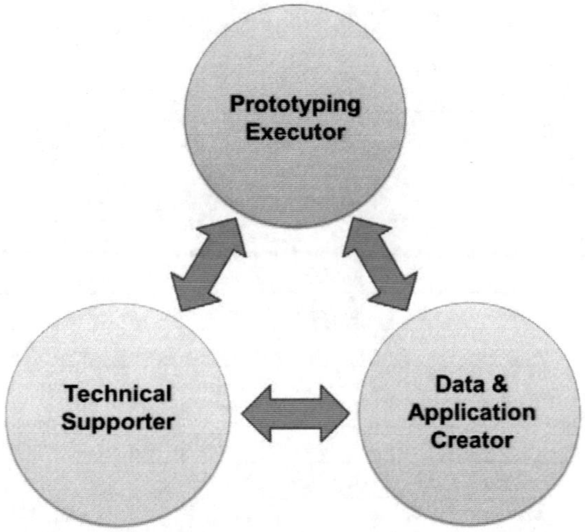

Figure 2. SAFE implementation structure.

It is expected that the Prototyping Executor applies the outcome of the prototype into their routine work related to environmental monitoring. The Technical Supporters / Data & Applications Creator validate their products/services within a practical framework, given that authority has been granted by the prototyping executor.

SAFE prototyping can provide some of the changing environmental parameters in land, forest, sea, coastal, agricultural products, and ecosystems. It can also provide a bridge to implementing operational and/or practical systems to contribute to solving national/regional climate change adaptation and mitigation issues by using space-based technologies in the Asia-Pacific region.

2.4. SAFE Project Design Matrix

Project Design Matrix (PDM) is a tool for management of project. PDM is adapted in some international cooperation project. Recently, SAFE secretariat is trying to connect the completed prototype to international donor project for practical sustainable use. Therefore, SAFE adapted PDM for management and evaluation of prototype. However, the SAFE PDM is not strictly same with commonly-used PDM. SAFE PDM is modified considering SAFE concepts; nature of prototype, voluntary manner and regarding practical or sustainable use of prototype results. In SAFE prototyping activities, each SAFE PDM is revised by SAFE secretariat after every SAFE workshop, and the evaluation of prototype is carried out based on the PDM.

2.5. SAFE prototyping activities

Through on-going SAFE prototyping, effects of climate change and related information by using space based technologies are provided to national/regional end users.

After launching the SAFE initiatives at APRSAF-15 in Hanoi in 2008, the workshops were held twice a year so far. The last workshop was held in early December, 2013, at Hanoi, Vietnam. in conjunction with APRSAF-20. The SAFE executers participated the Earth Observation Working Group (EOWG) in the APRSAF-20 and also gave a presentation about own prototype.

At the SAFE workshop in Hanoi last year, two prototyping proposals from Vietnam was endorsed by all of participants and adopted. This was reported at the EOWG in APRSAF-20. With the approval of two new proposals there are nine prototypes listed in Table 1 are on-going as of January 2014.

Table 1. SAFE prototype activities.

Status	Country / Executor	Theme
Completed	VIETNAM / NHMS,MONRE	Integrated water resource management
	VIETNAM / FIPI,MARD	Forest monitoring
	CAMBODIA / MOWRAM	Water Cycle and Agricultural Activities
	LAO PDR / WREA,WERI	Forest monitoring and management
	INDONESIA / LAPAN	Potential Drought Monitoring
	SRI LANKA / CCD	Risk of Sea Level Rise on Coastal Zone
	PAKISTAN / PMD	Monitoring Water Cycle Variations & Assessing Climate Change Impacts
	SRI LANKA / NARA	NARA Modeling ocean frontal zones using high resolution satellite and float data to locate tune fish aggregations
	THAILAND / DOF	Economic Fish Larvae Mapping and Monitoring
	VIETNAM / FIPI,MARD	Mangrove Forest Mapping and Carbon Stock Estimation
	SRI LANKA / GI,CEA	Mapping and Detecting Wetlands in River Basin
On-going	INDONESIA / ICALRD, IAARD, MOA	Assessment of drought impact on rice production in Indonesia by satellite remote sensing and dissemination with web-GIS
	CAMBODIA / MOWRAM	Water and Flood Security under the Climate Change
	BANGLADESH / LGED	Investigation of sedimentation process and stability of the area around the cross-dams in Meghna estuary
	INDONESIA / LAPAN	The assessment of Mangrove Forest Carbon Stock Monitoring of Indonesia using Remote Sensing Approach
	INDONESIA / ICALRD,IAARD,MoA	SAR Technology Application for Paddy Crop Monitoring in Center of Paddy Area, in Indonesia
	MALAYSIA / UPM	Monitoring of agricultural land abandonment using remote sensing
	VIETNAM / HCMIRG,VAST	Rice crop monitoring in the Mekong delta, Vietnam
	VIETNAM / VAWR	Assessment and Evaluation of Erosion and Sedimentation on the Coast from Hai Phong to Thanh Hoa Province, in Vietnam
	VIETNAM / NCHMF	Utilizing Satellite Data, Numerical Rainfall Forecasts, Combining with Ground Observations in Flood Forecasting for the Thai Binh River System

3. Accomplishments

Since the inception in 2008, SAFE has completed ten completed prototypes. The list of successfully completed prototypes is given in table 1. This paper describes two of the successfully completed prototypes in detail.

3.1. Use of spaced based technology characterize coastal changes and application in coastal zone management

3.1.1. Overview

As an Island Nation in the Indian ocean, Sri Lanka coastal zone is vulnerable for the severe impacts due to climatic change but mainly due to sea level rise; inundation, increased erosion, salinity intrusion, surges & wave over topping etc. Assessment of the hazards, vulnerabilities and risk are required for the identification of preparedness, adaptation and mitigation alternatives. One of the critical sources for these studies is access to reliable, continuous, and affordable data. The proposal by the Coast Conservation Department (CCD), Sri Lanka to use space based technologies to assess and monitor coastal changes of Sri Lanka, identify erosion and accretion and use of these information in formulating future coastal zone management program was favorably reviewed and adopted as a prototype of SAEF. The basic approach of the proposal was to use remote sensing and space based technology for assessing coastal dynamics and related changes objectively and cost effectively. The objective of this prototype is the followings;

1) Determine the availability, accessibility & reliability of data & tools for assessment of climate change impacts on coastal zone
2) Execution & streamlining observations, measurements, data collection & monitoring programs at national & regional level for enhancing the assessment of impacts on climate change
3) Assessment of risk on sea level rise spatially & temporally with available data

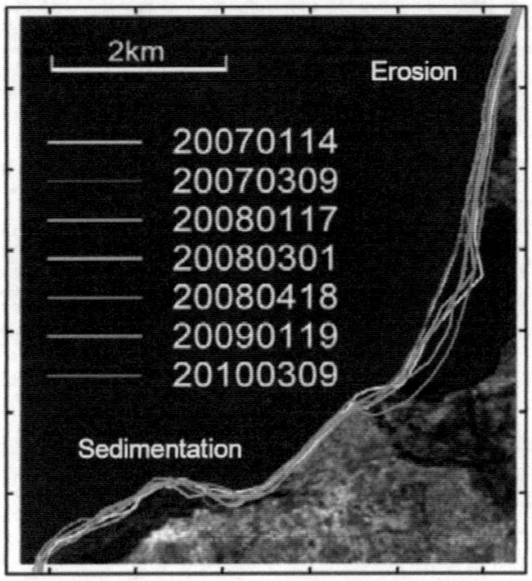

Figure 4. Trend of west coast shoreline changes in Kalpitiya area.

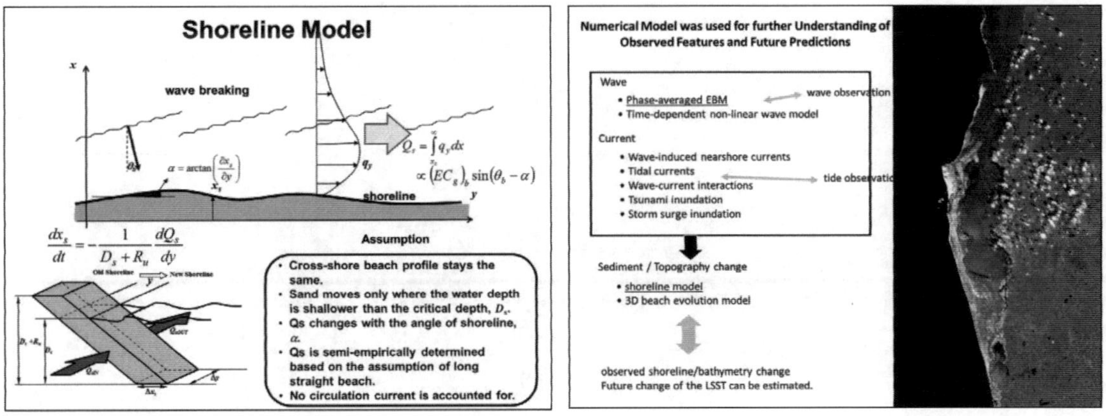

Figure 5. Elucidated show line model and future prediction.

3.1.2. Outcome
The outcome of activities of this very successful prototype can be summarized as below;

1) Contribution to Kalpitiya tourism project
Two meetings were conducted by the prototype executer with inviting coastal experts in the country & government high level officials involved in Kalpitiya tourism development project.
As the result, for the urgent decision making, interim report of the project was released on the request of the Sri Lanka government. Investors in the project area are officially instructed to follow the report for investment planning & EIA studies.

2) Contribute to national planning through stakeholder meeting
The meeting was held with 80 participants including state & local government, CBO, NGO & INGO and private developers from 45 agencies/institutions on August 25, 2011 as the stakeholder meeting, and following conclusions were confirmed;
- Thematic (coastal) standard for land use category mapping
- Provision of results as evidence for land ownership court cases
- Review of tourism & fisheries master development master plans in the area
- Continue monitoring of the coastal area

3.2. **Use of spaced based technology for drought monitoring**

3.2.1. Overview

This is a proposal submitted by National Institute of Aeronautics and Space (Lembaga Penerbangan dan Antariksa Nasional: LAPAN), Indonesia and successfully adopted as a prototype in 2010. Drought monitoring and providing early warning are major components of drought risk management. One of the goals of drought risk management is to increase society's coping capacity, leading to greater resilience and to reduce the need of government or donor interventions in the form of disaster assistance. Drought is the second major disaster in Indonesia. In the agricultural area, drought is the major disaster besides flood and landslide. The following are the objectives of this prototype;

1) To enhance the understanding on climate change-related drought in Indonesia

2) To develop the methods/techniques of drought monitoring over the agriculture areas in Java island accessible for users and public

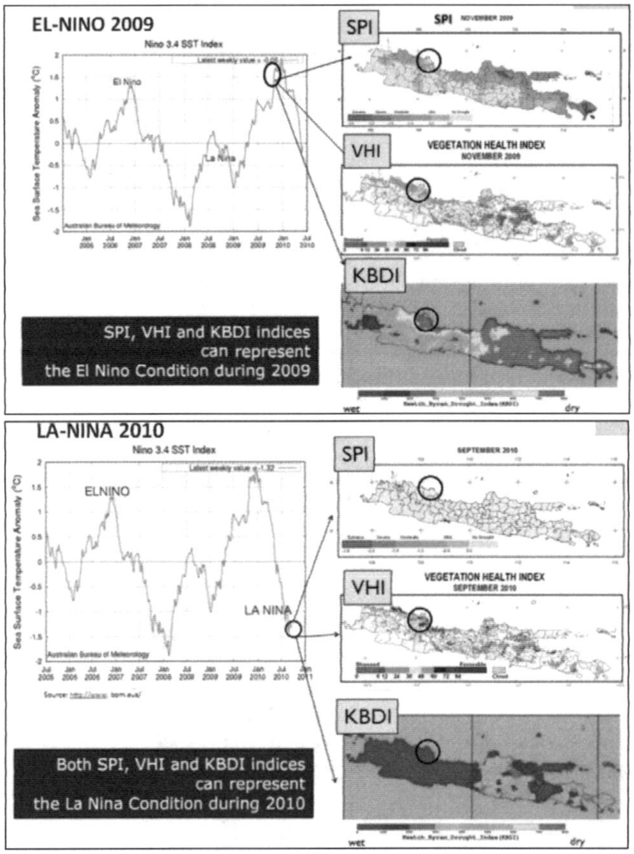

Figure 6. Drought comparison of the El-Nino and the La-Nina.

3.2.2. Outcome

1) The drought information and prediction is delivered every month to the end user through routine meeting activities, monthly bulletin, website and monthly report.
2) A system is developed and installed showing drought situation by the WEB-GIS
3) The outcome will be integrated in a couple countries under the international cooperative project funded by Asia Development Bank (ADB).
4) The derived information such as drought status and paddy cultivated area are provided to Asian Rice Crop Estimation & Monitoring (Asia-RiCE) initiative led by JAXA as one component of GEO Global Agricultural Monitoring (GEOGLAM).

4. Conclusion

SAFE prototyping is a regional activity under APRSAF assisting to bridge the gap between satellite based technology and application the technology in real-world operational work though the combination of experts from various agencies helping the end user. Since the inception of SAFE in 2008, prototypes have completed in several application areas providing opportunities to collaborating partners of number of countries to enhance the knowledge in the use of space based technologies in their own environment. Further, SAFE has developed a resource bank with successful application to share with the potential users in the region thanks to the collaborators under APRSAF. It expects that the outcome, such as utilization of developed system, gathered knowledge, potential use of remote sensing and approaches of applications can become a corner stone in contributing to find mitigation strategies against climate related hazards and risks to make the society a safer place.

Reference

[1] http://www.eorc.jaxa.jp/SAFE/index.html
[2] http://www.aprsaf.org/jp/about/

IOP Conf. Series: Earth and Environmental Science **20** (2014) 012047 doi:10.1088/1755-1315/20/1/012047

Recent coastline changes at the eastern part of the Meghna Estuary using PALSAR and Landsat images

M A Hussain[1], Y Tajima[2], K Gunasekara[3], S Rana[4] and R Hasan[5]
[1]Postdoc Researcher, Coastal Engg. Lab, The University of Tokyo, Japan
[2]Professor, Coastal Engg. Lab, The University of Tokyo, Japan
[3]Graduate Student, Coastal Engg. Lab, The University of Tokyo, Japan
[4]Senior Assistant Engineer, GIS Unit, LGED, Dhaka, Bangladesh
[5]Assistant Engineer, GIS Unit, LGED, Dhaka, Bangladesh

Email: asad@coastal.t.u-tokyo.ac.jp

Abstract: In the present paper, we have focused on the recent coastline changes at the eastern part of the Meghna Estuary of Bangladesh through the application of PALSAR (The Phased Array L-band Synthetic Aperture Radar) and Landsat images. The area under concern is very dynamic where the combined processes of erosion and accretion take place under the influence of river discharge with heavy sediment load, wind waves and tides. It falls under the macro-tidal zone with tidal range varying from 4m to 6m or even higher. Many parts of the area is covered with intertidal mudflats. They are inundated during the high tides and get exposed during the low tides. So for the detection coastline change consideration of tidal phase is quite important which has been done in the present study. Coastline change from the year 2007 to 2013 has been presented in this paper. From the analyses of satellite images the areas of erosion and accretion of the four major offshore islands have been calculated during the above mentioned period. It has been found that the annual rate of accretion of Urir Char island has decreased from 5.84 km^2 per year between 2007~2010 to 1.05 km^2 per year between 2010~2013. On the other hand, Sandwip island has been eroding at a higher rate of 3.15 km^2 per year between 2010~2013 compared to 0.34 km^2 per year between 2007~2010. Overall, the total area of Urir Char, Sandwip and Jahajir Char island increased by about 120 km^2 from 2007 to 2013, mainly contributed by the drastic expansion of Jahajir Char island during this period.

1. Introduction

The detection of coastline and its changes is an important constituent for the engineers, scientists and implementing agencies involved in coastal management. The position of the coastline and its historical change can provide important information for the design of coastal protection works, coastal

Content from this work may be used under the terms of the Creative Commons Attribution 3.0 licence. Any further distribution of this work must maintain attribution to the author(s) and the title of the work, journal citation and DOI.
Published under licence by IOP Publishing Ltd

IOP Conf. Series: Earth and Environmental Science **20** (2014) 012047 doi:10.1088/1755-1315/20/1/012047

development plans, and the calibration and verification of numerical models [1]. So, the overall management process has to be supported by a monitoring component of the shoreline which will require repeated surveys to take into account of the seasonal variability, storm events and other natural phenomena. The monitoring process must ensure a good compromise between accuracy, cost and time. This is why, during the recent past, for the purpose of coastline monitoring remote sensing technique is becoming increasingly popular (as shown by [2]~[7]) compared to GPS survey, aerial photography, airborne videography and video systems. It allows very quick data acquisition and processing compared to photogrammetry. At the same time satellite images with high spatial resolution provide comparable accuracy with other survey methods. The satellites with high temporal resolution have a short revisiting time (a few days) which allows to carry out monitoring studies of the area under examination, by means of images acquired at different times.

The coastlines around the Meghna Estuary of Bangladesh (figure 1) experiences dynamic morphological changes due to heavy sediment discharge from the combined flow from the Ganges, Brahmaputra and Upper Meghna Rivers with annual rate of nearly billion tons. Movement of coastlines at this scale cannot be found at any other parts of the world [8].

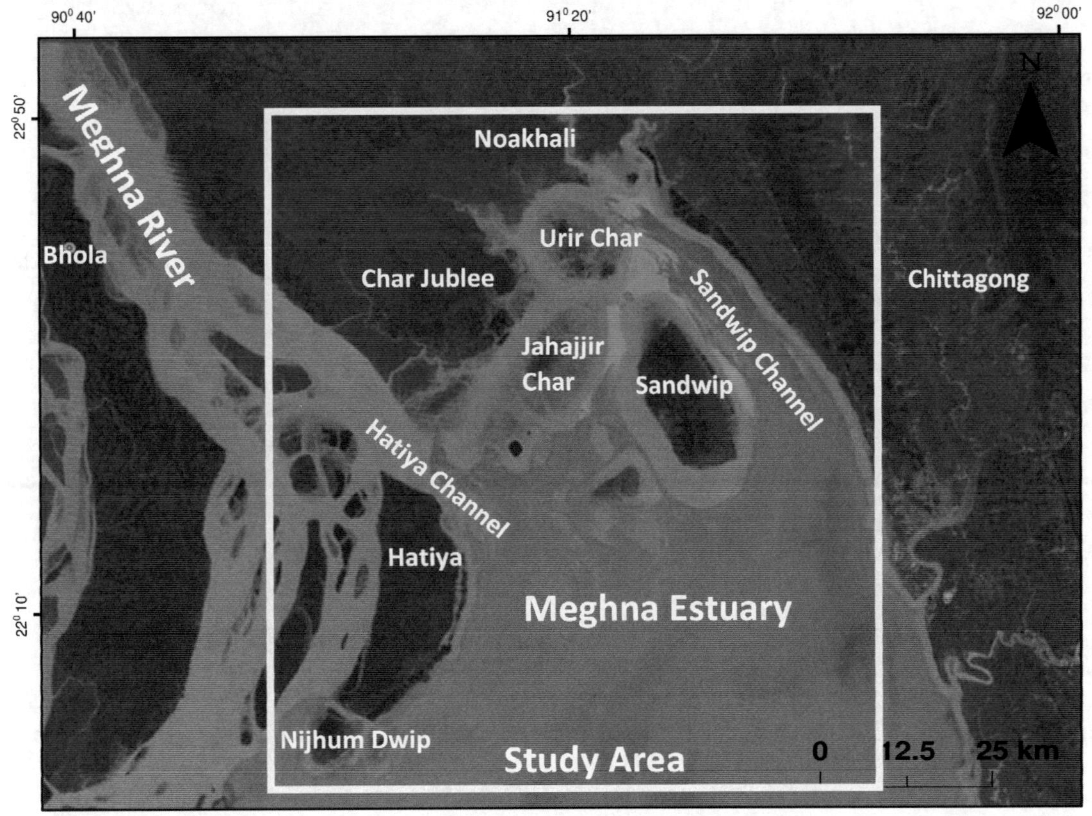

Figure 1. The Meghna Estuary and the study area (source: Google Earth).

Based on satellite images for the period of 1973~2000, Meghna Estuary Study [9] estimated the erosion and accretion in the region as 863.66 km^2 and 1371.68 km^2, respectively, which gives a net accretion of

508.02 km^2 area at a rate of 18.8 km^2 per year. As can be seen from figure 1, a number of small to large islands are located at the river mouth of the Meghna Estuary. Most of these islands are surrounded by intertidal mudflats. These mudflats are inundated during the high tide and get exposed during the low tide. In the satellite images acquired during the low tide the shallow areas appear as murky patches around the islands and adjacent to the coastlines. As already mentioned the study area under investigation receives huge amount of sediments from the rivers. That is why even sub-tidal areas in the Meghna Estuary exhibit similar murkiness when observed from the satellite images. So coastline detection becomes ambiguous. Again the very shallow water areas over the mudflats may appear as dark patches in some satellite images, for example for PALSAR images [10]. The process of land accretion is dominant in the region of the Meghna Estuary. About one fifth of the original supply of 1,100 million tons is retained in the Meghna Estuary and forms the material for land accretion in the central part of the coastal zone [1].

High-tide coastline method have been applied by [11] to study the erosion along the Kuakata coast located outside the Meghna Estuary area in the western side. Erosion and accretion along some of the coastal islands located in the central and western part of Meghna Estuary has also been studied by [12] for a period of 34 years (1977-2010) by using Landsat satellite images. Landsat images have also been used to study the morphological changes in the western and central part of the coastal areas of Bangladesh by [13] including the effect of mangrove forests for erosion control in these coastal belts. Another intriguing study was done by [14] involving remote sensing technology which focused on the transportation, distribution and deposition of suspended sediments in the Meghna Estuary through the application of Landsat TM images.

In this study, use of high-resolution image is important because it can help to distinguish the border between land and sea precisely. To find seasonal and yearly trend, comparing as many satellite images as possible is also essential. For this purpose, PALSAR images are suitable because it has not only a high-resolution mode but also its images are more frequently taken than other satellites of this kind. Its maximum ground resolution of PALSAR is 7m, and twenty-one images are available during the period 2007-2011. PALSAR is equipped with ALOS (The Advanced Land Observing Satellite), which was developed in Japan. As there is no more images available from PALSAR after 2011, we have used recent Landsat images to further update the erosion accretion scenario in the study area.

The overall goal of the present research is to develop an efficient method in order to assess the morphological changes in the Meghna Estuary area of Bangladesh. This involves processing and analyses of satellite images for coastline updating, validation of a numerical model for flow simulation in the estuary and also simulation of erosion-sedimentation processes in the estuary. The present paper focuses on the satellite data analyses to extract the amount of erosion and accretion at the eastern part of Meghna Estuary.

2. Methodology

As already mentioned, during the period of January 2007 to April 2011 a total of twenty-one PALSAR images were acquired under the present study. All the images are from Advanced Land Observing Satellite's Phased Array L-band Synthetic Aperture Radar (ALOS-PALSAR) and they have been collected from Japan Aerospace Exploration Agency (JAXA). All the products used for the study are

Level 1. Polarization of the images used for the present study are either HH or (HH+HV). The tidal phase at the time when these images were collected does not allow all these images to be compared directly. The existence of large intertidal mudflats and sediment laden estuarine flow makes the coastline detection process quite difficult. To overcome this problem only images collected near the peak of high tide period were analyzed.

A harmonic analysis was performed following a technique proposed by [15] to predict the tidal condition of the area at the time periods when the satellite images were available. In this analysis, the tidal constituents were calculated from observed data. The data were collected during Meghna Estuary Study during 1997 at the east of Hatiya island. The images which were acquired around the peak of high tides were then selected with the help of the results from harmonic analyses. The selected PALSAR images were acquired on 15 January 2007 at 16:29:50 LST and 10 March 2010 at 16:31:19 LST. The Landsat image at a recent date: 30 October 2013 at 10:20:29 LST was also selected fulfilling this criteria. Another Landsat image was selected on 12 September 2013 at 10:20:46 LST to identify the extent of intertidal mudflats around the major offshore islands at the eastern part of the Meghna Estuary. Figure 2 shows the simulated tidal water levels on 15 January 2007 and 10 March 2010 where the satellite data (PALSAR) acquisition times are marked by the dots. As can be seen both the images were acquired near the peaks of high tide periods.

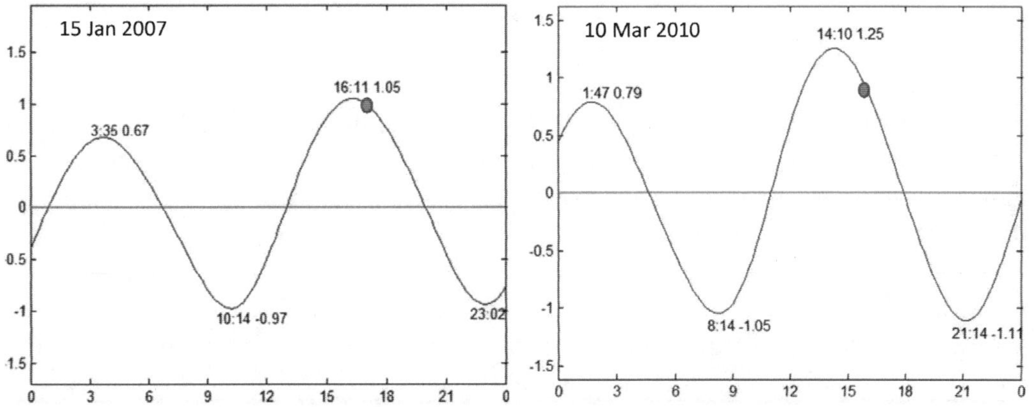

Figure 2. Predicted tidal water levels on 15 January 2007 and 10 March 2010; dots indicate the satellite image acquisition time.

Similarly, figure 3 shows simulated tidal water levels on 12 September 2013 and 30 October 2013 where the satellite data (Landsat) acquisition times are marked by the dots. As can be seen the image on 12 September 2013 was acquired near the lowest water level. So it should include maximum amount of intertidal mudflats around the study area. The other image on 30 October 2013 was acquired near the peak of high tide periods, so it should be comparable to the previously acquired PALSAR images to evaluate the erosion and accretion in the study area.

For PALSAR image processing, first each raw image file was geocoded to geographic (latitude-longitude) system by ASF MapReady, a freeware, at desired resolution and were converted to geo-tiff file format. Then adjacent images of the same time have been combined by using ARC-GIS software. Next the coastlines were identified through visual interpretation which involves careful image interpretation. Finally, for the purpose of calculating erosion and accretion desired images were overlaid to extract the area of erosion and accretion. Same procedures were repeated for each Landsat images except the use of ASF MapReady.

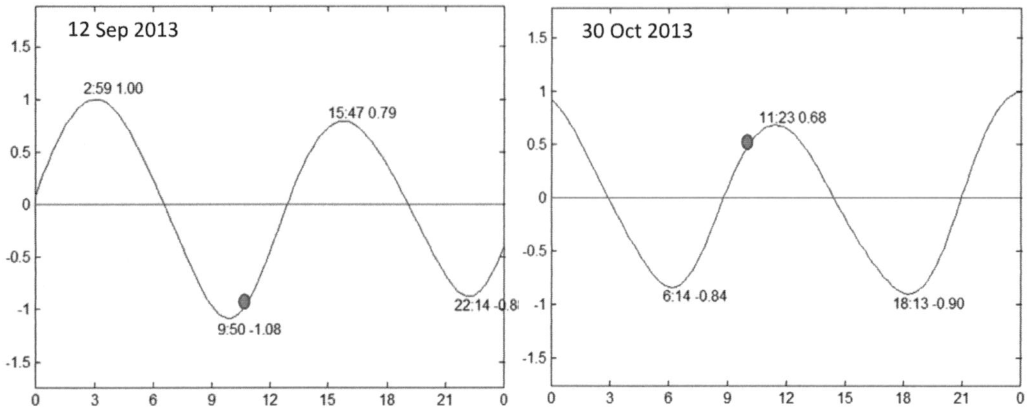

Figure 3. Predicted tidal water levels on 12 September 2013 and 30 October 2013; dots indicate the satellite image acquisition time.

3. Results and Discussion

In this section, first the areal extent of the four major islands are determined during the low and high tide periods of September and October, respectively, of 2013 to extract the inter-tidal mudflats around the islands. Next, the erosion and accretion around these islands are obtained from the comparison of high-tide coastlines from the years of 2007, 2010 and 2013.

3.1 High and low tide area and extent of mudflat detection

Table 1 shows the high and low tide areas and extent of intertidal mudflats of the four major offshore islands: Urir Char, Sandwip, Jahajir Char and Hatiya islands located at the eastern part of the Meghna Estuary. Clearly all the islands have significant proportion of intertidal mudflats, highest proportion being around Sandwip island (15%) and the lowest around Urir Char island (8%). Areal extent is highest around the Hatiya island (52km^2).

Table 1. Area in km^2 of four major islands and their intertidal mudflats.

Islands	Area (km^2) on 12 Sep 2013	Area (km^2) on 30 Oct 2013	Area (km^2) of Intertidal Mudflat
Urir Char	118.82	128.36	9.54
Sandwip	210.66	242.30	31.64
Jahajir Char	217.50	247.45	29.95
Hatiya	431.41	483.14	51.72

Figure 4 shows the coastlines extracted around Urir Char and Sandwip islands on 12 September 2013 and 30 October 2013 representing low tide shoreline and high tide shoreline, respectively. From the figure it is evident that Urir Char have intertidal mudflats around its three parts: east, north and west. Along the south part of Urir Char as the coastline is almost the same during high tide and low tide periods, indicating that there is no intertidal mudflats along the southern part of Urir Char. This was also verified during the field trip to Urir Char during December 2012, as shown in figure 5.

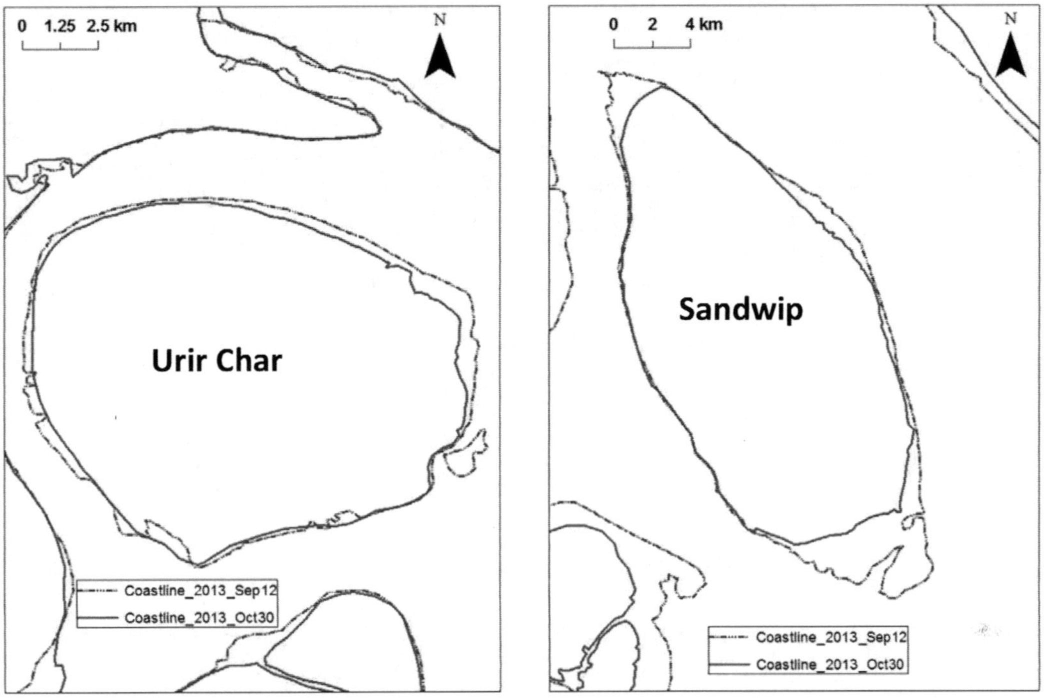

Figure 4. Coastlines extracted around Urir Char and Sandwip islands on 12 September 2013 (low tide) and 30 October 2013 (high tide).

Figure 5. Left: intertidal mudflats along the northern part of Urir Char and right: relatively steep slope along the southern part of Urir Char island.

Sandwip island, on the other hand, have large mudflats mainly at its north and south coasts. There is a small portion of mudflat along the middle part of its east coast. Along the rest of the Sandwip coast there is no further existence of mudflats.

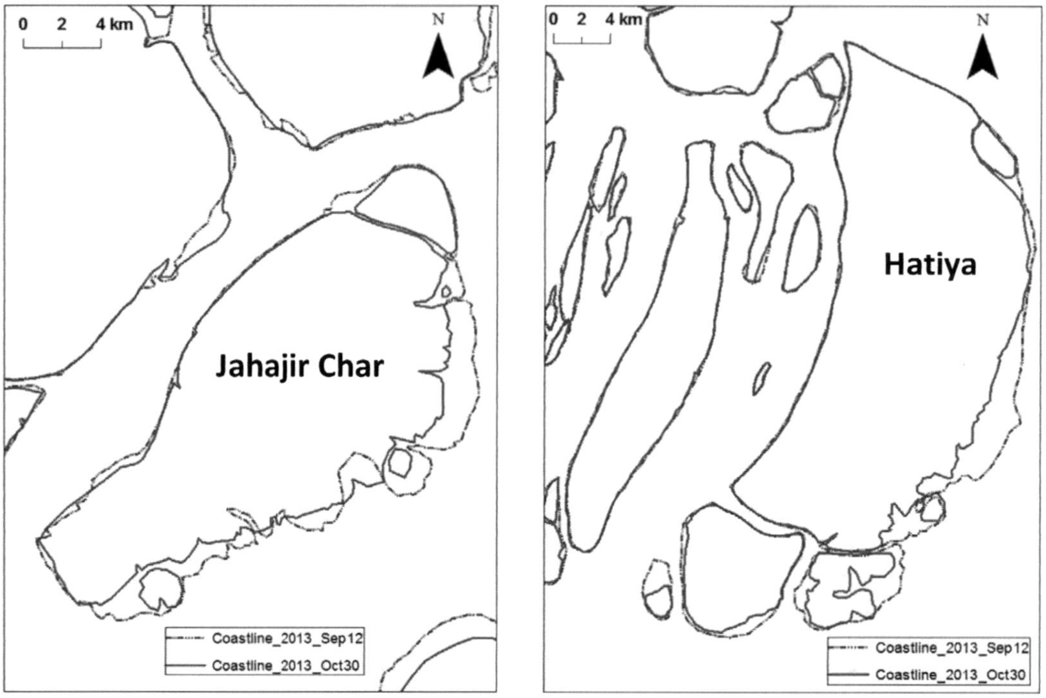

Figure 6. Coastlines extracted around Jahajir Char and Hatiya islands on 12 September 2013 (low tide) and 30 October 2013 (high tide).

Figure 6 shows the coastlines extracted around Jahajir Char and Hatiya islands on during low and high tide periods. From the figure it is evident that Jahajir Char island have most of its intertidal mudflats along its southern and south-eastern part. Along its north and north-west coasts there is no mudflats. Hatiya island also have large mudflats mainly at its south and south-east coasts and around the small islands near it. The small islands (including Nijhum Dwip island) at the south of Hatiya island are included as part of the Hatiya island as they almost get connected with the main Hatiya island during low tides. There is a small portion of mudflat along the middle part of its east coast. Along the rest of the Sandwip coast there is no further existence of mudflats.

3.2 Erosion and accretion around the major offshore islands

Table 2 shows the high tide areas of the four major islands at the eastern part of the Meghna Estuary during 2007, 2010 and 2013, extracted from PALSAR and Landsat images. These areas exclude the intertidal mudflat areas. The available PALSAR images do not cover the entire part of the Hatiya island. So the total area of the island is only obtained for 2013. Negative sign indicates erosion. The table shows that the overall rate of accretion of Urir Char island has slowed down to 1.05 km^2 per year during the recent three years, which was previously 5.84 km^2 per year between 2007~2010. On the other hand, Sandwip island has been facing erosion at a higher rate of 3.15 km^2 during the recent three years compared to the previous 0.34 km^2 per year. Jahajir Char island has been accreting at a slower rate in recent three years at a rate of 7.78 km^2 per year compared to the previous huge rate 40.55 km^2 per year mainly due to the erosion from the Char Jublee located at the northern part of Jahajir Char island.

Table 2. Total area and change of area in km^2 of four major islands.

Islands	Total area (km^2)			Area change (km^2)	
	Jan 2007	Mar 2010	Oct 2013	'07~'10	'10~'13
Urir Char	102.40	115.05	118.82	12.65	3.77
Sandwip	222.66	221.93	210.66	-0.73	-11.27
Jahajir Char	101.77	189.63	217.50	87.86	27.88
Hatiya	-	-	431.41	-	-

Figure 7 shows the coastlines extracted around the Urir Char and Sandwip islands during 2007, 2010 and 2013. It is evident that Urir Char is accreting along its three coasts: east, north and west and along the south, south-east part it is eroding. It appears that the intertidal mudflats are having an impact on the accreting process as accretion is taking place in those areas. The Sandwip island is primarily eroding along its south coast and north-east coast, while it is accreting along the north-west and middle part of east coast.

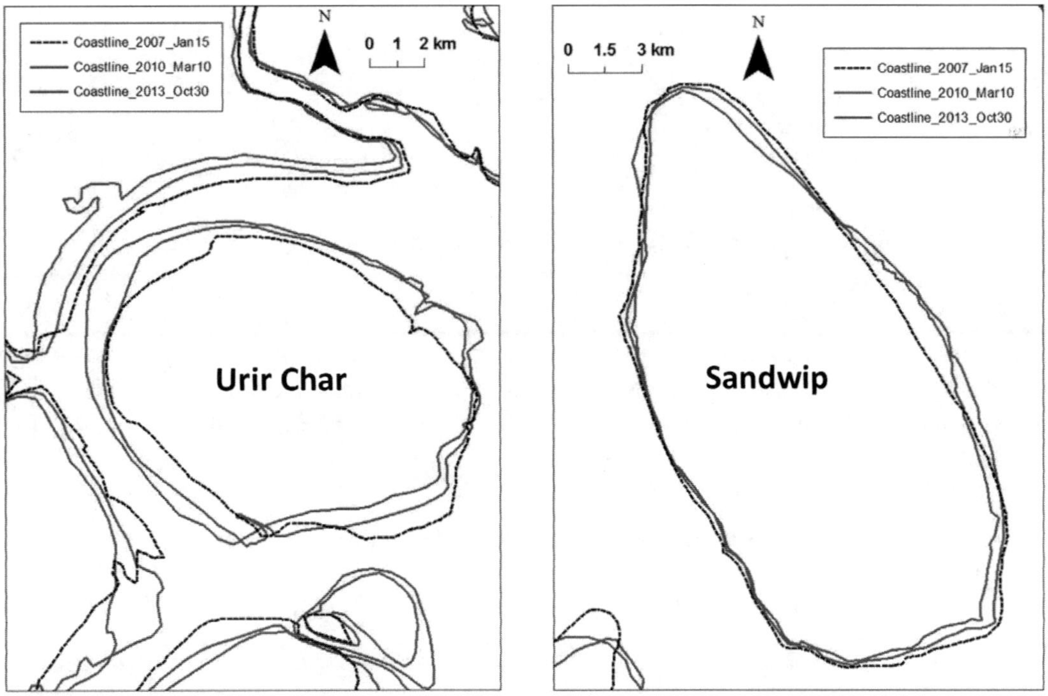

Figure 7. Coastlines extracted around Urir Char and Sandwip islands on 2007, 2010 and 2013.

Figure 8 shows the coastlines extracted around the Jahajir Char and north part of Hatiya islands during 2007, 2010 and 2013. It is evident that Jahajir Char island is accreting along south, south-east and south-west coasts. It appears that major accretion took place during 2007~2010 when parts of the Char Jublee from the northern part was attached to the Jahajir Char island. Jahajir Char is also facing some erosion along its northern part. Northern part of Hatiya island is clearly facing erosion at a rate of about half a km during 2007~2010 and 2010~2013.

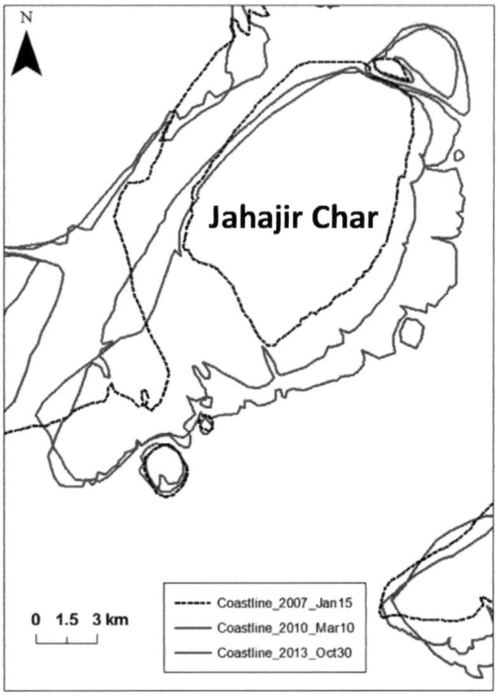

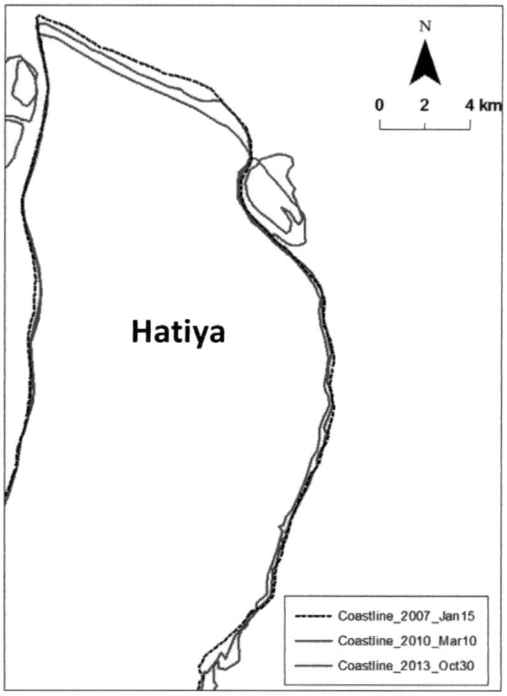

Figure 8. Coastlines extracted around Urir Char and Sandwip islands on 2007, 2010 and 2013.

Table 3 shows the total erosion and accretion during the study period at the four offshore islands at the eastern part of Meghna Estuary. It appears that in Urir Char island the rate of erosion has become higher in recent years while the rate of accretion has dropped. Similar trend is also observed at Sandwip and Jahajir Char islands. But still the rate of accretion in these four islands is much higher than the rate of erosion.

Table 3. Total erosion and accretion area in km^2 of four major islands.

Islands	Erosion area (km^2)		Accretion area (km^2)	
	2007~2010	2010~2013	2007~2010	2010~2013
Urir Char	3.31	5.63	15.92	10.26
Sandwip	9.08	11.49	8.62	1.46
Jahajir Char	4.16	16.36	90.77	45.90
North Hatiya	3.12	4.73	-	-

Overall, Meghna Estuary is known as an area where accretion dominates [16]. During the recent three years it appears that the rate of accretion has slowed down at the four major offshore islands at the eastern part of the Meghna Estuary. The newly accreted islands located south of Jahajir Char island between Sandwip and Hatiya islands may increase the overall rate of accretion in the eastern part of the Meghna Estuary.

4. Conclusions

High resolution satellite images from PALSAR and Landsat have been analyzed to study the erosion and accretion phenomena in the Meghna Estuary. The estuary is known as an area where accretion dominates. During the recent three years it appears that the rate of accretion has slowed down at the four major offshore islands at the eastern part of the Meghna Estuary. It has been found that the annual rate of accretion of Urir Char island has decreased from 5.84 km^2 per year between 2007~2010 to 1.05 km^2 per year between 2010~2013. On the other hand, Sandwip island has been eroding at a higher rate of 3.15 km^2 per year between 2010~2013 compared to 0.34 km^2 per year between 2007~2010. Overall, the total area of Urir Char, Sandwip and Jahajir Char island increased by about 120 km^2 from 2007 to 2013, mainly contributed by the drastic expansion of Jahajir Char island during this period.

Acknowledgement

This research has been carried out under a collaborative research project, Space Application for Environment (SAFE), one of international activities under the Asia-Pacific Regional Space Agency Forum (APRSAF). The subject of this specific project is 'Investigation of sedimentation process and stability of the area around the cross-dams in the Meghna Estuary', and is financially and technically supported by JAXA.

References

[1] Klein M and Lichter M 2006 Monitoring changes in shoreline position adjacent to the Hadera power station, Israel *Applied Geography* **26** pp 210–226 Elsevier

[2] Adegoke J O, Fageja M, James G, Agbaje G and Ologunorisa T E 2010 An Assessment of Recent Changes in the Niger Delta Coastline using Satellite Imagery *Journal of Sustainable Development* **3(4)**

[3] Boak E H and Turner IL 2005 Shoreline Definition and Detection: A Review *Journal of Coastal Research* **21(4)** pp. 688–703

[4] El-Asmar H M 2002 Short term coastal changes along Damiette – Port said coast northwest of the Nile Delta, Egypt *Journal of Coastal Research* **18(3)** pp. 433-441

[5] El-Raey M, El-Din S H S, Khafagy A A and Abo A I 1999 Remote sensing of beach erosion/accretion patterns along Damietta – Port Said shoreline, Egypt *International Journal of Remote Sensing* **20(6)** pp 1087-1106

[6] Liu H and Jezek K C 2004 Automated extraction of coastline from satellite imagery by integrating canny edge detection and locally adaptive thresholding methods *International Journal of Remote Sensing* **25(5)** pp 937-958

[7] Moore L J 2000 Shoreline mapping techniques *Journal of Coastal Research* **16(1)** pp. 111-124

[8] de Wilde K ed. 2011 *Moving Coastlines: Emergence and Use of Land in the Ganges-Brahmaputra-Meghna Estuary* University Press Limited

IOP Conf. Series: Earth and Environmental Science **20** (2014) 012047 doi:10.1088/1755-1315/20/1/012047

[9] Bangladesh Water Development Board (BWDB) 2001 *Hydro-morphological dynamics of the Meghna estuary* Meghna Estuary Survey-II

[10] Hussain M A, Tajima Y, Taguchi Y, Rana S, Hossain M A, Gunasekara K and Samarakoon L 2014 *Investigation of dynamic coastal morphological features around the Meghna Estuary using PALSAR images* Transactions of JSASS, Aerospace Technology Japan online Journal.

[11] Islam A Z M Z, Kabir S M H and Sharifee M N H 2013 High-tide Coastline Method to Study the Stability of Kuakata Coast of Bangladesh Using Remote Sensing Technique *Asian Journal of Geoinformatics* **13(1)**

[12] Alam M S and Uddin K 2013 A Study of Morphological Changes in the Coastal Areas and Offshore Islands of Bangladesh Using Remote Sensing, *American Journal of Geographic Information System* **2(1)** pp. 15-18

[13] Shibly A M and Takewaka S 2012 Morphological Changes along Bangladesh Coast Derived from Satellite Images, *Proceedings of Coastal Engineering*, JSCE **3**

[14] Islam M R, Begum S F, Yamaguchi Y and Ogawa K 2002 Distribution of suspended sediment in the coastal area off the Ganges-Brahmaputra River mouth: observation from TM data *Journal of Maritime Systems* **32** pp 307-321

[15] Boon J 2007 *Secrets of the Tide: Tide and Tidal Current Analysis and Applications, Storm surges and Sea Level Trends* Horwood Publishing Chichester UK

[16] Taguchi Y, Hussain M A, Tajima Y, Hossain M A, Rana S, Islam A K M S and Habib M A 2013 Detecting Recent Coastline Changes Around the Urir Char Island at the Eastern Part of Meghna Estuary Using PALSAR Images *Proceedings of the 4th International Conference on Water and Flood Management* Dhaka, Bangladesh

Indonesian drought monitoring from space. A report of SAFE activity: Assessment of drought impact on rice production in Indonesia by satellite remote sensing and dissemination with web-GIS

Rizatus Shofiyati[1*], Wataru Takeuchi[2], Parwati Sofan[3],Soni Darmawan[4], Awaluddin[5], andWahyu Supriatna[1]

[1.] *Indonesian Center for Agricultural Land Resources Research and Development (ICALRD)*
[2.] *University of Tokyo, Japan*
[3.] *Indonesian National Institute of Aeronautics and Space (LAPAN)*
[4.] *Bandung Institute of Technology (ITB), Indonesia*
[5.] *Agency for the Assessment and Application of Technology (BPPT), Indonesia*

*Corresponding author: rshofiyati@litbang.deptan.go.id or rshofiyati@yahoo.com
Jalan Tentara Pelajar 12, Cimanggu, Bogor, Indonesia

Abstract. Long droughts experienced in Indonesia in the past are identified as one of the main factors in the failure of rice production. In this regard, special attention to monitor the condition is encouraged to reduce the damage. Currently, various satellite data and approaches can withdraw valuable information for monitoring and anticipating drought hazards. Two types of drought, Meteorology and Agriculture, have been assessed. During the last 10 years, daily and monthly rainfall data derived from TRMM and GSMaP. MTSAT and AMSR-E data have been analyzed to identify meteorological drought. Agricultural drought has been studied by observing the character of some indices (EVI, VCI, VHI, LST, and NDVI) of sixteen-day and monthly MODIS data at a period of 5 years (2009 – 2013). Network for data transfer has been built between LAPAN (data provider), ICALRD (implementer), IAARD Cloud Computing, and University of Tokyo (technical supporter). A Web-GIS based Drought Monitoring Information System has been developed to disseminate the information to end users. This paper describes the implementation of remote sensing drought monitoring model and development of Web-GIS and satellite based information system.

1. Introduction

In Indonesia, the drought has occurred more frequently in recent years. Drought usually disrupts rice production. Sufficient data and information must be very important to anticipate this situation and would be strategic for agricultural development policy decision. Agricultural monitoring techniques to determine land surface information in near real time is needed and further analysis of data can be used in the current agricultural development planning.

Remote sensing has an ability to provide spatial information which can cover wide area and record multi - temporal information to help anticipate drought condition. Remote sensing has proven useful for large-area vegetation monitoring given the synoptic coverage, high temporal repeat cycle,

Content from this work may be used under the terms of the Creative Commons Attribution 3.0 licence. Any further distribution of this work must maintain attribution to the author(s) and the title of the work, journal citation and DOI.
Published under licence by IOP Publishing Ltd

continuity, and moderate resolution observation of satellite-based sensors. In particular, normalized difference vegetation index (NDVI) time-series data from the global imager, the Moderate Resolution Imaging Spectroradiometer (MODIS) and National Oceanic and Atmospheric Administration - Advanced Very-High-Resolution Radiometer (NOAA – AVHRR), have been widely used for vegetation and ecosystem monitoring [1,2,3]. Analysis of NDVI time-series data and NDVI-derived metrics have been an effective means in identifying vegetation condition anomalies (e.g., apparent declines in vegetation health). Operational efforts such as the U.S. Drought Monitor [4], National Drought Mitigation Center [5], and NOAA's National Weather Service [6] provide NDVI-derived products that describe the percentage or deviation of current vegetation conditions from the normal conditions are expressed historically in the NDVI data. The vegetation condition index (VCI) [7], which is based on a transformational of the AVHRR NDVI data, and the temperature condition index (TCI) calculated from AVHRR's thermal data [8] are operationally produced and commonly used for national- to global-scale drought monitoring [9,10,11]. Although these numerous operational products have been useful for vegetation monitoring, they are limited for effectively characterizing the impact of drought on vegetation because the anomalies caused by drought stress cannot be discriminated from anomalies produced by other environmental causes of vegetation stress (e.g., flooding, fire, pest infestation, and hail damage) and anthropogenic drivers (e.g., land cover/land use conversion). Additional information is required to discriminate the drought-impacted areas from locations where the vegetation is being influenced by these other environmental and anthropogenic factors [12].

Traditionally, climate-based drought indicators such as the palmer drought severity index (PDSI) and standardized precipitation index (SPI) have been used for drought monitoring. However, climate-based drought monitoring approaches have limited spatial precision at which drought patterns can be mapped since the indices are calculated from point-based meteorological measurements collected at weather station locations [13, 14]. In addition, weather stations are scarce in remote areas and they are not uniformly distributed. As a result, climate-based drought index maps depict broad-scale drought patterns that are produced from point-based data using statistical-based spatial interpolation techniques and the level of spatial detail in those patterns is highly dependent on the density and distribution of weather stations. Therefore, the spatial detail in climate-based drought index maps is limited due to the dependence on uneven and sparse weather station distributions, which limits drought planning and monitoring activities in areas not well covered by weather stations [12].

Several methods have been studied to identify drought level, such as standardized precipitation index (SPI) [15] and keetch-byram drought index (KBDI) for meteorological drought, normalized difference vegetation index (NDVI) anomalies, enhanced vegetation index (EVI) anomalies, vegetation condition index (VCI), temperature condition index (TCI), and vegetation health index (VHI). Satellite data used are MODIS, MTSAT, AMSR-E. Analysis shows that these methods can be used to identify the presence of anomalous levels of dryness and drought in the central region of rice.

This study is aimed to assess the use of satellite data to monitor drought conditions in paddy field in near real time, validate satellite -based drought models with field observation data, and disseminate the results by developing a Web-GIS satellite based system that is integrated with an existing crop calendar information system. The results of this study can be used to determine spatial location of drought area and its effect on rice production. Results are expected to help provide recommendations to prevent future agricultural drought events.

2. Methodology and data used

Human factors, such as water demand and water management can exacerbate the impact that drought has on a region. Because of the interplay between a natural drought event and various human factors, drought means different things to different people. In practice, drought is defined in a number of ways that reflect various perspectives and interests [16]. Wilhite and Glantz [17] categorized the definitions in terms of four basic approaches to measuring drought, namely meteorological, hydrological, agricultural, and socioeconomic. The first three approaches are dealt with ways to measure drought as a physical phenomenon. The last is dealing with drought in terms of supply and demand, tracking the effects of water shortfall as it ripples through socioeconomic systems.

Drought types having been analyzed for this study are meteorological and agriculture drought that are based on research conducted by Takeuchi and Gonzalez [8] and Roswintiati *et al.* [18, 19]. Several methods of meteorological drought index have been utilized. KBDI and the SPI of the daily and monthly rainfall data from TRMM and GSMap have been analyzed for several research areas in the last seven years. A 5-year time series (2009 – 2005) of 16-days composited of 250-m Terra MODIS data was used to process and analyze agricultural drought by observing some characters of some indices. The data used and performance analysis are presented in Table 1.

Table 1. Satellite data used and analyzed.

Satellite data	Description	Analysis	Data provider
MODIS	Resolution 250 m (16-day composite)	TCI, VCI, VHI	LAPAN
MTSAT	Resolution 4 km (daily)	KBDI	Univ of Tokyo
GSMaps	Resolution 0.1 deg (every hour)	Daily rainfall data	Univ of Tokyo

Classification of Kogan [7, 20] has been used to classify agricultural drought which has 5 classes (Table 2). The classification system is simpler than a more difficult to identify in the field of the 9th class of NEDIS NOAA.

Table 2. Drought classification.

SPI	Description
VHI > 40	No drought
30 > VHI ≥ 40	Mild drought
20 > VHI ≥ 30	Moderate drought
10 > VHI ≥ 20	Severe drought
VHI ≤ 10	Extreme drought

The following Figure 1 shows the flowchart of the study by integrating existing model using ground rainfall data and satellite data to obtain drought information. The information could facilitate farmers to plan their planting time, and local government to determine regional planning development, especially in agricultural field.

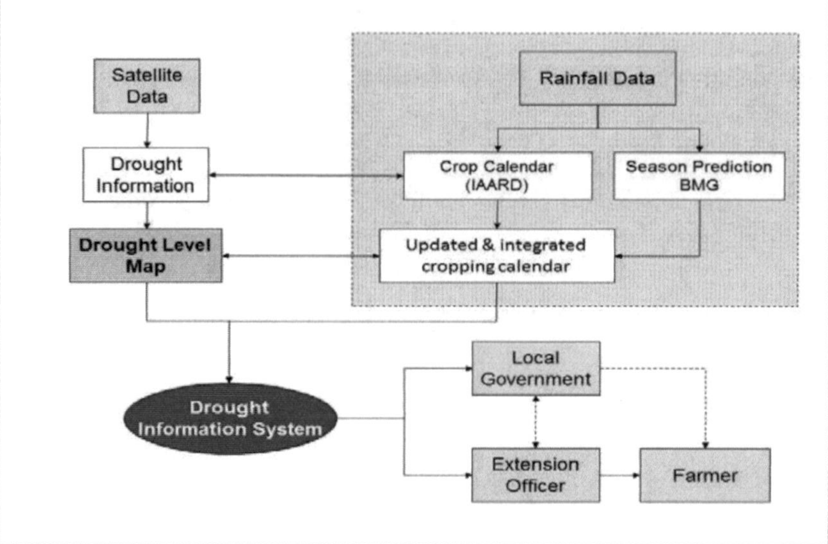

Figure 1. Flowchart of activity.

A number of institutions are involved in this project. The prototyping executor is the Indonesian Center for Agricultural Land Resources Research and Development (ICALRD) of Indonesian Agency for Agricultural Research and Development (IAARD), Ministry of Agriculture (MoA). Technical support is provided by the University of Tokyo, Geoinformatics Center (GIC) of the Asian Institute of Technology (AIT), Indonesian National Institute of Aeronautics and Space (LAPAN), and the Agency for the Assessment and Application of Technology (BPPT). JAXA provides MODIS, MTSAT, AMSR-E and ALOS data, the University of Tokyo supports the meteorological drought index, LAPAN provides agricultural drought index, BPPT develops Web-based Information System, and ICALRD-IAARD provides crop calendar data and rainfall data. End users are Indonesian local government services and farmers through extension officers. The results will be introduced and disseminated through workshops and seminars as well as through a web portal.

3. Results and discussions

Several methods have been studied to identify the index of drought in Indonesia, such as keetch-byram drought index (KBDI) for meteorological drought, normalized difference vegetation index (NDVI) anomaly, enhanced vegetation index (EVI) anomaly, vegetation condition index (VCI), temperature condition index (TCI), and vegetation health index (VHI). It reveals that these methods can be used to identify the level and anomaly of drought in rice production centers. The results of analysis shows that rainfall data derived from GSMap, KBDI of AMSR-E, and LSWC of MTSAT can be used to identify the level of meteorological drought. While NDVI, VCI, TCI, VHI from satellite data Terra MODIS resolution of 250 mx 250 m have been analyzed to identify agricultural drought level.

3.1. Drought analysis

Results at the district level show that MODIS data of 250 mx 250 m resolution can be applied and produce sufficient information to indicate drought conditions through the analysis of VCI, TCI and VHI. Terra MODIS resolution of 250 mx 250 m in the district of Indramayu is presented in Figure 2. The spatial resolution is sufficient to figure out detail information drought conditions at such district level. Based on the rainfall data obtained from satellite data GSMap, in 2009, the lowest amount of rainfall occurs during July to November, and the analysis indicates drought anomalies of KBDI. Different conditions occur in 2010, rainfall data and KBDI analysis shows that the region is not experiencing drought.

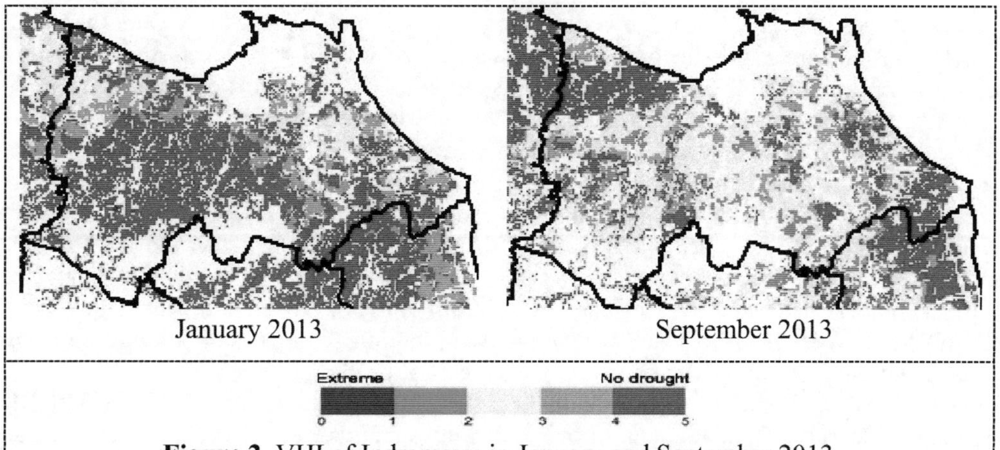

Figure 2. VHI of Indramayu in January and September 2013.

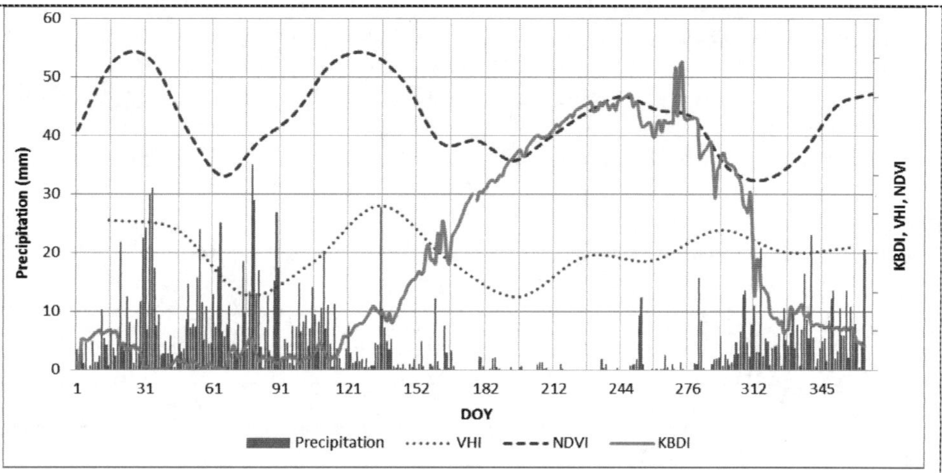

(a) Studi location of Ngawi District, East Java Province

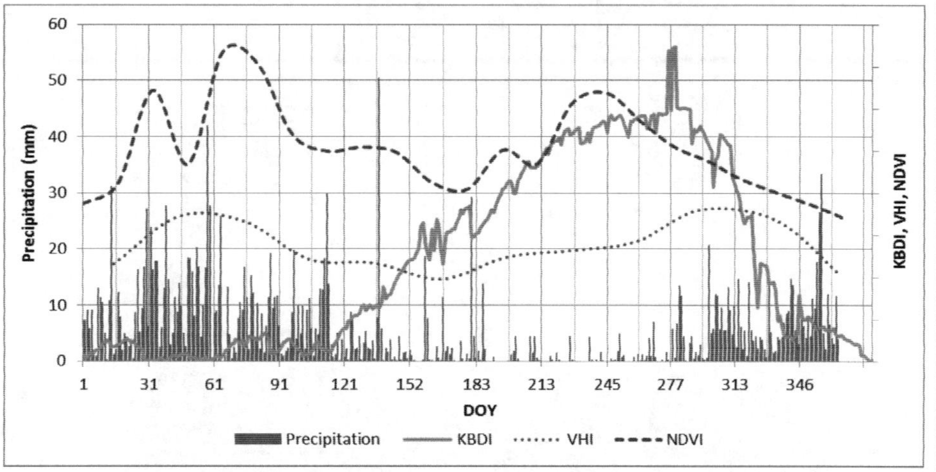

(b) Studi location of Indramayu District, West Java Province

Figure 3. Pattern of KBDI, VHI, NDVI, and rainfall condition

VHI characterizes vegetation health by combining estimation of moisture (VCI) and thermal (TCI) conditions [21]. KBDI measures the amount of moisture in the top layer of soil and covering leaves and vegetation [22]. The KBDI considers drought as an index on a scale from 0 to 2000, based on the moisture content of the soil [23]. It measures the amount of moisture in the top layer of soil and covers leaves and vegetation. Typical drought anomalies can be indicated by the increasing value of KBDI, as well as the decline value of VHI.

NDVI estimates photosynthetically absorbtion of radiation over the land surfaces [24, 25], at which it can gives a measure of the vegetative cover and displays greenness level on the land surface [26]. NDVI can show a number of paddy planting seasons, namely (a) at study location of Ngawi District, East Java Province indicates that the area has three planting time of paddy, and (b) at study location of Indramayu District, West Java Province which has two planting seasons. NDVI value indicates that the decrease is not directly affected by declining of precipitation.

Pattern of KBDI,VHI, NDVI, and TCI to precipitation are presented in Figure 3. KBDI can indicate drought anomaly. A small amount of rainfall shows an increase in KBDI that can be used as a meteorological drought parameters. However, in both study sites, drought anomalies indicated by KBDI does not necessarily affect the condition of vegetation. This means that the level of drought is not causing damage to rice crop. The VHI pattern is more sensitive to vegetation condition. At both locations, the decreasing of VHI value, which indicates drought, is followed by the decrease in the value of NDVI or greenness of vegetation. Similarly, on the opposite condition, the drought classification of VHI can be used to identify agricultural drought. In this respect, there is no correlation between KBDI and VHI and the increasing value of KBDI has no effect on VHI value.

3.2. Result Verification

Field verification was conducted during the dry season of August-September 2013 in Indramayu (Figure 4), West Java; DI Yogyakarta; Barito Kuala, South Kalimantan; Ngawi and Pasuruan, East Java, and Maros, South Sulawesi. The results show that the value of 500-600 KBDI obtained from MTSAT data indicates beginning of meteorological drought in paddy field. While agricultural drought, through VHI analysis of MODIS, has 60 % accuracy.

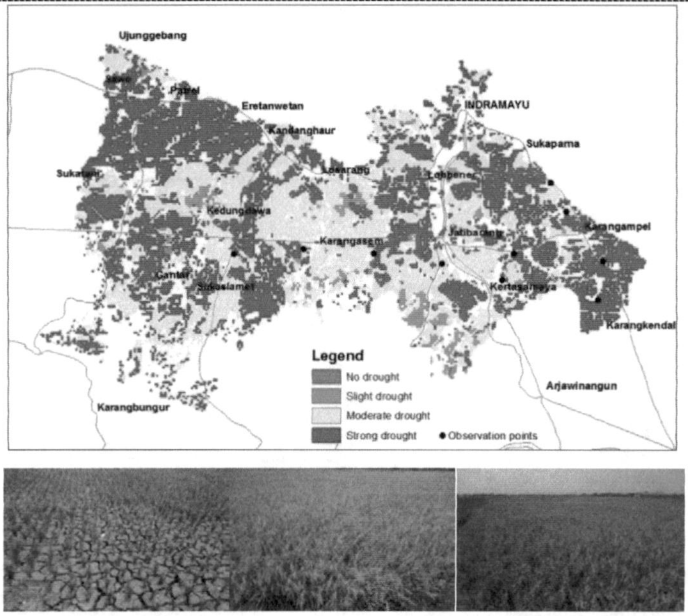

Figure 4. Drought map of VHI verification at Indramayu Distric of West Java.

In addition to field verification, several workshops and coordination meetings, such as Focus Group Discussion (FGD) and Workshops are conducted to enhance the output and to get feedback from end users. These meetings were attended by policy makers and other stakeholders such as farmers' coordinators, regional institutions' representatives, extension agents, and researchers.

3.3. Dissemination with Web-GIS

The result of the analysis of the level of agricultural drought is presented in printed form of drought levels distribution map of paddy fields as shown in Figure 5. This is also presented in Web-GIS based information system that can be accessed via internet for wider community. The Web-GIS based information system provides information on interactive and integrated drought condition by using Google API.

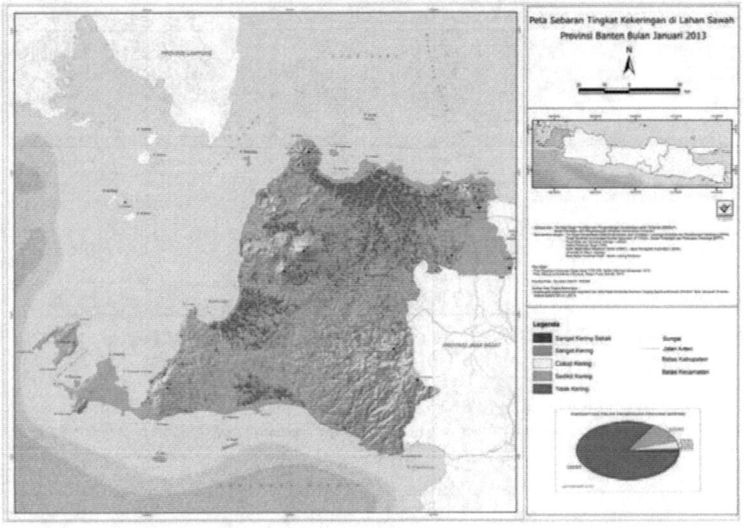

Figure 5. Map of drought levels in paddy field of Banten Province, January 2013.

The Indonesian Information System for Drought Monitoring is a daily and monthly product that provides a general summary of current meteorological and agricultural drought conditions. Multiple drought indicators, including various indices, outlooks, field reports, and news accounts are reviewed and synthesized. In addition, numerous experts from agencies and offices across the country are consulted and involved to generate the product. The result is the consensus assessment presented on the drought level map. Agricultural drought monitoring is presented in Indonesian Information System for Drought Monitoring as shown in Figure 6. Meterological drought monitoring and information can be searched at webgms.iis.u-tokyo.ac.jp [27] as shown in Figure 7.

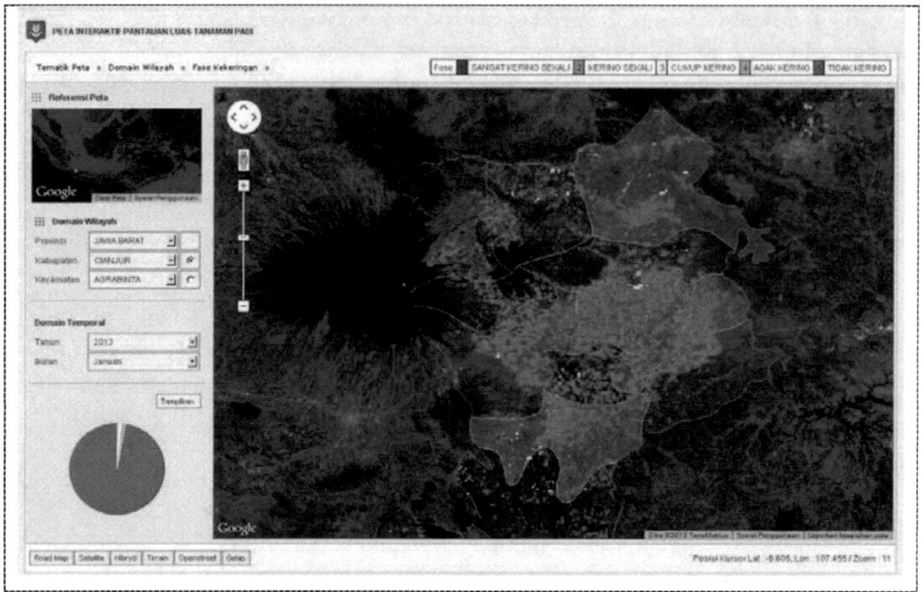

Figure 6. Web-GIS based information system of drought monitoring.

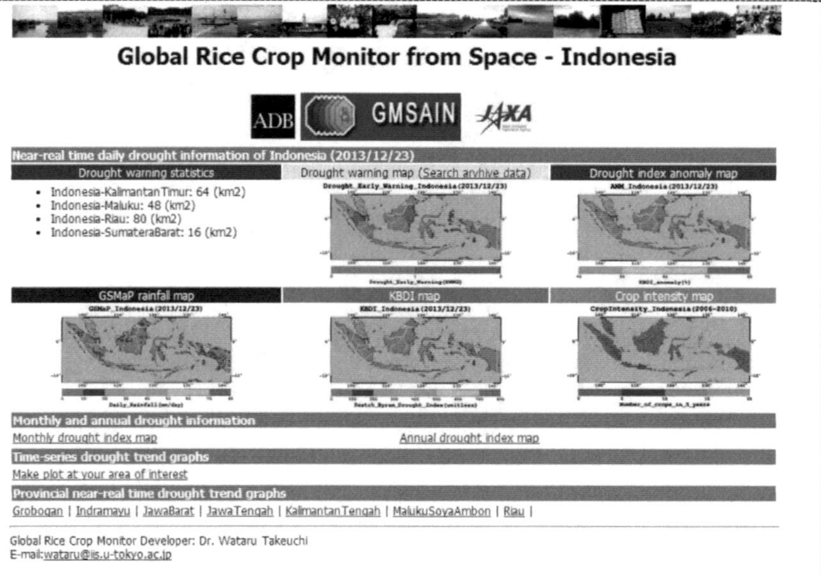

Figure 7. Meteorological drought monitoring information system.

4. Conclusions

In respect to the assessment of drought impact on rice production, several findings could be drawn from this study. These are related to the development models to create Web-GIS and satellite based information system.

The study reveals that the data obtained from NDVI and VHI from Terra MODIS resolution of 250 mx 250 m, rainfall data from GSMap, and KBDI from AMSR – E can be used to identify the presence of anomalous levels of dryness and drought in rice producing centers. Meanwhile, NDVI, VHI, KBDI can represent rainfall anomalies and also capture the onset of drought. VHI from MODIS, KBDI of MTSAT can be also be used to identify patterns of plant with its index to estimate the results.

Validation having been carried out in Indramayu district of West Java Province; Barito Kuala district of South Kalimantan Province, Ngawi and Pasuruan districts of East JavaProvince, and Yogyakarta of DI Yogyakarta Province indicate that (a) Meteorological drought for KBDI analysis with value ranging from 500 to 600 show the beginning of drought in paddy field, and (b) agricultural drought by analyzing VHI of 250 m x 250 m Terra MODIS has shown quite confident of 60 % accuracy.

Interactive and integrated Web-GIS based system using the Google API has been developed to disseminate information of drought condition for wider community. This is very important in preparing remote sensing based data for decision making process in the national agricultural development policies.

Acknowledgment

The authors are grateful to the Indonesian Center of Agricultural Land Resources Research and Development (ICALRD) of the Indonesian Agency of Agricultural Research and Development (IAARD), Ministry of Agriculture for opportunity to work on this study and to provide facilities, data, and encouragement to complete this study. Thanks are also conveyed to Japan Aerospace Exploration Agency (JAXA), Japan for providing satellite data and transfer of technology during the project activities. Many thanks are also delivered to University of Tokyo and to the Agency for Assessment and Application of Technology (BPPT) for providing technical support and facilities to develop Crop Monitoring System Information. Our thankful are also extended the Indonesian National Institute of Aeronautics and Space (LAPAN) for facilitating this collaboration.

References

[1] Tucker CJ, Townshend JRG, and Goff TE, 1985, African land cover classification using satellite data, *Science* 9227(**4685**):369-375.

[2] Reed BC, Brown JF, Vander Zee D, Loveland TR, Merchant JW, and Ohlen DO, 1994. Measuring phenological variability from satellite imagery, *Journal of Vegetation Science* 5:703-714.

[3] Jakubauskas ME, Peterson DL, Kastens JH, and Legates DR, 2002, Time series remote sensing of landscape-vegetation interactions in the southern Great Plains, *Photogrammetric Engineering and Remote Sensing* 68:1021-1030.

[4] Artusa A, 2012, *Drought Fact Sheet*. National Weather Service. http://droughtmonitor.unl.edu/

[5] National Drought Mitigation Center. 2013. http://www.drought.gov/drought/

[6] National Oceanic and Atmospheric Administration. *NOAA's National Weather Service*. http://www.weather.gov/

[7] Kogan FN, 1990, Remote sensing of weather impacts on vegetation in non-homogeneous areas. *International Journal of Remote Sensing* 11:1405–1419.

[8] Takeuchi W and Gonzalez L, 2009, Blending MODIS and AMSR-E to predict daily land surface water coverage, *Proceedings of International Remote Sensing Symposium (ISRS)*: Busan, South Korea.

[9] Liu WT and Kogan FN, 1996, Monitoring regional drought using the vegetation condition index, *International Journal of Remote Sensing* 17(**14**):2761-2782.

[10] Kogan FN, 1990, Remote Sensing of Weather Impacts on Vegetation in Non-Homogeneous Areas, *International Journal of Remote Sensing* 11 (1990), pp. 1405–1419.

[11] Unganai LS and Kogan FN, 1998, Drought monitoring and corn yield estimation in southern Africa from AVHRR data, *Remote Sensing of the Environment* 63:219-232.

[12] Brian DW, Tadess T, Brown, JF, and Gu Y, 2010, *The Vegetation Drought Response Index (VegDRI): A New Drought Monitoring Approach for Vegetation.* http://www.drought.gov/workshops/remotesensing/abstracts/brian_wardlow2.pdf.

[13] Palmer WC, 1965, *Meteorological Drought*. Research Paper No. 45, U.S. Department of Commerce Weather Bureau, Washington, D.C. 58 pp.
[14] Guttman NB, 1998, Comparing the Palmer Drought Index and the Standardized Precipitation Index, *J. Amer. Water Resources Assoc.* 34(**1**): 113-121.
[15] Guttman NB, 1999, Accepting the Standardized Precipitation Index: A calculation algorithm. *J. Amer. Water Resources Assoc.* 35(**2**):311-322.
[16] National Climatic Data Center. 2012. *Common types of drought*. National Weather Service.
[17] Wilhite DA and Glantz MH, 1985, Understanding the Drought Phenomenon: The Role of Definitions. *Water International 1* 10(**3**):111–120.
[18] Roswintiarti O, Parwati, and Anggraini N, 2010, Potential drought monitoring over agriculture area in Java Island, Indonesia. *Paper, presented in Meeting with Ministry of Agriculture and JAXA*: LAPAN, Indonesia.
[19] Shofiyati R, Takeuchi W, Samarakoon L, Sarwani M, and Sofan P, 2012, Assessment of Drought Impact on Rice Production in Java Island of Indonesia by Satellite Remote Sensing. *Proceedings of the Space Applications for Environment (SAFE) joint workshop with Climate R3and Mid-year EO-WG in the ISTS29*: Nagoya Japan.
[20] Kogan FN, 1987, Vegetation index for area analysis of crop conditions. *Proceedings of 18th Conference on Agricultural and Forest Meteorology*, pp. 103 – 106.
[21] NOAA STAR, 2010, STAR - Global Vegetation Health Products, Center for Satellite Applications and Research (STAR), NOAA Center for Weather and Climate Prediction (NCWCP).
[22] Buchholz G, and Weidemann D, 2000. The Use of simple Fire Danger Rating Systems as a Tool for Early Warning in Forestry. *International Forest Fire News* 23: 32-36.
[23] Keetch JJ, and Byram GA, 1968, *Drought index for forest fire control*, USDA Research Paper SE-38. Southeastern Forest Experiment Station.
[24] Mitchell T, 2002, *National Aeronautics and Space Administration (NASA) advanced very high resolution radiometer (AVHRR) normalized difference vegetation index (NDVI) 1981 - 2001*, Joint Institute for the Study of the Atmosphere and Ocean (JISAO), http://jisao.washington.edu/datasets/ndvi/#introduction.
[25] Piwowar JM, 2005, *Digital Image Analysis, Remote Sensing for GIS Managers*, Aronoff S (Ed.), pp. 287–335. ESRI Press, Redlands, California, 2005.
[26] Lillesand T, Kiefer RW, Chipman J, 2008, *Remote Sensing and Image Interpretation*, sixth edition. John Wiley & Sons, Inc., New York, New York.
[27] Takeuchi W, 2013, Global rice crop monitoring from space – Indonesia, webgms.iis.u-tokyo.ac.jp.

IOP Conf. Series: Earth and Environmental Science **20** (2014) 012049 doi:10.1088/1755-1315/20/1/012049

Seasonal analysis of precipitation, drought and vegetation index in Indonesian paddy field based on remote sensing data

S Darmawan[1,3,4], W Takeuchi[1], R Shofiyati[2], D K Sari[3] and K Wikantika[4]

[1]Institute of Industrial Science, University of Tokyo, 4-6-1 Komaba Meguro-ku, Tokyo, 153-8505, Japan.
[2]Indonesian Agency for Agricultural Research and Development Ministry of Agriculture, Jalan Tentara Pelajar No. 12 Bogor, West Java, Indonesia.
[3]Institut Teknologi Nasional (ITENAS), Jalan PHH. Mustafa No. 23, Bandung, West Java, Indonesia.
[4]Center for Remote Sensing, Institut Teknologi Bandung (ITB), Jalan Ganesha 10, Bandung, West Java, Indonesia.

Email : soni@iis.u-tokyo.ac.jp and soni_darmawan@yahoo.com

Abstract. Paddy field is important agriculture crop in Indonesia. Rice is a food staple for 237,6 million Indonesian people. Paddy field growth is strongly influenced by water, but the amount of precipitation is unpredictable. Annual and interannual climate variability in Indonesia is unusual. In recent years remote sensing data has been used for measurement and monitoring of precipitation, drought and vegetation index such as Global Satellite Mapping of Precipitation (GSMaP), Multi-purpose Transmission SATellite (MTSAT) and Moderate Resolution Imaging Spectroradiometer (MODIS). The objective of this research is to investigate seasonal variability of precipitation, drought and vegetation index in Indonesian paddy field based on remote sensing data. The methodology consists of collecting of enhanced vegetation index (EVI) from MODIS data, mosaicking of image, collecting of region of interest of paddy field, collecting of precipitation and drought index based on Keetch Bryam Drought Index (KBDI) from GSMaP and MTSAT, and seasonal analysis. The result of this research has showed seasonal variability of precipitation, KBDI and EVI on Indonesia paddy field from 2007 until 2012. Precipitation begins from January until May and October until December, and KBDI begins to increase from June and peak in September only in South Sumatera precipitation almost in all month. Seasonal analysis has showed precipitation and KBDI affect on EVI that can indicate variety phenology of Indonesian paddy field. Peak of EVI occurs before peak of KBDI occurs and increasing of KBDI followed by decreasing of EVI. In 2010 all province got higher precipitation and smaller KBDI so EVI has three peaks such as in West Java that can indicated increasing of rice production.

1. Introduction

Paddy field is one of the important agriculture crops in Indonesia. Large of Indonesian paddy field around 8.06 million ha and total production around 65 million tons/year (http://www.bps.go.id). Indonesia is third country in the world which large rice-producing after China and India [1]. Growth in rice production is popularly perceived as the most important indicator of agricultural development in

Content from this work may be used under the terms of the Creative Commons Attribution 3.0 licence. Any further distribution of this work must maintain attribution to the author(s) and the title of the work, journal citation and DOI.
Published under licence by IOP Publishing Ltd

Indonesia. Rice is the major crop for small farms, and the rice value chain is a key sector of the rural economy. Rice is also the food staple, and domestic demand for rice is very large. Historically, the thin world market for rice has meant that rice self-sufficiency was perceived as the key to ensuring national food security [2].

Phenology of paddy field can be characterized by three main periods: fisrt the flooding and paddy transplanting period; second the growing period (vegetative growth, reproductive and ripening stages); and third the fallow period after harves [3]. Many researcher mapping and monitor phenology of paddy field based on vegetation index derived from remote sensing data such as mapping paddy field in Southeast Asia [3], determining cropping type and cropping pattern of paddy field [4], multi sensor satellite data for rice production estimation [5], detecting rice phenology [6], global rice paddy field mapping [7] and Indonesian paddy field mapping [8]. Paddy field growing strongly influenced by water. In the fact amount of precipitation is unpredictable. Annual and interannual climate variability in Indonesia is unusual. It is not homogenous over the whole region and coherence of precipitation patterns varies seasonally [9]. According Quint et al.,[10] found that 93% of droughts in Indonesia between 1830 and 1953 occurred during El Niño years and four El Niño between 1973 and 1992. Information of precipitation and drought very important for transplanting period, harvest period and product estimation.

In recent years remote sensing has been used for measurement and monitoring precipitation and drought. According to Kubota et al.,[11] Global Satellite Mapping of Precipitation (GSMaP) has been started since November 2002. The aims of the GSMaP are the development of an advanced microwave radiometer algorithm based on the deterministic rain-retrieval algorithm of and the production of precise high-resolution global precipitation maps. And global Keetch Bryam Drought Index (KBDI) map based on precipitation derived from GSMaP and temperature from land surface temperature derived from Multi-purpose Transmission SATellite (MTSAT) has been generated [12]. The objective of this research to investigate seasonal variability of precipitation, drought and vegetation index in Indonesian paddy field based on remote sensing data. The methodology consists of collecting of enhanced vegetation index (EVI) from MODIS data, mosaicking of image, collecting of region of interest paddy field, collecting of precipitation and drought index based on Keetch Bryam Drought Index (KBDI) and seasonal analysis.

2. Methodology

Indonesia also called "Nusantara" is a country in Southeast Asia that is located on 6° N – 11° S and 95° E - 141° E, between the Asian and Australian continents and between the Pacific and Indian Oceans. Indonesia Consisting of 17,508 islands that is largest archipelagic country in the world. Indoensia have large area around 1.904.569 km^2 which population around 237,6 million people in 2012, total province is 33 province. (www.indonesia.go.id). For this study can see on the figure 1, we focus on six provinces that is West Java, East Java, North Sumatera, South Kalimantan, South Sulawesi and Papua. We choose their location because that location describes paddy field in each big island in Indonesia.

According to Indonesian statistical agency (http://www.bps.go.id) Indonesia paddy field in 2012 harvested area for whole Indonesia around 13 million ha with productivity 51.36 quintal/ha and total rice production around 69 million ton. The largest of harvested area in Java Island is 46% and the largest of rice production in Java Island is 53%. Out from Java such as North Sumatera, South Kalimantan and South Sulawesi have large harvested area and big rice production.

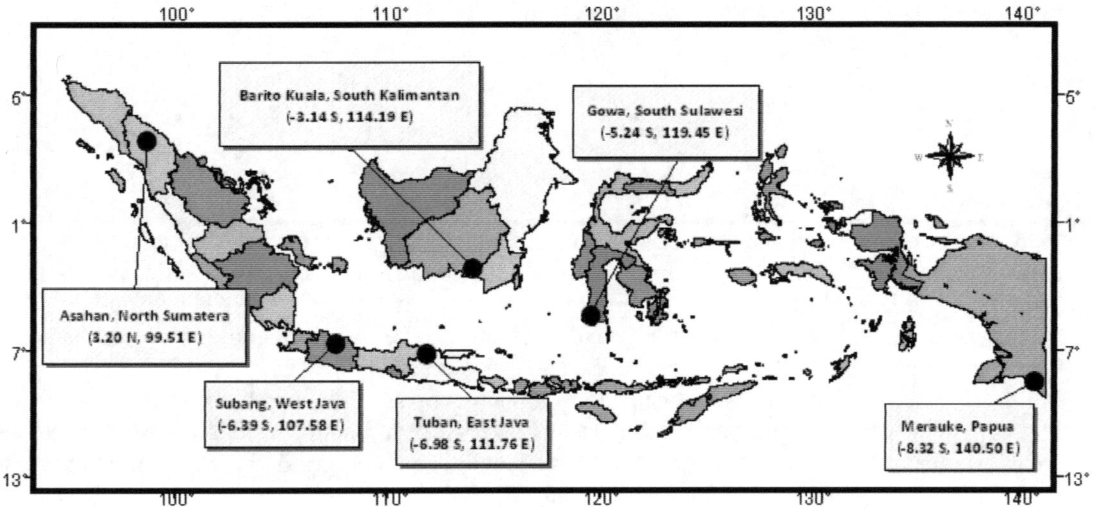

Figure 1. Case study focus on the six provinces in Indonesia.

In this research, methodology can see in the figure. 2. Methodology consists of collecting of enhanced vegetation index (EVI) from MODIS data, mosaicking of image, collecting of region of interest paddy field, collecting of precipitation and drought index based on Keetch Bryam Drought Index (KBDI) and seasonal analysis.

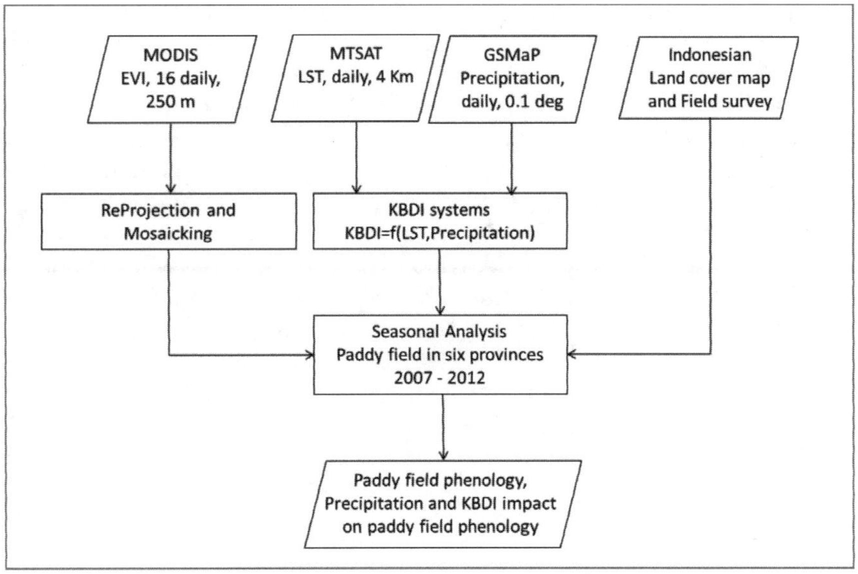

Figure 2. Methodology of seasonal analysis.

2.1 Enhanced Vegetation Index (EVI)

EVI was developed to optimize the vegetation signal with improved sensitivity in high biomass regions and improved vegetation monitoring through a de-coupling of the canopy background signal

IOP Conf. Series: Earth and Environmental Science **20** (2014) 012049 doi:10.1088/1755-1315/20/1/012049

and a reduction in atmosphere influences, the EVI equation follows [13]:

$$EVI=G\rho nir-\ \rho red\rho nir+C1\times\rho red-C2\times\rho blue+L \tag{1}$$

where ρ are atmospherically corrected or partially atmosphere corrected (Rayleigh and ozone absorption) surface reflectance, L is the canopy background adjustment that addresses nonlinear, differential NIR and red radiant transfer through a canopy, and C_1, C_2 are the coefficients of the aerosol resistance term, which uses the blue band to correct for aerosol influences in the red band. The coefficients adopted in the EVI algorithm are, L=1, C_1=6, C_2=7.5, and G (gain factor)=2.5 [14].

EVI derived from global MODIS vegetation indices (MOD13Q1 product) which designed to provide consistent spatial and temporal comparisons of vegetation conditions. Global MOD13Q1 data are provided every 16 days at 250-meter spatial resolution as a gridded level-3 product in the Sinusoidal projection. Lacking a 250m blue band, the EVI algorithm uses the 500m blue band to correct for residual atmospheric effects, with negligible spatial artifacts (https://lpdaac.usgs.gov/products/modis_products_table/mod13q1). MODIS products are organized in a tile system with the Sinusoidal (SIN) projection grid, and each tile covers an area of 1200 km by 1200 km (approximately 108 latitude by 108 longitude at equator). Twelve tiles (H27V08, H27V09, H28V08, H28V09, H29V08, H29V09, H30V08, H30V09, H31V08, H31V09, H32V08, H32V06) are needed to cover the study area.

2.2 Precipitation and Keetch-Byram drought index (KBDI)

In this study precipitation derived from GSMaP. The GSMaP project was promoted for a study "Production of a high-precision, high-resolution global precipitation map using satellite data," sponsored by Core Research for Evolutional Science and Technology (CREST) of the Japan Science and Technology Agency (JST) during 2002-2007. Since 2007, GSMaP project activities are promoted by the JAXA Precipitation Measuring Mission (PMM) Science Team (http://sharaku.eorc.jaxa.jp/GSMaP_crest/). According to Prasetya et al., [15] monthly spatial average rainfall showed very good agreement between satellite monthly rainfall data and rain gauge data across Indonesia.

The Keetch-Byram drought index (KBDI) is a continuous reference scale for estimating the dryness of the soil and duff layers. The index increases for each day without rain (the amount of increase depends on the daily high temperature) and decreases when it rains. The scale ranges from 0 (no moisture deficit) to 800. The range of the index is determined by assuming that there is 8 inches of moisture in a saturated soil that is readily available to the vegetation [16]. The equation for computing the incremental rate of change of the index, ΔQ, is:

$$\Delta Q=800-Q0.968\text{exp}0.0486T-8.30\Delta t\times 0.0011+10.88\text{exp}-0.0441M \tag{2}$$

Where t is the daily maximum temperature, M the mean annual rainfall, Q the current KBDI, and ΔQ is a time increment set equal to 1 day [16]. In this study KBDI is generated by Institute of Industrial Science, The University of Tokyo (http://webgms.iis.u-tokyo.ac.jp/KBDI/) [12]. KBDI generated from precipitation derived from Global satellite mapping of precipitation (GSMaP) and temperature from land surface temperature derived from Multi-purpose Transmission SATellite (MTSAT). KBDI systems can monitor agricultural and climatological drought in 10km grid size on a daily basis [17].

According to Xanthopoulos et al., [18] a relationship between KBDI and plant water potential exists, the relation varies depending on the plant species.

2.3 Seasonal Analysis

We have collected region of interest (ROI) of rice paddy field area from whole province level Indonesia, size of ROI around 4x4 pixels. ROI based on land cover map from Indonesian Ministry of Forestry and Indonesian base map from Indonesian Information Geospatial Agency (BIG). We have mosaicked MODIS vegetation index for whole Indonesia from 2007 until 2012, and we have collected statistical KBDI and precipitation data from http://webgms.iis.u-tokyo.ac.jp/DMEWS/plot.php 2007 until 2012. Finally we have seasonal variability of vegetation index, precipitations and KBDI from 2007 until 2012 for whole province in Indonesia.

3. Results and Discussion

EVI derived from reprojection and mosaicking MOD13Q1 product. We have reprojected and mosaicked twelve tiles (H27V08, H27V09, H28V08, H28V09, H29V08, H29V09, H30V08, H30V09, H31V08, H31V09, H32V08, H32V06) to cover overall study area. Precipitation derived from GSMaP and Drought has been calculated from MTSAT and GSMaP. Precipitation, EVI and KBDI can see in the figure 3. We have plotted ROI of rice paddy field in six provinces, which are West Java, East Java, North Sumatera, South Kalimantan, South Sulawesi and Papua. And we got statistical seasonal variability of EVI, precipitation and KBDI.

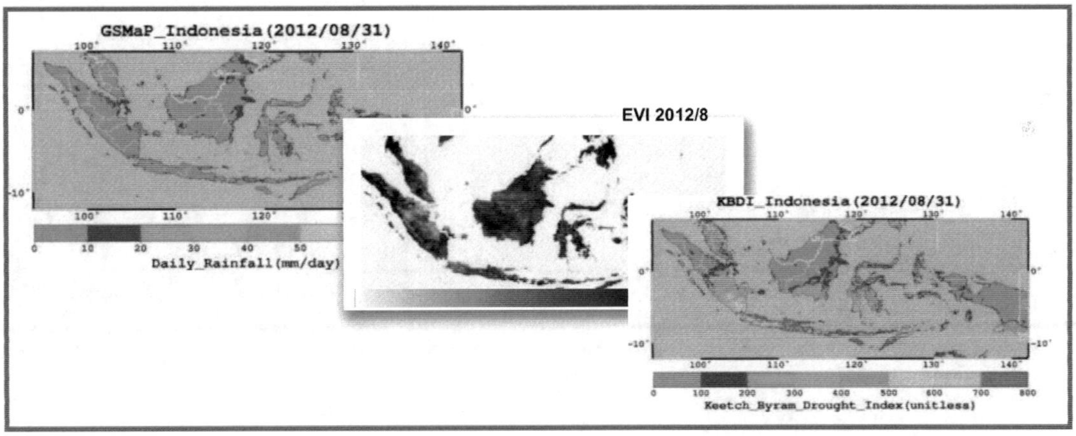

Figure 3. Precipitation, EVI and KBDI map in study area
(Derived from http://webgms.iis.u-tokyo.ac.jp/DMEWS/Indonesia/).

Statistical seasonal variability of EVI, precipitation and KBDI for six provinces in 2012 can see in the figure 4. According to figure 4 we can see seasonal variability of EVI in 2012 for six provinces. Seasonal variability of EVI shows of phenology of paddy field. EVI designed to enhance the contribution of vegetation properties and allow reliable spatial and temporal inter-comparisons of terrestrial photosynthetic activity and canopy structural variations [13]. Based on phenology of paddy field Lower EVI value that indicate flooding and rice transplanting period, and then EVI value increase until peak that indicate growing period (vegetative growth, reproductive and ripening stages), finally EVI value decrease that indicate fallow period after harvested. In this case EVI value have two peak except South Kalimantan only one peak, it means in 2012 got two harvested and only in South Kalimantan got one harvested.

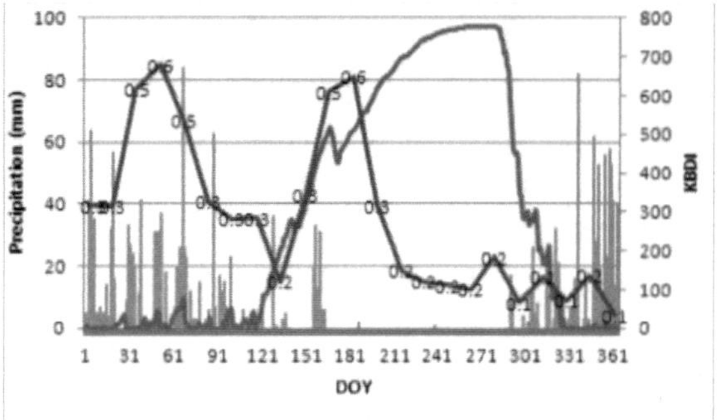

West Java (-6.39 S,107.58 E)

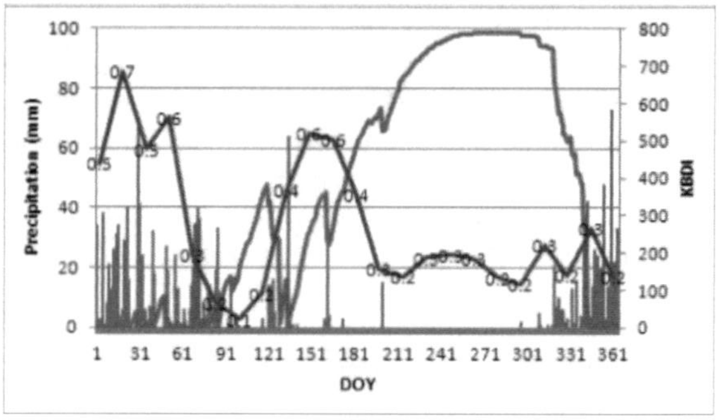

East Java (-6.98 S,111.75 E)

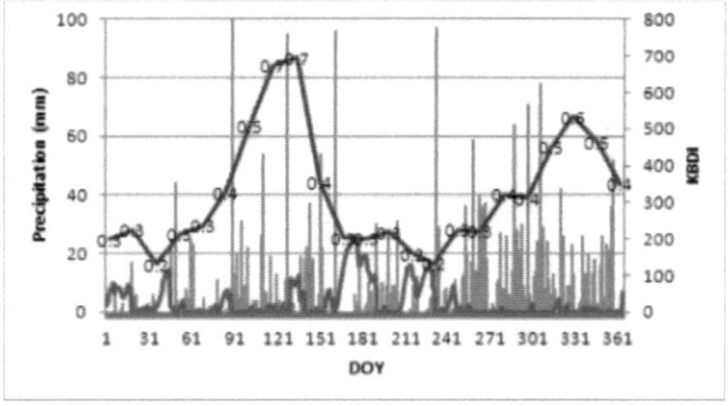

North Sumatera (3.20 N, 99.51 E)

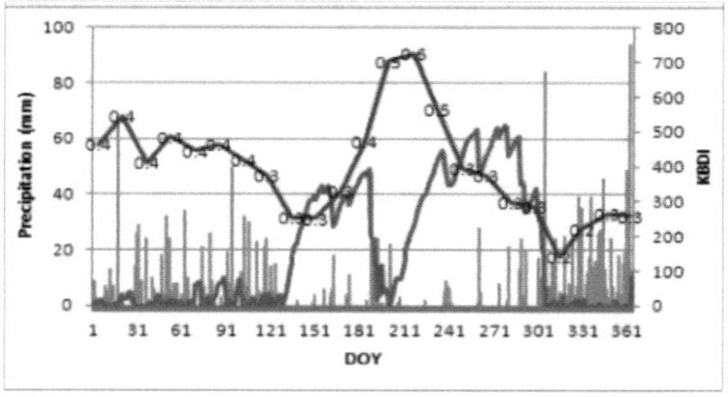

South Kalimantan (-0.10 S, 109.32 E)

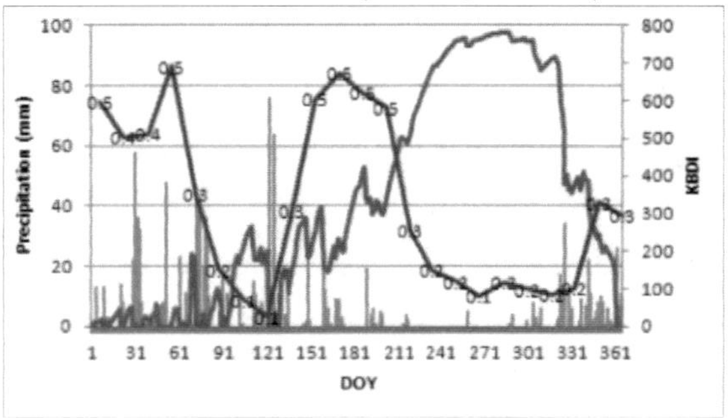

South Sulawesi (-5.24 N, 119.45 E)

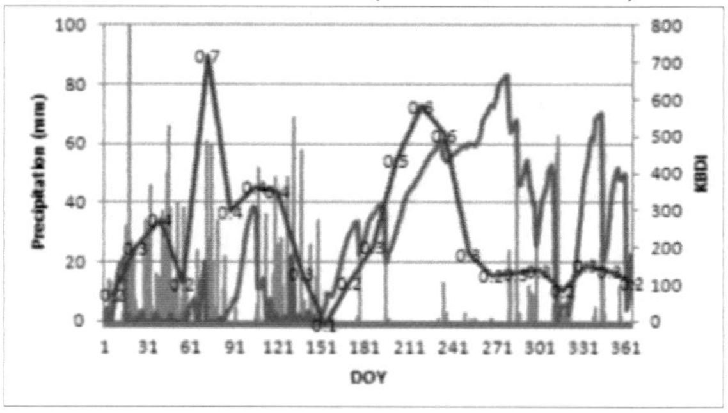

Papua (-8.32 S, 140.50 E)

Figure 4. Seasonal variability of EVI, Precipitation and KBDI in 2012
EVI (green line), precipitation (blue line) and KBDI (red line).

In West Java and East Java precipitation begins from January until May and October until December and from KBDI begins to increase from June and peak in September. In North Sumatera and South Kalimantan precipitation almost occurs all the month so KBDI value not height. In South Sulawesi precipitation begins from January until February and November until December so KBDI begins to increase from March and peak in October. In Papua precipitation begins from January until June and September until December so KBDI begins to increase from July and peak in the August. Sessional variability of precipitation and KBDI is different for each province. According to Aldrian and Susanto [19], in Indonesia seasonal variability of precipitation is categorized on three regions, that is region A

(West Java, East Java, South Kalimantan, South Sulawesi and Papua) has one peak and one valley rainfall and is influenced by two monsoon systems, the northwest monsoon (wet seasons) from November to March, and the southeast monsoon (dry seasons) from May to September. Region B (North Sumatera) has two peaks of rainfall in October–November and March–May. Those two peaks are associated with the southward and northward movement of the inter-tropical convergence zone. Region C has one peak and one valley rainfall, namely June–July for the wet season and November–February for the dry season.

We have calculated also EVI, precipitation and KBDI from 2007 until 2012, can see on figure 4. Mostly seasonal variability of EVI influenced by precipitation and KBDI. Peak of EVI occurs before peak of KBDI occurs and increasing of KBDI followed by decreasing of EVI.

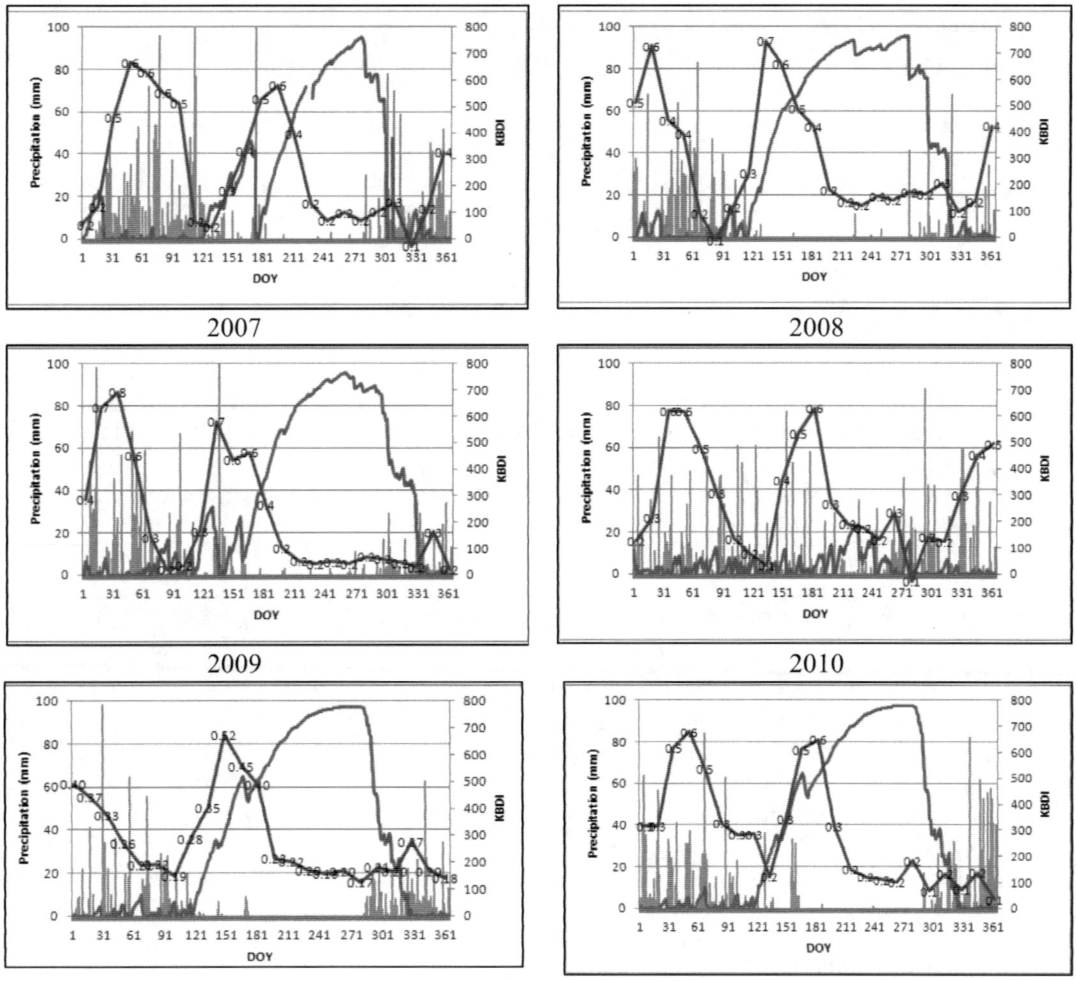

Figure 5. Seasonal variability of precipitation, KBDI and EVI 2007 – 2012 in West Java EVI (green line), precipitation (blue line) and KBDI (red line).

Basically In 2010, for all province got higher precipitation and smaller KBDI so that EVI has three peaks such as in West Java (Figure 4.) that can increase rice production. According to [20], paddy field production in Indonesia is subject to considerable climate variability that is mostly attributable to the El Niño-Southern Oscillation (ENSO) phenomenon.

4. Conclusion

We have showed seasonal variability of precipitation, KBDI and EVI. Precipitation begins from January until May and October until December, and KBDI begins to increase from June and peak in September only in South Sumatera precipitation almost in all month. Seasonal variability has showed precipitation and KBDI affect on EVI that can indicate variety phenology of Indonesia paddy field. Peak of EVI occurs before peak of KBDI occurs and increasing of KBDI followed by decreasing of EVI. In 2010 all province got higher precipitation and smaller KBDI so EVI has three peaks such as in West Java that can increase rice production. For future works, we will predict precipitation and drought and estimate rice production in study area.

Acknowledgment

I would like thanks to JAXA SAFE to support this study.

References

[1] Tobias A, Molina I, Valera H G, Mottaleb K A, and Mohanty S, 2012, *Handbook on rice policy in Asia*, International Rice Research Institute (IRRI), Philippines

[2] Simatupang P, and Timmer C P, 2008, Indonesian rice production: policies and Realities, *Bulletin of Indonesian Economic Studies*, **44:1**, 65-80

[3] Xiao X, Boles S, Frolking S, Li C, Babu J Y, Salas W, and Moore B, 2006, Mapping paddy rice agriculture in South and Shout East Asia using multi-temporal MODIS Image, *Remote Sensing of Environment* **100**, 95-113

[4] Shofiyati R, and Uchida S, 2011, Spatio Temporal Pattern Recognition of NDVI and TCT Wetness for Determining Cropping Type and Cropping Pattern of Paddy Fields, *Asian Journal of Geoinformatics*, Vol.**11**,No.3

[5] Shofiyati R, Ismullah I H, and Pawitan, 2011, Multi Sensor Satellite Data for Rice Production Estimation in an Effort to Support National Food Security, *J. Agric. Food. Tech.*, 1(**12**) 238-244

[6] Kani Sari D, Ismullah I H, Sulasdi W N, and Budi Harto A, 2010, Detecting Rice Phenology in Paddy Fields with Complex Cropping Pattern Using Time Series MODIS Data A Case study of Northern Part of West Java – Indonesia, *ITB J. Sci.* Vol. **42** A, No. 2, 91-106

[7] Jonai H, and W Takeuchi, 2012, Global rice paddy field mapping by integrating MODIS and AMSR-E measurementse, *33rd Asian conference on remote sensing (ACRS)*: Patthaya, Thailand

[8] Darmawan S, W Takeuchi, and R Shofiyati, 2013, Estimation of rice production and forecasting of agricultural drought in Indonesia based on modis data, *33rd Asian conference on remote sensing (ACRS)*: Bali, Indonesia

[9] Haylock M, and McBride J L, 2001, Spatial coherence and predictability of Indonesian wet season rainfall, *J Clim* **14**:3882–3887

[10] Quint W H, Zopf D O, Short K S, and Kuo Yang R T W, 1978, Historical trends and statistics of the Southern Oscillation, El Niño and Indonesian droughts, *Fish Bull* **76**:663–678

[11] Kubota T, S Shige, H Hashizume, K Aonashi, N Takahashi, S Seto, M Hirose, Y N Takayabu, K Nakagawa, K Iwanami, T Ushio, M Kachi, and K Okamoto, 2007, Global Precipitation Map using Satelliteborne Microwave Radiometers by the GSMaP Project : Production and Validation, *IEEE Trans. Geosci. Remote Sens.,* Vol. **45** No. 7 pp.2259-2275.

[12] Takeuchi W, Hirano T, Anggraini N, and Orbita R, 2010, Estimation of ground water table at forested peatland in Kalimantan using drought index towards wildfire control. *31st Asian conference on remote sensing(ACRS)*: Hanoi, Vietnam

[13] Huete A, Didan K, Miura T, Rodriguez E P, Gao X, and Ferreira L G, 2002, Overview of the radiometric and biophysical performance of the MODIS vegetation indices, *Remote Sensing of Environment,* **8**, 195– 213.

[14] Huete A, Justice C, and Liu H, 1994, Development of vegetation and soil indices for MODIS-EOS. *Remote Sensing of Environment,* **49**, 224–234

[15] Prasetia R, As-syakur A R, and Osawa T, 2013, Validation of TRMM Precipitation Radar satellite data over Indonesian region, *Theor Appl Climatol,* **112**:575–587

[16] Keetch J J, and Byram G M, 1968, A drought index for forest fire control, Res. Pap. SE-38. *Asheville NC: U.S. Department of Agriculture. Forest Service. Southeastern Forest Experiment Station*. 32 p. [Revised November 1988.]

[17] Oo K S. An V N, Takeuchi W, Lin S, and Win Z, 2012, Relating the kbdi with sea water intrusion to farm land, *33st Asian conference on remote sensing, (ACRS)*: Bangkok,Thailand, 26-30

[18] Xanthopoulos G, Maheras G, Gouma V, and Gouvas M, 2006, Is the Keetch-Byram drought index (KBDI) directly related to plant water stress ?, *V International Conference on Forest Fire Research D. X. Viegas*

[19] Aldrian E, and Susanto D R, 2003, Identification of three dominant rainfall regions within Indonesia and their relationship to sea surface temperature, *Int J Climatol* **23**:1435–1452

[20] Keil A, Zeller M, Wida A, Sanim B, and Birner R, 2008, What determines farmers' resilience towards ENSO related drought? An empirical assessment in Central Sulawesi, *Indonesia, Climatic Change,* **86**:291–307

Identification, evaluation and change detection of highly sensitive wetlands in South-Eastern Sri Lanka using ALOS (AVNIR2, PALSAR) and Landsat ETM+ data

Ajith Gunawardena[1], Tamasha Fernando[1], Wataru Takeuchi[2], Chathura H Wickramasinghe[3], and Lal Samarakoon[3]

[1] Geo-informatics Unit, Central Environmental Authority, Sri Lanka
[2] Institute of Industrial Science, University of Tokyo, Japan
[3]Geoinformatics Center, Asian Institute of Technology, Thailand

ajithr@cea.lk

Abstract. Sri Lanka is an island consists of numerous wetlands and many of these ecosystems have been indiscriminately exploited for a commercial, agricultural, residential and industrial development and waste dumping. Eastern River Basin Region in Sri Lanka is rapidly urbanizing, which leads more threats to the surrounding wetland ecosystems considerably. Therefore, it is important to identify and designated them as reserved areas where necessary in order to protect them under the National Environmental Act of Sri Lanka. Mapping and change detection of wetlands in the selected region is a key requirement to fulfill the above task. GIS and Remote Sensing techniques were used to identify and analyze the wetland eco systems. In this study Landsat ETM+, ALOS-AVNIR2, ALOS-PALSAR images were analyzed for identifying and change detection of wetlands. The secondary information and data were collected through a questionnaire survey to recognize the possible threats and benefits. The collected data and information were incorporated in identification, analyzing and ranking the wetlands. The final outcome of the project is to correlate the satellite data with the field observations to quantify the highly sensitive wetlands to declare as Environmental Protection Areas under the National Environment Act of Sri Lanka.

1. Introduction

Wetlands are the most important ecosystems all over the world it defined as areas of marsh, fen, peat land or water, whether natural or artificial, permanent or temporary, with water that is static or flowing, fresh, brackish or salt, including areas of marine water, the depth of which at low tides does not exceed six meters [9]. Wetlands provide many good services such as, reduce wave damage flood storage and conveyance, and sediment loading, erosion control, prevent and treat pollution, ground water recharge and discharge, produce crops and timber, habitat for wildlife and provide scenic beauty. Furthermore, it has historical, archaeological, heritage values and provides research opportunities too [3].

Currently many important wetlands have been overexploited for commercial, agricultural, residential and industrial development, directly affected to carrying capacity of water bodies and degradation of wildlife habitat. Not only that, natural phenomena such as spreading invasive plant on wetlands also becoming a major environmental problem [1].

Content from this work may be used under the terms of the Creative Commons Attribution 3.0 licence. Any further distribution of this work must maintain attribution to the author(s) and the title of the work, journal citation and DOI.
Published under licence by IOP Publishing Ltd

1.1. Current status and conservation strategies of wetlands in Sri Lanka

Coral reefs, Sand or salt flats, Marshes, Mangroves can be identified as marine and coastal wetland type in Sri Lanka while Fresh water swamp forest, Seasonal/intermittent freshwater ponds and marshes as inland wetlands [6]. Not only that, Sri Lanka is a classic example of "Hydraulic civilization" which had developed in 540 BC. Development of rice cultivation in the rolling planes of north and north central and the south east part of the country's in grand scale. Therefore, numerous large and small tanks linked with rice fields in a unique cascade system creating manmade wetlands. Sri Lanka is an island consists 65,610km^2 area having a 1340km long coastal belt with numerous types of wetland ecosystems rich with flora and fauna species, some are endemic and some listed as threatened in red list [5]. Most of the wetlands in North Western part are destructed for shrimp farming since late seventies. All wetlands comprise of a combination of soils, water, plants and animals. The interplay between these elements allow wetlands to perform several functions that are beneficial to human kind, while generating healthy wild life, fisheries and forest resources. The combination of these functions together with the fish biological diversity and cultural heritage of wetlands make these eco systems invaluable to people all over the country. The importance of wetlands can be elaborated into three major categories as uses, functions and attributes.

Majority of the wetlands in Sri Lanka are facing various threats that are posed by anthropological activities. According to the directory of Asian wetlands [8] the most frequently reported threats appears to be siltation. It should be realized that this is usually not caused by factors in the wetland itself by actions on land adjacent or away from wetlands. The development of aqua culture ponds is one of the serious threats to wetlands located in coastal areas. Habitat destruction, degradation direct loss, exploitation of species and spread of invasive alien species can be considered as other major threats on the wetlands.

Interest in conservation of wetlands can be traced back in the recent history to the late 19th century. The flora and fauna protection ordinance of 1937 can be considered as a major step in wetland conservation. Using this legislation, wetlands of importance to bird have been declared by the department of Wild life Conservation as sanctuaries and other protected areas. Sri Lanka signed the Ramsar Convention in 1971, but ratified it only in 1990. Interest in the development of the Muthurajawela Marsh was a major activity which led to a comprehensive study of the wetland and the preparation of an ecological profile and a management plan and the report was published in 1992. Since 1987 it was recognized that wetlands were falling under the responsibility of numerous agencies and such the need to introduce coordination was recognized. Accordingly by 1989, an interim wetland steering committee was established by the Central Environmental Authority.

During the past three decades the status of the wetlands in the Eastern and Northern provinces remained unknown and presently the areas are now rapidly populated and development projects are carried out in larger scales leads to the rapid exploitation of wetlands. Therefore, it needs to take immediate action to protect important wetlands under the National Environmental Act of Sri Lanka. To fulfill the above task, mapping and change detection of wetlands in a selected region is a key requirement.

2. Objectives

- To mapping and change detection of wetlands in South Eastern River Basin Region in Sri Lanka using Optical and PALSAR remote sensing techniques
- To Identification of important wetlands in the Eastern River Basin Region in Sri Lanka to declare as environmental protection areas under the National Environmental Act
- To Update the National Wetland Directory

The project is conducted by the Research and Development Unit (Geo-informatics) Unit of the Central Environmental Authority collaboration with the Japanese Aerospace Agency (JAXA) and the Geo-informatics Centre, Tokyo University, Japan, Asian Institute of Technology, Thailand. Survey Department of Sri Lanka, Department of Forest Conservation, Department of Wildlife

Conservation, Department of Irrigation, and National Wetland Steering Committee are the facilitators in this project.

2.1. Study area

Sri Lanka consists of 103 river basins according to the presence of major annual and perennial rivers. According to the "*Mahaweli*" management plan these basins are grouped in five (5) major river basins. The selected river basin is located in Eastern part of the country called "South- Eastern River basin" which consists different kinds of wetland systems including lagoon, estuary, marshland, man-made tanks, large reservoirs and streams. The total land area of selected river basin is abound 8816 km^2 and having 23 sub river basins.

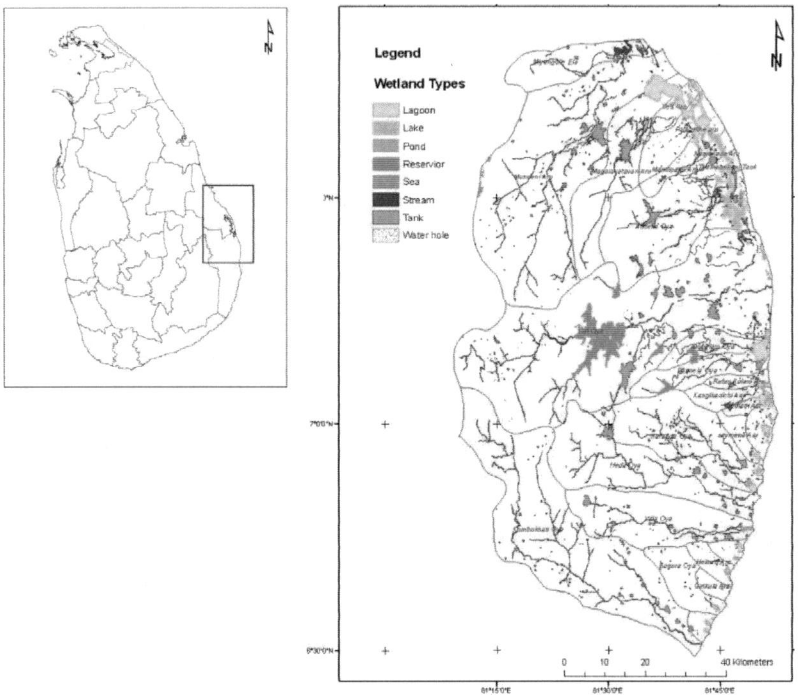

Figure 1. Wetland distribution of South-Easters river basin of Sri Lanka.

3. Methodology

Topographic maps (1:50,000) prepared by The Survey Department of Sri Lanka have been used to identification of sites; together with the River Basin map of the Mahaweli Authority of Sri Lanka. Accordingly research boundary which was South Eastern River basin of Sri Lanka is delineated. Preliminary land cover maps used in the field verification are prepared with the Landsat ETM+ images. Field survey carried out in order to obtain GPS locations of surrounding environment and obtain information from the community directly involved with wetlands for their living and that data were used to update land cover maps. While conducting field survey, identified wetlands were subjected to detail study in respect to wetland valuation method (table 2) developed by the Wetland Steering Committee, Sri Lanka. Field verified data were then in cooperated with the Landsat ETM+ and ALOS-AVNIR2 images and classified with supervised classification method to identification of wetlands and water bodies while ALOS-PALSAR were also used for identification of wetlands with supervised classification. Accuracy assessment of the classified three images has been done using field verified data. Areas of land cover types have been calculated separately for the identification of land cover changes in past decade. Statistical analysis was done for the identification of unique wetlands to be conserved through establishment of protection areas under the National Environmental Act (NEA) of Sri Lanka. According to the study identified 42wetland eco systems were subjected to detail study including biological, physical and socio-economic parameters collected in questionnaire survey during the field data collection. The significant importance of each wetland is obtained and ranked for further study and to be declared as Environmental Protection areas as per the NEA in future. Furthermore, identification of mangrove areas and calculation of above ground level biomass through remote sensing techniques has to be done in future [4].

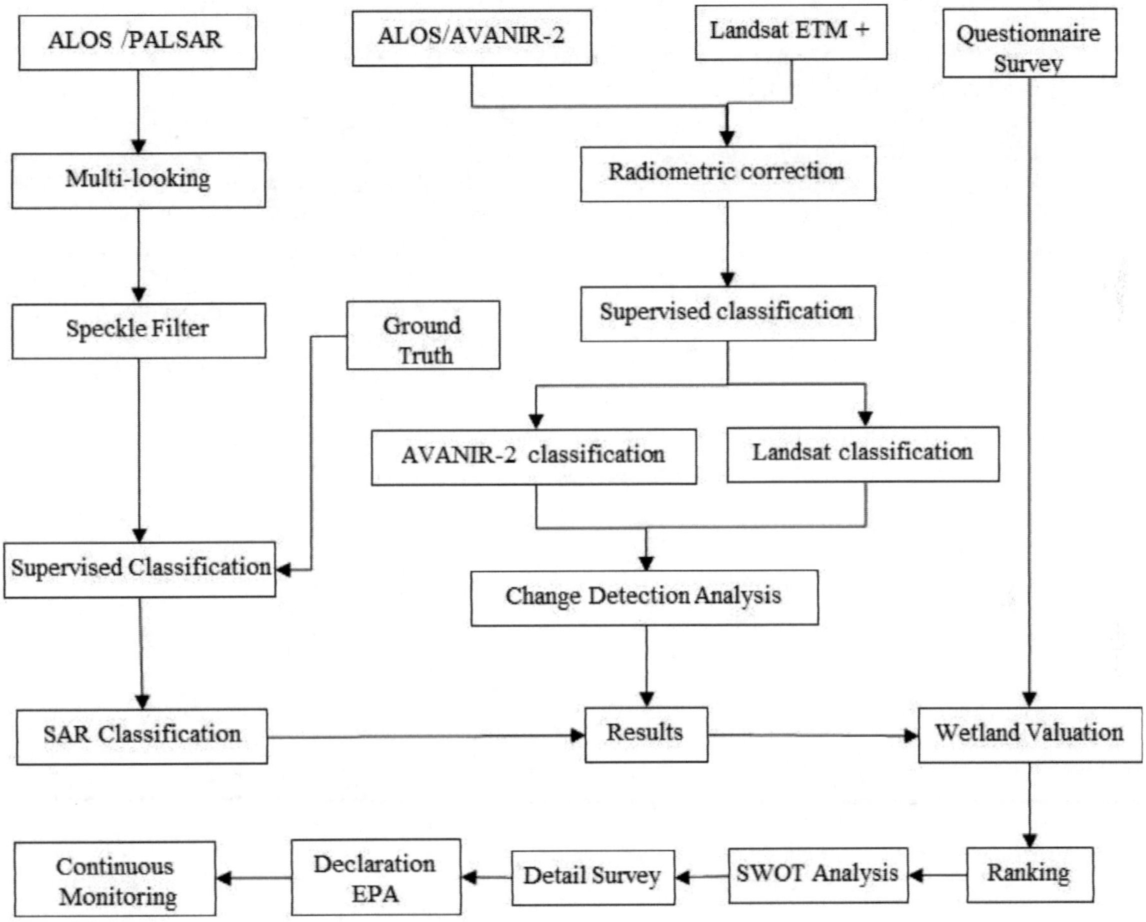

Figure 2. Image analysis methodology for change detection of wetlands and declaration of EPA process.

4. Results and discussion

The current and previous status, extent, conditions of the wetlands will be identified using past Landsat ETM+ (2005) and new satellite images (ALOS-AVNIR2 optical data from February/2005 and July/2009 respectively). Accuracy assessment was done using the GPS data and overall accuracy of wetlands and water bodies is over 85%. The current land use map created using the topographic maps and the major land cover type listed in the table1. The majority of the land cover type belongs to forest, paddy and homestead in selected study area.

Three remote sensing, sensors were used for land cover classification and change detection. These included Landsat ETM+, ALOS-AVNIR2 optical data from February/2005 and July/2009 respectively. ALOS-PALSAR data of March/2009 was also used to evaluate feasibility of SAR data for wetland mapping. As shown in the flow chart below supervised classification technique was used for both optical and SAR data. Ground truth data was vital for the classification of the PALSAR data as it's nearly impossible to visually distinguish land cover using SAR data. For optical data a combination of collected field data and visual analysis of the images were used.

4.1. Landsat data classification
Land cover classification in year 2005 showed the study area which is mostly covered by paddy fields and scrub. The Landsat data was highly accurate in identifying the wetland with an accuracy of 72%. Overall Landsat classification accuracy was 80%.

4.2. ALOS AVNIR2 classification
Accuracy assessment in the classification for ALOS-AVNIR2 showed 79% accuracy. Overall analysis shows some water bodies at the bottom left of the map is misclassified as mangrove. Lack of Short Wave Infrared band in ALOS-AVNIR2, which produces unique spectral signature that enables to distinguish mangroves from other vegetation types, makes it difficult to identify mangroves. However, mangrove areas are accurately demarcated using high resolution IKONOS images with the GPS ground sampling locations (Figure 5.)

Table 1. The extent of land cover type in the south eastern river basin derived from the topographic maps.

Land Cover	Area (ha)
Chena	59,133
Forest	324,923
Home garden	107,544
Paddy	139,500
Rubber	2,552
Stream	11,253
Tank	18,156
Tea	2,770
Other Plantation	25,671
Other Area	352,869
Total	1,044,370

4.3. Accuracy assessment of image classification
The accuracy of Landsat ETM+ data for mangrove mapping is 71% in contrast to the ALOS-AVNIR2 only showed an accuracy of 45%. However, ALOS-PALSAR can be used to identifying the shallow water body to an accuracy of 80% compared to the 70% accuracy in Landsat classification accurately. Other land cover classes had very similar accuracy level. According to the graph Landsat ETM+ was the most suitable for identification of different land use types with higher accuracy while the AVNIR2 was moderately suitable in this regard. As the island is within the tropical region the upper atmosphere was covered with clouds throughout the year and microwave images like ALOS-PALSAR was more suitable in delineation of forest and water logged area from other land cover types. It was observed, the dual polarization penetrates forest lands it was easily identifiable from

other features. Thermal band can be recommended for identification of rocky and highly urbanized areas as they emit more thermal energy during the day time.

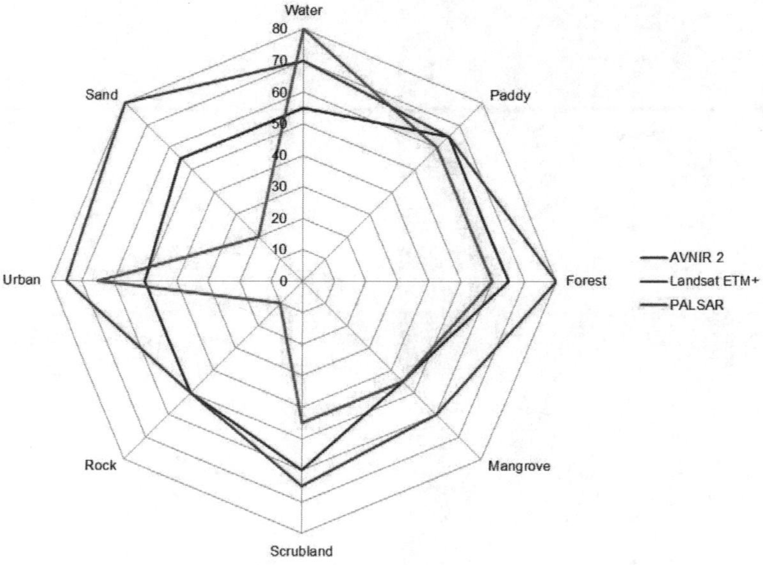

Figure 3. Accuracy comparisons of Landsat (ETM+), AVNIR2 and PALSAR land cover classification.

Detailed study was done for the evaluation of each wetland type using the standard valuation method developed by the National Wetland Steering committee of Sri Lanka. According to that, three major parameters identified as attributes, functions and uses with sub categories under the different weighted values. Using SPSS statistical software the data are analyzed and find the significant values. The highest value was recorded from *Nawagiriyawa* Tank (Value 75.54) due to very high biological value while providing a source of drinking water. There were 14 wetland systems selected for further study in identifying the potential to establishment as environmental protection areas with the value over 60, considering the vulnerability. Land development and urbanization are considered as the major threats of degradation of wetlands.

The feasibility study on selected wetlands is carrying out to obtain physical, biological and socio-economical parameters to propose at the National Wetland Steering Committee to establish EPA under the National Environmental Act of Sri Lanka. The strength, weakness, opportunities and threats have been identified (SWOT) during the field survey in the selected wetlands (annexed 01 and 02) areas [7]. At the final stage zoning maps for possible threats from hap hazardously utilization of the area for numerous activities have been developed together with the proposed Environmental Protection Area map (figure 5) . Also zoning maps for possible expansion with the increase of population for settlements, tourism, agriculture and infrastructure have been developed for sustainable environmental management.

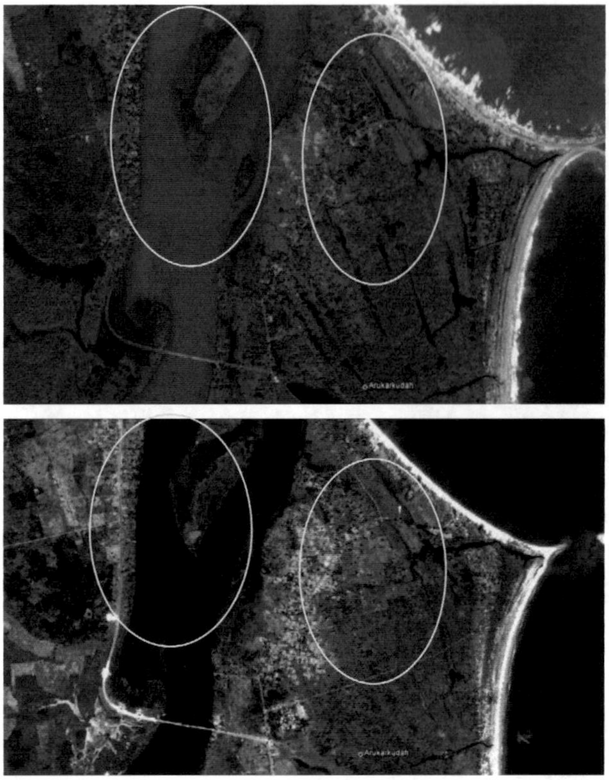

Figure 4. Land development around the *Panichchankerni* Lagoon over the past 6 years and it is one of the key threats to destruction of mangroves.

Table2. Parameters considered on wetlands valuation.

	Attributes
1	Biological Diversity
2	Historical and Cultural Value
3	Scientific Value
4	Uniqueness
	Function
5	Water Quantity Regulation
6	Water Quality Regulation
7	Habitat for Fish
8	Habitat for Wildlife
	Uses
9	Plant production
10	Animal
11	Mineral
12	Water Storage Supply / energy
13	Truism Recreation
14	Research Education
15	Waste Water Disposal
16	Land Development

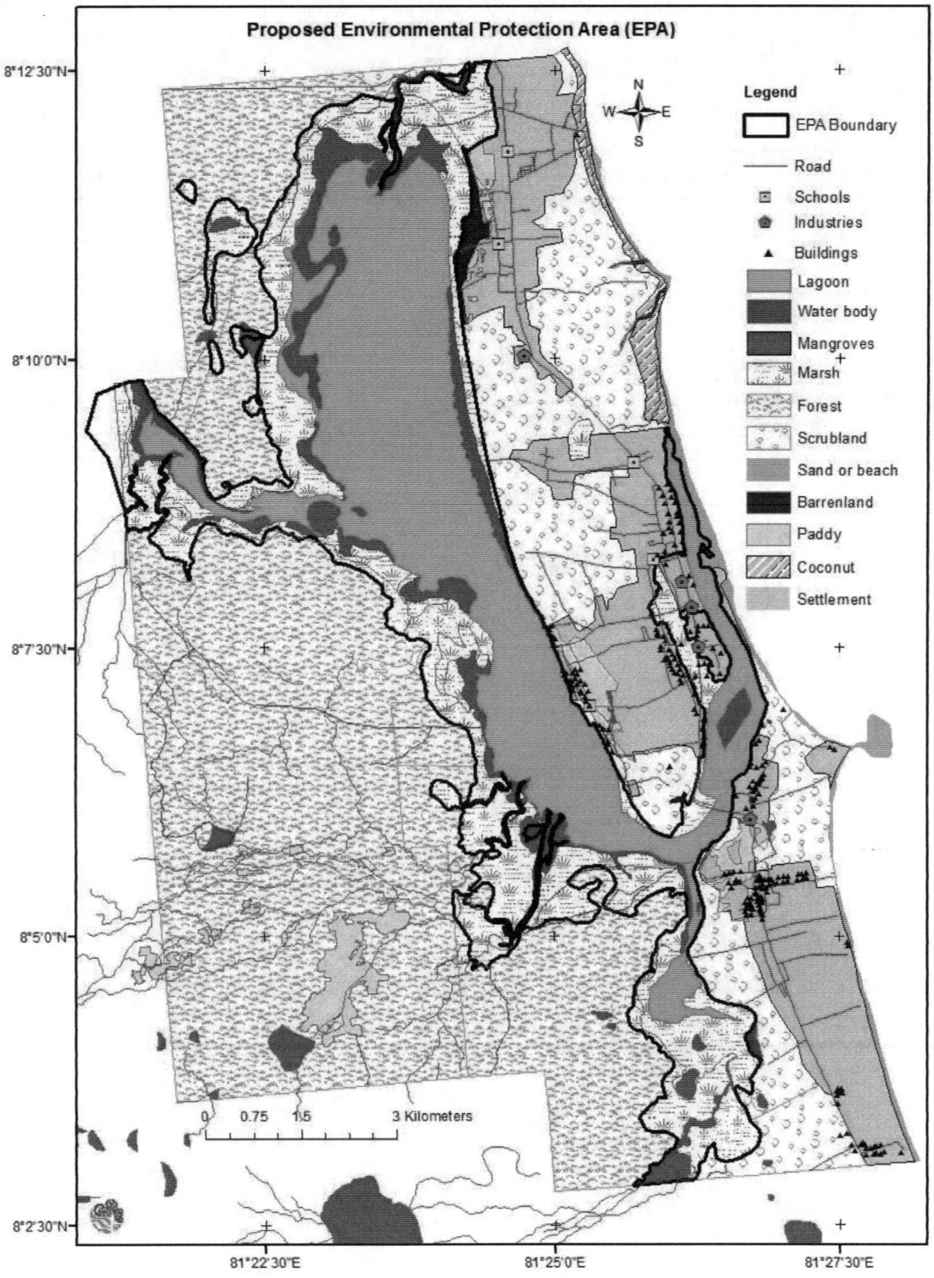

Figure 5. Proposed EPA for *Panichchankerni* Lagoon which is rich in mangroves and biological diversity.

5. Conclusion

There have been identified forty two (42) important wetland ecosystems during the field visits in the study area. The collected information through the questionnaire survey have been considered to identification of the most vital and threatened wetlands in the South eastern river basin. The selected seven wetlands are then analyzed using Landsat ETM+, AVNIR2 and PALSAR data and it has been revealed that the Landsat ETM+ is more eligible to spot out wetland eco systems and to avoid the

IOP Conf. Series: Earth and Environmental Science **20** (2014) 012050 doi:10.1088/1755-1315/20/1/012050

atmospheric disturbances ALOS - PALSAR have been used and it has been revealed that the polarized data are more suitable in delineation of forest and vegetative cover while AVNIR2 is moderately suitable in this methodology.

Finally, it has been done a statistical analysis to rank the identified wetlands according to the importance. At the final stage a SWOT analysis has been initiated to prepare guidelines in order to implement the areas as Environmental Protection Areas as per the National Environmental Act of Sri Lanka. Monitoring has been conducted by the CEA from time to time in order to prevent the violations of the act and further environmental implementations.

References

[1] CEA/Euro-consult (Netherlands), 1994a *Wetlands are no wastelands*, wetland conservation project, Sri Lanka

[2] CEA/IWMI/IUCN, 2006 National wetland directory of Sri Lanka, the central environmental authority, the world conservation union and the international water management institute, Sri Lanka

[3] IUCN, 2003 Wetland conservation in Sri Lanka. *Proceeding of the Symposium on wetland conservation and management. Ministry of environment , Colombo , Sri Lanka.*

[4] Gunawardena, A.R., Nissanka, S.P. and Dayawansa, N.D.K. 2006 *Relationship between above ground live biomass and satellite spectral responses. J. Trop. Agric. Res.* Vol. **18**:334-345

[5] MOE 2012. The National Red List 2012 of Sri Lanka; Conservation Status of the Fauna and Flora. Ministry of Environment, Colombo, Sri Lanka. viii + 476pp

[6] Karunathilake, K. M. B. C., 2003 *Status of mangroves in Sri Lanka. J. Co. Dev. ISSN*: 1410-5217, Vol. **7**, No 1

[7] Kustanti, A., Nugrobo, B. , Darusman, D .and Kusmana, C, 2009 *Integrated management of mangrove ecosystems in Lumpung mangrove centre* (LMC) East Lumpung regency, Indonesia. *J. Co. Dev.* Vol. **15**

[8] Scott DA 1989 A directory of asian wetlands, the world conservation union (IUCN), Cambridge, UK

[9] The convention on wetlands (Ramsar) 1971 *An active player in the fight against poverty*, http://ramsar.org/key_sgf_index.htm

Annexure 1: Internal and external strategy factors

	Weight	Rate	Scoring
Strengths			
1. Low population density	0.3	2	0.6
2. Less urbanization	0.2	1	0.2
3. Difficult to access some wetlands by community	0.3	2	0.6
4. Some areas are administrate under the security forces	0.4	4	1.6
5. Government and NGO support	0.4	3	1.2
6. Community support	0.5	4	2.0
7. Pollution control activities	0.4	4	1.6
Weaknesses			
1. Some important wetlands are privately owned	0.2	3	0.6
2. Exiting polluting Industries	0.3	2	0.6
3. Attitudes changing not easy	0.4	2	0.8
4. Lack of coordination	0.1	3	0.3
5. Not aware community well	0.2	3	0.6
6. Political influences	0.4	3	1.2
Total			11.9
Opportunities			
1. CEA implementation and WSC support	0.4	5	2.0
2. Other stakeholder support	0.3	4	1.2
3. Most of the wetlands are state owned	0.2	3	0.6
4. Allow Traditional fishing	0.1	4	0.4
5. Allow ecotourism	0.2	4	0.8
Threats			
1. Land filling	0.2	2	0.4
2. Encroachments	0.3	5	1.5
3. Waste disposal	0.2	5	1.0
4. Deforestation	0.2	2	0.4
5. Shrimp farming	0.4	2	0.8
6. Invasive species distribution	0.3	3	0.9
Total			10.0
Grand Total			22.1

Annexure 2: Proximity matrix of SWOT analysis

Proximity Matrix- Strength

Case	Squared Euclidean Distance						
	Low population	Less urbanization	Difficult to access	Security Control	Government and NGO	Community support	Pollution Control
Low population	0.00	1.17	0.00	5.00	1.40	6.00	5.00
Less urbanization	1.17	0.00	1.20	11.00	5.00	12.30	11.00
Difficult to access	0.00	1.17	0.00	5.00	1.40	6.00	5.00
Security Control	5.01	11.00	5.00	0.00	1.20	0.17	0.00
Government &NGO	1.37	5.04	1.40	1.20	0.00	1.65	1.20
Community support	6.00	12.30	6.00	0.20	1.70	0.00	0.20
Pollution Control	5.01	11.00	5.00	0.00	1.20	0.17	0.00

Proximity Matrix - Weakness

Case	Squared Euclidean Distance					
	Ownership	Polluting Industries	Attitudes	Lack of coordination	Not aware community	Political influences
Ownership	0.00	1.00	1.10	0.10	0.00	0.40
Polluting Industries	1.00	0.00	0.10	1.10	1.00	1.40
Attitudes	1.10	0.10	0.00	1.30	1.10	1.20
Lack of coordination	0.10	1.10	1.30	0.00	0.10	0.90
Not aware community	0.00	1.00	1.10	0.10	0.00	0.40

Proximity Matrix -Threats

Case	Squared Euclidean Distance					
	Land filling	Encroachments	Waste disposal	Deforestation	Shrimp farming	Invasive species
Land filling	0.00	10.22	9.36	0.00	0.20	1.26
Encroachments	10.22	0.00	0.26	10.22	9.50	4.36
Waste disposal	9.36	0.26	0.00	9.36	9.08	4.02
Deforestation	0.00	10.22	9.36	0.00	0.20	1.26
Shrimp farming	0.20	9.50	9.08	0.20	0.00	1.02
Invasive species	1.26	4.36	4.02	1.26	1.02	0.00
Political influences	0.40	1.40	1.20	0.90	0.40	0.00

Proximity Matrix

Case	Squared Euclidean Distance				
	CEA/WSC Support	Stakeholders Support	State ownership	Traditional Fishing	Eco-tourism
CEA/WSC Support	0.00	1.65	6.00	3.65	2.48
Stakeholders Support	1.65	0.00	1.37	0.68	0.17
State ownership	6.00	1.37	0.00	1.05	1.04
Eco-tourism	3.65	0.68	1.05	0.00	0.17
Eco-tourism	2.48	0.17	1.04	0.17	0.00

IOP Conf. Series: Earth and Environmental Science **20** (2014) 012051 doi:10.1088/1755-1315/20/1/012051

Characterization of mangrove forest types based on ALOS-PALSAR in overall Indonesian archipelago

S Darmawan[1,3,4], W Takeuchi[1], Y Vetrita[2], G Winarso[2], K Wikantika[3] and D K Sari[4]

[1]Institute of Industrial Science, University of Tokyo, 4-6-1 Komaba Meguro-ku, Tokyo, 153-8505, Japan.
[2]Indonesian National Institute of Aeronautics and Space (LAPAN), Jalan Kalisari No. 8 Pekayon Pasar Rebo, Jakarta Indonesia.
[3]Center for Remote Sensing, Institut Teknologi Bandung (ITB), Jalan Ganesha 10, Bandung, West Java, Indonesia.
[4]Institut Teknologi Nasional (ITENAS), Jalan PHH. Mustafa No. 23, Bandung, West Java, Indonesia.

Email : soni@iis.u-tokyo.ac.jp and soni_darmawan@yahoo.com

Abstract. Indonesia has largest mangrove forest in the world, total area around 3.5 million ha or 17% - 23% from mangrove forest in the world. Mangrove forest provides products and services, such as carbon balance of the coastal zone. Mapping and monitoring biomass of mangrove is very important but field survey of mangrove biomass and productivity in overall Indonesia is very difficult. Global-scale mosaics with HH and HV backscatter of Advanced Land Observing Satellite (ALOS) Phased Array L-band Synthetic Aperture Radar (PALSAR) which is 50-m spatial resolution has been generated. This image available for identification and monitoring mangrove forest. The Objective of this research to investigate characterization of mangrove forest types based on ALOS-PALSAR in overall Indonesian archipelago. Methodology consists of collecting ALOS-PALSAR image for overall Indonesian archipelago, preprocessing and mosaicking, collecting region of interest of mangrove forest, plotting, ground survey, characterization and classification. The result of this research has showed characterization of mangrove forest types based on ALOS-PALSAR. Indonesian mangrove forest types has HH value around -10 dB until -7 dB and HV value around -17 dB until -13 dB. Higher of HH and HV backscatters value indicated higher of level biomass. Based on scatter plot of HH and HV, Indonesian mangrove forest can be classified in three level biomass. Generally level biomass of mangrove forest in Indonesia archipelago is moderate.

1. Introduction

The term 'mangrove' is used to define both the plants that occur in tidal forests, and to describe the community itself [1],[2],[3]. According to Kuenzer et al., [4] mangrove ecosystems provide important products and services that can be divided into four categories ; regulating (e.g., shoreline protection), providing (e.g., fisheries, aquaculture, construction material, fuel, tannins, honey, traditional medicine, paper, and textiles), cultural (e.g. tourism and recreation, spiritual; and supporting (e.g., nursery habitats, nutrient cycling). Indonesia has largest mangrove area around 3.5 million ha or around 17% - 23% from large in the world [5]. In recent years global mangrove loss of more than 50%. Industrial lumber and wood chip operations, increasing human populations, industrialization, and agriculture have caused dramatic forest loss as well. In addition to the natural progression and succession stages

Content from this work may be used under the terms of the Creative Commons Attribution 3.0 licence. Any further distribution of this work must maintain attribution to the author(s) and the title of the work, journal citation and DOI.
Published under licence by IOP Publishing Ltd

IOP Conf. Series: Earth and Environmental Science **20** (2014) 012051 doi:10.1088/1755-1315/20/1/012051

of the ecosystem, a significant amount of the loss is triggered by natural forces, such as tsunamis, cyclones, and the threat of global warming. The related reduction in mangrove-related services and product delivery imposes serious limitations on the local residents [4].

Mapping and monitoring biomass of mangrove forest is very important but field survey of mangrove biomass and productivity in overall Indonesia is very difficult due to muddy soil condition, heavy weight of the wood [6] and very large area. According to Kuenzer et al. [4] remote sensing has been widely proven to be essential in monitoring and mapping highly threatened mangrove ecosystems. Many research studies on this subject have been carried out around the globe. Tropical and subtropical coastal mangroves are among the most threatened and vulnerable ecosystems worldwide. According to Lucas et al., [7] the launch of the Japanese Space Exploration Agency's (JAXA) Advanced Land Observing Satellite (ALOS) Phased Arrayed L-band SAR (PALSAR) in 2006 therefore represented a milestone in the global observation, characterization, mapping and monitoring or mangroves largely, because these provide more information on the three- dimensional structure and biomass of woody vegetation and the presence and extent of (primarily tidal) inundation. As data can be day or night regardless of weather conditions, mangroves can be observed more frequently, even in regions with prevalent cloud cover. L-band microwaves (wavelength approximating 25 cm) emitted by the ALOS-PALSAR penetrate through the foliage and interact primarily with the woody components of vegetation. Horizontally transmitted waves are either depolarized through volume scattering by branches in the canopy, with a proportion of vertically polarized microwaves returning to the sensor, or penetrate through the canopy and interact with the trunks, returning primarily through double bounce scattering as a horizontally polarized wave.

The Objective of this research to investigate characterization of mangrove forest types based on ALOS-PALSAR in overall Indonesia archipelago. Methodology consists of collecting ALOS-PALSAR image for overall Indonesia archipelago, preprocessing and mosaicking, collecting region of interest of mangrove forest, ground survey, plotting, characterization and classifications.

2. Methodology

Indonesia also called "Nusantara" is a country in Southeast Asia that is located on 6° N – 11° S and 95° E - 141° E, between the Asian and Australian continents and between the Pacific and Indian Oceans. Indonesia Consisting of 17,508 islands that is largest archipelagic country in the world. Indoensia have large area around 1.904.569 km^2 which population around 237,6 million people in 2012, total province is 33 provinces. (www.indonesia.go.id). For this study that can see in the figure 1, we focus on 20 sites mangroves forest in Indonesia that is (a) Langsa, Aceh, (b) Bengkalis, Riau,

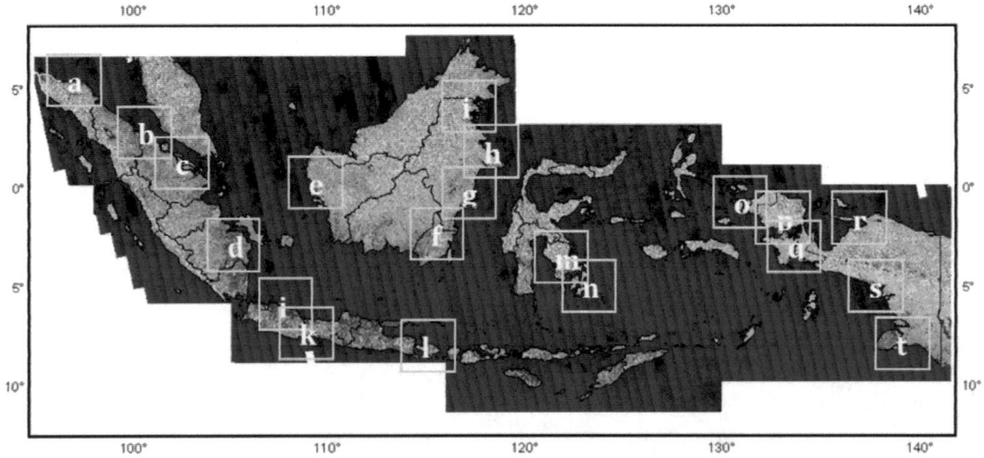

Figure 1. Study area dispersion of 20 sites (a - t).

(c) Indragiri Hilir, Riau, (d) Banyuasin, South Sumatera, (e) Pontianak, West Kalimantan, (f) Kotabaru, South Kalimantan, (g) Kutai Kartanegara, East Kalimantan, (h) Berau, East Kalimantan, (i) Nunukan, East Kalimatan, (j) Subang, West Java, (k) Cilacap, Central Java, (l) Badung, Bali (m) Bombana, Southeast Sulawesi (n) Muna, Southeast Sulawesi, (o) Sorong, West Papua, (p) Teluk Bintuni, Papua, (q) Teluk Bintuni, Papua, (r) Waropen, Papua, (s) Asmat, Papua, (t) Merauke Papua. We choose their location because their site has large area of mangrove forest that can describes mangrove forest types in overall Indonesia.

Generally mangrove can be found throughout in the Indonesian archipelago. Largest mangrove found in Papua around 1.3 million ha (38%), Kalimantan around 978 thousand ha (28%) and Sumatera 673 thousand ha (19%) [5]. In Indonesia archipelago mangroves grow and thrive well on the beach that has a large river and protected [5]. According to Rusila et al., [5] in Indonesia mangrove can be categorized in fourth zone; first open zone, which is mangrove zone front of in the sea and very effected by water from the sea. Majority of mangrove type is *Sonneratia alba.* Second middle zone, which is mangrove zone behind open zone. Majority of mangrove type is *Rhizophora.* Third payau zone, which is mangrove zones along the river. Majority of mangrove type is *Nypa* or *Sonneratia.* Fourth land zone, which is mangrove zone behind middle zone and payau zone, in this zone mangrove is very little affected by water from the sea. This zone has variety species than the other zones.

For this study, methodology consists of collecting ALOS-PALSAR image for overall Indonesia archipelago, preprocessing and mosaicking, collecting region of interest of mangrove forest, ground survey, plotting, characterization and classifications. The Methodology can see in the figure 2.

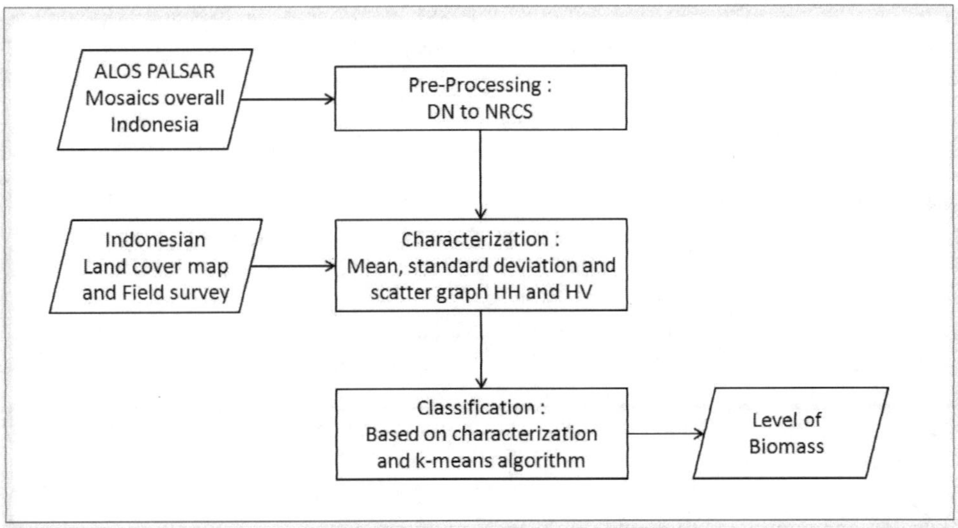

Figure 2. Methodology for characterization of mangrove forest on ALOS PALSAR.

The methodology in figure 2 can be described as follows :

2.1 Data collection

We used dual polarization ALOS PALSAR L-band HH and HV spatial resolution 50 m orthorectified mosaic product in 2008 from Kyoto and Carbon Initiative. We collected scene of Sumatra, Kalimantan, Sulawesi, Java and New Guinea to cover overall Indonesia archipelago. All data were acquired in fine beam dual mode at a viewing angle of 34.4 and delivered in single-look complex (SLC) as the normalized backscattering coefficient in slant-range geometry by JAXA. Secondary data we used land cover map from Indonesian ministry of forestry and Indonesian base map from Badan Informasi Geospasial (BIG).

2.2 Pre-processing

We collected ALOS PALSAR from http://www.eorc.jaxa.jp/ALOS/en/kc_mosaic/kc_map_50.htm, totally 28 images to cover overall Indonesia archipelago. All images are projected into geographical latitude and longitude coordinates. The digital number (DN) value of HH (DN_{HH}) and HV (DN_{HV}) backscatter intensities were converted to a normalized radar cross section (NRCS), (i.e., $\sigma°_{HH}$ and $\sigma°_{HV}$) by the following equations [8] :

$$\sigma°_{HH} = 10 \text{ x } \log_{10}(DN^2_{HH}) - 83.2 \quad (1)$$

$$\sigma°_{HV} = 10 \text{ x } \log_{10}(DN^2_{Hv}) - 80.2 \quad (2)$$

The 28 image were mosaicked into one continues image to cover overall Indonesia archipelago using the image analysis software. For visualization we used image composite with R=HH, G=HV and B=HH/HV.

2.3 Charactization and classification

We collected Region of Interest (ROI) of mangrove forest by land cover map from Indonesian ministry of forestry and Indonesian base map from Badan Informasi Geospasial (BIG). We collected 20 ROI of mangrove forest dispersion on overall Indonesia archipelago. And then calculate mean and standardeviation. After we known characterization of mangrove forest, we classify mangrove forest using k-means algorithm.

3. Results and discussion

Mangrove is plants that occur in tidal forests, and to describe the community itself. Mangrove can be found throughout in the Indonesian archipelago. We pre-processed 28 image ALOS-PALSAR include converted DN to NRCS, Mosaicked and then created image composite with R=HH, G=HV and B=HH/HV. The image cover overall Indonesia archipelago. Based on ALOS-PALSAR composite we can identify mangrove forest. In the figure 3 we show dispersion of mangrove forest from ROI in Indonesia archipelago.

Generally visual interpretation of mangrove forest based on ALOS-PALSAR mosaic very easy. In figure 3, we show also polygon of mangrove forest (white line) from land cover map derived from Indonesia ministry of forestry. Based on ALOS-PALSAR mosaic and land cover map as secondary data we can classify mangrove forest and non-mangrove forest.

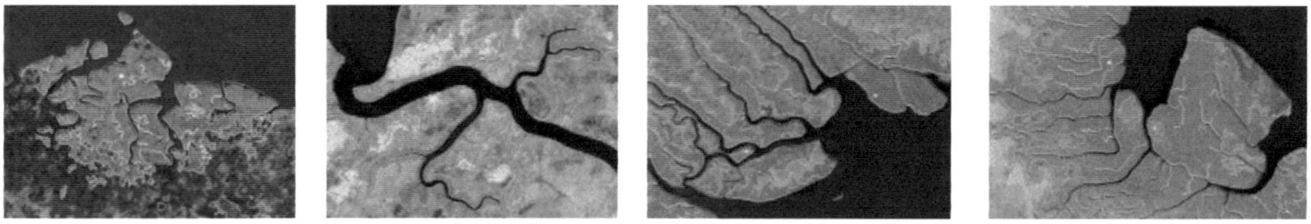

a.Langsa, Aceh **b**.Bengkalis Riau **c**.Indragiri, Riau **d**.Banyuasin, South Sumatera

IOP Conf. Series: Earth and Environmental Science **20** (2014) 012051 doi:10.1088/1755-1315/20/1/012051

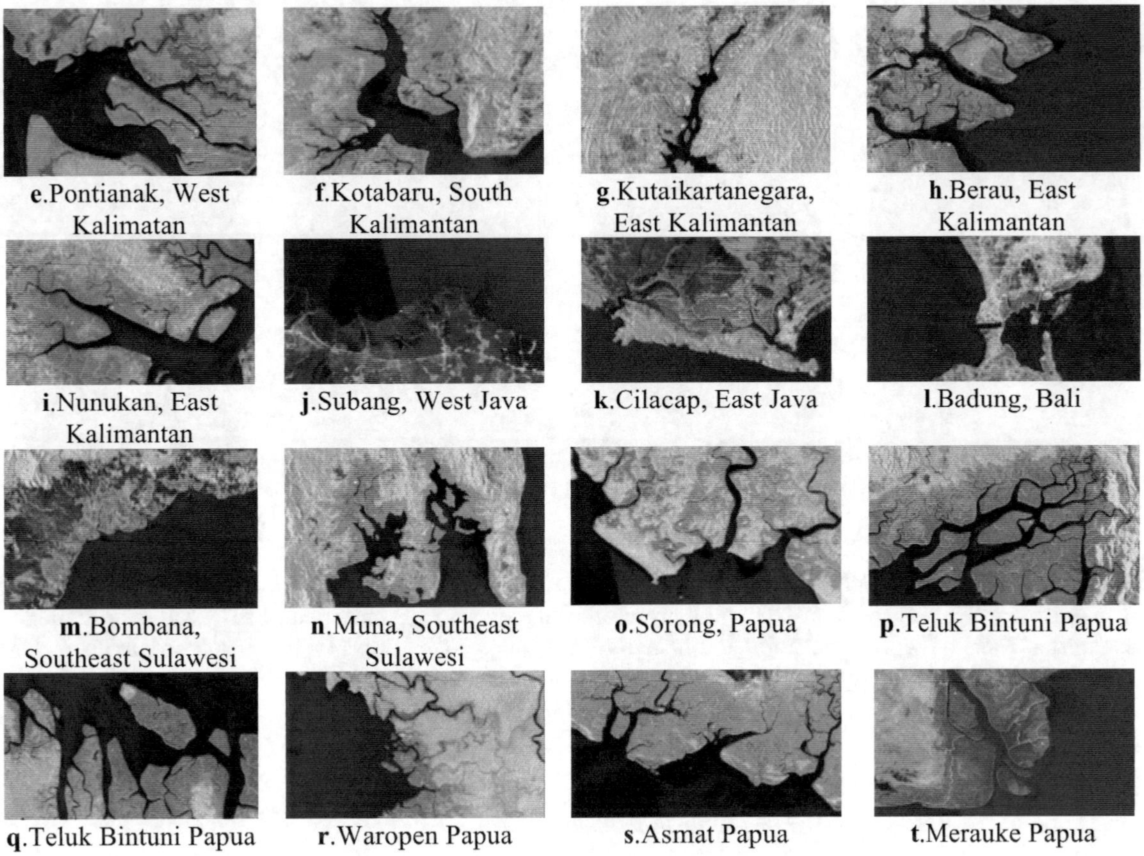

Figure 3. Composite of ALOS-PALSAR, R=HH, G=HV, B=HH/HV.

We collected ROI to know characterization of mangrove forest based ALOS-PALSAR. ROI spread overall Indonesia archipelago from (**a**) Langsa Aceh until (**t**) Merauke Papua. And then calculate mean and standard deviation from each ROI. Mean and standard deviation value for each ROI can see in the figure 4.

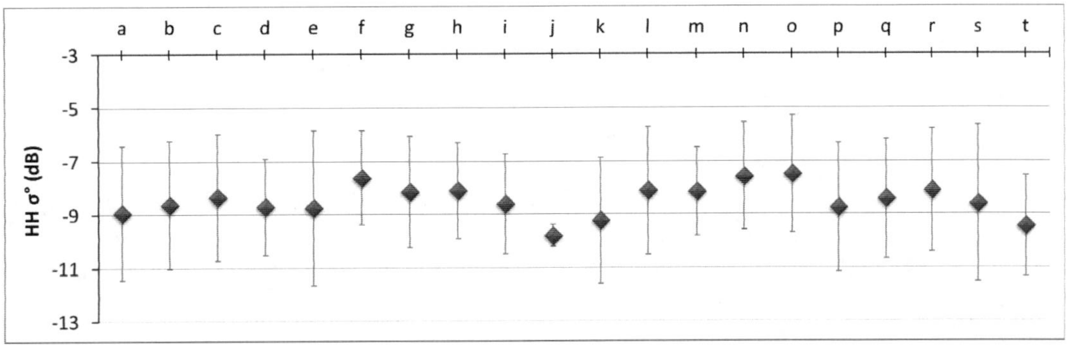

(a)

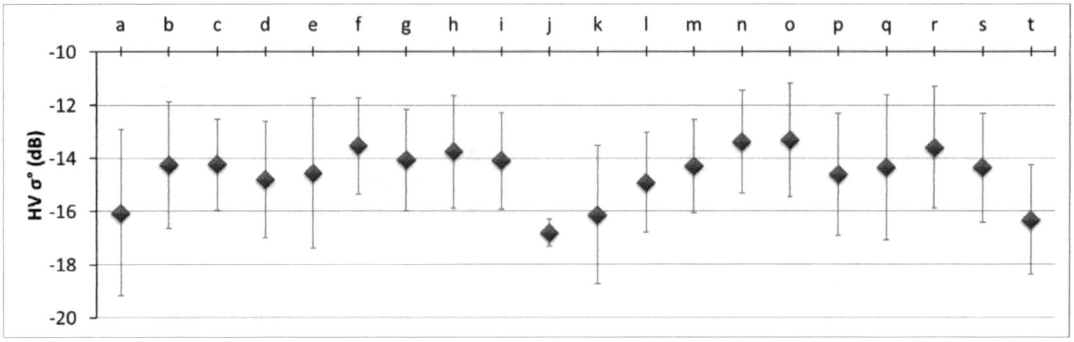

(b)

Figure 4. Mean and standard deviation of backscatter coefficients (a).HH and (b). HV.

In figure 4, we had seen characterization backscatter coefficient of mangrove forest. Characterization of mangrove forest for mean of HH value around -10 dB until -7 dB and mean of HV value around -17 dB until -13 dB. According scatter graph in figure 5, mangrove forest in Indonesia archipelago that can be classified on three classes. According to Takeuchi et al., [6] the higher of HH and HV value has the higher of tree and this condition can be indicated higher of biomass.

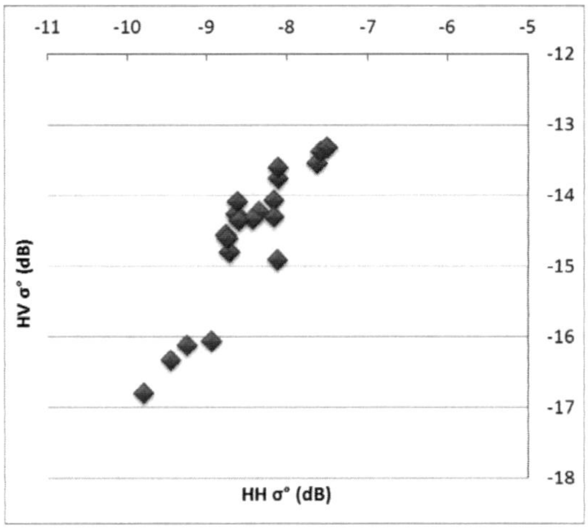

Figure 5. Scatter graph HH and HV.

Based on characterization, we classify of mangrove forest on three levels of biomass that are low biomass, moderate biomass and height biomass. According Rusila et al.,[5] type of mangrove majority In Indonesia in open zone and payau is *Sonneratia alba and Nypa* (low biomass), middle zone is *Rhizophora* (moderate biomass), and land zone tree mangrove (highest biomass) and According Kusmana et al.,[9] Indonesia has three species majority of mangrove that are *Rhizophora, Bruguiera,* and *Ceriops.* Classification of mangrove forest based on characterization of ALOS PALSAR on three level biomass can see in the figure 6.

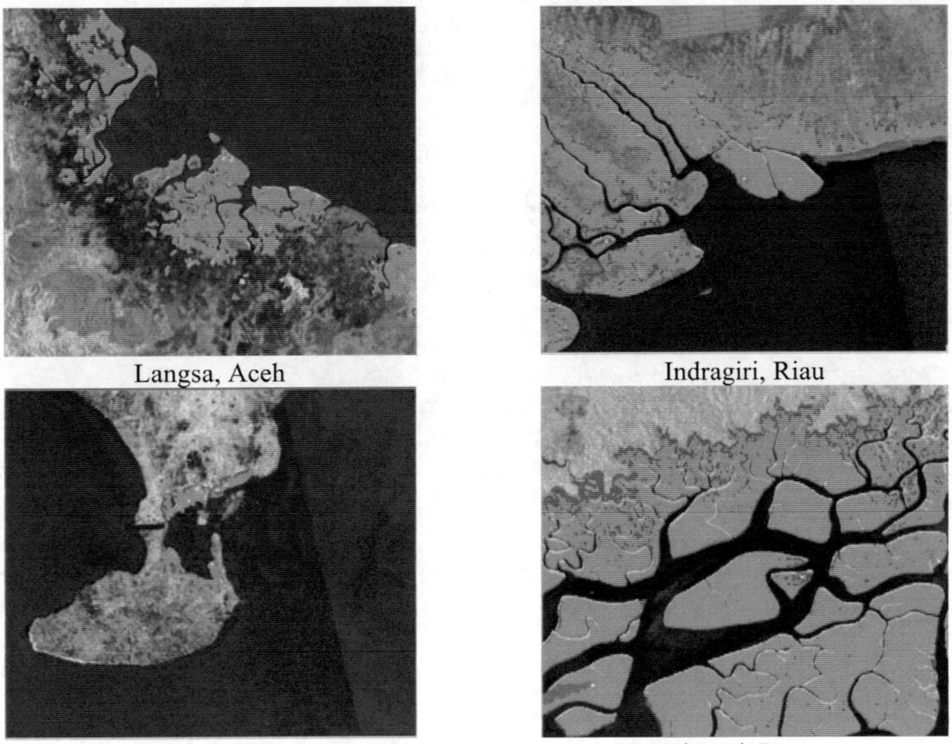

Langsa, Aceh Indragiri, Riau

Badung, Bali Bintuni, Papua

Figure 6. Spatial distribution of level biomass mangrove forest
Yellow=low biomass, green, moderate biomass, red=highest biomass.

4. Conclusion

We have showed characterization of mangrove forest types based on ALOS-PALSAR in overall Indonesia archipelago. Based on characterization Indonesia mangrove forest types has HH value around -10 dB until -7 dB and HV value around -17 dB until -13 dB. That can be classified into three level biomass of mangrove forest. Generally level biomass of mangrove forest in Indonesia archipelago is moderate. For future works we will validate image that has been classified and then if result of validation have accuracy assessment value very small we will try to classify using another parameter such as topography, geomorphology and tidal effect.

Acknowledgement

We would like thanks to JAXA SAFE to support this study.

References

[1] Tomlinson, P.B., 1986, *The Botany of Mangroves*, Cambridge University Press, Cambridge, U.K., 419 pp.

[2] Wightman, G.M., 1989, Mangroves of the Northern Territory. *Northern Territory Botanical Bulletin* No. 7. Conservation Commission of the Northern Territory, Palmerston, N.T., Australia.

[3] Giesen, W., Wulffraat, S., Zieren, M., and Scholten, L., 2007, *Mangrove Guidebook for South East Asia*, ISBN: 974-7946-85-8, FAO and Wetlands International, Bangkok.

[4] Kuenzer, C., Bluemel, A., Gebhardt, S., Tuan Vo Quoc and Dech, S., 2011, Remote Sensing of Mangrove Ecosystems: A Review, *Remote Sensing,* ISSN 2072-4292 www.mdpi.com/journal/remotesensing

[5] Rusila Noor, Y., M. Khazali, and I N.N. Suryadiputra, 1999, *Panduan Pengenalan Mangrove di Indonesia*. PHKA/WI-IP, Bogor.

[6] Takeuchi, W., Tien, D. V., Phuong, V. T., Van, A. N., Oo, K. S., 2011, Above ground biomass mapping of mangrove forest in vietnam by ALOS-PALSAR, ISBN 978-1-4577-1351-4 http://ieeexplore.ieee.org/stamp/stamp.jsp?arnumber=06087044&tag=1

[7] Lucas, R.M., Mitchell, A.L., Rosenqvist, A., Proisy, C., Melius, A., Ticehurst, C., 2007, The potential of L-band SAR for quantifying mangrove characteristics and change: Case studies from the tropics. *Aquat. Conserv.* , **17**, 245-264.

[8] Shimada, M., Isoguchi, O., Tadono, T., Isono, K., 2009, PALSAR radiometric and geometric calibration, *IEEE Transactions on Geoscience and Remote Sensing* **47**, 3915-3932.

[9] Kusmana, C., Sabiham, S., Abe, k., Watanabe, H., 1992, An Estimation of Above Ground Tree Biomass of a Mangrove Forest in East Sumatera, Indonesia, *Tropics,* Vol. **1** *(4).* 243-257.

Comparison of two classification methods (MLC and SVM) to extract land use and land cover in Johor Malaysia

B Rokni Deilmai[1], B Bin Ahmad[1], H Zabihi[2]
[1]Department of Remote sensing, Faculty of Geoinformation and Real Estate, Universiti Teknologi Malaysia, 81310 Johor, Malaysia
[2] Department of Geoinformation, Faculty of Geoinformation and Real Estate, Universiti Teknologi Malaysia, 81310 Johor, Malaysia

E-mail: bashirrokni@yahoo.com

Abstract. Mapping is essential for the analysis of the land use and land cover, which influence many environmental processes and properties. For the purpose of the creation of land cover maps, it is important to minimize error. These errors will propagate into later analyses based on these land cover maps. The reliability of land cover maps derived from remotely sensed data depends on an accurate classification. In this study, we have analyzed multispectral data using two different classifiers including Maximum Likelihood Classifier (MLC) and Support Vector Machine (SVM). To pursue this aim, Landsat Thematic Mapper data and identical field-based training sample datasets in Johor Malaysia used for each classification method, which results indicate in five land cover classes forest, oil palm, urban area, water, rubber. Classification results indicate that SVM was more accurate than MLC. With demonstrated capability to produce reliable cover results, the SVM methods should be especially useful for land cover classification.

1. Introduction

Land use and land cover are the basic assumption to identify the global ecology or environmental changes.Land use/ land cover mapping is a vital component, which has various parameters that are integrated on the basis of requirement. Land cover refers to a water body, cultivated land, built-up, natural vegetation, fallow land, glacial, rock/soil, artificial cover and others observed on the land. Global Information monitoring of land use and land cover is possible due to remote sensing technology in the form of spatial, spectral and temporal resolution. Remote sensing technology has many important roles, like reduction of survey time, latest map availability, more economic, digital image classification (pixels), spectral information etc. Image classification is an important part in many remote sensing applications. Classification of land cover is one of the most important task and one of the primary objectives in the analysis of remotely sensed data [16].
In this study, two different classification methods were compared for evaluating land cover features in the Johor Malaysia.

A Maximum Likelihood classifier, representing a conventional classification method and Support Vector Machines were applied. Maximum Likelihood is one of the most popular supervised

Content from this work may be used under the terms of the Creative Commons Attribution 3.0 licence. Any further distribution of this work must maintain attribution to the author(s) and the title of the work, journal citation and DOI.
Published under licence by IOP Publishing Ltd

classification method used with remote sensing image data. The Maximum Likelihood classification method is well known for the analysis of satellite images [5]. So far, satellite image interpretation using maximum likelihood approach was mostly applied for land cover classification [6] and monitoring of land use changes [17] showing overall high accuracies (mostly over 80%). MLC classification is based on parametric approach that involves assumption of the selected classes of signature in the normal distribution [1].

Improving land use/cover classification accuracy is an important issue in remote sensing literature. Moreover, advanced classification algorithm including Neural Network and Support Vector Machine approach instead of conventional classification method have been developed recently.

Support vector machines (SVMs) are a group of supervised classification algorithms that have been used recently in the remote sensing field. The classification accuracy produced by SVMs may show variation depending on the choice of the kernel function and its parameters [7].

Szuster et al. (2011) examined the performance of the SVM classification technique in the tropical coastal zone. He also compared this technique with the MLC and Artificial Neural Network techniques and has concluded that the SVM is better classifier for challenges like separating human infrastructures, for instance buildings from rocks and sandy beaches as they possess similar spectral signatures.

Yu et al. (2012) has applied the support vector machine (SVM) algorithm for the automated lithological classification in a part of northwestern India using ASTER imagery. He has also showed that SVM gives higher accuracy in comparison to MLC. The SVM algorithm has been used widely for pattern recognition applications. Many researchers in this field have found that a higher level of accuracy can be achieved by SVM than other processes of classification like MLC, artificial neural network (NN) etc. [11, 12, 14 and 15].

2. Data and methodology

2.1Study area

The study area considered in this study is the entire state of Johor. Johor is one of the developed states in Peninsular Malaysia In term of the area and population, Johor is the fifth largest and second populous state in Malaysia, With a total area of 19,210 km2 and population of 3,233,434 in 2010 [22]. The largest land uses in Johor are oil palm and forest (Figure 1).

Figure1.Studyarea showing the state of Johor [23]

2.2 Data used

Landsat TM satellite imagery has been selected for the classification of Land use by MLC and SVM algorithms. Date of acquisition of satellite imagery with spatial resolution of 30m has been 10

IOP Conf. Series: Earth and Environmental Science **20** (2014) 012052 doi:10.1088/1755-1315/20/1/012052

May, 2009 available from the Earth Explorer website [24]. It is corrected for geometric and topographic errors (level 1T).

2.3 Methods

Image classification is the stage of image analysis in which the multivariate quantitative measurement associated with each pixel is translated into a label from a pre-defined set (e.g., land use categories). Recall that the intent of the classification process is to assign pixels in an image to several classes of data based on their pixel value. Remote sensing classification is a complex process and requires consideration of many factors. The major steps of image classification may include determination of a suitable classification system, selection of training samples, image pre-processing, feature extraction, selection of suitable classification approaches, post-classification processing and accuracy assessment. Before classification, the land cover types in the study area were defined with the help of a land use map produced by the department of Agriculture Malaysia (year 2008). The main land cover types are forest, oil palm, urban area, rubber and water bodies.

Two types of classification techniques have been applied for analyzing the accuracy of land use map by using corrected satellite imagery. Training areas were selected according to the land use map of years 2006. Training stage is the most important part in supervised classification because it could influence the final classification results.

Maximum Likelihood classifier is the most common method used in the applications of remote sensing as a parametric statistical method where the analyst supervises the classification by identifying representative areas, called training zones. MLC is a process for determination of known class distribution as the maximum for a given statistic. This classification is a standard pixel based technique which is based on a multivariate probability density function of classes [9].

Early development of SVM started in 1970s and the popularity of SVM for pattern recognition and classification is actually surged in the late 1990s [19, 20]. The primary objective of the SVM method is the generation of a hyperplane that represents the optimal separation of linearly-separable classes in decision boundary space. In remote sensing, SVM was primarily used for the hyperspectral image classification and object detection [3, 10], although researchers have recently expanding its application for multispectral remote sensing data [2, 5 and 15]. The primary advantage of SVM was good generalization capability with limited training samples. The authors acknowledged SVM's limitations in parameter selection and computational requirements. However, SVM provided superior performance compared to most other image classification algorithms for both real-world remote sensing data and simulated experiments.

In general, the SVM represents a novel approach compared to conventional MLC classifiers. Additional SVM applications for regional scale land cover classification need to be conducted to better understand performance.

After classification, the accuracy of the classified images was assessed using reference data (land use map of year 2006). A total of 250 random points were selected from the images generating via a stratified random sampling method.The accuracy was assessed using error matrices (overall, user's and producer's accuracies and Kappa statistics.

3. Results and Analyisis

3.3 Landuse/landcover classification

The final LULC (Land use and land cover) maps presented in figure 2 show that the major classes are forest, oil palm, rubber, city and water. An evaluation of accuracy of the classified images (table 1) shows that the overall kappa and overall accuracy for Support Vector Machine (SVM) classification is 0.86 and 91.67%, and it is higher than the pixel based classification results with overall accuracy and Kappa 78.33% and 0.65.

A B

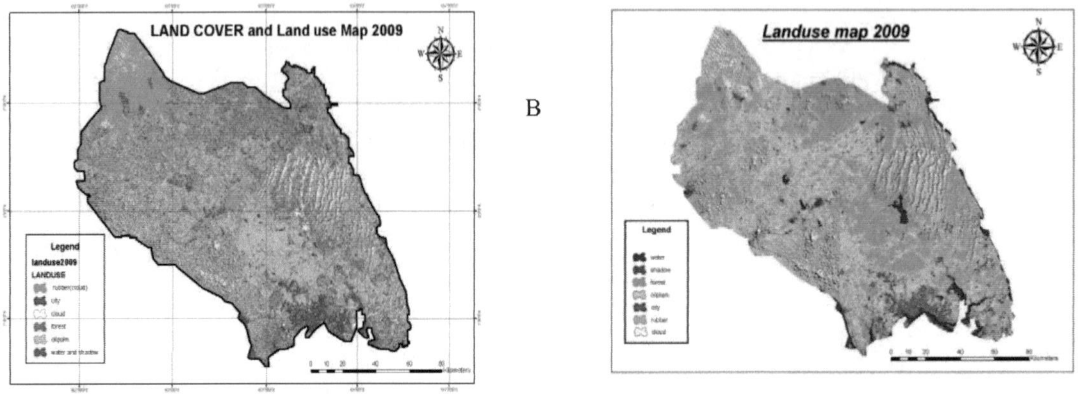

Figure 2. Land use/land cover maps produced using (A) maximum likelihood and (B) support vector machine for year 2009

Table 1. Accuracy assessment of maximum likelihood and support vector machine classifier

	Maximum likelihood		Support vector machine	
Land use - land cover	Producer's accuracy (%)	User's accuracy (%)	Producer's accuracy (%)	User's accuracy (%)
forest	65.5	71.43	87.5	87.5
oil palm	86.21	78.13	93.65	90.77
city	83.33	83.33	83.33	83.33
water	71.43	78.95	91.89	97.14
rubber	66.67	80.00	83.33	83.33
Overall accuracy	78.33%		91,67 (%)	
Kappa	0.65		0.86	

4. Conclusion

In this study, SVM and MLC classification techniques have been applied to identify the improved and accurate method for land use classification. Results of classifications from both the methods have been verified by land cover map for year 2006.Level of accuracy has also been measured through different proven parameters. It has been concluded that the accuracy level is comparatively better in SVM method than MLC method. The statistical Maximum Likelihood classifier cannot handle complex images so that many pixels cannot be classified correctly, but SVM also could overcome the mixed pixel problem in the image. According to SVM accuracy, it is better we use SVM for hyperspectral images.

References

[1] Al-Ahmadi, F. and Hames, A. (2009). Comparison of four classification methods to extract land use and land cover from raw satellite images for some remote arid areas, kingdom of Saudi Arabia. *Earth,* 20, 167-191.

[2] Foody, G. M. and Mathur, A. (2004). A relative evaluation of multiclass image classification by support vector machines. *Geoscience and Remote Sensing, IEEE Transactions on,* 42, 1335-1343.

[3] Gualtieri, J. A. and Cromp, R. F. (Year). Support vector machines for hyperspectral remote sensing classification. *The 27th AIPR Workshop: Advances in Computer-Assisted Recognition*, 1999. International Society for Optics and Photonics, 221-232.

[4] Hagner, O. and Reese, H. (2007). A method for calibrated maximum likelihood classification of forest types. *Remote Sensing of Environment,* 110, 438-444.

[5] Huang, C., Davis, L. and Townshend, J. (2002). An assessment of support vector machines for land cover classification. *International journal of remote sensing,* 23, 725-749.

[6] Kanellopoulos, I., Varfis, A., Wilkinson, G. and Megier, J. (1992). Land-cover discrimination in SPOT HRV imagery using an artificial neural network—a 20-class experiment. *International journal of remote sensing,* 13, 917-924.

[7] Kavzoglu, T. and Colkesen, I. (2009). A kernel functions analysis for support vector machines for land cover classification. *International Journal of Applied Earth Observation and Geoinformation,* 11, 352-359.

[8] Lewis, D., Corr, D., Gent, C. and Sheppard, C. (1992). Semi-supervised artificial neural networks for classification of remotely sensed images. *Remote Sensing from Research to Operation. Remote Sensing Society, Nottingham*, 489-497.

[9] Lillesand, T. M., Kiefer, R. W. and Chipman, J. W. (2008). *Remote sensing and image interpretation 6th edition (New York:John Wiley & Sons Ltd).* John Wiley & Sons Ltd.

[10] Melgani, F. and Bruzzone, L. (2004). Classification of hyperspectral remote sensing images with support vector machines. *Geoscience and Remote Sensing, IEEE Transactions on,* 42, 1778-1790.

[11] Mondal, A., Kundu, S., Chandniha, S. K., Shukla, R. and Mishra, P. (2012). Comparison of Support Vector Machine and Maximum Likelihood Classification Technique using Satellite Imagery.

[12] Naguib, A., Farag, M., Yahia, M., Ramadan, H. and Abd Elwahab, M. (2009). Comparative study between support vector machines and neural networks for lithological discrimination using hyperspectral data. *Egypt Journal of Remote Sensing and Space Science,* 12, 27-42.

[13] Nasrabadi, N. M. and Choo, C. Y. (1992). Hopfield network for stereo vision correspondence. *Neural Networks, IEEE Transactions on,* 3, 5-13.

[14] Oommen, T., Misra, D., Twarakavi, N. K., Prakash, A., Sahoo, B. and Bandopadhyay, S. (2008). An objective analysis of support vector machine based classification for remote sensing. *Mathematical Geosciences,* 40, 409-424.

[15] Pal, M. and Mather, P. (2005). Support vector machines for classification in remote sensing. *International journal of remote sensing,* 26, 1007-1011.

[16] Roy, P. and Giriraj, A. (2008). Land use and land cover analysis in Indian context. *Journal of Applied sciences,* 8, 1346-1353.

IOP Conf. Series: Earth and Environmental Science **20** (2014) 012052 doi:10.1088/1755-1315/20/1/012052

[17] Shalaby, A. and Tateishi, R. (2007). Remote sensing and GIS for mapping and monitoring land cover and land-use changes in the Northwestern coastal zone of Egypt. *Applied Geography,* 27**,** 28-41.

[18] Szuster, B. W., Chen, Q. and Borger, M. (2011). A comparison of classification techniques to support land cover and land use analysis in tropical coastal zones. *Applied Geography,* 31**,** 525-532.

[19] Vapnik, V. (2000). *The nature of statistical learning theory.* Springer.

[20] Vapnik, V. N. (1998). Statistical learning theory.

[21] Yu, L., Porwal, A., Holden, E.-J. and Dentith, M. C. (2012). Towards automatic lithological classification from remote sensing data using support vector machines. *Computers & Geosciences,* 45**,** 229-239.

[22] http://en.wikipedia.org/wiki/Johor_Bahru

[23] https://maps.google.com/

[24] http://earthexplorer.usgs.gov/

Modelling LULC for the period 2010-2030 using GIS and Remote sensing: a case study of Tikrit, Iraq

Sinan Jasim Hadi[1], **Helmi Z. M. Shafri**[2], **Mustafa D. Mahir**[3]

[1]Yer ve Uzay Bilimleri Enstitüsü, Anadolu üniversitesi, Eskişehir, Turkey
[2]Faculty of Engineering, University Putra Malaysia, Serdang, Malaysia
[3]School of Civil Engineering, Kuala Lumpur Infrastructure University College, Kajang, Malaysia
Email: sinan.jasim@yahoo.com

Abstract. This study extends upon the results of [1] to include the modeling of Land use/ Land cover (LULC). This study looks at the changes that occurred from 2010 to 2030 in Tikrit district, Iraq by predicting LULC for the year target 2030 by using the classified images for two points of time (2000 – 2010) as a foundation for the modeling process. The projected map, in comparison with 2010 LULC map, shows a significant decrease in vegetation area (45.11 km2) which must be regulated in order to maintain a green environment, and increase in the urban area (58.42 km2) which should be monitored to have sustainable development and control the eco-environment degradation. Also, in this study, it is shown clearly that the use of Geographic Information System (GIS) and remote sensing (specially IDRISI software) in modeling LULC is a suitable approach to understand the future pattern.

1. Introduction

One of the most important concerns of the world, nowadays, is the change in eco-environment that are caused by human exploitation and regional climate especially in arid and semiarid areas. Also, There are many ways of anthropogenic in changing the landscape of the world such as agriculture, deforestation, expanding farmland, and urban centres [1].

Degradation in the eco-environment has become one of the problems that affects Iraq, particularly the middle and southern part of it [1]. One million hectares is the estimated area that can be affected by this degradation [2]. Sand dunes, misuse of the cover of the plant, overgrazing were identified by Al-Janabi et al. (1988) as the causes of land degradation and cover changes. Important factors were added to these causes by Dougrameji (1999) these are the use of unsustainable practices continuously in addition to the deterioration and the poor maintenance of the infrastructure that was deteriorated because of the sanctions between 1990 and 2003. In last three to four decades, starting by the Iraq-Iran war and ending by the American invasion, Iraq has been involved in several wars. In particular, after 2003 the demography of Iraq has been changed due to the invasion dramatically, hence, the urban growth has been severely changed [1].

Modelling of land use and land cover is a scientific field that is growing rapidly because of its importance in identifying the effects of the humans on the environment. In view of this importance, scientists have constituted an international organization named Land use and Cover Change (LUCC) organization that is connected with the International Geosphere Biosphere Program and the International Human Dimensions of Global Change Program [5]. Furthermore, many algorithms and methods have been developed for modelling land use and cover.

One of the approaches that have been developed for forecasting Land use/ Land cover (LULC) is Cellular Automata (CA) which is defined as a dynamical discrete system in space and time that works by specific

Content from this work may be used under the terms of the Creative Commons Attribution 3.0 licence. Any further distribution of this work must maintain attribution to the author(s) and the title of the work, journal citation and DOI.
Published under licence by IOP Publishing Ltd

rules on a uniform grid based space [6]. CA involves cells and transition rules that are used to identify the state of a certain cell. It is especially interesting for land use and land cover modelling because of its ability to represent a complex system by a small set of rules and states with spatio-temporal behaviour (Figure 1). CA was successfully compiled in one of the models in the IDRISI software that, hence, gives this model power and easiness for performing modelling LULC.

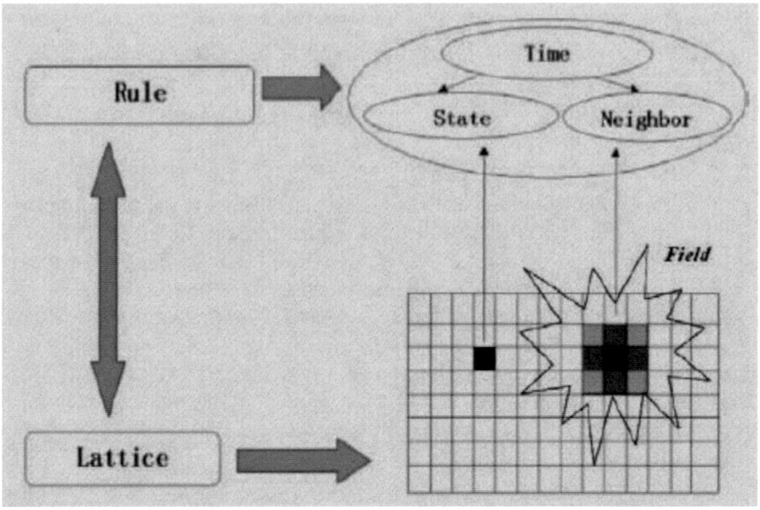

Figure 1. The components of CA and their relation (adopted from [7]).

CA_Markov is a model in the DRISI software. This model is a powerful tool for modelling and predicting land use and land cover change. It is a methodology that has been used widely in LULC modelling as it takes into consideration spatial interaction and stimulates multi LULC types [8]. In this research, an approach of detecting the change and predicting the change of a specific year is applied. This approach is an integrated method of remote sensing, GIS, and modelling (CA method), as the RS and GIS is used for detecting the change and providing basis data for CA model, the latter is used to predict the future LULC map.

In Iraq, there is a phenomenon called Desertification (i.e. A land losing its bodies of water till becomes dry) in addition to significant unmonitored Land use / Land cover (LULC) change due to several reasons such as the climate change around the world, the wars that Iraq has been involved, and the reduction in the amount of the water in the river that passes from Iraq [1]. The LULC change and desertification of the last four decades 1972-2010 and the factors that affected these changes were studied by [1].
This study aims at forecast in the future (i.e. 2030) pattern of land use and land cover in the area on the basis of the past and current situation and compare the result (by a change detection process) with the latest existing classified image (i.e. 2010).

2. Study area

Tikrit city that is located in Salahaldin province in Iraq was chosen due to the big number of immigrants that moved to Baghdad, in addition many people moved from Baghdad to Tikrit particularly because of the invasion in 2003. Another reason of choosing it is the importance of it as an administrative and economical centre of Salahaldin province. Tikrit is located 155 km from Baghdad the capital of Iraq. Also, it is the centre of Salahaldin province. Its population is 172,119 in 2007. The study area extends between

43°07'37.69" and 44°6'15.02" east and approximately between 34°26'20.41" and 35°4'1.96" north. Its area is about 2554.8 km^2 (Figure 2).

3. Data and methods

In this study, three types of software were used:

a) ArcGIS- was used for creating slope constraint and providing the administrative shapefile.
b) ENVI- was used for change detection analysis.
c) IDRISI Taiga- was used for presenting change detection graphs and for modeling land use and land cover.

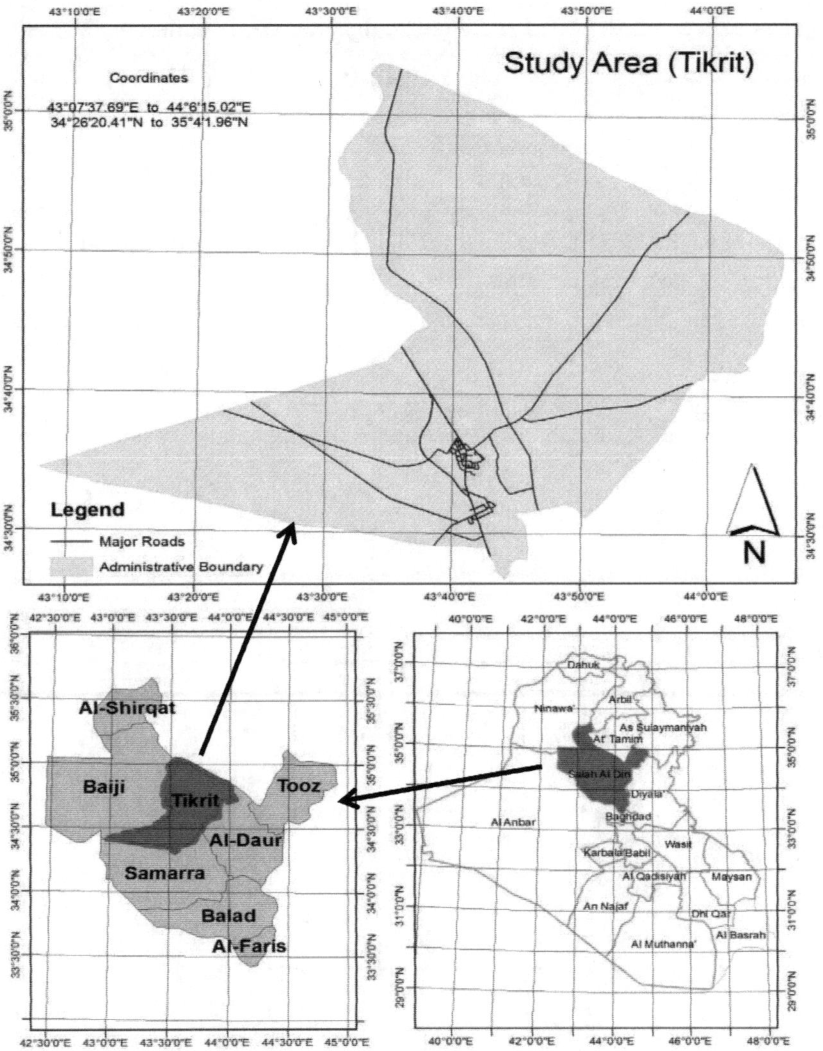

Figure 2. Study area.

The data that are used in the research are shown in table 1 and the detail of raw data are not going to be clarified as the detail can be seen in [1]. The studied data included thematic maps of 2000 and 2010 that were extracted from Landsat imagery for the two time epochs by using Maximum likelihood supervised classification. This step was done because for the process of modelling needs transition data that gives the pattern of changing to the model. This transition file was extracted by comparing the 2000 and 2010 classified maps. However, ancillary data were used in this research such as GIS map that were useful for deriving some of the restriction areas (i.e. areas cannot be developed or changed for example urban cannot be changed to water).

Table 1. Summary of data used for the study.

No.	Data Type	Acquisition Date	Source
1	Thematic Map	2000	[1].
2	Thematic Map	2010	[1].
3	Administrative and local government Map	2011	SAD Province
4	GIS Map for Iraq (UN)	2003	SAD Province
5	ASTER Global Digital Elevation Model V002 (DEM)	2011-03-16	NASA

The practical and theoretical aspects of land use and land cover change analysis and modeling implemented in this study involve several steps (Figure 3).

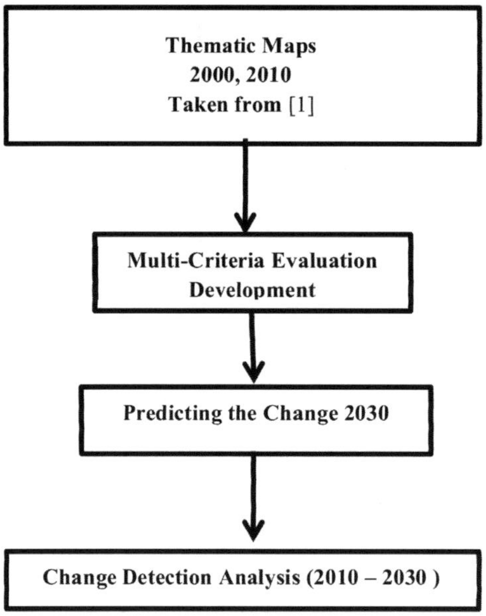

Figure 3. The methodology used in this study.

In order to predict land use and land cover changes for 2030, CA_Markov was performed in the environment of IDRISI Taiga software. As the name suggests, it is a combination of two components: Cellular Automata and Markov transition probabilities.

To perform CA_markov, preparing some inputs is required: Basic Land Cover map which was extracted by classifying a satellite image, Markov transition area file that was created by the use of Markov model in IDRISI and transition suitability image collection which is a collection of suitability images created by using MCE (Multi-Criteria Evaluation) model.

The first requirement was created in [1]. The second which is Markov Transition Area file was extracted by using Markov Module in IDRISI. The output will be a text file that contains the number of pixels, which are supposed and expected to change from one land cover type to the other through certain time units. The third requirement is the collection of suitability images that can be obtained by utilizing MCE module.

4. Criteria development

In the use of CA_Markov, criteria must be created, as they are the basis of making a measured and evaluated decision. Criteria consist of two types: Constraints and Factors.

4.1. Constraints

Constraints are used to limit the alternatives under consideration. Basically, it identifies the areas in a class that can be used for change from other classes to this class, and the areas that cannot be changed. They are represented as Boolean (Logical) maps. This means areas included in the consideration have the value 1 and the areas not included in the consideration have the value 0. These constraint images were created by using Reclass module to change them for 0 and 1 images. In this study, for each class there are constraints (Table 2).

It is worth mentioning that one of the constraints was created using ArcGIS. A DEM image which was obtained from NASA was first converted to a TIN (triangulated irregular network) image. Then the slope was created. Next, the areas with slope over 13° were extracted from the slope raster. This was used as a constraint for urban areas as they cannot be built in the areas with more than 13° slope.

4.2. Factors

Factors are commonly continuous images (measured in a continuous scale) defining the suitability area for the alternatives of the change. Simply put, they are images with measure of the change suitability by a continuous scale (e.g. 0 to 255). A fuzzy module is provided. The selection of the Fuzzy module parameters is based on the knowledge and the experience of the analyst. The standardization of a map is from 0 (least suitable) to 255 (most suitable).

After adding the constraints to the Decision wizard and creating the fuzzy map (considered as factor), the wizard asks to choose a method of weighting the factors and constraints. In this step, an equal weight method was chosen as there was no need for weighting the factors because they were weighted through the process of fuzzy module. The constraints were all given 0 values as they were not to be used for any change. So, the value of 1 was given to the factor layer which was created by the fuzzy module. Table 2 shows the factors and constraints and how they were considered.

Table 2. Classes' constraint and factors

Class	Constraints	Factors
Water	Non Water class	No factor
Vegetation	Urban, Water	Vegetation, Soil, Barren Land
Barren Land	Urban, Water	Barren Land, Soil, Vegetation
Soil	Urban, Water	Soil, Vegetation, Barren Land
Urban	Water, Slope	Urban, Soil, Vegetation, Barren Land

These steps were followed for each class based on 5 classes that were used (Table 2) in the maps which means they were done five times to create 5 suitability maps. Finally, gathering the five classes' suitability maps in one raster group file created the suitability map.

5. Modelling in CA_Markov

The requirements of the CA_Markov module were added to the module one by one: basic land cover which is the map of 2010 as the 2030 map had to be predicted, Markov transition areas file which was created using the 2000 and 2010 maps and a transition suitability image collection which was obtained by the decision support wizard. Then, the number of iterations was set to 20 with one iteration for each year (2010 to 2030). Finally, the module gave LULC image for 2030 as a predicted LULC.

6. Model validation

In order to ensure that the model is reliable in predicting an LULC for a specific project year, it must be validated using existing datasets. Therefore, the used model was validated by using two maps: the first one is the real land use and land cover of the year 2010, and the second map is a predicted (simulated) 2010 map. The latter was simulated by using the model and the same procedure that was implemented in predicting 2030 land use and land cover map, but, in this case, the 1990 and 2000 maps (Taken from [1]) were used for simulating 2010 maps. The validate module in IDRISI Taiga was used for validating the model by producing several parameters: Kno, Klocation, and Kstandard that are used to identify the accuracy of the model. Table 3 shows the result of the validation module.

Table 3. Kappa index result of the model validation.

Parameter	Value
Kno	0.7918
Klocation	1.0000
Kstandard	0.0157

For all of the Kappa statistics, 0% indicates that the level of agreement is equal to the agreement due to chance and 100% indicates perfect agreement. In comparing the reality map (classified) with the predicted map, the most important parameter is Kno which should be used to evaluate the overall success of the simulation or the model [9]. Klocation should be used to evaluate the ability of the simulation to identify location [9]. In this model, the value of Klocation indicates the model is perfect in specifying location. Kstandard shows almost no useful information because it compounds the location error with quantification error. So, it fails in being used for evaluating a model [9]. Hence, this value that is 0.0157 is neglected.

In addition, the CROSSTAB module was used for the same reason of validating the model that was used in simulating 2030 LULC map. It was utilized due its ability to produce Kappa index of agreement between the simulated and real map for the same time point for each class in addition to the overall Kappa coefficient.

IOP Conf. Series: Earth and Environmental Science **20** (2014) 012053 doi:10.1088/1755-1315/20/1/012053

CROSSTAB output provides many of the statistics that can be used by the user in order to separate between location error and quantification error. Location errors happen in case of there is a difference in the location of a category in a map from the location of the same category in the other map. Moreover, quantification errors happen in case of there is a difference in the quantity of cells in a specific category in a map from the quantity of the same category in the other map. Kappa Index of Agreement contains these two types of errors [9].

A real 2010 maps was taken from [1] as a classified map while a simulated 2010 was extracted in the same way of 2030. 1990 and 2000 classified maps were used in order to simulate (modelling) 2010; hence, real (classified) and simulated (modelled) were used in this module. Table 4 represents the values of the Kappa index agreement between the simulated and real 2010. In addition, the overall Kappa confident of the agreement was extracted and its value was 0.8404.

Table 4. Kappa index of agreement between the predicted and real 2010 maps.

Class	**Kappa Index of Agreement (KIA)**
NoData	0.9996
Water	0.8548
Vegetation	0.9175
Barren Land	0.7098
Soil	0.7253
Urban	0.4203

7. Result and discussion

Figure 4 shows the projected land cover/use for the year 2030. By visual examination and comparison with 2010 Land cover map (Figure 5), it can be seen that there is a significant increase in the urban areas, and this increase is of three types: one of them is in-fill urban expansion which occurs within existing urban areas; the second is extension of existing urban areas and the third is termed 'leap frog' expansion, which refers to dispersed urban area growth that is disconnected from existing urban areas.

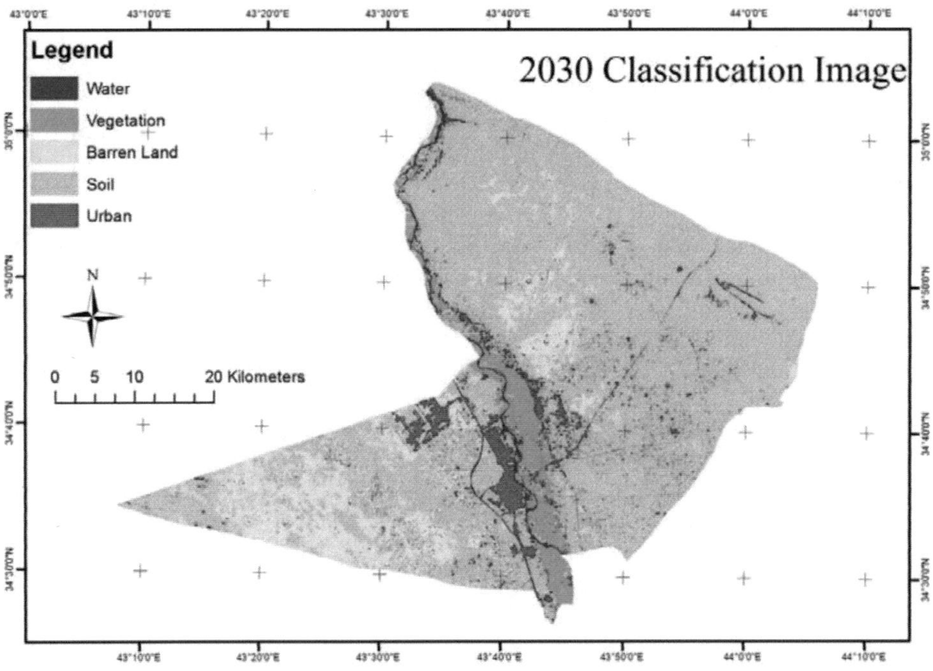

Figure 4. Projected Map of the year 2030.

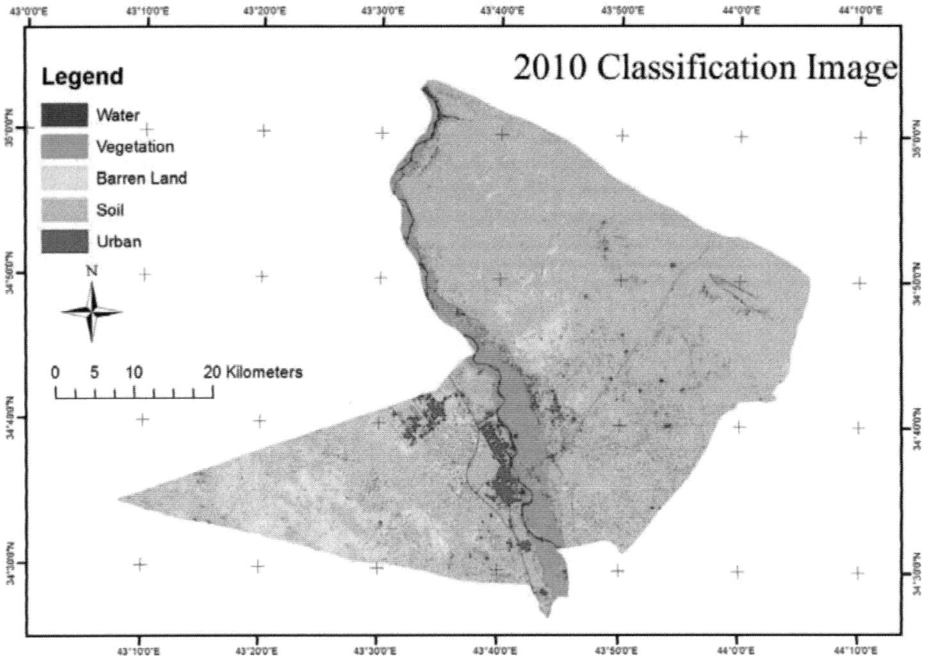

Figure 5. Projected Map of the year 2010.

Table 5 shows the areas of land covers that were calculated from the map that was predicted for the targeted year 2030.

Table 5. Areas of land covers in 2030 measured in km^2

No.	Land Cover	Area km2
1	Water	18.50
2	Vegetation	103.55
3	Barren Land	391.03
4	Soil	1866.37
5	Urban	175.24

Based on table 6 and figure 6, the most identifiable change that should be taken into consideration is the increase in the area of urban land cover by 58.42 km^2 with a percentage of 50.01% with no area loss through the period between 2010 and the prediction year 2030. As mentioned earlier, if a new planned base map is not used to regulate the growth, an urban sprawl threatens to appear in the city. Vegetation area is predicted to decrease in 2030 by 45.11 km^2 with a percentage of 30.34%, which is a large reduction that can significantly contribute to eco-environment degradation. A plan needs to be put in place to control this degradation by methods such as planting trees around the city to increase the green cover that can act as a dam for dust that comes from the west of Iraq to Tikrit city. The plan could incorporate irrigation projects in the soil area, which decreased by 111.51 km2 with a percentage of 5.64%, and the barren land which increased by 94.24 km^2 with a percentage of 31.75%. These projects can be used by farmers to increase their production and can be used by the government for starting new plantation projects.

Table 6. Losses and Gains in land covers/uses areas between 2010 and 2030 measured in km^2.

	Loss	Gain	Net Change	% of Change
Water	0	3.95	3.95	27.16
Vegetation	-45.44	0.33	-45.11	30.34
Barren Land	-8.62	102.86	94.24	31.75
Soil	-154.78	43.27	-111.51	5.64
Urban	0	58.42	58.42	50.01

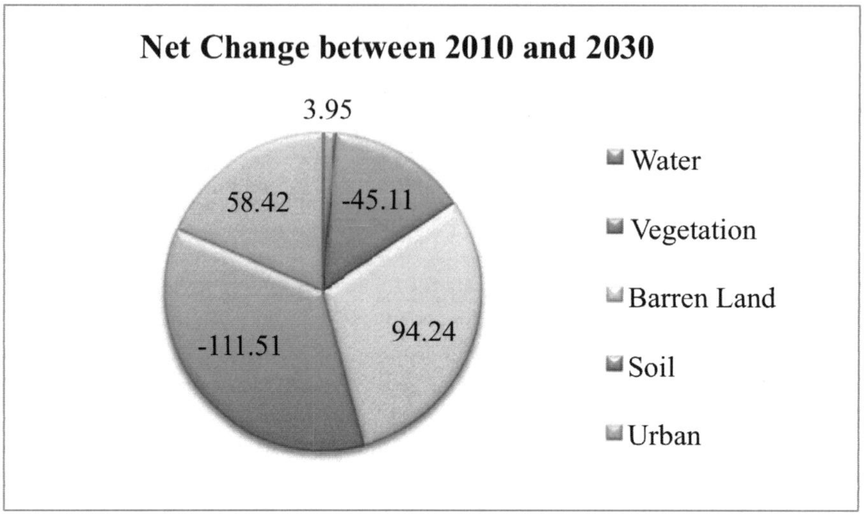

Figure 6. Net changes on the area of Land cover/use types in 2010-2030.

8. Conclusions

In the projected image for 2030, it is found that the vegetation will decrease by a great amount, which must be regulated in order to maintain a green environment. Also, three modes of urban expansion are expected to occur quite dramatically, which need to be monitored and controlled by a base map accompanied by strong laws to implement it. The most viable solution for such problems is building dams in the river in order to store the water, building some irrigation channels around the city and directing water to them so it reaches the largest possible area.

Moreover, by using different remote sensing datasets, satellite image, GIS vector maps, and hard copy maps, it is demonstrated that GIS and remote sensing are excellent ways of identifying the changes that take place in land cover and land use because of their ability to provide data for different times that facilitates decision making. Also, these tools are instrumental in modeling future land cover/ use.
Finally, predicting land use/land cover changes and identifying their consequences is based on the capability of understanding the drivers (past, current) of land use and land cover change.

References:

[1] S. J. Hadi, H. Z. M. Shafri, and M. D. Mahir, "Factors Affecting the Eco-Environment Identification Through Change Detection Analysis by Using Remote Sensing and GIS: A Case Study of Tikrit, Iraq," *Arab. J. Sci. Eng.*, vol. 39, no. 1, pp. 395–405, 2013.

[2] M. T. Jabbar and X. Zhou, "Eco-environmental change detection by using remote sensing and GIS techniques: a case study Basrah province, south part of Iraq," *Environ. Earth Sci.*, vol. 64, no. 5, pp. 1397–1407, Feb. 2011.

[3] K. Z. Al-Janabi, A. Jawad Ali, F. H. Al-Taie, and F. J. Jack, "Origin and nature of sand dunes in the alluvial plain of southern Iraq," *J. Arid Environ.*, vol. 14, no. 1, pp. 27–34, 1988.

[4] J. Dougrameji, "Aeolian Sediment Movements in Lower Alluvial Plain, Iraq," *Desertif. Control Bull.*, pp. 45–49, 1999.

[5] G. R. Pontius and H. Chen, "GEOMOD Modelling: Land use and cover change modeling tutorial," *Clark Univ.*, 2006.

[6] C. Hand, "Simple cellular automata on a spreadsheet," *Comput. High. Educ. Econ. Rev.*, vol. 17, 2005.

[7] Y. Xie, "Projecting Urban land use change: DUEM-dynamic urban evolution modeling, USGS-GLSC," 2002. .

[8] B. Assefa, "Land Use and Land Cover Analysis and Modeling in South Western Ethiopia: The Case of Selected Resettlement Kebeles in Gimbo Woreda," 2010.

[9] R. Pontius, "Quantification error versus location error in comparison of categorical maps," *Photogramm. Eng. Remote Sensing*, vol. 66, no. 8, pp. 1011–1016, 2000.

IOP Conf. Series: Earth and Environmental Science **20** (2014) 012054 doi:10.1088/1755-1315/20/1/012054

Monitoring temporal vegetation changes in Lao tropical forests

Chittana Phompila[1 2], Megan Lewis[2], Kenneth Clarke[2], Bertram Ostendorf[2]

[1]Faculty of Forestry, National University of Laos, Dongdok, Vientiane, Lao PDR
[2]School of Earth & Environmental Sciences, The University of Adelaide, Adelaide, SA 5005, Australia

E-mail: chittana.phompila@adelaide.edu.au

Abstract. Studies on changes in vegetation are essential for understanding the interaction between humans and the environment. These studies provide key information for land use assessment, terrestrial ecosystem monitoring, carbon flux modelling and impacts of global climate change. The primary purpose of this study was to detect temporal vegetation changes in tropical forests in the southern part of Lao PDR from 2001-2012. The study investigated the annual vegetation phenological response of dominant land cover types across the study area and relationships to seasonal precipitation and temperature. Improved understanding of intra-annual patterns of vegetation variation was useful to detect longer term changes in vegetation. The breaks for additive season and trend (BFAST) approach was implemented to detect changes in these land cover types throughout the 2001-2012 period. We used the enhanced vegetation index (EVI) data from the Moderate Resolution Imaging Spectroradiometer (MODIS) (MOD13Q1 products) and monthly rainfall and temperature data obtained from the Meteorology and Hydrology Department, Ministry of Agriculture-Forestry, published by Lao National Statistical Centre in this research. EVI well documented the annual seasonal growth of vegetation and clearly distinguished the characteristic phenology of four different land use types; native forest, plantation, agriculture and mixed wooded/cleared area. Native forests maintained high EVI throughout the year, while plantations, wooded/cleared areas and agriculture showed greater inter-annual variation, with minimum EVI at the end of the dry season in April and maximum EVI in September-October, around two months after the wet season peak in rainfall. The BFAST analysis detected abrupt temporal changes in vegetation in the tropical forests, especially in a large conversion of mixed wooded/cleared area into plantation. Within the study area from 2001-2012 there has been an overall decreasing trend of vegetation cover for native forests and mixed wooded/cleared lands, and by contrast an increase in EVI of plantations after 2008.

Content from this work may be used under the terms of the Creative Commons Attribution 3.0 licence. Any further distribution of this work must maintain attribution to the author(s) and the title of the work, journal citation and DOI.
Published under licence by IOP Publishing Ltd

IOP Conf. Series: Earth and Environmental Science **20** (2014) 012054 doi:10.1088/1755-1315/20/1/012054

1. Introduction

Information on spatial and temporal changes in vegetation at local, regional and global scales is important for improving our understanding of human - environment interactions [1, 2] including alterations in terrestrial ecosystems, carbon fluxes and global climate [3]. Satellite-based studies on vegetation phenology provide information for environmental risk assessment and land use change monitoring. This information is also useful for detecting and monitoring global environmental change as vegetation response is associated with temperature, soil moisture, and human activity [2]. Moreover, monitoring and forecasting changes in phenology is useful to understand the response of vegetation related to changing climatic conditions [4]. However, acquisition of such information from on-ground studies is difficult.

The use of information captured by earth-observing remote sensing instruments provides opportunities to overcome this challenge [5]. Information from space is essential for analysing the spatial and temporal patterns of vegetation from local to regional and continental scales [6-9]. Changes in vegetation can be analysed using remote sensing data time series, such as NDVI [7, 9, 10] and EVI [11, 12].

Several remote sensing approaches have been tested and successfully applied to study changes in vegetation composition, cover and structure. They have used a diversity of satellite datasets, with varying spatiotemporal resolutions, and have employed a range of statistical methods [9]. Amongst these, the BFAST approach is a powerful analysis to detect gradual and abrupt changes in satellite data time series [8, 13]. It has been successfully tested in south eastern Australia and South Somalia.

However, there is little known about the performance of BFAST in tropical forests. Monitoring vegetation in this region using satellite remote sensing is challenging due to atmospheric effects such as frequent cloud cover and high levels of aerosols. These factors have an influence on analysis of temporal signals of remote sensing data. As a result, vegetation phenology in tropical forests is not well documented [12]. Therefore, the primary purpose of this study was to attempt to detect intra and inter-annual changes in the vegetation of tropical forests across the study area, the southern part of Lao PDR, through analysis of MODIS satellite vegetation index data. The response of vegetation phenology to monthly average precipitation and temperature was investigated, providing understanding of intra-annual response, against which longer term changes in vegetation could be detected. The BFAST approach was used to detect vegetation changes in land cover types between 2001 and 2012.

2. Material and method

2.1. Study area

The study area was situated in Champasack Province, in the south of Lao PDR. It covers an area of 15,415 km^2, from 13°55'00"N to 15°22'00"N latitude and from 105°13'00" E to 106°55'00" E longitude (figure 1). It is bordered by Salavanh and Sekong Provinces in the north and northeast respectively, while the east is bordered by Attapeu Province. The province adjoins Thailand to the west and Cambodia to the south. About 58% of the Champasack Province is covered by native forests and it includes two of the largest Laotian national biodiversity conservation areas, Xepian and Dong Houa Sao. A range of natural ecosystems cover the region including savanna, dry, moist or mixed deciduous tropical forests, and semi-evergreen dry dipterocarpus forests. Champasack has relatively flat terrain comprising two different landscapes; 26% of the area is high altitude upland and 74% is low altitude, flat lowland. The altitude ranges from 75 - 1,284 m, but the majority of land is between 75 - 120 m above sea level. There are two distinct seasons: rainy (May-October) and dry (November-April). During the rainy season it is often windy, humidity is high and most of the 2,279 mm average annual rainfall occurs. In the dry season, conditions are mostly sunny, average temperatures are 28◦C - 31◦C and there is little rainfall.

2.2. Data

The research made use of the MODIS 16 days EVI composite product (MOD13Q1) time series from 2001-2012, with a spatial resolution of 250m. Both NDVI and EVI have been used to study changes in vegetation in tropical regions. In this study, we used EVI because it is better captures vegetation dynamics rather than NDVI [11], and it reduces sensitivity to soil and atmospheric effects but remains sensitive to a wide range of variation in canopy density [2]. The study area was covered by MODIS tile h28v07. Data was downloaded from the National Administration and Space Aeronautics (NASA) using a "MODISTools" package in R[1]. Data were reprojected to WGS84, UTM projection and zone 48 (MODIS reprojection tool version 4.01). Additionally, we also used monthly average rainfall and temperature data (2001-2012), from the Lao Meteorology and Hydrology Department, Ministry of Agriculture-Forestry, published by the Lao National Statistical Centre in this study.

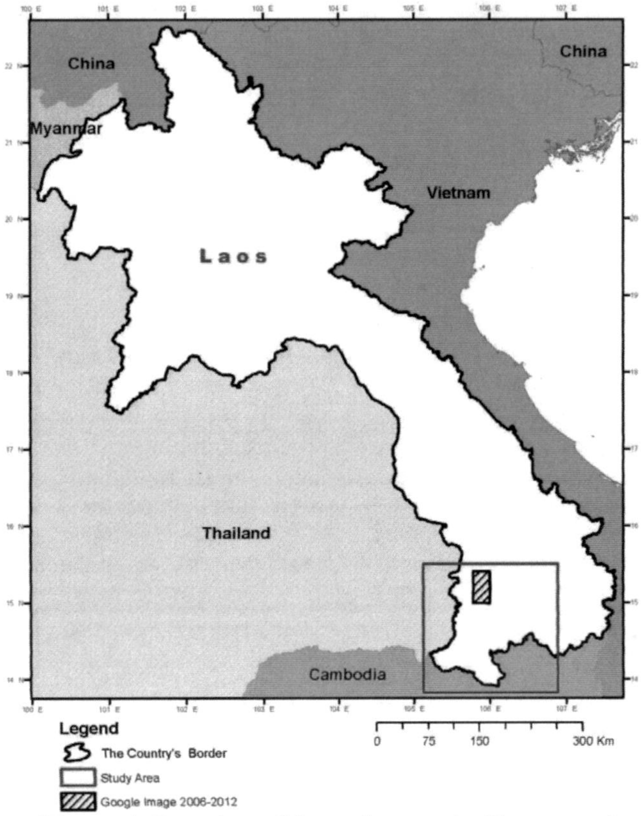

Figure 1. Location of the study area, in Champasack Province, Laos.

2.3. Data Analysis

Our aim was to investigate the dynamics of EVI in different land uses from 2001-2012. EVI for four dominant land uses was first compared and contrasted, and then examined in the context of the monthly average precipitation and temperature. We also applied the BFAST approach to detect temporal changes in tropical forests. EVI time series (2001-2012) were extracted for four different land use covers using a standard randomization approach. Samples were randomly computer generated within land use polygons digitized from high resolution satellite imagery available through Google Earth. The land uses represented native forest, plantation, mixed wooded/cleared area and agriculture. 250 samples for each land use were extracted and the average EVI for each 16-day composite date calculated from the 2001-2012 records. This provided average annual EVI profiles of the four major land uses. These were compared with monthly rainfall and temperature patterns for the same period. Secondly, inter-annual changes in vegetation of four different land uses were examined using the BFAST approach. In total, 1,000 random samples representing the four land uses were used in the model. The BFAST algorithm decomposes input time series datasets into three components: trend, seasonal and remainder components. Its formula is:

$$Yt = Tt + St + et \quad (1)$$

[1] http://cran.fhcrc.org/web/packages/MODISTools/index.html

Where ***Yt*** is the observed data at time t; ***Tt*** is the trend component; ***St*** is the seasonal component; and ***et*** is the remainder, or residual component.

3. Results and discussion

3.1. Seasonal patterns of vegetation

The seasonal pattern of vegetation response was well represented by the spatial averages of MODIS EVI for the four land use types (native forest, plantation, mixed wooded/cleared area and agriculture) (figure 2). Native forest had highest overall EVI with least variation throughout the year, but a minor peak in July and slightly higher EVI maintained though to December. Plantations also had high overall EVI, but with greater seasonal variation: lower than the native forests from January-June, and higher from August-November. The wooded/cleared areas showed even greater intra-annual range of EVI: much lower than the forests and plantations January-July, but with similarly high EVI in the latter half of the year. Both plantations and mixed wooded/cleared areas showed evidence of two peaks in EVI in August and October. By contrast, the agricultural lands had overall lower EVI, but with greater variation between seasons, with a maximum in September-November.

EVI dynamics exhibited systematic differences based on vegetation cover, species and management practices in the four land uses. A high canopy cover is maintained throughout the year for forests and plantations, with the plantation species showing more evidence of seasonal variation. In mixed wooded/cleared areas more deciduous trees and shrubs are prevalent, resulting in a pronounced seasonal contrast in EVI. Agricultural areas displayed a pronounced annual cycle of land clearance/preparation (Jan-April), followed by crop planting and growth (Jun-Oct) and harvest (Nov-Dec). During land preparation, the greater soil exposure results in very low EVI, which gradually increases during growth of the crops (predominantly rice), reaching a maximum in October.

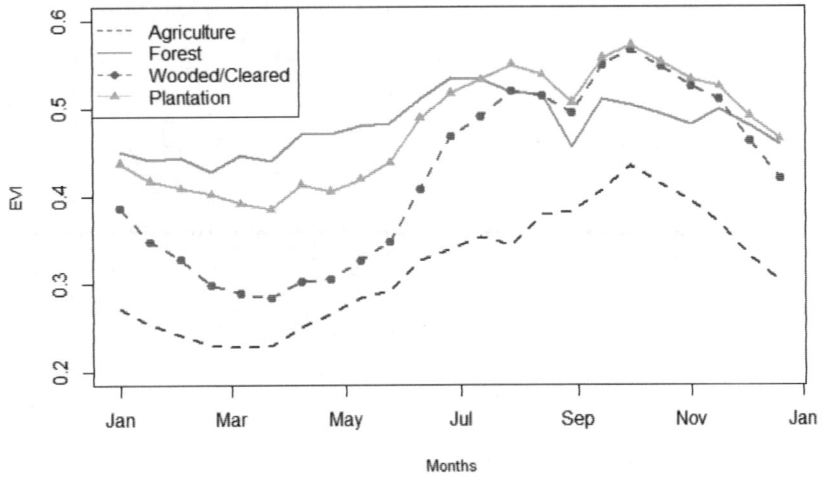

Figure 2. A long tern average of EVI profile across four land use types; native forest, plantation, agriculture and mixed wooded/cleared area, from 2001-2012.

3.2. EVI responses to monthly precipitation and temperature

The vegetation response in relation to monthly average rainfall and temperature (2001-2012) is shown in figure 3. The annual vegetation growth cycles for all vegetation/land cover types closely followed the precipitation pattern with a lag of two-three months but were inversely related to temperature. From January to April average rainfall was less than 250 mm (figure 3A), while average temperatures climbed from a low of 26◦C in January to the annual high of 31◦C in April (figure 3B). During this period, vegetation cover and/or photosynthesis was lower, resulting in lower EVI for all land uses. However, when rainfall increased from late May through September (300-400 mm), the greenness of vegetation started to increase and peaked in October, two months after the July peak of rainfall. This EVI signal still remained high almost two-three months after the end of rainfall, and then gradually declined in Nov-Dec. Soil moisture was still available to the deep-rooted perennial trees and shrubs of the forests, plantations and wooded/cleared areas during this period. It is evidence that intra-annual variations of average EVI are strongly influenced by seasonal rainfall.

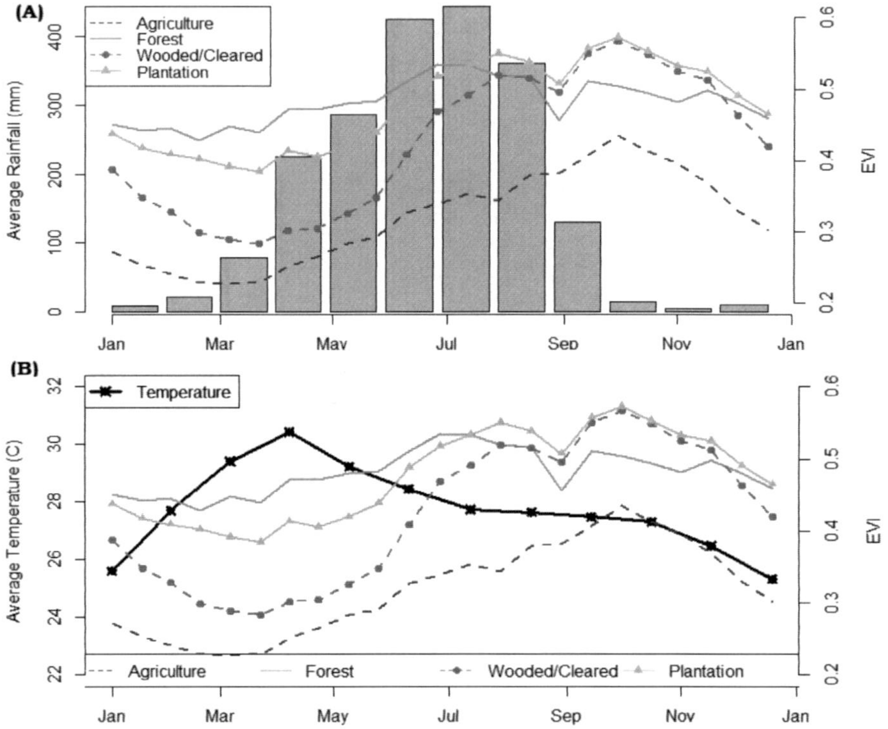

Figure 3. A comparison of the long term average of EVI (from 2001-2012) with; **(A)** 12-year monthly average precipitation; **(B)** 12-year monthly average temperature.

Figure 3B shows an inverse relationship between monthly average temperature and EVI within all four land-use covers. Although the minimum of EVI seemed to synchronize with higher temperatures in the dry period (Mar-April) and the different vegetation types appeared to actively response to a mid-range of temperature during June-Nov, temperature was not directly influencing EVI. From these two figures, it could be concluded that the rainy season (Jun-Nov) until the beginning of dry season is the best for vegetation growth in this region because of relatively suitable temperatures (26°C-28°C) and availability of soil moisture.

3.3. Detecting temporal changes of vegetation with BFAST model

Temporal changes in vegetation in the study area were detected using the BFAST model. Verbesselt, Hyndman (7) suggested that the trend component indicates gradual and abrupt change in vegetation, while the seasonal component indicates phenological changes. Results here showed that BFAST had a capability to detect abrupt changes in vegetation dynamics, especially when a large area of mixed wooded/cleared area was converted to plantation. This was confirmed by visual interpretation of two dates of Google Earth images (2006 and 2012). The 1,000 random samples were all examined, with combined samples for each land use illustrated and described here. Figure 4A shows the temporal changes in plantation area detected by the BFAST algorithm. The examination of the BFAST trend component for plantation suggests that clearing for plantations commenced from the beginning of 2004 to late 2007, followed by maturation and increase in plantation canopies from 2008 until 2011. Their EVI value dropped in early 2012, and then continued to increase.

The overall trend of vegetation in mixed wooded/cleared area was downward, with two abrupt changes detected by the BFAST algorithm: in 2005 and early 2010 (figure 4B). Although EVI in native forests remained stable from 2001-2011, there was an abrupt decrease in its response in 2012 (figure 4C). Figure 4D shows an overall gradual downward trend and the annual seasonality of vegetation of agriculture areas.

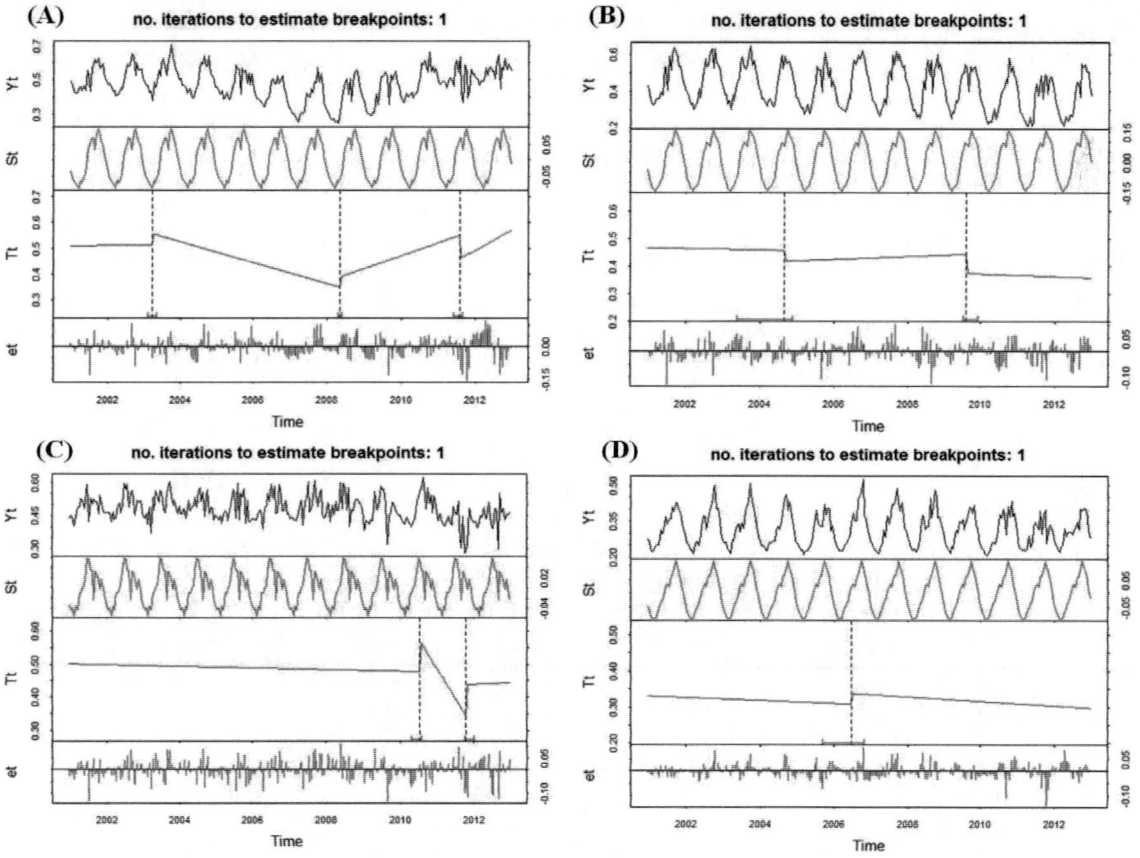

Figure 4. An example of temporal changes in vegetation across four different land use covers; (A) plantation, (B) mixed wooded/cleared area, (C) native forest, and (D) agriculture. *Yt=input data, St=seasonal pattern, Tt=overall trend or abrupt changes detected, et=remainder or noise.*

4. Conclusion

This study aimed to detect temporal changes in tropical forests across the study area in southern Lao PDR using enhanced vegetation index (EVI) time series data of MODIS. Firstly, the intra-annual response of vegetation in different land covers was investigated to provide an understanding of typical seasonal patterns. Secondly BFAST was applied to examine inter-annual changes in vegetation responses over the 2001-2012 periods. EVI well documented the annual seasonal growth of vegetation and well distinguished the characteristic phenological patterns of four different land use types; native forest, plantation, agriculture and mixed wooded/cleared areas. The maximum vegetation growth occurred two months after the peak of annual precipitation, coinciding with mid-range monthly temperatures. This suggests that typical seasonal patterns of vegetation growth appeared to be primarily determined by water availability, but not directly by temperature. This study has provided key information on vegetation seasonality, long term trends and land clearance and re-planting for the first time in a tropical forest of Lao PDR. The BFAST analysis revealed an overall trend of decreasing vigour for all land covers since 2001, with the exception of plantations. It demonstrated to be capable of detecting known clearance and revegetation, supported by independent evidence (Google Earth image interpretation), and a possible disturbance event in the native vegetation class in 2011. The BFAST analysis of MODIS EVI appeared as a promising tool to assess tropical land cover type changes.

5. Acknowledgement

This study was supported by the Australian Agency for International Development (AusAID) and the University of Adelaide. Acknowledgements are given to the National Aeronautics and Space Administration (NASA), Google Earth and the Lao National Statistical Centre for providing freely available satellite images and data which were used for this research.

6. References

[1] Chernetskiy M, Pasko I, Shevyrnogov A, Slyusar N, Khodyayev A. A study of forest vegetation dynamics in the south of the Krasnoyarskii Krai in spring. Advances in Space Research. 2011;48(5):819-25.

[2] Zhang X, Friedl MA, Schaaf CB, Strahler AH, Liu Z. Monitoring the response of vegetation phenology to precipitation in Africa by coupling MODIS and TRMM instruments. Journal of geophysical research. 2005;110(D12):D12103.

[3] Papes M, Tupayachi R, Martinez P, Peterson AT, Asner GP, Powell GVN. Seasonal Variation in Spectral Signatures of Five Genera of Rainforest Trees. Selected Topics in Applied Earth Observations and Remote Sensing, IEEE Journal of. 2013;6(2):339-50.

[4] Prabakaran C, Singh CP, Panigrahy S, Parihar JS. Retrieval of forest phenological parameters from remote sensing-based NDVI time-series data. Current Science. 2013;105(6):795-802.

[5] Ostendorf B, Hilbert DW, Hopkins MS. The effect of climate change on tropical rainforest vegetation pattern. Ecological modelling. 2001;145(2):211-224.

[6] Viña A, Tuanmu M-N, Xu W, Li Y, Qi J, Ouyang Z, et al. Relationship between floristic similarity and vegetated land surface phenology: Implications for the synoptic monitoring of species diversity at broad geographic regions. Remote Sensing of Environment. 2012;121(0):488-96.

[7] Mao D, Wang Z, Luo L, Ren C. Integrating AVHRR and MODIS data to monitor NDVI changes and their relationships with climatic parameters in Northeast China. International Journal of Applied Earth Observation and Geoinformation. 2011.

[8] Verbesselt J, Hyndman R, Zeileis A, Culvenor D. Phenological change detection while accounting for abrupt and gradual trends in satellite image time series. Remote Sensing of Environment. 2010;114(12):2970-80.

[9] Forkel M, Carvalhais N, Verbesselt J, Mahecha M, Neigh C, Reichstein M. Trend Change Detection in NDVI Time Series: Effects of Inter-Annual Variability and Methodology. Remote Sens-Basel. 2013;5(5):2113-44.

[10] Klein I, Gessner U, Kuenzer C. Regional land cover mapping and change detection in Central Asia using MODIS time-series. Appl Geogr. 2012;35(1-2):219-34.

[11] Coops NC, Wulder MA, Iwanicka D. Large area monitoring with a MODIS-based Disturbance Index (DI) sensitive to annual and seasonal variations. Remote Sensing of Environment.

IOP Conf. Series: Earth and Environmental Science **20** (2014) 012054 doi:10.1088/1755-1315/20/1/012054

2009;113(6):1250-61.
[12] Moreau I, Defourny P, editors. The vegetation phenology detection in Amazon tropical evergreen forests using SPOT-VEGETATION 11-y time series. Geoscience and Remote Sensing Symposium (IGARSS), 2012 IEEE International; 2012 22-27 July 2012.
[13] Verbesselt J, Zeileis A, Herold M, editors. Near real-time change monitoring in tropical forest using MODIS satellite image time series. 1st Forestry Workshop: Operational Remote Sensing in Forest Management, Prague, Czech Republic, 02 - 03 June, 2012; 2012; Prague, Czech Republic: EARSeL.

Land use and land cover change detection at Mirzapur Union of Gazipur District of Bangladesh using remote sensing and GIS technology

R Yesmin[1], A S M Mohiuddin[1], M J Uddin[1], and M A Shahid[2]
[1]Dept. of Soil, Water and Environment. University of Dhaka. Bangladesh.
[2]Bangladesh Space Research and Remote Sensing Organization (SPARRSO), Bangladesh.

Email: mjuddin66@yahoo.com

Abstract. A study was conducted with a view to identify and quantify the changes in land use and land cover occurred during the last 20 years at Mirzapur Union of Gazipur district of Bangladesh using remote sensing and GIS technologies. Two LANDSAT TM images of 1989 and 2009 with 30mx30m spatial resolution were used to determine the temporal land cover changes. Subsequently, a ground verification was done in the study site. The study revealed that forest cover was decreased by 20.29 % and settlement area was found to increase by 28.64% and water bodies was decreased by 6.25 %. In the same period of time, bare land was found to increase by 20.91 % due to the effect of clearing of forest area which is not replanted again. A lot of new infrastructures has been built by this time. Population pressure becomes double enhancing the deforestation of Sal (Shorea robusta). Most prominent features are the emergence of brick fields and various industries. Thus, the above study demonstrated the usefulness of RS and GIS technology regarding resource management and urban planning.

1. Introduction

Land is one of the most important non-renewable natural resources. The continuous growth of population exerts a serious pressure on land which is associated with agricultural demand and use, urbanization and economic development, science and technology etc. These all are accelerating the global warming and associated devastating impacts on limited land and other natural resources. So land use and land cover (LULC) change is a major issue of global environmental change and an important field in this area of research. Timely and precise information on land use and land cover change detection of earth's surface is extremely important for understanding relationships and interactions between human and natural phenomena for better management of natural resources and in decision making. So land-use/land-cover change is an important field in global environmental change research. Inventory and monitoring of land-use/land-cover changes are indispensable aspects for further understanding of change mechanism and modeling the impact of change on the environment and associated ecosystems at different scales [1-2]. To detect land use and land cover change, a comparison of two or more satellite images acquired at different times can be used to evaluate the temporal or spectral reflectance differences that have occurred between them [3]. The results quantify the land cover changes and demonstrate the potential of multi temporal Landsat data to provide an accurate, economical means to map and analyze changes in land cover in a spatio-temporal framework. It is also used as inputs to land type for management and policy decisions with regard to

Content from this work may be used under the terms of the Creative Commons Attribution 3.0 licence. Any further distribution of this work must maintain attribution to the author(s) and the title of the work, journal citation and DOI.
Published under licence by IOP Publishing Ltd

varied themes that has link with space such as urbanization, watershed management, deforestation, land degradation and so on.

The aim of this study is to monitor and analyze the spatio-temporal land use/land cover change patterns in Mirzapur union of Gazipur district in Bangladesh using multi-temporal Landsat imageries for the past 20 years (1989-2009).

2. Materials and methods

The study area, Mirzapur union of Gazipur district of Bangladesh (Figure 1) has been selected because migration of huge population is seen as because, it is very near to Dhaka mega city. It carries a good site for urban development center. As a result of population pressure, the local people are destroying forest, trees and valuable cultivable land for making various industries and brick fields etc. As a result, the forest area is gradually decreasing and the environment is changing significantly. Considering the above situations, two Landsat satellite image TM (Thematic Mapper) at resolution 30x30 m (projection: Bangladesh Transverse Mercator) were used in the present study. Some secondary data sources were also used such as topographical, soil and physiographic etc. The image related methodology adopted for this study involves both the Digital Image Processing (DIP) and GIS analysis. DIP was carried out in the computer system having the following software configuration; ERDAS IMAGINE- Image Processing Software was used for geo-referencing, classification, enhancement of the images and other related processing (Version 9.3). ARC Info Software was used for digitizing, displaying and analyzing all the vector layers. Landsat TM 1989 and 2009 were preprocessed to remove data errors and anomalies. Digital image processing system was executed to calculate the land cover (Figure 2).

2.1 Geo-referencing

Each of the images was geo-referenced. A sufficient number of Ground Control Points (GCPs), which were readily identifiable on the image and on pre-rectified image, were selected for calculations of a least-square fit, and the results were then used to adjust the image to image coordinates. Preferred GCPs were taken from recognizable, permanent features such as road intersection, pond, large building and the points were well dispersed for accurate rectification. The images were projected on to a plane, then rotated and scaled to a map projection system. For each image transformation, a root square means (RMS) error was calculated. The maximum RMS error was 1.5 pixels. This corresponds to a ground distance of 45 m for the TM images. Using this procedure, each raw satellite image was resampled, using the nearest neighbor algorithm, and transformed in to a file referenced to the Universal Transverse Mercator (UTM) projection.

2.2 Classification

A digital classification scheme was developed which categorized major land cover types (water bodies, forest, settlement, and bare land). Digital satellite images were classified using image processing techniques to enable the assignment of land cover classes to areas of similar spectral characteristics. Both supervised and unsupervised methods of classification were done. Unsupervised classification was based on ISODATA algorithm available in the ERDAS Imagine software. While the supervised classification method was based on maximum likelihood algorithm. The following chart depicts the techniques used to obtain land cover classes in vector form (Figure 3).

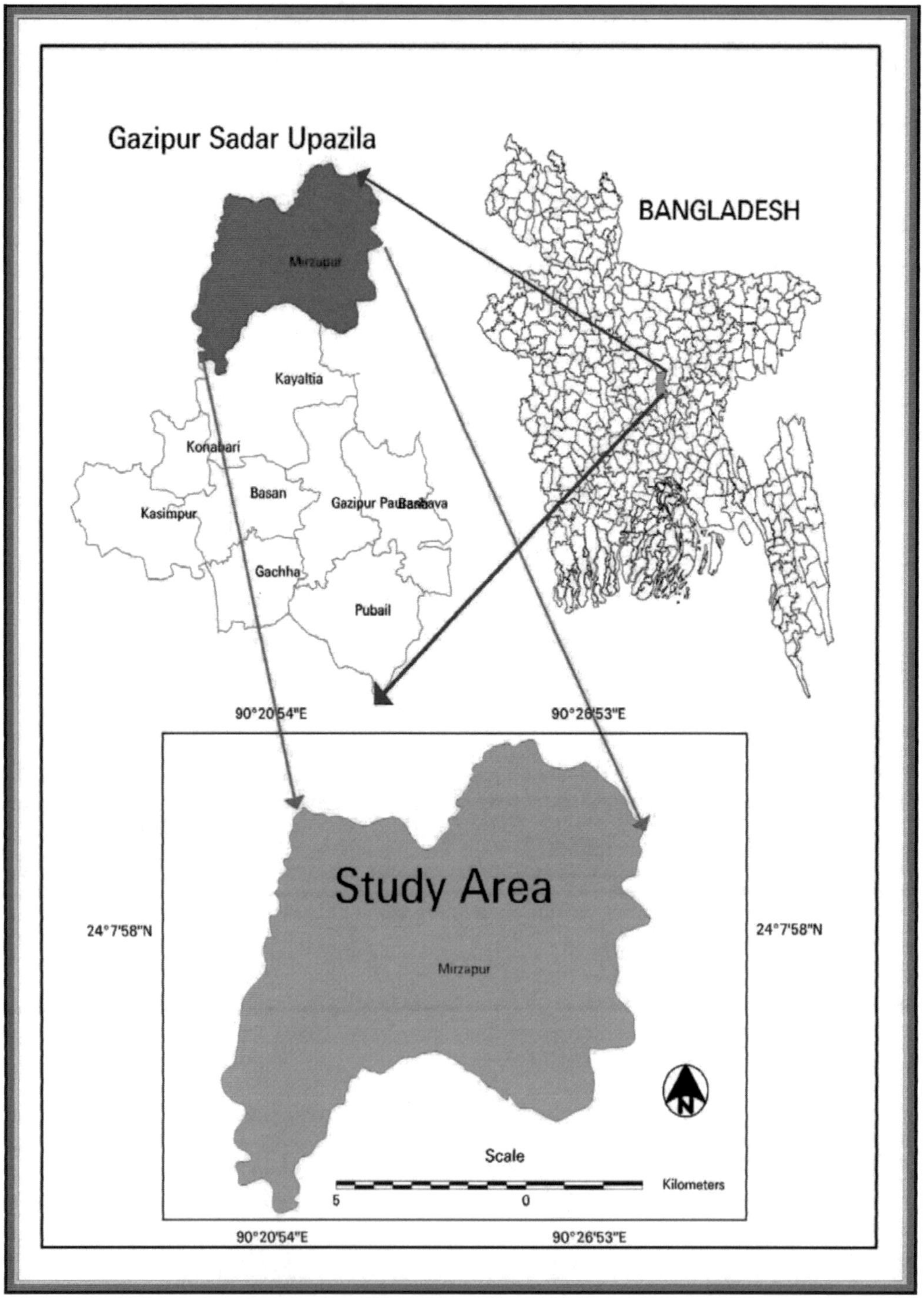

Figure 1. Location map of study site (Mirzapur Union of Gazipur District).

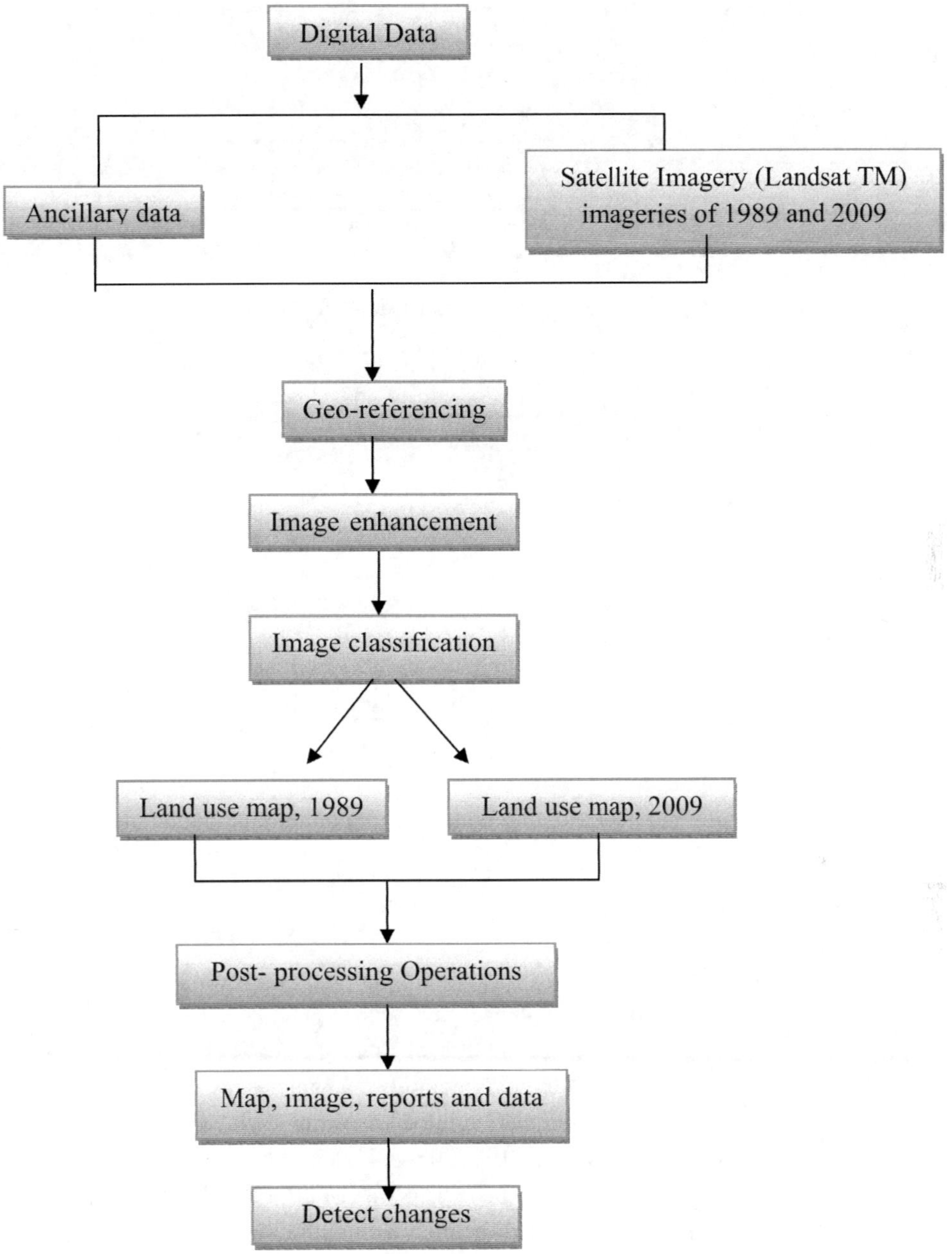

Figure 2. Depicting methods and procedures used in the study.

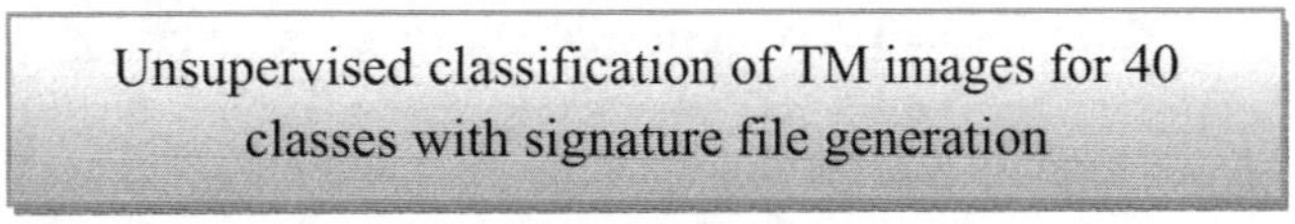

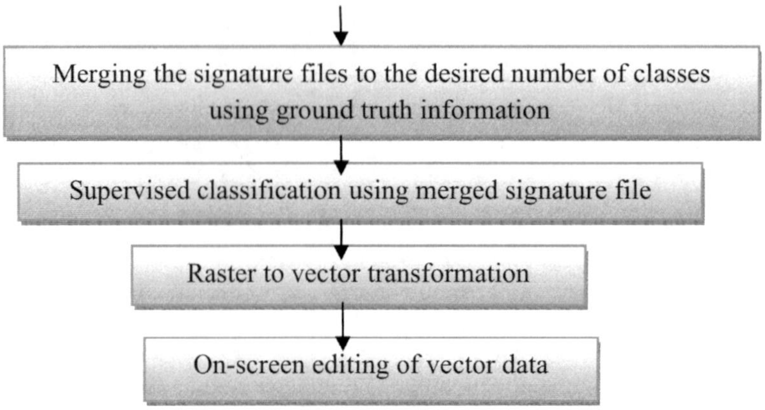

Figure 3. Techniques used to obtain land cover classes.

Four broad land cover classes were assigned each of the two images in time series: settlement, water body, forest, bare land (fallow land). To verify the land cover classes in the field the authenticity of the class boundary were checked. The location positions with latitude and longitude were noted down and geo-features characteristics were recorded in the field and then independently these records were examined whether the interpreted classes in the maps were right or wrong. Area calculation of the final four classes was done for both of the images.

3. Results and discussion

A classified image of the land cover 1989 (Figure 4) and 2009 (Figure 5) were done where four classes of land cover were identified and interpreted such as forest area, settlement, water bodies, bare (Fallow) land. The class forest was consisting of mainly Sal forest (Shorea Robusta). Adjacent crop fields and trees were included in settlement because of lower resolution of the images. The forest was identified in the satellite images due to its color in the images and by the knowledge of the position of the Sal forest in the study area. In 1989, about 2986.74 ha or 34.35% land was found to be occupied with forest area which was declined to 2380.59 ha or 27.38% in the year of 2009 (Figure 6). Within this 20 years period, the forest was decreased by 606.01 ha or 20.29%. The main cause of deforestation is the expansion of agricultural land which results clearing of forest cover especially in the plane areas where the highest population density exist. After deforestation some part of *Madhupur* forest has been converted to banana and pineapple plantation [4]. Some of the deforested area has been converted to rubber garden by the government. Population and economic pressures are two prominent factors leading to clearing of forests [4].

Settlement or homestead was appeared in all part of the study area and it was found to be lean, semi-compact, compact and semi-sprinkled pattern. The land cover class settlement includes houses, markets, shops, adjacent lands and associated trees, crop land, road and industrial area etc. Extensive areas of settlement were found to be mixed with vegetation. In the satellite images, settlement class was appeared as mixed color. In 1989, area occupied by settlement was 3959.55 ha or 45.54%. In 2009, the area occupied by settlement was 5093.10 ha or 58.58% of the total area. So, settlement area was increased by 28.64% between 1989 to 2009 (Figure 6). Population census report by BBS (2009) also indicates the increasing trends of population by those years [5].

Water bodies were identified in images by dark tone (blue, black color in classified images) that reflects very little amounts of light that varies upon turbidity (the higher the turbidity, the higher will be the light reflectance and lighter will be the color). Water bodies also identified by their shape, as rivers are appeared meandering and pond are rectangular in shape. In 1989, area of water bodies was occupied by 445.41 ha or 5.12% of total study area. In 2009, the areas occupied by water bodies was 413.46 ha (4.8%) So, water bodies were found to have decreased to 31.95 ha or 6.25% of the total area (Figure 6). The results were in well accordance with the findings of the authors [6, 7]. According to semi detailed soil survey data of SRDI [8]. Landsat TM image data revealed that the water bodies in terms of area coverage prevailed a decreasing trend in Savar Upazila of Bangladesh over time [9,10]. This change might take place due to sedimentation/siltation, shifting of water flow direction and rising of river beds.

Bare/Fallow land showed a white signature and fine grain size of the sands formed smooth surface. Low moisture content of bare soil (due to dry season) appeared to be in brighter tone than that of the other features. In 1989, bare land area was 669.33 ha or 7.7% and in 2009 it was 809.91 ha or 9.31% in 2009. The result indicated that the bare land was increased by 140.58 ha (20.91%) during the last 20 years (Figure 6). The causes of increasing bare land might be due to clearing of forest area for fuel and other purposes. Baid or lowland depression which was identified in the satellite images due to its shape. From the TM image analysis, the feature baid in 1989 was 1856.52 ha of total study area (21.35%) where in 2009, the area was 1156.86 ha (13.30%) which was found to be decreased by 699.66 ha (37.71%) (Figure 6). In most cases, this baid area was occupied by settlement which was due to increasing pressure of population. In Mirzapur union, brick fields were dominantly observed in both sides of the bank of the river Lubundaha where part of wetland turned into fallow lands. In these areas, brick fields are increasing day by day as the urbanization is going fast in the study site.

4. Conclusion

A significant change was detected in land use and land cover in the Mirzapur union of the Gazipur district of Bangladesh during the last 20 years (1989-2009). The changes occurred in different land covers like forest, water bodies were decreased by 20.29%, 6.25% respectively and where settlement (homestead), bare land (fallow land) were increased by 28.64% and 20.91% respectively. All of these changes were absolutely due to the increase of population or unplanned urbanization destroying the local forest resources.

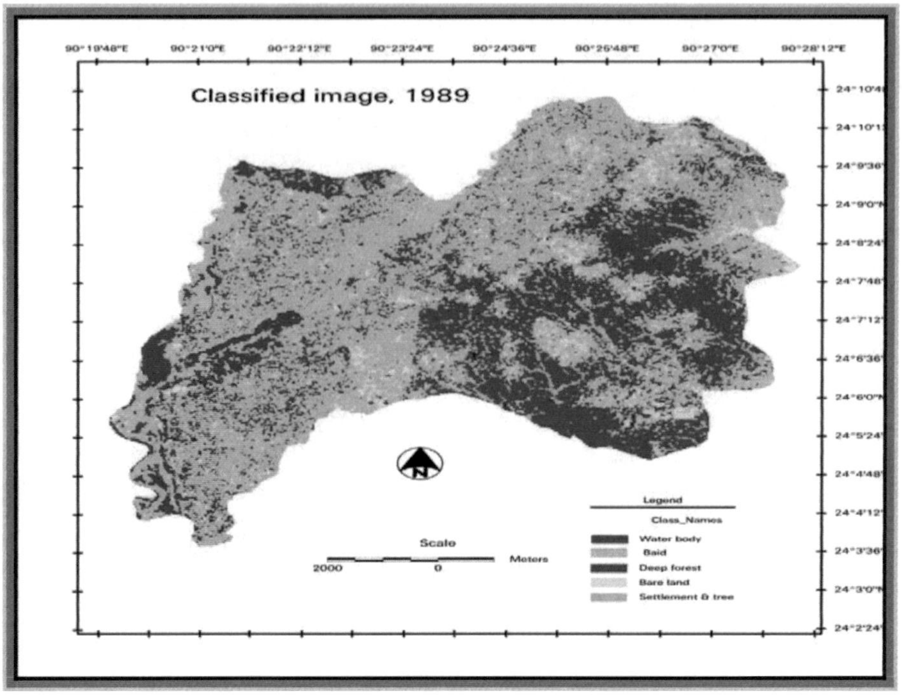

Figure 4. Land cover map of the year 1989.

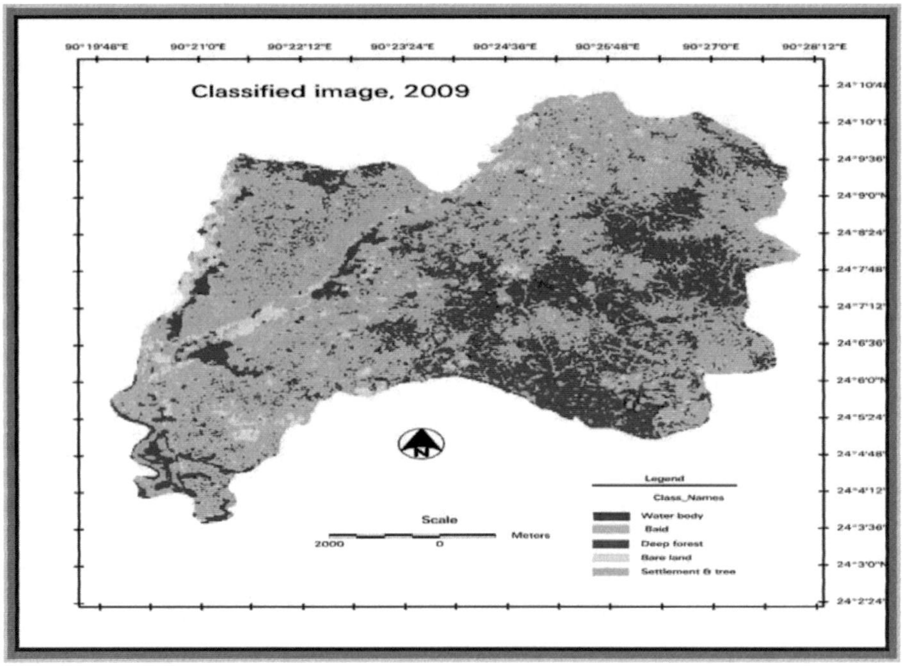

Figure 5. Land use and land cover map of the year 2009.

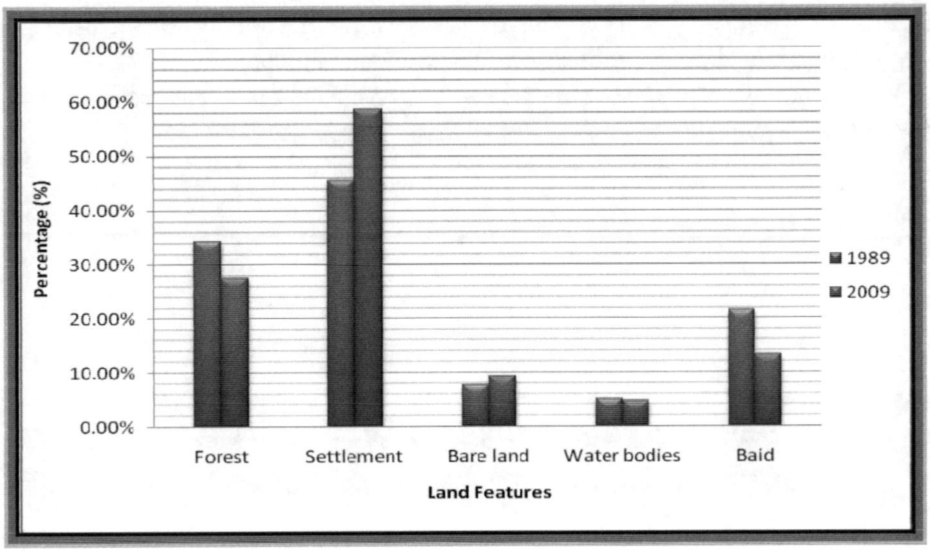

Figure 6. Land cover change in the Mirzapur Union Over the years 1989 and 2009.

5. References

[1] Tunner B L, II Skole D, Sanderson S, Fischer G, Fresco L and Leemans R 1995 *Land use and land-cover change; Science/Research Plan.* IGBP Report No. 35. HDP Report No. 7, International Geosphere-Biosphere Programme, Stockholm, Sweden, 132p.

[2] William E, R William B M and Turner B L 1994 *Modeling land use and land cover as part of global environmental change. Climatic Change.* **28**: 45–64.

[3] Yuan D and Elvidge C D 1998 Land cover change detection pilot study: Washington D.C area experiments. *Remote sensing of environment.* **66**: 166- 78.

[4] Uddin K and Gurung D R 2010 *Land cover change in Bangladesh: A knowledge based classification approach.* 10th International Symposium on High Mountain Remote Sensing Cartography, International Center for Integrated Mountain Development, Kathmandu (ICIMOD). Khumaltar, Lalitpur. Nepal. Pp.41-46.

[5] BBS 2009 Population Census of Bangladesh. Community Series: Zila-Dhaka. *Statistical Yearbook of Bangladesh. Bangladesh Bureau of Statistics (BBS).* Ministry of Planning, Government of the People's Republic of Bangladesh, Dhaka, Bangladesh.

[6] Momtaz F 2009 *Remote Sensing and GIS techniques in land use and land cover assessment of Savar Upazila with detailed classification of Amin Bazar union and Sarulia mouza. MS thesis*. Dept. of Soil, Water and Environment. University of Dhaka. Bangladesh.

[7] Akhter S 2009 *Assessment of land cover change in Savar Upazila with detailed land cover study at Kaundia union and Birulia mouza: A RS-GIS Approach. MS Thesis.* Dept. of Soil, Water and Environment. University of Dhaka. Bangladesh.

[8] SRDI 1992 *Land and Soil Resource Utilization Guide of Gazipur Sadar Upazila. Soil Resource Development Institute (SRDI).* Ministry of Agriculture. Bangladesh.

[9] SPARRSO 1997 *Annual report. Bangladesh Space Research and Remote Sensing Organization (SPARRSO).* Ministry of Defense. Bangladesh.

[10] SPARRSO 2006 *Annual report. Bangladesh Space Research and Remote Sensing Organization (SPARRSO).* Ministry of Defense. Bangladesh.

IOP Conf. Series: Earth and Environmental Science **20** (2014) 012056 doi:10.1088/1755-1315/20/1/012056

An analysis of LULC change detection using remotely sensed data; A Case study of Bauchi City

K M Kafi[1], H Z M Shafri[2], A B M Shariff[3]
Department of Civil Engineering, Faculty of Engineering, Universiti Putra Malaysia, 43400.Serdang, Selangor Malaysia

E-mail: kameelmk@yahoo.com

Abstract. Bauchi is one of the cities in the north eastern part of Nigeria that has witnessed tremendous expansion as a result of rapid urbanization. For the past decade, the city has been known to be relatively small both in size and population. But today, Bauchi is one of the fastest growing cities in northern Nigeria. Therefore, it is paramount to detect the nature and magnitude of these changes in order to determine its direction of future expansion. To achieve that, Remotely Sensed data from Landsat 7 ETM+ and Landsat 8 were utilized for the purpose of LULC change detection in Bauchi city. Landsat 7 image of 2003 and Landsat 8 image of 2013 were obtained for the purpose of the study. Three change detection algorithms were used to detect areas of change. They include supervised, unsupervised and post classification comparison. While the unsupervised classification was found to be less effective, the supervised classification produced good results with overall accuracies of 93.5% and 89.7% for the 2003 and 2013 respectively. Finally, "from-to" change was obtained using the post classification comparism.

1.0 Introduction

Cities are dynamic; this is because change is inevitable. These changes can be attributed to one factor or the other depending on socio economic, political and climatic condition of a given area. However, one fundamental factor behind city change both in terms of size and pattern remains the same for most cities, i.e "population growth". Nevertheless, other factors attributing to Land Use Land Cover (LULC) change are directly or indirectly dependent on population growth. As the population of a given area increase, the demand for land cover such as built up area also increase, while other land cover classes such as bare land, vegetation, decreases as a result of the increased demand for built up area. Consequently, these changes in most cases if not continuously checked comes with certain negativities which may affect not only the environment but also its inhabitants. However, Accurate and up-to-date land cover change information is necessary in understanding and assessing the environmental consequences of such changes [1]. Remote Sensing has the capability of capturing such changes, extracting the change information from satellite data requires effective and automated change detection techniques [2]. The importance of remote sensing was emphasized as a 'unique view' of the spatial and temporal dynamics of the processes in urban growth and land use change [3]. Satellite remote-sensing techniques have therefore, been widely used in detecting and monitoring land cover change at various scales with useful results [4]. As land becomes increasingly

Content from this work may be used under the terms of the Creative Commons Attribution 3.0 licence. Any further distribution of this work must maintain attribution to the author(s) and the title of the work, journal citation and DOI.
Published under licence by IOP Publishing Ltd

important and competition among alternative use intensifies, shift in land use may be affected by further institutional and environmental constraint [5]. Thus, it is imperative to consistently monitor the changes that are taking place so that planners can evaluate the impacts of ''from-to'' change and plan or suggest alternative land use for development purposes which will in turn help policy making and implementation process. Furthermore, monitoring urban changes leads to forecast of the amount of changes and the location of future built-up areas, which is extremely valuable [6]. This paper seeks to utilize remotely sensed data and GIS tool to analyse the LULC of Bauchi Town for the purpose of detecting change in the area by comparing between two dates images.

2.0 Study area

Bauchi is geographically located in the North Eastern part of Nigeria. It is regarded as the North Eastern gate because it is the only state in the North East that is bounded by at least one state in each of the Northern geo-political zones. Bauchiis located between latitudes 9° 3′ and 12° 3′ north of the equator, while longitudinally it is located between 8° 50′ and 11° east of the Greenwich meridian. It has an altitude of 785.2 meters. The vegetation covers of Bauchi are of two types namely; the Sahel Savanna and the Sudan savannah.

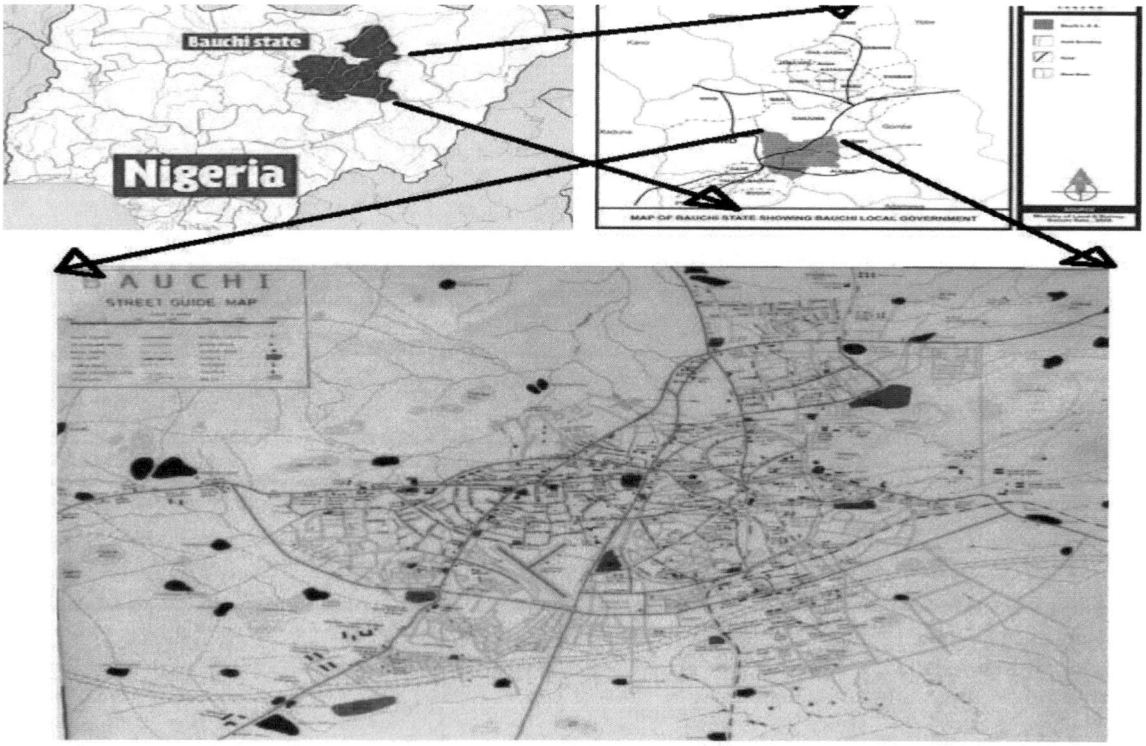

Figure 1. Map showing the study area.

Similarly Bauchi has two distinct types of terrain, namely rocky and flat terrain. The rocky terrain are mostly found in those areas that share boundary with plateau State i.e areas around the southern part of Bauchi, while the flat terrain regions are found mostly in the northern part of Bauchi where the soil type is generally sandy or semi desert soil. The flat surface of the terrain has been subjected to considerable human interference through farming, grazing and construction activities.

Bauchi experience it's raining season right from mid-June till mid – October, with August recording the highest amount of rainfall of 340 mm. the total mean annual rainfall stands at 1,091.4 mm. While the hottest temperature is experienced around the months of April and May with the highest temperature of $40.56^{O}C$. The coldest temperature (i.e during the harmattan season) is experienced around December and January with the minimum temperature of $6.11^{O}C$ and $7.22^{O}C$ respectively.

Bauchi city is the capital of Bauchi state and it is bounded by seven local governments which comprises of Kirfi, Ganjuwa, Alkaleri, TafawaBalewa, Dass and Toro respectively. While as a state, Bauchi share border with seven states that cut across the three different geo-political zones of the Northern Nigeria. These states include Kaduna, Jigawa and Kano State from the North West Zone, Plateau State from North Central Zone and Gombe, Taraba and Yobe State from the North East Zone.

3.0 Methodology

3.1 LULC classifications

The different LULC classes of the study area were grouped into five for easy analysis and assessment of change detection. The LULC classification includes Built up area, Wet land, Farm land, Shrub/grass and Other land. The built up area category includes high, medium and low densities; disperse settlements, and all other man-made structures such as schools, hospitals, industries, bridges and roads. The wet land category includes wet and dry river, river banks, water logged areas and small ponds. The farm land category includes lowland crop and highland crop fields with and without crops. Grass/shrubs were combined into a common class; this was due to the fact that grass and shrubs naturally grows together. Finally, other land category includes ditches, and rocks.

3.2 Image classification

3.2.1 Unsupervised. Unsupervised classification was carried out using the 6 bands of the multi date images in order to classify the image into clusters and to identify the different change classes by using both IsoData and K – Mean. This classification did not produce the needed result because some classes were merged and misclassified.

3.2.2 Supervised. The supervised classification was carried out using the Maximum Likelihood and a composition of band 1, 2 and 3 by applying the false color composite to select Region of Interest (ROI) for features like water body and shrubs/grass, while the true color composite was applied in choosing the ROI of built up area, farm land and other land. The spectral signature of each class was obtained from the images using ENVI. This was done through selection of ROI for each of the LULC category. The ROI helps in producing the map by defining an area in the map based on the color assigned to that category and the spectral homogeneity of the pixels of chosen area. This classification has yielded a good result after subjecting the classified maps to a confusion matrix. An overall accuracy of 93.50% and Kappa coefficient of 0.92 were obtained from the 2003 map, while for the 2013 map, the overall accuracy was found to be 89.66% and a Kappa coefficient of 0.848.

3.3 Change detection

3.3.1 Post classification. The two independently classified images where then ran for post classification comparison in order to produce a change detection analysis. By using the change detection statistical tool of the post classification, the matrix table of ''from – to'' change class was obtained. Finally, this classification proved to be effective because it presents the advantage of indicating the nature and magnitude of change that has taken place through pixel by pixel comparison.

3.4 Accuracy assessment

The accuracy assessment is a comparison of a classification with ROI or ground-truth data to evaluate how well the classification represents the real world. This is produced in a matrix table showing four different types of accuracies. However, Accuracy assessment requires that an adequate number of samples per map class be gathered when the classified results are compared with actual ground conditions [7].

3.5 Determination of magnitude of change

The magnitude of change is a degree of expansion or reduction in the LULC size. A negative value will present a decrease in LULC size while a positive value will indicate an increase in the size LULC class [5].

- The magnitude of change (K) is calculated by the simple equation

$$K = F - I \qquad \text{A.1}$$

- The percentage of change (A) is calculated by the formula

$$A = \frac{F-I}{I} x\ 100 \qquad \text{A.2}$$

Where: K = magnitude of change
A = percentage of change
F = first date (2003)
I = Reference date (2013)

4.0 Results

Change detection analysis is performed not only to detect changes that have occurred, but also to identify the nature of those changes and to determine the areal extent and spatial pattern of those changes [8].The result of the five categorized LULC of 2003 is shown in table 1. A total of 2756 pixels were randomly selected as ROI for the purpose of identifying the different LULC class, the selection of the ROI was aided by the use of aerial photo and the local knowledge of the study area. An overall accuracy of 93.50% and Kappa coefficient of 0.92 were obtained from the accuracy assessment. In terms of producer's accuracy, all the classes except farm land were over 96.0%,while for the user's accuracy all classes were above 92.50%.

Table 1. Confusion matrix table of LULC derived from satellite data (2003).

Producer's	**User's**						
Class	**Built up**	**Wet land**	**Farm land**	**Shrubs/Grass**	**Other land**	**accuracy (%)**	**accuracy(%)**
Built up	706	8	42	0	0	96.36	93.39
Wet land	6	493	16	0	16	95.17	92.84
Farm land	18	4	554	22	0	86.16	92.64
Shrub/Grass	3	0	26	361	0	94.26	92.56
Other land	0	13	5	0	463	96.66	96.26
Total	733	518	643	383	479		

Kappa accuracy: 0.92. **Overall accuracy:** 93.50%.

IOP Conf. Series: Earth and Environmental Science **20** (2014) 012056 doi:10.1088/1755-1315/20/1/012056

The table below shows the results obtained from the classified map of 2013. A total of 2581 pixels were randomly selected and checked using aerial photo and local knowledge of the area. An overall accuracy of 89.655% and Kappa coefficient of 0.85 were obtained. In terms of the producer's accuracy, all classes were above 83.0%, while for the user's accuracy all the classes except farm land were above 76.30%.

Table 2. Confusion matrix table of LULC derived from satellite data (2013).

Class	Built up	Wet land	Farm land	Shrubs/Grass	Other land	Producer Accuracy (%)	User Accuracy(%)
Built up	1215	10	2	5	4	89.21	98.30
Wet land	43	208	8	0	3	86.31	76.39
Farm land	62	19	211	39	3	88.44	63.17
Shrub/Grass	34	1	17	252	0	83.44	82.84
Other land	8	3	0	6	428	97.72	96.18
Total	1362	241	238	302	438		

Kappa accuracy: 0.85. **Overall accuracy:** 89.66%.

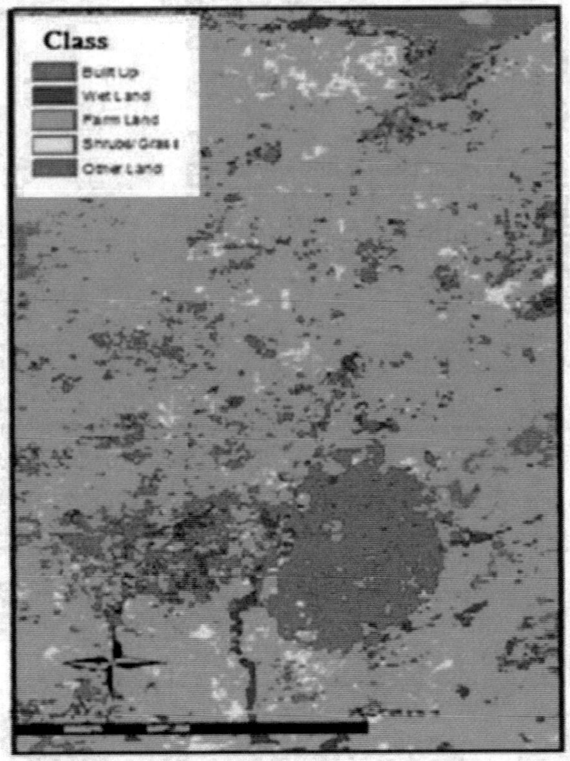

Figure 2. The classified map of Bauchi area in 2003, revealing a relatively smaller built up area surrounded by vast farm land.

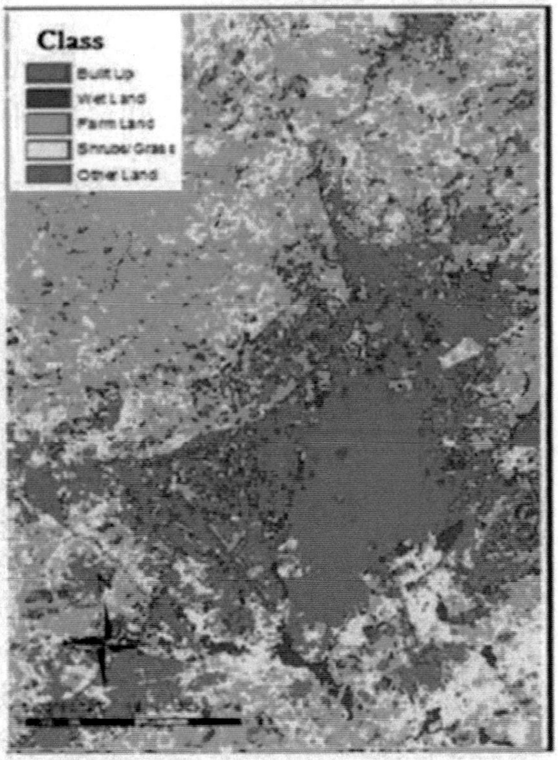

Figure 3. Change detection map of Bauchi in 2013, indicating the level of urban expansion that has taken place over the period of 10 years.

In the table below, shrub/grass has the highest positive value, indicating high increase of 696.7% from its initial 9.0 km^2, followed by built up area with a percentage increase of 138.45. On the contrary, farm land has the highest negative value, indicating a high level of conversion to other categories of land cover which reduces its size by 64.1%.

Table 3. Magnitude and percentage of LULC change from 2003 – 2013.

	2003		2013		Magnitude of change	Percentage of change
Class	Km^2	(%)	Km^2	(%)	Km^2	(%)
Built up	26.3	12.9	62.7	30.7	36.4	138.4
Wet land	5.1	2.5	9.7	4.8	4.6	90.2
Farm land	144.8	710	52	25.5	-92.8	-64.1
Shrub/Grass	9.0	4.4	71.7	35.1	62.7	696.7
Other land	18.8	9.2	7.9	3.9	-10.9	-58
Total	**204**	**100**	**204**	**100**		

The table below shows the amount of "from – to" change that has taken place over the period of Ten (10) years, where it revealed a great portion of the farm land has been converted to shrub/grass, built up area, wet land and other land respectively.

Table 4. Cross tabulation of LULC classes between 2003 – 2013 in (Km^2).

Class	Built up	Wet land	Farm land	Shrubs/Grass	Other land	Total
Built up	26.3	0	28.1	1.2	7.1	62.7
Wet land	0	5.1	4.4	0.2	0	9.7
Farm land	0	0	52	0	0	52
Shrub/Grass	0	0	58.3	7.6	5.8	71.7
Other land	0	0	2	0	5.9	7.9
Total	26.3	5.1	144.8	9.0	18.8	204

5.0 Discussion

5.1 Built up area

The built up area which comprises of the low, medium and high densities, road network and other man-made structures has witnessed a significant increase based on the analysis carried out. The total area covered by the built up area was estimated at 26.3 km^2 in 2003 has increased to 62.7 km^2 in 2013. This was as a result of the demand for shelter by both city inhabitants and emigrants from rural areas and some neighboring cities especially those who have been troubled by either religious crises in the Jos and TafawaBalewa town or the Boko Haram insurgency in some of the north eastern part of the country.

5.2 Wet land

The wet land which is made up of the river, water logged areas and small ponds has witnessed an increase from the initial 5.1 km^2 to about 9.7 km^2 resulting to a net increase of 4.6 km^2. This is because; over the past few years the amount of rainfall in Bauchi has drastically increased compared to five years back. For the past few successive years, flooding has been a serious issue both within and around Bauchi town (including peripheral settlements). This particular menace has made Bauchi to be categorized amongst the flood prone zones by the Nigerian Meteorological Agency (NIMET).

5.3 Farm Land

This constitutes the agricultural land which is mostly used for the cultivation of grains such as maize, sorghum and millet; also include cowpea such as beans and groundnut. Others include rice, sugar cane and beverages like tomatoes, onions and cabbage which are cultivated within the Fadama area. The study has shown a drastic decrease in the available agricultural land within the study area. The farm land which has an area of 144.8 km^2 in 2003 has decreased to 52.0 km^2 in 2013. This was attributed to the high demand for urban shelter and infrastructural development within the study area, another reason is that, quiet a number of farmers have abandoned farming for other petty businesses in the study area. This has resulted in conversion of some of the abandoned farm land to shrub/grass

5.4 Shrub / Grass

The shrub/grass which comprises of trees, shrubs and grasses has witnessed an increase over the last decade. This is however not unconnected to the fact that most farmers in the study area have abandoned farming activities and engages in other forms businesses. Even though some shrub/grass land were converted to built up area, still the conversion of farm land to shrubs/grass over the years, has significantly increase its size from the initial 9.0 km^2 in 2003 to 71.7 km^2 In 2013.

5.5 Other Land

This comprises of features such as rock outcrops and ditches that are found within Bauchi city and the environs. This study has also revealed a decrease in the amount of other land as a greater portion has been replaced by built up area while some of the ditches were used as waste disposal sites.

6.0 Conclusion

The utilization of Remote Sensing and GIS tools were helpful in detecting the magnitude of LULC change that has taken place in Bauchi over the span of 10 years. The study also revealed a significant expansion of Bauchi city size. All LULC classes except farm land and other land have increased over the years, with the built up area increased to 138% from 2003 to 2013. While farm land, has witnessed a drastic decrease of up to 178% over the period of 10 years. The study has revealed that the expansion of Bauchi city is more concentrated around the north eastern and South Western part of the city. This was as a result of the newly constructed estates that attracted other construction activities around the north eastern part, and also location of higher institutions such as Federal University, Federal Polytechnic and State Polytechnic around the south western part of the city. Meanwhile, expansion around the north western part of Bauchi was constrained by the location of military barracks in the area.

However, based on the LULC change detection algorithms used for the classification, the supervised classification was found to be very effective with overall accuracy of 93.5% and a Kappa coefficient of 0.94% for the year 2003, also an overall accuracy of 89.6% and Kappa coefficient of 0.848% for the year 2013. In general, the maps met the minimum accuracy of 85% as stipulated by the Anderson classification scheme, Anderson et al. [9]. The post classification was found to be helpful in producing the change detection statistics because it provided the matrix table of "from – to" change for the two dates images on the bases of pixel by pixel comparison.

Reference

[1] Giri, C, Zhu Z and Reed B 2005 "A comparative analysis of the Global Land Cover 2000 and MODIS land cover data sets". *Remote Sensing of Environment.**94*** 123-132.

[2] Roy D P, P E Lewis and C O Justice 2002 "Burned area mapping using multi-temporal moderate spatial resolution data—A bi-directional reflectance model-based expectation approach." *Remote Sensing of Environment* **83** 263-286.

[3] Batty M and Howes D 2001 Predicting temporal patterns in urban development from remote imagery. *Remote Sensing and urban analysis. S.* 185-204. London. Taylor and Francis.

[4] Martha S G, Emily H W and Nels B 2008 "Integrating muhi-temporal spectral and structural information to map wetland vegetation in a lower Connecticut River tidal marsh". *Remote Sensing of Environment, **112*** 4-048.

[5] Mahmud A &Achide, A. S 2012 "Analysis of Land Use/Land Cover Changes to Monitor Urban Sprawl in Keffi-Nigeria". *Environmental Research Journal, **6*** 129-134.

[6] Mundia, C N and M Aniya. 2005 "Analysis of land use/cover changes and urban expansion of Nairobi city using remote sensing and GIS." *Int. J. of Remote Sensing* **26**.13 2831-2849.

[7] Congalton, R G 1991 "A review of assessing the accuracy of classifications of remotely sensed data". *Remote sensing of environment, **37*** 35-46.

[8] Macleod R D and Congalton R G 1998 "A quantitative comparison of change detection algorithms for monitoring eelgrass from remotely sensed data". *Photogrammetric Engineering and Remote Sensing.***64** pp.207–216.

[9] Anderson R, Hardy E E, Roach J T and Witmer R E 1976 "A land use and land cover classification system for use with remote sensor data" *USGS Professional Paper***964**, Sioux Falls, SD, USA.

[10] Shalaby A and Tateishi, R 2007. "Remote sensing and GIS for mapping and monitoring land cover and land-use changes in the Northwestern coastal zone of Egypt". *Appl. Geography, **27*** 28-41.

[11] Rodriguez-Galiano, V and Chica-Olmo M 2012 "Land cover change analysis of a Mediterranean area in Spain using different sources of data: multi-seasonal Landsat images, land surface temperature, digital terrain models and texture". *Appl. Geography. **35*** 208-218.

[12] ZanganehShahraki S, Sauri, D, Serra P, Modugno S, Seifolddini F and Pourahmad A 2011 "Urban sprawl pattern and land-use change detection in Yazd, Iran". *Habitat Int. **35*** 521-528..

[13] Mallupattu P K. and Sreenivasula Reddy J R 2013 "Analysis of Land Use/Land Cover Changes Using Remote Sensing Data and GIS at an Urban Area, Tirupati, India". *The Scientific World J.*

[14] Walter V 2004. "Object-based classification of remote sensing data for change detection". *ISPRS J. of photogrammetry and remote sensing.**58*** 225-238.

[15] Shalaby A and Tateishi, R 2007 "Remote sensing and GIS for mapping and monitoring land cover and land-use changes in the Northwestern coastal zone of Egypt". *Appl. Geography. **27*** 28-41.

[16] Dewan A M and Yamaguchi Y 2009 "Land use and land cover change in Greater Dhaka, Bangladesh: using remote sensing to promote sustainable urbanization". *Appl. Geography, **29*** 390-401.

[17] Samir Kamh et al. 2012 "Evaluating urban land cover change in the Hurghada area, Egypt, by using GIS and remote sensing".*Int. J. of Remote Sensing*, **33** 41-68.

[18] Yang L, Xian G, Klaver J M and Deal B 2003 "Urban land-cover change detection through sub-pixel imperviousness mapping using remotely sensed data. *Photogrammetric engineering and remote sensing*". ***69*** 1003-1010.

[19] Singh A 1989 "Review Article Digital change detection techniques using remotely-sensed data". *International journal of remote sensing*, ***10*** 989-1003.

[20] Todd W J 1977 "Urban and regional land use change detected by using Landsat data". *J. of Research by the US Geological Survey*.**5** 527–534.

[21] Asselman N E and Middelkoop H 1995 "Floodplain sedimentation: quantities, patterns and processes". *Earth Surface Processes and Landforms*.***20*** 481-499.

[22] Gamba P, Aldrighi M and Stasolla M 2011 "Robust extraction of urban area extents in HR and VHR SAR images". *Selected Topics in Appl. Earth Observations and Remote Sensing, IEEE J.* **4** 27-34.

[23] Mortimore M, Ba M, Mahamane A, Rostom R S, del Pozo P S and Turner B 2005 "Changing systems and changing landscapes: Measuring and interpreting land use transformation in African drylands" *GeografiskTidsskrift-Danish Journal of Geography*, ***105*** 101-118.

[24] Woodcock Curtis E W and Alan H Strahler 1987 "The factor of scale in remote sensing". *Remote sensing of Environment* **21** 311-332.

[25] Coppin P, Jonckheere I, Nackaerts K, Muys B and Lambin E 2004 "Review ArticleDigital change detection methods in ecosystem monitoring: a review" *International journal of remote sensing **25*** 1565-1596.

[26] Petit C C and E F Lambin 2003 "Integration of multi-source remote sensing data for land cover change detection." *Int. J. of Geographical Information Sci.* **15**.8 (2001): 785-803.

[27] Mas J F 2010 "Monitoring land-cover changes: a comparison of change detection techniques." *Int. J. of remote sensing* **20** 139-152.

[28] Chen Zhaohua and Jinfei Wang 2010 "Land use and land cover change detection using satellite remote sensing techniques in the mountainous Three Gorges Area, China." *Int. J. of Remote Sensing* **31** 1519-1542.

[29] Yang X and C P Lo 2002 "Using a time series of satellite imagery to detect land use and land cover changes in the Atlanta, Georgia metropolitan area."*Int. J. of Remote Sensing* **23** 1775-1798.

[30] Bolstad Paul V and T M Lillesand 1992 "Rule-based classification models: flexible integration of satellite imagery and thematic spatial data."*Photogrammetric Engineering and Remote Sensing* **58** 965-971.

[31] Lo C P and Xiaojun Yang 2002 "Drivers of land-use/land-cover changes and dynamic modeling for the Atlanta, Georgia metropolitan area." *Photogrammetric Engineering and Remote Sensing* **68** 1073-1082.

[32] Rosin P L and Ioannidis E 2003 "Evaluation of global image thresholding for change detection". *Pattern Recognition Letters*. ***24*** 2345-2356.

Monitoring the variability of precipitable water vapor over the Klang Valley, Malaysia during flash flood

W Suparta[1,*], R Rahman[1] and M S J Singh[2]
[1] Space Science Centre (ANGKASA), Institute of Climate Change, Universiti Kebangsaan Malaysia, 43600 Bangi, Selangor Darul Ehsan, Malaysia
[2] Dept of Electrical, Electronic and Systems Engineering, Universiti Kebangsaan Malaysia, 43600 Bangi, Selangor Darul Ehsan, Malaysia

*E-mail: wayan@ukm.my

Abstract. Klang Valley is a focal area of Malaysian economic and business activities where the local weather condition is very important to maintain its reputation. Heavy rainfalls for more than an hour were reported up to 40 mm in September 2013 and 35 mm in October 2013. Both events are monitored as the first and second cases of flash flood, respectively. Based on these cases, we investigate the water vapor, rainfall, surface meteorological data (surface pressure, relative humidity, and temperature) and river water level. The precipitable water vapor (PWV) derived from Global Positioning System (GPS) is used to indicate the impact of flash flood on the rainfall. We found that PWV was dropped 4 mm in 2 hours before rainfall reached to 40 mm and dropped 3 mm in 3 hours before 35 mm of rainfall in respective cases. Variation of PWV was higher in September case compared to October case of about 2 mm. We suggest the rainfall phenomena can disturb the GPS propagation and therefore, the impact of PWV before, during and after the flash flood event at three selected GPS stations in Klang Valley is investigated for possible mitigation in the future.

1. Introduction

Klang Valley is a focal area of Malaysia where many activities are located and local weather condition plays an important role. Information on weather variability has a significant effect on economic activities, including communications, services and utilities as well as the local community [1]. As Klang Valley is the main economic area, it is very upsetting for the area to be affected by flash flood every year. Hence, more studies are needed to be carried out to provide basic information in setting up a mitigation plan against the disaster. National Oceanic and Atmospheric Administration (NOAA) has defined flash flood occurring less than 6 hours is a result of excessive rainfall within a short period of time while Department of Irrigation and Drainage (DID) Malaysia has defined flash flood as an unexpected flood due to heavy rainfall in local and its surrounding area. DID is monitoring flood based on daily total rainfall and river water level along nationwide river stream and have classified the water level of a particular station into four categories: normal, alert, warning and danger level depending on its cross sectional area. Daily total rainfall is also classified into 4 categories; light (1-10 mm), moderate (11-30 mm), heavy (31-60 mm) and very heavy (>60 mm). Information on local flood status is currently being spread online through an integrated system that monitors river water level and total daily rainfall.

Several studies on the hydrographical factor such as [2] were carried out but studies on the meteorological component need to be improvised [3]. Storm water Management and Road Tunnel

Content from this work may be used under the terms of the Creative Commons Attribution 3.0 licence. Any further distribution of this work must maintain attribution to the author(s) and the title of the work, journal citation and DOI.
Published under licence by IOP Publishing Ltd

(SMART) tunnel was developed to solve issues on flash flood and peak traffic jam hours in the Klang Valley. However, this system has become less effective due to uncontrolled development and economic activities in the area. Rapid development of building construction and public facilities may cause changes in meteorological and geological of a focal area [4]. As of now, Global Positioning System (GPS) Meteorology is widely used in predicting weather nowcasting and in the study extreme weather events such as severe flood, storm and typhoon. A research in China [5] during a 2-day observations found that the trend of precipitable water vapor (PWV) from GPS is high before heavy rainfall.

The main purpose of this study is to monitor the variability GPS PWV, total daily rainfall and surface meteorological data during short-term local weather nowcasting. According to [6] and [7], variation of GPS PWV in Peninsular Malaysia is highly correlated with daily total rainfall. Two incidents of flash flood in September and October 2013 which occurred during an inter-monsoon were investigated.

2. Methodology

2.1. Dataset and Location

The main parameters used in this study were GPS and meteorological data set. GPS data was supplied by Department of Survey and Mapping Malaysia (DSMM) via RTKnet website and meteorological dataset (pressure, mean surface temperature, relative humidity) were obtained from Malaysia Meteorological Department (MMD) and downloaded from the flight-meteorology website (http://www.wunderground.com). Supporting dataset consists of total daily rainfall and water level was obtained via *infobanjir* website developed by DID. This study considered all DID stations located along the Klang River within the Federal Territories of Kuala Lumpur. Figure 1 showed the location of GPS stations around incident area, which located less than 60 km distance and tabulated in details in Table 1.

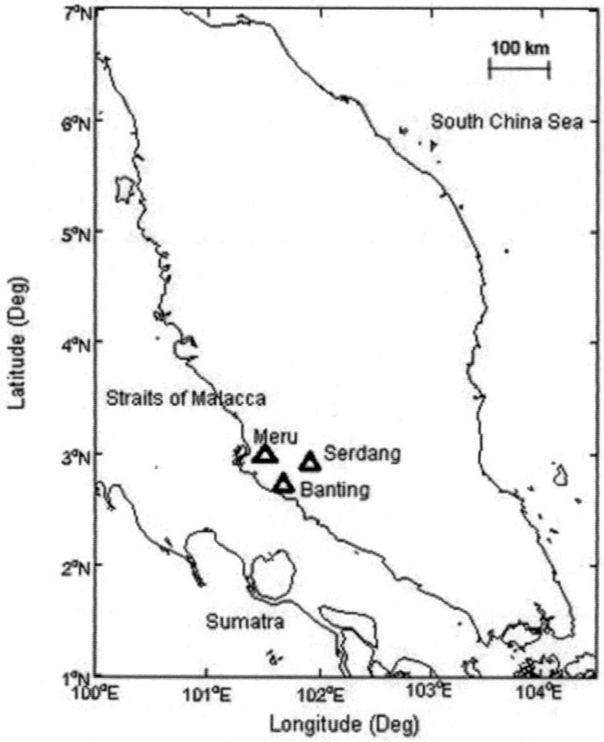

Figure 1. Location of GPS stations.

The Tropical Rainfall Measuring Mission (TRMM) precipitation data archive was used to find out the association of accumulated precipitation and rainfall. Data was freely obtained from http://gdata1.sci.gsfc.nasa.gov, which limited to 3 hours daily data measurement. Data was selected over Peninsular Malaysia within latitude of 0°N to 8°N and longitude of 99°N to 105°N.

Table 1. GPS coordinates and distance from point of incident.

GPS Station	Latitude (Deg)	Longitude (Deg)	Height (m)	Distance (km)
Banting (BANT)	2.83 N	101.54 E	8.83	57
Meru (MERU)	3.14 N	101.41 E	6.43	27
Serdang (UPMS)	2.99 N	101.72 E	100.38	28

2.2. Data Processing

GPS and meteorological data at every GPS station need to be cleaned before it can be processed to determine PWV. The useful noise of GPS reading as a result of neutral atmospheric delay in identifying PWV has been explained in detail by Suparta et al. [8]. The processing of PWV value is developed by using Matlab program called TroWav 2.0, which involved the regional weighted mean temperature, $T_m = 0.83663T_s + 48.103$ instead of global weighted mean temperature [9].

The main purpose of GPS station availability at particular places in Malaysia is for mapping and geodetic studies. Therefore, all GPS stations in the studied area are not equipped with meteorological sensors. As of that, meteorological parameter from interpolating technique has been used in this study [10] to determine PWV. Pattern of PWV from particular GPS station at 5 days duration during pre and post flash flood events were studied. This step also applied to study the variability of rainfall, water level and meteorological parameters. TRMM data was compared temporally and spatially to the heavy rainfall event to indicate the occurrence of flash flood.

3. Result and discussion

The daily variations of GPS PWV, total rainfall and river water level for flash flood events on 1 September and 10 October are depicted in Figure 2 and Figure 3, respectively. The 5 days mean of rainfall in September shows 2 mm higher compared in October. This September event also indicated double mean standard deviation (STD) compared to October. We also identified that the rainfall occurred every day for 5 days in September and 3 days in October.

A simple statistical analysis such as mean and STD of GPS PWV and rainfall was carried out. Both variations proved with river water level, meteorological parameters as well as TRMM precipitation image.

3.1. GPS PWV variation during flash flood in September 2013

Figure 2 showed the GPS PWV variability at three GPS stations with respect to total rainfall and water level changes. The incident of flash flood was indicated specifically at 18:00 LT that caused havoc in Klang Valley as shown by the shaded area. In the figure, phase 1, phase 2 and phase 3 are indicated for pre, during and post of flash flood events, respectively. Figure 2a showed the peak of PWV that correlated with peak of rainfall and water level. From September 2013 event, DID determined the water level into 4 levels: normal (14m), alert (17m), warning (18m) and danger (19m). By referring to the figure, the first light rainfall was recorded in late evening on 30 August due to a drop in PWV at all GPS stations (Figure 2b). It caused river water level to rise 0.5 m higher than the normal level for 2 hours (Figure 2c). A similar trend of PWV repeatedly occurred on the following day with moderate rainfall (11 mm) resulted in the rise of river water level 1.5 m over a normal level for few hours. This 9 hours of rainfall caused 3 mm drop of PWV at UPMS and BANT and 1 mm drop at MERU, but the drop in BANT station occurred earlier than in Klang and UPMS.

The next episode of rainfall that caused a flash flood event was in evening of 1 September. The rainfall was recorded to be 40 mm as a result of 6 mm PWV drop at UPMS and 2 mm PWV drop at Klang 2 hours before the heavy rainfall event. However, PWV at BANT did not show a similar trend as the PWV at UPMS and Klang. PWV at BANT was varied normally where it dropped early in the morning and became highest in the afternoon and later decreased in the evening. This phenomenon may be due to normal water evaporation from the earth's surface during sun rise and sunset. The heavy rainfall resulted in 1.5 meter rise of river water level that sustained until midnight of 1 September. This was also another factor contributed to the flash incident at flood-prone areas.

After the occurrence of heavy rainfall on 1 September, there was no rainfall recorded on 2 September except for an hour in the early morning. We found all GPS PWV reached their minimum peaks on 2 September followed by light and moderate rainfall on 3 September. It seemed the circulation of evaporation occurred on 2 September, which results in whole day rainfall on the following day. From the case of flash flood September 2013 analysis we found that the GPS PWV decreased almost 4 mm at 2 hours before heavy rainfall.

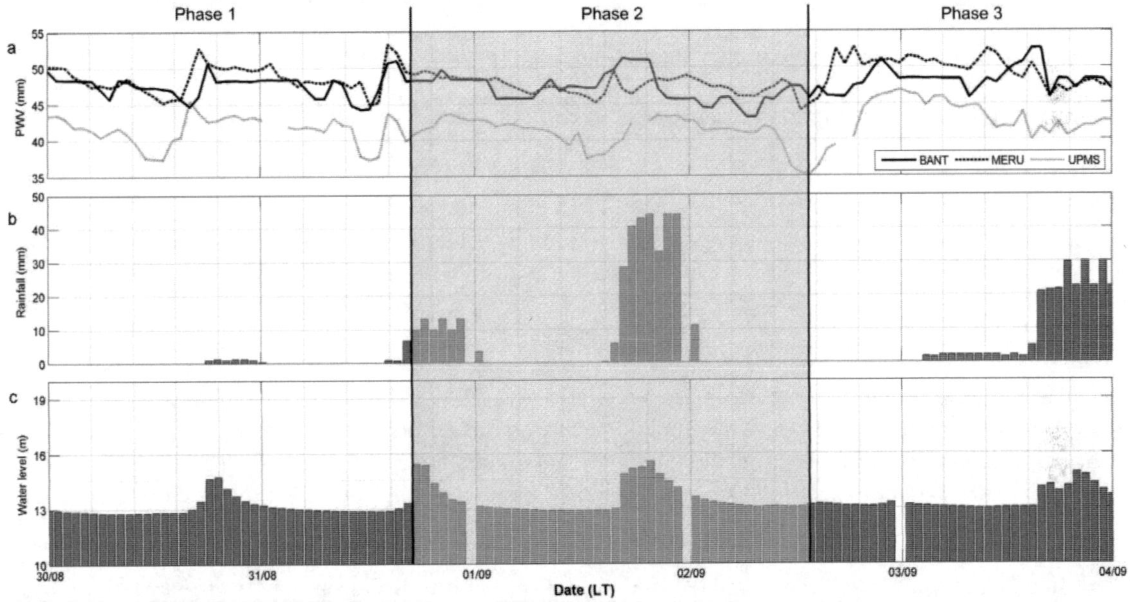

Figure 2. Variation of (a) PWV at three GPS stations, (b) rainfall and (c) river water level within 5 days from 30 August to 4 September where 1 September 2013 is flash flood event.

GPS PWV was found to be higher at MERU and UPMS within 40 hours of a flash flood event (shaded area). It was an increase of as much as 1 mm of mean GPS PWV at both stations. However, it did not happened in BANT where mean value of PWV at the same event was 1 mm lower as compared to phase 1. Mean of GPS PWV at MERU and UPMS decreased by 1 mm in phase 3 where it showed similarly before the flash flood event. It did not happened in BANT station where the mean value continuously increased by 1 mm in the following phase. Overall, GPS PWV was highest during phase of a flash flood event by 0.33 mm compared to pre and post events. Table 2 showed the mean and STD of PWV and rainfall during 5 days of pre and post flash flood events. The variation of GPS PWV was highest in UPMS during phase 1 and phase 2 followed by MERU and BANT. GPS PWV varied by less than 2 mm STD in phase 1. In phase 2, the PWV variation was recorded to be more than 2 mm and less than 3 mm compared to phase 1. It showed that GPS detected the atmospheric disturbances in phase 2 that could be a result of atmospheric water vapour and later on caused the flash flood. STD of GPS PWV decreased by 1 mm in phase 3. Individually, PWV varied the most in BANT and varied the least in UPMS.

IOP Conf. Series: Earth and Environmental Science **20** (2014) 012057 doi:10.1088/1755-1315/20/1/012057

Rainfall was recorded the highest during phase 2, which was 5 mm more in from pre event and 1 mm more in post event. The highest variation trend of rainfall was also similar to the trend of GPS PWV.

Table 2. Statistical value of GPS PWV and rainfall within 5 days observations from Figure 2.

Phase	Station		GPS		Rainfall	
	GPS	Rainfall*	Mean	STD	Mean	STD
	Banting		47.90	1.59		
1	Meru		48.00	1.70	2.09	4.21
	Serdang		41.10	1.86		
	Banting	Jambatan	47.02	2.04		
2	Meru	Petaling	48.67	2.16	7.42	15.30
	Serdang		42.32	3.16		
	Banting		48.03	1.98		
3	Meru		47.94	1.65	6.39	10.19
	Serdang		41.61	1.36		

*Rainfall data only recorded at Jambatan Petaling. All values are in mm.

3.2. GPS PWV variation during flash flood in October 2013

Another flash flood that hit Klang Valley was on 10 October after 3 hours of rainfall. Figure 3 showed the variation of GPS PWV at three GPS stations as well as total rainfall and changes of river water level as Figure 2. For October 2013 flash flood event, DID determine the water level into 4 levels: normal (24 m), alert (28 m), warning (29 m) and danger (30 m).

Similar to Figure 2, Figure 3a showed the variation of PWV at three nearest GPS stations that correlated to total rainfall and changes of river water level. PWV at UPMS recorded the lowest, while at BANT is the highest. During five consecutive days of monitoring, PWV reached a minimum peak in the afternoon with 2 minimum peaks at UPMS in the evening of 9 October. The first rainfall was recorded in the evening of 10 October as shown in Figure 3b that brought to flash flood at several parts in Klang Valley. Flash flood was reported hit over Klang Valley after 3 hours of rainfall after 1700 LT. From our observation, rainfall is occurred later after PWV dropped at all GPS stations. We identified 4 mm dropped of PWV at UPMS, while 4 mm and 9 mm dropped before rainfall at BANT and MERU, respectively. PWV at BANT and MERU dropped later than in UPMS with lead time 3 hours and 4 hours, respectively. The other 2 rainfall events were in the evening of 11 October and 12 October lead to drop the PWV at all GPS stations. The most PWV response with decreasing trend on 11 October before light rainfall can be seen at UPMS, and the least PWV was occurred at MERU. On 12 October, PWV at all GPS stations showed a decreasing trend before light rainfall with lead time of 2 hours. River water level in Figure 3c was drastically raised over warning level in the evening of 10 October, which believed causing a flash flood at the surrounding areas. The river water level then continued above the normal level on the following day and only declined to a normal level on 12 October. However, light rainfall in the evening of 12 October caused 1 m rise of river water level. Before that, we observed river water level drastically increased reaching alert level on 9 October even though no rainfall was recorded. It may be rainfall in the surrounding areas is not recorded at the Leboh Pasar DID station.

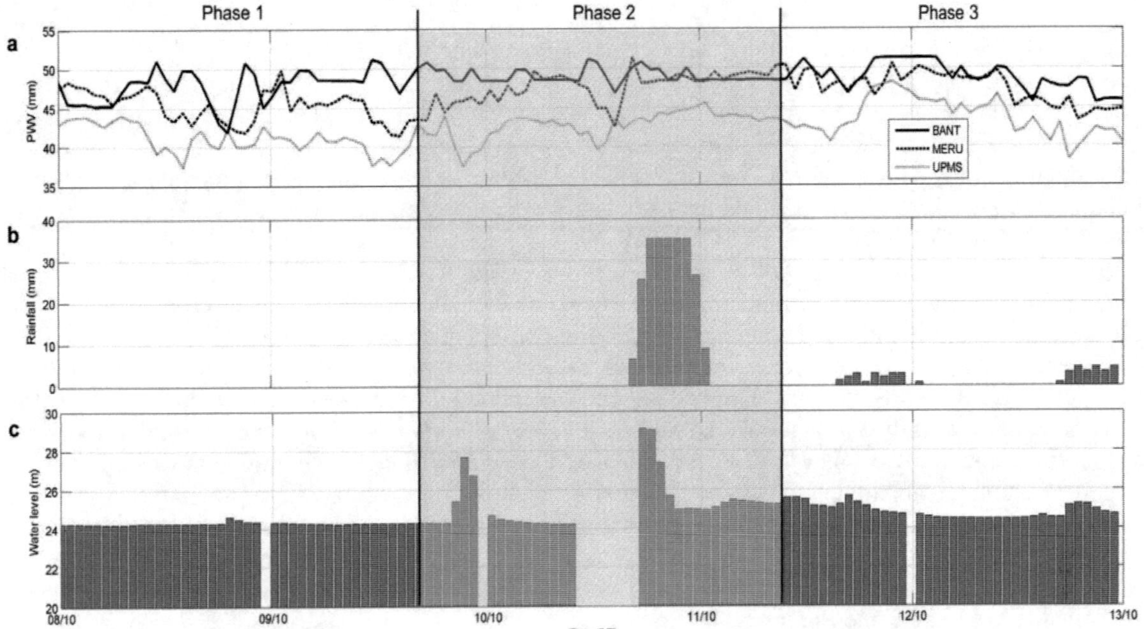

Figure 3. Variation of (a) PWV at three GPS stations, (b) rainfall and (c) river water level within 5 days of 10 October 2013 flash flood event.

As shown in Table 3, we found the mean of GPS PWV at all three stations were the highest within 40 hours (phase 2) during the event, which is 1 mm higher compared to phase 1 and phase 3, respectively. It showed phase 2 was wetter than phase 1 and phase 3 where the increasing of PWV trend (Figure 3a) in the evening of 9 October maintained at high level until morning of 12 October. The STD of PWV at every GPS station was almost 2 mm before and after the flash flood incident, but lowest in phase 2. The STD decreased 1 mm from phase 1 to phase 2 and increased 1 mm when entering phase 3. This pattern was opposite from the pattern of flash flood in September.

As mentioned before, there was no rainfall recorded in phase 1 that resulted in no variation of rainfall in this phase. The only first occurrence of rainfall in phase 2 was in the evening of 10 October, which recorded 35 mm with STD of about 13 mm. Light rainfall event repeatedly occurred in the evening of following day and another day which caused deviation of 1.52 mm.

Table 3. Statistical value of GPS PWV and rainfall within 5 days from Figure 3.

Phase	Station		GPS		Rainfall	
	GPS	Rainfall*	Mean	STD	Mean	STD
	Banting		48.11	1.92		
1	Meru		47.23	1.92	0.00	0.00
	Serdang		46.51	1.89		
	Banting	Leboh	48.79	0.89		
2	Meru	Pasar	48.80	0.89	6.11	12.57
	Serdang		48.02	0.88		
	Banting		47.52	1.53		
3	Meru		47.53	1.53	1.04	1.52
	Serdang		46.78	1.51		

*Rainfall data only recorded at Leboh Pasar. All values are in mm.

3.3. Association of PWV variation with accumulated precipitation from TRMM

From our analysis on the TRMM image specifically at latitude 2°N to 3°N and longitude 101.5°E to 200.0°E where three GPS stations are located (triangle), we found the accumulated precipitation was greatest in the evening during 1 September and 10 October flash flood event. As shown in Figure 4a, accumulated precipitation was less than 15 mm in the afternoon of 1 September. Precipitation >60 mm occurred above MERU. All GPS stations did not receive rainfall during this time until 14:00 LT. During this time, GPS stations in UPMS and BANT received 40 mm rainfall except in MERU where it received less than 20 mm rainfall or none as shown in Figure 4b. At this time (Figure 4b), North Malaysia (Kedah and Perlis) received more precipitation. For the three locations of GPS station, it is a proven that heavy rainfall was occurred in the evening on 1 September.

Figure 4c showed accumulated precipitation in the afternoon on 10 October where it was higher than 30 mm at all GPS stations. The accumulated precipitation increased more than 35 mm in the evening at almost all the GPS stations as shown in Figure 4d. It was similar to heavy rainfall recorded at this time amounting 35 mm hourly. As shown in the contour image, accumulation precipitation in October is higher compared in September. Overall in Figure 4, the accumulation precipitation within 6 hours in October was 15 mm greater compared to September. The accumulation of precipitation in September was less than 15 mm, which occurred in the afternoon. On the other hand, in these flash flood cases, the accumulated precipitation in Figure 4 was linearly correlated with the amount of GPS PWV value in Figure 2 and Figure 3.

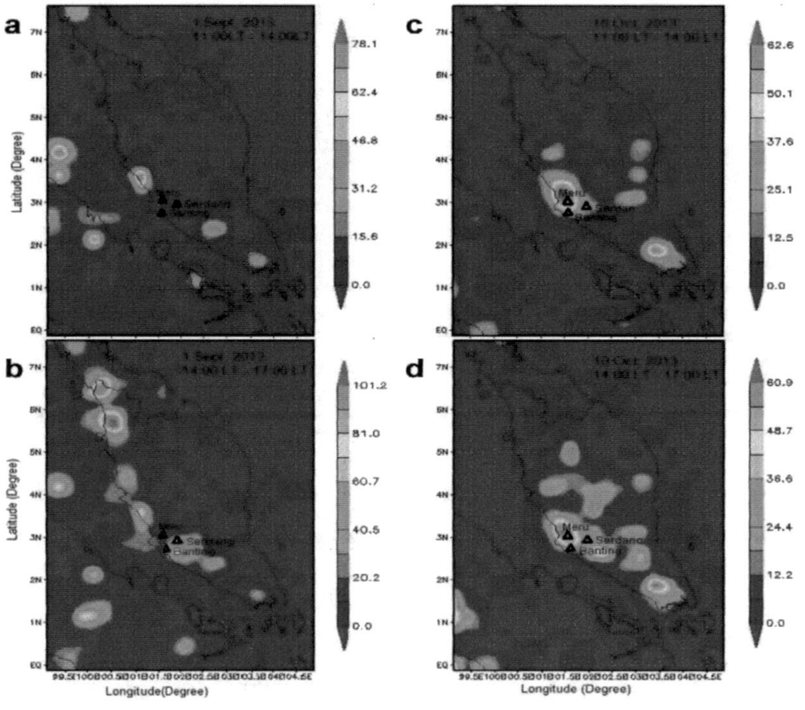

Figure 4. Accumulated precipitation (mm) from TRMM.

4. Conclusion

Monitoring the variation of GPS PWV during pre and post flash flood events over the Klang Valley in September and October 2013 flash flood was reported. These flash flood events during the inter-monsoon with heavy rainfall have been investigated by looking at the increasing trend of GPS PWV. Specifically, heavy rainfall in the evening during a flash flood can be detected earlier by looking at its decreasing trend of ~3 mm within 2 hours before the rainfall event. PWV variation was varied slowly in the afternoon of a flash flood event. During flash flood events, PWV value in October is higher with low STD than in September, which is contrasted with the amount of rainfall (precipitation is highest). This indicated that the variation of PWV is influenced by a winter monsoon. Detail of PWV variability for country mitigation will be investigated further by considering wind properties that associates with precipitation that possibly may help to improve weather nowcasting over Klang Valley.

Acknowledgements

This research is granted under FRGS/2/2013/TK03/UKM/02/3. The second author is MSc student supported by the Department of Public Service (JPA). The authors highly acknowledge Department of Survey and Mapping Malaysia (DSMM) for providing Malaysian GPS data, Malaysian Department of Irrigation and Drainage (DID) for supporting rainfall and water level data, Department of Meteorological Malaysia (DMM) for giving meteorological data.

References

[1] Lazo J K, Lawson M, Larsen P H and Waldman D M 2011 *Bulletin of the American Meteorological Society* **92** 709

[2] Toriman M E, Hassan A J, Gazim M B, Mokhtar M, Mastura S A S, Jaafar O, Karim O and Abdul Aziz N A 2009 *Research Journal of Earth Sciences* **1** 22

[3] Chan N W 1997 *Disaster Prevention and Management* **6** 72

[4] Low K C 2007 *Public Weather Services Workshop on Warning of Real-Time Hazards by Using Nowcasting Technology*

[5] Xie Y, Wei F, Zhang T and Hu L 2010 *Annales Geophysicae* **28** 1369

[6] Musa T A, Amir S, Othman R, Ses S, Omar K, Abdullah K, Sambsung L and Rizos C 2011 *Journal of Atmospheric and Solar-Terrestrial Physics* **73** pp 2410-2422

[7] Suparta W, Adnan J and Mohd Ali M A 2012 *American Journal of Applied Sciences* **9(6)** 825

[8] Suparta W, Abdul Rashid Z A, Mohd Ali M A, Yatim B and Fraser G J 2008 *Journal of Atmospheric and Solar-Terrestrial Physics* **70** 1419

[9] Suparta W and Iskandar A 2013 *Proceeding of the 2013 International Conference on Space Science and Communication* (IconSpace2013) p 190

[10] Bai Z and Feng Y 2003 *Journal of Global Positioning Systems* **2** 83

IOP Conf. Series: Earth and Environmental Science **20** (2014) 012058 doi:10.1088/1755-1315/20/1/012058

Comparison of Aerosol optical depth (AOD) derived from AERONET sunphotometer and Lidar system

Wei Ying Khor[1], Wan Shen Hee[1], Fuyi Tan[1], Hwee San Lim[1], Mohamad Zubir Mat Jafri[1] and Brent Holben[2]

[1] School of Physics, Universiti Sains Malaysia, 11800 Penang, Malaysia
[2] NASA/GSFC Code 923, Greenbelt Road Greenbelt, MD 20771 USA

Email: weiying326@hotmail.com

Abstract. Aerosol optical depth (AOD) is the measure of aerosols distributed within a column of air from the instrument or Earth's surface to the top of the atmosphere. In this paper, we compared the AOD measured by the Raymetrics Lidar system and AERONET sunphotometer. A total of 6 days data which was collected by both instruments were compiled and compared. Generally, AOD value calculated from Lidar data are higher than that calculated from AERONET data. Differences and similarities in the AOD data trend were observed and the corresponding explanations were done. Level 1.5 data of AERONET is estimated to have an accuracy of ±0.03, thus the Lidar data should follow the trend of the AERONET. But in this regards, this study was conducted less than one month and was very difficult to justify the differences and similarities between AOD measured by the Raymetrics Lidar system and AERONET sunphotometer. So further studies for an extended period will be needed and performed with more comprehensive LIDAR measurements. The slope of the best-fit straight line for the data points between the AOD values retrieved from LIDAR and the AERONET measurements is the closest to unity and the coefficient of determination is high (above 0.6692). Factors which affect AOD data were discussed. As a conclusion, the trends of the AOD of both systems are similar. Yet due to some external factors, the trend will be slightly different.

1. Introduction

In the atmosphere, there are atmospheric aerosols which play a very important role in most atmospheric processes, such as the air quality, visibility, clouds, precipitation and chemical processes. Aerosols originate from natural (sea salts, air-borne dust, volcanoes, and storms) and anthropogenic sources (fossil fuels combustion, biomass burning and gas-to-particle conversion) [1], [2], [3], [4]. They have different residence time, physical properties, chemical composition, refractive-index characteristics and climate-relevant properties due to the different sources and meteorological processes. In terms of Earth's radiative forcing, aerosols will affect the Earth's radiation balance directly by absorbing and scattering incoming and outgoing radiation and indirectly by acting as cloud condensation nuclei (CCN), which will then change the concentration of initial droplets, albedo, precipitation formation and lifetime of the clouds [1], [2], [3], [4]. Thus, measurements of these physical and optical properties of the particles (aerosols) are of great interest. To do so, we can use the Lidar [5] or AERONET sunphotometer [6].

Content from this work may be used under the terms of the Creative Commons Attribution 3.0 licence. Any further distribution of this work must maintain attribution to the author(s) and the title of the work, journal citation and DOI.
Published under licence by IOP Publishing Ltd

IOP Conf. Series: Earth and Environmental Science **20** (2014) 012058 doi:10.1088/1755-1315/20/1/012058

Light detection and ranging also known as Lidar is a technique whereby a beam of light is used to make range-resolved remote measurements. Lidar is used extensively in measuring the properties of the earth's atmosphere and also in ocean research [7]. Lidar also has a few military applications, including chemical [8] and biological agent detection [9]. For this study, we will only focus on atmospheric Lidar. Atmospheric Lidar makes use of the interaction, scattering and absorption of a beam of light with the particles or constituents in the atmosphere. A variety of atmospheric parameters can be measured, including aerosol and cloud properties, temperature and etc., which depends on the design of the Lidar. The paper entitled "Lidar observations of the vertical aerosol flux in the planetary boundary layer" shows two different Lidar systems i.e. Doppler wind Lidar for measuring the vertical wind velocity and a Raman Lidar to measure aerosol parameters [10].

AERONET which is also known as Aerosol Robotic NETwork program is a federation of ground-based remote sensing aerosol networks collaborated on by national agencies, institutes, universities, individual scientists and partners. This program provides a long term, continuous and readily accessible public domain database of aerosol optical depth, microphysical and radiative properties for aerosol research and characterisation, validation of satellite retrievals, and synergism with other databases [11]. Holben et.al described the concept and description of AERONET in detail in their paper [12].

Lidar uses a single wavelength to detect the optical depths of the atmosphere at different heights. Meanwhile, AERONET uses a sun photometer that delivers integrated optical depths of the atmospheric column at multiple wavelengths. The different measurement techniques of the two systems are as shown in Figure 1. The sunphotometer always points towards the sun using the incorporated sun-tracker, while the Lidar system points vertically into the atmosphere (Zenith angle of 90°). If a cloud is detected by the Lidar system, the sunphotometer which points in another direction may miss it.

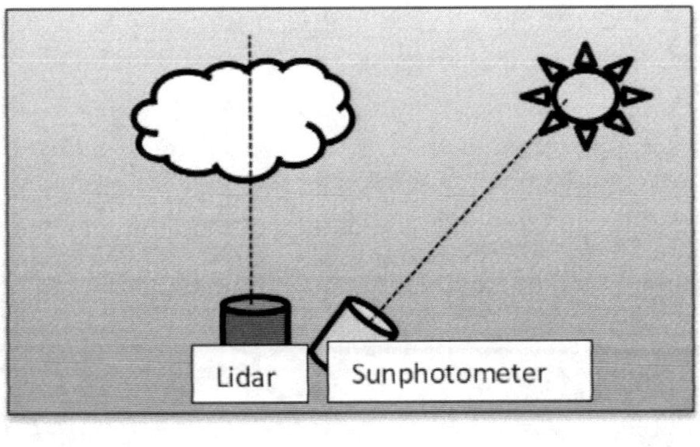

Figure 1: Simulation of the measurement techniques of the Lidar system and sunphotometer.

Although the two systems have different methods, the sunphotometer can be used to reinforce the aerosol profiling capabilities of a Lidar [13]. In the past, a few studies have been made to compare between the two systems [14], [15]. Most of the studies that were found did not compare the two systems in much detail. Therefore, this study will solely focus on comparing the two systems. And, the data of comparison is the aerosol optical depth (AOD). By definition, Aerosol Optical Depth (AOD) is the measure of aerosols (e.g. urban haze, smoke particles, desert dust, sea salt) which are distributed within a column of air from the instrument or Earth's surface to the top of the atmosphere.

2. Methodology and data collection

2.1. Lidar system

The Lidar system used is the Raymetrics LB100-ESS-D200 Eye-Safe Scanning Lidar System. Laser of wavelength 355nm was fired into the atmosphere at a zenith angle of 90° and azimuthal angle of 0°. Energy emitted per pulse for the lidar system is 6.50J. The laser pulse repetition rate is 20Hz. For this study, the laser pulse was set at 1200 shots per minute where data will be collected every minute throughout the whole operation of the Lidar system. The data retrieval was held throughout the weekdays of the month of December 2012. Each session was about 6 to 7 hours per day from 10.00 am till 5.00pm. All the data used in this study were only taken on sunny or cloudy days. In the case of rainy days, no data collection was done.

The laser system is placed on the rooftop of the School of Physics, USM, Penang, which is at a height of 50m above sea level. The position of the system is at longitude of 100.30 and latitude of 5.40. The temperature in the laser system is set to be 30 °C and the pressure is set as 1 atm. The temperature is set as so because the temperature in Penang, Malaysia has an average of 30 °C, where the temperature from morning (around 9am) till evening (6pm) would fluctuate from 23 to 32 °C (based on weather station data).

2.2. AERONET sunphotometer

For AERONET, the instrument used is the CIMEL Electronique 318A spectral radiometer (sunphotometer). It makes direct sun measurements in eight spectral bands between 340nm and 1640nm [12]. The radiometer is placed at the rooftop of School of Physics, USM, Penang. Its location is at a longitude of 100.302 and latitude of 5.358. Data is collected every day at a scheduled routine.

2.3. Method

Data of AERONET used are all of level 1.0 (unscreened data) and level 1.5 (cloud-screened data with an accuracy of the order of ±0.03 using the cloud-screening and quality control algorithm by Smirnov A. et al. [16]). However, the data are not quality assured as the final calibration is not applied. Level 2.0 cannot be downloaded, as the data is still in process and not available yet. Thus, for this study, data of level 1.0 and level 1.5 are used. AOD 355nm can be obtained from AOD 380nm and AOD 340nm using Equation 1 and Equation 2. We will first obtain the Angstrom coefficient, α by dividing the ln component of the AODs with the ln component of the corresponding wavelengths using Equation 1. Then, we get the AOD of 355nm through Equation 2. For Equation 2, either AOD 340 or AOD 380 can be used to calculate the final AOD 355.

$$\alpha = \left[\frac{ln\left(\frac{AOD_{340}}{AOD_{380}}\right)}{ln\left(\frac{\lambda_{340}}{\lambda_{380}}\right)} \right] \quad [1]$$

$$AOD_{355} = AOD_{340} \times \left(\frac{\lambda_{355}}{\lambda_{340}}\right)^{-\alpha} \quad [2]$$

Data obtained from the Raymetrics lidar system was analysed using the software provided, which is the Advanced Viewer. The Inver Klett-Fernald method is used by the software to obtain the value of the Extinction coefficient before solving it to get the AOD value for each specific time. For this method, we had to estimate the Lidar ratio. Since we do not have data for the Raman channel yet, we used the AERONET data as a guideline to choose the best Lidar ratio. As such, we chose Lidar ratio of 35 as this is the value where the Lidar AOD is almost the same as the AERONET AOD. This Lidar ratio also indicates that the atmosphere in Penang consists of mixed aerosols that are marine and anthropogenic aerosols.

Table 1: Comparison of Lidar AOD and AERONET AOD 355 (6th December 2012).

Time(hh:mm:ss)	AOD_380	AOD_340	Angstrom coeff, α	AOD_355	Lidar AOD
10:40:59	0.240795	0.28515	1.520057929	0.267038049	0.377
10:55:59	0.35596	0.394419	0.922408138	0.379020932	0.27

11:11:02	0.2991	0.335958	1.04479679	0.321140912	0.234
11:26:06	0.57342	0.606822	0.509028966	0.593632011	0.74
11:41:01	0.683612	0.719159	0.45575796	0.705147085	0.166
11:56:00	0.455541	0.496853	0.780471912	0.480390608	0.294
12:11:04	0.493709	0.537672	0.766931418	0.520161122	0.423
12:26:00	0.647435	0.696704	0.659400514	0.677150052	0.698
12:41:02	0.486539	0.530955	0.785432324	0.51325277	0.428
12:56:02	0.237743	0.282519	1.551400866	0.264216399	0.253
13:11:02	0.165109	0.203892	1.896906049	0.187859978	0.211
13:26:01	0.202725	0.245374	1.71662941	0.227846682	0.193
13:41:02	0.180132	0.218455	1.734223662	0.202696511	0.223
13:56:00	0.157009	0.198153	2.092470237	0.18103728	0.177
14:41:01	0.256351	0.298339	1.363739794	0.281281182	0.224
14:56:01	0.232074	0.271118	1.398039795	0.25523833	0.281
16:11:00	0.768126	0.824577	0.637593531	0.802188991	0.675
16:22:11	0.949568	0.99004	0.375257035	0.974129922	1.044

3. Results and Discussions

The AOD values for AERONET (the average of AOD 380nm and AOD 340nm) and Lidar (355nm) are compiled and the data of AERONET for Level 1.0 and Level 1.5 are compared with the data of Lidar respectively.

3.1. *Comparison of Lidar AOD and AERONET AOD (Level 1.0)*

Generally, the Lidar AOD is larger than the AERONET AOD. The difference is generally quite small except for some specific AOD points. The reason Lidar data is higher than the AERONET AOD may be due to the slight misalignment of the far-range sensing causing some signal noise in the lidar data. Noise present in the Lidar data are generally due to the large field-of-view of the Lidar system. The noise results from the interference from background light [17].

For AERONET AOD which is much higher than the Lidar AOD, as seen in Figure 2 (indicated by the circle), it is due to the presence of high clouds which are only detected by the sunphotometer but not the Raymetrics LB100-ESS-D200 Eye-Safe Scanning Lidar System. The sunphotometer measures the direct solar radiance up to the whole atmosphere column which might be contaminated with clouds while for the Lidar, it is only the integral of the extinction coefficient from the ground level till around 15km of height in the atmosphere. The presence of clouds is confirmed by the low or approximately 0 Angstrom coefficient obtained from AERONET.

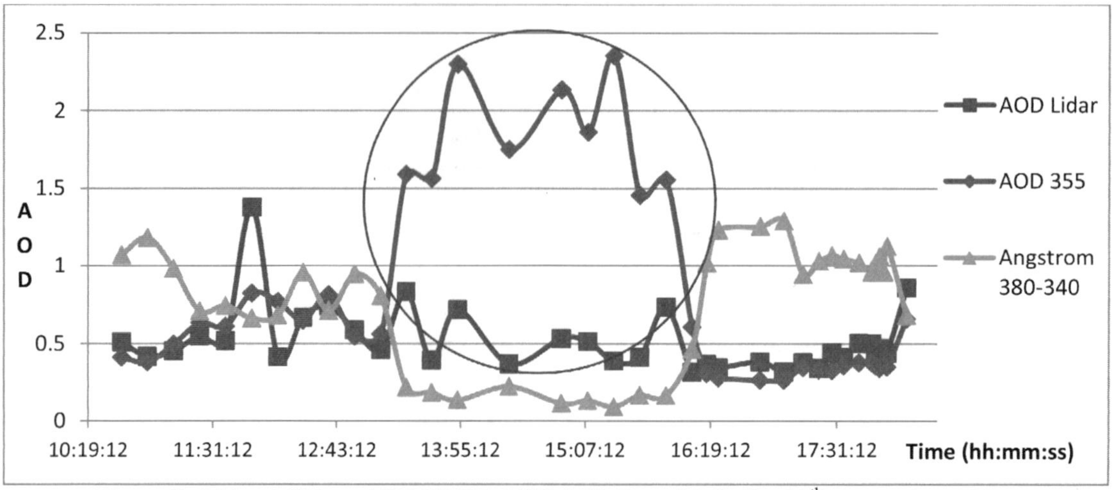

Figure 2: Lidar AOD vs AERONET AOD 355 (Level 1.0) for 30th November 2012.

For Level 1.0 data, which might be cloud-contaminated, most of the peaks shown in the AOD graphs for both AERONET and Lidar are represented by clouds (low and middle clouds) that exist in the sky. This was confirmed through our inspection of the time evolution graph for the total backscatter coefficient of the Lidar system. Clouds are the main factors of affecting the values of AOD. Thus there is an importance of comparing data which have undergone cloud-screening so that we can investigate the AOD value in the atmosphere.

3.2. Comparison of Lidar AOD and AERONET AOD (Level 1.5)

On average, the AOD for the atmosphere in Penang is less than 0.5. The peak of the AOD is mostly in the noon or afternoon as shown in Figure 3 and Figure 4. Peaks in the morning (11:31:12) and noon (13:26:24) may indicate the heavy traffic and activities around that particular time.

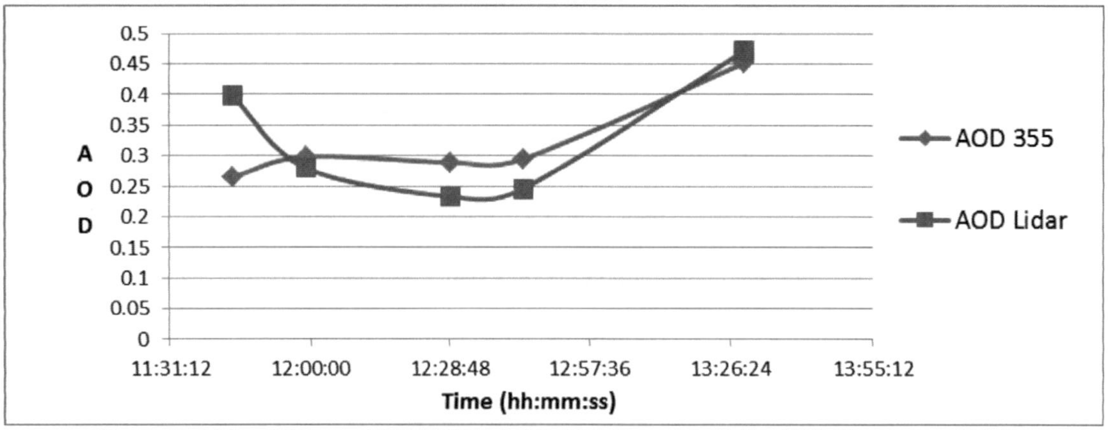

Figure 3: Lidar AOD vs AERONET AOD 355 (Level 1.5) for 13th December 2012.

IOP Conf. Series: Earth and Environmental Science **20** (2014) 012058 doi:10.1088/1755-1315/20/1/012058

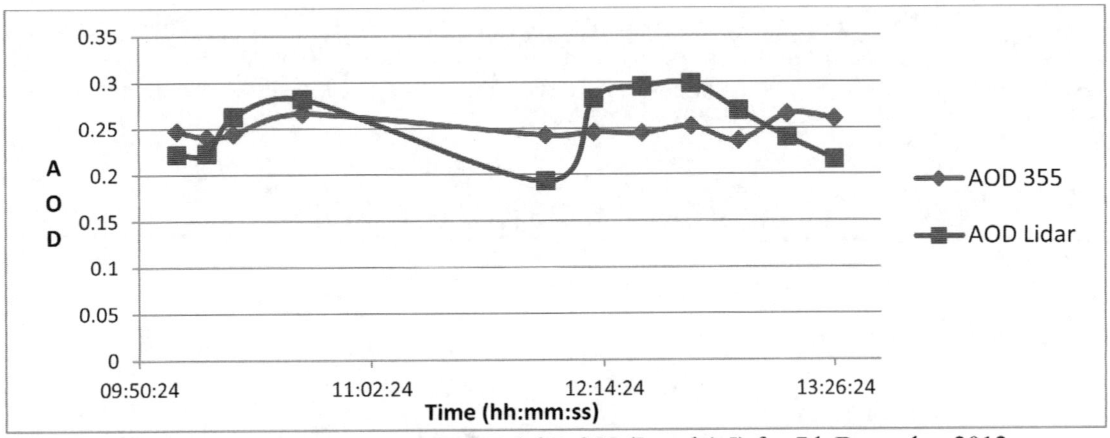

Figure 4: Lidar AOD vs AERONET AOD 355 (Level 1.5) for 7th December 2012.

On the 12th of December (Figure 5), we get a peak of around 0.71 of Lidar AOD. The sunphotometer also showed AOD of 0.6. This may be due to dust or aerosols coming from other parts of the country, which mix with the local anthropogenic aerosols and thus cause the AOD to increase more than normal. According to the weather station data, the weather at 1pm was cloudy while haze was occurring at 2pm. When analysing these data, we find a problem where the temporal variation between each data point is quite big. This is due to the cloud screening algorithm applied by AERONET [16]. The data which is suspected to be clouds will be completely screened off. Since Malaysia has a very cloudy weather, most of the time, the data for Level 1.5 and Level 2.0 will only be left with a few points after undergone cloud-screening. This will be a problem as the actual value of AOD at a particular time cannot be identified and is only estimated by connecting the data points, as for here, the data at 13:43:37 is connected with the data at 15:43:36. Although the Bayan Lepas weather station (~9km away from the Lidar and sunphotometer location) stated that there is haze, yet we cannot determine if it is true as there is only one data between 1pm to 3pm. We only have data at 13:43:37. The data for other time is removed due to contamination of clouds.

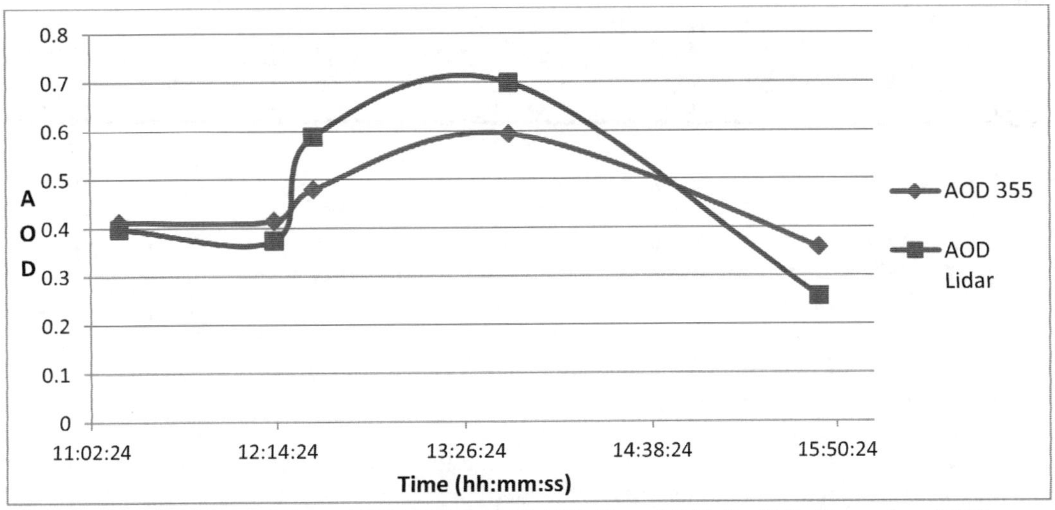

Figure 5: Lidar AOD vs AERONET AOD 355 (Level 1.5) for 12th December 2012.

As mentioned before, AERONET AOD has an accuracy of an order of ± 0.03. Thus, the trend of the AOD of Lidar should be similar to the AERONET. Although the values are slightly different, the trend is still quite similar. Referring to Figure 6, the slope of the best-fit straight line for the data points between the AOD values retrieved from LIDAR and the AERONET measurements is the closest to unity and the coefficient of determination is high (above 0.6692). This shows that the agreement between the two systems is good. However, the differences in the AOD should be noted and discussed.

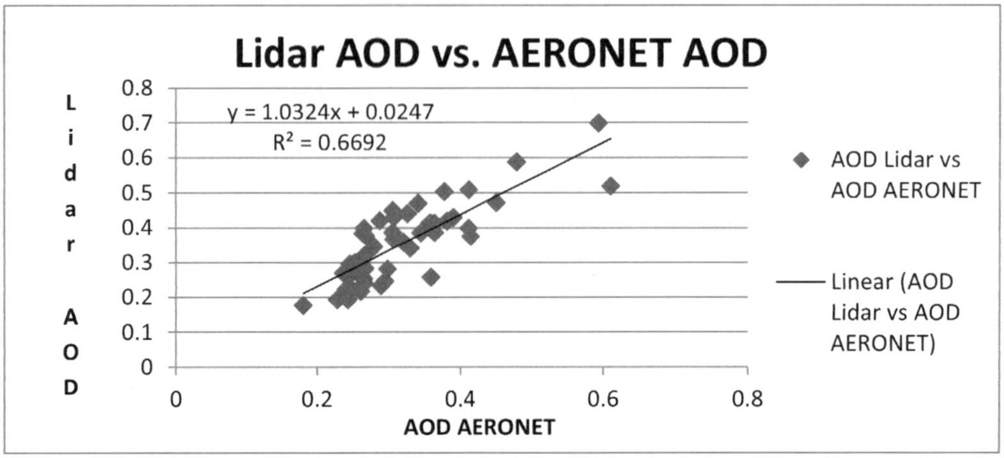

Figure 6: Best-fit straight line for Lidar AOD and AERONET AOD.

It is also found that the AOD data for both AERONET and Lidar are typically higher in the afternoon compared to in the morning. This suggests the presence of higher concentrations of aerosol in the afternoon. This is similar to the explanation of Yang Xun where as he investigated, the AOT value in the afternoon was mostly higher than in the morning [18]. The reason for the concentration of AOD in the afternoon is higher than in the morning may also suggest the effect of temperature on the aerosol distributions. Based on the weather conditions obtained from the nearby weather station, we can see that the temperature in the afternoon is higher than the temperature in the morning. On the whole, when the temperature increases at 12pm in the afternoon, the AOD graphs also show a trend of increasing in value after 12pm. This shows that temperature affects the concentration of AOD. This is consistent with the fact that, as the air in the planetary boundary layer (PBL) warms during the morning, the height at which thermal equilibrium occurs increases. Thus the depth of the PBL increases from dawn to several hours after noon. Since the PBL depth increases, there is a higher rate or possibility for aerosols to be distributed in the atmosphere.

4. Conclusion

All in all, the trend of the Lidar AOD and the AERONET AOD agree well and is almost similar. The differences in AOD of both systems may be due to presence of noise in Lidar data and presence of clouds in AERONET data. Due to the cloud-screening algorithm of AERONET, the points obtained for Level 1.5 is limited and thus the actual condition of the atmosphere may not be determined accurately. The AOD for Penang is normally less than 0.5 except on the 12th of December where there is haze. The slope of the best-fit straight line for the data points between the AOD values retrieved from LIDAR and the AERONET measurements is the closest to unity and the coefficient of determination is high (above 0.6692). The aerosol distribution may be affected by temperature. Further observations should be made and a more consistent and long term data collection should be done. This is so to more accurately observe the conditions of the atmosphere in Penang.

Acknowledgements
The authors gratefully acknowledge the financial support under the ERGS grant 203/PFIZIK/6730051, Universiti Sains Malaysia (USM) Short Term Grant 304/PFIZIK/6311082, RUI grant (Investigation

Of The Impacts Of Summertime Monsoon Circulation To The Aerosols Transportation And Distribution In Southeast Asia Which Can Lead To Global Climate Change) 1001/PFIZIK/811228 and ScienceFund, Environmental Effects And Its Influence of Increased Green House Gasses In Peninsular Malaysia, 305/PFIZIK/613615. We would like to thank the technical staff who participated in this project. Thanks are also extended to USM for support and encouragement.

References

[1] Balis D, Papayannis A, Galani E, Marenco F, Santacesaria V, Hamonou E, Chazette P, Ziomas I and Zerefos C 2000, Tropospheric LIDAR aerosol measurements and sun photometric observations at Thessaloniki, Greece, *Atmospheric Environment* **34**: 925 - 932.

[2] Landulfo E, Papayannis A, Artaxo P, Castanho A D A, Freitas A D, Souza R F & Moreira D S 2003, Synergetic measurements of aerosols over Sao Paulo, Brazil using LIDAR, sunphotometer and satellite data during the dry season, *Atmospheric Chemistry and Physics* **3**(5): 1523-1539.

[3] Georgoussis G C 2006, Monitoring of air pollution and atmospheric parameters using a mobile backscatter lidar system, *Óptica pura y aplicada*, **39**(1): 35-41.

[4] Chen W N, Chen Y W, Chou C C, Chang S Y, Lin P H, and Chen J P 2009, Columnar optical properties of tropospheric aerosol by combined lidar and sunphotometer measurements at Taipei, Taiwan, *Atmospheric Environment* **43**(17): 2700-2708.

[5] Welton E J, Voss K J, Gordon H R, Maring H, Smirnov A, Holben B N and Andreae M O 2000, Ground-based lidar measurements of aerosols during ACE-2: instrument description, results and comparisons with other ground-based and airborne measurements, *Tellus B* **52**(2): 636-651.

[6] Chin M, Diehl T, Dubovik O, Eck T F, Holben B N, Sinyuk A and Streets D G 2009, Light absorption by pollution, dust, and biomass burning aerosols: A global model study and evaluation with AERONET measurements, *Annales geophysicae: atmospheres, hydrospheres and space sciences* **27**(9): 3439.

[7] R. Barbini et al. 1998, *ICES J. Mar. Sci.* **55**: 793–802.

[8] N. S. Higdon et al. 1998, *Proc. Nineteenth ILRC, NASA, Hampton*Va. :651.

[9] W. B. Scott, Aviat 1995, *Week Space Technol.* 143: 44.

[10] Engelmann R, Wandinger U, Ansmann A, Müller D, Žeromskis E, Althausen D and Wehner B 2008, Lidar observations of the vertical aerosol flux in the planetary boundary layer, *Journal of Atmospheric and Oceanic Technology*, **25**(8): 1296-1306.

[11] AERONET website on System description, (March 2013) http://aeronet.gsfc.nasa.gov/new_web/system_descriptions.html (Viewed on March 2013)

[12] Holben B N, Eck T F, Slutsker I, Tanre D, Buis J P, Setzer A and Smirnov A 1998, AERONET—A federated instrument network and data archive for aerosol characterization, *Remote sensing of environment* **66**(1): 1-16.

[13] Otero L R 2005, Aerosol optical properties by means of a sunphotometer and LIDAR system in Buenos Aires Argentina, *Óptica pura y aplicada* **39**(1): 43-47.

[14] Mèndez J, Parsiani H, Sanchez E and RICO P 2009, Remote sensing of atmospheric particles using LIDAR, Calipso satellite, & AERONET: algorithm development, *WSEAS Transactions on Systems* **8**(1): 86-95.

[15] Schuster G L, Vaughan M, MacDonnell D, Su W, Winker D, Dubovik O and Trepte C 2012, Comparison of CALIPSO aerosol optical depth retrievals to AERONET measurements, and a climatology for the lidar ratio of dust, *Atmospheric Chemistry and Physics Discussions* **12**(5): 11641-11697.

[16] Smirnov A, Holben B N, Eck T F, Dubovik O and Slutsker I 2000, Cloud-screening and quality control algorithms for the AERONET database, *Remote Sensing of Environment* **73**(3): 337-349.

[17] Li J, Gong W and Ma Y Y 2012, Atmospheric lidar noise reduction based on ensemble

empirical mode decomposition, *Remote Sensing and Spatial Information Sciences.*
[18] Yang X 2008, Studies of urban atmospheric aerosols using lidar and sky radiometer.

IOP Conf. Series: Earth and Environmental Science **20** (2014) 012059 doi:10.1088/1755-1315/20/1/012059

Accuracy assessment of Terra-MODIS aerosol optical depth retrievals

Sahabeh Safarpour [1], Khiruddin Abdullah[1], Hwee San Lim[1], Mohsen Dadras[2]

[1] School of Physics, Universiti Sains Malaysia, Penang, Malaysia
[2] Faculty of Engineering, Department of Civil Engineering, Universiti Putra Malaysia, Serdang, Malaysia

E-mail: sahabeh.safarpour@yahoo.com

Abstract. Moderate Resolution Imaging Spectroradiometer (MODIS) aerosol products have been widely used to address environment and climate change subjects with daily global coverage. Aerosol optical depth (AOD) is retrieved by different algorithms based on the pixel surface, determining between land and ocean. MODIS-Terra and Global Aerosol Robotic Network (AERONET) products can be obtained from the Multi-sensor Aerosol Products Sampling System (MAPSS) for coastal regions during 2000–2010. Using data collected from 83 coastal stations worldwide from AERONET from 2000-2010, accuracy assessments are made for coastal aerosol optical depth (AOD) retrieved from MODIS aboard the Terra satellite. AOD retrieved from MODIS at 0.55μm wavelength has been compared With the AERONET derived AOD, because it is reliable with the major wavelength used by many chemistry transport and climate models as well as previous MODIS validation studies. After removing retrievals with quality flags below1 for Ocean algorithm and below 3 for Land algorithm, The accuracy of AOD retrieved from MODIS Dark Target Ocean algorithms (correlation coefficient R^2 is 0.844 and a regression equation of $\tau_M = 0.91 \cdot \tau_A + 0.02$ (where subscripts M and A represent MODIS and AERONET respectively), is the greater than the MODIS Dark Target Land algorithms (correlation coefficient R^2 is 0.764 and $\tau_M = 0.95 \cdot \tau_A + 0.03$) and the Deep Blue algorithm (correlation coefficient R^2 is 0.652 and $\tau_M = 0.81 \cdot \tau_A + 0.04$). The reasons of the retrieval error in AOD are found to be the various underlying surface reflectance. Therefore, the aerosol models and underlying surface reflectance are the dominant factors which influence the accuracy of MODIS retrieval performance. Generally the MODIS Land algorithm implements better than the Ocean algorithm for coastal sites.

1. Introduction

Aerosols, the suspended liquid, solid particles and small in the atmosphere, are significant ingredients of Earth's climate system. Between their many roles, they drive the hydrological cycle [10], force the worldwide energy budget [6] and in many concentrations are harmful to human health [14]. Determining aerosol worldwide distribution and changes over time are essential for realizing present and feasible future climate conditions [6]. Towards these aims; NASA has installed a collection of satellites known as the Earth Observation System (EOS) to display a number of main climate properties, including aerosols. one of these EOS-era satellite sensors are the MODerate resolution Imaging Spectroradiometers (MODIS, [17]), which have been flying in polar orbit on Terra since 2000 [17].

This paper focuses on the characterization of MODIS AOD uncertainty over the coastal regions because: (a) The MODIS AOD product over the coastal region is a simple union of the retrievals from algorithms that are planned for either over open ocean only or over land only, and neither algorithm has a separated plan to characterize the surface reflectance over the coastal region that is mostly influenced by a water reflectance and sand-water mixture contributed by the underlying seacoast and suspended matter in the coastal ocean; (b) the coastal region is mostly of high significance to its local economic development

Content from this work may be used under the terms of the Creative Commons Attribution 3.0 licence. Any further distribution of this work must maintain attribution to the author(s) and the title of the work, journal citation and DOI.
Published under licence by IOP Publishing Ltd

through either tourism or serving as a hub for freight transportation [20], Thus, the evaluation of the MODIS AOD product over the coastal region is decretive for studying the tendency of regional anthropogenic AOD and air pollution.

Several studies have found that the uncertainties in the instantaneous AOD retrievals from satellite sensors such as Moderate Resolution Imaging Spectroradiometer (MODIS) and Multi-Angle Imaging SpectroRadiometer (MISR) are generally within the (pre-launch) expected error (EE) envelope that is often characterized as a linear function of AOD itself. For example, in comparison with world-wide AOD measured from Aerosol Robotic Network (AERONET), MODIS AOD product is shown to have an EE envelope of $\pm(0.05+0.15AOD_{aeronet})$ over land and $\pm(0.03+0.05AOD_{aeronet})$ over the ocean ([15], [11], [12], [8]). This paper is designed to address the overall accuracy of MODIS AOD over coastal region with the Dark Target algorithms, over land and ocean ([15]), [11], [12]) and the Deep Blue algorithm over over bright-reflecting land surfaces, such as desert, semiarid, and urban regions ([3], [4]).

2. MODIS and AERONET AOD products

2.1. MODIS Data

MODIS Level 2 Collection 5.1 MOD04 aerosol data from 2 March 2000 through 28 December 2010 are used (http://giovanni.gsfc.nasa.gov/mapss/). The 550 nm wavelength is used for comparison with AERONET because it is consistent with the main wavelength used by many climate transports and chemistry models [9] also prior MODIS validation studies ([11], [12]). MODIS uses quality flags to characterize the accuracy of AOD retrievals. The quality flags range from 3 (high confidence) to 0 (low or no confidence) (Levy et al., 2010). The quality flags are determined to each MODIS AOD retrieval based on the quality and number of pixels used in the AOD algorithms ([15], [12]). Therefore, all MODIS data in this study is filtered by quality flag.

2.2. AERONET Data

AERONET AOD is derived from direct sun photometer measurements in certain or all of the following seven various spectral bands centered at 340, 380, 440, 500, 670, 940 and 1020 nm [2]. AERONET measures the extinction of direct beam solar radiation and applies the Beer_ Lambert_ Bouguer law to determine AOD [2] with uncertainties on the order of 0.01-0.02 [1]. Only cloud-screened and quality-assured AERONET Level 2 data are used in this study to evaluate the MODIS aerosol product [18].

3. Methodology

MODIS aerosol products in C5.1 are retrieved with two different algorithms, i.e. the Dark Target (DT) ((“DT” ocean and land; ([7], [15]) and the Deep Blue (DB) ([3], [4]).

3.1. Dark Target Algorithm

The MODIS aerosol algorithm is actually two totally autonomous algorithms, one for deriving aerosols over ocean and the second for aerosols over land. Both algorithms were developed and conceived before the Terra launch and are described in depth in [7] and [19]. In the MODIS Dark Target land algorithm, the 500 m spatial resolution reflectance at three bands (470, 650, and 2130 nm), after correction for ozone, water and carbon dioxide, are organized into nominal 10×10 km^2 boxes corresponding to 20×20 or 400 pixels per box. Following the deletion of cloud, snow/ice, and water pixels, the residual pixels are subjected to more process to retrieve AOD with four main steps: (1) recognizing dark pixels whose reflectance at 2130 nm falls in the typical value range of 0.01-0.25 and are about 20% to almost 50% of the corresponding reflectance at 650 nm (red band); (2) parameterizing the surface reflectance at visible (VIS) bands (470 nm and 650 nm) in term of the small wavelength infrared (SWIR, 2130 nm) surface and the corresponding “surface greenness”, the normalized ratio of measured reflectances at 1240 and 2130 nm channels [11]; (3) selecting the proper aerosol mode and type from a look-up-table, which is the function of geography and season; (4) deriving the optical thickness and mass concentration of the accumulation and coarse mode from the detected radiance. The details of dark target algorithm could be found in [7], [15] and [11].

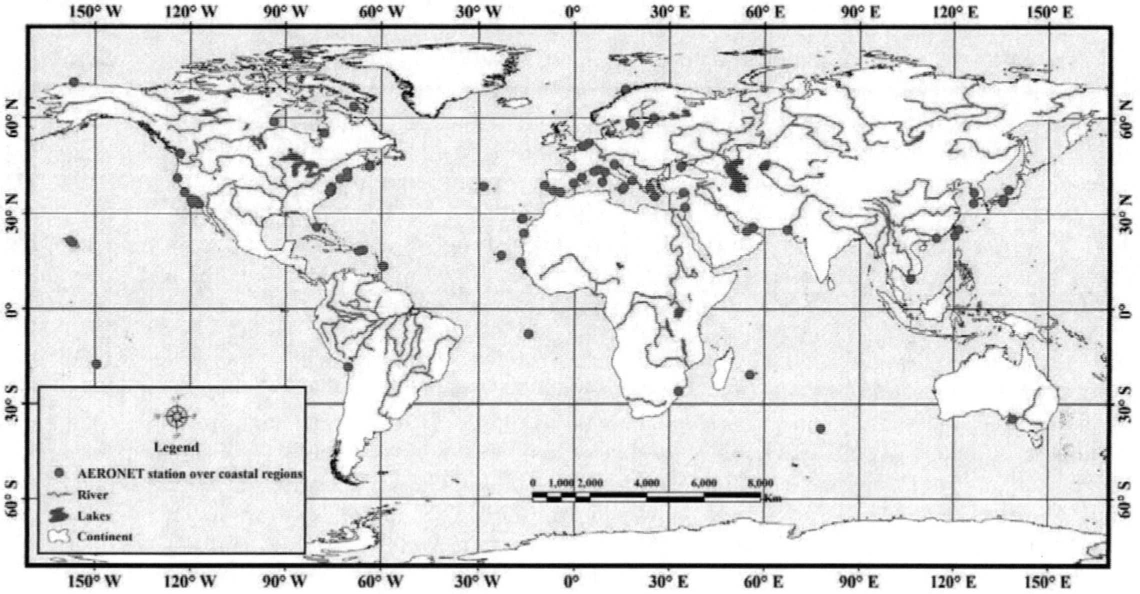

Figure1. Map of the location of all coastal AERONET sites

In the MODIS DT ocean algorithm, though the main inversion remains like the process described in [19], the masking of sediments and clouds, the special handling of heavy dust including dust retrievals over glint, and revisals of the lookup table are new. As in the land algorithm, after the ozone, carbon dioxide and water vapor corrections are applied, the primary phase in the ocean algorithm is to organize the reflectance of the six wavelengths used in the process ($\rho_{0.55}$, $\rho_{0.66}$, $\rho_{0.86}$, $\rho_{1.24}$, $\rho_{1.6}$, $\rho_{2.13}$) into nominal 10-km boxes of 20×20 pixels at 500-m resolution. This needs reducing the resolution of the 250-m channels ($\rho_{0.66}$ and $\rho_{0.86}$). The ocean algorithm needs all 400 pixels in the box to be recognized as ocean pixels by the MOD35 mask. This assists to decrease problems presented by shallow water near the coasts. If any land is encountered, the whole box is left for the land algorithm, but quality is reduced for coastal land retrievals.

3.2. Deep Blue Algorithm

The MODIS DT AOD retrieval is based on the relevance between the surface reflectance at SWIR and VIS, which is valid for most vegetated land surfaces but is generally invalid for desert, urban, arid, semi-arid areas [3]. The MODIS DB is offered to retrieve aerosol properties over brightly reflecting surfaces by using blue bands in which the surface reflectances are low enough to make such retrievals possible [4]. The main steps of the DB algorithm applied in the satellite processing stream are as follow: (1) apply the Rayleigh correction for terrain elevation to measured reflectances to account for variations in the surface pressure, and perform cloud screening by examining both the spatial homogeneity within a 3 × 3 pixel window and absorbing aerosol index values; (2) define the surface reflectance of a given pixel from a clear-scene database based upon its geolocation; (3) compare the surface reflectance at 412, 470, and 650 nm bands with the values recorded in a lookup- table in dimensions consisting of satellite zenith, solar zenith , and surface reflectance, relative azimuth angles, single scattering albedo and aerosol optical thickness [4]. The best match is obtained with the iterative method and maximum likelihood method; (4) relate the AOD and Angstrom exponent (α) after the mixing ratio and aerosol models are determined. The particularity of the deep blue satellite retrieval algorithm was described in [3].

2.3 MODIS-AERONET Collocation

Two methods are used for arranging the MODIS and AERONET AOD data. Initially, AERONET measurements within +/- 30 minutes of the MODIS overpass time are averaged and compared against MODIS AOD retrievals averaged within a 55 km diameter centered over the AERONET sites. Using the mean method MAPSS also saves the mode of the quality flags from each pixel within the averaging region (55 km) to represent the quality flag for the collocated MODIS retrieval [13]. The quality flag mode is then used to filter the AOD data for only high quality retrievals. Secondly, the MODIS AOD retrieval near to the AERONET site is paired with the AERONET measurement that is near to the MODIS overpass time (central method). [13], find that there is little difference between the mean and central methods, so, to be consistent with prior research and maximize data volume the mean method is used for of this study.

After getting the collocated MODIS retrievals at all AERONET sites a quality filter is run. AERONET sites must have at least 15 high quality (flag 3 for Land, or flags 1, 2, or 3 for Ocean) MODIS retrieval pairs to be used in this study. To be considered a coastal site, AOD measurements from AERONET must be paired with both the MODIS Ocean algorithm at minimum once and the MODIS Land algorithm at minimum once over the whole time period of this research (2000-2010). For the approximately 11 year record of Terra-MODIS and AERONET AOD pairs from MAPSS, the result from the mean collocation method, that is consistent with [5], shows that 83, or approximately 26%, of the AERONET stations have MODIS retrievals from both the Land and Ocean algorithms, and therefore those sites are designated as coastal

4. Result and Discussion

Although previous MODIS analysis, over a global average, was valuable for understanding MODIS error characteristics ([15], [11], [12]), an examination of coastal regions shows a reduction in MODIS accuracy. Note that the MODIS Land algorithm allows negative values when retrieving AOD, even though negative AOD values are not physically possible. From 2000-2010 approximately 500 retrievals from MODIS when paired with AERONET over the coastal regions resulted in negative AODs and those retrievals are not included in the analyses.

After removing retrievals with quality flags below 3 for Land and below 1 for Ocean, MODIS AODs are highly correlated with the paired AOD from AERONET with R^2 of 0.8, regardless if AODs are retrieved over coastal region, from Ocean algorithm, Land algorithm, and Deep blue algorithm (Figure 2). Compared with AERONET AOD, MODIS AOD retrievals from the Ocean algorithm have a correlation coefficient (R^2) of 0.844 and a regression equation of $\tau_M = 0.91 \cdot \tau_A + 0.02$ (where subscripts A and M represent AERONET and MODIS respectively) for all the coastal sites, MODIS AODs from the Dark Target Land algorithm have an R^2 of 0.764 and linear regression $\tau_M = 0.95 \cdot \tau_A + 0.03$ for coastal stations, and R^2 of 0.652 and $\tau_M = 0.81 \cdot \tau_A + 0.04$ for the Deep blue algorithm. However, the Ocean AOD correlation is consistently greater than the Land AOD correlation for coastal stations. The reasons of the retrieval error in AOD are found to be the different surface characteristics. So, the underlying surface reflectance and aerosol models are the main factors which impact the accuracy of MODIS retrieval performance. Generally the MODIS Land algorithm with R^2 0.744 and RMSD 0.1004 implements better than the Ocean algorithm with R^2 0.844 and RMSD 0.0596 for coastal sites.

Summary statistics for MODIS and AERONET values for the entire data set as well as stratified by algorithm are presented. Results showed that AERONET had a slightly wider range (0.002 – 2.0650) as compared to MODIS (0–3.1920). MODIS values were 0.0136 ± 0.00069 greater than AERONET; approximately 25.49% of the Total mean AOT (Table 1). The relative difference was larger in the dark target ocean algorithm (0.012, 27.79% relative error) and the dark target land algorithm (0.017, 27.30% relative error) as compared to the deep blue algorithm (0.020, 9.33% relative error).

Table 1. Descriptive Statistics for MODIS and AERONET AOT Variables by different algorithm

	Variables	N[1]	Mean	SD[2]	Minimum	Maximum	Difference (Standard Error)[3]	Relative Error[4], %
Dark Target Land Algorithm	MODIS	8214	.207	.181	.000	1.797	.017	27.30
	AERONET	8214	.189	.167	.002	2.044	(.0009)	
Dark Target Ocean Algorithm	MODIS	6200	.154	.144	.001	3.192	.012	27.79
	AERONET	6200	.141	.145	.004	2.023	(.0007)	
Deep Blue Algorithm	MODIS	839	.303	.272	.017	1.427	.020	9.33
	AERONET	839	.323	.271	.023	2.065	(.0058)	
Combination result	MODIS	15253	.191	.177	.000	3.192	.013	25.49
	AERONET	15253	.177	.171	.002	2.065	(.0006)	

[1]N refers to sample size.

[2]SD refers to arithmetic standard deviation.

[3]Difference is calculated as MODIS - AERONET.

[4]Relative error of MODIS is calculated as (MODIS- AERONET)/AERONET

Acknowledgment

This project was funded by University Science Malaysia short term grant no. No.304/PFIZIK/650049/E102. The Authors would like to acknowledge the NASA and AERONET for providing the data in this study.

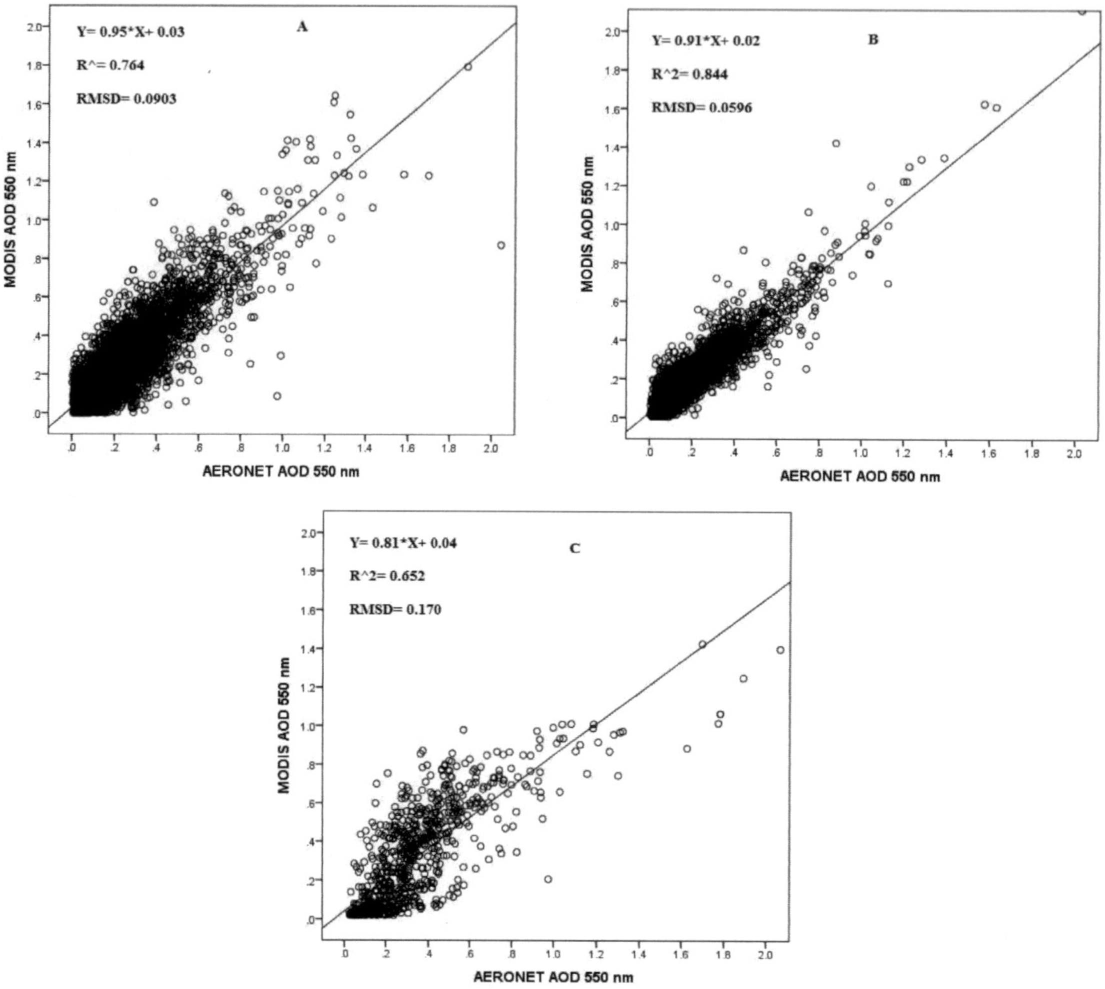

Figure 2. Scatter plot of AERONET AOD (x-axis) and the quality flag filtered Terra-MODIS AOD (y-axis) from 2000-2010. In (A) , (B) and (C) AODs in y-axis are respectively derived from MODIS dark target Land, dark target Ocean and deep blue products over the coastal AERONET station.

Reference

[1] Eck, T., Holben, B., Reid, J. and Dubovik, O. 1999. Wavelength dependence of the optical depth of biomass burning, urban, and desert dust aerosols. J. Geophys. Res. 104, 31333_31349.

[2] Holben, B., Eck, T., Slutsker, I., Tanre, D., Buis, J. P. and co-authors. 1998. AERONET _ a federated instrument network and data archive for aerosol characterization. Remote Sens. Environ. 66, 1_16.

[3] Hsu, N. C., Tsay, S. C., King, M. D., and Herman, J. R.: 2004. Aerosol Properties Over Bright-Reflecting Source Regions, IEEE T. Geosci. Remote, 42, 557–569, doi:10.1109/TGRS.2004.824067.

[4] Hsu, N. C., Tsay, S. C., King, M. D., and Herman, J. R.: 2006.Deep blue retrievals of Asian aerosol properties during ACE-Asia, IEEE T. Geosci. Remote., 44, 3180–3195.

[5] Ichoku, C., Chu, D., Mattoo, S., Kaufman, Y., Remer, L. and co-authors. 2002. A spatio-temporal approach for global validation and analysis of 1_4. DOI: 10.1029/2001GL013206.

[6] IPCC: Climate Change 2007: The physical science basis. Contribution of working group I to the Fourth Assessment Report of the Intergovernmental Panel on Climate Change [Solomon, S., Qin, D., Manning, M., Chen, Z., Marquis, M., Averyt, K. B., Tignor, M., Miller, H. L. (eds)]. Cambridge University Press, Cambridge, United Kingdom and New York, NY, USA, 996 pp., 2007.

[7] Kaufman, Y. J., Tanre, D., Remer, L. A., Vermote, E. F., Chu, A., Holben, B. N.: 1997. Operational remote sensing of tropospheric aerosol over land from EOS moderate resolution imaging Spectroradiometer, Journal of Geophysical Research, 102, 17051-17067.

[8] Kahn, R., Garay, M., Nelson, D., Levy, R., Bull, M. and co-authors. 2011. Response to ''Toward unified satellite climatology of aerosol properties. 3. MODIS versus MISR versus AERONET.'' J. Quant. Spectrosc. Radiat. Transfer. 112, 901_909. DOI: 10.1016/j.jqsrt.2010.11.001.

[9] Kinne, S. 2003. Monthly averages of aerosol properties: a global comparison among models, satellite data, and AERONET ground data. J. Geophys. Res. 108, 4634, D20. DOI: 10.1029/ 2001JD001253.

[10] Koren, I. and Feingold, G. 2011. Aerosol-cloud-precipitation system as a predator-prey problem, P. Natl. Acad. Sci. USA, 108, 12227–12232, doi:10.1073/pnas.1101777108.

[11] Levy, R., Remer, L. and Dubovik, O. 2007. Global aerosol optical properties and application to Moderate Resolution Imaging Spectroradiometer aerosol retrieval over land. J. Geophys. Res. 112, 1_15. DOI: 10.1029/2006JD007815.

[12] Levy, R., Remer, L., Kleidman, R., Mattoo, S., Ichoku, C. and co-authors. 2010. Global evaluation of the Collection 5 MODIS dark-target aerosol products over land. Atmos. Chem. Phys. 10, 10399_10420. DOI: 10.5194/acp-10-10399-2010.

[13] Petrenko, M., Ichoku, C., and Leptoukh, G. 2012. Multi-sensor Aerosol Products Sampling System (MAPSS), Atmospheric Measurement Techniques, 5, 913-926, doi:10.5194/amt-5-913-2012.

[14] Pope, III, C. A., Burnett, R. T., Thun, M. J., Calle, E. E., Krewski, D., Ito, K., and Thurston, G. D. 2002. Lung Cancer, Cardiopulmonary Mortality, and Long-term Exposure to Fine Particu late Air Pollution, JAMA, 287, 1132–1141, doi:10-1001/pubs.JAMA-ISSN-0098-7484-287-9-joc11435.

[15] Remer, L. A., Kaufman, Y. J., Tanre, D., Mattoo, S., Chu, D. A., Martins, J. V., Li, R. R., Ichoku, C., Levy, R. C., Kleidman, R. G., Eck, T. F., Vermote, E., and Holben, B. N. 2005. The MODIS aerosol algorithm, products, and validation, J. Atmos. Sci., 62, 947–973, doi:10.1175/JAS3385.1.

[16] Remer, L. A., Kleidman, R. G., Levy, R. C., Kaufman, Y. J., Tanr´ e, D., Mattoo, S., Martins, J. V., Ichoku, C., Koren, I., Yu, H., and Holben, B. N. 2008. Global aerosol climatology from the MODIS satellite sensors, J. Geophys. Res.-Atmos., 113, D14S07, doi: 10.1029/2007JD009661.

[17] Salomonsen, V., Barnes,W., Maymon, P., Montgomery, H., and Ostrow, H. 1989, MODIS – Advanced facility instrument for studies of the earth as a system, IEEE T. Geosci. Remote., 27, 145–153, 1989.

[18] Smirnov, A., Holben, B. N., Eck, T. F., Dubovik, O. and Slutsker, I. 2000. Cloud-screening and quality control algorithms for the AERONET database. Remote Sens. Environ. 73, 337_349. DOI: 10.1016/S0034-4257(00)00109-7.

[19] Tanré, D., Kaufman, Y.J., Herman, M., Mattoo, S., 1997. Remote sensing of aerosol properties over oceans using the MODIS/EOS spectral radiance. Journal of Geophysical Research 102, 16971e16988.

[20] Tibbetts, J. 2002. Coastal cities: living on the edge. Environ. Health Perspect. 110, 674_681.

IOP Conf. Series: Earth and Environmental Science **20** (2014) 012060 doi:10.1088/1755-1315/20/1/012060

Haze event monitoring and investigation in Penang Island, Malaysia using a ground-based backscatter Lidar

W S Hee[1,2], F Tan[1], H S Lim[1] and M Z Matjafri[1]
[1]School of Physics, Universiti Sains Malaysia, 11800 Penang, Malaysia.
Phone: 604-6535003 Fax: 04-6579150

E-mail: billy_manutd@hotmail.com

Abstract. During 24th July 2013 to 1st August 2013, a haze event struck Penang Island, causing the visibility to decrease and increase in Air Pollution Index (API). A ground-based backscatter Lidar, operate at 355nm which was setup at the roof top of the School of Physics, Universiti Sains Malaysia. It was used to monitor and investigate the haze event. For this work, we studied the daytime variation of the aerosol intensity, distribution, planetary boundary layer (PBL) height and the aerosol optical depth (AOD) values during these days. We found that the aerosol are very intense during the first two days of the haze event and slowly decline as time passed. Finally the haze event died off on 1st August 2013. As for daily aerosol distribution, aerosols are generally more intense during the afternoon. Its intensity is slightly lower in the morning and evening. Similar trends were observed for AOD values as they increase from morning to afternoon and slowly decrease in the evening. Most aerosols are found contained below the PBL which generally found at around 1000 - 2000 m in height.

1. Introduction

During the past 20-30 years, haze events caused by trans-boundary biomass burning smoke increased significantly in the Southeast Asia region. The haze event occurs recurrently every year during the dry season, typically around June to September. This biomass burning smoke usually comes from forest clearance using fire which goes rampant and developed into uncontrollably burning wild-fires [1]. The smoke is then transported by the Southwest monsoon across the boundary into other countries.

In June 2013, a serious haze event struck Peninsular Malaysia. The air pollution index (API) reported by the Department of Environment Malaysia (DoE) rose rapidly and reaches unhealthy and very unhealthy levels within a few days especially in the central and southern part of Peninsular Malaysia. It is worth to mentioning that in Muar, a city in the south of Peninsular Malaysia had an API reading on 23rd June 2013 that even exceeds 700, which is categorized in hazardous level according to the DoE official web site (http://apims.doe.gov.my/apims/). However, northern part of Peninsular Malaysia such as Penang was less affected by haze. The API generally maintained at healthy level until the mid of June. After that, the API readings in Penang Island start to fluctuate around healthy, moderate and unhealthy level. Generally, haze was found in the atmosphere when the API readings hike up to moderate and unhealthy level. This situation continued until the end of August.

Haze event are generally highlighted by the high concentration of aerosols suspended in the air. A collection of particles, either solid and/or liquid suspended in a gas is the simplest form of aerosol. Many daily events such as smoke, smog, fume, dust, haze, fog, mist, and cloud are considered as aerosol [2]. In terms of radiative forcing of Earth's climate, aerosol will affect the Earth's radiation balance directly by absorb and scatter incoming and outgoing radiation, depending on the aerosols'

[2] Corresponding author

Content from this work may be used under the terms of the Creative Commons Attribution 3.0 licence. Any further distribution of this work must maintain attribution to the author(s) and the title of the work, journal citation and DOI.
Published under licence by IOP Publishing Ltd

chemical composition. Indirectly, the aerosols will act as cloud condensation nuclei (CCN) and hence altering the concentration of initial droplets, albedo, precipitation formation and lifetime of the clouds. Aerosols may comes from both natural sources, such as volcanoes, sea salts, storms, and air-borne dust, as well as anthropogenic sources such as fossil fuels combustion, biomass burning and gas-to-particle conversion process. [3-6]. Aerosol generally shows high spatial and temporal variation, depending on the emission and dispersion process, chemical evolution, as well as meteorological process. Hence, high spatial and temporal resolution measurements, such as Lidar are very useful in air quality research [7].

Recently, more and more research has been carried out using Lidar to study the atmosphere. For example, Lidar and ceilometers were used to obtain aerosol and PBL profile over Athens, Greece. Good agreement with the results were obtained from both instruments as stated in [8]. Next, two micro-pulse Lidars were used to study the properties of atmospherics aerosol vertical distribution in Lanzhou China as stated by [9]. The aerosol was found to be distribute quite evenly in the lowermost two km of atmosphere. Then their concentration gradually decreased with increasing height. Lidar is also one of the popular instruments to study dust transportation in many area around the world. In Greece, Lidar were to study the Saharan dust layer over Athens [10]; while on the other side of the globe, a group of researchers did a case study on dust aerosol radiative properties in Lanzhou, China [11].

In this study, we use a ground based elastic backscatter Lidar to monitor and investigate the haze event. During the operation of the elastic backscatter Lidar, a collimated laser beam is emitted into the atmosphere. Then, the laser beam will undergo multiple scattering processes in the atmosphere before returning to the Lidar sysem. A telescope, usually set up in a coaxial or biaxial configuration with respect to the laser emitter will capture the return signal (or so called backscatter signal in Lidar terminology). Due to the high temporal (in terms of seconds) and spatial (3-15 m) resolution of the Lidar system, it is very useful in visualizing the vertical distribution of aerosols, as well as the instantaneous PBL structure by using aerosols as passive traces [3-5, 12]. Hence, the Lidar system is used during the haze event to monitor the changes of a few parameters such as aerosol backscatter coefficient (β_{aer}), AOD, and PBL height.

AOD was chosen as one of the parameter to monitor because AOD is a measure of the degree of light extinction by aerosol through absorption or scattering in the atmospheric column [13]. Light scattering and absorption by aerosols will diminish the solar radiation and limit the visual air quality. Consequently, visibility will be reduced and the air quality will decrease. This makes the AOD an important parameter to measure [6]. On the other hand, PBL is the lowermost sub-layer of the troposphere. It is directly affected by the earth and solar radiance, as well as the anthropogenic activities on the earth's surface [8]. Moreover, heat and moisture from the earth's surface must first be well mixed before circulating in the free troposphere. Thus, aerosol concentrations in the PBL are generally higher compared to that in the free troposphere [8, 14]. Generally, most aerosols are concentrated in the PBL, since the PBL is in direct contact with the ground. Nearly all aerosol sources are coming from ground level [15]. Hence, PBL height data are vital in atmospheric dynamics studies [8].

Since the 2013 haze event in Malaysia can generally consider as a recurrence event, it is thoroughly important to study and obtain data which can help further understanding of the aerosol emission and dispersion, as well as its effect on regional and global climatology. The main objective of this study is to monitor the life of the haze event and investigate it using parameters such as aerosol backscatter coefficients, AOD and PBL height obtained from a ground-based backscatter Lidar.

2. Instrumentation

2.1 Lidar System

An eye-safe ground based backscatter Lidar system, model no. LB100-ESS-D200, manufactured by Raymetrics SA was setup on the roof top of the School of Physics, Universiti Sains Malaysia, Penang.

IOP Conf. Series: Earth and Environmental Science **20** (2014) 012060 doi:10.1088/1755-1315/20/1/012060

The location are shown in Figure 1. This elastic backscatter Lidar system are capable of profiling aerosols and clouds by transmitting laser pulses into the atmosphere and measured the return time of the backscatter signal. It is operate with a Nd:YAG pulsed laser that emit 355nm laser beam into the atmosphere. The pulse duration is 5.04ns, the energy emitted in each pulse is 33.4mJ with 20Hz repetition rate. The returned signal is then collected by a Cassegrain telescope with a diameter of 200mm, 1mrad field-of-view and a complete overlap height of 180m. The captured signal is then spectrally analyzed, filtered and focused on a photon multiplier tube (PMT), with a corresponding spatial and temporal resolution of 7.5m and 1 minute respectively. Next, the current generated from the PMT will be detected by the Licel transcient recorder and the analog Lidar signal is obtained and recorded. By using a combination of a powerful A/D converter (12 Bit at 40 MHz) with a 250MHz fast photon counting system, the photon counting Lidar signal is obtained and recorded.

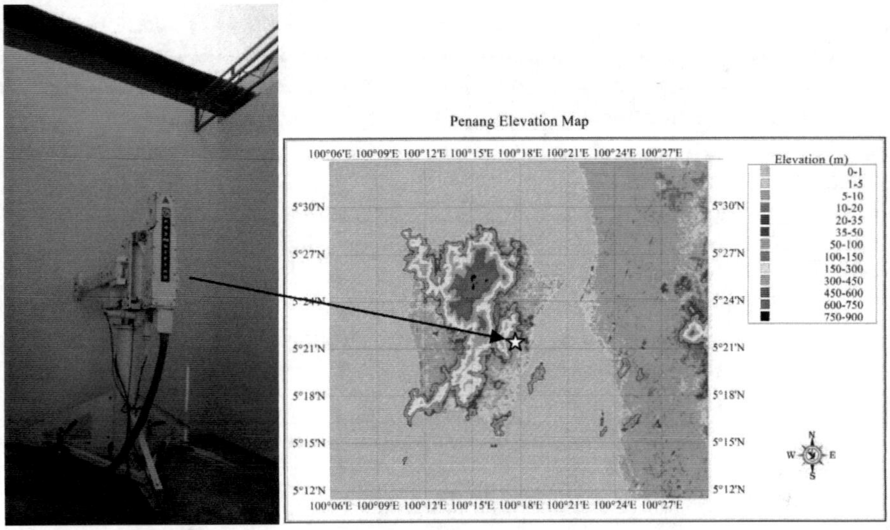

Figure 1: Topographic map produced from the global digital elevation model (GDEM) showing the high-mountain and low-plain regions in Penang. Five-edge star is point to the study site (Universiti Sains Malaysia, USM) where the Lidar system was deployed.

3. Methodology

3.1 Data Acquisition

The Lidar system was set to shoot at zenith on every weekday except on rainy day or public holiday. The acquisitions start from 0200 UTC to 1000 UTC, which correspond to 1000 to 1800 Malaysia local time. The Lidar system was set to an average of 1200 shots into a data file, which means that one data file will be produced every minute.

3.2 Data Pre-processing

The acquired data was being processed using Lidar Analysis.exe, which is the processing software provided by Raymatrics SA. Before starting the data processing, the usage of analog signal, photon counting signal or glued signal has to be determined, according to the conditions stated in [16]. Then, the background radiation is subtracted from the Lidar signal. If the photon counting signal or the glued signal is used, dead-time correction has to be further applied to the photon counting signal as given by [17]. Next, the range corrected signal (RCS) is obtained by applying the distance square law correction (z^2) to each data point to compensate for range-related attenuation from the atmosphere. After that, the temporal evolution of the RCS is plotted. This plot shows the general view of the atmospheric

IOP Conf. Series: Earth and Environmental Science **20** (2014) 012060 doi:10.1088/1755-1315/20/1/012060

conditions. Days with strong backscatter signals from the lower atmosphere, typically below 2000 m is selected and identified as hazy days. In this case, the hazy days lasted from 24/7/2013 until 1/8/2013.

3.3 Data Processing

When the pre-processing is completed, the aerosol backscatter coefficient (β_{aer}) is calculated with the aid of the Lidar Analysis.exe software. Here, the Klett inversion technique was used to retrieve the aerosol backscatter coefficient [18]. This was done using a simple numerical integration from the reference height until ground level (inward stepwise integration) since this method is more stable [19]. The reference height is a height where almost no aerosol backscatter signal is found, and the backscatter signal is mostly molecular [20]. To use the Klett inversion technique, an assumption has to be made on the Lidar ratio of the studied aerosol. In this case, the Lidar ratio was assumed to stand at 40 sr and remain constant throughout the haze periods. Although a Lidar ratio of 40 sr is slightly lower than the Lidar ratio for Southeast Asia aerosols and biomass burning aerosols, as stated by [21] and [22], but since Penang is a highly urbanized and industrialized city on an island, so that a certain degree of mixing between marine aerosols, Southeast Asia aerosols, as well as the biomass burning aerosols is expected, causing the Lidar ratio to be lower than the value suggested in the literature. After the aerosol backscatter coefficient was calculated, the temporal evolution of the aerosol backscatter coefficient was plotted.

Next, the hourly PBL height and hourly AOD were calculated. To determine the PBL height, the data files at each hour were selected. The Inflection Point Method (IPM) proposed by [14] was applied, where the minimum of the second derivative of the RCS gives the PBL height.

In addition for hourly AOD, the average light extinction effect by aerosols throughout a particular hour, assuming the aerosol loading in the atmosphere do not change very much throughout that hour was taken into account. Hence, the data files need to be averaged for that particular hour and AOD was calculated by formula below

$$AOD = \int_{R_0}^{R_{max}} \alpha_{aer}(r)dr \qquad (1)$$

where R_{max} is the reference height, R_0 is the height where the overlap is complete and α_{aer} is the aerosol extinction coefficient and α_{aer} is given by

$$\alpha_{aer} = L_{aer} * \beta_{aer} \qquad (2)$$

where L_{aer} is the aerosol Lidar ratio and β_{aer} is the aerosol backscatter coefficient.

4. Results and Discussion

The study period start from 24th July 2013 and end on 1st August 2013. A total of six daily measurements had been made during the study period, while the remaining days falls on weekends and public holiday where the Lidar is not operating. These measurements cover the whole life cycle of the haze event, starting from where the haze starting to be detect, right until the haze dies off. During these six days, significant differences in the aerosol distribution pattern are found during the haze event compared to the day where the haze dies off. This changes in aerosol distribution pattern will change the Earth radiative forcing and cloud formation in our atmospheric column, which ultimately will affect local and regional climate.

Aerosols are generally contained in the PBL, the first step was to determine the PBL height during the study period. Using the inflection point method suggested by [14], where the minimum of the second derivative of the RCS will provide information on the PBL height. The calculated PBL heights, in meter are shown in Table 1. Generally the PBL is located below 2000 m in height and will increase its height from morning to afternoon. Then its height will slowly decrease when approaching evening. This means that more aerosols will be loaded in the atmosphere in the afternoon compared to morning and evening.

Table 1: PBL height (m) evolution during the study period

Date/Time	02:00	03:00	04:00	05:00	06:00	07:00	08:00	09:00
24/07/2013	1450	1370	1760	1190	1100	1020	1400	1270
25/07/2013	600	960	960	1120	1240	1330	1180	1450
29/07/2013	540	580	920	820	990	900	840	640
30/07/2013	470	930	1040	1400	1140	1060	1150	1090
31/07/2013	580	1390	1110	810	1100	700	630	1330
01/08/2013	400	340	340	440	840	740	300	560

Figure 2 shows the temporal evolution of the aerosol backscatter coefficient for the study period and the black colour line indicates the PBL height. Each calculated PBL heights are plotted in respective profiles at respective time, while PBL heights at other times are extrapolated. From the profiles, the daytime aerosol distribution pattern and the intensity of the aerosol loading in the atmosphere can be observed. During the study period, aerosols were generally contained under the PBL, except for 30th July 2013, where a residual layer was found above the PBL around 0200 UTC to 0300 UTC. This residual layer had a very low concentration of aerosols, given by the low backscatter coefficient and was very short lived. Hence it was suspected that a convection effect was responsible for making the aerosol load higher than the PBL and it slowly decent and mixed together with other aerosol in the PBL. Next, on 24th and 25th July 2013, there was a thick aerosol layer loaded under the PBL. This thick aerosol layer indicated that the amount of aerosol loading in the atmosphere was very high, which was further supported by the high aerosol backscatter coefficient on these two days. This also shows that the haze event was serious on these two days. Then, it was followed by two slightly clearer days on 29th and 30th July 2013, where the aerosol backscatter coefficients were quite low under the PBL. After that, the haze returned again as the amount of aerosol in the atmosphere again increased on 31st July 2013. However, the amount of aerosol loading in the atmosphere at this day was not as high as the first two days since the aerosol backscatter coefficients were not as high as those obtained during the first two days. The aerosol layer was thinner compared to the first two days. Finally, on 1st August 2013, only a thin layer of aerosols was found loading in the atmosphere with low aerosol backscatter coefficient, which indicated that the haze had finally died off.

Moving on to daily aerosol distribution, aerosols were generally distributed under 2000 m in height, which was bounded by the PBL. Relatively heavily polluted days will have a higher PBL height compared to the mildly polluted days as shown in Figure 2. For example, 24th, 25th and 31st July 2013, which were heavily polluted days; its PBL height was generally maintained around 1000 to 1500m in height. As for mildly polluted days, the PBL height was maintained around 500 m to 1000 m in height, as shown in profiles for 29th and 30th July 2013. Compared to polluted days, a clear day has an even lower PBL height. For example, on 1st August 2013, after the haze died off, the PBL height was generally found at 500 m height which is very close to the Earth surface. As for daily aerosol distribution, it is much dependent on the PBL height. During the study period, most of the aerosols were contained under the PBL, except when there was a residual layer. Hence the aerosol layer generally will increase its height together with the PBL from morning to afternoon and decrease its height together when approaching evening, unless the residual layer exist and cause the aerosol layer load to be higher than the PBL, as shown in aerosol backscatter profiles for 30th July 2013 from 0200 to 0300 UTC.

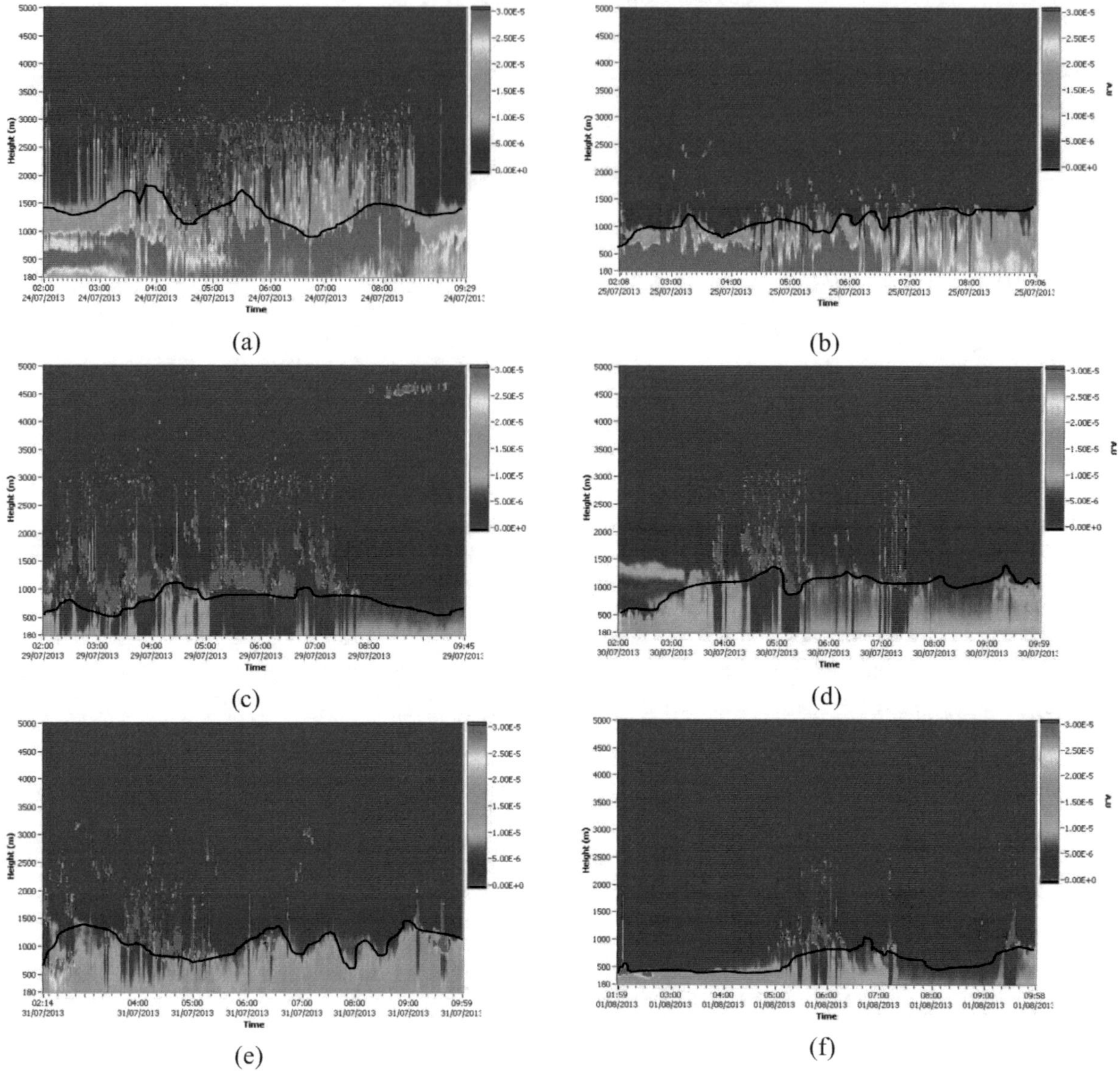

Figure 2: Temporal evolution of aerosol backscatter coefficient ($m^{-1}\ sr^{-1}$) on 24th July, 25th July, 29th July, 30th July 31st July and 1st August 2013. The black colour lines indicate the PBL height.

Figure 3 shows the vertical profiles of aerosol extinction coefficient. Based on the profiles, the aerosol extinction coefficient was decreasing very fast with increasing height. This proves that there was lesser and lesser aerosol with increasing height. It was found that the aerosol extinction coefficient is very low beyond 2000 m in height, which means that aerosols generally contained below 2000 m in height and only minor aerosols exist beyond that height. This suggests that most aerosols are concentrated under the PBL since the PBL height listed in Table 1 does not exceed 2000 m in height. As shown in Figure 3, the aerosol extinction coefficient for 25th July 2013 at 0235 UTC is the highest, followed by that of 24th July 2013 at 0305 UTC and 31st July 2013 at 0305 UTC. However, the aerosol extinction coefficient profile for 25th July 2013 at 0235 UTC is slightly different from that on 24th July 2013 at 0305 UTC and 31st July 2013 at 0305 UTC. The aerosol extinction coefficient for 25th July 2013 decreased very fast after 500 m in height and dropped even lower after exceeding 1000 m in height. For 24th July 2013 at 3.05 UTC and 31st July 2013 at 0305 UTC, the aerosol extinction coefficient started to drop around 1000 m in height and fell to a very low aerosol extinction coefficient value at 2000 m. This shows that on 24th and 31st July 2013, the aerosol layer had a thickness around 2000 m, but for 25th July 2013, the aerosol layer was thinner where its thickness was just around 1000m. Besides that, on 25th July 2013, aerosols were loaded very low at a height around 700 m, given by the high aerosol extinction coefficient up to this height. Compared to 25th and 31st July 2013, the aerosols were loaded higher in the atmosphere, up to around 2000 m in height, given by the relatively high aerosol extinction coefficient value up to this height. However, the light extinction capability of the aerosol at 25th July 2013 was much higher than that at 24th July 2013 and 31st July 2013, given by the relatively higher aerosol extinction coefficient in the lower atmosphere compared to the other two days. This suggested that the aerosols may undergo a characteristics change from 24th to 25th July 2013 or there might be an aerosol intrusion during night time on 24th July 2013 when the Lidar was not operating and the aerosols were completely mixed up with local aerosols by the next morning. On 30th July 2013 at 0220 UTC, the aerosol extinction coefficient profiles had two peaks, one at around 500 m in height, and the other one at around 1300 m in height. This demonstrates that there were two aerosol layers presented during that particular time, which agrees well with what is shown in Figure 2.

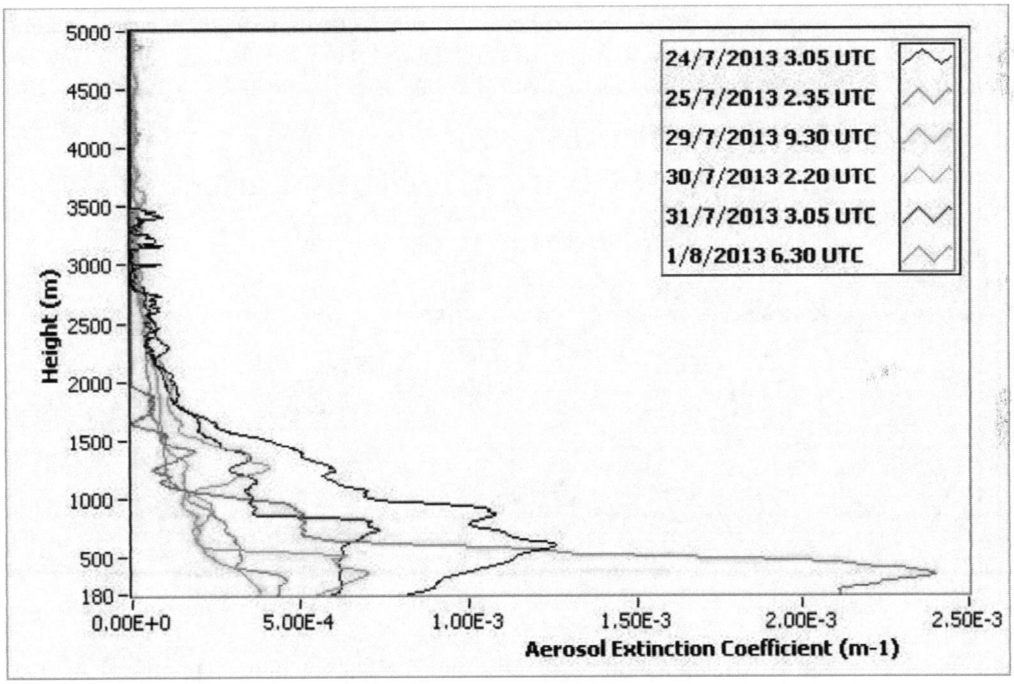

Figure 3: Aerosol extinction coefficient profiles (m^{-1}) for six different selected days and time during the study period.

Figure 4 displays the hourly AOD during the study period. Higher AOD generally means that there are more aerosols in the atmosphere, because these aerosols will obstruct the transmission of light either through absorption or scattering process. However, clouds are also considered as aerosols. Light will undergo a high degree of extinction when passing through clouds, causing an extremely high AOD. In the Southeast Asia region, AOD reading lower than 0.3 usually means that there is only minor aerosol contained in the atmosphere, while AOD reading higher than that represents polluted air. AOD reading higher than 2.0 is usually caused by cloud contamination. Based on Figure 3, 24th and 25th July were days with highly polluted air, marked by the high AOD reading. Then, 31st July 2013 was the day with moderately polluted air, as given by the lower AOD reading compared to 24th and 25th July 2013. Next, 29th and 30th July 2013 were days with slightly polluted air, stated by the low

AOD reading. Finally, 1st August 2013 was the day that the haze died off, as the AOD reading was very low. The hourly AOD distribution also showed a general view on how the aerosols were distributed in the atmosphere. Although the AOD distribution did not show a similar trend over the study period due to difference in aerosol concentration over the study period, but relatively high AOD was generally found at the afternoon, and lower AOD was found during the morning and evening, except that particular day was a highly polluted day, where high amount of aerosols were loaming around the sky all the time, causing the AOD to remain very high all the time. The AOD distribution trend means that more aerosols are distributed in the atmosphere during the afternoon compared to morning and evening. However, this pattern is similar during hazy days or a clear day. Although the AOD distribution trend might look similar to the PBL height evolution trend, but both trends are not identical. This is because AOD is a measure of the degree of light extinction in the atmospheric column by aerosols and it is much dependent on the types and composition of aerosol, as well as the age of the aerosol. However, PBL is more affected by the heat from the Earth and the sun, as well as the anthropogenic activity on the Earth surface. Direct heat from the sun during the afternoon together with peak human activity at that time will cause the PBL load higher than in morning and evening. Moreover, PBL is just a sublayer in the atmosphere where the aerosol concentration is a few magnitudes higher than that in the free troposphere, but the aerosols in the free troposphere is not negligible. These aerosols still contribute a significant value to the AOD reading causing the daily AOD trend to be different from the PBL height evolution trend.

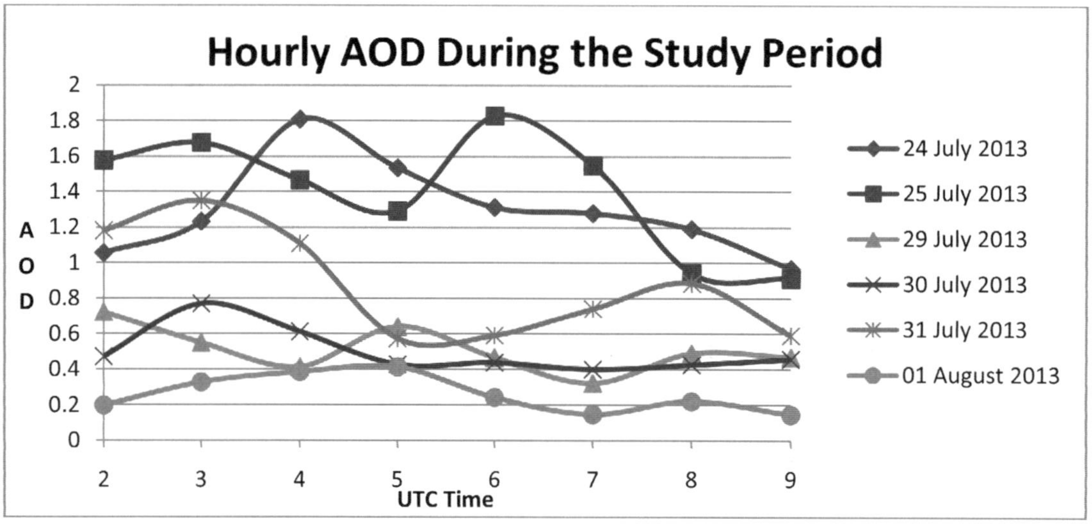

Figure 4: Hourly AOD Obtained using Lidar during the Study Period

5. Conclusion

In this study, a ground-based backscatter Lidar was used to study the haze event at Penang Island from 24th July to 1st August 2013. The PBL was found at a higher height in hazy days compared to the clear day after the haze dies off. Both the temporal evolution of aerosol backscatter coefficient and hourly AOD reading showed that high amount of aerosols were loamed under the PBL during the haze, particularly on 24th, 25th and 31st July 2013. These readings also showed that more aerosols were loaded in the atmosphere at morning and afternoon compared to evening, regardless of the atmospheric conditions. Overall, results from this study provide important results on aerosol variation and distribution during a haze event which is essential to future study on aerosol emission, growth, transportation and dispersion in Southeast Asia region. Future study should involve a synergy measurement which combines different in situ data such as particulate meter data, sunphotometer data, and satellite data into a more comprehensive result that allow us to study the haze event in details.

IOP Conf. Series: Earth and Environmental Science **20** (2014) 012060 doi:10.1088/1755-1315/20/1/012060

6. Acknowledgement

This project was carried out with financial support from RUI, Investigation Of The Impacts Of Summertime Monsoon Circulation To The Aerosols Transportation And Distribution In Southeast Asia Which Can Lead To Global Climate Change, 1001/PFIZIK/811228 and ScienceFund, Environmental Effects And Its Influence of Increased Green House Gasses In Peninsular Malaysia, 305/PFIZIK/613615.

7. References

[1] Salinas S V, Chew B N, Miettinen J, Campbell J R, Welton E J, Reid J S, Yu L E and Liew S C 2013 Physical and optical characteristics of the october 2010 haze event over Singapore: A photometric and Lidar analysis *Atmos. Res.* **122** 555 - 70.

[2] Hinds W C 1999 *Aerosol technology: Properties, behavior, and measurement of airborne particles* Wiley-Interscience 464.

[3] Balis D, Papayannis A, Galani E, Marenco F, Santacesaria V, Hamonou E, Chazette P, Ziomas I and Zerefos C 2000 Tropospheric Lidar aerosol measurements and sun photometric observations at Thessaloniki, Greece *Atmos. Environ.* **34** 925 - 32.

[4] Georgoussis G, Chourdakis G, Landulfo E, Hondidiadis K and Ikonomou A 2006 Monitoring of air pollution and atmospheric parameters using a mobile backscatter Lidar system *Optica Pura y Aplicada* **39** 35 - 41.

[5] Landulfo E, Papayannis A, Artaxo P, Castanho A D A, Freitas A Z d, Souza R F, Junior N D V, Jorge M P M P, Sanchez-Ccoyllo O R and Moreira D S 2003 Synergetic measurements of aerosols over Sao Paulo, Brazil using Lidar, sunphotometer and satellite data during the dry season *Atmos. Chem. Phys.* **3** 1523 - 39.

[6] Chen W-N, Chen Y-W, Chou C C K, Chang S-Y, Lin P-H and Chen J-P 2009 Columnar optical properties of tropospheric aerosol by combined Lidar and sunphotometer measurements at Taipei, Taiwan *Atmos. Environ.* **43** 2700–08.

[7] Zhang Y, Yu H, Eck T F, Smirnov A, Chin M, Remer L A, Bian H, Tan Q, Levy R, Holben B N and Piazzolla S 2012 Aerosol daytime variations over north and south America derived from multiyear aeronet measurements *J. Geophys. Res.* **117** D05211.

[8] Tsaknakis G, Papayannis A, Kokkalis P, Amiridis V, Kambezidis H D, Mamouri R E, Georgoussis G and Avdikos G 2011 Inter-comparison of Lidar and ceilometer retrievals for aerosol and planetary boundary layer profiling over Athens, Greece *Atmos. Meas. Tech.* **4** 1261 - 73.

[9] Cao X, Zhang L, Quan X, Zhou B, Bao J and Li X 2012 Lidar measurements for comparison of atmospheric aerosol extinction and scattering properties over Lanzhou, China *Appl. Mech. Mater.* **137** 256 - 61.

[10] Papayannis A, Mamouri R E, Amiridis V, Kazadzis S, Perez C, Tsaknakis G, Kokkalis P and Baldasano J M 2009 Systematic Lidar observations of saharan dust layers over Athens, Greece in the frame of Earlinet project (2004–2006) *Ann. Geophys.* **27** 3611 - 20

[11] Zhang L, Cao X, Bao J, Zhou B, Huang J, Shi J and Bi J 2010 A case study of dust aerosol radiative properties over Lanzhou, China *Atmos. Chem. Phys.* **10** 4283-93.

[12] Landulfo E, Papayannis A, Torres A S, Uehara S T, Pozzetti L M V, Matos C A d, Sawamura P, Nakaema W M and Jesus W d 2008 A four-year Lidar-sun photometer aerosol study at Sao Paulo, Brazil *J. Atmos. Ocean. Technol.* **25** 1463 - 68.

[13] Donkelaar A v, Martin R V, Brauer M, Kahn R, Levy R, Verduzco C and Villeneuve P J 2010 Global estimates of ambient fine particulate matter concentrations from satellite-based aerosol optical depth: Development and application *Environ. Health Perspect.* **188** 847 - 55.

[14] Menut L, Flamant C, Pelon J and Flamant P H 1999 Urban boundary-layer height determination from Lidar measurements over the Paris area *Appl. Optics* **38** 945 - 54.

IOP Conf. Series: Earth and Environmental Science **20** (2014) 012060 doi:10.1088/1755-1315/20/1/012060

[15] Matthias V and Bosenberg J 2002 Aerosol climatology for the planetary boundary layer derived from regular Lidar measurements *Atmos. Res.* **63** 221 - 45.

[16] Tan F, Beh B C, Tan C H, Lim H S, Abdullah K, Jafri M Z M, Welton E J and Lolli S 19 - 21 November 2012 *Preliminary analysis of ground based Lidar backscattered signal and performance evaluation in Penang Island* Ibrahim N B, Abd-Shukor R, Rahman I A, Ibarahim Z and Yap C C Colmar Tropicale, Bukit Tinggi, Pahang, Malaysia American Institute of Physics 1528 140 - 45.

[17] Donovan D P, Whiteway J A and Carswell A I 1993 Correction for nonlinear photon-counting effects in Lidar systems *Appl. Optics* **32** 6742 - 53.

[18] Klett J D 1981 Stable analytical inversion solution for processing Lidar returns *Appl. Optics* **20** 211-20.

[19] Fernald F G 1984 Analysis of atmospheric Lidar observations: Some comments *Appl. Optics* **23** 652 - 53.

[20] Ansmann A, Wandinger U, Riebesell M, Weitkamp C and Michaelis W 1992 Independent measurement of extinction and backscatter profiles in cirrus clouds by using a combined raman elastic-backscatter Lidar *Appl. Optics* **31** 7113 - 31.

[21] Cattrall C, Reagan J, Thome K and Dubovik O 2005 Variability of aerosol and spectral Lidar and

backscatter and extinction ratios of key aerosol types derived from selected aerosol robotic network locations *J. Geophys. Res.* **110**.

[22] Muller D, Ansmann A, Mattis I, Tesche M, Wandinger U, Althausen D and Pisani G 2007 Aerosol-type-dependent Lidar ratios observed with raman Lidar *J. Geophys. Res.* **112**.

AUTHOR INDEX

AUTHOR INDEX